International Conference on Architecture, Civil and Hydraulic Engineering (ACHE2019)

AA007687

IOP Conference Series: Materials Science and Engineering
Volume 794

Guangzhou, China
20 - 22 December 2019

ISBN: 978-1-7138-4721-2
ISSN: 1757-8981

Printed from e-media with permission by:

Curran Associates, Inc.
57 Morehouse Lane
Red Hook, NY 12571

Some format issues inherent in the e-media version may also appear in this print version.

Institute of Physics
Dirac House, Temple Back
Bristol BS1 6BE UK

Phone: 44 1 17 929 7481
Fax: 44 1 17 920 0979

techtracking@iop.org

Additional copies of this publication are available from:

Curran Associates, Inc.
57 Morehouse Lane
Red Hook, NY 12571 USA
Phone: 845-758-0400
Fax: 845-758-2633
Email: curran@proceedings.com
Web: www.proceedings.com

TABLE OF CONTENTS

IOP Conf. Series: Materials Science and Engineering **794** (2020) 011001 doi:10.1088/1757-899X/794/1/011001

Preface

The IAECST-International Conference on Architecture, Civil and Hydraulic Engineering (ACHE 2019), jointly organized by Hohai University and Southeast University, was held in Guangzhou, China on December 20-22, 2019. With the aim of enhancing the academic exchange between China and other foreign countries and boosting the development of Architecture, Civil and Hydraulic Engineering, ACHE 2019 was held to bring experts and scholars around the word together to share the latest research, discuss the existing problems and challenges, and explore the frontier technologies in the fields of Architecture, Civil and Hydraulic Engineering.

During the conference, Prof. Liaojun Zhang from Hohai University served as our conference Chairman, who also gave a marvelous speech on site. Then we had Prof. Wang Jingquan, Dean of School of Civil Engineering at Southeast University to share his insightful speech on Bridge and Tunnel Engineering. Followed by them, Prof. Zou Degao, representing Dalian University, gave a speech on anti-seismic design of underground structure engineering, and Prof. Wu Chengqing, representing University of Technology Sydney, talked about his analysis and design of structures under blast loading.

We are happy to present the proceedings of ACHE 2019 covered a wide variety of topics including geotechnical engineering, structural engineering, disaster prevention & reduction and protection engineering, water resources and hydrological research, and other related areas. All paper have been through rigid review process to comply with the requirements of international publication.

We want to thank the conference Chairman, the excellent keynote speakers and all attendees for making this conference a great success. Sincere gratitude also goes to the generous sponsors for their great supports. We hope the readers around the world can enjoy the proceedings of ACHE 2019 and find some reference in it. We wish the International Conference on Architecture, Civil and Hydraulic Engineering can be held better in the future and we would like to see you next time. Thank you.

The Committee of ACHE 2019

IOP Conf. Series: Materials Science and Engineering **794** (2020) 011001 doi:10.1088/1757-899X/794/1/011001

Committee member

Chairman of the Seminar
Liaojun Zhang, Hohai University

Speaker at the seminar
Jingquan Wang, Southeast University
Chengqing Wu, University of Technology Sydney
Liaojun Zhang, Hohai University
Degao Zou, Dalian University of Technology

Academic Committee
Yingjian Chen, Changjiang Institute of Survey, Planning, Design and Research (CISPDR)
Guanghu Jin, Yanbian University
Qingying Lui, Xi'an Peihua University
Zheng Li, China Architecture Design & Research Group
Xin Li, Beijing General Research Institute of Mining &Metallurgy
Yiwen Qian, Zhejiang Institute of Hydraulics & Estuary
Genquan Qin, Sichuang Technology Co., Ltd
Ji Ren, Zhejiang Design Institure of Water Conservancy & Hydro-electric Pow
Junying Rao, Guizhou University
Shoufeng Tang, China University of Mining and Technology
Hua Wang, Yellow River Institute of Hydraulic Research
Wei Xu, Yellow River Institute of Hydraulic Research
Haoming Yang, Yellow River Institute of Hydraulic Research
Lanlin Zou, Wuhan University of Science and Technology
Yingjie Zhao, Zhejiang Design Institute of Water Conservancy & Hydro-electric Pow
Cheng Zhang, China University of Geosciences

IOP Conf. Series: Materials Science and Engineering **794** (2020) 011002 doi:10.1088/1757-899X/794/1/011002

Peer review statement

All papers published in this volume of *IOP Conference Series: Materials Science and Engineering* have been peer reviewed through processes administered by the proceedings Editors. Reviews were conducted by expert referees to the professional and scientific standards expected of a proceedings journal published by IOP Publishing.

IOP Conf. Series: Materials Science and Engineering **794** (2020) 012001 doi:10.1088/1757-899X/794/1/012001

A construction method of water conservancy model library based on Microservice

Kuaile Feng[1,2], Jianzhong Zhou[1,2*], Yi Liu[1,2], Feifei He[1,2], Zhongzheng he[1,2] and Benjun Jia[1,2]

[1]School of Hydropower and Information Engineering, Huazhong University of Science and Technology, Wuhan, Hubei, 430074, China

[2]Hubei Key Laboratory of Digital Valley Science and Technology, Wuhan, Hubei, 430074, China

*Corresponding author's e-mail: jz.zhou@hust.edu.cn

Abstract. With the development of network information technology, the water conservancy industry is faced with the new situation of big data and informatization, and the water resources informatization is getting more and more attention. The water conservancy model library is the core part of water conservancy informatization, which provides many functions, including runoff forecast, reservoir operation, accuracy assessment and so on. This paper makes full use of mainstream software science and technology to construct the architecture of water conservancy model library based on microservice, uses Json as the interactive data format of the model, and encapsulates the water resources model as a service. The model library has the advantages of scalability and high availability.

1. Introduction

With the development of network information technology, the water conservancy industry is faced with the new situation of big data[1] and informatization, and the water resources informatization is getting more and more attention[2]. The traditional decentralized water conservancy model management has been unable to meet the new needs of informatization. Water resources informatization has various contents, including information collection, data processing, data mining[3], water conservancy calculation, decision support and so on.

Among them, the water conservancy model library is the core part of water conservancy informatization. Water conservancy model library is an organic combination of various water conservancy related models, including runoff prediction model[4], reservoir operation model, forecast evaluation model, flood operation[5] and so on. The water conservancy model library publishes models as resources for external use. And at the same time, models can be invoked to support more complex business functions. All kinds of water conservancy models simulate the collected data, obtain corresponding model results, and help and guide decision-makers in water resources management.

It is necessary to make full use of advanced software scientific achievements to build water conservancy model library. By adopting the model architecture of microservice, the lightweight communication data format and model interaction method, the water conservancy model library can have good scalability and high availability.

2. Model library architecture

There are many kinds of traditional professional water conservancy models, and each model has specific application scenarios, which makes it difficult to organize and manage these models. The construction of model library mainly faces the following problems:

- Provides common internal data access interface
- Provides common external model access interface
- Provides method of communication between models

In order to solve the above problems, we need to make full use of the latest achievements of network information technology. We build high performance water conservancy model library with microservice architecture, implement the encapsulation of water conservancy professional model with Restful (Representational State Transfer) interface[6]. And Json (JavaScript Object Notation) data format[7] is used to realize the communication between water conservancy models. The architecture of water conservancy model library is shown in the figure below:

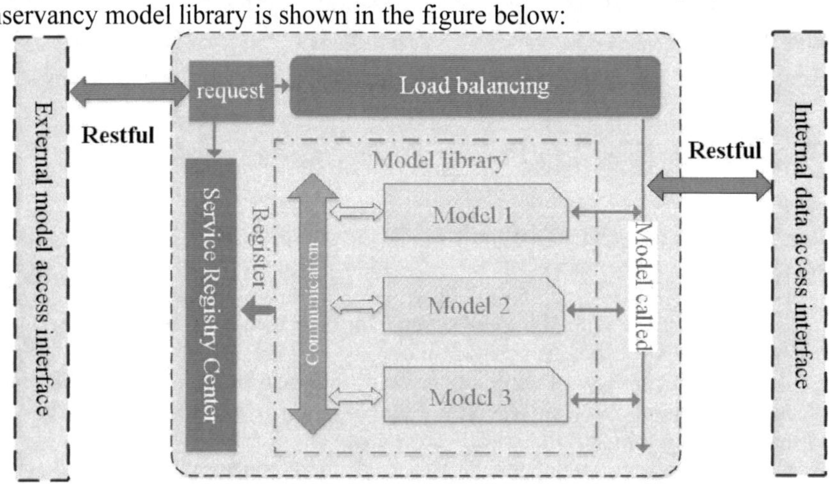

Figure 1. The architecture of water conservancy model library.

We encapsulate the water model as a service and provide Restful interfaces, including external call interfaces and data service interfaces. To ensure high availability and stability, each model service is published as a small application cluster with an accessible application interface.

3. Microservice

Microservice architecture is a way to develop a single application as a set of small services, each running in its own process and communicating with lightweight mechanisms (usually HTTP resources). Microservice architecture is a promising implementation of service-oriented architecture with recognized advantages in terms of modularity and continuous delivery[8]. Services are built around internal business functions and can be deployed independently, which is the mainstream technology to develop models with high reusability, high scalability and high maintainability. Therefore, microservice is an ideal architecture for constructing water conservancy model library. The basic microservice architecture is shown in figure 2.

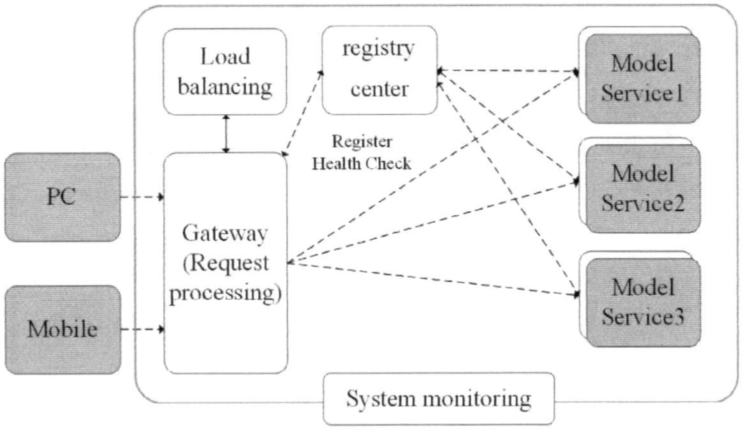

Figure 2. Basic architecture of microservice.

The microservice architecture manages water conservancy model services as interface resources, and each water model service can be accessed through a URL (Uniform Resource Locator) path[9]. As shown in figure 2, the basic microservice architecture consists of multiple modules. Where the gateway forwards the external request to the water model service and returns the model result. The load balancing module distributes the external requests to the water conservancy model service evenly and disperses the pressure of each model service. The registry center is responsible for maintaining and managing running water conservancy models. The system monitoring module is responsible for system flow monitoring, performance monitoring, error handling, etc.

4. Model communication

Traditional Web services, with SOAP protocol[10] as the core, package each model into a linkable component that conforms to a certain interface standard and serve as an analog computing service. As the standards have become more numerous, SOAP has become more bloated and more complex to use. REST technology makes up for the shortcomings of SOAP. Restful Web services are an increasingly popular way to interact with data[11]. Web services based on REST architecture style have the advantages of addressable, standard universality and connectivity. The restful style of interfaces in shown in the figure below.

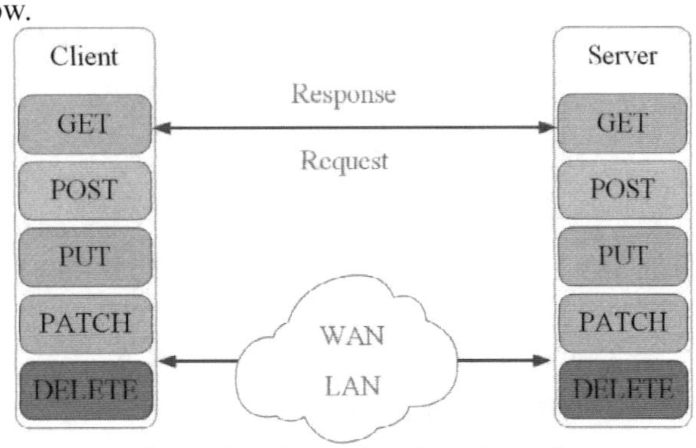

Figure 3. Schematic of restful style.

Rest-based Web service is a kind of lightweight implementation of Web service. Compared with traditional single software architecture, it has excellent compatibility to various heterogeneous platforms. It includes the following set of specifications:

(1) A Uniform Resource Identifier has an ID defined for all resources.
(2) Link all resources together.
(3) REST emphasizes a uniform interface between components
(4) Stateless communication

IOP Conf. Series: Materials Science and Engineering **794** (2020) 012001 doi:10.1088/1757-899X/794/1/012001

(5) Rest interface is scalable

Compared with the traditional system integration scheme, the water conservancy model library based on REST architecture is more highly available and reliable. The URI defined by REST style can not only realize the mutual invocation of each module of the system, but also improve the development efficiency of new functional modules.

5. Model interface

The core of establishing the interface of water conservancy model library is to define the interactive data format of the model. Json is a lightweight data interchange format[12]. It stores and represents data in a completely programming language-independent text format. It is easy to read and write, but also easy to machine analysis and generation, and effectively improve the efficiency of network transmission. The basic Json format specification is shown in figure 4.

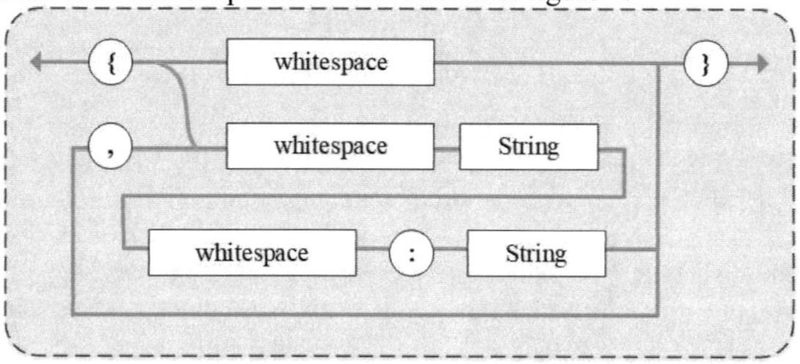

Figure 4. Json format specification.

Water conservancy models need to receive input and return results, which require a common data format to define interfaces. Json is used to define input and output parameter formats of various hydraulic models, which can quickly and efficiently encapsulate and interact the models.

6. Data access interface

Data resource is the basis of the model library. All water conservancy models need to input data, and there are many kinds of water conservancy data, including terrain, vegetation, climate, runoff, economy and other data. At the same time, different data formats are also different. In order to provide a common data sharing method, data resources need to be encapsulated as data services. On the one hand, data resources are provided for various water conservancy models, and on the other hand, models and data are decoupled. The unified data interface is shown in figure 5.

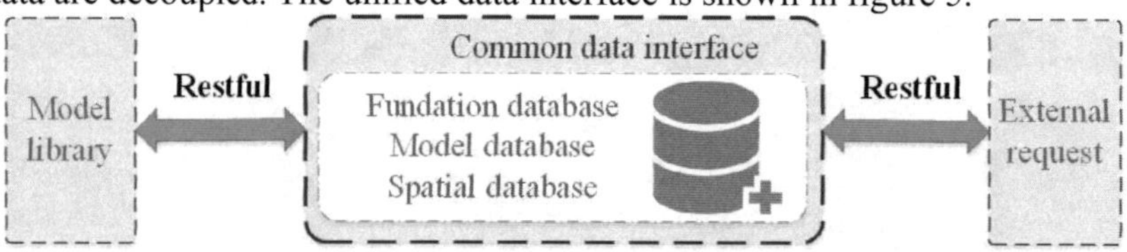

Figure 5. Common data access interface.

Data service is the main factor limiting the performance of model library, so it is necessary to conduct clustering deployment to improve the concurrency performance and the data processing efficiency of data service.

7. Summary

The water conservancy model library is the core part of water conservancy informatization. It is necessary to improve the performance of the model library. In this paper, we adopt microservice architecture, which makes different water conservancy models run and maintain independently. We adopt water model service interfaces of restful style, which provide a simple and unified method of resource access. We provide an independent data service interface to decouple data from models and improve the development and maintenance efficiency of the model library.

Acknowledgements
This research was financially supported by the National Key R&D Program of China (2016YFC0402210); the National Key R&D Program of China(2016YFC0402205); the National Natural Science Foundation Key Project of China (No. 91547208, No. U1865202) and the National Natural Science Foundation of China (No. 51579107).

References

[1] Min C, Mao S, Liu Y. (2014) Big Data: a Survey. J. Mobile Networks & Applications. 19(2): 171-209.
[2] Wang X, Zhou X. (2006) Application and Integration of Information Technology in Water Resources Informatization. In: Asia-pacific Web Conference. Harbin. pp. 1004-1009.
[3] Savic DA, Walters GA. Hydroinformatics, (2013) Data Mining and Maintenance of Uk Water Networks. J. Anti-corrosion Methods and Materials. 46(6): 415-425.
[4] Feng K, Zhou J, Liu Y, et al. (2019) Hydrological Uncertainty Processor (HUP) with Estimation of the Marginal Distribution by a Gaussian Mixture Model. J. Water Resources Management. 1-16.
[5] Liu Y, Qin H, Mo L, et al. (2019) Hierarchical flood operation rules optimization using multi-objective cultured evolutionary algorithm based on decomposition. J. Water resources management. 33(1): 337-354.
[6] Lablans M, Borg A, Ückert F. (2015) A Restful Interface to Pseudonymization Services in Modern Web Applications. J. Bmc Medical Informatics & Decision Making. 15(1): 2.
[7] Wang G. (2011) Improving Data Transmission in Web Applications Via the Translation Between Xml and Json. In: Third International Conference on Communications & Mobile Computing. pp. 182-185.
[8] Shami A. (2019) Exploring Microservices as the Architecture of Choice for Network Function Virtualization Platforms. J. IEEE Network. 33(2): 1-9.
[9] Ducut E, Liu F, Fontelo P. (2008) An Update on Uniform Resource Locator (url) Decay in Medline Abstracts and Measures for Its Mitigation. J. Bmc Medical Informatics and Decision Making. 8(1): 23.
[10] Tsenov M. (2002) Application of Soap Protocol in E-commerce Solution. In: Intelligent Systems, First International IEEE Symposium. Bulgaria. pp. 59-62.
[11] Lanthaler M, Gütl C. (2011) A Semantic Description Language for Restful Data Services to Combat Semaphobia. In: 5th IEEE International Conference on Digital Ecosystems and Technologies. Daejeon, Korea. pp. 47-53.
[12] Rodrigues C, Afonso J, (2011) Tomé P. Mobile Application Webservice Performance Analysis: Restful Services with Json and Xml. In: International Conference on ENTERprise Information Systems. Beijing. pp. 162-169.

The Optimal Utilization of Flood in Small and Medium magnitude for Cascade Reservoirs in Jinsha River

Chen Fang[1], Wang Xuemin[2]

[1]Bureau of Hydrology, Changjiang Water Resources commission, Wuhan 430010, China

[2]Changjiang Survey, Planning, design and Research CO., LTD., Wuhan 430010, China

[*]Corresponding author's e-mail: 1103448274@qq.com

Abstract. The large reservoirs generally undertake flood control, power generation, navigation and many other scheduling tasks. In flood season, flood control is the primary task for reservoirs. Then, how to realize effective utilization of the flood control capacity for flood interception and take into account other comprehensive benefits simultaneously, has become an urgent problem to be solved. In order to solve the problem, a multi-objective optimization model which can give full play to the comprehensive benefits under the premise of guaranteeing the safety of flood control is proposed. With the case study of Xiluodu and Xiangjiaba cascade reservoirs (XXCR), a practical scheme of small and medium-sized floods' utilization (SMFU) is presented. Finally, the SMFU scheme is applied to the operation in comparison with the original design scheme to verify its effectiveness. The results indicate that the SMFU scheme can improve power generation and navigation guaranteed rate while ensuring flood control safety, which adequately indicates that SMFU is an effective approach to make full use the flood resources in flood season.

1. Introduction

With the growing energy demand, more and more key reservoirs are being built in the world, especially in developing countries. These reservoirs not only contribute to flood control, but also play important roles in power generation, water supply and navigation, etc. Some researchers have devoted themselves to the economic dispatching of cascade reservoirs[1]-[4], and some have spent lots of time on flood control which is extremely important for the safety of people's life and property[5]-[8]. Besides, lots of attentions have been paid to water supply, navigation and ecological environment[9]-[12].

Most of key reservoirs generally undertake all the tasks above. In the design schemes of the reservoirs, the highest priority is assigned to flood control for safety, while power generation and navigation comes after. Their operation water levels can't exceed flood limited water level (FLWL) unless devastating flood happens, and they must be reduced to FLWL as soon as possible once the flood peak passed. The design scheme pays too much attention to flood control, reserving too much storage for rare devastating flood. The flood control storage can't be utilized effectively.

In order to manage flood reasonably, the flood is generally classified by flood peak or flood volume. According to the Chinese Standard GB/T 22482-2008 (AQSIQ and SAC 2008), the flood whose return period is less than 5 years is defined as small flood. The flood whose return period is in the range of 5 years to 20 years is defined as medium flood. The flood whose return period is more than 20 years is defined as extreme flood. When dealing with a flood, different measures will be adopted according to the magnitude of the flood.

Considering flood seasonal variation, a seasonal FLWL method giving multiple FLWLs in periods has been proposed to supply more storage for beneficial utilization[13]-[14]. For making full use of limited flood control storage, some operation strategies[15]-[17] as well as optimization models[18]-[20] have also been developed. These researches have already received various degrees of success. However, they mainly focused on flood control for devastating flood, not suitable for daily operation when small and medium-sized floods happen.

As 3~7 days runoff forecasting technology matures, its application on reservoir operation becomes more and more meaningful[21]-[23]. Instead of the traditional fixed FLWL, some dynamic control of FLWL methods have been studied and proposed[24]-[26]. The operation water level of reservoir changes real-time with the current inflow and forecasting inflow. The operation process of reservoir will be optimized with more time and information provided by 3~7 days runoff forecasting.

Xiluodu and Xiangjiaba cascade reservoirs (XXCR), located in the lower reaches of the Jinsha River, constitute one of the largest cascade systems. As the main power sources of "West Power to East" project, XXCR supply plenty of electricity for many provinces such as Hubei, Guangdong and Shanghai. Meanwhile, they also have to reserve enough storage for the safety of projects and downstream large cities in flood season. Moreover, the downstream navigation requires that Xiangjiaba should have a reasonable water discharge. Though XXCR undertake these tasks at the same time, the current design scheme mainly focuses on flood control without considering beneficial utilization in flood season.

Compared with rare devastating flood, small and medium-sized floods frequently happen in flood season. If part of flood control storage is supplied for beneficial utilization in normal flood, the integrated benefits will be sharply increased while flood control safety is insured. Therefore, a multi-objective optimization model is proposed in this paper to balance flood control and beneficial utilization. With Xiluodu and Xiangjiaba cascade reservoirs (XXCR), a practical scheme of small and medium-sized floods' utilization (SMFU) is presented. Finally, the proposed SMFU scheme is applied to the operation in comparison with the design scheme to verify its feasibility and effectiveness.

2. Problem formulation

SMFU is an effective approach to realize the tradeoff between flood control and beneficial utilization in flood season. Because of multiple objectives and various constraints, SMFU is a complex multi-objective optimization problem which can be formulated as follows:

2.1Objective function

Most of key reservoirs generally undertake the tasks of comprehensive utilization such as flood control, power generation and navigation. The objective of flood control is to determine water level and water discharge for the safety of the project and protective zones. The runoff in flood season is often much bigger than annual runoff, which can be utilized fully for power generation. Moreover, navigation should be taken into account if the downstream reach plays an important role in transportation. The objectives are described as follows:

(1) Upstream flood control safety. The water level of reservoir should be kept as low as possible to ensure the safety of project and upstream areas.

$$\min F_1 = \max\{Z_t\} \qquad t \in [1,T] \tag{1}$$

where F_1 denotes the maximum water level over all periods, the minimum of F_1 is to ensure upstream safety; Z_t denotes the water level in the t-th period; T is the number of periods.

(2) Downstream flood control safety. The water discharge of reservoir should be kept as small as possible to prevent downstream protective zones from drowning.

$$\min F_2 = \max\{Q_t\} \qquad t \in [1,T] \tag{2}$$

where F_2 denotes the maximum water discharge over all periods, and the minimum of F_2 is to ensure downstream safety; Q_t denotes the water discharge in the t-th period.

(3) Maximize total power generation. With plenty of flood resources in flood season, the reservoir can store certain flood to raise its water level for more power generation.

$$\max F_3 = \sum_{t=1}^{T} N_t \Delta t = \sum_{t=1}^{T} A H_t q_t \Delta t \tag{3}$$

where F_3 denotes the total power generation over all periods; A denotes output coefficient; Δt is the interval of scheduling; N_t, H_t and q_t denote the output, water head and power discharge in the t-th period, respectively.

(4) Maximize navigation guaranteed rate. Downstream navigation requires an appropriate range of water discharge. If the water discharge is outside the range, the reach will be not suitable for navigation. So it should be kept in the range as far as possible.

$$\max F_4 = \frac{1}{T}\sum_{t=1}^{T} K_t \quad \text{and} \quad K_t = \begin{cases} 1 & Q_t \in [Q_{\min}^{nav}, Q_{\max}^{nav}] \\ 0 & Q_t \notin [Q_{\min}^{nav}, Q_{\max}^{nav}] \end{cases} \tag{4}$$

where F_4 denotes the guaranteed rate over all periods; K_t is an intermediate variable; $Q_{\min}^{nav}$ and $Q_{\max}^{nav}$ denote the minimum and maximum water discharges for navigation, respectively.

2.2Constraints

There are kinds of equality and inequality constraints to insure the feasibility of results during the optimization process. These constraints are presented as follows:

(1) Water level constraints

$$Z_{t,\min} \le Z_t \le Z_{t,\max} \tag{5}$$

where $Z_{t,\min}$ and $Z_{t,\max}$ denote the lower and upper water level limits, respectively.

(2) Water discharge constraints

$$Q_{t,\min} \le Q_t \le Q_{t,\max} \tag{6}$$

where $Q_{t,\min}$, $Q_{t,\max}$ denote the minimum and maximum water discharge limits, respectively.

(3) Output constraints

$$N_{t,\min} \le N_t \le N_{t,\max} \tag{7}$$

where $N_{t,\min}$ and $N_{t,\max}$ denote the minimum and maximum output limits, respectively.

(4) Water balance equation

$$V_{t+1} = V_t + [I_t - Q_t]\Delta t \tag{8}$$

where I_t and V_t denote inflow and water storage, respectively.

(5) Water head equation

$$H_t = (Z_t + Z_{t+1})/2 - f_{zd}(Q_t) \tag{9}$$

where f_{zd} denotes the relation function between water discharge and downstream water level.

(6) Water spillage equation

$$\begin{cases} q_t = \min\{Q_t, q_{\max}\} \\ S_t = \max\{Q_t - q_{\max}, 0\} \end{cases} \tag{10}$$

where S_t denotes the water spillage in the t-th period; $q_{\max}$ denotes power discharge capacity.

(7) Initial water level

$$Z_0 = Z_{begin} \tag{11}$$

3. Operation demands of XXCR

Located in the lower reaches of Jinsha River, Xiluodu is an important reservoir with annual regulation ability. Xiangjiaba located downstream from Xiluodu about 150km is another key project. Considering the closely hydraulic connection, XXCR will obtain more benefits if the two reservoirs operate as a complex by joint dispatching. Moreover, some large cities such as Yibin and Luzhou are located in the downstream reaches of XXCR. For the safety of these cities, the safe water discharges of XXCR are required in flood season. The basic parameters and topological structure of XXCR are shown as Table 1 and Fig.1.

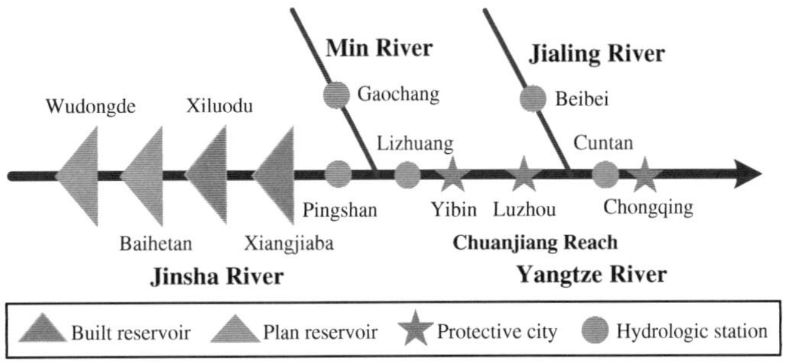

Figure 1. The topological structure of rivers, reservoirs, protective cities and hydrologic stations in the lower reaches of Jinsha River

Table 1. The basic parameters of XXCR

Reservoir	Normal water level (m)	Dead water level (m)	Flood limited water level (m)	Regulation storage (m^3)	Flood control storage (m^3)	Installed capacity
Xiluodu	600	540	560	64.6×10^8	46.5×10^8	13500 MW
Xiangjiaba	380	370	370	9.03×10^8	9.03×10^8	6400

3.1Operation rules of XXCR

Xiluodu is a key project privileged to power generation, incorporating with flood control and navigation. Xiangjiaba is a project with undertaking the tasks of power generation, navigation and flood control. As the re-regulation reservoir of Xiluodu, it should cooperate with Xiluodu during operation. According to the "Reservoir Dispatching Rules of Xiluodu" and "Reservoir Dispatching Rules of Xiangjiaba" approved by government departments, XXCR must follow some basic rules during normal operation.

(1) Flood control rules

From July 1 to September 10, XXCR undertake flood control task to ensure the safety of Chuanjiang Reach, and cooperate with the Three Gorges for the safety of Jingjiang Reach. XXCR have to supply $10\times10^8 m^3$ as reserved storage for devastating flood although they have $55.53\times10^8 m^3$ flood control storage in flood season. In general, XXCR keep their water levels under 560/370m unless they execute flood control task to reduce flood. Considering the actual operation factors, their water levels can rise to 561/372.5m within a short time, but they must be reduced to FLWL as soon as possible before flood comes. Once the flood peak passed, XXCR have to reduce their water levels timely while not increasing flood control pressure. These flood control rules can be considered when the SMFU scheme is drafted.

(2) Power generation rules

XXCR also undertake power generation and peaking modulation besides flood control. Considering the safety of projects and protective zones, power generation has lower priority than flood control in flood season. The water levels of XXCR are determined by flood control rules rather than power generation rules in flood season. However, they can vary from their dead water levels 540/370m to their normal water levels 600/380m in other seasons. XXCR start to impound water in early September, and their water levels will reach to 600/380m in late September. Because of water supply, their water levels gradually drop to 540/370m in dry season, and they will rise to FLWL 560/370m again in late June as flood season comes.

(3) Navigation rules

The Jinsha River is an important water corridor to transport regional resources. As the key projects on the river, XXCR have the responsibility to meet the navigation requirements of the Lower Jinsha River. According to the dispatching rules, navigation also submits to flood control in flood season, and it should be coordinated with power generation. Therefore, the rules for flood control and power generation can't be neglected when the navigation scheme is drafted. For the navigation safety of Lower Jinsha River, the tail water level of Xiangjiaba should be kept in the range of 265.8~277.25m relating to the water discharge of 1200~12000 m^3/s. Moreover, the daily variation of the tail water level is required less than 4.5m/d for avoiding strong fluctuation, and the hourly change should be less than 1.5m/h.

3.2Operation constraints of XXCR

Because of different requirements of water level, flood control and beneficial utilization are conflicting in flood season. Therefore, XXCR have to satisfy various constraints for normal operation. With the above operation rules, some important constraints are summarized.

(1) Upstream flood control constraints

According to the flood control rules, XXCR must supply 10×10^8 m^3 as reserved storage for devastating flood. Considering their different flood control storages, Xiluodu is suitable to supply the whole reserved storage for maximizing the total power generation. Therefore, the operation water level of Xiluodu can't exceed 593m in normal flood, and the storage between 593m and 600m is reserved for devastating flood. Meanwhile, the operation water level of Xiangjiaba is kept in the range of 370m to 380m.

(2) Downstream flood control constraints

Some large cities such as Yibin and Luzhou are located in Chuanjiang Reach. Their flood control safety requires the reasonable water discharges of XXCR. The warning water level of these cities is 300m, and the corresponding runoff is 24500m^3/s. The safe water level of these cities is 303m, and the corresponding runoff is 29000m^3/s. Thus, the final water discharge of XXCR (that is, the water discharge of Xiangjiaba) can't exceed 24500m^3/s in normal flood, and it can't exceed 29000m^3/s in devastating flood.

(3) Power generation constraints

Most of the inflows exceed the power discharge capacities of XXCR in flood season. The water discharge makes little sense to the improvement of power generation. In other words, the power generation can't be increased unless the operation water level is raised. However, upstream flood control requires low operation water level for safety. To balance flood control and power generation, the SMFU scheme won't be taken unless the current inflow exceeds power discharge capacity. The power discharge capacity of Xiluodu is 7500 m^3/s. The power discharge capacity of Xiangjiaba is 6500m^3/s.

(4) Navigation constraints

Besides flood control and power generation, XXCR also have to ensure the navigation safety of Chuanjiang Reach (that is, Lower Jinsha River). Therefore, they have to keep the final water discharge in the range of 1200m^3/s to 12000m^3/s. Moreover, the daily variation of the tail water level of Xiangjiaba can't exceed 4.5m/d, which requires that the daily variation of the final water discharge should be kept under 4500m^3/s.

4. The SMFU scheme for XXCR

Because of practical demands and limited leading time, the SMFU model in Section 2 can't be optimized effectively during actual operation. Compared to an unpractical optimal scheme, a reasonable scheme which takes all objectives and various constraints into account is more meaningful. Here, a practical SMFU scheme for XXCR is presented.

4.1Inflow characteristics of XXCR

The flood of Jinsha River not only happens frequently, but also lasts for a long time with huge volume. SMFU is an effective approach to make full use of flood resources, but it is relative to flood characteristics closely. We can't draft the SMFU scheme for XXCR without the detailed analysis of flood characteristics. According to the historical runoff data in flood season from 1940 to 2010, the probabilities of different inflows are listed in Table 2:

Table 2. The interval and cumulative probabilities of different inflows in flood season

Inflow range (m^3/s)	Probability	Cumulative	Inflow range (m^3/s)	Probability	Cumulative
0~6500	18.76%	18.76%	16000~17000	1.82%	94.60%
6500~7500	12.52%	31.28%	17000~18000	1.76%	96.36%
7500~9000	17.62%	48.90%	18000~19000	1.25%	97.61%
9000~10000	9.84%	58.74%	19000~20000	0.96%	98.57%
10000~11000	8.96%	67.70%	20000~21000	0.49%	99.06%
11000~12000	7.03%	74.73%	21000~22000	0.41%	99.47%
12000~13000	6.31%	81.04%	22000~23000	0.24%	99.71%
13000~14000	5.54%	86.58%	23000~24500	0.13%	99.84%
14000~15000	3.72%	90.30%	24500~29000	0.16%	100.00%
15000~16000	2.48%	92.78%	>29000	0%	100.00%

As the results listed in Table 2, 81.24% of the inflows exceed 6500m^3/s, and 68.72% of the inflows exceed 7500m^3/s. Most of the inflows can meet the requirements for the expected outputs of XXCR. In other words, power generation is mainly affected by water level instead of water discharge in flood season. Thus, the SMFU scheme can improve power generation by raising water level. Moreover, 74.73% of the inflows are smaller than 12000m^3/s, which indicates that the navigation requirements can be met most of the time.

With 3~7 days runoff forecasting, A pre-impound & pre-release scheme can be taken to balance power generation and flood control in small flood. XXCR will pre-impound water to raise their water levels when the inflow is smaller than 12000m^3/s. On the contrary, they will pre-release stored water for flood control when the forecasting inflows exceed 12000m^3/s. For further analysis, the durations of the floods exceeding 12000m^3/s are listed in Table 3:

Table 3. The durations of different historical floods in flood season

Flood peak discharge (m^3/s)	Flood duration (days)			
	<10 days	10~20 days	20~30 days	>30 days
12000~15000	6	28	16	13
15000~18000	1	14	10	5
18000~21000	0	8	6	4
21000~24500	1	0	2	7
>24500	0	1	1	0
Total times	8	51	35	29

As it can be seen from the results in Table 3, 51 floods last about 10~20 days, and 64 floods last more than 20 days. The flood duration of XXCR is very long, and it will be longer as the flood peak flow increases. However, the available storage of XXCR is limited in flood season, which prevents from reducing flood peak adequately. The SMFU scheme for XXCR has to balance the current utilization and later operation during flood control stage.

4.2Availability of inflow forecasting

Inflow forecasting is an important factor for the pre-impound & pre-release scheme. Xiluodu started the first impounding in May 2013. Xiangjiaba started the first impounding in October 2012. XXCR have been in normal operation for a short time, few researches on the inflow forecasting were presented. The leading time of the inflow forecasting for XXCR is 12~48h instead of 3~7 days at present. According to the 12~48h forecasting results from May 2013, the accuracies of the forecasting inflows for XXCR in flood season are listed in Table 4.

Table 4. The accuracy of 12-48h inflow forecasting for XXCR in flood season (%)

Leading time	2013			2014			Average
	July	August	September	July	August	September	
12 h	97.79	98.1	98.45	97.55	97.19	98.81	97.98
24 h	96.62	98.33	98.11	96.72	97.65	98.36	97.63
48 h	93.65	97.26	95.46	92.73	95.15	96.44	95.12

As Table 4 shows, XXCR have excellent 12~48h inflow forecasting results. The average accuracy of the 48h forecasting inflows reaches to 95.12%, which can satisfy the accuracy demand of inflow forecasting for the SMFU scheme completely. As the owner of XXCR, the China Three Gorges Corporation has been trying to prolong leading time although 3~7 days inflow forecasting for XXCR hasn't been done at present. The SMFU scheme for XXCR will be more reasonable and effective with 3~7 days forecasting inflows.

The accurate 3~7 days inflow forecasting is a hotspot issue, which has attracted lots of researchers in recent years. Bao[22] have coupled ensemble weather predictions with Grid-Xinanjiang model for the flood forecasting of Xixian catchment, obtaining excellent 6-day flood forecasting results. Webster[27] have proposed a new ensemble flood prediction scheme to realize 10-day forecasting floods in Ganges and Brahmaputra. Bürger[21] and Gouweleeuw[23] have obtained perfect forecasting results within 5-day leading time by ECMWF Ensemble Prediction System. According to these researches, ensemble prediction is an effective approach to realize the 3~7 days inflow forecasting for XXCR. Therefore, the SMFU scheme for XXCR will be reasonable and effective.

4.3Pre-impound & pre-release scheme

The pre-impound & pre-release scheme will be taken when the current inflow is smaller than 12000m^3/s. The upper water level limits of XXCR are key factors to balance flood control and power generation in the scheme. If they are too high, the operation water levels won't be reduced timely without affecting navigation. On the contrary, the benefits of power genera- tion won't be realized adequately. For determining the upper water level limits, three schemes 565/375m, 570/375m and 575/375m are selected as candidates. Based on 3~7 days leading time, the extra water discharges of XXCR in the three schemes are listed in Table 5:

Table 5. The extra water discharges of XXCR with different leading times and upper water level limits

Upper water level limits (m)	Pre-release water volume (10^8 m^3)	Extra water discharge (m^3/s)				
		3 days	4 days	5 days	6 days	7 days
565/375	4.92/4.38	1898/1690	1424/1267	1139/1014	949/845	813/724
570/375	10.30/4.38	3974/1690	2980/1267	2384/1014	1987/845	1703/724
575/375	15.91/4.38	6138/1690	4604/1267	3683/1014	3069/845	2631/724

As Table 5 shows, if the upper water level limits are set to 575/375m, the total extra water discharge of XXCR will reach to 3350m^3/s within 7 days leading time, and it will increase to 7800m^3/s within 3 days leading time. The extra water discharges are so large that the water levels can't be reduced timely. The extra water discharges are small within all leading times in 565/375m scheme, but low water levels are inefficient for power generation. Compared to other schemes, the extra water discharges and operation water levels of XXCR are acceptable in 570/375m scheme. The scheme can balance power generation and flood control, thus the upper water level limits of XXCR are selected to 570/375m.

Analyzed in above description, XXCR won't impound water unless their power discharge capacities are satisfied, and they will release stored water if any forecasting inflow exceeds 12000m^3/s. The inflow is classified by the grades of 6500, 7500 and 12000 in pre-impound & pre-release scheme. Besides, a grade of 10000 is added to satisfy the daily variation of water discharge. The detailed pre-impound & pre-release scheme is summarized as follows:

a) When the current inflow I<6500m^3/s, XXCR keep their water levels 560/370m, and their water discharges equal the inflow; b) When 6500≤ I<7500m^3/s, XXCR keep their water levels under 560/375m, and their water discharges equal inflow/6500 m^3/s; c) When 7500≤ I<10000m^3/s, XXCR keep their water levels under 570/375m, and their water discharges equal 7500/6500 m^3/s; d) When 10000≤ I<12000m^3/s, XXCR keep their water levels under 570/375m, and their final water discharge equals 10000m^3/s; e) When the forecasting inflow within leading time I_{fore}≥12000m^3/s, XXCR reduce their water levels to FLWL 560/370m as soon as possible while keeping their final water discharge under 12000m^3/s.

4.4Detailed flood control scheme

According to the design schemes of XXCR, they will not impound flood unless the current flood exceeds 24500m^3/s. In such a way, the flood control storage can't be utilized adequately. Their water levels are kept FLWL 560/370m most of the time, which leads to poor power generation. Their water discharges are very big in most floods, the flood control safety of downstream protective zones can't be ensured effectively. Therefore, a detailed flood control scheme is proposed to realize the trade off among different demands.

In this flood control scheme, XXCR not only prevent the flood exceeding 24500 m^3/s, but also utilize the flood in the range of 12000~24500m^3/s. With limited flood control storage, XXCR can't reduce all floods to 12000m^3/s for navigation, but reduce flood peak moderately. So the range of 12000~24500 is divided into several small ranges by the grades of 15000, 18000 and 21000. According to these small ranges, XXCR will impound flood by the lower limit of range, and release flood by the upper limit of range.

XXCR can't supply all storage for the current flood during the impounding stage. They have to reserve enough storage for later more serious flood. Their operation water levels must be limited to balance the current utilization and later operation. With the historical data and the impounding rules above, the required storage for each flood in the range of [12000, 24500] is calculated to determine the upper water level limits of XXCR. These required storages are classified by the small ranges and presented as Fig.2.

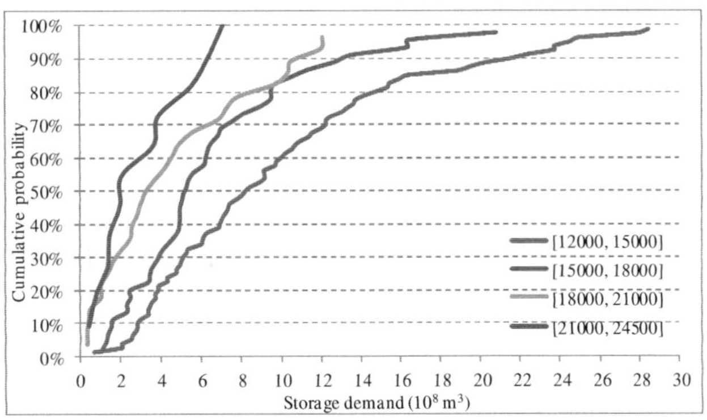

Figure 2. The cumulative probability curve of required storage for different floods in the range of [12000, 24500] m^3/s

As Fig.2 shows, some of the required storages are too big to be met with limited flood control storage. Especially, the maximum required storage for the floods in [12000, 15000] is nearly 30×10^{8}m^3, and that for the floods in [15000, 18000] exceeds 20×10^{8}m^3. The floods in the two ranges require about 50×10^{8}m^3 storage, while the total available storage of XXCR is 55.53×10^{8}m^3 in flood season. Moreover, part of the available storage must be supplied for the flood exceeding 24500m^3/s. Thus, the available storage of XXCR must be divided into several parts to reduce the floods in different ranges reasonably.

Besides the required storages above, the maximum required storage for the floods in the range of [24500, 29000] is 15.3×10^{8}m^3, which must be supplied fully for downstream flood control safety. The available storage for small and medium-sized floods will be only $30.23\times10^8 m^3$ if the reserved storage for devastating flood is deducted. Considering the navigation demands, more storage will be supplied for the floods in [12000, 15000]. Here, an allocation scheme of water level and available storage is presented as Table 6.

Table 6. The practical allocation scheme of water level and useable storage of XXCR for different floods

Flood grade (m^3/s)	Storage allocation (10^8 m^3)	Cumulative storage (10^8 m^3)	Upper water level limits (m)		Useable storage (10^8 m^3)		Total useable storage (10^8 m^3)
			Xiluodu	Xiangjiaba	Xiluodu	Xiangjiaba	
[12000, 15000]	14	14	570	375	10.30	4.38	14.68
[15000, 18000]	7	21	575	375	15.91	4.38	20.29
[18000, 21000]	5	26	580	375	21.52	4.38	25.90
[21000, 24500]	4	30	583	375	25.14	4.38	29.52
[24500, 29000]	15.3	45.3	593	380	36.50	9.03	45.53
>29000	–	–	600	380	46.50	9.03	55.53

As Table 6 shows, the required storages for the floods in [24500, 29000] are both satisfied, about 75% of the required storages for the floods in [12000, 15000] are satisfied, and more than 60% of the required storages for the floods in other ranges are satisfied. The scheme can satisfy most of the required storages for the floods in other ranges, and supply enough storage for devastating flood. In such a way, XXCR will impound flood under the maximum water level limits while reserving adequate storage for later more serious flood.

4.5The SMFU scheme for XXCR

In the above sections, the pre-impound & pre-release scheme and the detailed flood control scheme are proposed for different floods. They can be summarized as the SMFU scheme for XXCR, which considers flood control, power generation and navigation simultaneously. The SMFU scheme is presented as Table 7 (When the flood exceeds $10000m^3/s$, XXCR impound or release flood by determining the final water discharge.).

Table 7. The SMFU scheme for XXCR in flood season which contains the water discharges and upper water level limits of XXCR corresponding to different floods

Flood grade (m^3/s)	The water discharges of XXCR (m^3/s)		Upper water level limits (m)		Notes
	Xiluodu	Xiangjiaba	Xiluodu	Xiangjiaba	
<6500	Inflow	Inflow	560	370	–
[6500, 7500]	Inflow	6500	560	375	Pre-impound
[7500, 10000]	7500	6500	570	375	Pre-impound
[10000, 12000]	10000		570	375	Pre-release
[12000, 15000]	12000		570	375	Navigation
[15000, 18000]	12000-15000-12000		575	375	Flood control
[18000, 21000]	12000-15000-18000-15000-12000		580	375	Flood control
[21000, 24500]	12000-15000-18000-21000-18000-15000-12000		583	375	Flood control
[24500, 29000]	12000-15000-18000-21000-24500-21000-18000-15000-12000		593	380	Flood control
>29000	12000-15000-18000-21000-24500-29000-24500-21000-18000-15000-12000		600	380	Devastating flood

The following notes can't be neglected during the practical dispatching of SMFU:

a) XXCR impound flood based on their usable storages, and release flood based on their stored water volumes; b) XXCR pre-impound water to raise the water levels when the current inflow and forecasting inflows are smaller than 12000m^3/s; c) XXCR pre-release the stored water to reduce their water levels when any forecasting inflow exceeds 12000m^3/s; d) When the current flood exceeds 12000m^3/s, XXCR impound flood by the lower limit of flood range during rising stage while keeping their water levels under the limits, release the stored flood by the upper limit of flood range during falling stage while keeping their water levels above 570/375m; e) The water discharges will be reduced if the flood peak lasts less than 3 days.

5. Application benefits analysis

The proposed SMFU scheme is applied to the dispatching of XXCR in comparison with the design scheme to verify its effectiveness. With the current runoff forecasting technology, the inflows in the next 7 days are supposed as forecasting inflows during simulation.

5.1Risk and benefits of flood control

The safety of projects and protective zones must be ensured during SMFU. Otherwise, no matter how much the benefits of power generation and navigation will make no sense. To verify the risk and benefits of flood control, the typical adverse floods in 1966 and 1998 are selected. Considering all of the historical floods are normal floods, 1% design floods are also obtained to verify flood control risk further. According to the flood characteristics, the 1% design flood of 1966 is obtained by Peak and Volume-Amplitude Method, but the 1% design flood of 1998 is obtained by Volume-Amplitude Method with 30 days flood volume. With these floods, the simulation results are presented as Fig.3 and Fig.4.

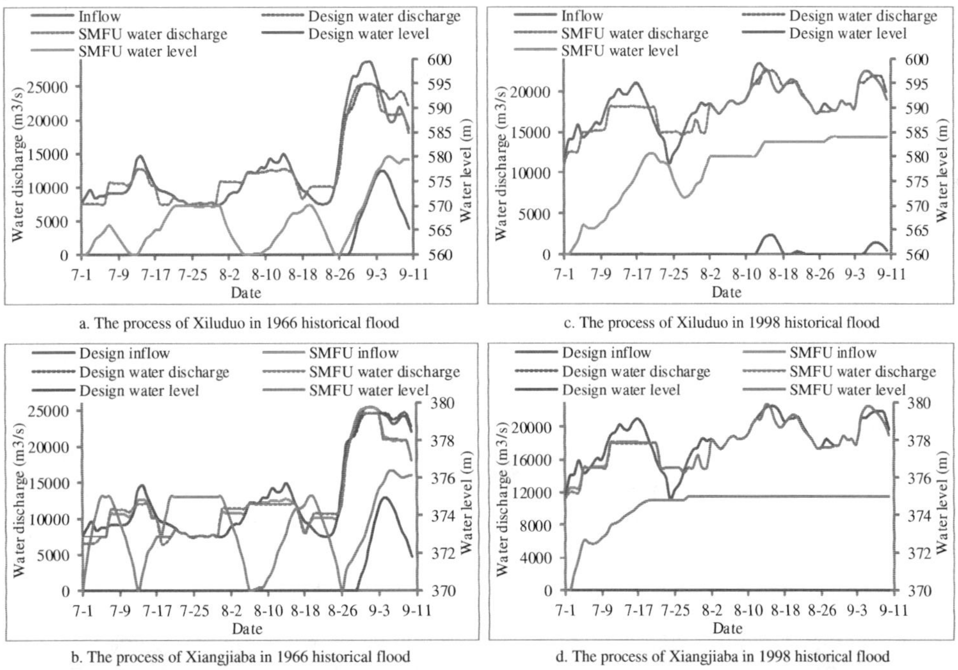

a. The process of Xiluduo in 1966 historical flood

c. The process of Xiluduo in 1998 historical flood

b. The process of Xiangjiaba in 1966 historical flood

d. The process of Xiangjiaba in 1998 historical flood

Figure 3. The operational processes (inflow, water discharge and water level) of XXCR in different historical floods in design scheme and SMFU scheme; (a) Xiluodu in 1966 historical flood, (b) Xiangjiaba in 1966 historical flood, (c) Xiluodu in 1998 historical flood, (d) Xiangjiaba in 1998 historical flood.

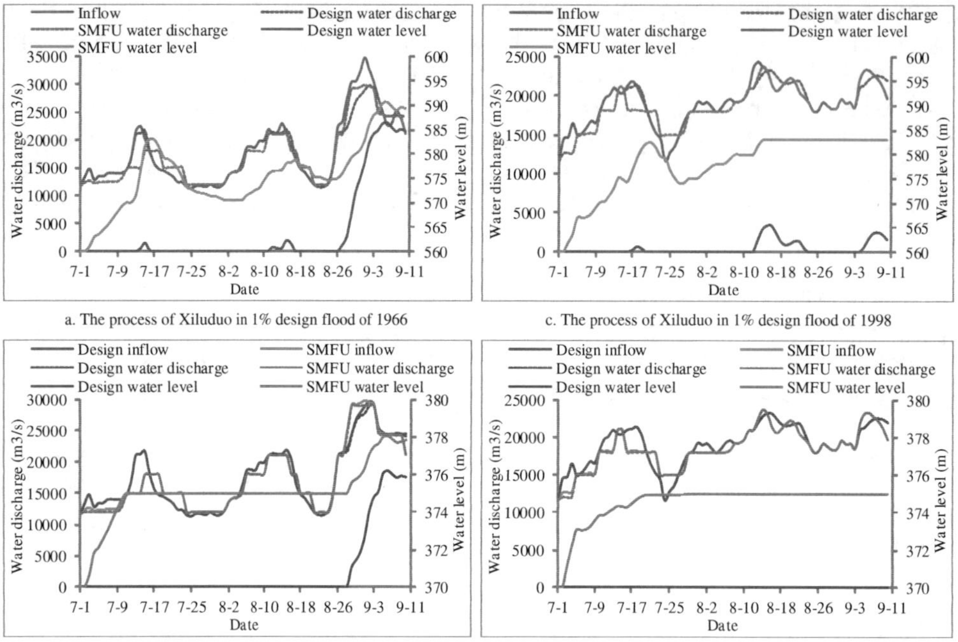

a. The process of Xiluduo in 1% design flood of 1966

c. The process of Xiluduo in 1% design flood of 1998

b. The process of Xiangjiaba in 1% design flood of 1966

b. The process of Xiangjiaba in 1% design flood of 1998

Figure 4. The operational processes (inflow, water discharge and water level) of XXCR in different design floods in design scheme and SMFU scheme; (a) Xiluodu in 1% design flood of 1966, (b) Xiangjiaba in 1% design flood of 1966, (c) Xiluodu in 1% design flood of 1998, (d) Xiangjiaba in 1% design flood of 1998.

As it can be seen from Fig.3 and Fig.4, the proposed SMFU scheme can ensure the safety of projects and protective zones in normal and devastating floods, which indicates that the scheme is feasible. In the 1966 historical flood, XXCR can reduce the water levels to FLWL 560/370m timely before serious flood comes while keeping their final water discharge under 12000m^3/s. All the inflows of Xiangjiaba exceed 12000m^3/s during the whole season of other floods. According to the detailed flood control scheme, Xiangjiaba will impound flood during the rising stage and release flood during the falling stage. Thus, its water levels rise with the increasing of flood peak before serious flood comes.

As the figures show, Xiluodu plays the main role in runoff regulation during SMFU. With different flood characteristics, the benefits of flood control are different in the floods of 1966 and 1998. XXCR can reduce each flood peak effectively in the floods of 1966, obtaining perfect effects of flood control. However, they can't reduce later flood peaks in the floods of 1998. Due to the long durations of these floods, they have to reserve enough storage for more serious flood. Moreover, their water discharges increase or decrease step by step during the flood control stage. The water levels of XXCR can be reduced to 570/375m quickly after the peak flood passed.

Compared with the design scheme, the SMFU scheme can obtain higher water levels and smaller water discharges while subjecting to kinds of constraints. In such a way, upstream projects may take certain flood control pressure, but the flood control pressure of downstream protective zones is sharply reduced. In other words, upstream safety and downstream safety are considered and balanced. Moreover, the smaller water discharges bring better navigation guaranteed rate, the higher water levels bring more power generation.

The inflow forecasting is a vital important factor for pre-impound & pre-release scheme. The accuracy will be decreased with the increasing of leading time. Therefore, the forecasting schemes with different leading times are applied to the SMFU for XXCR. The processes of Xiluodu in the 1966 historical flood with different leading times are shown as Fig.5 (Because the processes of Xiangjiaba are similar, they are not presented in the paper.):

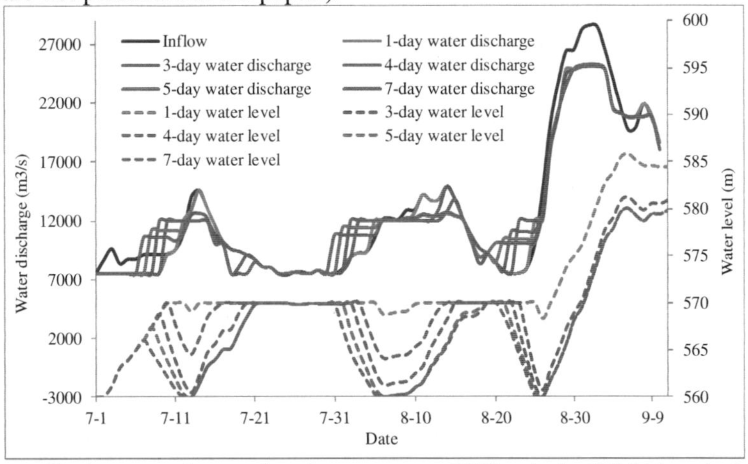

Figure 5. The water discharge and water level processes of Xiluodu in 1996 historical flood with 1-day, 3-day, 4-day, 5-day and 7-day inflow forecasting.

As Fig.5 shows, the water levels of Xiluodu can be reduced to FLWL 560m timely within 4~7 days, and they are also sharply reduced within 3 days. However, the reduction of water level is very small within 1 day. The water discharges of Xiluodu are smaller than 12000m^3/s during pre-release stage with different leading times, and they will be decreased as leading time increases. In a word, the SMFU scheme can reduce the water levels of XXCR within 3~7 days while not affecting navigation safety.

5.2Benefits of power generation and navigation

The SMFU scheme also considers power generation and navigation besides flood control. It can improve the benefits of power generation and navigation by the analysis above. In order to further quantify the benefits, the total power generation and navigation guaranteed rate of XXCR are obtained with the historical floods. The results are shown as Fig.6 (The navigation guaranteed rate is calculated by judging the water discharge of Xiangjiaba.):

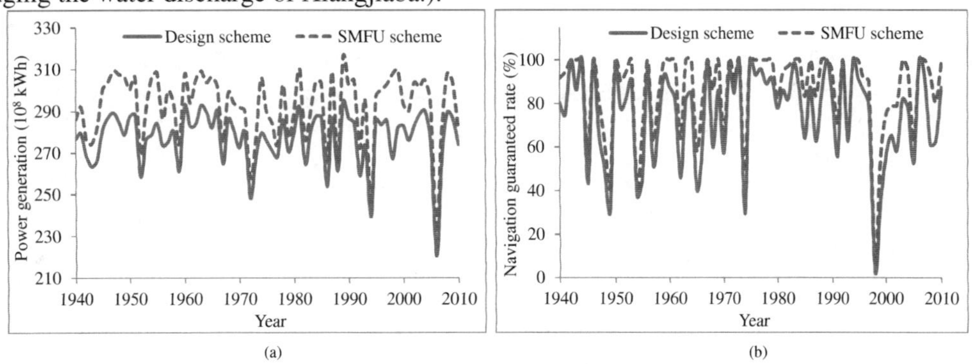

Figure 6. The results of XXCR in design scheme and SMFU scheme in the historical floods from 1940 to 2010; (a) The annual total power generation of XXCR; (b) The annual navigation guaranteed rate of XXCR.

Compared to the design scheme, the proposed SMFU scheme can improve the total power generation and navigation guaranteed rate of XXCR. According to the detailed results from 1940 to 2010, the average power generation is 276.96×10^8 kwh in the design scheme, while that is 292.54×10^8

kwh in the SMFU scheme. The power generation is improved by 15.58×10^8 kwh on average, which takes about 5.62% of that in design scheme.

Because of different inflow characteristics, the navigation guaranteed rate varies from rear to year. All the inflows are less than $12000m^3/s$ during the whole flood season in some years, the rate can reach to 100%. On the contrary, the rate is very small in other years due to devastating flood. The SMFU scheme can improve navigation guaranteed rate no matter how the rate varies every year. Compared with 75.33% in the design scheme, the mean navigation guaranteed rate is 86.38% in the SMFU scheme, which is improved by 11.05%. In a word, the SMFU scheme is benefit for the power generation and navigation of XXCR.

6. Conclusions

In order to make full use of flood resources, a multi-objective optimization model which takes flood control, power generation and navigation into account is proposed in this paper. A pre-impound & pre-release module is proposed to balance power generation and flood control. A detailed flood control module is proposed for flood control and navigation. With the model and two modules, a practical scheme of small and medium-sized floods' utilization (SMFU) for cascade reservoirs is presented.

Focusing on Xiluodu and Xiangjiaba cascade reservoirs (XXCR), the practical demands of comprehensive utilization are summarized according their dispatching rules. Based on the detailed analysis of inflow characteristics and inflow forecasting, the flood is classified by several ranges, the available storage of XXCR is divided to several parts. The parameters of pre-impound & pre-release module and the detailed flood control module are determined. The SMFU scheme for XXCR is drafted and applied to the operation.

According to the simulation results, XXCR pre-impound water to raise their water levels in small flood, and lower their water levels to FLWL 560/370m timely before serious flood comes while keeping the final water discharge under $12000m^3/s$. When facing medium flood or devastating flood, XXCR raise their water levels and increase their water discharges step by step, upstream safety and downstream safety are both ensured effectively. The total power generation is improved by 5.62%. The navigation guaranteed rate is improved by 11.05%.

The SMFU scheme can improve power generation and navigation guaranteed rate while ensuring the safety of upstream projects and downstream zones. SMFU which can realize the tradeoff among different tasks is an excellent approach to make full use flood resources.

Acknowledgments

This work is supported by the National Key Research and Development Program of China (Item Nos.2018YFC1508002). The authors also appreciate the insightful comments and suggestions from anonymous reviewers.

References

[1] Barros, M. T., Tsai, F. T., Yang, S.-l., Lopes, J. E., and Yeh, W. W. (2003). "Optimization of large-scale hydropower system operations." *Journal of Water Resources Planning and Management*, 129(3), 178-188.

[2] Li, C., Zhou, J., Ouyang, S., Ding, X., and Chen, L. (2014). "Improved decomposition–coordination and discrete differential dynamic programming for optimization of large-scale hydropower system." *Energy Conversion and Management*, 84, 363-373.

[3] Wang, J., Liu, S., and Zhang, Y. (2014). "Quarter-hourly operation of large-scale hydropower reservoir systems with prioritized constraints." *Journal of Water Resources Planning and Management.*

[4] Yeh, W. W. G. (1985). "Reservoir management and operations models: A state - of - the - art review." *Water resources research*, 21(12), 1797-1818.

[5] Ahmed, E.-S. M. S., and Mays, L. W. (2013). "Model for determining real-time optimal dam releases during flooding conditions." *Natural hazards*, 65(3), 1849-1861.

[6] Kumar, D. N., Baliarsingh, F., and Raju, K. S. (2010). "Optimal reservoir operation for flood control using folded dynamic programming." *Water resources management*, 24(6), 1045-1064.
[7] Wei, C.-C., and Hsu, N.-S. (2008). "Multireservoir real-time operations for flood control using balanced water level index method." *Journal of environmental management*, 88(4), 1624-1639.
[8] Yazdi, J., and Neyshabouri, S. S. (2012). "Optimal design of flood-control multi-reservoir system on a watershed scale." *Natural hazards*, 63(2), 629-646.
[9] Chang, J.-x., Bai, T., Huang, Q., and Yang, D.-w. (2013). "Optimization of water resources utilization by PSO-GA." *Water resources management*, 27(10), 3525-3540.
[10] Vieira, J., Cunha, M. C., Nunes, L., Monteiro, J. P., Ribeiro, L., Stigter, T., Nascimento, J., and Lucas, H. (2010). "Optimization of the operation of large-scale multisource water-supply systems." *Journal of Water Resources Planning and Management*, 137(2), 150-161.
[11] Wu, B., Chen, Y.-B., Zeng, Y., Zhao, Y., and Yuan, C. (2011). "Evaluation on effectiveness of carbon emission reduction of the power generation and shipping functions of the Three Gorges Reservoir." *Resources and Environment in the Yangtze Basin*, 20(3), 257-261.
[12] Yang, N., Mei, Y., and Zhou, C. (2012). "An optimal reservoir operation model based on ecological requirement and its effect on electricity generation." *Water resources management*, 26(14), 4019-4028.
[13] Liu, P., Guo, S., and Li, W. (2008). "Optimal design of seasonal flood control water levels for the Three Gorges Reservoir." *IAHS-AISH publication*, 270-277.
[14] Yun, R., and Singh, V. P. (2008). "Multiple duration limited water level and dynamic limited water level for flood control, with implications on water supply." *Journal of hydrology*, 354(1), 160-170.
[15] Marien, J., Damazio, J., and Costa, F. (1994). "Building flood control rule curves for multipurpose multireservoir systems using controllability conditions." *Water Resources Research*, 30(4), 1135-1144.
[16] Needham, J. T., Watkins Jr, D. W., Lund, J. R., and Nanda, S. (2000). "Linear programming for flood control in the Iowa and Des Moines rivers." *Journal of Water Resources Planning and Management*, 126(3), 118-127.
[17] Wasimi, S. A., and Kitanidis, P. K. (1983). "Real‐time forecasting and daily operation of a multireservoir system during floods by linear quadratic Gaussian control." *Water Resources Research*, 19(6), 1511-1522.
[18] Li, X., Guo, S., Liu, P., and Chen, G. (2010). "Dynamic control of flood limited water level for reservoir operation by considering inflow uncertainty." *Journal of hydrology*, 391(1), 124-132.
[19] Qin, H., Zhou, J., Lu, Y., Li, Y., and Zhang, Y. (2010). "Multi-objective cultured differential evolution for generating optimal trade-offs in reservoir flood control operation." *Water resources management*, 24(11), 2611-2632.
[20] Tilmant, A., and Kelman, R. (2007). "A stochastic approach to analyze trade‐offs and risks associated with large‐scale water resources systems." *Water resources research*, 43(6).
[21] Bürger, G., Reusser, D., and Kneis, D. (2009). "Early flood warnings from empirical (expanded) downscaling of the full ECMWF Ensemble Prediction System." *Water Resources Research*, 45(10).
[22] Bao, H., Zhao, L., He, Y., Li, Z., Wetterhall, F., Cloke, H., Pappenberger, F., and Manful, D. (2011). "Coupling ensemble weather predictions based on TIGGE database with Grid-Xinanjiang model for flood forecast." *Advances in Geosciences (ADGEO)*, 29, 61-67.
[23] Gouweleeuw, B., Thielen, J., Franchello, G., De Roo, A., and Buizza, R. (2005). "Flood forecasting using medium-range probabilistic weather prediction." *Hydrology and Earth System Sciences*, 9(4), 365-380.
[24] Chen, J., Guo, S., Li, Y., Liu, P., and Zhou, Y. (2013). "Joint operation and dynamic control of

flood limiting water levels for cascade reservoirs." *Water resources management*, 27(3), 749-763.

[25] Li, Y., Zhou, J., Zhang, Y., Qin, H., and Liu, L. (2010). "Novel multiobjective shuffled frog leaping algorithm with application to reservoir flood control operation." *Journal of Water Resources Planning and Management*, 136(2), 217-226.

[26] Zhou, Y., Guo, S., Liu, P., and Xu, C. (2014). "Joint operation and dynamic control of flood limiting water levels for mixed cascade reservoir systems." *Journal of Hydrology*.

[27] Webster, P. J., Jian, J., Hopson, T. M., Hoyos, C. D., Agudelo, P. A., Chang, H.-R., Curry, J. A., Grossman, R. L., Palmer, T. N., and Subbiah, A. (2010). "Extended-range probabilistic forecasts of Ganges and Brahmaputra floods in Bangladesh." *Bulletin of the American Meteorological Society*, 91(11), 1493-1514.

The Idea and Key Technical Prospect on Integration between Underground Reservoir and Surface Water System

Li Zhou[1, 2], Lingxue Liu[1], Xiaodong Li[1], Tianqi Ao[1*]

[1]State Key Laboratory of Hydraulic & Mountain River Engineering, College of Water Resources & Hydropower, Sichuan University, Chengdu, 610065, China

[2]Nanjing Hydraulic Research Institute, Nanjing, 210029, China

[*]Corresponding author's e-mail: aotianqi@scu.edu.cn

Abstract. The problem of global water security is extremely serious due to the increasingly poor and uneven distribution of water resources and severe water pollution in some regions. Water security is a major long-term strategic demand of China, involving water resource supply, the prevention and control of flood, drought and water pollution, etc. In order to solve the prominent problems of global surface water shortage and the groundwater level drops caused by the over exploited of the traditional groundwater reservoir, the underground reservoir is proposed under the guidance of national strategy "Deep Into the Earth". The protection and early warning of water security would be better through the interaction between underground reservoir and surface water system. This paper provides solutions for water security issues. In which, the preliminary concept of underground reservoir is defined, and the theoretical system of water security for underground reservoirs and surface water is put forward. Moreover, several key technologies are also proposed, including the distributed rainfall-runoff and water environment simulation with high precision, site selection and design of underground reservoir, the early warning and protection of water environment in underground reservoir, the joint operation of water quantity and quality between underground reservoir and surface water system, etc.

1. Introduction

Water is the source of life, the basis of production and ecology and the fundamental resource on which human beings depend for survival. The fresh water that is available to humans accounts for 2.5% of global water, but only 0.26% of which can be directly utilized after excluding the hard-to-exploit water in the Antarctic ice sheets, Greenland ice sheets and some mountain glaciers. Worse, the global freshwater is unevenly distributed, and 60% of which belongs to the top nine countries(Brazil, Russia, Canada, China, the United States, Indonesia, India, Colombia, and Congo), while 80 countries and regions with 1.5 billion people are short of fresh water[1]. Moreover, with the development of social economy, the problem of water environmental pollution is increasingly aggravated. More than 420 billion m^3 of sewage are discharged into rivers, lakes and seas every year, polluting 5.5 trillion m^3 of fresh water which is equivalent to 14% of global runoff[2]. It not only aggravates the shortage of available irrigation water to become an important constraint to food production, but also directly affects the safety of water drinking, food production with resulting in huge economic losses. According to the "United Nations World Water Development Report 2018"[3], the global demand for water resources is growing at a rate of 1% per year. About 5 billion people will live in water shortage areas by 2050 if there is no action been taken.

With the constant extension and expansion of definition, "water safety" has been replaced by "water security". Generally, "safety" refers to the safe of human health and production technology activities, covering the natural attributes of "water safety", such as water resource safety, water environment safety, etc[4]. However, "security" contains social and political security issues that can guarantee its stability and economic development[5], and it covers the social attributes of "water security". Thus, "security" that has comprehensive implications has been used by a large number of scholars[6][7][8], such as the internationally recognized "food security". In 2009, UNESCO(United Nations Educational Scientific and Cultural Organization) defined water security as the water resources that can ensure living and development of humans in quantity and quality, which can maintain the sustainable health of human and ecological environment, protect lives and properties from water disasters[8]. And the implication of "water security" has been extended to the resources, environment, ecology, society, politics and economy. Karen Bakker[9] discussed the challenges and opportunities of water security research, indicating that the problem of water security had received a great attention in the world.

In recent years, surface water conservancy projects have been built all over the world to solve the water security problems including water resource shortage and uneven distribution in time and space[10], such as the California Water Diversion Project, China South-to-North Water Diversion Project[11] and China Three Gorges Project[12]. However, the problems of water security cannot be completely solved by relying only on surface water systems. As early as 1998, Qian[13] predicted the prospect of underground space development, and considered the planning of which in recent years[14][15]. Chen[16] proposed the strategic concept of building a large-scale deep earth science and engineering laboratory in China, and showed that the need to develop underground space is becoming more and more urgent. Xie[17] also presented several ideas of deep earth science in hydraulic engineering. At the China National Science and Technology Innovation Conference in 2016, it was pointed out that "Deep Into the Earth" is a problem of science and technology strategy that must be solved in China[18], and water is the first factor for deep earth space utilization and deep resource development[19]. Therefore, the construction of the underground reservoir integrated with the surface water system would gradually form the early warning theoretical system of water security, which is a new way to solve the water security problems.

2. Development trend of underground reservoir

At the beginning of the 20th century, the practice of artificial groundwater recharge has been conducted in Japan, America, Soviet Union, Netherlands, France and other countries. In the 1950s, the freshwater storage experiments in the salt water layer were started in America[20] and the artificial recharge of groundwater was carried out by using natural aeolian dunes In Netherlands[21]. After the 1970s, ASR(Aquifer Storage and Recovery) was gradually formed in America, and until July 2002, there were 56 ASR in operation and more than 100 systems has been built[22]. Now, the ASR project has become an important part of CERP(the Comprehensive Wetland Restoration Plan). In order to solve the problems of water shortage and seawater intrusion, an artificial groundwater reservoir(the catchment area is 0.6 km^2 and the total storage capacity is 9000 m^3) was built in 1972 in Nagasaki, Japan[23], which is the first one in the world. And the groundwater reservoirs in Miyako and Sunakawa of Okinawa[24] were successively constructed. In 1975, the Nangong Groundwater Reservoir(the total storage capacity is 480 million m^3)[25], the first one in China, was built in Hebei Province. And then, the Jiahe Groundwater Reservoir and the Dagu River Groundwater Reservoir in Shandong[26], the Maguan Groundwater Reservoir in Guizhou[27] were gradually built.

As an effective, economic and environmental friendly underground development project, the groundwater reservoir has been developed rapidly, and it is known abroad as "aquifer recharge" or "aquifer storage and recovery". Most scholars believe that the groundwater reservoir is a groundwater development project that uses the natural water storage space within the earth's crust to store water resources[28]. However, the underground reservoir proposed in this paper is different from the traditional groundwater reservoir. It is preliminarily defined as the reservoir that built artificially at a certain depth underground, with surface water dispatching or natural recharge as the main water source. And the project which is similar to this concept is the "Outside Drainage System of the Capital Area" that was built between 1992 and 2006 in Tokyo[29], which is the most advanced sewer drainage system in the world. The Tokyo drainage system is located about 50 meters underground along the Sixteenth National Highway in Kasugabe City, and the total length of which is 6.4 kilometers. There are five giant shafts with a height of 65m and a width of 32m, and an artificial underground reservoir with a length of 177m, a width of 77m and a height of about 20m, which is supported by 59 large columns with a height of 18m and a weight of 500 tons[30].

The underground reservoir has many advantages, including no land occupation, no immigration, no damage to the ecological environment, no risk of flooding and dam failure, the reduction of water loss caused by evaporation and the provision of foundation for deep water networks. Especially, it can avoid the problem of groundwater over-exploitation and effectively solve the problem of uneven spatial and temporal distribution of water resources. The underground reservoir can be used as a source of water supply to solve the problem of water shortage, which improves the existing traditional mode of water resources development, flood prevention and mitigation, and water environmental protection. And through the comprehensive operation with surface water, it can improve the utilization of water resources and protect the surface water environment. In the flood season, it can also be used as a flood discharging pool to achieve flood control and disaster mitigation. Furthermore, the underground reservoir group, which is constituted by a number of independent or connected underground reservoirs based on regional hydrometeorology, topographical geology, water supply and water environmental protection requirements, is linked and coordinated with the surface water system to solve the water security issues jointly.

3. The prospect on theory and key technology of integration between underground reservoir and surface water system

The study of underground reservoir and its linkage with surface water system needs interdisciplinary integration. This research has the dual characteristics of theoretical science and technical science. The basic theories involve mathematics, physics, geography, geology, ecology, environmental science, material science, control science, management science, etc, and the basic technologies include measurement technology, control technology, computer technology, information technology, etc. The theoretical study of the integration between underground reservoir and surface water system mainly involves three aspects: (1) The basic theory research of underground reservoir and underground reservoir group, such as the precisely definition of underground reservoir, the main function of underground reservoir, and the construction area that need to be studied. (2) The design theory of underground reservoir. The design concept, design principles, design standards and design methods should be studied from the perspectives of hydrometeorology, water environment and water security. (3) The scientific basis and method of the integration between the underground reservoir and surface water system in water quantity and quality, and the early warning system of water security by integrating underground reservoirs with surface water system should be formed eventually.

3.1Key technology of distributed rainfall - runoff-water environment simulation and early warning with high precision

The distributed meteorological, hydrological and water environment integrated simulation model is developed to provide high-precision hydrological and water environment forecast information for the coordinated and optimized operation of underground reservoirs and the linkage with surface water system.. The main research contains four parts. (1) To develop a watershed scale rainfall forecasting model with higher resolution and precision. (2) To develop a physically distributed hydrological model with more comprehensive and scientific structural system. (3) To build a distributed coupled forecast model with integration of meteorology, hydrology and water environment. (4)To build the parameter transfer function that indicates the quantitative relationship between parameters and the

IOP Conf. Series: Materials Science and Engineering **794** (2020) 012003 doi:10.1088/1757-899X/794/1/012003

physical characteristics of the underlying surface, it can effectively solve the key problem of hydrological forecasting in the widespread ungauged basins.

3.2Key technology of site selection, storage capacity design and early warning of underground reservoir

Considering the basin composition characteristics, rainfall, water quantity, topography, water resource demand and allocation and water environment endowment, the site selection of underground reservoir should be carried out by using remote sensing and GIS technology, and many analysis methods including fuzzy comprehensive analysis, analytic hierarchy process, etc. To study the system and method of early warning index of underground reservoir, the design of capacity, elevation and water function should be based on the simulation results of regional water resources and water environment conditions by using integrated forecasting model of meteorology, hydrology and water environment.

3.3Key technology of early warning and protection of water environment simulation for underground reservoir

Underground reservoirs are in the special environmental conditions, such as no sunlight, low velocity of flow and wind, long storage time of water body with weak self-purification ability. Thus, the environmental hydraulics, microbiology and other interdisciplinary scientific theories and technical methods need to be applied to research the water quality, evolution rule, early warning index system as well as the water environmental protection technology of the underground reservoirs.

4. Conclusions

There is a shortage of fresh water resources available for humans, although the amount of water on the surface of the earth is large. In recent years, with the rapid development of global economy and population, water security has been seriously challenged. However, the problems of water environment security, water resources security, water ecological security and water disaster avoidance security cannot be effectively solved by relying only on the surface water system. And the traditional ground reservoir cannot fully ensure the region water security due to the geographical limitations, the over-exploitation of groundwater and so on. In order to realize the efficient utilization of water resources and to solve the problem of water security, the idea of improving the water security and early warning system by using deep earth space and establishing the integration system of underground reservoirs and surface water is put forward. In this paper, the underground reservoir is preliminarily defined, and three scientific theoretical problems are summarized: the basic theory of underground reservoir, the design theory of underground reservoir, the integration theory of underground reservoir and surface water system in terms of water quantity and quality. In addition, several key technologies are proposed including the distributed high precision rainfall-runoff-water environment simulation, site selection and design of underground reservoir, the early warning and protection of water environment in underground reservoir, the joint operation of underground reservoir and surface water system, etc. In a word, this paper provides solutions for water security problems, and it is expected to effectively solve the uneven spatial and temporal distribution of water resources, and to promote the development and utilization of deep earth space.

Acknowledgments

This research is supported by the International Cooperation Project of Sichuan Province, China (grant number 2010HH0005). It is very appreciated for the valuable suggestions and comments from all the reviewers, and for all the contributors of the references cited in this paper.

References

[1] Li, Z.F. (2018) Carrying forward green infrastructure and optimizing grey infrastructure for solving water problems in an integrated manner-a review of UN World Water Development Report in 2018. China Water Resources, 07: 1-3 (in Chinese).

[2] Zhu, P. (2013) Water pollution and prevention. Sciences & Wealth, 12: 320 (in Chinese).
[3] WWAP (United Nations World Water Assessment Programme). 2018. The United Nations World Water Development Report 2018: Nature-based Solutions. UNESCO, Paris.
[4] Chen, E.J., Shi, G.Q., Gu, Q.Y. (2005) Theoretical framework of water safety system, 03: 9-11 (in Chinese).
[5] Xia, J., Shi, W. (2016) Perspective on water security issue of changing environment in China. Journal of Hydraulic Engineering, 03: 292-301 (in Chinese).
[6] Wang, X.J., Zhang, J.Y., Gao, J., Shahid, S., Xia, X.H., Geng, Z., Tang, L. (2018) The new concept of water resources management in China: ensuring water security in changing environment. Environment, Development and Sustainability, 20: 897-909.
[7] Vaseashta, A., Braman, E., et al. (2011) Sensors for water safety and security. In: Sensors Applications Symposium IEEE. San Antonio. pp. 478-486.
[8] United Nations Environment Program. (2009) Water Security and Ecosystem Services: The Critical Connection. UNEP, Nairobi.
[9] Bakker, K. (2012) Water security: research challenges and opportunities. Science, 337(6097): 914-915.
[10] Wang, X.L. (2004) Famous water diversion projects in foreign river basins and regions. Water Resources and Electric Power, 01: 1-25 (in Chinese).
[11] Wang, Y.S., Yang, Y.Y. (2005) South-to-North Water Transfer project of China. Yangtze River, 07: 2-5, 71(in Chinese).
[12] Dai, H.C., Pang, Y.X. (2005) Study on relationship between TGP and the eco-environment of the middle & lower Yangtze. Journal of Hydroelectric Engineering, 04: 26-30.
[13] Qian, Q.H. (1998) To meet the climax of urban underground space development in China. Chinese Journal of Geotechnical Engineering, 01: 112-113 (in Chinese).
[14] Qian, Q.H. (2017) To Transform Way of Urban Development by Constructing Underground Utility Tunnel. Tunnel Construction, 06: 647-654 (in Chinese).
[15] Qian, Q.H. (2017) The consideration of promoting urban underground space planning and construction. Urban and Rural Development, 18: 60-65 (in Chinese).
[16] Chen, H.S. (2010) Deep underground science in China: status and prospects. Bulletin of National Natural Science Foundation of China, 02: 65-69 (in Chinese).
[17] Xie, H.P., Gao, F., Ju, Y., Zhang, R., Gao, Z.M., Deng, J.H. (2017) Novel Idea and Disruptive Technologies for the Exploration and Research of Deep Earth. Advanced Engineering Sciences, 01: 1-8 (in Chinese).
[18] Xie, H.P., Gao, M.Z., Zhang, R., Xu, H., Wang, Y.H., Deng, J.H. (2017) The subversive idea and its key technical prospect on underground ecological city and ecosystem. Chinese Journal of Rock Mechanics and Engineering, 06: 1301-1313 (in Chinese).
[19] Xie, H.P., Xu, W.L., Liu, C., Yang, X.G. (2018) The subversive idea and key technical prospects on underground hydraulic engineering. Chinese Journal of Rock Mechanics and Engineering, 04: 781-791 (in Chinese).
[20] Getchell, F., Wiley, D. (1995) Artificial recharge enhances aquifer capacity. Water Engineering and Management, 142(11): 24-27.
[21] Jonoski, A., Zhou, Y.X., Nonner, J., Meijer, S. (1997) Model-aided design and optimization of artificial recharge-pumping systems. Hydrological Sciences Journal, 42(6): 937-953.
[22] Zhu, S.Y., Tian, J.C., Li, Q.D. (2008) Current Situation and Development Trend of Research on Groundwater Reservoir. Water Saving Irrigation, 04: 23-27 (in Chinese).
[23] Li, L.W., Shu, L.C., Li Q.D. (2006) Concept and design theory of groundwater reservoir. Journal of Hydraulic Engineering, 05: 613-618 (in Chinese).
[24] Wang, Y.J., Lei, J. (2005) Introduction to underground reservoirs in Okinawa, Japan. China Rural Water and Hydropower, 03: 78-80 (in Chinese).

IOP Conf. Series: Materials Science and Engineering **794** (2020) 012003 doi:10.1088/1757-899X/794/1/012003

[25] Li, W.G., Guo, W.D., Fang, M. (2002) Research for Utilization of the Store Water Layer to Relieve the Lack of Water. Journal of Drainage and Irrigation Machinery Engineering, 06: 29-31 (in Chinese).
[26] Li, X.R., Zhao, C.F. (2003) Application of high-pressure jet grouting to underground reservoir with pebble bed. Journal of Hydraulic Engineering, 03: 80-85, 92 (in Chinese).
[27] Wang, W.F. (1999) Study on the construction conditions of Maguan underground reservoir and its benefit at Puding, Guizhou. Carsologica Sinica, 01: 48-49, 51-56 (in Chinese).
[28] Zhao, T.S. (2002) Discussion of problems for groundwater reservoir. Hydrogeology & Engineering Geology, 05: 65-67 (in Chinese).
[29] Xin, D., Xu, X.Y., et al. (2012) Comparative Research on Water Resource Management for Capital Area of China and Japan— A Case Study of Beijing and Tokyo. South-to-North Water Transfers and Water Science & Technology, 01: 60-64 (in Chinese).
[30] Li, Z.D. (2011) Japan's efficient "Capital perimeter drainage system". Life and Disaster, 8: 20 -21 (in Chinese).

IOP Conf. Series: Materials Science and Engineering **794** (2020) 012004 doi:10.1088/1757-899X/794/1/012004

Research and Application of Reservoir Safety Monitoring System for Intelligent Water Conservancy

Xiuwen Li[1], Shunfu Zhang[1,2]* Hao Wu[1], Dongcheng Tian[1],

[1]China Institute of Water Resources and Hydropower Research, Beijing, 100038, China

[2]Research Center on Flood & Drought Disaster Reduction of the Ministry of Water Resources, Beijing, 100038, China

*Corresponding author's e-mail: 553592704@qq.com

Abstract: Safety monitoring systems are an important part of intelligent water conservancy projects. Based on the overall scheme requirements of intelligent water conservancy in the new era, we analyzed the new requirements of the current safety monitoring system. Intelligent perception, infrastructure, data resources, intelligent analysis and application display of reservoir safety monitoring system under the background of intelligent water conservancy are described respectively. Combined with the exploration of the intelligent safety monitoring system of Hekou Village reservoir, the functions of the system are introduced for further improvement. References and suggestions are provided for the construction of safety monitoring systems for intelligent water conservancy projects.

1.Introduction

Intelligent water conservancy [1] refers to the application of cloud computing, big data, Internet, mobile Internet, artificial intelligence and other newly developed information technologies in water conservancy projects, such as rivers, lakes, groundwater reservoir, hydropower station and water gates, as well as water storage, drainage and other water management activities through perception, networking, information sharing and intelligent analysis. These efforts would serve as basis for the supervision and management of modern water conservancy, inundation and drought prevention, as well as intelligent processing, decision-making and associated services of water conservancy.

Reservoir safety monitoring system for intelligent water conservancy has a wide range of applications in business with high technical demands. The underlying data can be obtained through thorough perception of automation equipment, and the monitoring results can be extracted, analyzed and timely transmitted and published by using the new generation of information technologies, so as to ensure safe operation and management of intelligent, scientific and economically-efficient reservoirs [2]. Compared with the traditional digital security monitoring, which carries out only post-processing and analysis of the monitoring data, intelligent monitoring combines with theories and methods of intelligent analysis, and proposes predictions based on historical data, and gives early warning in time.

2.New Requirements for Intelligent Water Conservancy for Safety Monitoring

Intelligent water conservancy has put forward new requirements on the construction of reservoir safety monitoring system. The safety monitoring system is supposed to be three-dimensional, one-stop and intelligent. To achieve this, a strategic system should be adopted to solve newly-emerging problems and realize new leapfrogs [3].

(1) Full three-dimensional coverage

Reservoirs that needed to be monitored in an intelligent way usually feature a large area, and complex geological and topographic conditions. The comprehensive perception of safety monitoring system should integrate various channels for monitoring. On the basis of the original monitoring techniques, the earth space information technology should be fully integrated to cover the whole reservoir and river basin, so as to construct a multi-dimensional safety monitoring system to collect data from both the sky and the earth. Through the construction of intelligent sensor information network, it provides comprehensive, accurate, dynamic and timely safety monitoring spatio-temporal data for intelligent water conservancy.

(2) Requirements for one-stop production capacity

Safety monitoring and supply service needs to improve the top-level design, and establish one-stop supply and service capabilities integrating monitoring design, system construction, data collection, data sorting, data analysis and data mining.

(3) Requirements for specialized intelligent services

Intelligent application is the ultimate goal of intelligent water conservancy. The security monitoring system should be based on the spatio-temporal information platform framework of intelligent water conservancy, and with the help of cloud computing and big data technology, an in-depth mining on the massive monitoring data of heterogeneous multi-sources should be conducted. Finally, comprehensive monitoring data sources, real-time collection, automatic collation, intelligent analysis, knowledge services, multiple expressions and network collaboration will be achieved. It provides convenient, efficient and intelligent safety monitoring services for intelligent water conservancy [3].

3.General characteristics and framework of intelligent safety monitoring

3.1 General characteristics of intelligent monitoring

Intelligent monitoring system can add, delete and modify monitoring points and parameters as needed, which is an extensible and improvable system. It is compatible, inclusive, unified and coordinated. The system uses the current information technology while taking into account the trend of future technologies to accommodate the need for update. It meets the demands for real-time monitoring of the reservoir to detect abnormal conditions in a timely manner.

3.2 The overall framework of intelligent monitoring system

As an important subsystem of intelligent water conservancy, intelligent safety monitoring system has no construction standard. Based on the specific construction planning objectives of the intelligent water conservancy, the design of the technical scheme is carried out according to relevant norms and standards. The construction of intelligent safety monitoring system relies on the intelligent water conservancy infrastructure as well as software and hardware, thus promoting efficiency and lowering cost. The intelligent safety monitoring system of reservoirs, assisted by advanced information technologies such as intelligent perception, big data analysis, cloud computing, heterogeneous network integration, BIM+GIS and artificial intelligence, could fully perceive the structured and unstructured data of reservoirs, ensuring rapid response, scientific decision-making and collaborative management.

IOP Conf. Series: Materials Science and Engineering **794** (2020) 012004 doi:10.1088/1757-899X/794/1/012004

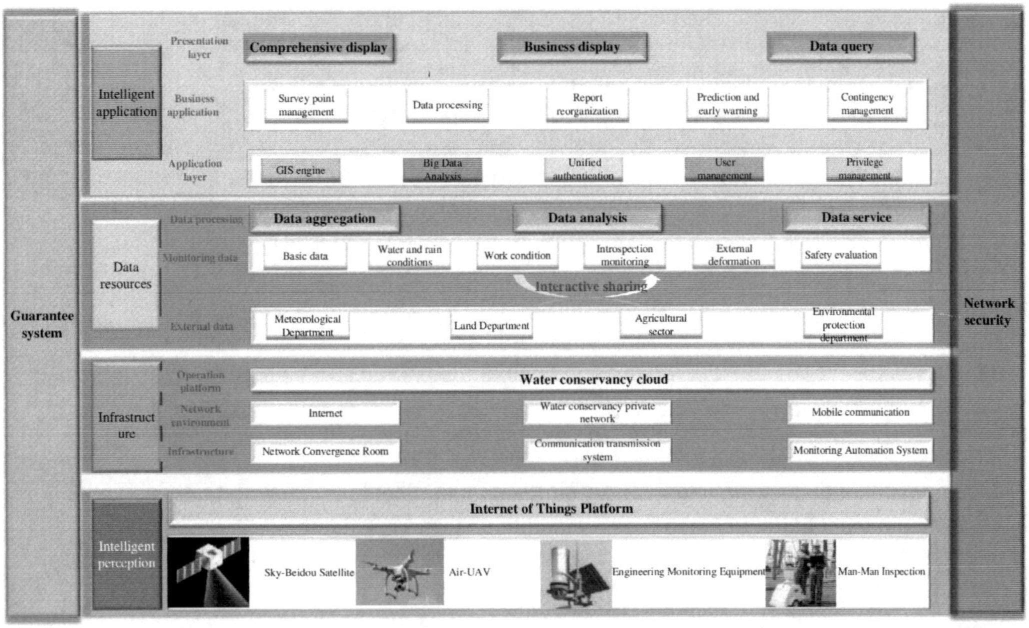

Figure 1 The overall framework of intelligent monitoring system

4.Examples of intelligent monitoring construction of reservoirs

4.1 Project overview

Hekou Village reservoir is located at the exit of the gorge in the last section of Qinhe river, a first-grade tributary of the Yellow River. It is a key project to control the flood and runoff of Qinhe river and an important part of the flood control system in the lower reaches of the Yellow River. Hekou Village dam covers a basin area of 9223km2. Hekou Village reservoir is rated Level 2 in terms of scale and grade-II as an engineering project, consisting of concrete rockfill dam, No.1 spillway, No.2 spillway, diversion power tunnel, spillway and hydropower station.

4.2 Key technologies of intelligent monitoring system

(1)Full perception

Reservoir safety monitoring mainly include external deformation, internal deformation, seepage and seepage pressure, stress and strain, environmental quantity and so on. The monitoring system integrates Beidou satellite, unmanned aerial vehicle, monitoring sensor and manual inspection. The data is collected by layers. Monitoring parameters, such as displacement, settlement, stress and strain and seepage, are collected in a real-time manner that go from the sensor, data collection module, data automatic collection device to the central monitoring station. The database in the central station server allows sharing of resources such as data, parameters and other information. The administrative host or the superior authority can access the database remotely on the Internet. Field management personnel can process parameters, databases and other operations through the software, which, by giving different levels of permission, facilitates the utilization of the database by personnel from different levels.

(2)Infrastructure

The automatic safety monitoring system of Hekou Village reservoir adopts a fully distributed and open network with intelligent node control while the automatic data collection system is based on the Ethernet fiber network [6].

The standard RS232/RS485 interface of each field data acquisition unit (MCU) is converted to RJ45 Ethernet interface through serial port server, and independent and unique static IP address is set on the serial port server so that each MCU can conduct network addressing through TCP/IP protocol. In order to ensure reliable long-distance transmission and improve the ability of the network against interference, single-mode optical fiber is adopted as the communication network medium between test stations and between test station and sub-station. The computer in the centralized control center and the MCU of each test station allows the mutual conversion between Ethernet electrical signals and optical fiber signals through the optical fiber transceiver. The safety monitoring system of Hekou Village reservoir adopts a three-layer C/S and B/S hybrid structure, namely the three-layer structure consisting of data platform, business service and human-machine interface.

(3)Data resources

Data resource layer [7] mainly includes three parts: 1) data aggregation platform. Middleware technology is applied to realize heterogeneous information aggregation of multi-system and data management based on distributed protocol. 2) comprehensive database. Integrated management of safety monitoring, water condition, water quality and other data along with geographical data as well as documentation work together to form a unified information management database. 3) database management and operation maintenance. The database allows remote backup to enable mass data storage and ensure data security; by integrating the system and data resources, it can not only reduce the cost of data management, but also ensure the integrity and consistency of the data.

(4)Intelligent analysis

Detailed monitoring indicators are set up for each monitoring items and monitoring points for abnormality evaluation of real-time monitoring data through the background data analysis. Combined with the existing model, the system compares real-time monitoring data with historical data for intelligent analysis, which would automatically inform corresponding management personnel of any abnormalities, thus allowing intelligent analysis and notification.

(5)Application display

The measurement points of the project including safety, water condition, gate, water quality and ecological flow and other monitoring indicators were classified by GIS to establish vector layers, which were superimposed on the map layers and connected with the spatial information management database to from an overall picture of the reservoir. Through an electronic map that can be scaled and shifted with various parameters for selection, the reservoir could be monitored in a real-time manner along with descriptions of the overall evaluation of the reservoir.

4.3 Main functions of intelligent monitoring

(1)Basic service

① Automatic data collection and storage: functions including real-time information collection, distributed remote measurement control, fault diagnosis of sensors and data collection devices, data inspection in the process of data collection, alarm for out-of-limit data and other functions.

② Manual data entry and storage: keyboard entry, file import, special procedures loading and other ways allowing artificial data collection and storage in the database.

③ Results calculation and processing: Through the design formula in the system, the original measurement values can be directly converted into the monitoring results, and error or error data can be identified and eliminated automatically and manually.

④ Graph drawing and qualitative analysis: the system designed the duration curve, spatial distribution diagram, correlation analysis diagram and statistical table for characteristic values, which can directly show the structure of the basic nature and safety of the measured objects.

⑤ Basic information management: the system provides the display of basic information, such as engineering attributes, events, drawings of main buildings, management organization and personnel information, GIS maps and so on.

⑥User rights management: the system sets permissions according to the user right policy, allowing users at different levels to access the system.

(2)Intelligent analysis and management

① Analysis and quantitative analysis: modeling based on multiple linear stepwise regression, partial least-squares regression, the statistical model, hybrid model and neural network to study the influence of various factors (such as water level, temperature, rainfall, etc.) on the hydraulic structures, including deformation, seepage and stress strain as well as the scale of the influence, providing basis for structure safety assessment and prediction.

② Evaluation of the safety of structures: based on the theories and expertise of dam engineering and rock-soil mechanics, a three-level evaluation system for local anomalies of structures and the overall safety of the project is established, and the analysis and evaluation of the monitoring results obtained are drawn up[7].

③ Prediction of structure safety: based on numerical simulation and historical monitoring result model stored in the system model database, the trend of structure property and safety under specific conditions are predicted, and early warnings will be given according to the prediction, providing decision support for project operation and maintenance.

(3)Intelligent application

Mobile application: it enables relevant personnel to obtain key monitoring data and deliver early warning in a prompt manner so as to facilitate timely investigation, early warning release and emergency disposal by engineering personnel. It can also use the mobile Internet terminal system for emergency training.

4.4 For further improvement

Data mining and fusion analysis can be introduced to determine the authenticity of monitoring data. For example, covariance crossover algorithm can be used for data fusion and evaluation of the same index measured in different time periods and places in safety monitoring. For abnormal data, support vector machine and D-S evidence theory can be used for this purpose. The matrix decomposition technique can be used for heterogeneous data fusion of multi-level monitoring data from different departments. In addition, it is necessary to make assessments based on water conservancy expertise. By integrating information technology, data mining and expertise for analysis and decision making, the internal rules of water resource data will be better understood so as to identify the authenticity of data and improve its accuracy [4]

5.Conclusion

Intelligent monitoring involves every link of the safety monitoring system. Only by adopting advanced information technology, full perception, intelligent analysis and scientific decision-making can modern reservoir safety management be realized. The article first briefly introduces the function and requirements of the safety monitoring system of the intelligent water reservoir. Then based on the Hekou Village reservoir in Henan Province, the paper introduces the design of the system and its application, the modeling analysis and quantitative analysis, safety evaluation of the structure, safety forecast, early warning and emergency management, mobile applications as well as information release. The paper then points out defects of the system in need of further improvement. The system can effectively manage the safety monitoring of large reservoirs and provide powerful technical support for further analysis and decision-making.

Acknowledgments

This paper was jointly supported by National Key R&D Program of China (Mechanism of Landslide due to Seepage in expansive soil bank slope and dam;2017YFC1501201)

References

[1] Working Group on Intelligent Water Conservancy Overall Programming. Intelligent Water Conservancy Overall Programming [M]. Beijing, 2019.

[2] Yang Mingxiang,Jiang Yunzhong,Tian Yu,etc.Demand analysis of smart water resource[J].Journal of Tsinghua University(Science and Technology),2014,54(01):133-136+144.

IOP Conf. Series: Materials Science and Engineering **794** (2020) 012004 doi:10.1088/1757-899X/794/1/012004

[3] Liu Feng, Li Dahong, Huang Zhangyu, etc. Research on data acquisition technology of "land, water, air and sky" safety monitoring for intelligent watershed [J]. Sichuan Hydropower, 2017, 36 (01): 13-17.
[4] Du Hongyan, Xue Huifeng, Hou Junjie, etc. Studies on integration of water resources data with water smart technologies[J]. China Water Resources, 2018 (23): 61-64.
[5] Huang Jianming. Research on key technology of integrated management information intelligentialize system in Agang Reservoir [J]. Water Resources Informatization, 2018 (06): 62-68.
[6] Yi Guangjun, He Hu, Sun Jianhui. Dam safety monitoring information system based on domain-driven design [J]. Dam and safety, 2016 (01): 37-41.
[7] Hu Bo, Chen Jun, Li Xin, etc. Design and application of dam safety monitoring information management system [J]. Hydropower and pumped storage, 2017,3(06): 25-29+24.

IOP Conf. Series: Materials Science and Engineering **794** (2020) 012005 doi:10.1088/1757-899X/794/1/012005

River Classification in Line with China's New Requirements of Water Resources Management

Huojian Huang[1], Yong Yuan[1*], Ziqiang Xing[1], and Ding Wang[1]

[1]General Institute of Water Resources and Hydropower Planning and Design, Beijing, 100120, China

[*]Corresponding author's e-mail: yuany@giwp.org.cn

Abstract. River classification facilitates efficient management of rivers. Due to the complexity and variety of river morphology in China, classification of Chinese rivers has never developed maturely. The current river classification system used in China becomes even less qualified for new management requirements aiming to environmental protection and restoration as a response to the worsening environmental conditions represented by the deterioration of river functions and the scarcity of water resources, which have been greatly exacerbated by the impressive growth of Chinese economy. Moreover, some well-known classification methods discussed internationally cannot adapt to Chinese specific conditions appropriately, although they have been developed beyond the Chinese ones. This thus highlights the necessity to establish an updated classification system that is adequate for the emerging river management requirements of China. In these circumstances, this paper summarizes the currently representative river classification methods and proposes a pathway to improve river classification methodology for China.

1. Introduction

River system is an outcome of the Earth's evolution. Rivers flow by gravity on the surface or underground regularly or intermittently. Rivers play an important role in the Earth's water cycle and a major freshwater resource. They are not only a main type of habitats for aquatic organisms, but also a critical foundation of the development of human society and economy, being considered as the birthplace of human civilization[1]. Rivers provide water resources for human life and production, meanwhile transport, dilute and degrade pollutants discharged. Another important component of terrestrial water resources, lake water, also depends on water supply from rivers. It denotes that river health is closely related to the development and prosperity of society. However, due to the large population and rapid economic development in China, many rivers have been irrationally and unscientifically overexploited and utilized. River health has been jeopardized coinciding with deteriorated ecological function and intensified water pollution. Unfortunately, river managers have not developed a perfect approach to fully responding to these issues. It is not uncommon to observe frequent violation of water policies, elevated flood risk, mismatches between water resources supply and demand in upper and lower reaches, left and right banks and main and tributaries of the rivers, and increased disputes over water rights.

In recent years, China has started to emphasize environmental protection, advocated sustainable development and harmonious coexistence between humans and nature, proposed ecological civilization, and strengthened water resources protection through rational development and utilization. River classification is an important means of understanding rivers. It is the basis of conducting river health assessment, the foundation of river pollution control and ecological restoration, and the premise and assurance of scientific river management and comprehensive development and utilization of river basins[2]. Based on the above requirements, this paper discusses typical river classification methods. The discussion is made through the lens of the current Chinese river situations, associated with the requirements of integrated river basin management. It puts forward the ideas to improve river classification methods suitable for China, which can provide reference for the follow-up research.

2. Importance of river classification to China

In recent years, China has paid more attention to the protection and restoration of river environment, the harmonious coexistence of humans and water, and the sustainable development of river basins. Promulgation and implementation of water function zoning is a new means of river management, which is helpful to the improvement of river water quality and the protection of ecological environment[3]. In the report of the Eighteenth National Congress, the Central Committee pointed out that the construction of ecological civilization should be placed at the same important position as the construction of political, economic, cultural and social civilizations. In 2013, the Ministry of Water Resources proposed to accelerate the construction of water ecological civilization. Water ecological civilization advocated the concept of human-water harmony and realizes the sustainable utilization of water resources, which is embodied in strengthening the protection of water sources and the control of total water use, promoting water recycling, and building a water-saving society. Nowadays, the idea of environmental water demand and the study of ecological restoration have also been emphasized. The environmental water demand is to protect the long-term stability and balance of ecosystems, with to the greatest extent meeting the water demand of economy and society, and realizing the rational allocation and sustainable utilization of water resources[4]. River water ecological restoration aims to restore the damaged river ecosystem structures and functions, making it sustainable to provide ecological services, and realizing the harmonious coexistence of people and rivers[5].

Based on China's new viewpoints and requirements for river water resources management in the new era, classifying rivers scientifically becomes very important. Classification is the process of grouping or categorizing a series of complex objects according to similarity or relationship using established criteria[2]. River classification is the process of grouping rivers according to their features. It is the first step to understand the complexity of rivers and the basis of scientific and effective management of rivers. Its significance lies in: (1) predicting the behaviour of rivers according to the morphology of rivers; (2) understanding the relationship between important parameters of rivers according to the monitoring data of typical river sections; (3) extrapolating the known river data according to the mechanisms to estimate similar river conditions; and (4) providing a unified and consistent information platform for the exchange of professionals in different disciplines. The ultimate goal of river classification is to understand rivers scientifically and to manage them efficiently by dividing rivers into sections with similar structures and functions[3,6]. Therefore, river classification is the first step to realize rational development, utilization and protection of rivers. At this stage, it should be in line with the requirements of ecological protection and sustainable development.

3. Typical river classification methods

The earliest and most well-known classification of rivers was proposed by Davis in 1899. He divided the river channels into three stages according to the geomorphological evolution: youth stage, maturity stage and old age stage, which were equivalent to the highland, lowland and coastal areas in space [7-8]. Since then, during the research and development for more than a century, scholars have proposed dozens of classification methods, based on geomorphology, river bed sediment, characteristic species, ecology, landscape, protection and utilization, etc. Among them, classification methods based on the theory of geomorphology and river dynamics are the most systematic and well-developed[2-3]. Since river morphology and sediment process are inseparable, many scholars consider these two factors together to classify rivers. This paper considers the classification methods based on geomorphology and sediment characteristics into one category. This category can be subdivided into many different types. The representative methods are summarized in Table 1. The methods have developed in a trend from simple to complex, incorporating multiple factors, levels, and disciplines, and become more quantitative[3]. Due to the limitation of space in this paper, it is impossible to enumerate each classification method. Here, only a few representative methods are outlined.

Table 1. Typical river classification methods based on geomorphological and sediment characteristics

Classification basis	Indicator	Contributor	Classification outcome	Notes
River network link	Stream order	Horton[9]; Strahler[12]	First-, second-, third orders, etc.	
Sediment	Sediment transfer, morphology	Schumm[11]	Erosion, transport, and deposition zones	
	Sediment transfer, morphology	Montgomery [12]	Colluvial, confined, and floodplain channels	
	Sediment transfer, morphology	Schumm[15]	Suspended, mixed, and bed loads	For alluvial rivers
	Sediment transfer, morphology	Church[16]	Boulders, cobbles, gravel, sand, fine sand, and silt	
	Sediment, morphology, and topography	Whiting and Bradley[17]	35 types, denoted using letters and numbers	Headwater rivers
Channel pattern	Planform geometry, slopes, and flow rate	Leopold and Wolman[16]	Straight, meandering, and braided patterns	For alluvial rivers with floodplains
Floodplain morphology	Water energy, grain size, and geometry	Nanson and Croke[17]	High-energy non-cohesive, medium-energy non-cohesive, and low-energy cohesive floodplains, then subdivided into 13 types	For alluvial rivers with floodplains
Hierarchical classification	Reach-scale morphology, slope, and channel material	Rosgen[6]	First classified into 8 groups, according to entrenchment ratio, width/depth ratio, and sinuosity; then subdivided into 94 types based on slope and channel material	
	Channel morphology and geometry	Montgomery and Buffington [20]	First classified into colluvial, bedrock, and alluvial types; the alluvial type were then divided into cascade, step pool, plane bed, pool riffle, and dune ripple	For small-scale mountain streams
	Geometry, channel morphology, sediment	Brierley and Fryirs [19]; Brierley et al. [20]	Channels divided into a series of river styles linking to cross-section and floodplain evolution.	Mainly focusing on Australian rivers

Classifying rivers into different orders according to the relationship between main and tributaries is the most commonly used qualitative classification method, put forward by Horton[9] and Strahler[10]. Similar methods are often used in describing the relationships between the main and tributaries of China's rivers. The river order can be combined with physical parameters such as the length of a river sections, the area of a river basin, the slope and size of a river course. Although this method visually and concisely represents the river network structure, it does not reflect the different morphological characteristics and evolution processes of rivers of the same order[21].

IOP Conf. Series: Materials Science and Engineering **794** (2020) 012005 doi:10.1088/1757-899X/794/1/012005

Sediment transport is an important natural function of rivers. Sediment transport mode and characteristics play a key role in the evolution of river morphology, and become the leading factor and entry point of river classification. Schumm proposed in 1977 that rivers can be divided into erosion, transport, and deposition zones according to sediment transport processes[21]. Based on this, Montgomery[12] indicated that rivers could be further divided into colluvial channels, confined valleys with larger slopes, and unconfined floodplains with smaller slopes. This classification system roughly correlates the morphological characteristics of river reaches with aquatic habitats. For example, habitats in unconfined floodplains are more diversified than confined valley area. On the other hand, the sediment factor can be combined with other river factors for classification. For example, Schumm[13], proposed a conceptual framework for classifying alluvial rivers that related channel pattern and stability to sediment particle sizes, sediment transport mode, ratio of bed load to total load, channel slope, and width-to-depth ratio. Church[14] improved Schumm's classification system and introduced it into mountain rivers with large slopes. According to the grain size of sediment, the rivers were divided into six types. The morphological characteristics, stability and sediment transport dynamic characteristics of each type of rivers were given and sketched in schematics. Whiting and Bradley[15] developed a process-based classification protocol for headwater rivers based on bed mobility characteristics. Associating sediment particle size with channel size and valley stability, the headwater rivers were sorted into 35 types. However, this method has some limitations and can only be applied to rivers in source areas[21].

Most river classifications have been developed involving channel patterns (i.e., planform geometry) and combined qualitative and quantitative analyses. It is because to obtain river morphological parameters is relatively easy. For example, Leopold and Wolman[16] divided rivers into straight, meandering and braided categories, and proposed thresholds to distinguish meandering rivers from braided rivers based on discharge and channel slope. Subsequent investigations modified this method to distinguish anastomosing and wandering channels by considering many parameters, such as grain size, sediment load, riparian vegetation, channel roughness, width, and depth. In recent years, the application of new technologies such as GIS, such as Beechie et al.[22], has also promoted the progress of river morphological classification development. Both qualitative and quantitative analyses of river morphological characteristics are found to be used. However, most of quantitative data are empirical and can be only used in limited circumstances[21]. On the other hand, the planform geometry of rivers is closely related to the floodplain morphology, which also links to the energy of river water. For example, Nanson and Roke[17] classified rivers into three types according to river water energy – high, medium and low – and further divided the river into 13 types according to the grain size of floodplain sediments and the topography of riparian zones. Kellerhals et al.[23] predicted the evolution process of large rivers (bankfull width > 20 – 30 m) by combining the planform morphology of mountain rivers with channel pattern/sinuosity, frequency of channel islands, bar type, and lateral activity of the channel and floodplain. The Forestry Department of British Columbia (FPC) of Canada, on the other hand, classified the small to intermediate channels (bankfull < 20 – 30 m) into three types: riffle-pool, cascade-pool, and step-pool[21]. It can be seen that the combination of the river morphology and the surrounding zones can predict the river evolution trend and provide a reference for human engineering work along the river. However, most of these methods are limited to specific types of river sections, and their applications are relatively narrow[21].

Hierarchical classifications are probably the most comprehensive and widely accepted classification methods. Hierarchical classification theory was first proposed by Frissell et al.[24]. Large-scale temporal and spatial systems (primarily for stream habitats) were divided into many small-scale elements and associated with watershed geomorphic features and events. The most representative and widely used hierarchical classification method was proposed by Rosgen[6]. In this method, rivers were first divided into 8 major stream types based on entrenchment, width-to-depth ratio, and sinuosity, followed by subdivision into 94 minor channel types for different landscapes. Another widely accepted hierarchical method was proposed by Montgomery and Buffington[18]. According to field observations, small-scale mountain rivers were divided into seven types: colluvial, bedrock, cascade, step pool, plane bed, pool riffle, and dune ripple. The stability of the river channel and its response to environmental changes were predicted by the surrounding topography and river bed sediments. Brierley and Fryirs[19] and Brierley[20] modified Rosgen's method to make it more suitable for Australian rivers. They developed a River Style Framework, according to land type and degree of confinement, river character, and river behavior, which can be used to establish a relationship between river evolution process and ecological change process, and to preliminarily predict the evolution trend of rivers and the potential of ecosystem restoration. However, this method is mainly based on the case study of Bega River in Australia, and is doubtful about the adaptability in other areas.

In terms of China, the river types are complex and diverse, because of great regional disparity in geography, climate, environment, economy, culture and management measures. However, the research on river classification in China is relatively insufficient with slow development[2-3]. Most of the classification methods only consider single factor and scale and qualitative descriptive indicators, for example, based on destination, flowrate, flow velocity, source of water, etc. This can only give a general idea about China's river system, but cannot meet the increasingly strict river management requirements[3].

4. Recommendations of future river classification development for China

The river classifications proposed by other countries have been developed ahead of Chinese ones, whereas they are found to be difficult to apply in China. This is mainly because China's geography and river characteristics are out of the scope of these studies. Some classification indicators working well in the original regions are not suitable for Chinese cases. Secondly, foreign research is supported by detailed historical data and a large number of field observations, while similar data in China are relatively scarce, which makes it difficult to be used. Thirdly, most of these classifications are based on the characteristics of rivers from the perspective of a certain discipline, such as geology, river dynamics, hydrology, hydraulics, ecology and so on. However, under the condition that China is stressing ecological protection and sustainable development, these classification methods are not a perfect fit. It is required to classify rivers for the sake of river protection and restoration. This is what the endeavor should be made on, but still short of discussion.

As mentioned above, China is paying more attention to the protection of ecological environment, sustainable development and harmonious coexistence between humans and nature. When we study the river classification, we should take this goal into account and find an appropriate river classification approach to facilitating river water resources protection and sustainable development and utilization. Specific recommendations are as follows.

4.1 River classification should be in alignment with river management requirements

According to the needs of environmental protection and sustainable development, there are two major tasks of river management for China: fulfilling environmental flow requirements and conducting ecological restoration. Ensuring environmental flows can maintain long-term stability of river ecosystem and achieve the goal of sustainable development. The ecological restoration of damaged rivers aims to improve the structure and function of river systems. Therefore, we should explore how river classification can help us to complete these tasks. This requires a clear understanding of important river indicators for environmental flows and ecological restoration, from which rivers can be classified accordingly.

4.2 River classification should be based on field observation and monitoring data

Due to the lack of historical river data in China, it is impossible to develop classification methods using a large number of data, while data-based classifying could be the most rational approach. Therefore, striving for long-term collection of river data across multiple disciplines is desired, including hydrology, hydraulics, water quality, geomorphology, ecology, etc. Scientific classification cannot be achieved without sufficient data. In addition, social and economic data should be also included. The statuses of social and economic development, production and living conditions, and future plans directly determine water consumption, hence affects the achievement of environmental flow and ecological restoration goals. Among river data, ecological data may be the scarcest. Scientifically and continuously collecting ecological data of rivers is highly demanded.

5. Conclusions

The complexity and diversity of natural, geographical and human factors in China make China's river system very complicated. Degradation of river function and shortage of water resources, accompanied by rapid economic development, urge China to improve the environment and protect ecosystems. However, Chinese river classification is not suitable enough to meet the requirements, while the methods of other countries do not well adapt to Chinese cases. Therefore, it is imperative to find a scientific and appropriate way to classify the rivers to meet our current river management requirements. We should proceed from the key factors that need to be considered for river environmental flow demand and ecological restoration, and try to establish a data-based classification system. In this sense, river classification should be simultaneously developed with the theory and practice of environmental flows and ecological restoration.

Acknowledgments

This work was supported by the National Key Research and Development Program: Technical System for Comprehensive Guarantee of Ecological Water Demand in River, Lake and Wetland Systems (grant number: 2017YFC0404506).

References

[1] Ni, J.R., Gao, X.W. (2011) Comprehensive approach for classification of river systems and identification of ecological characteristics II. Applications. Shuili Xuebao, 42 (10): 1177–1184.
[2] He, P., Shi, P.J., Liu, S.K., Gao, J.X., Li, X.R. (2008) Classification of river systems. Advances in Water Science, 19 (3): 434–442.
[3] Zhao, Y.J. (2016) Progress and prospects of river classification research. Water Resources and Power, 34 (2): 26–30.
[4] Zhang, L., Li, L.J., Liang, L.Q., Li, J.Y., Jiang, D.J., Xu, M.X., Song, W.X. (2008) Progress on the research of theory and calculation method of ecological water requirement. Transactions of the CSAE, 24 (7): 307–312.
[5] Yu, L.J., Lv, X.Y., Song, S.Y., Liang, J., Wang, X.Q. (2013) The preliminary structure of definition index system for river water ecological restoration threshold. Ecology and Environmental Sciences, 22 (1): 170–175.
[6] Rosgen, D.L. (1994) A classification of natural rivers. Catena, 22: 169–199.
[7] Davis, W.M. (1899) The geographical cycle. The Geographical Journal, 14 (5): 481–504.
[8] Newson, M.D., Clark, M.J., Sear, D.A., Brookes, A. (1998) The geomorphological basis for classifying rivers. Aquatic Conservation: Marine and Freshwater Ecosystems, 8: 415–430.
[9] Horton, R.E. (1945) Erosion development of streams and their drainage basins: hydrophysical approach to quantitative morphology. Bulletin of the Geological Society of America, 56: 275–370.
[10] Strahler, A.N. (1957) Quantitative analysis of watershed geomorphology. Transactions, American Geophysical Union, 38 (6): 913–920.
[11] Schumm, S.A. (1977) The Fluvial System. Blackburn Press, Caldwell, NJ.
[12] Montgomery, D.R. (1999) Process domains and the river continuum. Journal of the American Water Resources Association, 35 (2): 397–410.
[13] Schumm, S.A. (1985) Patterns of alluvial rivers. Annual Review of Earth and Planetary, 13: 5–27.

[14] Church, M. (2006) Bed material transport and the morphology of alluvial rivers. Annual Review of Earth and Planetary Sciences, 34: 325–354.
[15] Whiting, P.J., Bradley, J.B. (1993) A process-based classification system for headwater streams. Earth Surface Processes and Landforms, 18: 603–612.
[16] Leopold, L.B., Wolman, M.G. (1957) River channel patterns: braided, meandering, and straight. U.S. Geological Survey Professional Paper 282 – B, Washington DC. pp. 39–84.
[17] Nanson, G.C., Croke, J.C. (1992) A genetic classification of floodplains. Geomorphology, 4 (6): 459–486.
[18] Montgomery, D.R., Buffington, J.M. (1997) Channel-reach morphology in mountain drainage basins. Geological Society of America Bulletin, 109 (5): 596–611.
[19] Brierley, G.J., Fryirs, K. (2000) River styles, a geomorphic approach to catchment characterization: implication for river rehabilitation in Bega Catchment, New South Wales, Australia. Environmental Management, 25 (6): 661–679.
[20] Brierley, G., Fryirs, K., Outhet, D., Massey, C. (2002) Application of the River Style framework as a basis for river management in New South Wales, Australia. Applied Geography, 22: 91–122.
[21] Buffington, J.M., Montgomery, D.R. (2013) Geomorphic classification of rivers. In: Shroder, J., Wohl, E. (Eds.), Treatise on Geomorphology. Academic Press, San Diego, CA. pp. 730–767.
[22] Beechie, T.J., Liermann, M., Pollock, M.M., Baker, S., Davies, J. (2006) Channel pattern and river-floodplain dynamics in forested mountain river systems. Geomorphology, 7: 124–141.
[23] Kellerhals, R., Church, M., Bray, D.I. (1976) Classification and analysis of river processes. Journal of the Hydraulics Division, American Society of Civil Engineers, 102: 813–829.
[24] Frissell, C.A., Liss, W.J., Warren, C.E., Hurley, M.D. (1986) A hierarchical framework for stream habitat classification: viewing streams in a watershed context. Environmental Management, 10 (2): 199–214.

Risk Assessment of Debris Flow in Ya'an City Based on BP Neural Network

Xiaoyu Liu, Xiaolong Shu, Xing Liu*, Zhe Duan and Zhiyi Ran

College of Water Resources and Hydropower, Sichuan Agricultural University, Ya'an 625014

*Xing Liu's e-mail: hydro_lx@sicau.edu.cn

Abstract. Since the occurrence of debris flow hazards is related to many factors, this paper selects five evaluation indicators: rainfall, lithology level, slope gradient, undulation degree and distance from faultages based on the three scopes of geomorphology, meteorology and geology. In this paper, the data extracted from the occurrence point of historical debris flow disaster is used, and BP neural network is used for training, then the weights of evaluation indicators are obtained. Finally, the GIS software is used to superimpose and analyze these indicators, and the risk of debris flow in Ya'an city is divided into five levels, thus the danger distribution map of debris flow in Ya'an city is obtained. And the map is compared with historical debris flow disaster occurrence points, the results show that, three districts of Tiancheng County, Lushan County and Baoxing County have the highest overall risk; the overall risk of Yingjing County, Yucheng District and Shimian County is high; and the overall risk of Hanyuan County and Mingshan District is relatively low. The evaluation results reflect the actual situation of the spatial distribution of debris flow in Ya'an city, which can provide a scientific basis for the development of reasonable flood control and disaster against planning in Ya'an city.

1.Introduction

Debris flow is one of the most common mountain geological disasters. It is a rapid surface runoff formed by saturated dilution of soft soil hills, which is caused by short-term heavy rainfall and other bad weather. Debris flow carries a large amount of solid detrital materials, such as sediment and rocks, flows to low-lying areas and causes loss of life and property, which is characterized by sudden developed, quickly formed, serious harm and difficulty in monitoring [1]. China is a mountainous country with nearly 70% of the country's land area is mountainous. The number of mudslides recorded alone has reached 111,000, which is spread over mountainous areas of 26 cities and is a large number and a wide range. As a whole, the average annual direct economic loss is huge. Among the areas mentioned above, Yunnan, Sichuan and Gansu are serious disaster areas for landslides and debris flows. These areas are mostly located in the seismic belt on the edge of the Qinghai-Tibet Plateau. In addition, the precipitation of the mountains on the surface of the Tibetan Plateau leads to the loosening of the surface structure, which forms a large amount of loose deposit strata. Moreover, human improper activities damage the environment, causing frequent mudslide occurrences.

In 1982, Chinese scholar Wang Lixian proposed a method for dividing the risk of debris flow valleys [2]; in 1986, Tan Bingyan proposed that 15 basic indicators should be selected from the direction of determining the severity of debris flow gully and scored according to the requirements of quantitative theory [3]; in 1999, Tang Jiafa used GIS software to classify the debris flow in the upper reaches of the Minjiang River [4]; in 2002, Li Huixia evaluated and classified the risk degree of soil and water loss in Ya 'an area from two aspects of the status and driving force of soil and water loss, and classified the risk degree of soil and water loss in Ya 'an city [5]; in 2004, Liu Xilin selected a number of indicators to the debris flow through a comparative analysis of the vulnerability of mudslides in Sichuan Province in 1990 and 2000. The quantitative calculation of disaster risk shows that Ya'an city is a high-risk area for mudslide disasters in Sichuan Province [6]; in 2011, Wang Xueliang used analytic hierarchy process (AHP) to select 10 index factors to construct a hierarchy index system for debris flow risk evaluation [7]; in 2013, Zhao Xin used multiple regression analysis to establish a single-ditch debris flow risk assessment model [8]; in 14 years, based on GIS spatial modeling, Jin Rong selected four factors combined with analytic hierarchy process to evaluate the risk of debris flow in Shimian County and Hanyuan County of Ya'an city [9]; in 2015, taking the debris flow in Dayyelong gully, Tianquan county, Ya 'an city as the research object, Wang Jiazhu calculated the dynamic characteristics of debris flow and made a quantitative evaluation of its risk [10].

With the development of artificial intelligence technology, object evaluation based on machine deep learning has become a future development trend. This paper will combine BP neural network and GIS software to analyze various debris occurrences of debris flow disasters, and evaluate the risk of debris flow in Ya'an city, which can provide a scientific basis for scientifically, reasonably predicting and coping with the occurrence of debris flow in this area.

2.Materials and Methods

2.1 Overview of the study area

Ya'an city is located at the western edge of the Sichuan Basin. It spans north latitude 28°51′10″~30°56′40″and east longitude 101°56′26″~103°23′28″ , which is the transition zone from the Sichuan Basin to the Qinghai-Tibet Plateau. The terrain is higher in the north, west and south, and lower in the middle and east. The mountain ranges in the city, the land surface is rugged, and the types of landforms are complex and diverse. There are many mountains, few hills and flat dams. The climate type is subtropical monsoon humid climate, which has large difference in north and south. The annual average temperature is between 14.1 °C to 17.9 °C, the average annual rainfall is about 1800 mm, and the resident population is about 1,537,800.

The causes of debris flow hazards are complex, and there are many factors affecting and controlling the risk of debris flow. Therefore, the selection of evaluation indicators should follow the principles of scientific, operational, hierarchical, objective, comparability, qualitative and quantitative [11]. The establishment of the evaluation index system should meet the principles of applicability, completeness, comparability, comprehensiveness and systematicness [12]. Therefore, according to the previous research results [13], and combined with the actual situation of the study area and GIS spatial analysis tools, the selected evaluation indicators are: lithology level, undulation degree, slope gradient, distance from faultage and rainfall.

2.2 Data processing

The terrain topographic evaluation indicators are extracted from “ASTGTM 30m resolution data”, and the slope gradient and the undulation degree are obtained from research area DEM by the GIS software. The original data of the geological evaluation indicators are derived from the “2.5 million Chinese geological map”, the lithology level of the stratum is obtained from the lithology data in the geological map through GIS software, and then the solidity coefficient is divided into five grades according to the "Rock Classification and Hardness Level", the distance from the faultage is obtained through the fault file to establish a grading buffer; hydrometeorological evaluation factor uses the monthly rainfall data of 0.25°×0.25° for the “TRMM 3B43 satellite” from 1998 to 2016. The average rainfall is calculated by MATLAB software. The measured data of the debris flow originated from “Sichuan Province in 1997”. The spatial distribution data of the 1:10 000 000 debris flow in Ya'an city and the “Handheld Query System for the Hidden Danger Points of Sichuan Province”. Among the five indicators, the slope and rainfall are positive correlation indicators, that is, the larger the index value, the higher the risk; the undulation, lithology level and distance from the fault are negative correlation indicators, that is, the more the index, the smaller the risk.

Because the indicators are not uniform and the range of values varies greatly, which is not conducive to neural network training. It is necessary to standardize the data and eliminate its dimensions, and the formula is as follows [14]:

$$X = \frac{X_i - X_{\min}}{X_{\max} - X_{\min}} \tag{1}$$

where X is the standard value of the indicator; X_i is the measured value of a certain indicator; X_{max} and X_{min} are the maximum and minimum values of the indicator respectively.

If the indicator is negatively correlated with risk, the reverse indicator is normalized to:

$$X = 1 - \frac{X_i - X_{\min}}{X_{\max} - X_{\min}} \tag{2}$$

After normalizing, the indicators are divided into five grades according to the risk according to the degree of influence on the risk of debris flow. The higher the grade, the greater the risk (Table 1):

Table 1 Range of each indicators

Level	I	II	III	IV	V
Lithology level	0	0-0.247	0.247-0.498	0.498-0.749	0.749-1
Undulation degree	0-0.353	0.353-0.512	0.512-0.641	0.641-0.786	0.786-1
Slope gradient	0-0.164	0.164-0.297	0.297-0.420	0.420-0.582	0.582-1
Distance from faultage	0-0.200	0.200-0.400	0.400-0.600	0.600-0.800	0.800-1
Rainfall	0-0.165	0.165-0.293	0.293-0.455	0.455-0.672	0.672-1

2.3 Research methods

Artificial Neural Networks (ANN) is a nonlinear dynamic system composed of a large number of simple processing units (neurons), which has learning, memory, calculation, recognition, prediction and many other intelligent processing functions. It is different from the modern electronic computers called "computers" and the artificial intelligence systems based on symbolic deduction, but another artificial intelligence system that mimics the human brain nervous system at different levels and degrees [15].

BP neural network is an error back-transfer learning algorithm. The principle is to transform an input vector into an output vector through an implicit layer to realize mapping from input space to output space. The forward mapping is implemented by the weight, and the output of the network under the current weight is compared with the expected output of the mapping requirement that is desired to be implemented [16]. The BP algorithm has not only input layer nodes, output layer nodes, but also one or more hidden layer nodes. For the input signal, it must be propagated forward to the hidden layer node. After the action function, the output signal of the hidden node is propagated to the output node, and finally the output result is given.

This paper chooses a 3-layer neural network model as a debris flow risk assessment model. The specific steps are shown as follow[17]:

(1) Operation initialization: the three-layer structure weight W_{ij}, W_{jk} and a threshold value Q_j of the network model to be established are respectively assigned a small enough random value.

(2) Each model pair (input value and expected output value) of the sample set (known sample) is input to the network.

(3) Calculate the output values of each layer node through forward propagation between networks:

$$X_j^{(s)} = f\left(\sum_i W_{ij}^{(s)} X_i^{(s-1)} + Q_j\right) \tag{3}$$

where: s is the input and output known sample number: i, j is the serial number of each corresponding layer; $X_j^{(s)}$ is the connection weight of the j_{th} neuron on the s_{th} layer; $W_{ij}^{(s)}$ is the connection weight of the i_{th} neuron to the j_{th} neuron on s_{th} layer; f is the action function, which generally is the sigmoid function, as $f = 1/(1+\exp(-x))$.

(4) Calculate the error δ_k of the output layer node k and the error σ_j of the intermediate layer node j based on the actual output o_k of the output layer and the expected output t_k and the output y_j of the intermediate layer node j:

$$\sigma_k = (o_k - t_k)o_k(1-o_k) \tag{4}$$

$$\sigma_j = \sum_s \delta_k W_{jk}(1-y_j) \tag{5}$$

where: s is the total number of learning modes.

(5) Adjust the connection weight of each layer according to the error of each layer

$$W_{jk} = W_{jk} + \alpha\delta_k y_j \tag{6}$$

$$W_{ij} = W_{ij} + \alpha\sigma_j I_i \tag{7}$$

where: α is the learning indicator and I_i is the output value of the input layer node i.

(6) Repeat step (4) to step (7). If the output layer cannot obtain the desired output value, the error signal between the actual output value and the expected output value will be circulated along the original link path until the accuracy requirement is reached, that is, the actual output. The difference between the value and the expected value is within a certain acceptable range, and the network training ends. After the training, the network found the influence of each evaluation indicator on the risk of debris flow from the sample example, and fixed it in the form of network connection weight and other structural parameters, and remained stable. At this point, the trained neural network can be used to input the unknown sample parameters to be predicted, and the corresponding prediction results can be directly obtained.

3.Results and discussion

This paper takes 270 point position data of actual debris flow disasters in Ya'an as input data, 200 data as training samples and 70 data as test samples. The results are shown in Table 2. Then use the grid calculator of GIS spatial analysis to superimpose each index factor according to formula 8 to obtain the comprehensive distribution map of debris flow hazard in Ya'an city, as shown in Figure 1.

Table 2 Weights of indicators

Factor	Lithology level	Undulation degree	Slope gradient	Distance from faultage	Rainfall
Weight	0.140	0.094	0.450	0.197	0.118

$$P = L\times\alpha + U\times\beta + S\times\chi + D\times\delta + R\times\varepsilon \tag{8}$$

In the formula, P is the flood hazard index; L is the lithology index; U is the undulation index; S is the slope index; D is the distance from the faultage; R is the rainfall index; α, β, γ, δ, ε are the weight of each indicator.

IOP Conf. Series: Materials Science and Engineering **794** (2020) 012006 doi:10.1088/1757-899X/794/1/012006

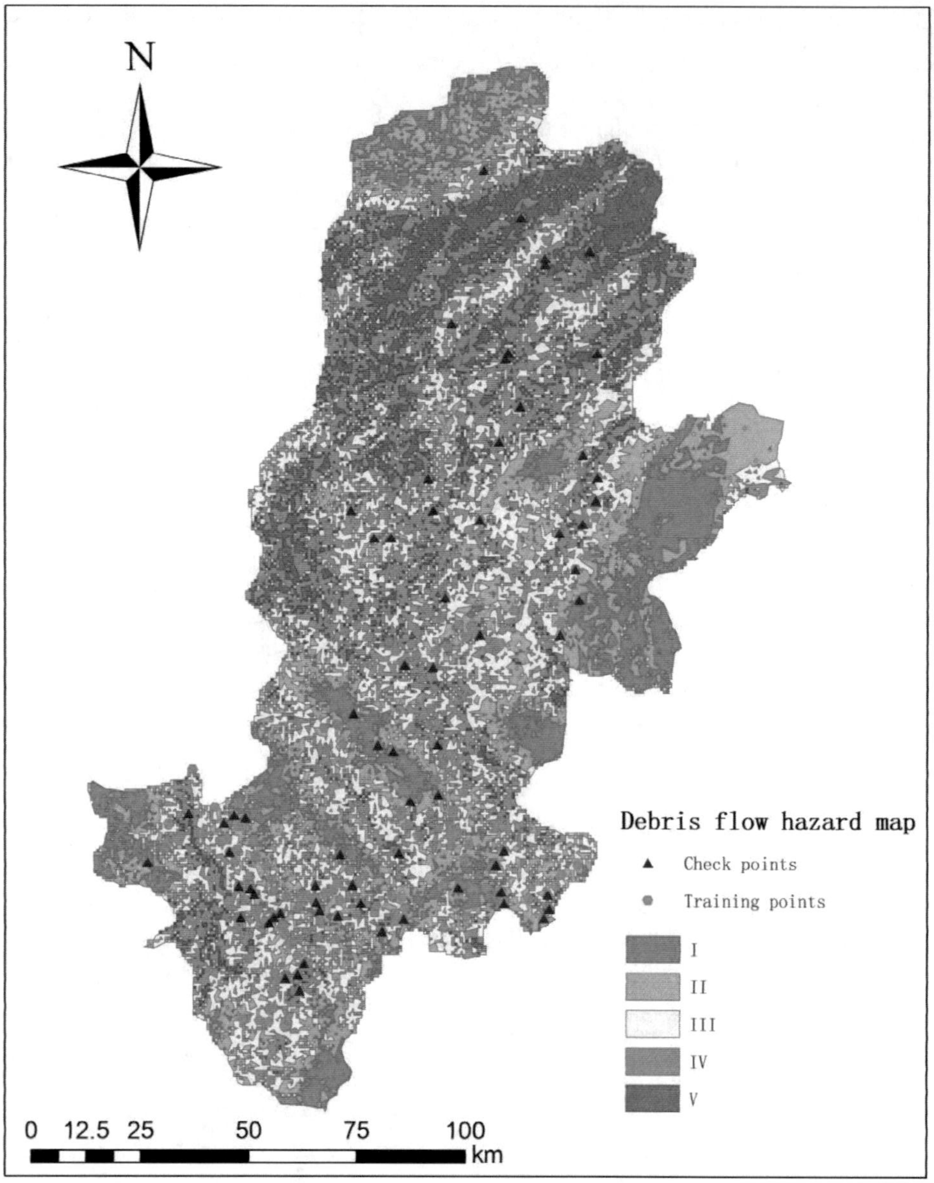

Fig.1 Debris flow hazard map

It can be seen from the figure that the risk of debris flow in the northern part of Ya'an city is significantly higher. Through the analysis of the test samples, 70% is located in the medium and above areas of debris flow hazard, and the results reflect the actual situation of the spatial distribution of debris flow in Ya'an. The extremely high-risk area accounts for 15.3% of the total area. The central and southern parts of Baoxing County, the north-central part of Lushan County, the western part of Tianquan County, and the central part of the central part of Shimian County are extremely dangerous. These areas belong to the area with large and strong slope fluctuation, which are distributed on the fault zone, and are prone to debris flow disaster; the high-risk areas account for 25.2%, including southern Baoxing County, Lushan County, Tianquan County, middle of Yingjing County, and north-central Shimian County, these areas have relatively large undulations and less vegetation cover.

If heavy rains or flash floods occur, debris flows are prone to occur; the moderately dangerous areas account for 29.7%, mainly distributed in the central and southern parts of Ya'an city. Most of these areas are not distributed on the fault zone and the geological conditions are good, the possibility of debris flow is small; the low-risk, extremely low-occupancy areas accounts for 29.8%, mainly located in the southeast of Yucheng District, north of Baoxing County, west of Shimian County and Mingshan District. Most of these areas have good vegetation coverage and flat terrain, and are not prone to debris flow disaster.

Statistical analysis was performed using GIS software, and the results as shown in Table 3 were obtained. It can be seen from the table that the V-level areas of Tianquan County, Lushan County and Baoxing County account for the largest proportion, and the proportion of Grade IV areas is above 25%. It is necessary to be wary that these IV-level areas may break through the critical value. It is converted into a V-class dangerous zone, so these three administrative districts need to do scientific disaster prevention work; Yingjing County, Yucheng District and Shimian County III and above account for more than 50%, and it is also necessary to be alert to the possibility of mudslides. . The areas of Grade IV and Grade V in Hanyuan County and Mingshan District are relatively small, and the overall risk of debris flow is the lowest, but it is also necessary to pay attention to the prevention of debris flow.

Table 3 Hazard ratio and the proportion of different hazard levels in each district (county)

District (county) risk ranking	Hazard levels				
	I	II	III	IV	V
Tianquan County (1)	3.894	14.873	30.557	29.313	21.363
Lushan County (2)	1.995	9.726	31.671	35.411	21.197
Baoxing County (3)	6.885	14.541	29.947	28.069	20.558
Yingjing County (4)	5.581	20.155	31.550	25.969	16.745
Yucheng District (5)	13.090	35.622	24.036	16.308	10.944
Shimian County (6)	11.202	30.995	28.058	21.588	8.157
Hanyuan County (7)	14.824	28.283	30.689	18.661	7.543
Mingshan District (8)	24.219	39.063	25.000	10.156	1.562

4.Conclusions

(1) On the basis of summarizing the research results of predecessors, select rainfall, lithology level, distance from faultage, undulation degree and slope gradient as evaluation indicators, construct an evaluation index system, and each index factor is superimposed with its respective weights by GIS software. Then the risk distribution map of the debris flow disaster in Ya'an city is obtained. Finally, the analysis results are compared with the location of the debris flow disaster at the checkpoint, and the research results show good consistency.

(2) Using BP neural network and combining with the actual occurrence location point of debris flow disaster, the evaluation indicators are extracted and standardized for learning and training, then the weight of each evaluation index is obtained, and the accuracy of the research results is improved.

(3) According to the risk distribution map, the areas with high risk of debris flow disasters are mainly concentrated in the central part of Baoxing County and the western part of Tianquan County and most areas of Lushan County. The population of these three administrative areas is relatively densely distributed, which will aggravate the impact of mudslide disasters. Therefore, appropriate

disaster prevention work should be done.

(4) This paper uses the 0.25°×0.25° TRMM satellite multi-year average rainfall index, which has a large spatial scale, but is still insufficient spatial differentiation of mountain rainfall. And there is also a lack of description of local short-term heavy rainfall. The surface rainfall on the hourly and empty scales is the average maximum three-day rainfall for many years, and is greater than the average annual rainstorm days of the 50 mm flood season. The above shortcomings are expected to be improved in future research.

Reference:

[1] Wang J Z, Ren G M, Yu T B, Gao B. (2015) Development characteristics and risk assessment of debris flow in Dayelonggou, Lushan earthquake area, Sichuan. J. Chinese Journal of Geological Hazard and Control, 26(04): 1-5.

[2] Zhao X, Cheng Z L, Liu J K, Liu D X, Shi L. (2013) Study on Risk Assessment of Single Gully Debris Flow in Dongchuan District of Yunnan Province. J. Journal of Catastrophology, 28(01): 102-106.

[3] Wang X L, Li J Y. (2011) Research on debris flow risk assessment system based on analytic hierarchy process. J. China Mining, 20(10): 113-117.

[4] Han Y S, Cui P, Liu H J, Yang M Q, Li C K. (2008) Risk assessment method and application of debris flow disaster. J. Chinese Journal of Safety Science, 18(12): 140-147+26.

[5] Liu X L, Su P C. (2004) Risk Assessment of Debris Flow in Sichuan Province. J. Disaster Science, (02): 25-30.

[6] Li H X. (2002) Analysis on Risk Assessment, Zoning and Development Trend of Soil and Water Loss in Ya'an Area. J. Journal of Soil and Water Conservation, (06): 17-19.

[7] Tang J F, Xie H. (1999) Study on debris flow risk zoning under the support of gis technology——Taking the upper reaches of Minjiang River as an example. J. Sichuan Surveying and Mapping, (03): 120-122.

[8] Tan B Y. (1986) Quantitative comprehensive evaluation of the severity of debris flow ditch. J. Bulletin of Soil and Water Conservation, (01): 51-57+44.

[9] Yu J R, Yan L H, Li X. (2014) Evaluation of debris flow risk based on gis space modeling. J. Science Chinese, (4):30-30.

[10] Wang L X. (1982) About the classification of wild rivers. J. Journal of Beijing Forestry College, (03): 94-107.

[11] Fu X G. (2008) Research on Benefit Evaluation System of Slope Runoff Distribution Project in Earth and Rock Mountain Area. D. Chang'an University.

[12] Cui P, Yang K, Wei F Q, Chen X Q. (2001) Index system for debris flow disaster assessment. J. Journal of Natural Disasters, (04): 36-41.

[13] Li X Y, Liu Q S, Bai S Y. (2017) Risk Assessment of Debris Flow Disaster in Ya'an City, Sichuan Province. J. Bulletin of Soil and Water Conservation, 37(02): 278-283+288.

[14] Lai C G, Wang Z L, Song H J. (2011) Risk assessment of flood disaster in Beijiang River Basin based on bp neural network. J. Hydropower Energy Science, 29(03): 57-59+161.

[15] Mao J, Zhao H D, Yao W. (2011) Development and application of artificial neural network. J. Electronic Design Engineering, 19(24): 62-65.

[16] Wang L. (2014) Principle, classification and application of artificial neural network. J. Science and Technology Information, (03): 240-241.

[17] Liu Y J, Hu H T, Bai Z Y. (2001) Neural network method for risk assessment of debris flow. J. Geology and Exploration, (02): 84-87.

Spatial downscaling of monthly precipitation based on TRMM satellite in Ya'an

Ting Chen[1,2], Xiaoyu Liu[1], Xing Liu[1*], Xiaodong Li[3] and Jie Liu[1]

[1]College of Water Conservancy and Hydropower Engineering, Sichuan Agricultural University, Ya'an, Sichuan, 625014, China

[2]College of Hydrology and Water Resources, Hohai University, Nanjing, Jiangsu, 210098, China

[3]College of Water Conservancy and Hydropower Engineering, Sichuan University, Chengdu, Sichuan, 610065, China

[*]Corresponding author's e-mail: hydro_lx@sicau.edu.cn

Abstract. In the study of precipitation in ya'an with complex terrain, a statistical downscaling algorithm based on the monthly precipitation data of TRMM 3B43 from 2000 to 2008, digital elevation model (DEM) and normalized difference vegetation index (NDVI) was presented. The precipitation fields from TRMM 3B43 were downscaled to 1 km × 1 km resolution simulated precipitation by constructing multiple linear regression equation and using regression model and residuals method. Combined with the observation values obtained from 8 weather stations in the study area, the downscaled precipitation was tested in the two aspects of overall accuracy and singel station. On this basis, the spatial and temporal distribution characteristics of monthly precipitation in recent 9 years in the study area were analyzed. The results showed that: The correlation coefficients of TRMM before and after downscaling and observation values were 0.863 and 0.854, and the biases were -0.08 and -0.12. The correlation and accuracy of the two types of data were basically consistent on the whole. 75% of the stations had a high correlation with the downscaling results, and the absolute value of the bias of each station was no more than 0.35, which showed that the data accuracy was high, the data was suitable for the precipitation analysis in the research area. The distribution of monthly precipitation in 8 stations over the years showed that the precipitation in the middle of ya'an was concentrated in January and November to December. From February to April and July to August the north received more precipitation than the south, while the opposite was true from May to June and September to October, and the August in the main flood season reached the heaviest rainfall of the year.

1.Preface

Precipitation, as the most active meteorological element in the process of hydrological cycle, climate change and ecosystem balance, has a strong driving force, and its amount plays a key role in the water vapor distribution and heat balance of the region and even the whole world[1-2]. Atmospheric movement, underlying surface conditions and geographical location act on precipitation, which makes precipitation have strong spatial and temporal heterogeneity[3-4]. Therefore, high spatial resolution continuous distributed precipitation data plays an important role in constructing regional hydrological and meteorological models and analyzing regional precipitation characteristics[5-6]. In the past, researchers spatially interpolated based on the data from the dot-form meteorological observatory to obtain regional precipitation data. This traditional method is usually limited by the observation range and topography of monitoring instruments[5], and has advantages in areas with high-density meteorological observatory network. Although many observation stations have been set up in ya'an area, which are located in the western Sichuan plateau, the number of stations is still very limited. The precipitation is greatly affected by the local special topography and complex climate. Therefore, the density of stations can not meet the needs of regional precipitation research. Different from point rainfall, area rainfall is based on the whole research area or basin, which has the advantages of high data continuity and wide coverage. Therefore, other methods with high feasibility and accuracy can be used to obtain surface distributed precipitation data for precipitation research.

The rapid development of RS and GIS provides a new way for precipitation research[7]. In the late 1990s, TRMM (Tropical Rainfall Measuring Mission), a remote sensing satellite developed jointly by the United States and Japan, carried a detector with the function of precipitation measurement. Through space-time detection and inversion of tropical and subtropical rainfall[8], it was obtained that the spatial resolution of precipitation data with strong spatial-temporal continuity, large coverage area and small geographical constraints is as high as $0.25° \times 0.25°$[9]. At present, a large number of studies have shown that TRMM data and meteorological observation data have good consistency and high applicability[10-11]. However, when constructing hydrological and meteorological models, the accuracy of data often can't meet the demand, and regional precipitation analysis is limited to some extent. Therefore, it is necessary to downscale TRMM precipitation data in order to obtain higher precision spatial distribution precipitation.

Many domestic and oversea scholars have used statistical downscaling method to construct the linear function relationship between TRMM 3B43 precipitation data and other underlying surface elements to study the regional precipitation distribution characteristics. The two types of underlying surface factors that are often taken into account are terrain and normalized difference vegetation index (NDVI). Topography is an important factor affecting local climate. Considering the influence of topography on precipitation, we can start with specific factors such as elevation, slope, aspect and so on, and concretize complex elements. However, there are some limitations for the lack of consideration of other underlying surface factors. For example, Ma JH[5] and Yang WY[6], taking into account the complex topographic conditions, extracted five types of topographic elements from DEM, including elevation, aspect, slope, longitude and latitude, and established a multiple linear regression equation between TRMM 3B43 annual average precipitation data and extracted topographic elements. Satellite precipitation data in the study area in the past 10 years was obtained from 0.25° (the low resolution) to 1 km (the high resolution), which reflected the basic characteristics of local precipitation better. Based on the principal component-stepwise regression method, Fan XW[12] reduced the dimension and eliminated the insignificance of seven types of topographic factors extracted from the study area, deepened the multivariate linear regression analysis method, established the annual and seasonal downscaling model, and found that the precision of precipitation data with 1 km resolution was higher, which provided a methodological reference for the study of spatial precipitation in arid areas. There will be some limitations if only the influence of NDVI is considered. Immerzeel[13] only considered NDVI as a factor, and constructed a polynomial regression equation between TRMM precipitation and NDVI. Because of the absence of relevant geographical factors, the 1 km high resolution data could not meet the needs of research. Some scholars took the NDVI factor into account while considering the topographic impact. Based on Immerzeel's research and considering NDVI and DEM factors simultaneously, Park and Jia SF[14-15] obtained the simulated precipitation distribution with a high resolution of 1 km in the study area. Li J[16] took the middle part of Tianshan Mountains as the research area. Based on the correlation among TRMM 3B43 precipitation and NDVI, DEM, aspect, longitude and latitude, the multivariate linear regression equation was constructed to obtain higher resolution spatial distribution data 250 m × 250 m, which was suitable for precipitation research in Tianshan Mountains. Zheng J[17] obtained high-precision monthly precipitation data with resolution of 1 km by further analyzing the lag of

IOP Conf. Series: Materials Science and Engineering **794** (2020) 012007 doi:10.1088/1757-899X/794/1/012007

NDVI factor to precipitation response and eliminating the data with weak correlation. The addition of NDVI factor on the basis of considering topographic influence not only improved the spatial variability of regression model, but also improved the coupling degree between variables. However, it had the shortcoming of imperfect analysis of the spatial characteristics of NDVI itself.

Most of the previous studies focued on downscaling study and spatiotemporal analysis of annual precipitation data, but the time scale has not been significantly improved, and there was no analysis of the spatial characteristics of the introduced NDVI factor. Considering the complex topography and vegetation coverage in ya'an, based on the analysis of the spatial correlation of NDVI, combined with longitude latitude, elevation, aspect and NDVI factors, this paper studied the spatial downscaling of monthly precipitation data from 2000 to 2008 in the study area, used the measured data of eight meteorological stations in the study area to test the accuracy, and then carried out the precision test. The spatial and temporal distribution of monthly precipitation in the study area was analyzed to provide an effective data basis for the construction of meteorological and hydrological models in ya'an.

2.Survey and data processing of research areas

2.1Survey of research areas

Ya'an, located in the western margin of Sichuan Basin, belongs to the upper reaches of the Yangtze River. Its geographic coordinates range from 28°51′-30°56′ north latitude, and 101°56′- 103°23′ east longitude. The study area is adjacent to the western Sichuan plain in the East and the western Sichuan plateau which is the eastern part of the Qinghai-Tibet Plateau in the east. It is located at the southeastern edge of the western Sichuan plateau. As a transitional zone from Sichuan-Chongqing Basin to the western Sichuan Plateau, the mountain ranges are crisscrossed and the relief degree is relatively large (Figure 1 (a)). The overall topography is higher in the north, West and south, lower in the middle and East (Figure 1 (b)), and uneven distribution of precipitation in the north and south. The unique topography of "windward slope" and "trumpet mouth" in ya'an makes it easy for water vapor to rise and condense in the circulation process, forming precipitation. The local climate type is subtropical monsoon humid climate, with more rainfall and high humidity. The annual rainfall is about 1000-1800 mm, it is the most rainfall area in Sichuan province[18]. In ya'an, there are two main river systems, Dadu River and Qingyi River. The river network has high density, large drop, rapid flow and abundant hydropower resources. Ya'an forest is widely distributed, and its vegetation coverage is as high as 50.97%. Influenced by the special topography and landform, complex climatic characteristics, dense river network distribution and high vegetation coverage in ya'an, the interaction between the air flow from the western plateau and the warm and wet air flow from the basin has formed the precipitation characteristics of ya'an with more rainy days, longer rainy hours and larger rainfall. The spatial distribution of 8 basic meteorological stations selected in this study is shown in figure 1 (b).

IOP Conf. Series: Materials Science and Engineering **794** (2020) 012007 doi:10.1088/1757-899X/794/1/012007

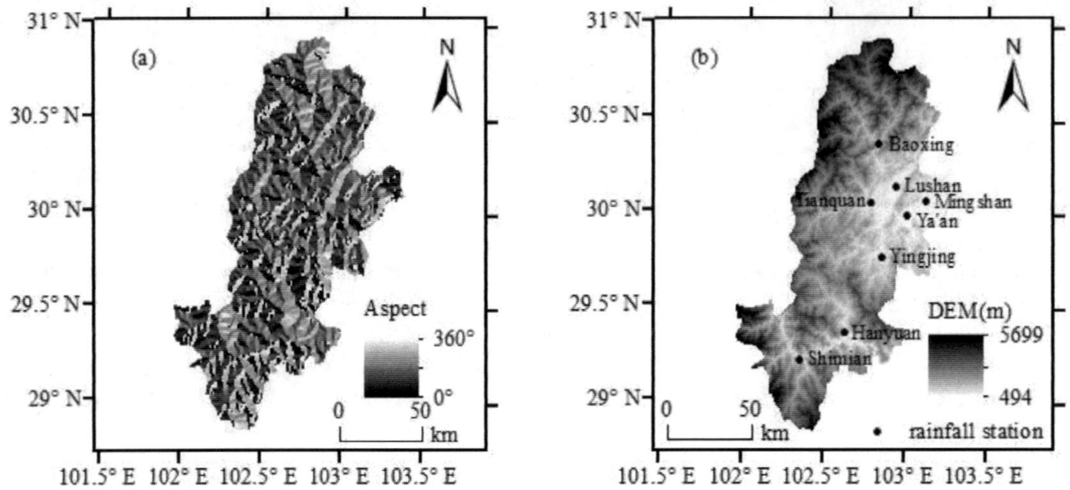

Figure 1. (a) Aspect of research area (b) Topographic elevation and rainfall Station distribution in the study area.

2.2Data source and preprocessing

2.2.1Remote sensing data. The remote sensing data products used in this paper include TRMM 3B43 satellite precipitation data set, MODIS data product NDVI data, data of digital elevation model.

The spatial resolution of TRMM 3B43 data is 0.25°, the temporal resolution is 1 month, and the precipitation unit is mm/h. The data comes from the DAAC (Distributed Active Archive Center) of NASA Goddard in the United States, which is available free of charge through the NASA website (http://trmm.gsfc.nasa.gov/). In this study, we used MATLAB program to read the precipitation matrix data of TRMM 3B43 in HDF format from 2000 to 2008, and converted it into point and surface files in vector format through ARCGIS.

The NDVI data of the same period is MODND1M, which is synthesized in China. It comes from MODIS synthesized products in China. It is calculated by monthly maximum synthesis method. The spatial resolution is 500 m and the temporal resolution is 1 month. Considering the lag of vegetation response to precipitation, this study covered the period from February 2000 to January 2009. A total of 108 NDVI images were downloaded, which could be obtained free of charge through the geospatial data cloud website (http://www.gscloud.cn/). Through projection conversion, resampling and mask processing of ARCGIS, NDVI data with spatial geographic information at resolutions of 0.25° and 1 km were obtained, which served as the data basis for vegetation precipitation hysteresis analysis, multiple linear regression model construction and downscaling precipitation data calculation.

DEM data comes from the GDEMDEM data set, which is processed from the data of the first edition of ASTER GDEM (V1). It is a digital elevation data product with global spatial resolution of 30 m × 30 m. According to the latitude and longitude of the study area, 12 DEM data covering the study area were downloaded. The geospatial data cloud website (http://www.gscloud.cn/) could download the data for free. Preprosessing of DEM was the same as NDVI, and aspect of corresponding resolution were extracted at the same time.

2.2.2Meteorological station data. Meteorological station data includes daily precipitation data of 8 rainfall stations in Baoxing, Lushan, Ya'an, Xingjing, Hanyuan, Shimian, Mingshan and Tianquan. This data comes from Sichuan Meteorological Bureau. After strict quality control and inspection, the quality is good. In this study, the daily precipitation (unit: 0.1mm/day) of the stations was calculated, and the monthly precipitation (unit: mm/month) was obtained. The average monthly precipitation of each station was calculated year by year, which served as the test data basis for the downscaling precipitation results.

3.Research method

3.1Introduction and analysis of NDVI factor

3.1.1Time lag analysis of vegetation response to precipitation. The vegetation growth status and coverage area are greatly affected by precipitation. It takes a long time from rainfall to the ground, through vegetation interception and absorption, vegetation growth and coverage expansion. During this period, NDVI showed a lag to precipitation[19]. Zheng J studied the correlation between the monthly NDVI value and the precipitation of the same month, the previous month and the previous two months in the western Sichuan Plateau from 2001 to 2013, and obtained that the lag period of vegetation relative precipitation in the study area was about one month[17]. Ya'an is in the western Sichuan plateau, which satisfies this conclusion. Therefore, the NDVI value of vegetation in the next month was selected as the vegetation factor corresponding to the monthly precipitation when constructing the multi-linear regression downscaling model.

3.1.2Spatial autocorrelation analysis of vegetation. (1) Gloable Moran's index
The spatial distribution of vegetation is affected to some extent by the spatial distribution of precipitation. In this part, the global Moran's index was introduced to illustrate the spatial autocorrelation of the normalized vegetation index NDVI. The formula is as follow:

$$I = \frac{\sum_{i=1}^{n}\sum_{j=1}^{n} w_{ij}(x_i - \bar{x})(x_j - \bar{x})}{S^2 \sum_{i=1}^{n}\sum_{j=1}^{n} w_{ij}} \tag{1}$$

In the formula, w_{ij} is a spatial weight matrix (used to measure the degree of influence between pixels I and j), x_i and x_j are the NDVI values of spatial pixels i and j, and S^2 is the variance of NDVI samples. $S^2 = \frac{\sum_{i=1}^{n}(x_i - \bar{x})^2}{n}$, $\sum_{i=1}^{n}\sum_{j=1}^{n} w_{ij}$ is the sum of all spatial weights, n is the number of spatial pixels.

The value range of global Moran index is [-1, 1].When the global Moran's I is greater than 0, the data shows positive spatial correlation, and the greater the spatial correlation is, the more obvious the spatial correlation is; when the global Moran's I is less than 0, the data shows negative spatial correlation, and the smaller the value is, the greater the spatial difference is, then the NDVI factor is not used as the influencing factor to construct the downscaling model. When equal to 0, the space is random. In this paper, the global Moran's I index was calculated by GeoDa software.

(2) Local Moran's index

Influenced by human factors, there was an unreasonable correlation between regional precipitation and vegetation cover. In order to eliminate the abnormal values in NDVI, local Moran's index was introduced on the basis of global Moran's index, which could reveal the correlation between current spatial pixels and their adjacent spatial pixels, identify the degree of spatial dispersion and reflect the spatial variability of vegetation. The concrete calculation formulas are as follow:

$$I = \frac{n^2}{\sum_{i=1}^{n}\sum_{j=1}^{n} w_{ij}} \frac{(x_i - \bar{x})\sum_{j=1}^{n}(x_j - \bar{x})}{\sum_{j=1}^{n}(x_j - \bar{x})^2} \tag{2}$$

When the local Moran's I is greater than 0, the NDVI aggregation is relatively concentrated, and the positive spatial correlation is relatively strong; when the local Moran's I is less than 0, the NDVI aggregation is relatively discrete and the positive spatial correlation is relatively weak. If the index is less than a certain value, the part of NDVI value should be excluded in the construction of downscaling model. In this paper, ARCGIS software was used to calculate the local Moran's I.

3.2Construction of spatial downscaling model

3.2.1Principle of statistical downscaling. Statistical downscaling is a common empirical downscaling method. The linear or non-linear function relationship between large-scale and low-resolution meteorological change elements and regional small-scale and high-resolution meteorological change elements is constructed by using the statistical empirical data collected by satellites, so as to realize the transformation of spatial scale and obtain small-scale. On the basis of statistical downscaling method, this paper chose multivariate linear regression method to construct a downscaling model suitable for the study area.

3.2.2Downscaling steps. (1) Taking the study area as the boundary, a buffer with a distance of 25 km was established by using the neighborhood analysis method, which included and superimposed the original area. The monthly precipitation, NDVI and Dem data areas of TRMM 3B43 at 0.25° resolution in the study area were obtained by using the established buffer as the clipping boundary. The central positions of each data pixel correspond one by one.

(2) The monthly precipitation and five kinds of independent variables were extracted from the grid center at 0.25° resolution, and the multiple linear regression model was constructed. Namely:

$$P_{LR} = aX + bY + cN + dD + eA + C \tag{3}$$

In the formula, P_{LR} is the predicted monthly precipitation with resolution of 0.25° × 0.25°, mm; X and Y are longitude and latitude respectively, °; N is normalized difference vegetation index; D is elevation, m; A is aspect, m; C is constant; a, b, c, d and e are linear regression simulation fitting coefficients.

(3) Calculated the $\Delta TRMM_{LR}$ of precipitation residual caused by various topographic factors and vegetation coverage at a resolution of 0.25°.

$$\Delta TRMM_{LR} = TRMM_{LR} - P_{LR} \tag{4}$$

In the formula, $TRMM_{LR}$ is the original monthly precipitation value at a resolution of 0.25°, mm. The precipitation residual $TRMM_{HR}$ at 1 km high spatial resolution is extracted by spline interpolation.

(4) The regression coefficients at 0.25° × 0.25° and the corresponding factors at 1 km resolution were substituted into equation (3) to obtain the predicted monthly precipitation value (P_{HR}) at high resolution. Using the method of "regression model + residual", the monthly simulated precipitation ($TRMM_{HR}$) of each year in the study area under 1 km × 1 km was obtained. Namely:

$$TRMM_{HR} = P_{HR} + \Delta TRMM_{HR} \tag{5}$$

3.2.3Precision testing index of downscaling result. For the scale-down results of 108 months from 2000 to 2008, the measured data of time and space at 8 rainfall stations in the study area were tested. The results were validated by the following three indicators: correlation coefficient (R), bias (BIAS) and root mean square error (RMSE):

$$R = \frac{\sum_i^n (H_i - \overline{H})(M_i - \overline{M})}{[\sum_i^n (H_i - \overline{H})^2 (M_i - \overline{M})^2]^{1/2}} \tag{6}$$

$$B = \frac{\sum_i^n M_i}{\sum_i^n H_i} - 1 \tag{7}$$

$$RMSE = [\sum_i^n (H_i - M_i)^2 / n]^{1/2} \tag{8}$$

Among them, H_i is the measured precipitation at the rainfall station, M_i is TRMM 3B43 precipitation or precipitation at downscales, i is the rainfall station, and n is the total number of data. In practice, the coordinates of TRMM monthly precipitation grid unit at 0.25° and 1 km resolution were calculated according to the longitude and latitude position of the meteorological station, and then the monthly precipitation at the center of the pixel was extracted as the value of M_i. R is used to measure the linear relationship between H_i and M_i. The value range is [-1, 1]. The closer R is to -1 or 1, the higher the consistency of data is. Bias and RMSE are used to describe the deviation or deviation between TRMM monthly precipitation and downscaling results and observed monthly precipitation values. The closer Bias approaches 0 or RMSE is, the higher the accuracy of TRMM monthly precipitation or downscaling results is.

4.Result and discussions

4.1Overall accuracy test

TRMM 3B43 precipitation data was downscaled to obtain simulated precipitation with high resolution. To verify its effectiveness, monthly measured precipitation at eight precipitation stations in ya'an from 2000 to 2008 was used to verify the downscaling results from January to December. The monthly precipitation from the corresponding location of each rainfall station and the monthly precipitation from TRMM 3B43 data were extracted separately. The measured precipitation from each rainfall station was taken as an independent variable, and a linear regression analysis was made. The correlation coefficient R was used to reflect the correlation between variables. Bias (B) and RMSE were calculated to test the accuracy of the data. Figure 2 shows the scatter plots and error statistics of the observed monthly precipitation and TRMM 3B43 monthly precipitation (a) and downscaled monthly precipitation (b) in ya'an.

From figure 2, it can be seen that: (1) The correlation coefficient between downscaling simulated precipitation and observed monthly precipitation at rainfall station is 0.85, which is close to that of TRMM 3B43. The correlation degree between the two types of data and observed monthly precipitation data is basically the same. (2) Compared with the original data, the deviation value of the downscaled simulated precipitation data is slightly higher, which indicates that the downscaling results are slightly lower as a whole. However, the deviation values of the two types of data are low, indicating that the deviation degree has not changed significantly. (3) RMSE is sensitive to very large or very small outliers in predicted data. When there is a big difference between predicted and measured values, RMSE will be very large. By comparing the RMSE values of the two types of data in the graph, we can find that the RMSE values of the two types of data are close, which shows that the accuracy of the two types of data is basically the same. The data correlation and accuracy of downscaling simulated precipitation results are slightly lower than that of TRMM 3B43 data. The main reason is that TRMM satellite is affected by atmospheric circulation, complex terrain and underlying surface coverage differences in the study area, which results in some errors in data inversion results. This study combines five factors with TRMM 3B43 data to reduce the scale. Degree model construction, considering factors have certain representativeness and limitations, will produce certain errors in the process of model construction. But on the whole, the correlation and accuracy of the two types of data are not very different. The downscaling model in this paper still has certain reliability, and can reflect the spatial and temporal characteristics of precipitation distribution in ya'an more comprehensively.

IOP Conf. Series: Materials Science and Engineering **794** (2020) 012007 doi:10.1088/1757-899X/794/1/012007

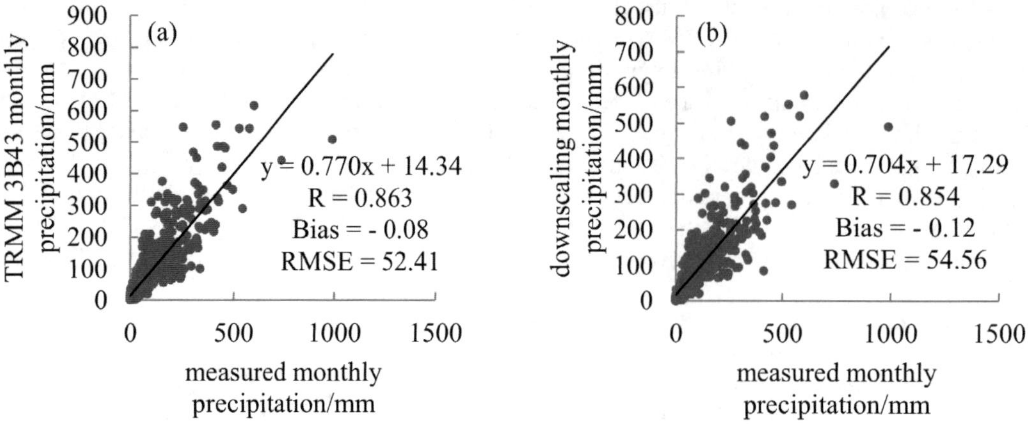

Figure 2. The scatter plots of the measured monthly precipitation and TRMM 3B43 monthly precipitation (a) and downscaling monthly precipitation (b) in ya'an.

4.2Accuracy test of single station

Based on the measured precipitation data and downscaling simulated precipitation data from 2000 to 2008, taking the monthly precipitation at 8 stations as independent variable, and the corresponding downscaling monthly precipitation at the corresponding location as the dependent variable, the linear regression analysis was carried out. The linear relationship and error statistics between the monthly precipitation measured at each rainfall station and the simulated monthly precipitation are shown in figure 3. From figure 3, it can be seen that: (1) The correlation coefficient between the downscaling simulated values and the measured values of Lushan, Mingshan, Ya'an and Tianquan stations is more than 0.90, and the discrete points are highly concentrated and evenly distributed near the regression line. The correlation between the downscaling simulated values and the measured values is significant and the data consistency is high. (2) At Baoxing and Xingjing stations, the R value of the downscaling simulated values and the measured values is more than 0.85, and the distribution of discrete points is relatively concentrated near the trend line, and the data consistency is relatively high. (3) At Hanyuan station and Shimian station, the R value between the downscaling simulated values and the measured values is more than 0.80 and the data consistency is low, but Bias and RMSE are low, which indicates the data accuracy is high. Among them, Shimian station has the lowest data consistency among the eight stations. The main reason is that Shimian station is located in the valley area, with an elevation of 891 m, surrounded by the middle mountain area (elevation 1000-3500 m), and the elevation difference is large, which is mainly manifested by complex and diverse mountain geomorphological characteristics. So that great differences can be seen between data of rainfall stations and TRMM satellite precipitation, leading to low consistency of data. (4) Among all rainfall stations, Lushan station has the most significant correlation between the two types of data, R= 0.941, Bias= -0.07, RMSE= 33.83. The data deviation is the lowest, and it is the most reliable station with the highest consistency and accuracy among the eight rainfall stations. The accuracy of downscaling simulated data and the consistency between the measured data and the downscaling simulated data of each station are obviously different in space, which is mainly related to the accuracy of TRMM 3B43 detection data itself, the complex topography of each station and the accuracy of multiple linear regression model.

IOP Conf. Series: Materials Science and Engineering **794** (2020) 012007 doi:10.1088/1757-899X/794/1/012007

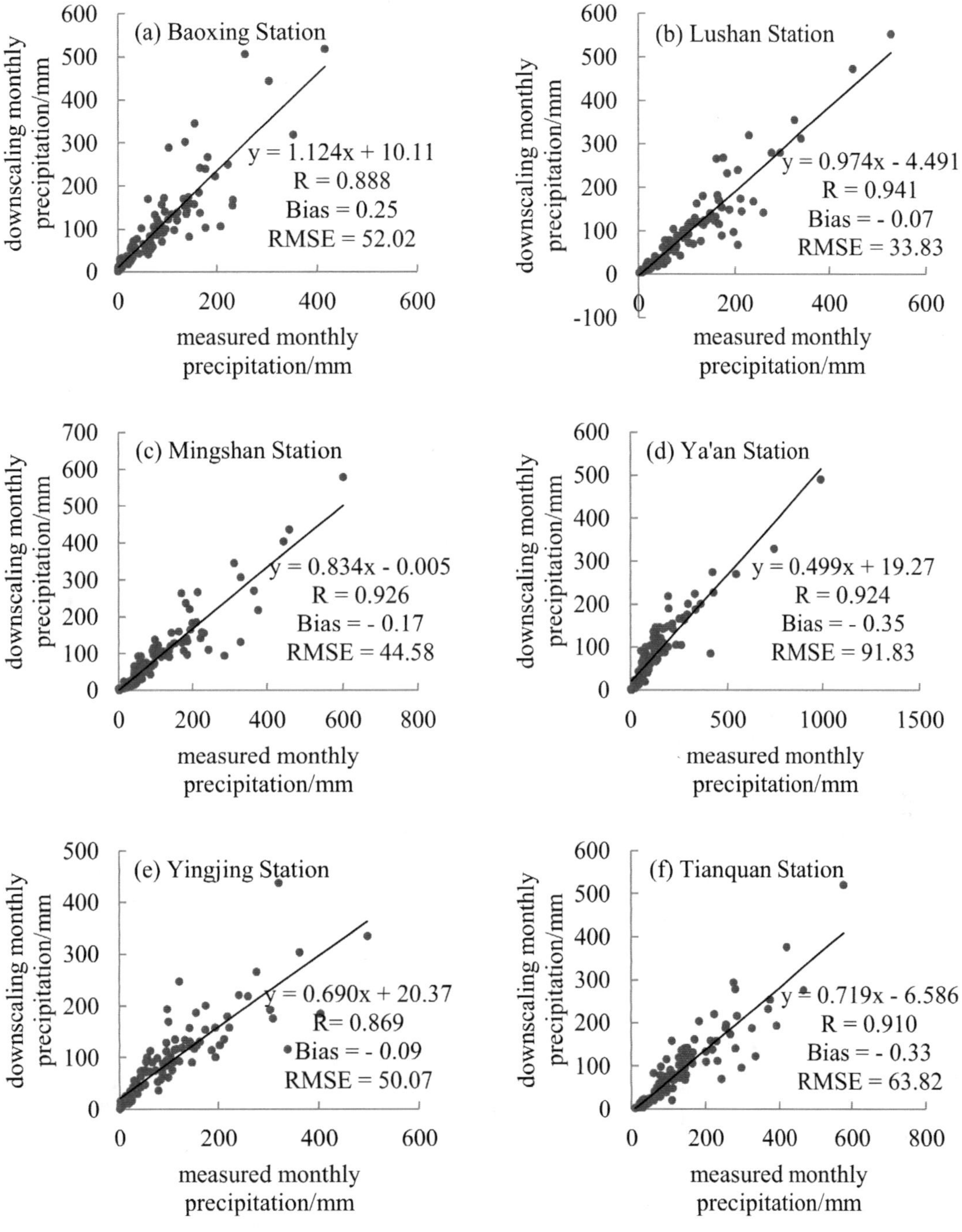

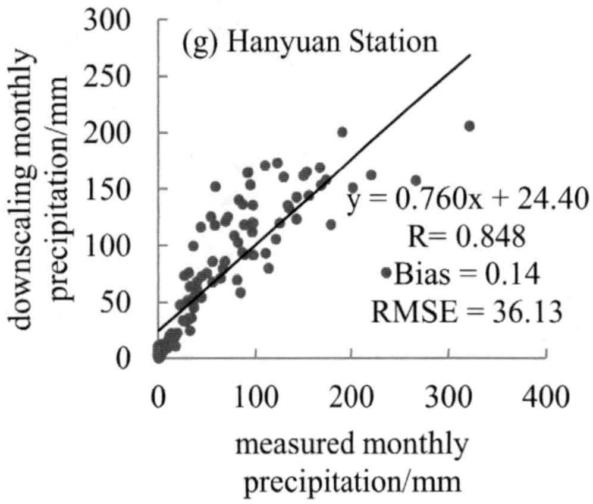

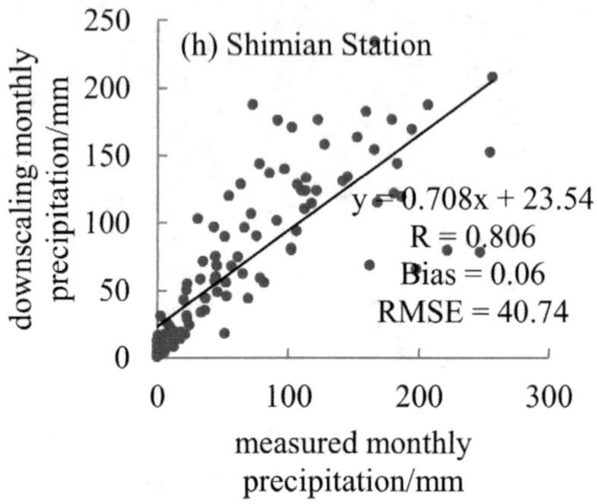

Figure 3. Linear relation and error statistics between monthly precipitation at downscales and monthly precipitation at various rainfall stations.

4.3Spatial and temporal distribution characteristics of monthly precipitation in the study area

The monthly downscaling simulated precipitation of each station from 2000 to 2008 was analyzed by eigenvalue statistics. The statistical results are shown in table 1. From table 1, it can be seen that the minimum monthly precipitation of 8 stations in the whole region is 0 mm, which occurs in January and December. The standard deviations of precipitation data series are 3.98 mm and 4.19 mm, respectively. The maximum monthly precipitation is 578.55 mm, which occurs in August, and the corresponding standard deviation is 132.25 mm. The average value shows that the annual precipitation is mainly concentrated from May to September, and the precipitation is over 100mm, the total precipitation from May to September accounts for about 77% of the annual precipitation.

Table 1. Statistics of annual monthly precipitation characteristic value in ya'an.

month	Precipitation (mm)			standard
	minimum value	maximum value	average value	deviation(mm)
1	0.00	17.47	9.93	3.98
2	1.70	44.78	16.29	9.59
3	4.67	75.56	37.69	19.25
4	35.73	169.74	76.68	21.44
5	54.06	165.02	104.07	26.26
6	76.56	218.26	144.45	29.80
7	78.64	345.44	177.97	58.87
8	55.93	578.55	239.72	132.25
9	50.68	199.80	117.76	34.43
10	19.69	116.07	64.41	21.99
11	2.08	77.89	25.07	16.87
12	0.00	20.42	7.49	4.19

The distribution of monthly average precipitation in 2000-2008 at 1 km spatial resolution in the study area is shown in figure 4, and the variation trend of monthly precipitation in the main flood season is shown in figure 5. From figure 4, it can be seen that the central ya'an has concentrated precipitation in January, November and December. The northern precipitation in February, March, April, July and August is more than the southern precipitation. The the southern precipitation in May, June, September and October is more than the northern precipitation. The precipitation in August is the most abundant, and the maximum monthly average precipitation is 319.13 mm, which appears at

Baoxing Station in the north. Secondly, in July, the maximum precipitation is 215.81 mm, which appears at Lushan Station in the north. Compared with the monthly average precipitation of nine years in the whole region, the maximum monthly precipitation of Baoxing Station in August and Lushan Station in July are higher than the average of the whole region in corresponding period. The minimum precipitation in January is 5.12 mm, which appears in Hanyuan station in the south. From January onwards, the monthly average precipitation in ya'an increases gradually. The monthly average precipitation from May to September exceeds 100 mm. In July and August, ya'an enters the main flood season. The monthly average precipitation is higher than 200 mm. After September, the monthly average precipitation gradually decreases, which is consistent with the trend of the monthly average precipitation in Table 1 for nine years.

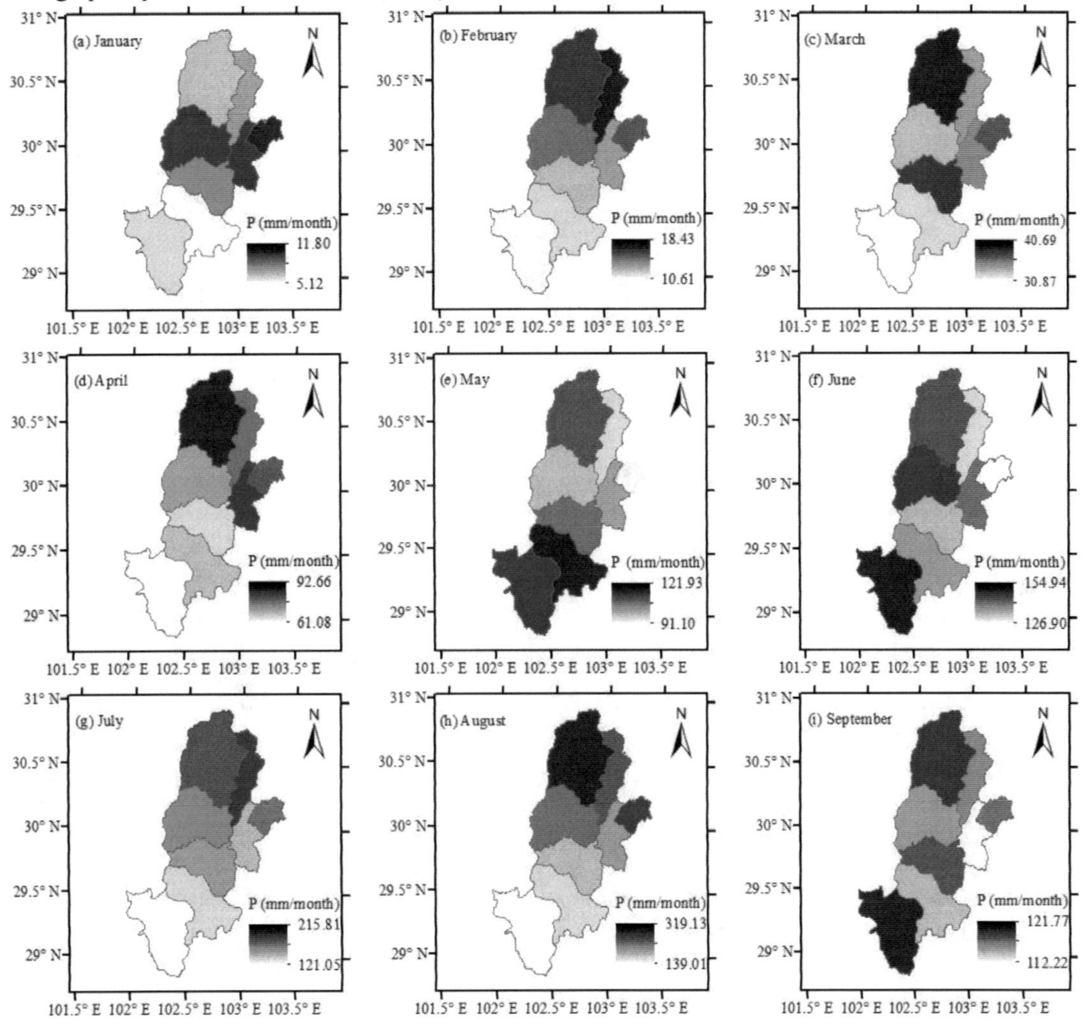

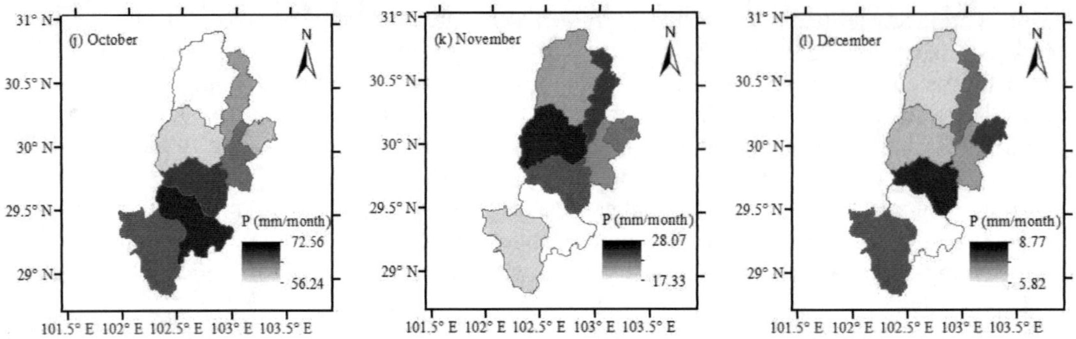

Figure 4. Distribution of annual monthly average precipitation in ya'an.

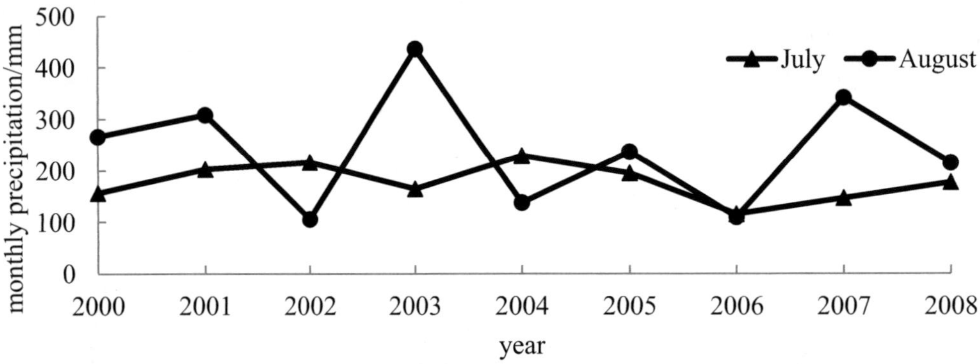

Figure 5. Change trend of monthly precipitation in the main flood season of ya'an from 2000 to 2008.

The main flood season in ya'an is July to August. As can be seen from figure 5, the precipitation in ya'an has been decreasing gradually within the nine years, with a minimum of 115.51 mm in 2006. The precipitation in August fluctuates greatly, with a minimum of 105.60 mm and a maximum of 436.33 mm in 2002 and 2003, respectively.

5.Conclusion and Prospect

Based on the TRMM 3B43 monthly precipitation data in ya'an from 2000 to 2008, the monthly precipitation data at 1 km resolution in the study area were obtained by constructing the multiple linear regression equation between longitude latitude, elevation, vegetation normalization index, aspect and satellite monthly precipitation data, and employing the method of "regression model + residual" to reduce the scale. The reliability of the downscaling model was verified by the measured data of 8 rainfall stations, and then the spatial and temporal distribution characteristics of monthly precipitation data in the study area were analyzed. The conclusions are as follows:

(1) The overall accuracy test shows that the R values of TRMM 3B43 data before and after downscaling are 0.863 and 0.854, and Bias values of them are -0.08 and -0.12, respectively. The correlation and accuracy between the original data and the downscaling results are basically the same. Accuracy test of single station shows that the measured data from about 75% rainfall stations are highly correlated with downscaling simulated precipitation, and the data deviation of each station is small, with the absolute value is no more than 0.35. And the root mean square error of 75% rainfall stations is relatively small. Among them, the correlation coefficient between predictive values and measured values in Lushan station is as high as 0.941, Bias is -0.07 and RMSE is 33.83. The test results show that the 1 km downscaling model in this paper has certain credibility on the regional scale, and is suitable for the precipitation study in ya'an.

(2) Monthly downscaling simulation results show that the precipitation in ya'an is concentrated in

the middle region in January, November and December, but the amount of precipitation is relatively small, the low-value area is located in the southern Hanyuan; the north is more than the south from February to April the same in July and August. The high-value area is located in the northern Baoxing and Lushan; the south is more than the north in May, June, September and October. The precipitation in ya'an increases significantly from May to October, and enters the main flood season in July and August. The precipitation in August reaches the maximum in the year.

After downscaling, the spatial resolution of TRMM precipitation data is improved and its continuous spatial distribution is stronger, which reduces the limitation of analyzing the spatial and temporal variation characteristics of precipitation in ya'an. However, there are still some errors between the downscaling data and the measured data, which are mainly related to the factors affecting the accuracy of TRMM data and the accuracy of the constructed multivariate linear regression model. Therefore, in the future research, we can consider using the measured precipitation to revise the processing, in order to improve the reliability of the data, or using GPM satellite precipitation, which has higher spatial and temporal accuracy[20], to make it more suitable for the analysis of spatial and temporal variation characteristics of precipitation in the study area.

References

[1] Liu, Q., Fu, Y.F. (2007). Research on Asian summer precipitation based on TRMM/TMI. Scientia Sinica(Terrae). 37(1): 111-122.

[2] Langella, G., Basile, A., Bonfante, A., et al. (2010). High-resolution space–time rainfall analysis using integrated ANN inference systems. Journal of Hydrology. 387(3): 328-342.

[3] Zhang, T., Li, B.L., He, Y.Q., et al. (2015). Spatial and Temporal Distribution of Precipitation Based on Corrected TRMM Data in Hengduan Moutains. Journal of Natural Resources. 30(2): 260-270.

[4] Bohnenstengel, S.I., Schlünzen, K.H., Beyrich, F. (2011). Representativity of in situ precipitation measurements - A case study for the LITFASS area in North-Eastern Germany. Journal of Hydrology. 400(3/4): 387-395.

[5] Ma, J.H., Qu, C., Zhang, H.X., et al. (2013). Spatial downscaling of TRMM precipitation data based on DEM in the upstream of Shiyang River Basin during 2001-2010. Progress in Geography,32(9):1423-1432.

[6] Yang, W.Y., Ma, J.H., Yang, W.K. (2014). Variation Characteristics of Precipitation Based on TRMM Data During 2001-2010 in Linxia of Gansu Province. Journal of Arid Meteorology, 32(6): 934-939.

[7] Michaelides, S., Levizzani, V., Anagnostou, E., et al. (2009). Precipitation: Measurement, remote sensing, climatology and modeling. Atmospheric Research, 94(4): 512-533.

[8] Kummerow, C., Barnes, W., Kozu, T., et al. (1998). The Tropical Rainfall Measuring Mission (TRMM) sensor package. Journal of Atmospheric and Oceanic Technology, 15(3): 809-817.

[9] Hao, Z.C., Tong, K., Zhang, L.L., et al. (2011). Applicability Analysis of TRMM Precipitation Estimates in Tibetan Plateau. Journal of China Hydrology, 31(5): 18-23.

[10] Lü, Y., Yang, S.T., Cai, M.Y., et al. (2013). The Applicability Analysis of TRMM Precipitation Data in the Yarlung Zangbo River Baisin. Journal of Natural Resources, 28(8): 1414-1425.

[11] Liao, H.T., Xiao, T.G. (2017). Applicability test of TRMM precipitation data in Sichuan and Chongqing region . FUJIAN ZHILIANG GUANLI, (14): 232-233.

[12] Fan, X.W., Liu, H.L. (2018). Downscaling Method of TRMM Satellite Precipitation Data over the Tianshan Mountains. Journal of Natural Resources, 33(3): 478-488.

[13] Immerzeel, W.W., Rutton, M.M., Droogers, P. (2009). Spatial downscaling of TRMM precipitation using vegetative response on the Iberian Peninsula. Remote Sensing of Environment, 113(2): 362-370.

[14] Park, N.W. (2013). Spatial Downscaling of TRMM Precipitation Using Geostatistics and Fine Scale Environmental Variables. Advances in Meteorology, 2013(Pt.2): 237126-1-237126-9.
[15] Jia, S.F., Zhu, W.B., Lü, A.F., et al. (2011). A statistical spatial downscaling algorithm of TRMM precipitation based on NDVI and DEM in the Qaidam Basin of China. Remote Sensing of Environment: An Interdisciplinary Journal, 115(12): 3069-3079.
[16] Li, J., Zhang, X. (2015). Downscaling Method of TRMM Satellite Precipitation Data. Scientia Geographica Sinica, 35(9): 1164-1169.
[17] Zheng, J., Lü, L., Feng, W.L., et al. (2016). Spatial Downscaling Simulation of Monthly Precipitation Based on TRMM 3B43 Data in the Western Sichuan Plateau. Chinese Journal of Agrometeorology, 37(2): 245-254.
[18] Peng, G.K., Li, Z.Y., Chai, F.X. (1985). The Relationship between Topography and Precipitation in Ya'an District. Plateau Meteorology, (3): 230-240.
[19] Jiang, L., Feng, W.L., Guo, B. (2014). Analysis of Dynamic Monitoring of Vegetation Change and the Correlation with Precipitation Factor in Yalu Tsangpo River Basin During the Past 13 years. Resources and Environment in the Yangtze Basin, 23(11): 1610-1619.
[20] Tang, G.Q., Wan, W., Zeng, Z.Y., et al. (2015). An Overview of the Global Precipitation Measurement (GPM) Mission and It's Latest Development. Remote Sensing Technology and Application, 30(4): 607-615.

Hazard evaluation of mountain flood disaster based on multi-source remote sensing information—a case study in city of rain, Ya'an

Juan Gong[1,2], Xing Liu[1*], Jie Liu[1], Yuanlin Li[1] and Shaoxing Zhao[1]

[1]The College of Water Resources and Hydropower Engineering, Sichuan Agricultural University, Ya'an City, Sichuan Province, 625000, China

[2]The College of Water Resources and Civil Engineering, China Agricultural University, Beijing City, Beijing, 100083, China

[*]Corresponding author's e-mail: hydro_lx@sicau.edu.cn

Abstract. Because the hazard of flood disaster is mainly related to the natural attribute of flood disaster, this paper makes full use of multi-source remote sensing information to reflect the meteorological and underlying surface conditions in Ya'an City, constitutes an evaluation index system according to selecting the rainfall, runoff depth, the basin storage capacity, elevation, slope and water system, from the two aspects of disaster-inducing factors and disaster-pregnant environment; trains historical flood occurrence sites by using BP neural network, obtains the weights of each research index; overlays analysis each index according to the index model, based on the GIS technology; and obtains the torrential flood hazard distribution map of Ya'an City as well as comparing it with the occurrence sites of historical flood disasters. The results show that the flood hazard level is the highest in the eastern part of Ya'an, the lowest level area of the flood hazard is the southern part. The evaluation results well reflect the actual situation of regional flood spatial distribution, which can provide a scientific basis for the development of reasonable flood control and disaster mitigation planning in Ya'an City.

1. Introduction

Flood disaster is current one of the natural disasters that have brought huge losses to human. The Fifth Assessment Report of the Intergovernmental Panel on Climate Change (IPCC AR5) indicates that global warming is occurring and will continue in the future due to the increased human emissions [1]. Resulting in natural system variation and frequent extreme climate, this will greatly affect the frequency, intensity, spatial extent, duration of flood disasters and increase the flood risk in some areas. In China, the mountain basin area covers for about two-thirds of the national land area, and the population accounts for about one third of the total population of the country. The losses caused by mountain torrents are particularly serious, which has become a significant factor restricting the sustainable development of mountain economy and society. In order to avoid and reduce the disaster loss from the source, it is necessary to evaluate the risk of mountain flood disaster scientifically.

From the perspective of disaster system theory, the risk of flood disaster refers to the damage caused by flood to the natural environment and human society to a certain extent, with two attributes of nature and society. Hazard evaluation is the basis of risk assessment, which involves the natural attributes of flood disasters. In brief, it analyzes flood disasters from two aspects: disaster-inducing factors (causes of flood) and disaster-pregnant environment (natural environment for flood disasters)

[2]. Research on regional flood risk assessment has always been a hotspot, and methods of historical flood, Hydro-hydraulic model and index are often adopted [3]. Historic flood method is simple in calculation, but difficult in data acquisition, and the consistency of data statistics is difficult to ensure. Therefore, hydrological frequency calculation method is often adopted to calculate the recurrence period [4-5]. The hydro-hydraulic method can calculate the flow, possible flood inundation scope, depth and other elements through mathematical simulation of flood by means of hydrological and hydraulic models, and has high calculation accuracy, but needs to observe precipitation, flow, water level and other data, limiting its application in mountain watershed [6-7]. With the development of GIS and RS technology, index method is more commonly used evaluation methods in recent years, which has low data requirements and is suitable for flood assessment in non-data regions. The most intuitive method of index method is to extract flood inundation scope and depth indicators based on remote sensing images, usually used for rapid extraction of hazard areas and assessment of post-disaster losses in sudden flood disasters [8-9]. Huang Shifeng et al. [10] used the natural geomorphology of the Liaohe River Basin as the research object, and used the river network density index to characterize the underlying surface conditions of the basin, and analyzed its relationship with flood hazard. As the occurrence of flood is affected by many factors, it is more reasonable to use a variety of indicators to obtain comprehensive indicators through weighting. Thus, analytic hierarchy process, fuzzy comprehensive evaluation, grey clustering, entropy weight method, random forest, artificial neural network [11-13] and other methods are often used to judge the weight. However, there are great uncertainties in analytic hierarchy process, fuzzy comprehensive evaluation and grey clustering. Moreover, when using entropy weight method, even small differences will lead to entropy weight change exponentially when the entropy value tends to 1, so that certain indexes are endowed with unreasonable weight. With the development of artificial intelligence technology, object evaluation based on machine learning algorithm, such as random forest, artificial neural network, etc. has become the trend of future development.

In this paper, GIS technology and multi-source remote sensing information such as satellite precipitation, digital elevation, land cover and soil type are applied to the Ya'an area where the data is lacking. Taking the natural attribute characteristics of flood disaster as the target and the actual flood occurrence sites as the training data of BP neural network, hazard evaluation of mountain flood disaster is carried out in Ya'an administrative region to improve the reliability of hazard map. In order to provide a scientific basis for the flood control and disaster reduction and disaster assessment in Ya'an City.

2. The research methods

2.1 BP neural network

Back Propagation Neural Network [14] is a multi-level feedforward Network trained according to the algorithm of error reverse transportation, which is one of the most widely used Neural networks nowadays. BP neural network has strong mirroring capability with sigmoid hidden layer and linear output layer. The main operational steps are :(1) provide training examples to the network, i.e. learning samples, including input and expected output; (2) determine the allowable error between the actual output and the expected output of the network. Change all the connection weights in the network so that the output produced by the network is closer to the expected output until a certain allowable error is met.

Through calculation, the calculation formula for the weight adjustment of the BP learning algorithm with the three-layer feedforward network is obtained as follows:

$$\nabla w_{jk} = \eta(d_k - o_k)o_k(1-o_k)y_j \tag{1}$$

$$\nabla v_{ji} = \eta(\sum_{k=1}^{l} \delta_k^0 w_{jk})(1-y_j)x_i \tag{2}$$

In the formula, W is weight matrix between the hidden layer and the output layer; V is weight matrix between the input layer and the hidden layer; the constant $\eta \in (0,1)$ indicates the proportional coefficient; d_k, o_k are the expected output vector and the output layer output vector; x_i, y_j are the input vector and the hidden layer output vector; For the output layer: $j=1,2,\ldots,m$; $k=1,2,\ldots,l$; For the hidden layer: $i=0,1,2,\ldots,n$; $j=1,2,\ldots,m$.

2.2 Construction of evaluation index system

It is difficult for flood disaster to make unified quantitative analysis due to its complicated formation conditions and numerous influencing indexes. This paper considers that each impact index can be analyzed separately and classified according to its influence degree on flood disaster. Then, each index is superimposed with corresponding weight value, using the spatial overlay function of GIS system, analyzing comprehensively the influence of each factor on flood hazard, so as to make the analysis result more reliable.

The basic conditions for the formation of flood disasters include disaster-inducing factors and disaster-pregnant environment, reflecting the natural properties of floods. Disaster-inducing factors are the causes of flood, such as rainfall, water conditions of river, etc. Disaster-pregnant environment refers to the natural factors that form floods, such as topography and landforms, soil vegetation and so on. According to local conditions and the actual situation of flood in Ya'an City, the following indicators are selected:

(1) Disaster-inducing factors

Rainfall and runoff depth are selected as the indicators of disaster-inducing factors: Ya'an City is a mountainous city, and its floods belong to mountain torrents. Mountain torrents, as a strongly destructive surface runoff, are mainly caused by heavy rainfall. At the same time, the runoff depth is commonly used to measure the size of surface runoff. The greater the runoff depth is, the greater the possibility of flood disasters is.

(2) Disaster-pregnant environment

The basin storage capacity, elevation, slope and water system are selected as disaster-pregnant environmental indicators: the basin storage capacity represents maximum soil moisture content in the basin, and the smaller the value is, the lower the upper limit of soil moisture content is, the easier surface runoff is to form, and the greater the impact on flood hazard is. Topography and landforms are important factors for the formation of flood disasters. The probability of flood disasters is higher in low-lying and Middle-Mountain areas than in high-lying areas such as hills. However, in areas with relatively flat mountains, once torrential rain occurs, the accumulated water will wash down along the mountains to plain areas due to lack of obstacles, easily outbreak of mountain torrents. River network system is another important factor affecting flood disasters. The closer to the river system, the stronger the water-collecting capacity is, the weaker the seepage capacity is, the higher the hazard of mountain torrent disasters is.

3. Survey of research areas

Ya'an City is located in the western margin of the Sichuan Basin, in the middle reaches of the Qingyi River, in the upper reaches of the Yangtze River, east to Chengdu, west to Ganzi, south to Liangshan, and north to Aba. It is known as "Western Sichuan Throat", "Tibet Gateway" and "National Corridor". The topography is high in the southwest and low in the northeast, belonging to low-lying and middle-mountain zone (as shown in Figure 1). The city covers an area of 15,000 square kilometers and administers 6 counties and 2 districts with a total population of 1.53 million. Known as "Rain City" and "Sky Leak", most of the county's annual rainfall of 1000-1800 millimeters above, an average annual rainfall of about 1800 millimeters, is relatively more rainfall area in Sichuan province. In addition, its main rivers: both Dadu River and Qingyi River belong to Minjiang River system, with nearly 100 tributaries, and are located near Longmenshan fault zone, prone to geological disasters and rainstorm floods and other disasters. In the past five years, Ya'an City has suffered several major floods. In July 2013, after the Lushan earthquake, the Longxi River basin in Yucheng district suffered extraordinary flood in more than 50 years. On July 7 and 8 of the same year, the accumulated rainfall of Shangli town in Yucheng district reached 221.8 millimeters, causing flash floods. On August 27, 2017, Ya'an City was hit by torrential rain and mountain torrents, flooding many parts of the country.

4. Data and processing

4.1 Data sources

The specific description of each data is shown in Table 1.

Table 1. Data introduction.

Category	Data name	Data source	Data description
Disaster-inducing factors	TRMM 3B43 satellite rainfall	NASA in USA	Global grid data of monthly rainfall of 0.25°×0.25° from 1998 to 2016
Disaster-pregnant environment	Isogram of annual average runoff depth	Sichuan hydrological bureau	Synchronous series from 1956 to 2000
	Land cover UMD	University of Maryland, USA	Global land cover grid data of 1 km
	Soil type	International Food and Agriculture Organization, FAO	Global soil type grid data of 1 km
	Digital elevation model, DEM	Geospatial data cloud http://www.gscloud.cn/	30m DEM based on the latest STRM V4.1 data mosaic and collation
Disaster condition	Historic flood sites in Ya'an City	Flood and drought control command of Sichuan Provincial People's Government /Network information collection	79 points for flood and drought control command and 10 points for network collection, specific to villages.

4.2 Data processing

4.2.1 Standardized processing. Due to the different data download sources, different measurement units and large variation range, it is not conducive to neural network learning and training, so it is necessary to standardize data to eliminate its dimension. The specific algorithm is as follows:

$$X = \frac{x_i - x_{\min}}{x_{\max} - x_{\min}} \tag{3}$$

In the formula, X is the standard value of the index; x_i is the actual value of a certain index; $x_{\max}$, $x_{\min}$ is the maximum and minimum value of the index respectively.

4.2.2 Natural breakpoint classification method. In order to reflect the degree of impact of various indicators on flood hazard, this paper uses the natural breakpoint classification method to classify each index. The classification principle is to reduce the difference in the level and increase the difference among the levels. The calculation formula is as follows:

$$SSD_{i-j} = \sum_{k=i}^{j} (A[k] - mean_{i-j})^2, (1 \le i < j \le N) \tag{4}$$

In the formula, A is an array (the length of the array is N); $mean_{i-j}$ is the average in each level.

The standardized indicators are divided into five levels using equation (4), and the classification is as shown in Table 2:

Table 2. Range of indexes.

Level	Very low	Low	Medium	High	Very high
Rainfall	90-100	100-106	106-117	117-126	126-146
Runoff depth	254.28-690.15	690.15-1029.98	1029.98-1369.82	1369.82-1753.97	1753.97-2138.13
The basin storage capacity	0-0.04	0.04-0.12	0.12-0.18	0.18-0.28	0.28-0.45
Elevation	488-1250	1250-1926	1926-2624	2624-3461	3461-5700
Slope	0-12.69	12.69-23.11	23.11-32.55	32.55-42.96	42.96-82.99
Hazard	0.239-0.400	0.400-0.467	0.467-0.529	0.529-0.607	0.607-0.826

4.2.3 Index processing. (1) Rainfall index. Torrential rain is usually an important cause of mountain torrents in Ya'an City. At present, ground precipitation stations and radar are the main methods to obtain rainfall data. Precipitation station observation, as one of the most direct precipitation measurement method, is relatively mature, the most widely applied, in terms of a single measurement point, it has strong accuracy. However, because of the limitations of station distribution and observation scope, especially influenced by the complex mountainous terrain, it can not reflect the large-area spatial distribution and intensity change of rainfall. Moreover, most rainfall interpolation methods tend to produce too smooth rainfall curves, which underestimate the impact of spatial variability, such as the estimation of rainstorm value. The products of TRMM satellite precipitation are characterized by wide coverage (50°S-50°N), good accuracy and high spatial resolution (0.25°×0.25°). Among them, the products of 3B43 month precipitation are significantly correlated with the observation data of ground meteorological stations. Therefore, this paper selects the monthly data of TRMM satellite precipitation during the period of 1998-2016 for 19 years, and uses MATLAB software to calculate the average annual rainfall. According to the principle that the greater the rainfall is, the higher the flood hazard is, the layer is divided into five levels by the natural breakpoint classification method, and the distribution map of the impact degree of rainfall on flood hazard is shown in Figure 2.

(2) Runoff depth index. In this paper, the runoff depth contour of Ya'an City is drawn on the basis of the multi-year average annual runoff depth contour map of Sichuan Province. The iterative finite difference interpolation technology is used to interpolate it into the grid surface, which is divided into 5 levels after standardized processing, the distribution map of the impact degree of runoff depth on flood hazard is shown in Figure 3.

(3) The basin storage capacity index. According to statistics in 2019, the vegetation coverage rate in Ya'an City reachs 64.77%, and runoff process of the whole city is greatly affected by vegetation. The basin storage capacity can be calculated by the plant root depth and field capacity listed in Formula (5). The plant root depth is determined by determining vegetation types in the study area, according to the standard table of root depth of different land cover types [15] (as shown in Table 3). There are 8 kinds of FAO soil numbers in this study area: 4351, 4287, 4329, 3085, 3963, 4270, 3967, 4269. According to the proportion of FAO soil type particles provided by Yamanashi University (as shown in Table 4) and the soil triangle provided by USDA [16], the soil types in this study area can be divided into three types, namely, sandy clay loam, clay and loam. The field capacity corresponding to different soil types is found in Table 5 [15]. Then, the basin storage capacity is obtained. As it is inversely proportional to the flood hazard, the distribution map of the basin storage capacity is obtained by subtracting the grid value from the maximum value and dividing levels, as shown in Figure 4.

$$M = W_m \times F_r \tag{5}$$

In the formula, M is the basin storage capacity; W_m is the field capacity; F_r is the plant root depth.

Table 3. Root depth for various land cover types.

UMD land cover classification	Evergreen coniferous forest	Evergreen broad-leaved forest	Mixed forest	Woodland	Forest steppe	Dense thickets	thickets	Grass land	Arable land
Root depth/m	1.0	1.25	1.125	0.997	0.872	0.651	0.578	0.75	0.75

Table 4. Soil particle proportion of each FAO soil type in the study area.

	Soil particle ratio			
FAO soil number	Clay%	Sand%	Silt%	USDA soil type
4351	23.3	55.628	21.072	sandy clay loam
4287	41.3	39.91	18.89	clay
4329	30.62	46.42	24.9	sandy clay loam
3085	39.699	38.059	22.275	clay
3963	25.245	45.022	29.733	loam
4270	40.96	44.32	14.8	clay
3967	26.75	51	22.25	sandy clay loam
4269	43.03	42.6	14.44	clay

Table 5. Field capacity for various USDA soil type.

USDA soil type	Field capacity	USDA soil type	Field capacity
Sand	0.08	Sandy clay loam	0.27
Loamy sand	0.15	Silt clay loam	0.36
Sandy loam	0.21	Clay loam	0.34
Silt loam	0.32	Sandy clay	0.31
Silty sand	0.28	Silty clay	0.37
Loam	0.29	Clay	0.36

(4) Elevation and slope indexes. In GIS, topographic elevation is expressed by DEM, and topographic variation degree is often expressed by slope. The influence of elevation and slope on flood hazard is negatively correlated, so it is necessary to convert it into positive correlation, and then classify it to obtain the distribution map of its influence on flood hazard, as shown in Figure 5 and 6.

(5) Water system index. The spatial analysis function of GIS is used to hydrologic analysis DEM to get the flow accumulation, and the water system with the flow accumulation of no less than 15000 is taken as the main water system in the study area. Then, the buffer analysis of the water system is conducted according to the buffer width listed in Table 6 after comprehensively considering the different impacts of river grade and topographic variation on flood hazard. This paper establishes the two-level buffer. According to the principle of buffer formation, the impact degree of buffers on flood hazard at all levels is quantified: the magnitude of the first-level buffer is 0.9, the magnitude of the

second-level buffer is 0.7, and the magnitude of the non-buffer is 0.5. The distribution map of the influence of water system on flood hazard is shown in Figure 7.

Table 6. River buffer width values at all levels.

River rank		Absolute elevation (m) 488-1249	1249-1925	1925-2623	2623-3459	3459-5700
First-level buffer	Trunk stream	3000	2500	2000	1500	1000
	First-level river	2500	2000	1500	1000	500
	Second-level and others	1500	1000	750	500	300
Second-level buffer	Trunk stream	6000	5000	4000	3000	2000
	First-level river	5000	4000	3000	2000	1000
	Second-level and others	3000	2000	1500	1000	600

5. Comprehensive analysis of flood hazard

In this paper, 89 points of actual location data of flood in Ya'an City are used as input data, 54% of training samples and 46% of test samples are randomly selected from districts and counties. The weights of each index are calculated by using the BP neural network implementation steps given above, and the results are shown in Table 8. Afterwards, using the raster calculator to superimpose each index according to formula (6), the comprehensive distribution map of flood hazard in Ya'an City is shown in Figure 8, which is 0.01°×0.01°.

Table 7. Weights of indexes.

Index	Rainfall	Runoff depth	The basin storage capacity	Elevation	Slope	Water system
Weight	0.321	0.114	0.125	0.163	0.134	0.143

$$P = R\times\alpha + E\times\beta + H\times\gamma + S\times\varepsilon + W\times\delta + M\times\lambda \tag{6}$$

In the formula, P is flood hazard index; R is rainfall index; E is runoff depth index; H is elevation index; S is slope index; W is water system index; M is the basin storage capacity index; α, β, γ, ε, δ, λ are weights of each index respectively.

From Figure 8, it can be seen that the flood hazard in the eastern region of Ya'an City is higher than that in the western region, while the flood hazard in the northern and central regions is higher than that in the southern region. Through the analysis of the test samples, 75.6% is located in the area with high and very high flood hazard, and 92.7% is located in the area with medium or above flood hazard, which better reflects the actual spatial distribution of flood in Ya'an City. The very high-hazard areas account for 10.7% of the total area, including Mingshan District in the east of Ya'an City, Dachuan Town, Baosheng Township, Yuquan Township and Longmen Township in Lushan County, as well as Shangli Town, Zhongli Town, Bifengxia Town, Fengming Township, Beijiao Town, and Daxing Town in the north of Yucheng District. These areas belong to low and middle mountain zone, located in the middle reaches of the Qingyi River, with gentle topography, high rainfall and developed water system. The basin storage capacity in Mingshan District is low, which is more likely to form surface runoff, about 22.5% of the historical flood occurrence points concentration distribution in the area. The high-hazard areas account for 20.7%, including the southern of Lushan County, the southern of Yucheng District, the Zishi Township and surrounding townships of Tianquan County, the surrounding areas of Yingjing County, Lingguan Town of Baoxing County, etc., which concentrate 42.7% of historical occurrence flood points. The medium-hazard areas account for 33.2%, mainly located in the northern and central regions of Ya'an City. Low and very low hazard areas account for 35.4%, mainly located in Hanyuan County and Shimian County in the south of Ya'an City, most regions of which are in the high mountain, belonging to the Dadu River Basin, with low rainfall and the lowest flood hazard compared to other districts and countys from Figure 8. However, the urban section of trunk stream of Dadu River and the lower reaches of first-level rivers which are Tianwan River, Nanya River and Liusha River, are low-lying, have medium flood hazard and distribute many historical flood occurrence points. The results shown in Table 8 are obtained by using zonal statistics of GIS.

Table 8. Hazard ranking and the proportion of different hazard levels in each district (county) (%).

Hazard ranking of district (county)	Hazard levels				
	Very low	Low	Medium	High	Very high
Mingshan District (1)	-	-	-	5.166	94.834
Lushan County (2)	-	0.176	17.414	31.750	50.660
Yucheng District (3)	0.104	0.519	16.390	54.979	28.008
Tianquan County (4)	-	3.045	47.182	46.364	3.409
Yingjing County (5)	1.587	17.949	54.884	24.481	1.099
Baoxing County (6)	0.139	16.011	65.889	17.090	0.870
Hanyuan County (7)	39.379	50.100	10.220	0.301	-
Shimian County (8)	45.553	47.091	7.107	0.249	-

6. Conclusion

(1) On the basis of summarizing previous research results, selecting rainfall, runoff depth, the basin storage capacity, elevation, slope and water system factors are as evaluation indexes to build an evaluation index system. BP neural network is used to learn and train samples combined with the actual flood occurrence points to get the weight of each evaluation index, according to which, superimposing the index factors on the GIS platform, so as to obtain the hazard distribution map of flood in Ya'an City. Finally, the analysis results are compared with historical flood sites, and the research shows that the consistency is good.

(2) In terms of the hazard distribution map, high-hazard areas of flood are mainly concentrated in the east of Ya'an City. The highest-hazard areas are in the east of Lushan County and Yucheng District and most of Mingshan District. Some of these three administrative regions belongs to post-disaster reconstruction areas, which will aggravate the impact of flood disasters. So corresponding countermeasures for flood control and disaster reduction should be done well.

(3) TRMM satellite rainfall of 0.25°×0.25° in this paper has a large spatial scale, and is still insufficient spatial difference description of rainfall in mountainous areas, as well as lacking of description of topo-short-term heavy rainfall. The surface rainfall of smaller spatial scale should be considered. In addition, this article also ignores the impact of hydraulic structures such as reservoir, dam, dike and so on. The above shortcomings hope to be improved in future research.

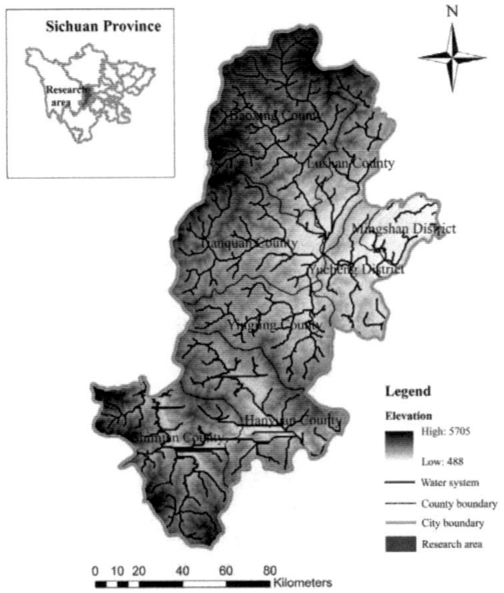

Figure 1. Geographical location of Ya'an City.

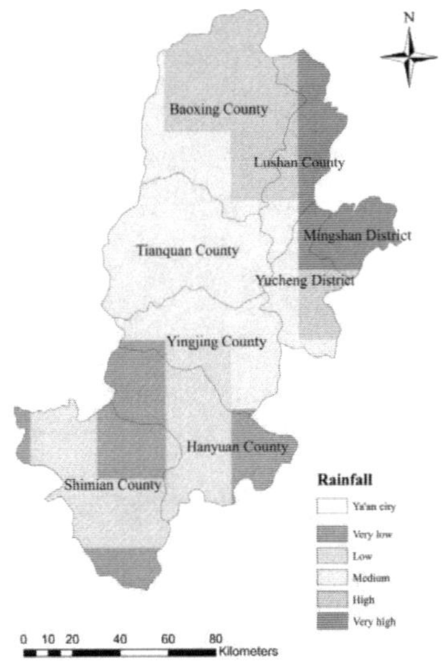

Figure 2. Rainfall impact map.

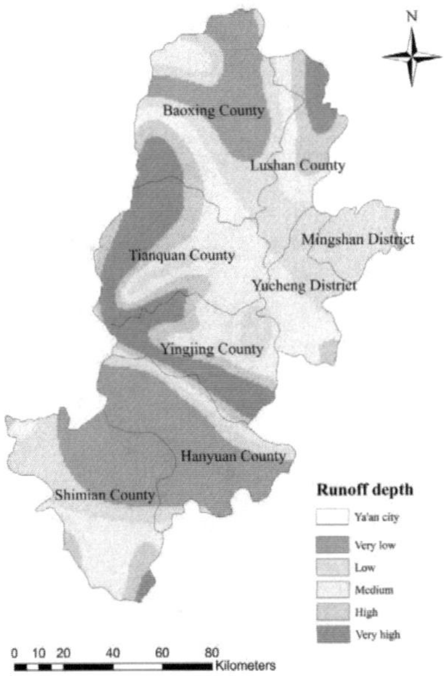

Figure 3. Runoff depth impact map.

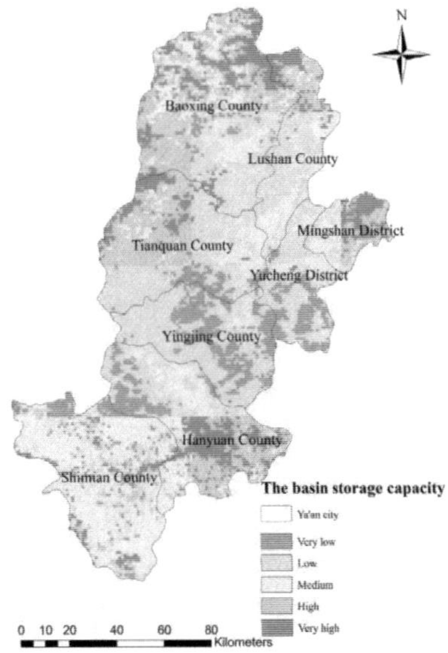

Figure 4. The basin storage capacity impact map.

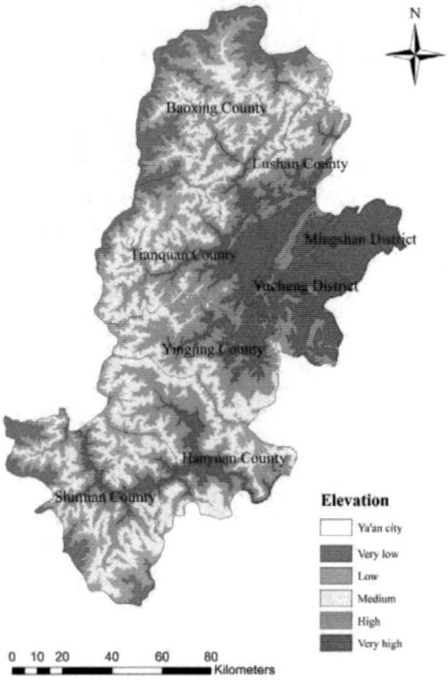

Figure 5. Elevation impact map.

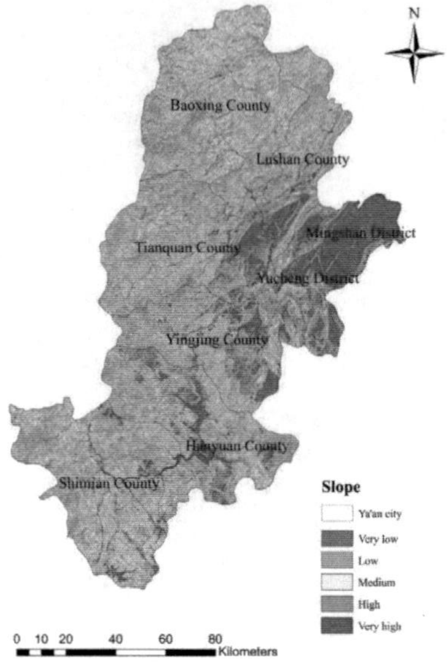

Figure 6. Slope impact map.

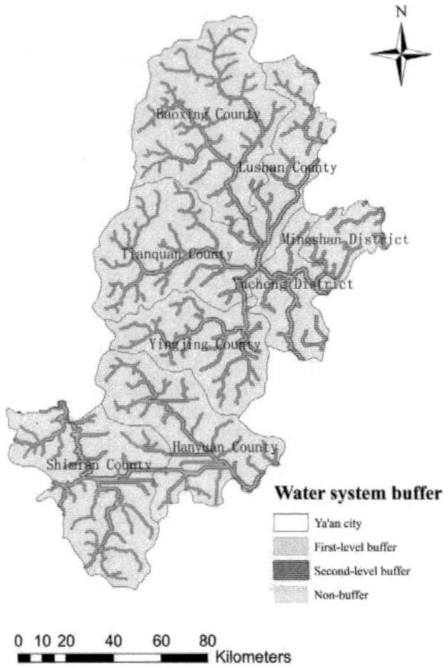

Figure 7. Water system buffer map.

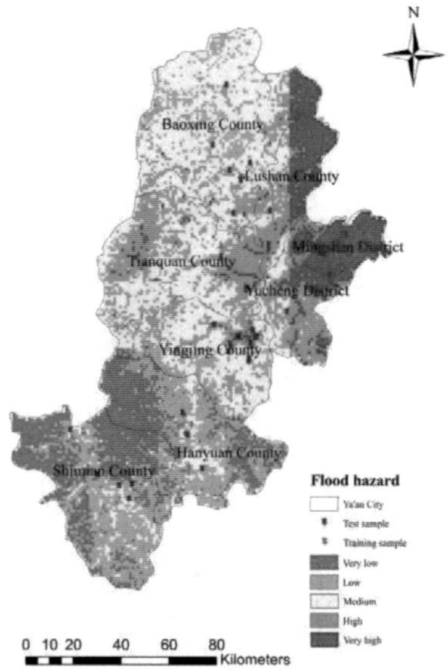

Figure 8. Flood hazard map.

Reference

[1] ZHAO Zongci, LUO Yong, WANG Shaowu, et al. (2015) Science issues on global warming. J. Journal of Meteorology and Environment, 31(01):1-5.

[2] WANG Shaoyu, LIU Jia. (2012) Multi-attribute dynamic evaluation of urban flood disaster vulnerability. J. Advances in Water Science, 23(3): 334-340.

[3] MAO Dehua, HE Zilin, HE Xiguang, et al. (2009) Review and prospect of research on flood risk analysis at home and abroad(I): status quo of research on risk analysis of flood hazard. J.

Journal of Natural Disasters, 18(01): 139-149.

[4] Archer D R, Parkin G, Fowler H J. (2017) Assessing long term flash flooding frequency using historical information. J. Hydrology Research, 48(1): 1-16.

[5] SUN Zhangli, ZHU Xiufang, PAN Yaozhong, et al. (2017) Flood Risk Analysis: Progress, Challenges and Prospect. J. Journal of Catastrophology, 32(3): 125-130, 136.

[6] Manfreda S, Samela C, Gioia A, et al. (2015) Flood-prone areas assessment using linear binary classifiers based on flood maps obtained from 1D and 2D hydraulic models. J. Natural Hazards, 79(2): 735-754.

[7] Gusyev M, Gädeke A, Cullmann J, et al. (2016) Connecting global - and local - scale flood risk assessment: a case study of the Rhine River basin flood hazard. J. Journal of Flood Risk Management, 9(4): 343-354.

[8] QI Shuhua, SHU Xiaobo, Daniel Brown, et al. (2015) Flooding hazard mapping for Poyang Lake Region with remote sensing and water level records. J. Journal of Lake Sciences, 21(5): 720-724.

[9] Rosser J F, Leibovici D G, Jackson M J. (2017) Rapid flood inundation mapping using social media, remote sensing and topographic data. J. Natural Hazards, 87(1): 103-120.

[10] HUANG Shifeng, XU Mei, CHEN Deqing. (2001) GIS-based extraction of drainage network density and it's application to flood hazard analysis. J. Natural Hazards, 2001(04): 129-132.

[11] LIU Rui. (2016) A Study on Modeling of Flood Risk Evaluation Based on Bayesian Networks. D. East China Normal University.

[12] LAI Chengguang, CHEN Xiaohong, ZHAO Shiwei, et al. (2015) A flood risk assessment model based on Random Forest and its application. J. Journal of Hydraulic Engineering, 46(1): 58-66.

[13] Kourgialas N N, Karatzas G P. (2017) A national scale flood hazard mapping methodology: The case of Greece–Protection and adaptation policy approaches. J. Science of the Total Environment, 601: 441-452.

[14] WANG Lei. (2014) Principle, classification and application of artificial neural network. J. Science & Technology Information, 2014(03): 240-241.

[15] YUAN Fei. (2006) Hydrological Processes Modeling Considering the Effect of Vegetation. D. Hohai University.

[16] Rawls W L, Ahuja L R, Brakensiek D L. (1993) Infiltration and soil water movement, In: Maidment, D.R.(Eds)handbook of hydrology, New York: McGraw-Hill Inc, 5.1-5.51.

IOP Conf. Series: Materials Science and Engineering **794** (2020) 012009 doi:10.1088/1757-899X/794/1/012009

Accuracy and Grade Demonstration and Research of Elevation Control Network for Long distance Water Diversion Projects in Plateau

Xin Zhang*, Zhe Yang, Muxia He

Changjiang Institute of Survey, Planning, Design and Research, Wuhan, Hubei, 430010, China

*Corresponding author's e-mail: zsimba@163.com

Abstract: According to the characteristics of the Water Diversion Project in Central Yunnan, the application of error theory analysis and simulation calculation method, the influence value of the elevation error of the channel line on the longitudinal slope is obtained. It is deduced that the first-level elevation control network of the project should use the second-level level control network. Combined with the analysis and calculation of the error in the weakest point elevation, it is analyzed that the project should use the third-level encryption high-speed control network. Finally, through the valuation calculation, it is determined that the Water Diversion Project in Central Yunnan does not need to correct the lunisolar attraction in the elevation control network.

1. Introduction

The Water Diversion Project in Central Yunnan is the first of the key water conservancy landmark projects under construction in China. The project involves 35 cities and 6 districts. The average annual water intake is 3.403 billion cubic meters, and the water receiving area is 36,900 square kilometers[1]. The scale of the project is huge, and the magnificent project must have first-class surveying and mapping support, so that the design, construction and supervision and management units along the line can work under the control of the unified precision and standard framework[2,3]. In meanwhile, the construction control network is the basis of the surveying and mapping work, which can provide unified plane and elevation control benchmarks and precision guarantee for all parties involved in the project.

The length of the Water Diversion Project in Central Yunnan is more than 660km, and the water diversion line is from 2035m to 1400m above sea level. With the characteristics of long span of survey area, high average altitude and large variation of height difference, it is the largest plateau long-distance water transfer project in China at this stage. Among them, the study of the layout of its elevation control network is of great salience. Therefore, this paper systematically carries out the accuracy and grade demonstration of the elevation control network for the characteristics of the Water Diversion Project in Central Yunnan.

2. Layout Analysis of the First-level Elevation Control Network

The elevation control network is divided into a first-level elevation control network and an encrypted elevation control network. By applying the error theory to analyze the influence of the error of the channel elevation measurement on the longitudinal slope, the elevation measurement accuracy and measurement level can be selected scientifically, reasonably and feasibly[4~6].

Let i be the slope of the channel line of the canal, h is the height difference between the two ends of the channel, and L is the length of the channel. Then:

$$\mathrm{i} = \frac{h}{L} \tag{1}$$

Expressed in logarithmic form as:

$$\ln i = \ln h - \ln L \tag{2}$$

Differentiate equation (2):

$$\frac{di}{i} = \frac{dh}{h} - \frac{dL}{L} \tag{3}$$

According to the law of error propagation, convert equation (3) to medium error:

$$\left(\frac{m_i}{i}\right)^2 = \left(\frac{m_h}{h}\right)^2 + \left(\frac{m_L}{L}\right)^2 \tag{4}$$

Equation (4) shows the relationship between the slope of the channel line, the height difference, and the relative error of the length. It is assumed that the error caused by the measurement affects the slope of the channel line by no more than 1/10 of the slope, and that is:

$$\frac{m_i}{i} \le \frac{1}{10} \quad \text{or} \quad \left[\left(\frac{m_h}{h}\right)^2 + \left(\frac{m_L}{L}\right)^2\right]^{1/2} \le \frac{1}{10} \tag{5}$$

The length measurement error is generally up to 1/100,000, so the error affecting the slope of the channel line is mainly the height difference measurement error m_h. Therefore, equation (5) can be approximated as:

$$\frac{m_h}{h} \le \frac{1}{10} \tag{6}$$

Substituting equation (1) into equation (6), the error allowed by the height difference measurement of the channel line is:

$$m_h \le \frac{1}{10} \times i \times L \tag{7}$$

The height difference error of the channel line is composed of basic elevation control error m_1, station elevation control error m_2, and elevation stakeout error m_3, which is:

$$m_h = \left(m_1^2 + m_2^2 + m_3^2\right)^{1/2} \tag{8}$$

If the elevation stake-out error is strictly required, it can reach ±5mm, and the precision gradient $m_1/m_2=1/3$, which can be obtained by formula (7) and formula (8):

$$m_1 \le \left(\frac{(iL)^2}{1000} - 2.5\right)^{1/2} \tag{9}$$

Equation (9) reflects the mathematical relationship between the measurement error of the first stage high control network and the length of the channel section and the length of the channel section. The longer the segment L considers, the greater the measurement error allows.

In addition, the medium error of the height difference of the level measurement is obtained as follows:

$$m = M_w L^{1/2} \tag{10}$$

In the equation, M_w is the total error measured per kilometer level. The first value is 1.0mm, the second is 2.0mm, the third is 6.0mm, and the fourth is 10.0mm. It can be seen from equation (10) that the value of the error m in the height difference of the segment is also related to the value of L.

Therefore, the determination of the leveling level is closely related to the value of L. According to the requirements of the national leveling specification, a leveling stone should be buried every 4~8km, that is, 4km≤L≤8km. The L value is selected within this range and the corresponding level of measurement is selected. If it meets the needs of engineering construction and is easy to implement, it means that the length of the section considered is appropriate. Otherwise, the length of the canal can be recalculated and analyzed until a reasonable length of the canal is selected.

The slope i of the Water Diversion Project in Central Yunnan has a maximum of 1/1000 and a minimum of 1/15000. Therefore, i=1/15000, 1/6000, 1/1000; L=4km; M_w=2.0, 6.0, 10.0 are respectively substituted into equations (9) and (10), and the calculation results are shown in Table 1. In the table, m_1 is the error in the basic elevation control point, m_2 is the error in the station elevation control, and m_{II}, m_{III}, and m_{IV} are the medium errors in the height difference of the second, third, and fourth level measurement.

Table 1. Measurement error table of L=4km channel height difference Unit: mm

i	m_1	m_2	m_{II}	m_{III}	m_{IV}
1/15000	8.3	24.9			
1/6000	21.0	63.0	4	12	20
1/1000	126.5	379.5			

It can be seen from Table 1 that when the distance between the elevation points is about 4 km, the construction control network of the channel line should select the second-level accuracy to meet the requirements of the construction of the water diversion project in the middle. However, the difference between m_1 and m_{II} is large, indicating that the value of L can be appropriately adjusted.

Further, the value of L is changed to 8 km, and the equations (9) and (10) are substituted. The results are shown in Table 2.

Table 2. Measurement error table of L=8km channel height difference Unit: mm

i	m_1	m_2	m_{II}	m_{III}	m_{IV}
1/15000	16.8	50.4			
1/6000	42.1	126.3	5.7	17.0	28.3
1/1000	253.0	759.0			

It can be seen from Table 2 that when the distance between the elevation points is about 8 km, the third-level leveling control accuracy of the channel line elevation construction control network can basically meet the requirements of the construction of the water diversion project in the middle.

The above is an analysis based on theoretical values. In the actual elevation measurement of the water diversion project in the middle of the river, due to the location of the plateau, the elevation point spacing should not be too large. Therefore, it is recommended to use 4 to 8km spacing and set the second-class standard as the first-level elevation control network for the main canal.

3. Layout Analysis of Encrypted Elevation Control Network

The purpose of the encryption elevation control network setting is to measure the elevation of the measurement station. Generally, an attachment level route is arranged between two adjacent basic level points. The weakest point in the route is in the middle of the level. The error in the weakest elevation is:

$$m_{weakest} = \frac{1}{2} M_w S^{1/2} \tag{11}$$

In the equation, S is the level line length and the unit is km.

Another $m_{weakest} \geq m_2$, take $m_2 = m_{weakest}$, substituting it into equation (8) and taking into account equation (7):

$$M_w \le 2\left(\frac{i^2 \times L^2 \times 10^{10} - m_1^2 - m_3^2}{S}\right)^{1/2} \tag{12}$$

In the equation, L is the length of the channel, and S is the length of the level line, and the unit is km.

When the first-level elevation control adopts the second-class level measurement, m_1=4mm, and L=4km, S=8km. Take the elevation stakeout error m3=±5mm, ±10mm and ±20mm respectively, substituting into equation (12), calculate the M_w value, and list the results in Table 3.

Table 3. Calculation of the error in the measurement of the level of each kilometer Unit: mm

i	M_w		
	m_3=±5mm	m_3=±10mm	m_3=±20mm
1/15000	18.3	17.2	12.1
1/6000	46.9	46.5	44.9
1/1000	282.8	282.7	282.5

In general, when the value of M_w is 6~10mm, the third-level encryption is used. When the value of M_w is 10~15mm, it is encrypted with four levels. When the value of M_w is above 15mm, it is encrypted with five levels. According to the numerical values in Table 3, when the error value in the elevation stakeout is controlled within 10 mm, only the fifth-level encryption is used. However, when the error value in the elevation stakeout is higher than 10 mm, especially to a value of 20 mm, a fourth level of encryption is required. Therefore, the level of four or more can meet the accuracy requirements of the encrypted elevation control network.

On the other hand, due to the numerous buildings in the Water Diversion Project in Central Yunnan, the layout of its encrypted elevation network is mainly to meet the needs of building construction control. The tunnel project accounts for about 92% of the total length of the water diversion project. Therefore, meeting the tunnel elevation control network layout requirements, basically meet the requirements of all building elevation control network deployment. The average length of the tunnel in the Water Diversion Project in Central Yunnan is about 9.6km. According to the *Specification for construction survey of water and hydropower projects*[7], the elevation construction control network related to the long tunnel of 5km~20km should meet the second-class or third-class precision requirement. Besides, considering the demand for hierarchical network, it is recommended to design the encryption elevation control network, that is, the design level of the building's first-level elevation construction control network, to meet the elevation accuracy requirements of the ground control network for tunnel construction and other building construction measurements.

4. Error Analysis of the lunisolar attraction in the Elevation Control Network

The direction of gravity at any point on the earth, in addition to the distribution of matter within the Earth and the centrifugal force generated by the Earth's rotation, is also affected by the sun, the moon, and other celestial bodies, as well as the Earth's outer atmosphere. Due to the effect of the lunisolar attraction, the vertical direction of the earth changes instantaneously, that is, the vertical line deviates from the effect, so the same tilt change occurs in the horizontal plane passing through the point. In addition, due to the influence of the lunisolar attraction, there will be ground deformation. In precision elevation leveling, the corrections that estimate the effects of lunisolar attraction are called tidal corrections. The data show that in extreme cases, the influence of lunisolar attraction does not exceed 0.1 mm per kilometer. In most cases, the error per kilometer is about 0.01 mm to 0.02 mm[8,9]. This error is not reflected in the segment closure difference, the route closure difference, and the loop closure difference, but this kind of error will accumulate in the north and south. For China, the cumulative value in the north-south direction can reach 12cm.

The water diversion project in the middle reaches about 370km. Under normal circumstances, according to the error of 0.01mm~0.02mm per kilometer of gravity, the cumulative error of the project can reach 3.7mm~7.4mm. This is much smaller than the accuracy requirements of the first-level elevation control network and the encrypted elevation control network analyzed in the previous section, so it can be ignored. Therefore, the tidal correction is not a must in the Water Diversion Project in Central Yunnan.

5. Conclusion

In this paper, the accuracy and grade demonstration of the elevation control network are systematically carried out on the basis of the characteristics of the Water Diversion Project in Central Yunnan. By analyzing the influence of the channel height measurement error on the longitudinal slope and the simulation results, it is recommended that the first-level elevation control network of the main channel use the second-level level control network. Combined with the analysis and calculation of the error in the weakest point elevation, it is recommended to use the third-class level to encrypt the elevation control network. Finally, through the valuation calculation, it is determined that the Water Diversion Project in Central Yunnan does not need tidal correction of the elevation control network.

Generally speaking, it provides theoretical support for the design and establishment of the elevation control network of the Water Diversion Project in Central Yunnan, and serves the construction of major national water conservancy projects with high precision.

References

[1] Zhang X., Yang A.M., Xu Q.F., et al. (2014) Key Technology on Establishment of an Independent Coordinate System for Water Diversion Project in Central Yunnan. Geomatics and Information Science of Wuhan University, 39(9): 1047-1051.

[2] Blewitt G., Lavallee D., Clarke P. (2001) A New Global Mode of Earth Deformation: Seasonal Cycle Detected. Science, 294: 2342-2345.

[3] Zhang X., Luo H.B., Xiang W., et al. (2016) Research on selection of ellipsoid transformation methods for measurement of Water Diversion Project in Central Yunnan. Yangtze River, 47(3): 45-48.

[4] Giovannetti, V., Lloyd,S., Maccone, L. (2001) Quantum enhanced positioning and clock synchronization. Nature, 412: 417-419.

[5] Li Z.H., Li P., Ding D., et al. (2018) Research Progress of Global High Resolution Digital Elevation Models. Geomatics and Information Science of Wuhan University, 43(12): 1927-1942.

[6] Dach R., Schildknecht T., Hugentobler U., et al. (2006) Continuous Geodetic Time Transfer Analysis Methods. IEEE transactions on ultrasonics, ferroelectrics, and frequency control, 53(7):1250-1259.

[7] Specification for construction survey of water and hydropower projects. SL52-2015.

[8] Yang A.M., Yan J.G., Yang C.H., et al. (2010) Study on construction surveying control network of South-to-North Water Diversion Middle Route Project. 41(19):30-33.

[9] Fang W.H., Fan L.Z., Du Z.H. (2017) Issues related to evaluation of safety monitoring system for hydraniic engineering. China Water Resources, 12: 40-43.

IOP Conf. Series: Materials Science and Engineering **794** (2020) 012010 doi:10.1088/1757-899X/794/1/012010

Analysis on Feature of Soil_River_Ground_Plants Water Based on Stable Oxygen Isotope in Poyang Lake Wetland

Yuru LIN[1*], Zhimin DENG[2], Ying XIONG[3], Jijun XU[1]

[1]Changjiang River Scientific Research Institute, Wuhan, Hubei, 430010, China;

[2]Changjiang Water Resources Protection Institute, Wuhan, Hubei, 430051, China;

[3]Bureau of Hydrology, Changjiang Water Resources Commission, Wuhan, Hubei, 430010, China;

*Corresponding author's e-mail: lin_yuru@126.com

Abstract. Poyang Lake is the largest freshwater lake in China, the water level of Poyang Lake was declined continuous, and extreme droughts have frequently occurred in recent years, the structure and function of the wetland ecosystem were destroyed. In this paper, Stable oxygen isotopes was used as a tracer to identify composition of soil water, river water, ground water, and plants water. The focus was on characterize the variation with depth of the stable oxygen isotope composition and identifying the main sources of oxygen in soil water, ground water, river water and plants water, understanding the mechanisms of water movement among them. Soil water sampled at 10 cm intervals over the top 100 cm at two sections in the Poyang Lake wetland, and water samples collected from the plants, river water and ground water, were analysed for stable oxygen isotope composition, exploring the recharge and discharge of relationship between them. The results shown that the values of $\delta^{18}O$ at depth of 0~30cm at section I were declined with increasing soil depth, the value of $\delta^{18}O$ at depth of 0~40cm at section II had rare variation with increasing soil depth. The groundwater supply at section I was mainly influenced by precipitation, at section II was by precipitation and river water. The average value of $\delta^{18}O$ in the domain species plants leaves of phalaris arundinacea, crex cinerascens, and phragmites communis was -0.9‰, -4.23‰, -5.25‰ respectively.

1. Introduction

Poyang Lake is the largest shallow freshwater lake in China, as well as an important international wetland. In recent years, high water level in Poyang Lake is gradually decreasing, while the time of low water level duration is increasing [1], and annual minimum water level has been on a decrease trend [1-2]. Extreme drought events have occurred frequently in Poyang Lake [3-6], changing of water level had significant influences on soil water content of bottomland. The low water level has represented more important reason for degradation of the wetland plants [7]. Therefore, study on the moisture source, migration and dynamic rules in circulation of soil-river-ground-plants by features of stable oxygen isotopes, had great significances on protection and recovery of wetland plants and sustainable development of wetlands ecosystem.

Analyzation of water stable isotope played an increasingly important role in studying moisture source, migration, dynamic rules and water resource utilization of plants in circulation of soil-river-ground-plants by analysed oxygen isotope ratio. Dawson and Ehleringer [8] shown that the water demand of plants growing along the river were rarely from the river water. By using the mix model of oxygen isotopes, Shen et al [9] applied the direct inference method and iso-source model to evaluate ponded water, soil water and ground water's contributions to rice crop by using stable isotopes of hydrogen and oxygen. Zhan et al used [10] the stable isotopes and hydrological data to analyse the interactions between precipitation, river water, lake water and shallow ground water in Poyang Lake. Zhang [11] analysed water supply sources of soil in Poyang Lake wetland by using stable hydrogen and oxygen isotopes. In this paper, we taken the Poyang Lake as study area, based on field acquisition of soil, river water, ground water and plants water, by useing stable isotope to analyse features of oxygen isotope and moisture sources in soil of wetland, and features of oxygen isotope with plants.

2. Research area and data

2.1Research area

The Poyang Lake located in 115°49′~116°46′E and 28°21′~29°52′N, in administrative level it belong to Jiangxi Province, and downstream of Yangtze river. National Nature Reserve of Poyang Lake in Jiangxi have jurisdiction over 9 lakes, including Dahu Pool, Sha Lake, Beng Lake, Zhushi Lake, Meixi Lake, Zhonghu Pool, Dacha Lake, Xiang Lake and Changhu Pool, was an important international wetland with abundant biological diversity. The annual average temperature is 16.5~17.8℃. The annual precipitation is 1,368.7~1,633.8mm, from April to June accounts for 46%. The annual evaporation is 800~1,200mm, from July to September accounts for 50%. Featured as the flood in summer and drought in autumn. Major flood period of Poyang Lake lasted from April to July, and the high-water level were kept from August to September due to the flow jacking and backward with Yangtze River. The development soil type of wetland in Poyang Lake included red soil, paddy soil, yellow-cinnamon soil, alluvial soil and moisture soil. The moisture soil was the typical wetland soil mainly distributed in the embankment and the ecotone of the lake basin. Wetland phytocoenosium in Poyang Lake distributed in annuluses along moisture gradient, due to lakeshore terrain slope, soil structure and water depth. Based on dominant community radio, carex cinerascens were the most widely distributed plants and covered the biggest area.

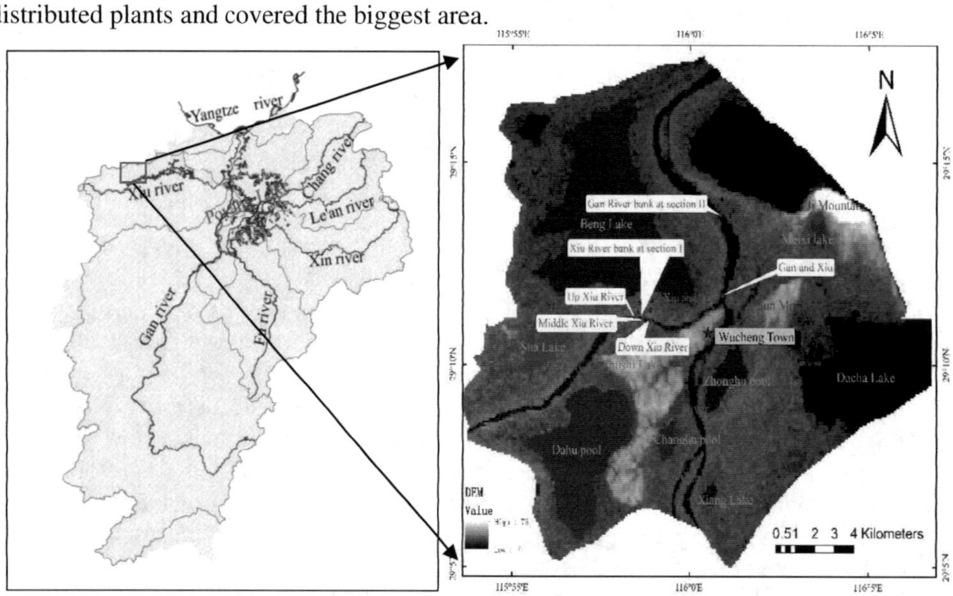

Fig. 1 Study area and the location of sections

2.2Data collecting and processing

Since the fluctuation of water level of Banghu affected a little by low water level period of Poyang Lake [12], two sections were arranged, section I was arranged at upstream of Xiu River along the Beng Lake, section II was arranged at downstream of Gan River along the Beng Lake, and each section was laid out four sampling points every 150m. 99 samples are taken from the river water, ground water, soil and plant, shown as Table 1, including 77 soil samples. Soil samples are taken by cylindrical auger boring with the depth of 100 cm, sampling is taken every 10 cm deep, however, at the sampling point 4 (the deepest layer, about 90~100cm deep underground) at Section II, soil sample cannot be taken due to soil texture and soil water content. Meanwhile, soil volumetric moisture content of each layer was measured by TDR instrument. 8 dominant species plants samples were gathered around soil sample points, dominant species at section I was phragmites, and at Section II was carex cinerascens. 8 ground water samples were gathered at soil sample points. 7 river water samples were gathered at the upper, middle and lower position of the sections, shown as Figure1.

All water samples were filled in 60ml with polyethylene plastic bottles, the bottle covers were sealed by seal membrane or plastic wrap to avoid bubble in the bottles and prevent evaporation of water sample, then the bottles were kept in incubator with ice bag before cold storage in the laboratory. Soil samples were placed in 8ml glass bottles, and the bottlenecks were sealed by plastic wrap before storage in the incubator. Plants samples were wrapped by freezer bags, and bag mouths were sealed by plastic wrap before storage in the incubator. Soil samples and plants samples were frozen in the refrigerator after back to the laboratory. Soil moisture and plants moisture were extracted by vacuum extraction device in Isotopic Laboratory of Chinese Academy of Forestry [13]. Stable oxygen isotope contained in water samples were measured in Stable Isotope Laboratory of State Key Laboratory of Water Resource and Hydropower Engineering Science in Wuhan University, $\delta^{18}O$ content in water samples were measured by MAT253 isotope ratio mass spectrograph connected with Flash EA/HT, and the analytical precision of $\delta^{18}O$ instrument was 0.3‰. Measure results of all water samples presented by micrometer on the basis of V-SMOW standard.

$$\delta^{18}O = \left[\frac{(^{18}O/^{16}O)_{sample}}{(^{18}O/^{16}O)_{V-SMOW}} - 1 \right] \times 1000 \tag{1}$$

Tab.1 The types and number of samples

sample types	river water	ground water	soil water		plants water	
			section I	section II	section I	section II
numbers	7	8	39	38	4	4

3. Results and Discussions

3.1 $\delta^{18}O$ features of soil water

The composition features of the oxygen isotope of soil water at section I shown as Figure 2(a). The values of oxygen isotope in the soil water were in a range of -10.48‰~-5.23‰, and the mean value was -8.36‰. The values of oxygen isotope in soil water were decreased with increasing depth. The values of $\delta^{18}O$ in soil water were concentrated at the surface of soil in the depth of 0~30cm, and were decreased with the increasing depth, due to the isotope fractionation caused by strongly evaporation of the moisture in surface soil. Oxygen isotope values of the deep soil below the depth in 70cm had rarely variation. While the value of oxygen isotope was little at oil surface in the depth of 0~40cm at sampling point 2, and values were increased with the increasing depth, this trend was in keeping with results by Tian Richang [14], caused by the different plants covered.

The composition features of the oxygen isotope of soil water at section II shown as Figure 2(b). The values of oxygen isotope in the soil water were in a range of -12.39‰ ~ -6.55‰, and the mean value was -8.63‰. The value of $\delta^{18}O$ at the depth of 0~40cm in surface soil water changed on the small side at each sampling point, standard deviation was 0.39, 0.50, 0.77 and 0.24 respectively. Clark's [15] research shown that the composition of oxygen isotope almost had no change when the standard deviation of the $\delta^{18}O$ value in the soil water was less than twice of analytical error. In this experiment, measurement accuracy of $\delta^{18}O$ was 0.3‰, the composition of $\delta^{18}O$ in the surface soil water were relatively stable, the reason might be that the dominant species of plants at sampling points in section II were carex cinerascens, which had large coverage [16, 17] and had small surface evaporation. Bristow and Horton's [18] research shown that much of soil had great influenced on physical environment of surface soil, especially at the depth of above 40cm of soil. With the increased depth of soil, $\delta^{18}O$ value changed in disorder. At the depth of 0~40cm, at sampling point 1, the mean value of $\delta^{18}O$ of soil water was -9.71‰, at sampling points 2~4, the values of $\delta^{18}O$ were similar, the mean value was -7.67‰. It's shown that soil permeability at sampling point 1 was better than other three sampling points, which may relate to plants coverage. The oxygen isotope value at the depth of 60cm of sampling point 1 and at depth of 80cm of sampling point 3 were -11.29‰ and -12.39‰ respectively. The reason for small oxygen isotope value at the depth of 60cm of sampling point 1 needs to be verified further, but effected from precipitation could be excluded. Figure 3 shown that the soil water content at the depth of 60cm was lowest, it's impossible that precipitation flows to this area through the preferential flow form. At sampling point 3, the precipitation infiltrates into soil deeply through the preferential flow form due to soil pores.

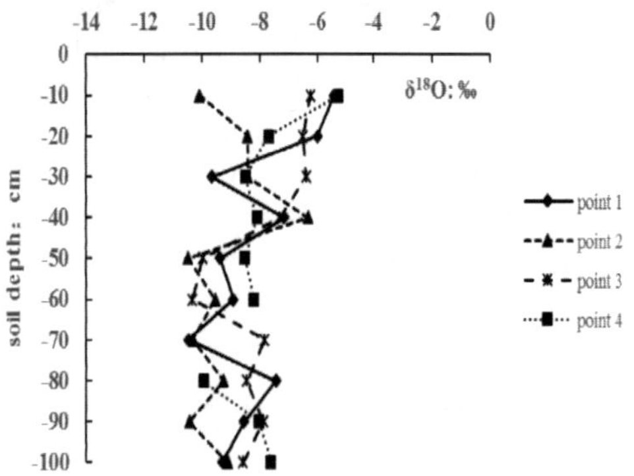

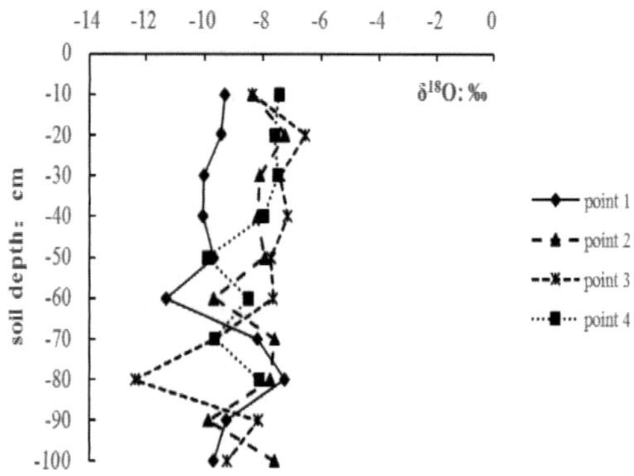

section I section II

Fig.2 $\delta^{18}O$ value of soil water at different sections

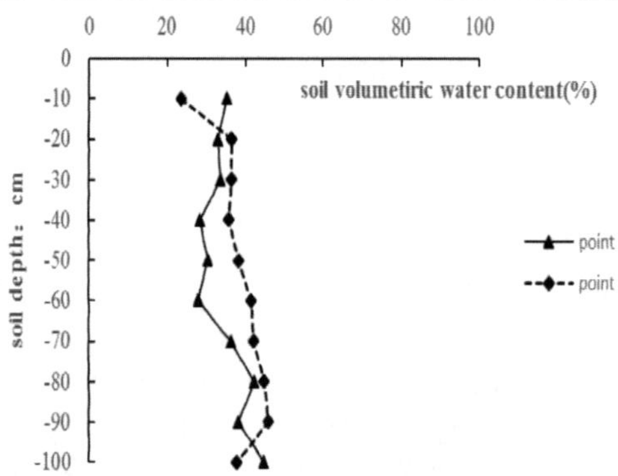

Fig.3 The volumetric water content of soil at point 1 and point 3 in section II

3.2 $\delta^{18}O$ features of ground water and river water

The composition features of the oxygen isotope of ground water at section I and II shown as Figure 4.The values of oxygen isotope in the ground water were respectively in a range of -8.15‰ ~ -6.64‰ and -5.33‰ ~ -3.14‰, and the mean value were -7.13‰ and -4.14‰ respectively. Oxygen isotope value of the ground water of section I was less than section II, four sampling points at each section, the oxygen isotope values of the ground water presented the same trends that at middle sampling points were greater than two sides, which might related to their respectively supply source and buried depth.

Table 2 Shown that the buried depth of the ground water at section I was greater than that at section II, so the ground water at Section I influenced by evaporation was weaker and the isotope value was small than that at section II. Figure 5 shown that the oxygen isotope values in the Xiu River that flows through section I were in a range of -7.24‰ ~ -3.43‰. Gan River flows through section II, at two sampling points in section II the oxygen isotope values were -3.35‰ and -4.64‰ respectively, expected "Up Xiu River" and "down Xiu River" these two points, the oxygen isotope value from river water were lower, others sampling points were basically similar. In general, the oxygen isotope value from river water in Gan River was greater than that in Xiu River, which is consistent with the results from Liang Yue[19]. Figure 4 and 5 shown that the oxygen isotope value of ground water from sampling points at section I were less than that from river water in Xiu River, while the oxygen isotope value of ground water from sampling points at section II were almost similar with that from river water in Gan River. According to the isotope value of local atmospheric precipitation [20], ground water at section I was mainly from precipitation, and at section II was from precipitation and the water of river and lake [11].

The depth of ground water shown as Table 2, combined with oxygen isotope values of soil water and ground water, we known that oxygen isotope value of the ground water at sampling point 1 of section I was similar to that in soil water of corresponding underground depth, while others sampling points' oxygen isotope values of the ground water were greater than that in soil water of corresponding underground depth, and all oxygen isotope values of ground water were greater than that in surface soil water (except that the values of oxygen isotope in ground water at sampling point 3 and 4 of section I were similar with that in the depth of 20cm for surface soil water).These conclusions might be related to the borehole bored at field, plants types and its coverage, shallowed water lever of ground water and exposed heel without sealing, caused the strongly evaporation and $\delta^{18}O$ enrichment.

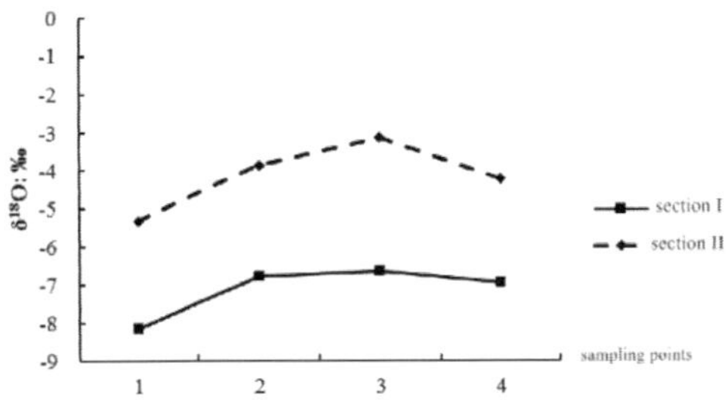

Fig.4 $\delta^{18}O$ value of ground water

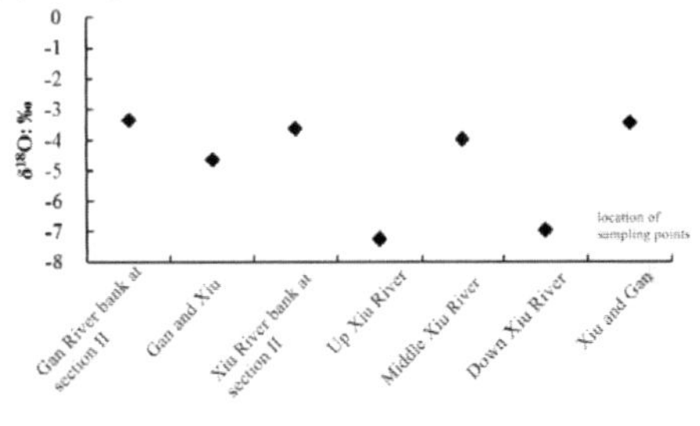

Fig.5 $\delta^{18}O$ value of river water[1]

Tab. 2 the depth of groundwater of sampling points

sampling points		1	2	3	4
depth of ground water(m)	section I	0.65	0.7	0.97	0.85
	section II	uncanned	0.7	0.44	0.28

3.3 $\delta^{18}O$ values of plants leaf water

Oxygen isotope value in leaf water of surface dominant species plants at section I and section II shown as Figure 6. The research shown that the dominant species plants at section I and Section II were phalaris arundinacea, phragmites and carex cinerascens respectively, the oxygen isotope values in leaves water were in a range of -6.52‰ ~ -0.9‰ and -5.04‰ ~ -3.58‰ respectively, the mean values were -4.16‰ and -4.23‰ respectively. Oxygen isotope value in the leaves water of phalaris

[1] "Xiu River bank at section I" means the river water from the Xiu River and located at the lines linked with the sampling points at section I with the Xiu River. "Up Xiu River" means the river water frome the Xiu River and located at the upstream of lines linked with the sampling points at section I with the Xiu River. "Xiu and Gan" means the river water from the river confluence of Xiu River and Gan River. "Gan and Xiu" means the river water from the Gan River which were newly mixed of Xiu River and Gan River. Other points represented the similar meaning, all sampling points shown as Figure 1.

arundinacea was the largest, the mean value was -0.9‰; and that of phragmites was the smallest, the mean value was -5.25‰; and that of Carex cinerascens was located in middle, the mean value was -4.23‰. There was a little difference in oxygen isotope values in plants leaf water of same dominant species plants, and at the same section with the different sampling points, the reason might be that the influence on plant transpiration varies from migration features of main element (Ca, Mg, etc.) in different landform sections (sloping land, depression, etc.) [21].

Distribution and depth of plants root system were important factors to determine the moisture source used by plants [22]. Hair root of xeromorphic phragmites communis mostly distributed in the depth of 0~30cm underground [23], root system of carex cinerascens generally distributed in the depth of 0~20cm underground [24], and root system of phalaris arundinacea distributed in the depth of 0~30cm underground. The oxygen isotope value of surface soil water corresponding to plants root system depth at different sampling points of two sections shown as Table 3. Only at sampling point 3 of section I the oxygen isotope value of soil water was similar with corresponding oxygen isotope value of plants leaf water, at others sampling points the oxygen isotope values of soil water were less than corresponding oxygen isotope values of plants leaf water, the reasons shown as: (1) the isotope fractionation was not occurred when water absorbed by plants root system and conveyed to leaves [25, 26], but when the leaf water directly contacted with the air, heavy isotope in leaf water will enriched with plant transpiration [27].(2) the distribution depth of plants root systems and the isotope value of absorbed water from soil were different, means the moisture source of plants were different.

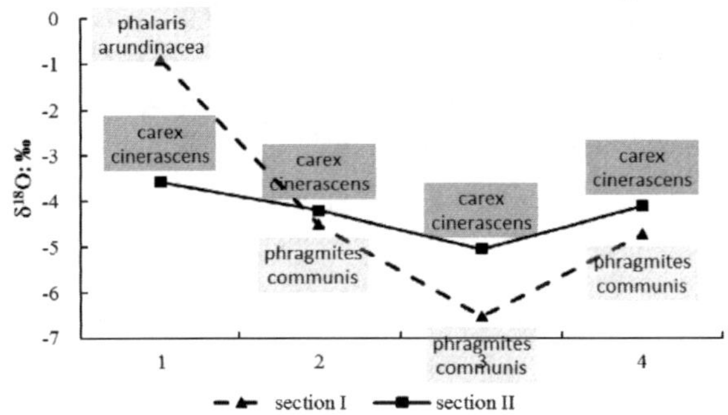

Fig.6 $\delta^{18}O$ value of plants water

Tab.3 The average $\delta^{18}O$ value of surface soil at two sections

section I	sampling points	1	2	3	4
	soil depth(cm)	0~30	0~30	0~30	0~30
	$\delta^{18}O$(‰)	-7.02	-8.98	-6.39	-7.14
section II	sampling points	1	2	3	4
	soil depth(cm)	0~20	0~20	0~20	0~20
	$\delta^{18}O$(‰)	-9.37	-7.82	-7.47	-7.54

4. Conclusion

Based on the field acquisition, the data included river water, ground water, soil water and plants water at two sections, by using stable isotopic technique, analysed the features of stable isotope in soil water, plants water, river water, and ground water, explored the moisture source relation between them, the following conclusions were drawn:

(1) The oxygen isotope in the soil water of section I was in a range of -10.48‰~-5.23‰, the mean value was -8.36‰. Oxygen isotope in soil water at the depth of 0~30cm of section I were enriched, and were decreased with the increasing depth. At sampling point 2, due to the plants coverage, the oxygen heavy isotope in soil water at the depth of 0~40cm were increased with the increasing depth. The oxygen isotope in the soil water of section II was in a range of -12.39‰~-6.55‰, the mean value

was -8.63‰, which was similar with the mean value of oxygen isotope in soil water of section I. $\delta^{18}O$ composition in surface soil water at depth of 0~40cm of section II were relatively stabled, due to the carex cinerascens large coverage of plant dominant species in the section and little influences on surface by evaporation.

(2) The oxygen isotope in ground water at section I and section II were in a range of -8.15‰~-6.64‰ and -5.33‰~-3.14‰ respectively, the mean value were -7.13‰ and -4.14‰ respectively. Based on oxygen stable isotope value in ground water, river water and lake water at two sections, and with buried depth of ground water, shown that the ground water in section I was mainly from precipitation, and in section II was mainly from precipitation, the river water and lake water.

(3) In the average value of oxygen isotope in the dominant species plants water of the Poyang Lake, phalaris arundinacea was the largest, the mean value was -0.9‰; the second was carex cinerascens, the mean value was -4.23‰; and phragmites communis was the smallest, the mean value was -5.25‰.

Acknowledgments
This study was financially supported by: National Key Research and Development Program (2017YFC0405304); National Natural Science Foundation of China (41601043); The Fundamental Research Funds for Central Public Welfare Research Institutes (CKSF2019499 and CKSF2017061).

References

[1] DENG Z M, ZHANG X, LUO W, et al. Response of Poyang Lake wetland carex to the water level change[J], Journal of Basic Science and Engineering, 2014,22(5), 865-876.
[2] OUYANG Q L, LIU W L. Variation characteristic of water level in Poyang Lake over 50 years[J], Resources and Environment in the Yangtze Basin, 2014.23(11), 1545-1550.
[3] LUO W, ZHANG X, DENG Z M, et al. Variation of the total runoff into Poyang Lake and drought-flood abrupt alternation during the past 50 years[J], Journal of Basic Science and Engineering, 2013,21(5), 845-856.
[4] Li X, Ye X. Spatiotemporal characteristics of dry-wet abrupt transition based on precipitation in Poyang Lake basin [J], Water, 2015, 7(5), 1943-1958.
[5] You H, Xu L, Liu G, et al. Effects of Inter-Annual Water Level Fluctuations on Vegetation Evolution in Typical Wetlands of Poyang Lake [J], Wetlands, 2015, 35(5), 931-943.
[6] Li M, Zhang Q, Li Y, et al. Inter-annual variations of Poyang Lake area during dry seasons: characteristics and implications[J], Hydrology Research, 2016, 47.S1:40-50.
[7] HU Z P, GE G, LIU C L. Cause analysis and early warning for wetland vegetation degradation in Poyang Lake[J], Resources and Environment in the Yangtze Basin, 2015,24(03), 381-386.
[8] Dawson T E, Ehleringer J R. Streamside trees that do not use stream water[J], Nature, 1991, 350(6316), 335-337.
[9] Shen Y J, Zhang Z B, Gao L, et al. Evaluating contribution of soil water to paddy rice by stable isotopes of hydrogen and oxygen[J], Paddy and Water Environment, 2015, 13(1), 125-133.
[10] Zhan L, Chen J, Zhang S, et al. Isotopic signatures of precipitation, surface water, and groundwater interactions, Poyang Lake Basin[J], Environmental Earth Sciences, 2016, 75(19), 1307.
[11] ZHANG X, DENG Z M, PAN G Y, et al. Variation in stable isotope composition in soil water in Poyang Lake wetland[J], Acta Ecologica Sinica, 2015, 35(22), 7580-7589.
[12] HU C H, JIANG J h, ZHU H H. Analysis on water-level relationships between Banghu depression and Poyang Lake and its submersion and emersion of bottomland[J], Oceanologia et limnologia sinica,1997, (28),623-630.
[13] West A G, Patrickson S J, Ehleringer J R. Water extraction times for plant and soil materials used in stable isotope analysis[J], Rapid Communications in Mass Spectrometry, 2006, 20(8), 1317-1321.

[14] TIAN R C, CHENG H S, SONG X F, et al. Characteristics of soil water movement using stable isotopes in red soil hilly region of northwest Hunan[J], Envirionmental Science, 2009, 30(9), 2747-2754.
[15] Clark I D, Fritz P. Environmental isotopes in hydrogeology [M], New York: CRC Press LLC, 1997.
[16] WU J D, LIUG H, JIN J F, et al. Structure analysis of beach vegetation in Poyang Lake in autumn[J], Jiangxi Science, 2010,28(4),549-554.
[17] HU D D, OUYANG K H, DAI Z H, et al. Investigation on community characteristics and α-diversity of Carex cinierascens meadow steppe in Poyang Lake wetland[J], Pratacultural Science, 2013,30(6),844-848.
[18] Bristow K L, Horton R. Modeling the impact of partial surface much on soil heat and water flow[J], Theoretical and applied climatology, 1996, 54(1-2), 85-98.
[19] Liao Y. Biogeochemistry of heavy metal and nitrogen and isotopic tracing in Bang Lake(Poyang Lake) [D], Nanchang University, 2014.
[20] ZHAO H P, ZHANG X, DENG Z M, et al. The characteristics of stable hydrogen and oxygen isotopes of atmospheric precipitation in Poyang Lake wetland[J], Journal of Water Resources Research, 2015, 4, 257- 264.
[21] LI W, LIU Y, YU L J, et al. Characteristics of primary elements migration at different geomorphological positions and their correlation to plant transpiration in Guilin karst experimental section[J], Journal of Huazhong Agricultural University, 2007, 26(1), 55-58.
[22] DUAN D Y, OUYANG H. Application of stable hydrogen and oxygen isotope in analyzing plant water use sources[J], Ecology and Environment, 2007, 16(2), 655-660.
[23] LI X C, HU S J, LI Y T, et al. Study on the root distribution and soil water dynamics under Phragmites arid area [J], Acta pratacultural sinica, 2008, 17(2), 97-101.
[24] ZHU H H, ZHANG B, et al. Poyang Lake-Hydrology Biology Deposit Wetland Development and rehabilitation [M]. China Science and Technology University Press, 1997.
[25] Brunel J P, Walker G R, Walker C D, et al. Using stable isotopes of water to trace plant water uptake//Stewart G R. Stable Isotopes in Plant Nutrition[J], Soil Fertility and Environmental Studies. Vienna: International Atomic Energy Agency, 1991, 543-551.
[26] Alessio G A, De Lillis M, Brugnoli E, et al. Water sources and water-use efficiency in Mediterranean coastal dune vegetation[J], Plant Biology,2004, (6),350-357.
[27] Rundel P W, Ehleringer J R, Nagy K A et al. Stable isotopes in ecological research[C], ecological studies. Springer, New York, 1989, 68: 142-162.

Design and Implementation of River Data Acquisition and Storage System Based on Cloud Platform

Rui Long[1*], Chunyan Deng[1] and Min Wang[1]

[1]Changjiang River Scientific Research Institute, Changjiang Water Resource Commission, Wuhan, Hubei, 430010, China

[*]Corresponding author's e-mail: longrui113@outlook.com

Abstract. In view of the huge geospatial information data in the Yangtze River Basin, messy data storage and difficulty in searching, a cloud platform for river scientific research was established based on Browser/Server Architecture. The system design comprehensively considers and balances the advancement and maturity, enabling the system to mass-store and manage data from various units or departments, multiple types and multiple channels, and to provide advanced data management methods to ensure data security and data unified management, thus providing researchers with fast and reliable data information. The system realizes a set of standards, a map, a library and a platform for the whole basin, which is of great significance to the standardized management and application of data in the Yangtze River Basin.

1. Introduction

With the rapid development of computer technology and Internet technology, the world has entered the digital age. Technologies such as big data, Internet of Things, cloud computing, virtualization, and artificial intelligence have gradually matured and are being applied in various industries. Water conservancy informatization refers to the process of making full use of modern information technology, in-depth development and utilization of water conservancy information resources, modernization of collection, transportation, storage, processing and service of water conservancy information, and comprehensive improvement of the efficiency and effectiveness of water conservancy business activities[1]. Water conservancy informatization is of great significance for water regime monitoring, flood situation monitoring, water resources management, automated office, information service and other system construction[2]. The Yangtze River Basin is China's largest watershed, which spans the three major economic zones of Southwest China, Central China, and East China. Its natural characteristics are closely related to its socio-economic characteristics. Due to the large area of the Yangtze River Basin, across the Qinghai-Tibet high-altitude zone, the southwestern tropical monsoon zone and the Central China subtropical monsoon zone[3-5], the hydrometeorological conditions are very complicated, and the relevant data are extremely large. In order to strengthen the standardized management of water resources information in the Yangtze River Basin, comprehensively improve the data management level of the river basin, and give full play to the comprehensive utilization of engineering research and scientific research, CRSRI (Changjiang River Scientific Research Institute of Changjiang Water Resources Commission) of River research institute have established a river research cloud platform based on the .Net technology platform, which improved the management and application level of water conservancy information.

2. Overall System Design

In order to achieve comprehensive supervision of all kinds of data in the Yangtze River Basin, it is necessary to sort and classify the data itself to achieve unified data standards, real-time data updates, and accurate keyword search. The system consists of five parts: disk array and storage management module, remote data transmission module, distributed data storage and management module, and data display module. The system design comprehensively considers and balances the advanced and mature aspects, enabling the system to mass-store and manage data from various units or departments, multiple types and multiple channels, and provide advanced data management methods to ensure data security and data unified management to provide researchers with fast and reliable data information.

3. Hardware Design

The hardware part of the system mainly includes: 2U rack-mounted integrated controller control system, Intel 64-bit high-performance processor, 10Gb/s network port, and disk array. The hardware system provides two copies of exactly the same hardware, and the standby hardware is always on standby to ensure that in case of system failure, redundant configuration components step in and assume the work of the failed parts, so as to keep the server running. The distributed storage architecture of river research cloud is shown in **Figure 1**.

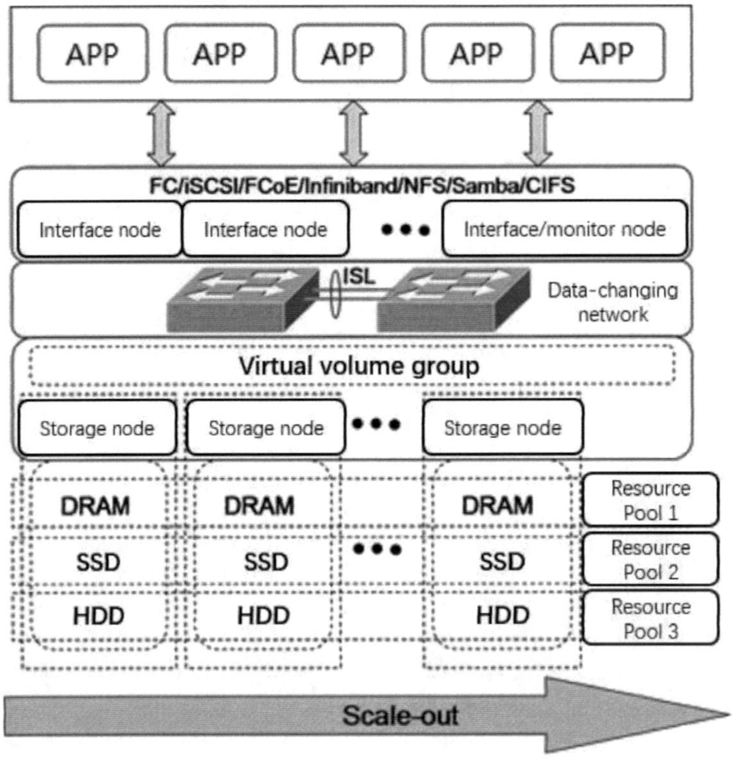

Figure 1. Distributed storage architecture of river research cloud

4. Software System Design

The River Research Cloud Platform System focuses on providing detailed geospatial information services to researchers. The system uses a multi-layer implementation mechanism, namely the Browser/Server Architecture (B/S Architecture). In the B/S Architecture, each node is distributed on the network. These network nodes can be divided into browser end, server end and middleware, and the functional tasks of the system are completed through the links and interactions between them. The three levels of division are logically divided. In practical applications, different physical divisions are performed according to the actual physical network[6].

The overall architecture of the system consists of front-end UI, presentation layer, application layer, data layer, and data interaction. The logical architecture diagram of the system is shown in **Figure 2**. In terms of external (horizontal) level, security assurance system, standard specification system and operation and maintenance system run through all levels of the platform to ensure that the internal (vertical) platforms conform to standards, safety and reliability, thus ensuring the system's operational quality. In terms of internal (vertical) level, each platform provides technical support from bottom to top, and finally provides a display platform for users to ensure the effective operation of the system.

The system platform is mainly composed of five modules: map data point, data entry, data search, data display and data application.

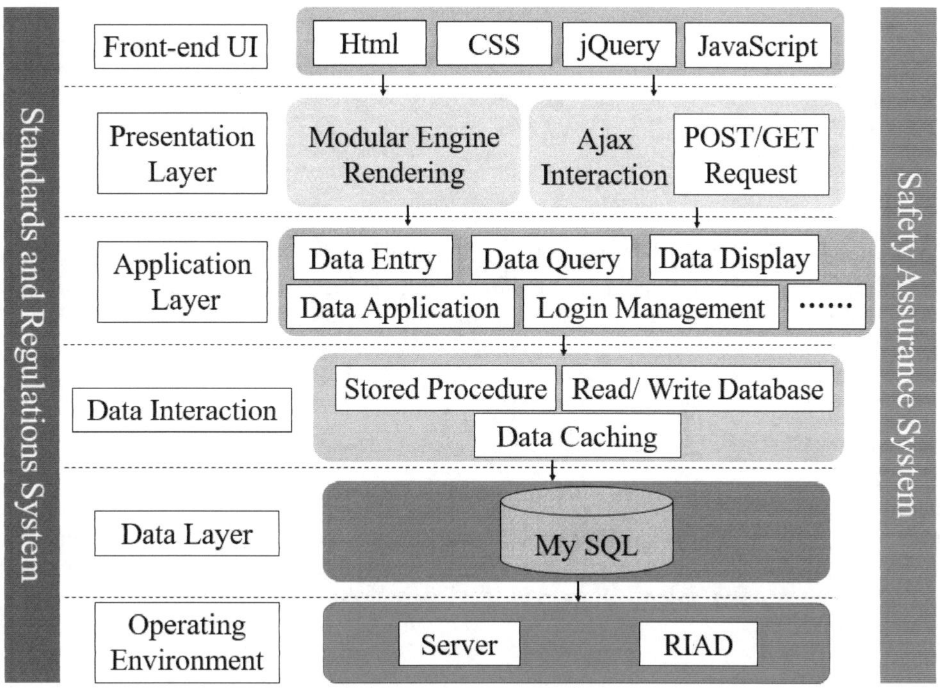

Figure 2. Logical architecture diagram of the system

4.1Map data points

The main page of this module displays the map of the Yangtze River Basin. Users can zoom in and out on the map. The module provides satellite map, road condition map, traffic map and other layers. Data points can be edited along the river basin. You can click on the data points to view all the measured data in the corresponding location.

4.2Data entry

This module mainly relies on users to input the latest watershed data, and can be used for single or batch data uploading. The uploaded data backstage will automatically detect and classify the types and upload them to disk array. The storage folder structure is: /Upload/ data type/experiment type/time/file, and the information is saved to the database.

4.3Data Search

Users can perform fuzzy or precise search under this module.

Fuzzy query: the users can input the corresponding conditions in the search input box and click the search and send the search parameters request to the backstage through methods of *ui* and *ajax*. Then the backstage gets the parameters to the database search and returns the results to the list.

Precise search: the users fill in the contents in the detailed search box and search it. Similarly, sending the search parameters request to the backstage through methods of *ajax*. Then the backstage gets the parameters to the database search and returns the results to the list.

4.4Data display

The function of this module is to display all picture and text video data stored on the platform. The page displays the data classified by the upper, middle and lower reaches of the Yangtze River. Click the corresponding data to display the backstage data to the data list through *ajax*.

4.5Data Application

At this stage, the module can draw three-dimensional topographic maps of before and after channel erosion and deposition, and its contrasted situation according to the *.dat* data, and extract the sections from the comparison results. In the future, new functions will be introduced to provide researchers with a convenient cloud-based office environment.

5. Conclusion

This paper proposes a solution of river big data management based on computer cloud platform. This system is designed for river geographic information data, and its focus is to sort and classify the data itself, so as to achieve unified data standards, real-time update of data and accurate search of keywords. The platform has basically completed the above functions and put into use, realizing the design objectives of a set of standards, a map, a database and a platform for the whole basin. In the future, we will continue to improve the platform and realize the goal of cloud office for scientific researchers as soon as possible.

Acknowledgments

The authors gratefully acknowledge Hawk soft Co., Ltd. for their assistance in the field. This paper was supported by the National Key Research and Development Project (2017YFC0405306, 2016YFC0402309), Central Research Institutes of Basic Research and Public Service Operations (CKSF2015049/HL).

References

[1] Baidu baike. Water conservancy informatization. https://baike.baidu.com/item/%E6%B0%B4%E5%88%A9%E4%BF%A1%E6%81%AF%E5%8C%96/4886850?fr=aladdin.

[2] Li Wenxiong. (2019) Research on the construction of liaoning water conservancy geographic information service platform based on GIS cloud technology. Heilongjiang Science and Technology of Water Conservancy, 47(6): 91-93.

[3] GEMMER M, JIANG Tong, SU Buda, et al. Seasonal precipitation changes in the wet season and their influence on flood drought hazards in the Yangtze River Basin, China [J]. Quaternary International, 2008, 186: 12-21.

[4] JIANG Tong, ZHANG Qiang, ZHU Deming, et al. Yangtze floods and droughts(China) and teleconnections with ENSO activities(1470-2003) [J]. Quaternary International, 2006, 144: 29-37.

[5] LU Er, LIU Siyuan, LUO Yali, et al. The atmospheric anomalies associated with the drought over the Yangtze River Basin during spring 2011 [J]. Journal of Geophysical Research-Atmospheres, 2014, 119(10): 5881-5894.

[6] Chen Wei, Fang Peipei. Contrast Analysis on Information System of C/S and B/S Structure 2003[j], INFORMATION SCIENCE 2003, 21(3): 313-315.

IOP Conf. Series: Materials Science and Engineering **794** (2020) 012012 doi:10.1088/1757-899X/794/1/012012

Study on the hydraulic characteristics of the gradually expanding and gradually contracting watercourse with different width-narrow ratios

Wang Xueting[1], Huang Lingyu[2], Qi Lijian[1], Zeng Yun[1]*

[1]Sichuan Agricultural University Institute of Water Conservancy and Hydro-Electric Power, Sichuan, 625014, Yaan

[2]The people's Government of Chongqing town in the market area, Chongqing, 401420, Qijiang

*Corresponding author's e-mail: yunzeng@sicau.edu.cn

Abstract. The characteristics of the water flow in the gradually expanding and gradually contracting watercourse are very complicated, and the width-to-narrowness ratio has a significant influence on the hydraulic characteristics of the gradually expanding and gradually contracting watercourse. Five three-dimensional hydraulic calculation models with different width-to-narrowness ratios were established to study its hydraulic characteristics. The results show that: (1) river flow gradually appears as the recirculation zone with the increase of the width-to-narrowness ratio. The recirculation zone increases gradually with the increase of the width-to-narrowness ratio and the shapes are different. The flow velocity in the recirculation zone is negatively correlated with the width-to-narrowness ratio. (2) When the width-to-narrowness ratio is small, the shearing force distribution of gradually expanding and gradually contracting sections are basically in “Hump-shape” along the Y-axis; When the width-to-narrowness ratio is large, the section basically shows a trend of "parabola-like" change that increases first and then decreases near the shore, and the average shearing force of the expanding section is larger than that of the contracting section. (3) As the width-to-narrowness ratio of the watercourse increases, the shearing force at the widest part of the river section gradually decreases, and the shearing force of the side wall at the contracting section is larger than that of the expanding section.

1. Introduction

The topography of southwestern China is mainly mountainous. Narrow channels formed by geological processes in the canyon area, and after a long riverbed succession, it is easy to form a wide and narrow watercourse. It is of engineering significance to study its water flow characteristics for river regulation. Nguyen Thanh Hoan [1] used LDV method to observe and analyze the flow velocity distribution of the river's gradually expanding open channel and the related laws of water flow turbulence. Singha et al. [2] found that changes in the geometry of the river channel will change the distribution of turbulent energy in the water flow. Through a large number of investigations, Lucy et al. [3] have shown that the important factor affecting the evolution of riverbeds is the change of river width. Armellini et al. [4] studied and analyzed that the bank shape is also an important factor affecting the structure of the separation zone of the river flow. Wu Huali et al. [5] introduced in detail the research results of the characteristics of water and sediment movement in wide and narrow rivers at home and abroad, which pointed out that,in the river wide variation, the flow velocity distribution, turbulence structure, resistance characteristics and scouring and silting characteristics mechanism of river flow can be considered in the future research. Zhou Sufen et al. [6] carried out numerical simulation of plane flow in wide and narrow rivers, and simulated two-dimensional flow characteristics such as the width gradient influence of wide and narrow channel on water depth and flow velocity. Wang Shuying et al [7] based on the indoor generalization experiment, tested and analyzed the characteristics of partial water flow movement such as the water level of gradually expanding and gradually contracting watercourse, the section circulation, and the water flow separation area. In recent years, many scholars started from numerical simulations on different watercourses, analyzed and studied the characteristics of the water flow. Ma Fuxi et al. [8] used the VOF method to track the free surface, and used the modified κ-ε model to solve the Rcynolds equation and performed a three-dimensional numerical simulation of the water flow with irregular boundaries, which proved that it can simulate complex water flow phenomena well. ; Yan Xufeng et al [9] proposed a significant influence on the water flow structure through the tank model test, and the variation law of the local head loss coefficient along the path was analyzed based on the SMS hydrodynamic numerical simulation. Zhang Guangbi et al. [10] used the VOF method to track the simulated free surface to conduct numerical simulation of the velocity field before and after the construction of the embankment within a range of 2500m on the upper and lower reaches of the embankment, which provided technical support for design and decision-making. Miao Shuting et al [11] established a three-dimensional water flow calculation model for wide and narrow rivers in mountainous areas, and explored the characteristics of water flow structure such as average cross-section flow velocity and bed shearing force under different flow rates, and then derived the evolution law of river channels.

In order to obtain a deeper understanding of the water flow characteristics of the expanding and contracting watercourses under different width-to-narrowness ratios, based on the previous studies, this paper uses the three-dimensional numerical model of the gradually expanding and gradually contracting water courses to explore the water flow structure characteristics, and provides a theory support for flood control projects of wide and narrow rivers during the mega-flood period.

2. Model establishment

2.1 Model governing equation

Numerical experiments were carried out by using a Standard *k-ε* two-equation model and VOF method for tracing simulated free surface. The Standard K-ε model is a turbulence model governing equation for jointly solving the kinetic energy-*k* equation and the dissipation rate-ε equation to obtain the turbulent viscosity coefficient, which is discretized using the finite volume method. The dissociation of each parameter adopt the first-order precision upwind precision format. The coupling of speed and pressure is processed using the SIMPLEC algorithm. The governing equations for the model are as follows:

Continuous equation:

$$\frac{\partial \rho}{\partial t}+\frac{\partial \rho u_i}{\partial x_i}=0 \tag{1}$$

Momentum equation:

$$\frac{\partial \rho u_i}{\partial t}+\frac{\partial}{\partial xj}(\rho u_i u_j)=-\frac{\partial P}{\partial x_i}+\frac{\partial}{\partial x_j}[(\mu+\mu_t)[\frac{\partial u_i}{\partial x_j}+\frac{\partial u_j}{\partial x_i}]] \tag{2}$$

k equation:

$$\frac{\partial \rho k}{\partial t}+\frac{\partial(\rho u_i k)}{\partial x_i}=\frac{\partial}{\partial x_i}[[\mu+\frac{\mu_t}{\sigma_k}]\frac{\partial k}{\partial xi}]+G+\rho\varepsilon \tag{3}$$

ε equation:

$$\frac{\partial(\rho\varepsilon)}{\partial t}+\frac{\partial(\rho u_i\varepsilon)}{\partial x_i}=\frac{\partial}{\partial x_i}[[\mu+\frac{\mu_t}{\sigma_\varepsilon}]\frac{\partial\varepsilon}{\partial x_i}]+C_{1\varepsilon}\frac{\varepsilon}{k}G-C_{2\varepsilon}\rho\frac{\varepsilon^2}{k} \tag{4}$$

$$\mu_t=\rho C_\mu\frac{k^2}{\varepsilon} \tag{5}$$

$$G=\mu_t\left[\frac{\partial u_i}{\partial x_j}+\frac{\partial u_j}{\partial x_i}\right]\frac{\partial u_i}{\partial x_j} \tag{6}$$

$$\frac{\partial\alpha_w}{\partial t}+u_i\frac{\partial\alpha_w}{\partial x_i}=0 \tag{7}$$

In the above formula: P is the pressure term considering gravity; ρ and μ are density and molecular viscosity coefficients respectively; δ_κ and δ_ε are turbulent Prandtl numbers of k and ε, respectively, taking δ_κ=1.0, δ_ε=1.3; $C_{1\varepsilon}$ and $C_{2\varepsilon}$ are the constants of the ε equation, taking $C_{1\varepsilon}$=1.44, $C_{2\varepsilon}$==1.92; C_μ is the empirical constant in equation (5), taking C_μ=0.09, turbulent viscosity coefficient μ_t from turbulent energy k and turbulent dissipation rate ε It is also obtained according to equation (5); the turbulent energy generation term can be defined by equation (6), G is the turbulent energy generation term caused by the average velocity gradient; and equation (7) is the control differential equation of the water volume fraction α_w, and the solution is solved. Controlling the differential equation is to simulate the tracking of the water-air interface, μ_i and χ_i are the velocity component and the coordinate component, respectively, and t represents time.

2.2 Model size and boundary conditions

The different width-to-narrowness ratios of the gradually expanding and decreasing watercourse sections will lead to different water flow structures and have a great influence on the stability of watercourses and bank slopes [12-13].In order to better study the water flow characteristics of the gradually expanding and gradually contracting watercourses under different width-to-narrowness ratios, the experimental study model of the gradual expansion and contraction of river flow characteristics in Sichuan Key Laboratory [7] was simulated, as shown in Figure 1. The model simulates a river with a total length of 225m, a slope of 10‰ and a height of 6m. The distance between the two narrowest sections is 40m, and the narrowest section is 10m. The working condition is the widest part of the working condition. It is 20m, 30m, 40m, 50m, 60m, and the side line is a sinusoid of Y=±(B sin0.15708x+0.5). The value of working condition 5 to B is 5, 10, 15, 20, 25 in order. The control flow rate of the model is 120m^3/s, and the width-to-narrowness ratio of working condition 1 to working condition 5 is 2:1, 3:1, 4:1, 5:1, 6:1, and there are 5 typical cross sections in the middle. (The figure is indicated by a broken line and is numbered 1-9 from upstream to downstream), and the water flow direction is as shown in Fig. 1.

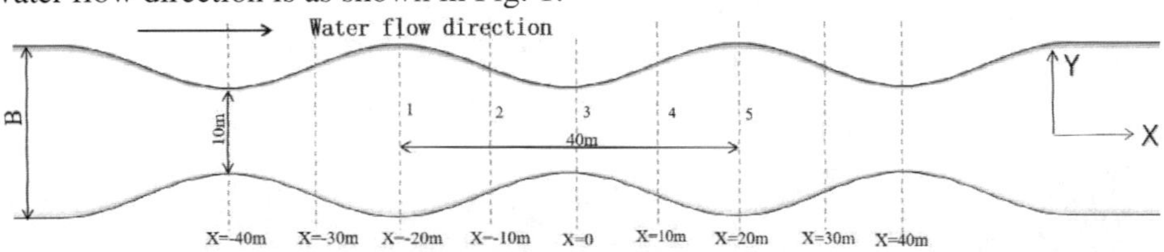

Figure 1 Typical section and dimensions of the model

Boundary condition setting: the bottom end of the water tank is set as the water flow inlet, the type is the speed inlet boundary; the outlet is set as the mixed outlet of the water flow and the atmosphere, the type is the pressure outlet boundary; the side wall is the condition of no slip solid wall surface; the upper part of the model is set as atmospheric imports, the type is the pressure import boundary.

IOP Conf. Series: Materials Science and Engineering **794** (2020) 012012 doi:10.1088/1757-899X/794/1/012012

3. Modeling results and analysis

3.1 Water level changes along the river

The variation of the water level change of this gradually expanding and gradually contracting watercourse is shown in Figure 2. The variation of the water surface line under the five working conditions is basically gradually increasing to the gradually decreasing water level, and the water level is falling near the narrowest section. The downstream is from the rapid flow to the slow flow, and the narrow section and the widest section are formed. The water level between the two is quite different. The water surface line of the upstream divergent section X=-36m to X=-20m is relatively smooth. The water level is increased along the downstream tapered section X=-20m to X=-6m, and the water level is about 3.70m. In the tapered section X = -6 m, the minimum value of 2.49 m appears near the narrow section X = 2 m. In the section X=-6m, the water falls from the slow flow to the rapid flow, and the water surface line is reduced to the lowest at the section X=2m, which is about 2.66m; the flow state of the downstream transition section is slow flow, and a hydraulic jump occurs from section X=2m to section X=6m.

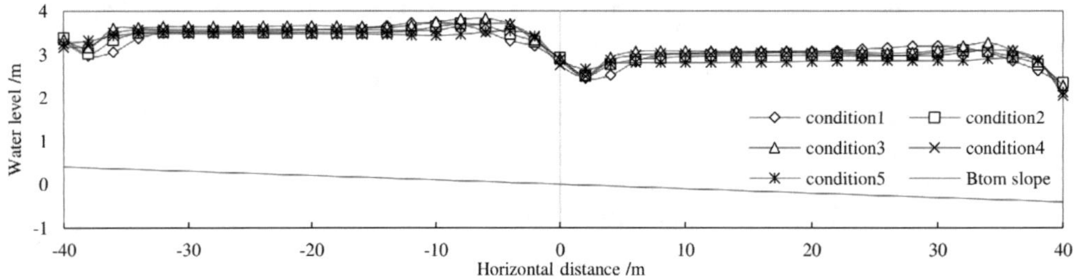

Figure 2 Water level distribution along the watercourse

3.2 Average flow rate of different sections

Different width-to-narrowness ratios have a great influence on the average flow velocity of the divergent and tapered sections. The average flow velocity of the sections under the five conditions shown in Figure 3 is distributed along the path. With the increase of the width-to-narrowness ratio of the watercourse, the average flow velocity of each section is gradually reduced, and the average section velocity of the gradient section is more susceptible to the aspect ratio than the widest section. The average flow velocity of the narrowest section X=0 is about 1.86m/s, and the average flow velocity of the section is larger than that of the transition section. The average flow velocity of the channel under this condition is the same as the average flow velocity of the same section. The maximum cross-sectional average flow rate is about 32%, and the average flow velocity of the section from the working condition 2 to the working condition 5 is 53%, 64%, 74%, and 75%, respectively. It can be seen that as the width-narrow ratio increases. The average flow velocity of the section changes gradually. The working conditions are as follows: the five-segment expansion section x=-30m average flow rate is 1.72m/s, 1.14m/s, 0.98m/s, 0.82m/s, 0.65m/s, tapered section x=-10m The average flow rate is 1.55m/s, 1.06m/s, 0.96m/s, 0.85m/s, and 0.81m/s. It can be seen from the data that the working condition 1, 2, and 3 are due to the tapered section X=-10m. The action of muddy water hinders the movement of water flow, which causes the average flow velocity of the section to decrease. The average flow velocity of the tapered section is smaller than that of the gradually expanding section. Under the working condition 4and 5, the slope of the tapered section X=-10m is larger, resulting in the flow velocity. The flow rate is larger than that of the diverging section, that is, there is a special channel width between working condition 3 and 4, which determines the relative size of the average section flow velocity of the diverging section and the tapered section.

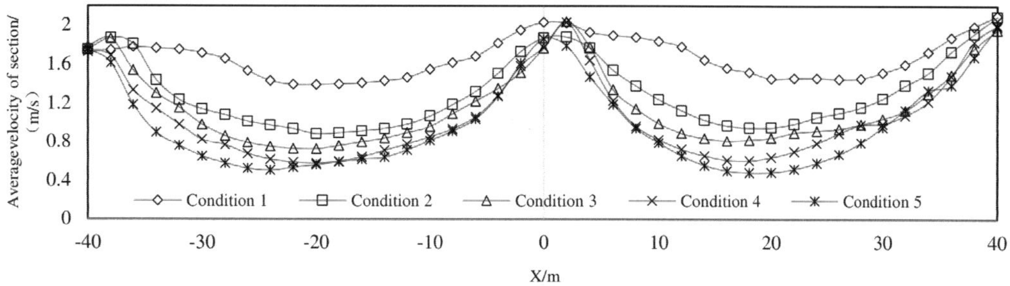

Figure 3 Flow velocity distribution along the central axis of the section

3.3 Typical watercourse speed field

The flow velocity vector of different working conditions in the transition section is shown in Figure 4 (the solid line represents the flow line of the recirculation zone). It can be seen from the figure that the larger the width-to-narrowness ratio, the smaller the flow rate variation of the narrowest section, and the recirculation zone. The larger the range. The width of the widest part is 20m, and the flow velocity distribution of the wide section and the narrow section is relatively uniform. According to the streamline, it can be seen that the streamline near the widest point is straight and there is no reflow; as the width-to-narrowness ratio increases, the reflow gradually occurs. The working area 2 to the working condition 5 recirculation area accounted for 36%, 50%, 56%, and 63% of the water surface of the typical river section. As the width ratio increases, the range of the recirculation zone also increases. The shape of the recirculation zone is different; the condition of the recirculation zone has obvious reflow, and the shape of the recirculation zone of the upstream wide section is elliptical. The condition of the recirculation zone of the upper section of the working condition is rounded, and the flow velocity changes uniformly. When the width-to-narrowness ratio of the recirculation zone at the downstream wide section is relatively small, the flow velocity in the recirculation zone is significantly reduced; under the working condition, the recirculation zone shape like a sharp triangular, and the sharp portion appears at the end of the tapered section.

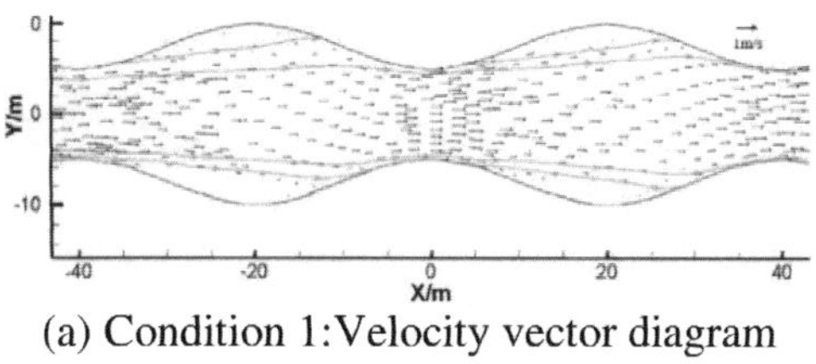

(a) Condition 1:Velocity vector diagram

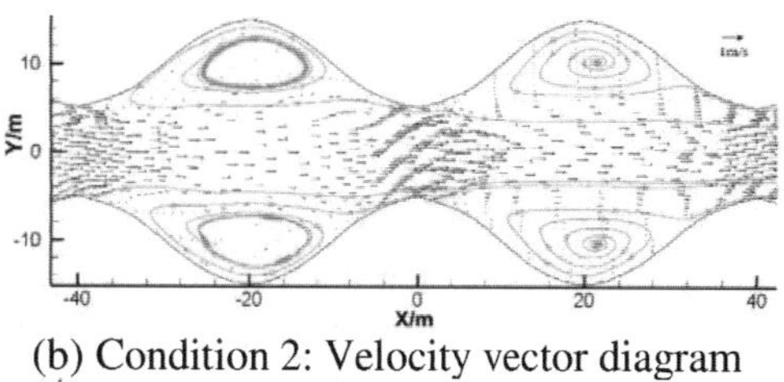

(b) Condition 2: Velocity vector diagram

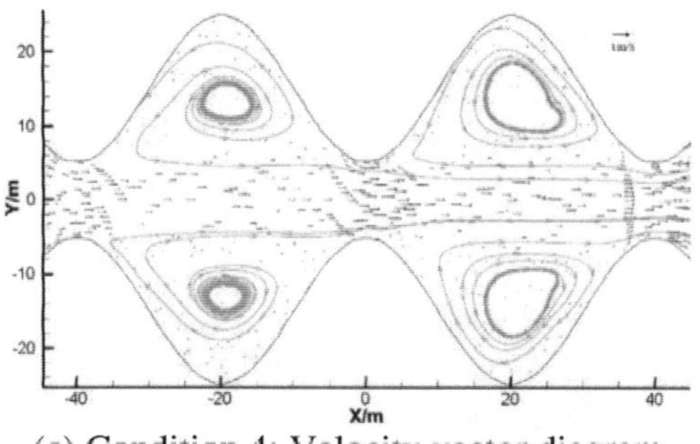

(c) Condition 4: Velocity vector diagram

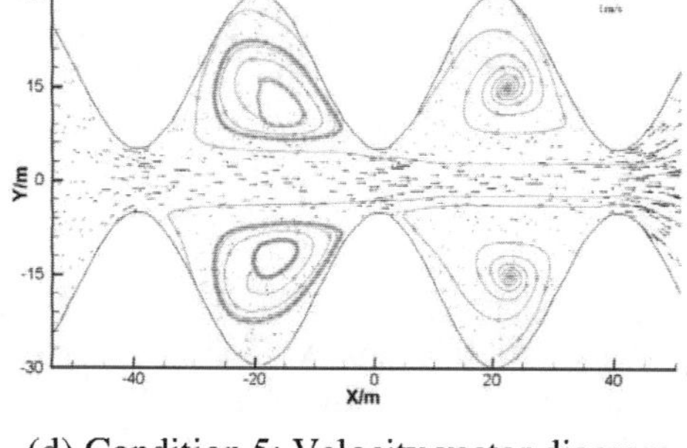

(d) Condition 5: Velocity vector diagram

Figure 4 Flow rate near the surface of a typical river section under different conditions

Under the same flow rate, as the width-to-narrowness ratio increases, the average flow velocity variation in the narrow section of the channel is small, and the average flow velocity in the recirculation zone decreases relatively slowly; when the width-to-narrowness ratio of the channel is 2:1, there is basically no recirculation zone. As the width-to-narrowness ratio increases, the recirculation zone gradually appears. The range of the recirculation zone gradually increases as the aspect ratio increases, and the shape of the recirculation zone changes continuously with the shape of the side wall.

3.4 Shearing force of river bed

The shearing force of the bed surface under typical working conditions is shown in Fig. 5. When the width and width are relatively small, the shearing force distribution of the progressively expanding and tapered section is basically "hump-like" along the Y-axis; when the width-to-narrowness ratio is large, the section is basically It shows the trend of "parabola" which decreases first and then decreases near the shore, and has a "hump" shape in the middle, and changes sharply near the 1/4 or 3/4 position near the water bank. Under the working condition, the shearing force of the bed with the widest section X=-20m presents a "hump-like" distribution. As the width-to-narrowness ratio increases, the bed shearing force of the widest section presents a "parabolic" distribution near the shore. The maximum shearing force of the bed surface from the working condition 1 to the working condition five section X=0 is 20.02Pa, 18.84Pa, 19.15Pa, 18.41Pa and 18.80Pa, which is less affected by the change of the width-to-narrowness ratio. The average bed shearing force of the divergent section X=10m and the tapered section X=-10m of the working condition 1 to the working condition 5 are 9.60 Pa and 7.30 Pa, respectively, and the water level of the tapered section is larger than the divergent section due to the hydrophobicity. The shearing force of the bed surface of the dilating section of the working condition 1 to the working condition 5 X=10 m is larger than that of the tapered section X=-10 m.

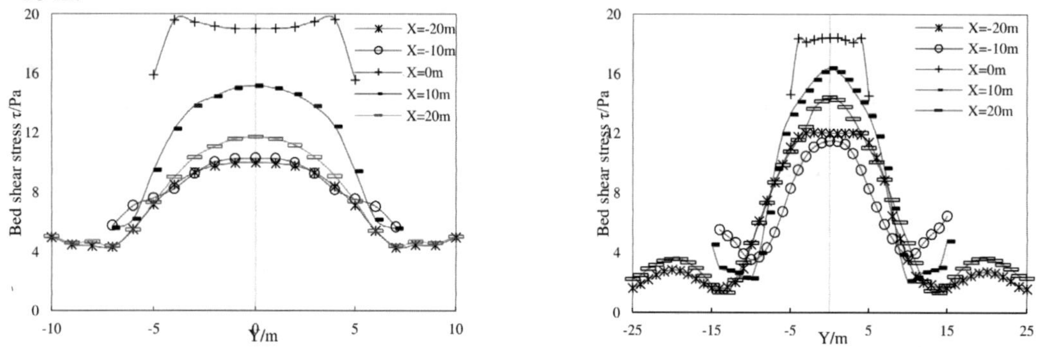

(a) Condition 1 (b) Condition 4

Figure 5 Bed shearing force under typical conditions

3.5 Shearing force of the side wall

Figure 6 as the shearing force distribution of the side wall of each section of the transition section under typical working conditions. The shearing force of the side wall changes with height: the shearing force of the side wall decreases with the increase of height at the near water surface. Small, basically showing a vertical trend distribution. When the width-to-narrowness ratio is 2:1, the shearing force distribution of the edge of the gradient section is concentrated, and the minimum is 3.69Pa and the maximum is 9.96Pa. When the width-to-narrowness ratio is 3:1, the shearing force of the side wall of the narrowest section X=0 reaches the maximum at the bottom of the channel, which is about 13.47Pa, while the maximum value of the shearing force of the side wall of the section is generally located under other conditions. 83% water depth. The average shearing force of the divergent section X=10m and the tapered section X=-10m is 7.31Pa and 4.32Pa, respectively. The trend of the two changes is the same: first increase with height and slowly increase, near the water surface The height is increased and decreased, and the shearing force of the side wall of the diverging section is greater than the shearing force of the side wall of the tapered section. As the width-to-narrowness ratio increases, the shear strength of the side wall of the tapered section X=-10m increases with respect to the width and width, and the width of the diverging section X=10m and the shear wall of the widest section decreases relative to the hour.

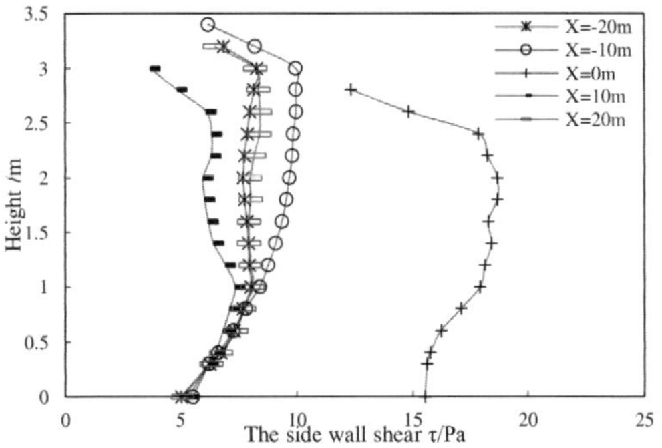

(a) Condition 1

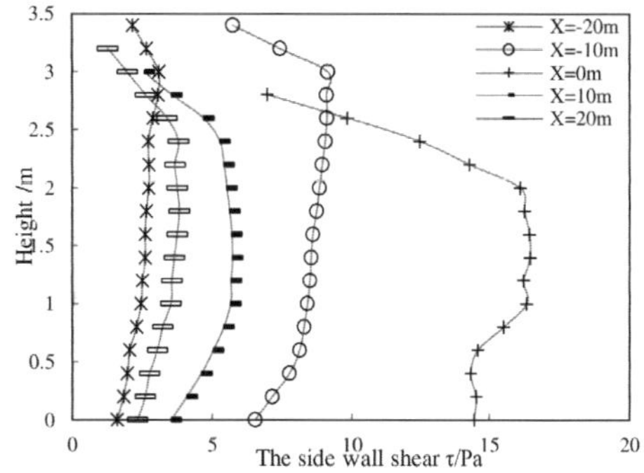

(b) Condition 4

Figure 6 Sidewall shearing force under typical conditions

The distribution of the average side wall shearing force of a typical section under different operating conditions can be seen from Figure 7. With the increase of the width-to-narrowness ratio, the average side wall shearing force of the widest section of the channel X=-20m and X=20m gradually decreases with the increase of the width-to-narrowness ratio of the channel, and the average side wall of the two sections under the working condition The shearing force is the largest, 7.47Pa and 7.65Pa respectively; the average side wall shearing force of section X=0, X=-10m, and X=10m shows a decreasing trend when the aspect ratio is less than 3:1, and the aspect ratio is from The average side wall shearing force increases gradually from 3:1 to 4:1. When the width to width ratio is 4:1, the average side wall shearing force changes substantially. Under normal conditions, the average side wall shearing force of each section is small, the average side wall shearing force of the widest section X=0 is 10.16Pa, and the average side wall shear of the widest section X=-20 and X=20m. The shearing force is greater than the gradient section X=-10m and X=10m. The main reason is that the recirculation zone has an effect on the flow velocity of the water flow, and the flow gradient of the gradient section is smaller.

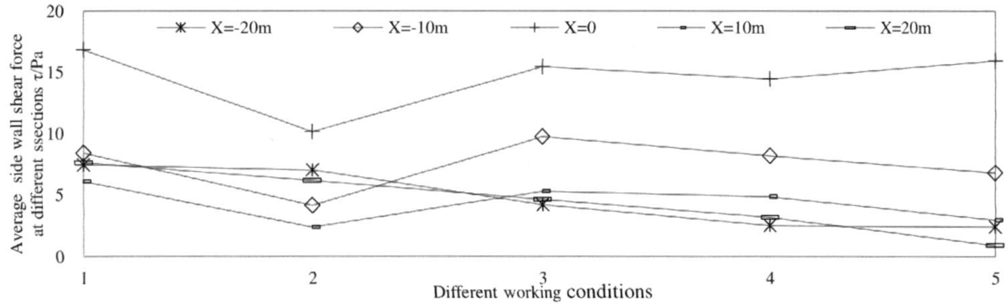

Figure 7 Average side wall shearing force under five working conditions

4. Conclusion

(1) Under the same flow rate, as the width-to-narrowness ratio increases, the flow velocity of the whole watercourse gradually decreases. With the increase of the width-narrow ratio, the recirculation zone gradually appears on the widest part of the river channel, and the shape of the recirculation zone increases with the width-narrow ratio. The gradual change gradually, the cross-sectional flow velocity in the recirculation zone is negatively correlated with the width-to-narrowness ratio.

(2) When the width-to-narrowness ratio is small, the shearing force distribution of gradually expanding and gradually contracting sections are basically in "Hump-shape" along the Y-axis; When the width-to-narrowness ratio is large, the section basically shows a trend of "parabola-like" change that increases first and then decreases near the shore. The maximum shearing force of the bed surface

appears in the narrowest section, the minimum value appears in the gradually contracting section, and the river bed surface shearing force of the expanding section of the river section is larger than that of the contracting section.

(3) As the width-to-narrowness ratio of the watercourse increases gradually, the shearing force of the side wall at the widest part of the river section gradually decreases; the shearing force of the side wall of the gradually expanding and gradually contracting section and its changing trend are basically the same, and the side wall shearing force of expanding section is greater than the contracting section.

Acknowledgement
Foundation item: project of Education Department of Sichuan Province (18ZB0468)

Reference

[1] Nguyen Thanh Hoan, Rob Booij, Marcel J.F. Stive, et al, (2007) Decelerating open channel flow in a gradual expansion[C]. Asian and Pacific Coast2007, September 21-24, 2007, Nanjing, China: 902-915.
[2] Singha A, Balachandar R. (2011) Structure of wake of a sharp edged bluff body in a shallow channel flow[J]. Journal of Fluid sand Structures, 27(2):233-249.
[3] Lucy C, Quine T A, Nicholas A. (2010) An experimental investigation of autogenic be haviour during alluvial fan evolution[J]. Geomorphology, 115(3/4):278-285.
[4] Armellini A, Casarsa L, Giannattasio P. (2009) Separated flow structures around a cylindrical obstacle in a narrow channel[J]. Experimental Thermal and Fluid Science, 33(4):604-619.
[5] Wu Huali, Zhang Xiaofeng, JinZhongwu. (2015) Research advances of characteristics of flow and sediment transportation in wide and narrow alternat[J].Yangtze River, 08:14-18.
[6] Zhou Sufen, Yi Zijing, Yan Xunfeng, et al, (2013) Two-dimensional numerical simulation of flows in wide and narrow alternated channels in mountainous [J]. Advances in Science and Technology of Water Resources, 33(1):22-26.
[7] Wang Suying, Zhou Sufen, Zhao Xiaoe, et al. (2013) Experimental Study on the Flow Characteristics at Local Diverging-converging Sections in Mountain Lotus Root Shape Channel [J]. Advanced Engineering Sciences, 45(supplement 2):51-54.
[8] Ma Fuxi, Wang Jinrui. (1996) Three-dimensional numerical simulation of flow [J]. Journal of Hydraulic Engineering, 8:39-44.
[9] Yan Xufeng, Yi Zijing, Liu Tonghuan, et al. (2011) Flow Structure and Characteristics of Local Head Loss in Transition Channel[J].Journal of Yangtze River Scientific Research Institute,28(9):1-5.
[10] Zhang Guangbi, Dengjun, Liu Chao,et al. (2007) Numerical Simulation Study on Three-dimensional Velocity Field of Fluvial Channel Current[J].Journal of Hydraulic Engineering ,01:58-62.
[11] Miao Shuting, Lin Hu, Qi Lijian, et,al. (2016) Flow Characteristics in Wide and Narrow Alternated Channels in Mountainous Areas under Different Discharges [J].Journal of Yangtze River Scientific Research Institute,33(08):64-68.
[12] Lu Jinyou, JinZhongwu, Yao Shiming,et al. (2010) Discussion on bed load transport in the backwater reach with a canyon, Journal of Sediment Research, 02:70-74.
[13] YaoShiming, Lu Jinyou, Xu Haitao. (2005) Study of Vertical Velocity Distribution at Huang ling miao Hydrologic Section, Journal of Yangtze River Scientific Research Institute, 04:8-11.

IOP Conf. Series: Materials Science and Engineering **794** (2020) 012013 doi:10.1088/1757-899X/794/1/012013

Study on Comprehensive Evaluation of Protection and Utilization of Artificial canal in Yichang City

Mingyuan Zhou, Jijun Xu*, Xiaofeng Hong and Junlong Liu

Water Resources Comprehensive Utilization Research Institute, Changjiang River Scientific Research Institute of Changjiang Water Resources Commission, Wuhan, Hubei, 430010, China

*Corresponding author's e-mail: xujj07@163.com

Abstract. In order to effectively evaluate the protection and utilization of artificial canals. This article takes the Dongshan Canal as an example, combining the characteristics of Dongshan Canal, and constructing artificial canal protection and utilization evaluation index system based on the conceptual model of "driving force-pressure-state-impact-response-management". At the same time, the comprehensive evaluation method of the protection and utilization of the artificial canal is established by combining the analytic hierarchy process. The results show that: The comprehensive evaluation result of Dongshan canal protection and utilization is 0.80 (I grade), and the river has a high degree of protection and utilization. The evaluation results are consistent with the actual situation of Dongshan canal, which verifies the feasibility of the index system constructed in this paper.

1. Introduction

With the acceleration of industrialization, urbanization and modernization, the uncertainty and complexity of artificial canals in the large-scale development and utilization of human beings are stimulated, which leads to serious threats to the river ecosystem, and then to different degrees of degradation[1]. The issue of the protection and utilization of artificial canals has attracted the attention of the government and the community[2-4]. At present, ensuring the health of artificial canals has become an important management goal of the government.

The Dongshan Canal (DSC) built in 1960, also known as the Yichang canal, is an inter-basin and inter-regional water transfer project. DSC has played an important role in the urban development and hydropower industry in Yichang city since it was built and operated. It is a milestone of Yichang and even Hubei hydropower. In recent years, under the opportunity of the strategic background of the Yangtze river economic belt and the rise of central China, the social economy of Yichang city has developed rapidly and the urbanization process has also accelerated significantly. In the meanwhile, the pressure of DSC has also increased, and the status of river protection is worrying. Therefore, the DSC protection and utilization status is evaluated, the problems of river protection and utilization are analyzed, and corresponding river management measures are formulated to ensure the healthy development rivers in Yichang city and realize the DSC as the target of residents' leisure, entertainment, popular science and sightseeing places.

In view of this, this paper takes the DSC as the research area, combines the actual conditions and the characteristics of the DSC, based on the representative principles, operability, functional requirements, harmonious coexistence and long-term development and other index system construction principles, draws on the research results of river protection and utilization. [5-7], and

IOP Conf. Series: Materials Science and Engineering **794** (2020) 012013 doi:10.1088/1757-899X/794/1/012013

constructs the DSC protection and utilization evaluation system. At the same time, the analytic hierarchy process is used to establish a comprehensive evaluation model of river protection and utilization, and the protection and utilization of DSC is quantitatively evaluated, in order to provide a favorable reference for the control of artificial canal.

2. Research Area

2.1Basic Situation of Watershed

DSC(111°18′～111°21′, 30°41′～30°51′), also known as the Yichang canal, is a inter-basin and inter-regional water transfer project. It is located on the left bank of the Yangtze river and north of Yichang City. It flows through Yiling district economic development zone, Xiaoxita street, Xiyaowan township, Xiling street, Yichang development zone, Dongshan Park, Dagong Bridgein Xiling district and Wanshouqiao street, involving 5 township streets and 3 developments parks, and returns to the Yangze rive at Wanshouqiao street in Wujiagang district. The section from Tangduhe reservoir to Dongshan power station is an artificial river section with a river length of about 27 km and a drainage area of 11.2 km^2. The Dongshan power station to Wanshouqiao is a natural river channel with a river length of about 3.5 km and a drainage area of 10.7 km^2.

2.2Current status of water resources development and utilization

The development and utilization of water resources in the DSC is relatively high. Its water intake is mainly for power generation and water intake, industrial water intake, urban life water intake (including domestic water, public water), agricultural water intake and other water use (landscape entertainment, ecological water use). According to the survey, there are currently two power generation and water supply facilities in the Dongshan Canal Basin, namely Dongshan Hydropower Station and Yaowan Water Plant. There is a reservoir in the basin, which is the Meiziya reservoir, which plays a more obvious role in regulating the water resources utilization of the basin and greatly improves the utilization of water resources. DSC, Tangduhe reservoir and Meiziyu reservoir are the spare water sources of Yichang city water supply plants.

2.3 Featured positioning

From the completion of the Yaowan water plant during 1962 to 2002, the DSC has been the main source of water for urban use in the past 40 years. In recent years, with the development of the city, the water environment and ecology of the DSC have been seriously damaged. After the "Dongyi project" was officially put into operation, the water source of Yichang Yishui plant was changed from the canal to the Dongfengqu Guanzhuang reservoir. The status of the canal was reduced from the original "main source of Yichang urban water supply" to "important standby living water source". Combined with the current situation of water resources development, environmental protection, ecological landscape, and related planning requirements of DSC, the characteristics of DSC are: urban leisure, ecological corridor, standby water source, taking into account power generation and irrigation functions.

IOP Conf. Series: Materials Science and Engineering **794** (2020) 012013 doi:10.1088/1757-899X/794/1/012013

Figure1 DSC location and its administrative area

2.4Source of information

The basic data used in this paper comes from the "one river and one river policy", "procedures for the planning of river basin planning" (SL201-2015), "guidelines for the compilation of river and lake shoreline protection and utilization planning" (trial), "national ecological protection red line-technical guide for delineation of ecological function red lines (trial)", (circular [2014] No. 10) and other policy plans. Taking into account the integrity and reliability of the data, the water quality data is derived from the actual monitoring of the Yichang water resources bureau.

3. Construction of indicator system

3.1Indicator system

The construction of index system is the basis of river protection and utilization evaluation. However, due to the difference of natural conditions and the influence of human activities, the river ecosystem has regional characteristics. Based on the existing research and field investigation of DSC, referring to DPSIRM conceptual framework, and following five principles of scientificity, integrity, operability, functional needs, harmonious coexistence and long-term development, this paper constructs an evaluation index system for DSC protection and utilization. The key indicators are listed in Table 1.

3.1.1Driving force index. The DSC covers a drainage area of 21.9 km^2, with a population of 0.88 million along the coast, and its affected area involves nearly 200, 000 residents. The DSC mainly provides water for power generation and water supply along the canal enterprises of Dongshan power plant, and provides ecological environment water for the canal city section. Runoff and population are important driving factors for the protection and utilization of DSC. Therefore, this paper chooses the runoff of dry-flow period to account for the annual average runoff ratio and population density as the driving force indicator.

IOP Conf. Series: Materials Science and Engineering **794** (2020) 012013 doi:10.1088/1757-899X/794/1/012013

3.1.2Pressure indicators. Combined with the social and economic development of DSC and the development and utilization of the basin, this paper selects the protection and utilization pressure of DSC from three aspects: water resources, water environment and riparian zone:

- Water resources pressure index. At present, the development and utilization of water resources in DSC is relatively high. The water intake is mainly for power generation and water intake, industrial water intake, urban life water intake (including domestic water, public water), agricultural water intake and other water use (landscape entertainment, ecological water use). Therefore, this paper selects urban residents' water quota, rural residents' water quota, total water use, irrigation water utilization coefficient as water resource pressure.
- Water environmental pressure index. The water pollution in the basin is dominated by urban (township) domestic pollution and agricultural planting pollution. The downstream section of the Meiziya reservoir is densely populated, and a large amount of domestic sewage is discharged into the river. At the same time, the sewage collection pipeline along the DSC is still not perfect, and the downstream urban section has not fully realized the distribution of rain and sewage. There are many domestic sewage and mixed sewage discharges along the line. Therefore, this paper chooses the waste water discharge as the water environment pressure.
- The riparian belt uses the pressure index. Riparian vegetation plays an important role in maintaining the health of river ecosystems[8]. Deforestation and destruction of vegetation planted on river banks will have a direct impact on the stability and health of river ecosystems, and lead to threats to human health and safety that depend on river ecosystems[9]. Therefore, this paper selects the vegetation coverage rate of forest grass as an indicator to reflect the utilization status of the riparian zone.

3.1.3Status indicators. Considering the influence of the pressure index and the driving force index, the state indicators are divided into three aspects: water resources, water environment, and riparian zone:

- Water resource status indicator. According to the statistical information of the DSC on the national economic and social development statistics, the population growth rate of the DSC is calculated. Considering that the urbanization rate of the DSC will be improved in the future, combined with the alternate water source and irrigation function of the DSC, this paper chooses the water supply guarantee rate as the water resource status indicator.
- Water environment status indicators. The water environment status indicator is based on the "Planning of the water intake, sewage outlet and emergency water source along the Yangtze River Economic Belt along the Yangtze River", the water quality target meets the Class III or better than the Class III water standard as the "suitable water intake zone" and "not suitable water intake zone". The proportion of water quality conforming to or better than Class III is selected as the water environment state.
- Riverside status indicators. Combined with the functional orientation of the canal city leisure and ecological corridor, this paper selects the natural degree of the bank slope as the riparian state indicator.

3.1.4Impact indicators. The DSC has comprehensive benefits such as water supply and irrigation. However, according to the water quality monitoring data, the water quality of DSC is poor, which can not meet the requirements of class II water quality target in water function area. In order to visually reflect the impact of the DSC, this paper selects the water availability and drinking water safety ratio as the impact indicators.

3.1.5Response indicators. In view of the functional characteristics and existing problems of the DSC, considering the current situation of water resources development, environmental protection, ecological landscape in the DSC basin and the relevant planning requirements of Yichang city, this paper chooses the sewage treatment capacity and the degree of ecological water security as the response indicators.

3.1.6Management indicators. In the full interpretation of the existing ecological function red line zoning, water environment quality red line zoning, water environment function zoning, water function zoning, livestock breeding area, other industry development plans, national and local laws and regulations on river protection in the DSC basin, this paper chooses the soundness of river protection and management system, the adaptability of flood control standards, the degree of river and cross-strait cleaning, and the establishment of environmental supervision system around rivers as management indicators.

Table 1. River protection and utilization indicators and implications.

Target layer	Factor layer	Indicator layer	Indicator meaning	Indicator type
River protection and utilization	Driving force	Runoff accounted for annual average runoff ratio during dry season	Characterizing natural factors driving water resources	+
		The population density (people/km^2)	Characterizing population density to drive river ecosystems	-
	Pressure	Urban residents' water quota (L/people/day)	The pressure of urban economic development on water resources	-
		Rural residents' water quota (L/people/day)	The pressure of rural economic development on water resources	-
		Total water use (L. $(people.d)^{-1}$)	Characterizing the pressure of regional socioeconomic development on water consumption	-
		Irrigation water utilization fact(%)	Characterizing the degree of protection of agricultural water use by farmland water conservancy facilities	+
		Waste water discharge	Representing the pressure of social development on the water environment	-
		Forest and grass vegetation coverage(%)	Indicates the percentage of the vertical projected area of the plant per unit area	+
	Status	Water supply guarantee rate(%)	Reflecting the carrying capacity of the water environment in the basin	+
		Water quality meets or exceeds class III ratio(%)	Reflecting the carrying capacity of the water environment in the basin	+
		Shore slope natural degree(%)	Reflecting the degree of human interference in the riparian zone	+
	Influences	Water availability ($108m^3$)	Indicates the impact of water resources	+
		Drinking water safety ratio(%)	Indicates the impact of the water supply	+
	Response	Sewage treatment capacity(%)	Representing a response to water environmental governance, direct means	+
		Ecological water security(%)	Indicates the level of ecological water demand protection	+

Management	The soundness of river protection and management system	Indicates the soundness of the river system	+
	Flood protection standard adaptability	Indicates the validity of management standards	+
	River and cross-strait cleaning	Indicates the effectiveness of the river system	+
	Establishment of environmental supervision system around the river	Indicates the implementation status of the river system	+

3.2Grading standards

According to the established DSC protection and utilization evaluation index system, combined with domestic and foreign river health index standards, relevant policies and planning objectives, DSC protection and utilization evaluation grades are divided. The specific grading standards are as follows:

Table 2. DSC Protection and Utilization Evaluation Grade

Target layer	Factor layer	Indicator layer	Grading				
			1-0.8	0.8-0.6	0.6-0.4	0.4-0.2	0-0.2
River protection and utilization	Driving force	Runoff accounted for annual average runoff ratio during dry season(%)	1.3-2	1.1-1.3	0.9-1.1	0.7-0.9	≥2 or<0.4
		The population density（people/km^2）	>400	400-600	600-1400	1400-3500	3500-5000
	Pressure	Urban residents' water quota（L/people/day）	<190	190-210	210-230	230-400	>400
		Rural residents' water quota（L/people/day）	<50	50-70	70-90	90-110	>110
		Total water use（L.（people.d）$^{-1}$）	1.5-4.5	4.5-7.5	7.5-10.5	10.5-13.5	13.5-16.5
		Irrigation water utilization fact (%)	80-100	60-80	40-60	20-40	<20
		Waste water discharge	<0.05	0.05-0.075	0.075-0.1	0.1-0.125	>0.125
		Forest and grass vegetation coverage (%)	<10	10-20	20-30	30-50	>50
	Status	Water supply guarantee rate（%）	>80	70-80	60-70	50-60	<50
		Water quality meets or exceeds class III ratio (%)	100	70-100	60-70	50-60	<50
		Shore slope natural degree (%)	>95	80-95	65-80	50-65	<50
	Influences	Water availability（108m^3）	>170	140-170	110-140	80-110	<80
		Drinking water safety ratio (%)	<5	5-10	10-30	30-50	>50
	Response	Sewage treatment capacity (%)	>95	80-95	65-80	50-65	<50
		Ecological water security (%)	≥90	70-90	60-70	40-60	<40

	The soundness of river protection and management system	high	Higher	medium	Lower	low
Management	Flood protection standard adaptability	Adapt	-	Basic adaptation	-	Not adapted
	River and cross-strait cleaning	high	Higher	medium	Lower	low
	Establishment of environmental supervision system around the river	excellent	well	general	qualified	Failed

3.3Evaluation method
The Analytic Hierarchy Process (AHP) was first proposed by American operations researcher T.L. Saaty et al. in the 1970s. It is a hierarchical and structured multiple target decision analysis method that combines quantitative and qualitative methods[8], has the characteristics of system, flexibility and conciseness, and can better solve the problem of multilevel and Multiple structure in the river system. In general, after determining the evaluation indicators for the study of the river section, the evaluation of each layer by the analytic hierarchy process can be divided into the following steps:

3.3.1Determination of indicator weight. (1) Structural judgment matrix. After establishing an index system with hierarchical institutions, the factors in the same layer are compared in pairs, and the relative merits and demerits of the indicators are judged by the scoring method. Finally, judgment results results form a judgment matrix. The specific scale is defined as shown in Table 3 below.

Table 3. Judgment matrix scale definition

Scaling	meaning
1	Two elements are of equal importance
3	One is slightly more important than the other compared to the other
5	One is more important than the other compared to the two elements
7	One is stronger than the other compared to the two elements
9	One element is more important than the other compared to the other
2, 4, 6, 8	Intermediate value of the above two adjacent judgments
reciprocal	If the factors i and j must have the importance ratio of the judgment matrix to be a_{ij}, then the ratio of the factors j to i is $a_{ij}=1/a_{ij}$.

- Hierarchical single sorting and consistency test. The feature vector corresponding to the largest eigenvalue of the matrix is normalized, which is the ordering of the corresponding elements of the same level corresponding to the relative importance of the upper layer, which is the weight. The specific calculation steps are as follows.First calculate the maximum eigenvalue of the judgment matrix $\lambda_{\max}$;
- Then calculate the consistency indicator CI

$$CI = \frac{\lambda_{\max} - n}{n-1} \tag{1}$$

In the formula, n is the order of the decision matrix

- Calculate the average random consistency ratio CR

$$CR = \frac{CI}{RI} \tag{2}$$

- In the formula, RI is the average consistency indicator and its value is determined as shown in the table.
- When CR < 0.10, The results of the analytic hierarchy process are considered to be satisfactory.

Table 4. Judgment matrix random consistency index RI value table

n	1	2	3	4	5	6	7	8	9
RI	0	0	0.58	0.9	1.12	1.24	1.32	1.41	1.45

- Hierarchical total ordering and consistency check
- The hierarchical total ordering is to appropriately combine the calculation results of step (2), and then obtain the relative weights of all the elements of each layer in the hierarchical structure to the total target.
- For the total ordering of the hierarchy, a consistency check is also required, and the total ranking result of the hierarchy has satisfactory consistency (when CR < 0.10). Otherwise, the judgment matrix needs to be re-adjusted.

3.3.2Establishment of evaluation model. The evaluation model adopts the analytic hierarchy process combining qualitative and quantitative analysis. Considering the simplicity and ease of operation of the evaluation model, the evaluation index of river protection and utilization control planning is calculated according to the following formula:

$$E = \sum_{i=1}^{n} \lambda_i \times \{\sum_{j=1}^{mi} \lambda_{ij} M_{ij}\} \tag{3}$$

In the formula, λ_i is the weight of the *i*-th criterion layer, n criteria layers in total; λ_{ij} is the weight of the *j*-th indicator selected for the *i*-th criterion layer in the criterion layer; M_{ji} is the score of the *j*-th indicator selected for the *i*-th criterion layer.

3.3.3Determination of evaluation level. According to the comprehensive index obtained from the evaluation model, the status of river protection and utilization control planning can be evaluated. The larger the comprehensive index is, the higher the degree of river protection and utilization is. The evaluation results are now divided into 5 levels, as shown in the following table.

Table 5. River protection and utilization scale

River protection and utilization	grade 1 (high)	grade 2 (Higher)	grade 3 (general)	grade 4 (Lower)	grade 5 (low)
composite index	>0.8	0.8-0.7	0.7-0.6	0.6-0.5	<0.5

4. DSC evaluation

4.1River protection and utilization evaluation

According to the requirements of AHP, through expert consultation and combined with a large number of river protection and utilization research judgment, the indicators of each layer are discriminated and tested for consistency according to table 3, and the index weight is calculated. Similarly, the weight of each criterion layer is calculated, and the results are as shown in Table 6.

Table 6. DSC Protection and Utilization Evaluation Index Weight

Factor layer	Indicator layer	Index Weight	Indicator value	Scale value	Evaluation index
Driving force	Runoff accounted for annual average runoff ratio during dry season	0.06	<0.4	0.20	0.01
	The population density (people/km^2)	0.06	400-600	0.80	0.05

	Urban residents' water quota（L/people/day）	0.03	<190	1.00	0.03
	Rural residents' water quota（L/people/day）	0.03	90-110	0.40	0.01
	Total water use（L.（people.d）$^{-1}$）	0.05	1.5-4.5	1.00	0.05
Pressure	Irrigation water utilization fact（%）	0.04	40-60	0.60	0.02
	Wastewater discharge	0.04	0.1-0.125	0.40	0.02
	Forest and grass vegetation coverage (%)	0.04	<10	1.00	0.04
	Water supply guarantee rate (%)	0.06	70-80	0.80	0.05
Status	Water quality meets or exceeds Class III ratio (%)	0.06	70-100	0.80	0.05
	Shore slope natural degree (%)	0.06	80-0.95	0.80	0.05
Influences	Water availability（10^8m^3）	0.06	<5	1.00	0.06
	Drinking water safety ratio (%)	0.11	40-70	0.40	0.05
Response	Sewage treatment capacity (%)	0.06	65-80	0.60	0.04
	Ecological water security (%)	0.06	70-90	0.80	0.05
	The soundness of river protection and management system	0.06	Basic adaptation	0.80	0.05
Manageme nt	Flood protection standard adaptability	0.03	Basic adaptation	0.60	0.02
	River and cross-strait cleaning	0.03	Basic adaptation	0.80	0.02
	Establishment of environmental supervision system around the river	0.05	well	0.80	0.04
	composite index	1	-	-	0.70

4.2Comprehensive evaluation of DPSIR river protection and utilization in DSC

According to the above method, the comprehensive evaluation index of DSC is 0.7, and the degree of river protection and utilization is high. From the weight of the criterion layer, the impact factors of the DSC protection and utilization are as follows: stress factor (0.17) >management factor (0.13) >impact factor (0.11) >state factor (0.1) >response factor (0.09) >kinetic force factor (0.06). Pressure factor is the main pressure on river protection and utilization caused by various social and natural driving forces [11]. The reduction of water supply guarantee rate and the increase of total water consumption are the direct causes of water pressure on the DSC. Therefore, the stress factor has the greatest impact on river protection and utilization, and the weight in the protection and utilization evaluation system is also the largest. The management factor is the emphasis and refinement of the active intervention and recovery measures of human response in the response indicators [12]. The large management factor indicates that management policies and means are of great significance for river protection and utilization, and should be highly valued by relevant departments. The influence factors refer to the feedback results and influence degree of the change of river ecosystem status on the social and economic development, human life, resources and environment under the pressure and driving force . The large weight indicates the high development and utilization of the river ecosystem. The natural ecological environment and human society of the DSC are greatly affected, so effective measures need to be taken for timely control. The state factor and response factor are important indicators of river protection and utilization, and the two factors are equally important for river protection and utilization. The driving force factor is the fundamental driving force and potential driving force for the change of river ecosystem [13], and its weight is large, indicating that the degree of development and utilization is high. The natural ecological environment of DSC is greatly affected by human society, so effective measures should be taken to control it in time. State factor and response factor are important indicators of river protection and utilization, and they play an equally important role in river protection and utilization. The weights of the two factors are small and similar, indicating that the DSC has a good protection and utilization status at present, and the control is relatively small. The driving force factor is the fundamental driving force and potential incentive for the change of river ecosystem[14]. Because DSC is an artificial canal, it is mainly affected by human disturbance activities, so the driving factor weight is small. Judging from the scoring index of the criterion layer, the ecological driving and social driving of the DSC basin are not large. The pressure on the water resources, water environment and riparian zone in the basin is small, and the pressure on the river is also small. Therefore, the DSC is in good condition and can basically meet the control requirements for river protection and utilization. However, the low response and management values indicate that the local awareness of the protection and utilization of DSC is relatively weak, and the management level needs to be improved. It is necessary to enhance the management and control capabilities of river protection and utilization.

From the index level, in the comprehensive evaluation index system of DSC protection and utilization, there are 12 indexes whose scores are more than 0.8 or above, accounting for 63.16%, and 7 indexes whose scores are less than 0.6 or below, accounting for 36.84%. This shows that although the protection and utilization degree of DSC is relatively high, there are still some abnormal changes in its structure, which are mainly manifested in the following aspects: large rural water quota, low efficiency of irrigation water utilization, large amount of sewage discharge, small proportion of drinking water safety, and low sewage treatment capacity. According to the weight value of each indicator in Table 6 and the evaluation index of each index, the safety ratio of drinking water, the guarantee rate of water supply, the natural degree of bank slope, the water quality meet or exceed the proportion of class III, the natural degree of bank slope, and the river cutoff throughout the year can be found. The frequency, the degree of ecological water security, and the soundness of the river protection and management system are the main influencing factors for the protection and utilization of the DSC. Therefore, the next river protection work of the DSC should focus on the red line of water resources, the red line of sewage discharge, and the red line of shoreline protection.

5. Conclusion

The evaluation index system of the control planning for the protection and utilization of the artificial canal involves a wide range of contents, including technical level and management level. The value of evaluation index and weight is also restricted by many factors, so it is difficult to value and quantify. Based on the DPSIRM model, the DSC protection and utilization comprehensive evaluation index system is used to calculate the river protection and utilization in the study area by using the analytic hierarchy process, which can better reflect the actual situation of the DSC protection and utilization. The analysis shows that:

- The degree of protection and utilization of DSC is relatively high, but there are still some abnormal changes in some structures. Combined with the weights of various indicators and the index of various indicators, it can be seen that water resources, water environment and riparian zone are the main influencing factors of DSC. Therefore, the next river protection work of the DSC should focus on the red line of water resources, the red line of sewage discharge, and the red line of shoreline protection.
- The river protection and utilization index system established in this study has broadened the ideas and theories of river protection and utilization research to a certain extent, but there are still many deficiencies, such as the selection and analysis of some indicators is too subjective, and the DSC is not analyzed. The applicability of the research results is also limited by the dynamic changes of different years, but the research results are reliable for the analysis of river protection and utilization in the study area, which can better meet the research objectives.

Acknowledgments

This work is funded by the National Natural Science Foundation of China (No. 51779013), National Public Research Institutes for Basic R&D Operating Expenses Special Project (No. CKSF2019478.), Water Conservancy Science and Technology Innovation project of GuangDong Province (2017-03),

Natural Science Foundation of Hubei Province(2017CFB606). Special thanks are given to the anonymous reviewers and editors for their constructive comments.

references

[1] ZHAI J., XU GB., GUO SY.,etal. (2016) Research on River Health Assessment Method Based on Coordinated Development Degree. J. Journal of Hydraulic Engineering., 47: 1465–1471.

[2] CUI LH., LiQ., ZHOU XH., etal. (2009) Purification effect of two kinds of composite constructed wetland systems on canal sewage in Dongguan. J. Ecology and Environment., 18: 1688–1692.

[3] YU KJ., L DH., L W. (2004) On the Ecological Infrastructure Strategy and Implementation Approach of the Grand Canal Area. J. Progress in Geography., 1:1–12.

[4] YU KJ., XI MS. (2010) Composition of the Grand Canal Heritage Corridor from the Perspective of OccurrenceComposition of the Grand Canal Heritage Corridor from the Perspective of Occurrence. J. Progress in Geography., 29: 975– 986.

[5] ZHU C., JIA HY., ZHOU Q. (2019) Comprehensive Evaluation of River Health in the Middle and Lower Reaches of Hanjiang River. J. Journal of Hydroecology., 40: 1–8.

[6] ZHENG BAO., LUO WS. (2019) Study on Index System and Weight of River Ecosystem Health Assessment. J. Hydropower and New Energy., 8: 60–65.

[7] SHU CK., YANG K., WANG QMM., etal. (2017) Study on the method of weighting in river health assessment. J. International Journal Hydroelectric Energy., 35: 61–65.

[8] BeschtaRL. (1991) Stream habit at management for fish in the northwestern United States: the role ofrip ari anvegetation. J. American Fisheries Society Symposium., 10:53–58.

[9] WANG JH, L XG, TIAN HJ. (2008) Theory and Application Technology of Riverbank Wetland Research J. Wetland Science., 2: 97–104.

[10] YUAN Y., HUANG JG., ZHANG XB., etal. (2017) Preliminary study on the flood control safety evaluation index system of reservoir bank rectification project. J. Express Water Resources & Hydropower Information., 38: 114–118

[11] ZHOU LL., WANG L., YU J. (2014) Water Resources Sustainable Utilization Evaluation System Based on Water Footprint Theory-Taking Jimo City as an Example. J. Resources Science., 36: 913–921.

[12] ZHANG F., YANG J., XI JIAN., etal. (2014) Health Assessment of Nansi Lake Lake Ecosystem Based on DPSIRM Healthy Distance Method. J. Resources Science,36:831-839.

[13] WANG JH., LU XG., TIAN JH. (2008) Theory and Application Technology of Riverbank Wetland Research. J. Wetland Science., 2: 97–104.

[14] WANG Y., ZHOU LH., WEI XUAN. (2018) Vulnerability evaluation index system based on social-ecological system for desertification reversal process. J. Acta Ecologica Sinica., 38: 829–840.

IOP Conf. Series: Materials Science and Engineering **794** (2020) 012014 doi:10.1088/1757-899X/794/1/012014

Development of Tianwan river system and characteristics of river geomorphology

Tang Di, Tan Long, Wu Xuefeng and Qi Lijiang*

College of Water Resources and Hydropower, Sichuan Agricultural University, Ya'an 625014

*Qi Lijiang's e-mail: qilij7@163. com

Abstract. The uplift of Gongga Mountain has greatly affected the development of water system and topography of Tianwan River Basin. In this paper, ArcGIS software is used to extract the water system and main channel profile of Tianwan River Basin, and Horton theory is used to analyze the water system development and river network characteristics of Tianwan River Basin. The results show that the water system development of Tianwan River Basin basically conforms to Horton's law , but the Horton ratio is larger at some levels, and the Horton ratio of the right bank is larger than that on the left bank. There are many river fissures in the basin, the largest of which are the Bawanghai Sea, which raises the river erosion benchmark and effectively suppresses the upstream disasters.

1. Introduction

Gongga Mountain is a kind of geomorphic and geological landscape formed under the influence of the geotectonic structure. Because of the squeeze between the Qinghai-Tibet Plate and the Yangzi Plate and the distinct difference in the uplifting due to neotectonic movement, the inner mountain rivers lie toward each other, running from north to south[1]. The Tianwan River Basin originates from the western slope of Gongga Mountain, with the left bank developing around the Mountain, which gives it the unique geological and geomorphological features and river network characteristics. The geomorphological characteristics of the basin influence the structure of the river system. At the same time, geological structures and the natural environment also have an impact on the patterns of river network. The features of the geomorphology and water system structure determines the water ecology and water environment[2]. In recent years, as the DEM data becomes more accurate, many scholars have studied the effects of the river network morphology and geomorphology on river systems using Horton's law. For example, Liu Le et al. [3] compared Qinghai-Tibet Plateau river network with the general river networks described by Horton's law so as to analyze its development law and the influence of plateau uplift on river network and topography. Lei Xue et al. [4] based on the digital terrain analysis method, compared the differences in the morphology of the small-and medium-sized mountain river networks in the hilly, plateau and plain, respectively. Yu Guoan et al. [5] analyzed the morphology of Yajiang River network and the characteristics of its geomorphology through field investigation, using SRTM and Google Earth. There are also scholars who have studied the relationship between the characteristics of the water system and that of the geological and geomorphological development by calculating the fractal dimension of basin systems[6-9]. This paper analyzes the characteristics of the water system and geomorphological development of the Tianwan River Basin based on the Horton's law.

2. Overview of the study area

Tianwan River originates from the west slope of Gongga Mountain. It is located at 101°41′~102°11′ east longitude and 29°12′~29°47′ north latitude. As the first-level tributary of the right bank of the middle reaches of Dadu River, it flows through Kangding County of Ganzi Prefecture and Shimian County of Ya'an City. The basin is located in the transition zone from the Sichuan Basin to the Qinghai-Tibet Plateau, with the subtropical monsoon climate. The vertical and horizontal gullies in the basin are mostly in the formation and development stages of geomorphological development, with a large bed ratio, strong lateral erosion and traceability erosion, and a complex geological structure. The north-south Caoke fault and Moxi fault run cross this area, which are mainly composed of P2 marble, metamorphic rocks and D2 slate. The drainage area totals 1441km^2, with the total length of the main stream of 89km. The average annual flow of the estuary is 42.3m^3/s, the elevation ranges from 1015m to 7375m, and the average river channel gradient is 44.2‰.

3. Research method

This paper extracts the map of water system and the main channel longitudinal section of the Tianwan River Basin based on the GDEMDEM-30 resolution digital elevation data (2009) using ArcGIS software, and obtains the number, average length, and average area of the channels. Then, obtain the Horton ratio between the water systems at different levels using Horton topology theory. For the convenience of the study, the threshold of river length is taken as 0. 5km and the water system level is determined according to the Horton-Strahler system[10-11] (Level 1 is the lowest; two Level 1 rivers merge into a Level 2 river; two Level 2 rivers merge into a Level 3 river, and so on). The Level 6 river is the main stream of the Tianwan River, based on which the water system development and river network characteristics of the Tianwan River Basin is analyzed. At the same time, combined with field investigation and measurement of the Tianwan River Basin, this paper obtains the features of the topographical geology in the basin and the characteristics of the main fracture points.

4. Water system development and river network morphology

The channels in the upper reaches of the Tianwan River Basin (above Bawanghai) is mainly in the shape of a Chinese character called "非", and the middle and lower ones are mostly in the shape of branches (Fig. 1). The left and right banks of the basin are quite different: the area of the left bank is 499.2km^2, while that of the right bank area of 941.8km^2 (the ratio of the left bank area to that of right bank is 0.53); the total length of the left bank is 420.6km, and that of the right bank is 805.3km (the ratio of total length of left bank to that of the right bank is 0.52).

Figure 1 Water System of Tianwan River Basin

Based on Horton-Strahler classification, the following ratios are calculated using N_ω, the number of rivers of level ω, L_ω, the average length, and A_ω, the average basin area. Then, the topological properties of the river network mechanism can be analyzed macroscopically.

$$\frac{N_\omega}{N_{\omega+1}} = R_{B\omega}; \frac{L_\omega}{L_{\omega-1}} = R_{L\omega}; \frac{A_\omega}{A_{\omega-1}} = R_{A\omega} \quad (1)$$

$R_{B\omega}$, $R_{L\omega}$, and $R_{A\omega}$ represent the branch ratio, length ratio and area ratio, respectively, which are collectively referred to as the Horton ratios.

It is suggested that most natural river networks are in line with the Horton's law, with the ratio varying within a small range (the river network branch ratio ranges between 3 and 5; the river network length ratio: 1. 5 and 3; area ratio: 3 and 6[11]. With the increase of river level, the branch ratio, length ratio, and area ratio tend to reach 4, 2, 4[12]. However, when the rivers are influenced by geological structures, the development of river networks tends to deviate. According to Table 1, there are 835 Level 1 rivers, 183 Level 2 rivers, 37 Level 3 rivers, 7 Level 4 rivers, 2 Level 5 rivers, and 1 Level 6 river. The overall branch ratio, length ratio and area ratio of the basin are calculated to be 4.56, 2.36, and 5.18, respectively, indicating that the water system development in the Tianwan River Basin is generally consistent with Horton's law.

Table 1 Basic parameters of rivers at all levels in the Tianwan River Basin

River level	Number of rivers / strips	Average length /km	Average watershed area/km^2
1	835	0.80	0.99
2	183	1.57	6.30
3	37	4.05	39.96
4	7	9.91	186.67
5	2	24.49	679.56
6	1	/	/

Figure 2 shows the change of Horton ratios of rivers of different levels in the entire basin and on the left and right banks. Affected by geology, the Horton ratios of the left and right banks are significantly different. The branch ratio of the entire basin is between 3.50 and 5.29, that of the left bank is between 2 and 4.56, and that of the right banks is between 2 and 5.50. Although the ratio of the maximum and the minimum level of the left and right banks are different to a small extent, that of the right bank of the other levels are significantly larger than that of the left bank (Fig. 2(a)). The length ratio of each level of the whole basin is between 1.96 and 2.56, that of the left bank is between 1.65 and 2.19, and that of the right bank is between 1.31 and 3.35. The length ratio of the right bank of Level 2 and Level 3 is 2.01 times that of the left bank, while there is little difference between that of the other levels (Figure 2(b)). The area ratio of the whole basin is between 3.64 and 6.37, that of the left bank is between 3.67 and 6.38, and that of the right bank is between 3.62 and 6.47. The area ratio of the entire basin is not much different from that of two sides, but that of the right bank is slightly larger than that of the left bank (Fig. 2(c)).

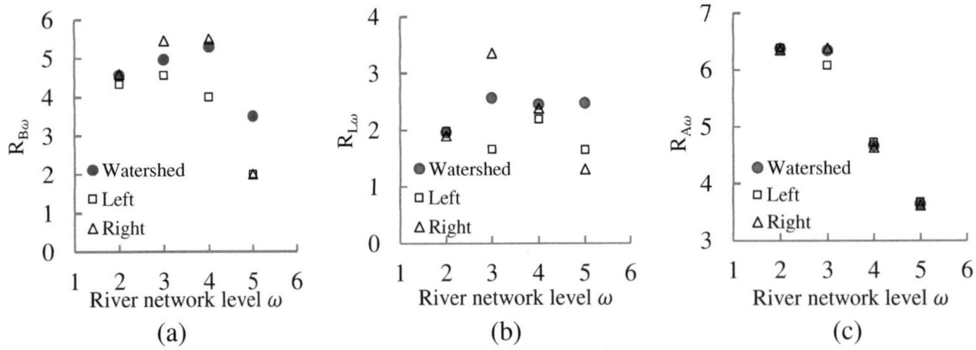

Figure 2 Horton ratios of different water systems

There also exist some situations where the Horton ratio of the Tianwan River Basin is not within the scope of Horton's law. For example, the branch ratio between Level 4 to Level 5 rivers is 5.29, and that between Level 3 and Level 4 of the right bank is 5.45. The length ratio between Level 3 and Level 4 is 3.35. However, the area ratio between the entire basin and the left and right banks of Level 1 and Level 2, Level 2 and Level 3 exceeds the range of Horton's law. Due to the strong uplift of the Gongga Mountain crust, the characteristics of the river network deviate from the general law. Since the geological uplifting effect on the right bank is smaller than that of the left bank, the Horton ratio of the right bank is generally larger than that of the left bank.

5. River landscape

The difference of geological structure and lithology of the Tianwan River Basin are mainly reflected in the huge difference of geological development of the left and right banks. The basin originates from the western slope of Gongga Mountain, which is located at the junction of the Qinghai-Tibet Plate and the Yangtze Plate, where the frequent geological tectonic activities result in many folds and faults. New tectonic movement of the Gongga Mountain has been active since the Quaternary Period, leading to strong and differentiated block uplifting[13]. As the landform develops, the neotectonic movement has a greater impact on the marginal zone. Therefore, the rising crust of the Gongga Mountain has a greater impact on the topography of the left bank of Tianwan River, resulting in the huge difference in the distribution and the features of water system in the left and right banks.

Under normal circumstances, the longitudinal section of the river is in the shape of a concave, but that of the main stream of the Tianwan River is in the shape of a convex. There is a relatively obvious change in the slopes of Bawanghai (left branch) and Nichanggou (left branch). The average longitudinal slope above the Bawanghai is 35‰, that from Bawanghai to Nichanggou is 50‰, and that below Nichanggou is 37‰. Therefore, the area above Bawanghai is upstream, with a total length of 41km, the area between Bawanghai to Nichanggou is midstream, totaling 33km, and the area below Nichanggou is downstream, totaling 15km (Fig. 3).

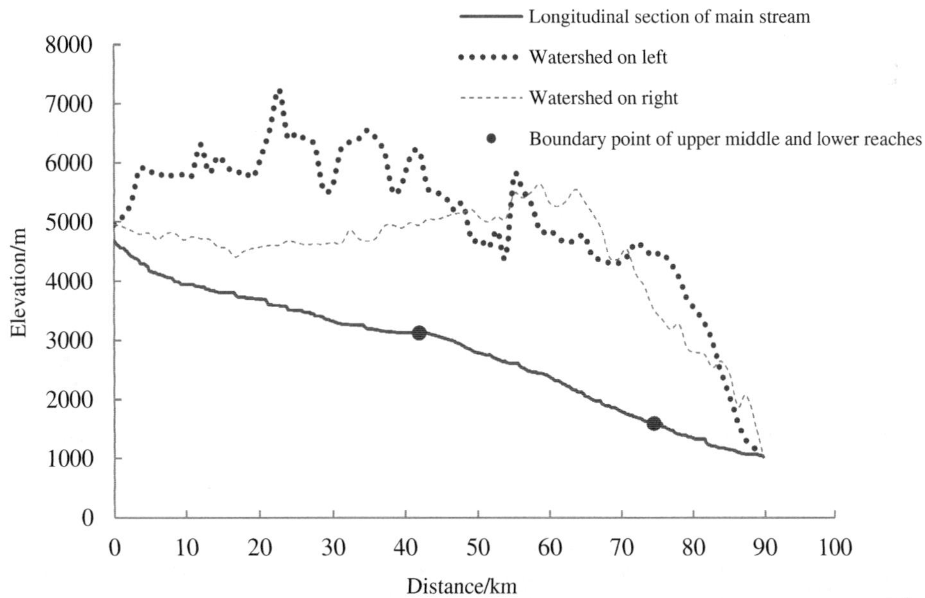

Figure 3 The longitudinal section of the Tianwan River main stream and the watershed of both sides

The transition point on the longitudinal section of the river is called the "river crack" [14]. It can be seen from Fig. 3 that the elevation difference between the left bank watershed and the channel shows basically the same trend in the whole basin. The altitude of the right bank watershed continues to increase, that is, the elevation difference of the middle and lower channel and the left and right banks are relatively large. The cracks in the Bawanghai are mainly formed due to the accumulation of solid matter during mudslides. According to the 2015 survey, its formation and influence range is about 12km upstream and about 9km downstream. Based on the trend of the longitudinal profile of the river, the thickness of the crack deposit is about 178m. The upstream channel is relatively wide, with a maximum width of 455m. In addition, there are many turbulences and shoals, and the average slope is 5.2‰. The downstream is narrow, with the average slope decreasing to 39.1‰ (Fig. 4(a)). There are also many cracks in the middle and lower reaches of the Tianwan River. In the middle reach of the Gongjiagou, the average slope of the upper reaches of the crack is 29.7‰; the average slope of the downstream is 77‰ (Fig. 4(b)); at the junction of the middle and lower reaches in the vicinity of Nichanggou, the average slope of the upper reaches of the crack is 54.6‰; the average slope of the downstream reaches is 64.5‰ (Fig. 4(c)). As can be seen from Figure 4, the upper reaches of these cracks are relatively broader and more gentle, while the downstream ones are relatively narrower (Figure 4).

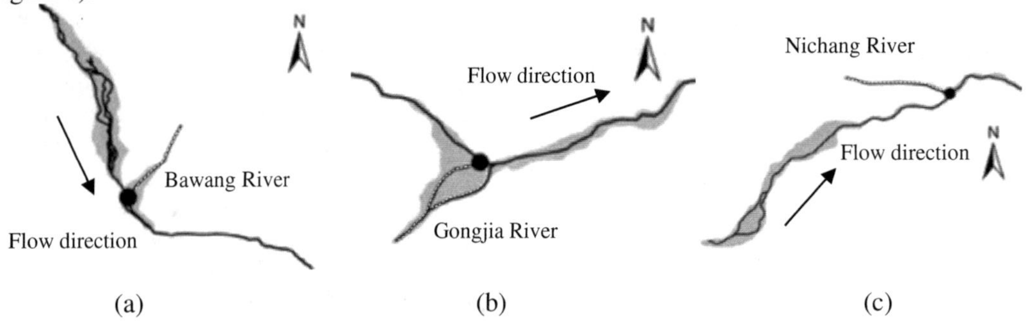

Figure 4 Figure of the river channel near some cracks

The formation of cracks lifted the erosion datum of the riverbed, which slowed down the erosion rate of the upstream channel, and even prevented undercutting or turned into siltation, thus widened the riverbed. The formation of the Bawanhai barrier has changed the erosion datum point, slowed down the undercutting of the downstream channel, and greatly reduced the occurrence of upstream disasters. Therefore, there has been no mudslide and collapse landslide in the upper reaches of the Bawanghai; Gongjiagou for a long time. In addition, within the influence range of cracks at Gongjiagou and Nichanggou, there is a small number of geological disasters. In the other areas of the middle and lower reaches, the main river and the branch ditch continued to undercut, making the mountain body less stable. In addition, the larger height difference between the two sides also leads to more geological disasters such as collapse, landslide and mudslide every year.

6. Conclusion

The water system of the Tianwan River Basin has the characteristics of a branched river network. The branch ratio, length ratio and area ratio of the river network are basically in accordance with Horton's law. However, due to the geological tectonic activities of the Gongga Mountain, there are some situations where the ratios are deviated from the Horton's law. Generally speaking, Horton ratios of the right bank are bigger than those of the left bank. River cracks in the basin slowed down the undercutting in the upper main river and branches, which significantly prevented the occurrence of upstream geological disasters.

Acknowledgement

Fund projects: Sichuan Province Education Department subsidized research projects (16ZB0046)

Reference

[1] Yang G. (2005) Characteristics of geological and geomorphic tourism resources in Gongga Mountain area [A]. China Geological Society Tourism Geosciences and National Geopark Research Branch. Proceedings of the 19th National Conference on Tourism Geosciences and

Shaoguan Tourism Development Strategy[C]. China Geological Society Tourism Geosciences and National Geopark Research Branch: Chinese Geological Society, :4.
[2] Zhang J F, Wu Y Q, Pan M H, et al. (2013) Analysis of the relationship between landform features and water system structure in the upper reaches of Dongjiang River [J]. Resources Science, 35 (03): 496-504.
[3] Liu L, Wang Z Y, Yu G A, et al. (2015) Statistical law of river network in Qinghai-Tibet Plateau and the influence of plateau uplift [J]. Journal of Tsinghua University (Natural Science Edition), 55(09): 964-970.
[4] Lei X, Luo M L, Zhou Y, et al. (2018) Analysis of river morphological characteristics in typical geomorphic areas based on dem [J]. Hydrology, 38(04): 48-54.
[5] Yu G A, Wang Z Y, Liu L, et al. (2012) Development of the Yarlung Zangbo River system and river geomorphology under the influence of neotectonic movement [J]. Advances in Water Science, 23(02): 163-169.
[6] Liang C L, Liang H Q, Zhang Z L. (2006) Analysis on the fractal dimension and geomorphological development stage of the Zulu River Basin [J]. Soil and Water Conservation Research, (03): 187-188+191.
[7] Barbera P L, Rosso R. (1989) On the fractal dimentions of stream networks[J]. J. Water Resources Res. 25: 735-741
[8] Li W, Wang Y L, Xu Z, et al. (2017) Fractal characteristics of river network in Yanhe River Basin based on dem[J]. Anhui Agricultural Science Bulletin, 23(05): 121-124.
[9] An Q, He Z H, Zhao C W, et al. (2019) Fractal dimension estimation and geomorphological characteristics of water system in the karst dam construction area in central Guizhou [J]. Water Resources and Hydropower Technology, 50(04): 73-81.
[10] Horton R E. (1945) Erosional development of streams and their drainage basins: Hydrophysical approach to quantitative morphology[J]. Geological Society of America Bulletin, 56: 275-370.
[11] Strahler A N. (1957) Quantitative analysis of watershed geomorphology[J]. EOS Transactions, 38(6): 913-920.
[12] Liu H X, Wang Z Y. (2007) Morphological characteristics and distribution of typical river networks [J]. Journal of Hydraulic Engineering, (11): 1354-1357.
[13] Zheng L M. (2017) Analysis of grain size and surface characteristics of quartz particles in the eastern slope of Gongga Mountain slope [D]. Lanzhou University.
[14] Lu Z C, Jia S F. (1991) Watershed geomorphology system [M]. Dalian: Dalian Publishing House.

Influence of parameters valuing on calculation results in reservoir sediment mathematical model

Chunyan Deng[1,2*], Junqiang Xia[2], Rui Long[1], Yuan yuan[1]

[1]River Department, Changjiang River Scientific Research Institute, Wuhan, Hubei, 430010, China

[2]State Key Laboratory of Water Resources and Hydropower Engineering Science, Wuhan University, Wuhan, Hubei, 430072, China

[*]Corresponding author's e-mail: yaoye0121@163.com

Abstract. Sediment deposition in the reservoir not only has a serious impact on the service life of the reservoir and the existence value of the hydropower station, but also profoundly affects the erosion and siltation of the downstream riverbed and the safety of the dam. Being an important method to the research of sedimentation problem, calculation accuracy of mathematical model is a subject worthy of continuous discussion. In this paper, influence of some key parameters of the model are analysed. The calculation results show that the changes of dry density, water flow sedimentation parameters and saturation recovery coefficient have significant influence on the calculation results, and the change of dry density has the greatest influence, while the effect of flocculation is the weakest. It is worth noting that when considering these influencing factors, this paper adopts a simplified processing method, and the mechanism of each parameter is complex, and further research is needed.

1. Introduction

After the construction of reservoirs on the river, due to the elevation of water level and reduction of the flow velocity, a huge number of sediment from the upstream is silted into the reservoir, which is a direct threat to the service life of the reservoir and the safety of the dam. To reduce the impact of the reservoir sediment deposition, and fully bring out the comprehensive benefits of the reservoir, it's necessary to carry out quantitative analysis on reservoirs silting. Over the years, China has accomplished a lot of work on reservoir siltation [1], mainly through measured data analysis, flume experiment, and physical and mathematical modelling.

This paper mainly focuses on the calculation accuracy issue of mathematical models. In previous work, the author studied the siltation of Yantan Reservoir, and adjusted the relevant calculation parameters in the model by comparing with the measured data, and improved the calculation results of the model. The calculation results of sedimentation in the reservoir area are close to the measured values [2]. Based on the preliminary work, this paper takes Yantan Reservoir as an example to study and analyze the influence of various parameters in the reservoir water sediment mathematical model on the calculation results.

2. Analysis of Related Parameters in Reservoir Sediment Mathematical Model

The calculation accuracy of the one-dimensional mathematical model is closely related to the valuing of the parameters in the function. The parameters that need to be analysed mainly consists of dry density, water flow sediment-carrying force factor, saturation recovery coefficient, roughness and flocculation. The mechanism of each influencing parameter is as follows.

(1) Influence of dry density

The dry density of reservoir sediments varies widely. It affects the conversion of silt weight and volume, the variation of silt volume after compaction of fine-grained sediments, and the variation of sediment starting velocity and sediment-carrying capacity. It is an important parameter for reservoir siltation calculation, and needs to be carefully chosen [1]. Some existing studies have shown that: the dry density of sediment in the reservoir area increases along with the increasing of the particle size, and the initial dry density of the fine-grained sediment (the initial dry density refers to the dry bulk density during the initial stabilization of the sediment at the begging of siltation). When d is between 0.005 to 0.010 mm, the value of dry density adopts 778 kg/m^3 [1]. If the sedimentation period is very long, usually stabilized dry density (the dry bulk density of the sludge after long-term compaction) is directly adopted. According to Han Qiwei et al. [1], when d is between 0.005 to 0.010mm, the dry density adopts 1000~1300kg/m^3.

(2) Influence of sediment carrying force factor

The calculation of sediment carrying force is key module of the sediment mathematical model, whose calculation accuracy directly affects the suspended sediment concentration and riverbed deformation amplitude. The formula of Zhang Ruiqi is adopted to calculate the water flow sediment carrying force in this paper. The formula is as follows:

$$S_* = K\left(\frac{U^3}{gh\omega}\right)^m$$

Where: K, m are the coefficient and index of sediment carrying force of flow, U and h are the average flow velocity and average water depth of the section, respectively, and g is the acceleration of gravity, ω is the sedimentation speed of the particle.

In the formula of water flow and sedimentation proposed by Zhang Ruiqi et al. [1], K and m are hard to determinate, and need to be calibrated by site measuring. Therefore, the parameters used in the calculation of the mathematical sediment model varies in each project. For example, according to the results of the Gongzui station data, K=0.124, m=1.05, however, according to water tank experiment result by Liu Jianjun et al. [3], According to the results of the flume experiment, the parameters of the suspended sediments on the Hongshui River where the Yantan Reservoir are located, it's more reasonable to adopt 0.185 and 0.666 for K and m.

(3) Influence of saturation coefficient

The saturation recovery coefficient is an important parameter reflecting the recovery rate of sediment concentration towards saturated sediment concentration, or sediment carrying capacity during unbalanced sediment transportation. Saturation recovery coefficient directly affects the amount of riverbed erosion and siltation, which has an important impact on the calculation of sediment erosion and deposition [4]. Some scholars have studied the valuing of the saturation recovery coefficient [5~8], and basically have the following consensus: ① Different particle size groups have different saturation recovery coefficient. ② Value of saturation recovery coefficients decrease as the sediment particle size increase, ③Value of saturation recovery coefficient varies with space and time. In the one-dimensional water and sediment model, empirical results are generally used, that is, the saturation recovery coefficient of sediment is 0.25. When flushing, the saturation recovery coefficient of sediment is 1.0 [3]. The greater the saturation recovery coefficient is, the greater the ratio of sediment concentration along the river is, and the faster the sediment concentration recovers to the sediment carrying capacity.

(4) Influence of fine particle sediment flocculation

Finer viscous particles will condense together in the water, and flocculation will occur, which will affect sediment transport and sedimentation to some extent. Sediment particle size is one of the important factors affecting flocculation. In a specific water environment, only sediment particles below the threshold particle size will flocculate [9]. In addition, according to Wang Dangwei et al. [10], fine-grained sediment provides the basic conditions for flocculation, but whether it will flocculate also closely relates to the water environment factors, especially to the species and concentration of ions in the water. For the convenience of model calculation, this paper only considers the influence of sediment particle size on flocculation.

(5) Influence of roughness

The influence of water flow friction in the one-dimensional mathematical model is usually reflected by the roughness (*n*). The rougher the boundary surface of the river, the greater the roughness; the smoother the boundary surface, the smaller the roughness. In the modelling, it is usually obtained through the measured values of the river section or referring to similar engineering experience. In this paper, the value of roughness is determined by the site measurement in the reservoir area.

3. Introduction to the preliminary calculation results of Yantan Reservoir

(1) Model introduction

The sedimentation and siltation calculation of Yantan Reservoir adopts the one-dimensional constant non-uniform unbalanced sediment transport mathematical model "SUSBED-2" developed by Yang Guolu and Wu Weimin [11]. This model has been widely promoted in domestic design institutes, and widely used in the research and design of water conservancy and hydropower projects in small and medium rivers [11].

Yantan Reservoir locates on the Hongshui River, which is the upper reaches of the Xijiang River in Guangxi Province. The calculated river section in the model ranges from the dam site of Yantan Hydropower Station to the upper bank of Longtan Hydropower Station upstream, with a total length of 166.6km. SUSBED-2 is adopted to recalculate the water and series of Yantan Reservoir from April 1994 to December 1999. The calculation results are detailed in the reference [2].

(2) Value of relevant parameters

The relevant parameters in the modelling are obtained based on the measured data of the watershed the Yantan Reservoir locates and with reference to similar engineering experience. The values of the calculated parameters are as follows: the dry density is 800 kg/m^3; the saturation recovery coefficient of the suspended sediment is 0.7; the coefficient and index of the suspended sediment carrying force are 0.185 and 0.666; the flocculation of the fine sediment is closely related to the particle size. The threshold particle size is 0.025mm; the initial roughness adopts the roughness during river flood period, which is 0.0422.

4. Influence of parameter adjustment on the model calculation results

In order to analyze the influence of each parameter on the model calculation results, this paper mainly adopts the single factor adjustment analysis method, that is, when sensitivity analysis is performed on a certain parameter, the other parameters are kept unchanged. Adjust the parameters by ±20% to compare the effect of this factor on the calculation model.

Table 1 shows the effect of each parameter adjustment on model calculation results. The table indicates that the larger the dry density, the smaller the cumulative sedimentation amount, and the sediment discharge ratio is almost unchanged; the change of the K of the water flow sedimentation parameter has almost no effect on the cumulative sedimentation amount and the sediment discharge ratio, however, the change of parameter m will cause variation in the cumulative sedimentation amount, and they are proportional; when the saturation recovery coefficient in siltation condition increases or decreases by 20% on the basis of 0.7, the corresponding cumulative sedimentation amount decreases; Considering the effect of flocculation of fine-grained on sediment when the threshold particle size increases or decreases by 20%, the calculation result of the cumulative sedimentation amount does not change.

5. Conclusion

Through the analysis above, following conclusion can be drawn:

(1) According to the sensitivity analysis of the parameters, the influence of dry density, water flow sedimentation parameters and recovery saturation coefficient have great affection on the calculation results, among which, the variation of dry density has the greatest influence.

(2) In the calculated river section, the size of the sand is rather small, where the sediment with a particle size of less than 0.01 mm accounts for 57% of the total sand. In this paper, the flocculation effect of fine-grained sediment is studied briefly. The results show that the effect of flocculation is not significant. Changing the value of the demarcation particle size has little effect on the calculation results.

(3) The natural sedimentation pattern of the reservoir is complex. In the calculation, considering the relevant influencing factors and the complexity of the model itself, the influencing parameters are simplified. In this paper, the effects of sediment carrying capacity change caused by sediment thickness variation and unbalanced sediment transportation is not thoroughly considered. Moreover, with the time passes, the change of reservoir topography will also cause the change in sediment carrying capacity and sediment concentration. They have increased the difficulty researching the sedimentation in reservoirs and will be the key in the future researching.

Table 1. Sensitivity analysis of various parameters.

Influencing parameter	Before adjustment			Variation /%	After adjustment		
	value	cumulative sedimentation /($10^8 m^3$)	Sediment discharge ratio/%		value	cumulative sedimentation /(10^8 m^3)	Sediment discharge ratio /%
Dry density /(kg/m^3)	800	1.15	71.7	+20	960	0.94	71.8
				-20	640	1.46	71.7
Flow sediment carrying force factor	*K*=0.185	0.77	67.5	+20	0.222	0.77	67.5
				-20	0.148	0.77	67.5
	m=0.666	0.77	67.5	+20	0.799	0.72	69.2
				-20	0.533	0.81	65.9
saturation recovery coefficient	0.7	0.82	66.3	+20	0.84	0.65	72.1
				-20	0.56	0.78	67.5
flocculation (threshold particle size/mm)	0.025	0.86	64.7	+20	0.024	0.86	64.7
				-20	0.016	0.86	64.7

Acknowledgments

This research was financially supported by the Strategic Priority Research Program of the Chinese Academy of Sciences (Grant No. XDA23040502), National Key R&D Program of China (2017YFC0405306, 2016YFC0402309, 2017YFC0405304), National Natural Science Foundation of China (Grant No. 51779014), the Special fund for basic scientific research business of central public research institutes (Grant No. CKSF2015049/HL).

References

[1] Han Qiwei. (2003) Reservoir siltation. Beijing. Science Press.

[2] DENG Chunyan, WU Weimin, XIA Junqiang. (2013) Numerical modeling of sedimentation process in Yantan Reservoir, Engineering Journal of Wuhan University, 46(1): 19-25.

[3] Liu Jianjun. (2003) Experimental Report on Sediment Problems in the Reservoir Area of Longtan Hydropower Station. Changsha: China Southern Survey and Design Institute, State Power Corporation, 8-42.

[4] HAN Qiwei, CHEN Xujian. (2008) Theoretical method for calculating the saturation recovery coefficient. Journal of Sediment Research, 6:8-11.

[5] HAN Qiwei. (2007) Theoretical study of nonequiliirium transportation of nonuniform suspended load. Water Resources and Hydropower Engineering, 1:14-23.

[6] YANG Jinying. (2005) Study on the recovery saturation coefficient of over-saturation sediment transport in settling tanks. Journal of Sediment Research, 3:42-47.
[7] SHI Chuanwen, LUO Quansheng. (2003) Study on the Restoration and Saturation Coefficient α in 1-D Over saturation Sediment Transport. Journal of Sediment Research, 1:59-63.
[8] WANG Xinhong, CAO Ruxuan, SHEN Jin. (2003) Investigation on coefficient of saturation recovery of non-uniform suspended load. Journal of Hydraulic Engineering, 3: 120-128.
[9] WANG Dangwei, YANG Guolu, YU Minghui. (2003) Method to determine the maximal diameter of fine sediment aggregation and influence factors analysis. Journal of Sediment Research, 1:74-80.
[10] WANG Dangwei, JI Zuwen, SHENG Yumin. (2019) Study on sediment flocculation in the Three Gorges Reservoir. Journal of Sediment Research, 44(3): 1-6.
[11] Yang Guolu, Wu Weiming. (1994) SUSBED-1 movable bed modeling of graded sediments. Journal of Hydraulic Engineering, 4:1-10.

IOP Conf. Series: Materials Science and Engineering **794** (2020) 012016 doi:10.1088/1757-899X/794/1/012016

Analysis and Determination of early warning index for Mountain Flood Disasters—A Case Study of Yalin River Small Watershed

Shi Zheng[1, 2], Hu Lin-lin[1, 2], Chen Huan-Bao[1, 2] and Tang Hong-xia[1, 2]

[1]Zhejiang Institute of Hydraulics & Estuary, Hangzhou 310020, Zhejiang Province, China

[2]Zhejiang key laboratory of disaster prevention and mitigation for hydraulics, Hangzhou 310020, Zhejiang Province, China

E-mail: 13858051984@126.com

Abstract: Determination of the critical rainfall is very important for the influence by mountain torrents with disaster prevention and reduction in significance, which is the priority among priorities of mountain flood early warning work. In this paper, the small watershed of Yalin River as the example, has carried on the analysis and the calculation of flood disaster early warning index. The results show that, the using method is simple and practical, with the certain reference value to the mountain torrent disaster warning.

1. Introduction

Zhejiang Province is mountainous, with hilly area accounting for more than 70% of the province's land area. Mountain flood caused by hard rainfall is suddenly, and it has become one of the major disasters causing casualties in flood disasters. Mountain flood is natural disaster caused by persistent heavy rainstorms in mountainous areas. It seriously threatens the safety of people's lives and property because of its wide range, sudden strong and destructive power[1].

This paper chooses Yalin River Basin in Xiangshan County to study the mountainous flood disaster prevention. Through calculation and analysis, the critical rainfall and critical water level are determined, which can provide some reference for the prevention of mountain torrent disasters.

2. Watershed General Situation

Yalin river rise in Shejian Mountain, and flows northwest from Southeast through the village of Huxiaopu, Yalin river and Andong. Yalin River Basin covers an area of 6.68 km^2 and the main stream is 5.02 km long, with ratio descending of 13.4‰.

The topography of the basin is mainly low mountains and hills. The mountain elevation is generally between 300~ 600 m, and the peak is 545m. The terrain inclines from southwest to north. Yalin river is steeper, and the source is short and the current is fast. The watershed has good vegetation and soil and water conservation. The main soils are red clay and loess.

Yalin river is located in the subtropical monsoon climate. The general climate characteristics are as follows: significant alternations of winter and summer monsoon, abundant distinct rainfall in four seasons, and uneven distribution of precipitation in a year. The flood of this river basin is caused by heavy rain. The flood in April to June is Mei-xun rain, and the flood in July to October is caused by typhoon and heavy rain.

3. Hydrological Calculation

3.1 Design rainstorm

(1) Design rainstorm calculation

There are no rainfall gauging stations in Yalin river watershed, so the design rainstorm is calculated by referring-to-diagram method. According to the usage regulation of "Short-duration Rainstorm in Zhejiang Province", the average point rainfall and Cv values of 10 min, 1 h, 6 h and 24 h were obtained by mapping the geographical location of Yalin river watershed, and then the Kp value was calculated from the parameters of each duration rainstorm using line fitting method (Cs/Cv=3.5). After conversion of the point-surface coefficients, the design rainstorm of each duration and frequency in Yalin river watershed was obtained.

(2) Design rain pattern

Designed rainstorm process distribution includes daytime distribution and hourtime distribution, and they are determined according to "Short-duration Rainstorm in Zhejiang Province".

Daytime distribution: The maximum rainfall day is the second day, the rainfall of the first day is 60% of H_{3d}-H_{24h} and the rainfall of the third day is 40% of H_{3d}-H_{24h}, as shown in table 1.

Table 1. Daytime Distribution of Designed Rainstorm

Daytime distribution	First day	Second day	Third day
Percentage of $H_{3d}-H_{24h}$	60		
Percentage of H_{24h}		100	
Percentage of $H_{3d}-H_{24h}$			40

Hourtime distribution: The design rainstorm in each period is calculated according to the rainstorm formulas. The formulas for calculating the design rainstorm (H_i) and the rainstorm attenuation index (n) in each period are as follows:

If t_i is between 10 and 60 minutes, then $H_i = H_{10} \times \left(\frac{t_i}{10}\right)^{1-n_{10,60}}$,$n_{10,60} = 1 + 1.285 \times lg\left(\frac{H_{10}}{H_{60}}\right)$;

If t_i is between 1 and 6 hours, then $H_i = H_6 \times \left(\frac{t_i}{6}\right)^{1-n_{1,6}}$,$n_{1,6} = 1 + 1.285 \times lg\left(\frac{H_1}{H_6}\right)$;

If t_i is between 6 and 24 hours, then $H_i = H_{24} \times \left(\frac{t_i}{24}\right)^{1-n_{6,24}}$,$n_{6,24} = 1 + 1.661 \times lg\left(\frac{H_6}{H_{24}}\right)$.

The difference value between rainfall at adjacent periods (ti) is the interval rainfall from bigness to smallness. The rain pattern of 24 hours in the maximum rainfall day is arranged according to the following rules: the maximum hourly rainfall is at 21:00, and the second maximum rainfall is on the left of the maximum hourly rainfall. The remaining items ranged from large to small odd Numbers on the left and even Numbers on the right. After 24:00 on the right, the remaining periods of rainfall ranged from large to small on the left. The other two days are arranged according to the 24-hour rain pattern of the maximum rainfall day.

3.2 Design flood

(1) Runoff yield calculation

Zhejiang Province belongs to humid areas in the south China. The main pattern of runoff yield is stored-full runoff, which is mean that the rainfall first satisfies the shortage of soil water, and after the soil is full (the soil moisture content reaches the field capacity), the surplus rainfall will produce runoff. The calculation of net rain is based on the simple deduction method, i.e. the initial loss is 25 mm, the later loss is 1 mm/h during the maximum rainfall day, and the loss is 0.5 mm/h in the remaining two days.

(2)Confluence calculation

Analyzing the watershed characters by one-thousandth topographic map. The calculation of design flood uses the rational formula method. The calculation results are shown in table 2.

Table 2. Design Flood Calculation Results

River Basin Control Section	Flood factor	Design flood values at different frequencies (m^3/s)				
		$H_{1\%}$	$H_{2\%}$	$H_{5\%}$	$H_{10\%}$	$H_{20\%}$
Control Section of Yalin river small Watershed	Flood peak discharge	139	119	92	73	54
	Flood level	17.13	16.89	16.55	16.27	15.93

4. Determination of Early Warning Index

The critical rainfall is determined by the "water level-discharge backstepping method". According to the specific conditions of river embankments in small watershed, select appropriate number of control sections, analyse critical water levels of each section, determine the discharge of control section under critical water level by hydraulic calculation. Using hydrological calculation and iteration method to solve the problem, there must be a rainfall to match flood peak formed after runoff yielding and confluence, at which time the rainfall is the critical rainfall. Then, on the basis of critical rainfall, the flood warning index is synthetically determined by considering the river section shape, flood rising and submergence speed[2].. The results of early warning index determination are shown in table 3.

Table 3. Results of Early Warning Index Calculation

Administr-ative area	Small watershed	categ-ory	soil moisture content	time interval (h)	Early warning index (mm)	
					Ready to transfer of index	Immediate transfer of index
Yalinxi Village, Daxu Town	Yalin river watershed	raina-ll	75mm	1	78	91
				2	84	101
				2.5	86	105
			90mm	1	63	77
				2	70	88
				2.5	72	92

Based on the basic theory of hydrostatistics, the method of water level-discharge backstepping is used to determine the critical rainfall in this paper. The data used in this method are relatively complete, and the calculated index has a scientific basis, but there are also some weaknesses. The main points are as follows: (1) there is often a shortage of hydrological data in flood-prone areas, which can not meet the calculation requirements[3]; (2) rainfall station layout density in mountainous areas is too small to affect the accuracy of calculation; (3) The index determination requires a lot of investigation and review work. It is suggested that different methods for determining index should be adopted in practical work, such as analogy method and interpolation method[4]. Meanwhile, the conversion between critical rainfall and critical water level should be further explored so as to calculate and analyze early warning index of mountain torrents more conveniently.

5. Conclusions

This paper takes Yalin river small watershed in Xiangshan County as an example, analyses and calculates the early warning index of mountain flood, which has certain reference value for areas with mountain torrent prevention tasks. The preliminary early warning index need to be revised according to the actual rainfall and disaster situation in the future, so as to make the initial value closer to the true value, in order to provide more scientific and technical reference for decision makers in the future mountain flood prevention work.

Acknowledgements
This research was financially supported by Zhejiang Hydraulic Science and Technology Research Project (Grant No. RC1702).

References
[1] Ye Y.; Wang Z.Y.; Fang B.Q. An analysis method for ascertain critical rainfall of mountain flood disaster of small watershed in ZheJiang Province. *Journal of China Hydrology* **2008**, *28(1)*, 56-58.
[2] Li D.; Chen G.C.; Xie P.; Chen X.; Yang D.G. Qian J. Estimation of the Critieal Rainfall Resulting in outburst of Floods and Debris Flows in the Ungauged Areas in Urumqi , Xinjiang. *Arid Land Geography* **2005**, *28(4)*, 441-444.
[3] Liu Z.Y.; Yang D.; Xie P.; Hu J.W. Dynamic Critical Rainfall-based Torrential Flood Early Warning for Medium-small Rivers. *Journal of Beijing Normal University(Natural Science)* **2010**, *46(3)*, 317-321.
[4] Li C.Z.; Sun Y.D. Determination of flood warning index for mountain flood. *China Water Resources* **2012**, 54-56.
[5] Guo L.; Wang Y.L.; Li Q.; Zhai X.Y.; Zhang X.L.; Liu R.H. Application Composition Warning Index in Flash Flood Early Warning. *Yellow River* **2018**, *40(7)*, 38-41.
[6] Liu G.; Yuan P.; Huang K.W.; Li J.H. The Design Flood of Ungauged Areas in Sichuan Province Based on HEC-HMS. *China Rural Water and Hydropower* **2014**, *4*, 65-68.

IOP Conf. Series: Materials Science and Engineering **794** (2020) 012017 doi:10.1088/1757-899X/794/1/012017

Monthly runoff forecasting via an improved extreme learning machine

Wen-jing Niu[1]; Yu-bin Chen[1]; Yao-wu Min[1]; Yu-rong Li[1]; Xiao Zhang[1]; Zhong-kai Feng[2, *]

[1]Bureau of Hydrology, ChangJiang Water Resources Commission, Wuhan 430010, China; Email: dgniuwenjing@163.com

[2]School of Hydropower and Information Engineering, Huazhong University of Science and Technology, Wuhan 430074, China; Email: myfellow@163.com

*Correspondence: myfellow@163.com; Tel.: +86-15842496712

Abstract: Generally, monthly runoff prediction is of great importance for effective water resource planning and management. Extreme learning machine (ELM) is a novel training tool for the famous single layer feed-forward neural network. Due to the satisfying performance, ELM is chosen for monthly runoff forecasting in this research. Nevertheless, it is unfortunately found that ELM easily falls into local optima in practice because the randomly-determined computational parameters remain unchanged during the learning process. Thus, this paper tries to develop an improved extreme learning algorithm (IELM) where the evolutionary algorithm is used to search for satisfying computational parameters while the Moore-Penrose generalized inverse method is used to determine the output weights. The IELM method is applied to forecast the monthly runoff of Hongjiadu reservoir in southwest China. The results show that the proposed method outperforms several traditional algorithms with respect to the performance indicators. Thus, this paper provides a new and effective artificial intelligence approach for the monthly runoff forecasting.

1. Introduction

Long-term runoff prediction is an important precondition for effective water resources planning, management and operation [1-3]. Therefore, in the past several decades, hydrologic time series prediction has received tremendous attention of researchers, and models from different perspectives have been proposed to improve the quality of forecasting accuracy [4-6]. In general, these models can be broadly classified into two different groups: physical based models and data based models [2, 3]. For physical based models, although they can help us understand the physical mechanism between rainfall and runoff, due to the complex structures and numerous parameters of models, the predictive ability of physical based models are limited, especially when it is lack of long-series data in monthly runoff prediction. On the other hand, data based models, having the ability of self-adaptation and self-study, can provide accurate prediction results with fewer data and have become increasingly popular in the long-term streamflow prediction among engineers and researchers [1, 7-9].

Since 1980s, plenty of data-based streamflow prediction methods were introduced into the field of hydrological prediction, such as artificial neural network models (ANN) [10, 11], support vector machine [12, 13] and hybrid models [9, 14]. Among them, as one of the most classical methods, ANN was widely used in the field of hydrological forecasting. However, ANN was usually trained by the conventional gradient-based learning algorithms represented by back-propagation (BP), leading to relatively slow learning rate and easily be trapped in a local minimum [11]. Recently, a novel algorithm, called extreme learning machine (ELM), has been developed for feedforward neural networks with single hidden layer [15, 16]. In ELM, the weights between the input and hidden layers and hidden biases are first randomly, then the weights linking hidden layer to output layers are calculated by the Moore-Penrose (MP) generalized inverse [17, 18]. In this way, compared with other traditional algorithms, ELM not only avoids the calculating parameters like learning rate and stop criteria, but also has faster training speed and higher generalization ability [17]. Now, ELM is at the forefront of nonlinear regression and classification, and has been used for wind forecasting, system identification and machine learning. Therefore, in this paper, ELM is introduced for monthly hydrological series forecasting. However, it was found that, in some cases, ELM may be stuck in local optimum because the randomly preset parameters, such as the hidden neurons biases, may tend to contribute less in minimizing the cost function. In other words, there are certain promotion spaces to enhance the performance of ELM [16, 19, 20]. Evolutionary algorithms provide powerful tools for optimizing the neural network parameters [21, 22]. Thus, in this research, the particle swarm optimization (PSO), a popular population-based stochastic method [23, 24], was used to select the ELM network parameters. Moreover, to ensure better searching precision, chaos was employed to generate the initial population of PSO. Then, we propose a novel learning algorithm called improved extreme learning machine (IELM) where PSO and chaos were used in an effort to cooperate with ELM for monthly runoff prediction. Simulation results show that the proposed IELM method outperforms the ELM method.

The rest of this paper is organized as follows. The IELM method is first proposed after the brief introduction of ELM. Then, the study area, data sets and the measures of accuracy are introduced. In the next sections, the prediction results and discussions are given. Finally, the conclusions are presented followed by acknowledgments.

2. Extreme Learning Machine (ELM)

Suppose the number of nodes in input layer, hidden layer and output layer of the ELM networks are *M*, *L* and *N*, respectively. The input and output layers in the ELM network employ the linear activation function, while the hidden layer uses the nonlinear activation function, such as sigmoid function and radial basis function. In this paper, the sigmoid function is chosen as the transfer function of the hidden layer because it behaves well in many different fields, which can be given as the following formula:

$$g(x)=\frac{1}{1+e^{-x}} \tag{1}$$

where x and $g(x)$ denote the input value and output value of the hidden node, respectively.

The optimization objective of ELM is to minimize the total empirical errors of the training data sets with J samples $(\boldsymbol{x}_j,\boldsymbol{t}_j)$ [15, 25], where $\boldsymbol{x}_j=\left[x_{j,1},x_{j,2},\cdots,x_{j,M}\right]\in\boldsymbol{R}^M$, $\boldsymbol{t}_j=\left[t_{j,1},t_{j,2},\cdots,t_{j,N}\right]\in\boldsymbol{R}^N$, and $j=1,2\cdots,J$. The objective function can be expressed as

$$\min E(\boldsymbol{w},\boldsymbol{b},\beta)=\sum_{j=1}^{J}\left\|\boldsymbol{t}_j-\boldsymbol{o}_j\right\| \tag{2}$$

where $\boldsymbol{w}$ is the connection weight matrix between the input layer and the hidden layer; $\boldsymbol{b}$ is the bias of the hidden layer; β is the connection weight matrix between the hidden layer and the output layer; $\boldsymbol{o}_j$ is the output vector of the ELM network after given inputs $\boldsymbol{x}_j$.

Once weights $\boldsymbol{w}$ and bias $\boldsymbol{b}$ are randomly assigned, the optimization problem in Eq. (2) is equivalent to obtain the least square solution of a linear system showed in Eq. (3). Using the Moore-Penrose generalized inverse matrix theory [17], the weight matrix β between hidden layer and output layer can be estimated as $\beta=\boldsymbol{H}^{+}\boldsymbol{T}$, where $\boldsymbol{H}^{+}$ is the generalized inverse matrix of the hidden layer outputs matrix $\boldsymbol{H}$.

$$\boldsymbol{H}\beta=\boldsymbol{T} \tag{3}$$

$$\boldsymbol{H}=\begin{bmatrix}\boldsymbol{H}_1\\ \boldsymbol{H}_2\\ \vdots\\ \boldsymbol{H}_J\end{bmatrix}_{J\times L}=\begin{bmatrix}g(\boldsymbol{w}_1\cdot\boldsymbol{x}_1+b_1) & g(\boldsymbol{w}_2\cdot\boldsymbol{x}_1+b_2) & \cdots & g(\boldsymbol{w}_L\cdot\boldsymbol{x}_1+b_L)\\ g(\boldsymbol{w}_1\cdot\boldsymbol{x}_2+b_1) & g(\boldsymbol{w}_2\cdot\boldsymbol{x}_2+b_2) & \cdots & g(\boldsymbol{w}_L\cdot\boldsymbol{x}_2+b_L)\\ \cdots & \cdots & \cdots & \cdots\\ g(\boldsymbol{w}_1\cdot\boldsymbol{x}_J+b_1) & g(\boldsymbol{w}_2\cdot\boldsymbol{x}_J+b_2) & \cdots & g(\boldsymbol{w}_L\cdot\boldsymbol{x}_J+b_L)\end{bmatrix}_{J\times L} \quad (4)$$

$$\boldsymbol{\beta}=\begin{bmatrix}\boldsymbol{\beta}_1^{\mathrm{T}}\\ \boldsymbol{\beta}_2^{\mathrm{T}}\\ \vdots\\ \boldsymbol{\beta}_L^{\mathrm{T}}\end{bmatrix}_{L\times N} \text{ and } \boldsymbol{T}=\begin{bmatrix}\boldsymbol{t}_1\\ \boldsymbol{t}_2\\ \vdots\\ \boldsymbol{t}_J\end{bmatrix}_{J\times N} \quad (5)$$

where $\boldsymbol{w}_l$ is the connection weight vector between the input layer and the lth node in the hidden layer; $\boldsymbol{w}_i\cdot\boldsymbol{x}_j$ denote the inner product of the weight vector $\boldsymbol{w}_i$ and input vector $\boldsymbol{x}_j$; b_l is the bias of the lth node in the hidden layer; $\boldsymbol{\beta}_l^{\mathrm{T}}=\left[\beta_{l,1},\beta_{l,2},\cdots,\beta_{l,N}\right]\in\boldsymbol{R}^N$ is the weight vector connecting the lth hidden node and all output nodes.

3. Improved Extreme Learning Machine (IELM)

Since ELM can directly obtain the weights between hidden and output layers by solving the linear system analytically, it can save much of time spent on tuning parameters. However, the output weights of ELM are based on the pre-decided input-hidden weights and biases of hidden layer, the model trained by ELM is random and may be easy to be run into local optimum [19, 22]. Therefore, in this section, a novel method called IELM is proposed, where PSO is employed for estimating optimal ELM parameters. Given a set of training data and L hidden nodes with an activation function $g(x)$, the IELM algorithm can be summarized in the following steps:

Step 1. Preparation for optimization. Set maximize iterations and population size in PSO, and divide data into training and testing sets.

Step 2. Set k=1, and initialize all particles with chaos mapping.

A set of I individuals, where each one is composed of all input weights and hidden bias of a ELM network, are initialized using chaos mapping as the first population. The particle $\boldsymbol{X}_i^k$ can be expressed as follows:

$$\boldsymbol{X}_i^k=\left[w_{1,1}^{i,k},w_{2,1}^{i,k},\cdots,w_{M,1}^{i,k},\cdots,w_{1,l}^{i,k},\cdots,w_{m,l}^{i,k},\cdots,w_{M,L}^{i,k},b_1^{i,k},\cdots,b_l^{i,k},\cdots,b_L^{i,k}\right] \quad (6)$$

Where $w_{m,l}^{i,k}$ in particle $\boldsymbol{X}_i^k$ represents the weight between the mth input node and the lth hidden node ; $b_l^{i,k}$ in particle $\boldsymbol{X}_i^k$ represents the bias of the lth hidden node.

Step 3. Calculate the output weights and fitness values of all individuals.

Decode the individual vector to an ELM network, and calculate the ELM network output weight matrix $\boldsymbol{\beta}_i^k$ with the following equations.

$$\boldsymbol{\beta}_i^k=\left(\boldsymbol{H}_i^k\right)^{+}\boldsymbol{T} \quad (7)$$

Where $\boldsymbol{H}_i^k$ is the hidden layer outputs matrix of the ith particle in the kth generation.

Then the fitness value $f(x)$ of each individual is calculated using the following equations.

$$f\left(\boldsymbol{X}_i^k\right)=\sqrt{\frac{\sum_{j=1}^{J}\left\|\sum_{i=1}^{L}\beta_i g\left(\boldsymbol{w}_i\cdot\boldsymbol{x}_j+b_i\right)-\boldsymbol{t}_j\right\|_2^2}{N\times J}} \quad (8)$$

Step 4. Update the best-known position of each particle and the best known position of the whole population according to the following two formulas:

$$\boldsymbol{PBest}_i^k=\begin{cases}\boldsymbol{X}_i^k & if\,(k=1)\\ \arg\min\left\{f\left(\boldsymbol{PBest}_i^k\right),f\left(\boldsymbol{X}_i^k\right)\right\} & otherwise\end{cases} \quad (9)$$

$$GBest_i^k = \begin{cases} \arg\min\left\{\min_{1\le i\le I}\left\{f\left(PBest_i^k\right)\right\}, f\left(GBest_i^{k-1}\right)\right\} & if\ (k>1) \\ \arg\min\left\{\min_{1\le i\le I}\left\{f\left(PBest_i^k\right)\right\}\right\} & otherwise \end{cases} \tag{10}$$

Step 5. Calculate the inertia factor, and update current position of each particle.

Step 6. Set k=k+1, and check whether the calculation can be stopped or not. Output the optimal ELM network if the maximum iterations are reached; otherwise, go back to Step 3.

4. Measures of Accuracy

The performances of forecasting models can be evaluated by different statistical measure criteria. In this paper, four commonly used metrics, respectively defined from Eqs. (11)-(14), are selected to evaluate the forecast results: the root mean squared error (*RMSE*), mean absolute percentage error (*MAPE*), Nash-Sutcliffe efficiency coefficient (*E*), and coefficient of correlation (*R*). The *RMSE* measures the precision of the model by comparing the deviation between the observed and forecasted data; the *MAPE* evaluates the absolute percentage error of the prediction; the *E* reveals the capability of the model in predicting values away from the mean; the *R* shows the linear relationship between the forecasted and measured data.

$$RMSE = \sqrt{\frac{1}{n}\sum_{i=1}^{n}\left(y_i - \tilde{y}_i\right)^2} \tag{11}$$

$$MAPE = \frac{1}{n}\sum_{i=1}^{n}\left|\frac{y_i - \tilde{y}_i}{y_i}\right| \times 100 \tag{12}$$

$$R = \frac{\sum_{i=1}^{n}\left[\left(y_i - y_{avg}\right)\left(\tilde{y}_i - \tilde{y}_{avg}\right)\right]}{\sqrt{\sum_{i=1}^{n}\left(y_i - y_{avg}\right)^2\left(\tilde{y}_i - \tilde{y}_{avg}\right)^2}} \tag{13}$$

$$E = 1 - \frac{\sum_{i=1}^{n}\left(y_i - \tilde{y}_i\right)^2}{\sum_{i=1}^{n}\left(y_i - y_{avg}\right)^2} \tag{14}$$

where y_i and $\tilde{y}_i$ are the observed value and model output value of ith data, respectively. y_{avg} and $\tilde{y}_{avg}$ represent the mean value of observed and modeled values, respectively. n is the total number of observed data values.

5. Study Area and Data Set

In this study, the monthly streamflow data from Hongjiadu reservoir located in Wu River was used to test the effectiveness of the proposed method. The monthly streamflow data sets from January 1951 to December 2015 are studied. The first 54 years from 1951 to 2005 were used for training while the rest were used for validation. To avoid the numerical problems caused by larger attribute values dominating smaller ones, the data sets are usually normalized before applied for prediction. In ELM and IELM modeling process, the runoff data sets were scaled to the range between 0 and 1 by Eq. (15). After simulation, the normalized output data should be rescaled to the original forecasted data following the contrary procedure of Eq. (15).

$$q_t' = \frac{q_t - q_{\min}}{q_{\max} - q_{\min}} \tag{15}$$

where q_t' and q_t are the scaled and original flow value at period t, respectively; $q_{\max}$ and $q_{\min}$ are the maximum and minimum values of original streamflow data series, respectively.

6. Prediction Modeling and Input Selection

In general, forecasting models that predict future runoff value based on past records are built. The object using antecedent values to forecast inflow values can be generalized in the following form $q=\text{Model}(\boldsymbol{x})$, which denotes the nonlinear relationship between input vector $\boldsymbol{x}$ and the target value q. In this research, q is the flow value in the next period, while, for the sake of simplicity, $\boldsymbol{x}$ is usually the antecedent flow values with various time lags.

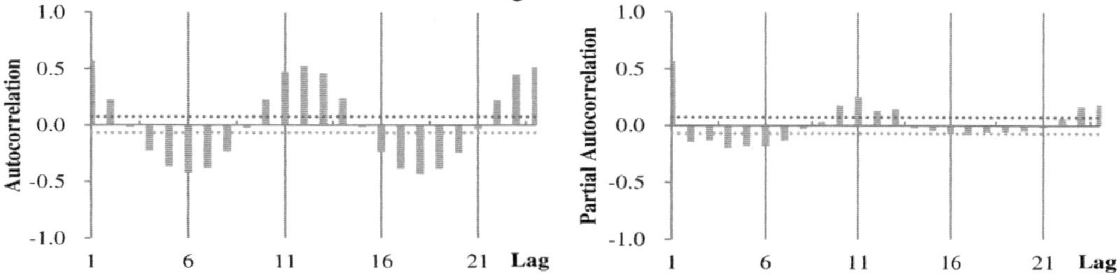

Figure 1. The (left) ACF and (right) PACF value of Hongjiadu monthly inflow series.

Input variables occupy an important place in model training phase and implementation phase because it can reflect the basic information about the system being studied. Therefore, the input variables should be selected carefully to enhance the precision of the model. In this study, the autocorrelation function (ACF) and the partial autocorrelation function (PACF) are two classic indicators used to determine the number of input vectors [26]. Figure 1 shows ACF and PACF values for lag 1-24 in Hongjiadu, respectively. It can be found that, at lag 12, ACF exhibits peak value while PACF shows a significant correlation. Hence, on the basis of the above-mentioned analysis, the variables $q_{t-1}, q_{t-2}, \cdots, q_{t-12}$ possess abundant information for future flow q_t and are selected as input vector for monthly forecasting model.

7. Results and Discussion

In order to verify the effectiveness of the proposed method, a typical three-layer feed-forward ANN model using traditional back-propagation algorithm is selected as the benchmark method. For the sake of comparison, the same training and validation data sets, respectively, are used for three forecasting models, and the four quantitative indexes are employed to evaluate the performances of all the developed models. For ANN model optimized by back-propagation algorithm, the optimal number of ANN hidden nodes was identified with a trial-error procedure by varying the number of hidden nodes. Similarly, the number of hidden neurons in ELM and IELM models is also determined by searching for the optimal performance. For IELM, the maximal iteration is 200 and the number of individuals is 50.

Table 1 shows the statistics results of Hongjiadu reservoir in terms of four performance indicators. It can be seen from Table 1 that three models developed have satisfactory performances during training and validation periods, and the IELM method are superior to those in ANN and ELM. In the training period, the IELM model can improve the ANN model with 6.25% and 12.63% reduction in *RMSE* and *MAPE*, while the improvements regarding *R* and *E* were 6.51% and 3.21%, respectively. In the validation period, when compared with ELM model, the *R* and *E* values of the IELM method increase by 2.85% and 1.59% respectively, while the values of the forecasting results regarding *RMSE* and *MAPE* decrease by 4.05% and 0.60%. Thus, it can be concluded that the IELM model can obtain better results than ANN and ELM models with obvious improvements in four statistical indexes for monthly streamflow prediction.

Table 1. Performance indices of various models for Hongjiadu reservoir.

Models	Training Period				Validation Period			
	RMSE	*MAPE*(%)	*E*	*R*	*RMSE*	*MAPE*(%)	*E*	*R*

ANN	79.546	45.682	0.650	0.806	62.811	50.935	0.705	0.840
ELM	76.746	41.818	0.674	0.821	59.431	49.406	0.736	0.859
IELM	74.576	39.915	0.693	0.832	57.026	49.710	0.757	0.873

Table 2 shows the observed and forecasted peak flow values using three models. In term of the maximum peak flow in the validation period, the forecasted value of ANN, ELM and IELM are 413.1 m^3/s, 404.4 m^3/s, and 429.4 m^3/s, corresponding to about 40.1%, 41.3% and 37.7% underestimation, respectively. The main reason was that, from 2011 to 2012, it was dry years where runoff was mainly recharged by base flow, and this kind of situation was a fairly rare occurrence in the history of Hongjiadu reservoir, making the performances of all models poorer than that in other years. On the whole, the absolute average relative error of IELM is 25.7%, while ANN has 26.2% error and ELM predicts peak with 20.3% error. Thus, for peak flow estimation, the IELM model has better nonlinear mapping ability than ELM and ANN.

Table 2. Observed vs. forecasted peak flow by various models in Hongjiadu reservoir.

Peak NO.	Date	Observed peak	Forecast Peak			Relative error (%)		
			ANN	ELM	IELM	ANN	ELM	IELM
1	2006/7	173.9	241.57	278.70	202.82	38.9	60.3	16.6
2	2007/7	329.7	261.97	275.84	203.91	-20.5	-16.3	-38.2
3	2008/6	422.7	368.8	284.7	323.6	-12.8	-32.7	-23.4
4	2009/7	201.8	190.6	224.6	200.0	-5.5	11.3	-0.9
5	2010/7	335.0	318.7	273.7	309.2	-4.9	-18.3	-7.7
6	2011/6	111.6	52.0	66.6	78.5	-53.4	-40.3	-29.7
7	2012/7	689.1	413.1	404.4	429.4	-40.1	-41.3	-37.7
8	2013/6	288.4	263.9	267.9	287.2	-8.5	-7.1	-0.4
9	2014/7	509.2	199.3	373.0	387.4	-60.9	-26.7	-23.9
10	2015/6	206.5	229.7	222.0	256.6	11.2	7.5	24.2
Average (absolute)						25.7	26.2	20.3

Figures 2~4 show the scatter plots of observed data versus forecasted data using various models developed for Hongjiadu reservoir during training and validation periods. The observed and calculated monthly streamflow by three models in the training period and validation period are shown in Figure 5. It can be clearly seen that all three models can give close approximations for observed runoff data, and the IELM model mimics flow data better than that by ELM and ANN models. It means that the IELM can overcome drawbacks of both ELM and ANN models, and effectively capture the dynamic changing processes of Hongjiadu reservoir.

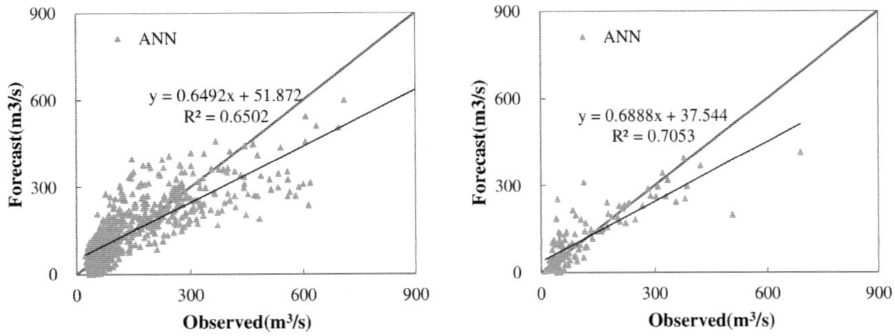

Figure 2. The scatter plots of observed vs. forecasted value using ANN model during the (left) training and (right) validation periods.

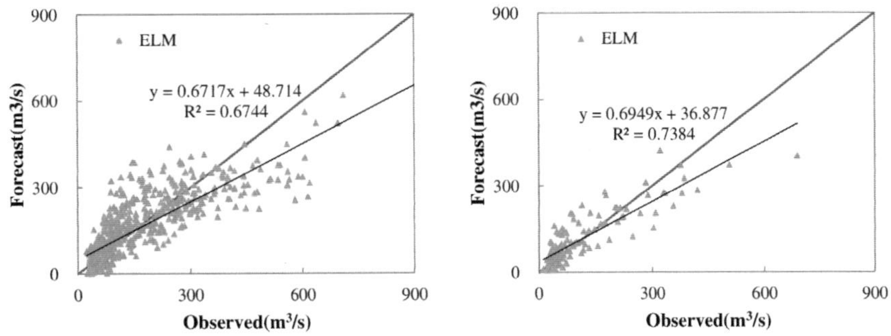

Figure 3. The scatter plots of observed vs. forecasted value using ELM model during the (left) training and (right) validation periods.

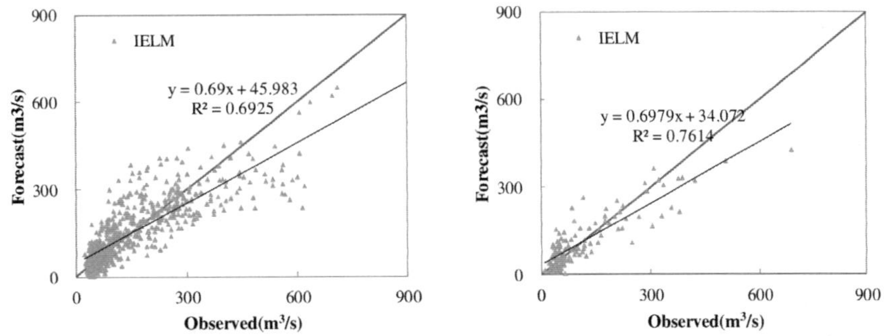

Figure 4. The scatter plots of observed vs. forecasted value using IELM model during the (left) training and (right) validation periods.

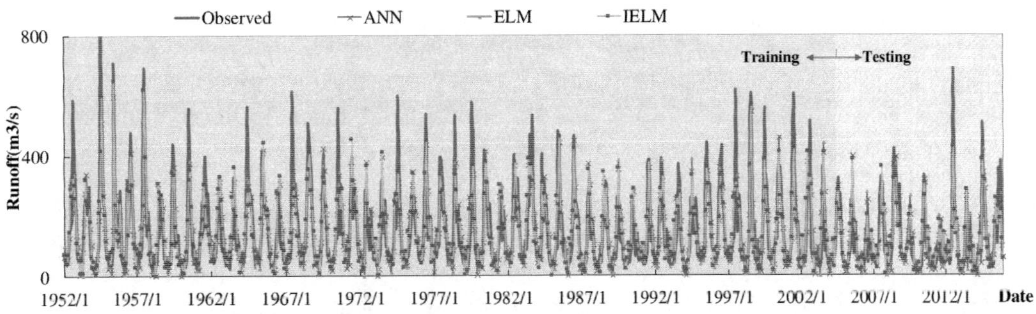

Figure 5. Observed and forecasted values in training and validation phases by various models.

To summary, the abovementioned results show that, in the monthly streamflow prediction modeling, the IELM model using population-based algorithms for parameter selection can enhance the generalization ability of the ELM method in some degrees. In addition, due to the complexity of hydrological system, the models for prediction should be developed according to the hydrological characteristics of the river basins being studied.

8. Conclusions

In the present study, the accuracy of extreme learning machine (ELM), one method with growing popularity for regression and classification, has been investigated for forecasting reservoir monthly streamflow. In order to enhance the performance of ELM, the improved extreme learning machine (IELM) using particle swarm optimization and chaos to select the parameters of ELM network is proposed. The results in Hongjiadu reservoir indicate that the IELM method outperforms the ANN and ELM models in terms of four different statistics measures, increasing the R and E by 2.85% and 1.59% respectively compared to ANN. Thus, with the measurable improvements on the precision of monthly runoff prediction, the IELM model can act as a superior tool to ANN and ELM, and can help reservoir operate in a more sustainable manner. Moreover, the IELM model considering the effects of other hydrological variables in different temporal scales needs to be studied in the future.

Acknowledgments

This study is supported by the National Key R&D Program of China (2018YFC1508002), National Natural Science Foundation of China (51709119), Natural Science Foundation of Hubei Province (2018CFB573), and Fundamental Research Funds for the Central Universities (HUST: 2017KFYXJJ193).

References

[1] Wang, W.C.; Chau, K.W.; Cheng, C.T.; Qiu, L., A comparison of performance of several artificial intelligence methods for forecasting monthly discharge time series. *J Hydrol* **2009**, 374, (3-4), 294-306.

[2] Valipour, M.; Banihabib, M.E.; Behbahani, S.M.R., Comparison of the ARMA, ARIMA, and the autoregressive artificial neural network models in forecasting the monthly inflow of Dez dam reservoir. *J Hydrol* **2013**, 476, (2), 433-441.

[3] Chau, K.W.; Wu, C.L.; Li, Y.S., Comparison of Several Flood Forecasting Models in Yangtze River. *J Hydrol Eng* **2005**, 10, (6), 485-491.

[4] Zhao, T.; Zhao, J., Joint and respective effects of long and short-term forecast uncertainties On reservoir operations. *J Hydrol* **2014**, 517, (2), 83-94.

[5] Wu, C.L.; Chau, K.W.; Li, Y.S., Methods to improve neural network performance in daily flows prediction. *J Hydrol* **2009**, 372, (1-4), 80-93.

[6] Huang, S.; Chang, J.; Huang, Q.; Chen, Y., Monthly streamflow prediction using modified EMD-based support vector machine. *J Hydrol* **2014**, 511, (7), 764–775.

[7] Liu, P.; Lin, K.; Wei, X., A two-stage method of quantitative flood risk analysis for reservoir real-time operation using ensemble-based hydrologic forecasts. *Stochastic Environmental Research & Risk Assessment* **2015**, 29, (3), 803-813.

[8] Lund, J.R., Flood Management in California. *Water-Sui* **2012**, 4, (1), 157-169.

[9] Cheng, C.T.; Feng, Z.K.; Niu, W.J.; Liao, S.L., Heuristic Methods for Reservoir Monthly Inflow Forecasting: A Case Study of Xinfengjiang Reservoir in Pearl River, China. *Water-Sui* **2015**, 7, (8), 4477-4495.

[10] 10.Wang, W.C.; Chau, K.W.; Lin, Q.; Chen, Y.B., Improving forecasting accuracy of medium and long-term runoff using artificial neural network based on EEMD decomposition. *Environ Res* **2015**, 139, 46-54.

[11] Chen, Y.H.; Chang, F.J., Evolutionary artificial neural networks for hydrological systems forecasting. *J Hydrol* **2009**, 367, (1-2), 125-137.

[12] Lin, G.F.; Chen, G.R.; Huang, P.Y.; Chou, Y.C., Support vector machine-based models for hourly reservoir inflow forecasting during typhoon-warning periods. *J Hydrol* **2009**, 372, (1-4), 17-29.

[13] Lin, G.F.; Chou, Y.C.; Wu, M.C., Typhoon flood forecasting using integrated two-stage Support Vector Machine approach. *J Hydrol* **2013**, 486, 334-342.
[14] Wu, C.L.; Chau, K.W.; Li, Y.S., Predicting Monthly Streamflow Using Data-Driven Models Coupled with Data-Preprocessing Techniques. *Water Resour Res* **2009**, 45, (8), 2263-2289.
[15] Huang, G.B.; Zhou, H.; Ding, X.; Zhang, R., Extreme learning machine for regression and multiclass classification. *IEEE Transactions on Systems, Man, and Cybernetics, Part B (Cybernetics)* **2012**, 42, (2), 513-29.
[16] Huang, G.B.; Zhu, Q.Y.; Siew, C.K., Extreme learning machine: Theory and applications. *Neurocomputing* **2006**, 70, (1-3), 489-501.
[17] Huang, G.B.; Chen, L.; Siew, C.K., Universal approximation using incremental constructive feedforward networks with random hidden nodes. *IEEE Transactions on Neural Networks* **2006**, 17, (4), 879-92.
[18] Yin, Y.; Zhao, Y.; Li, C.; Zhang, B., Improving Multi-Instance Multi-Label Learning by Extreme Learning Machine. *Applied Sciences* **2016**, 6, (6), 160.
[19] Zhu, Q.Y.; Qin, A.K.; Suganthan, P.N.; Huang, G.B., Evolutionary extreme learning machine. *Pattern Recogn* **2005**, 38, (10), 1759-1763.
[20] Cao, J.; Lin, Z.; Huang, G.B., Self-Adaptive Evolutionary Extreme Learning Machine. *Neural Process Lett* **2012**, 36, (3), 285-305.
[21] Ch, S.; Anand, N.; Panigrahi, B.K.; Mathur, S., Streamflow forecasting by SVM with quantum behaved particle swarm optimization. *Neurocomputing* **2013**, 101, (3), 18-23.
[22] Nahvi, B.; Habibi, J.; Mohammadi, K.; Shamshirband, S.; Razgan, O.S.A., Using self-adaptive evolutionary algorithm to improve the performance of an extreme learning machine for estimating soil temperature. *Computers & Electronics in Agriculture* **2016**, 124, 150-160.
[23] Chau, K.W., Particle swarm optimization training algorithm for ANNs in stage prediction of Shing Mun River. *J Hydrol* **2006**, 329, (3-4), 363-367.
[24] Kennedy, J.; Eberhart, R., In *Particle swarm optimization*, IEEE International Conference on Neural Networks, 1995. Proceedings, 1995; 1995; pp. 1942-1948 vol.4.
[25] Huang, G.; Huang, G.B.; Song, S.; You, K., Trends in extreme learning machines: A review. *Neural Networks* **2015**, 61, 32-48.
[26] Lin, J.; Chuntian, C.; Chau, K., Using support vector machines for long-term discharge prediction. *Hydrological Sciences Journal* **2010**, 51, (4), 599-612.

IOP Conf. Series: Materials Science and Engineering **794** (2020) 012018 doi:10.1088/1757-899X/794/1/012018

Study of the Thickness of Curtain Grouting Ring for the life-cycle of Tunnel in Water-rich Fault Fracture Zone

Yanling Jia[*1,3,4], Ouyang Lu[2,3,4], Shaowen Zhou[3,4] and Gang Wang[3,4]

[1]School of Highway, Chang'an University, Xi'an, Shanxi, 710064, China

[2]College of Civil Engineering, Changsha University of Science and Technology, Changsha, Hunan, 410004, China

[3]Guangxi Key Laboratory of Road Structure and Materials, Nanning, Guangxi, 530007, China

[4]Survey and Design Division, Guangxi Transportation Research & Consulting Co. Ltd, Nanning, Guangxi 530007, China

[*]Corresponding author's e-mail: 3859442@163.com

Abstract. Junchang Tunnel passes through fault fracture zone where joints and fracture developed, the water-rich zone is wide and the geological condition is extremely complex, which is likely to induce the water gushing and mud gushing. Based on the project, the proper thickness of curtain grouting ring for the life-cycle of tunnel in water-rich fault fracture zone is analyzed, and the main conclusions are as follows: (1) Comprehensive advance geological forecast including ultra-high density resistance method, land sonar method, transient electromagnetic method and drilling method are adopted to achieve good prediction results in the fault fracture zone of DK7+812 to DK7+862; (2) The full-face curtain grouting method can effectively reduce the formation deformation caused by tunnel excavation in the tunnel construction in water-rich fault fracture zone, and the effect of water plugging is great; (3) According to the formula of thick wall tube, the experience of grouting and water plugging in existing tunnels, and the comparison of finite element method, the thickness of curtain grouting ring should be greater than 5m to meet the safety requirements. Considering the safety requirements and the reduction of curtain grouting ring in the operation period, the thickness of curtain grouting ring needs to be increased to 7m.

1. Introduction

Against the backdrop of the Western Development in China and the expansion of transportation networks, more highway and railway tunnels inevitably have to pass across the water-rich fault fracture zones [1~3] with poor self-stability, high water pressure, and large water volume. This kind of zone is difficult to block, and the tunnel lining structure is often cracked with water and mud gushing, even leading to landslide disasters [4~6], thereby causing a huge threat for tunnel construction and operation. In recent years, domestic and foreign scholars have been carrying out research on the tunnels in the water-rich layer, mainly focusing on value rationalization and stability analysis of grouting reinforcement ring parameters of the tunnel curtain [1] [2], prevention and treatment of water and mud gushing during tunnel construction [3~6], key techniques of construction in unfavorable geological sections [8] [9], and assumption of external water pressure value of the rich-water lining [9~12]. The research has played an active role in the aspect of tunnel construction.

However, the current research focuses mainly on the stability analysis during tunnel construction, but less on geological prediction before construction and lining safety analysis during life-cycle operation of the water-rich fault fracture zone. For example, in the early stage of tunnel construction, it is often found that the geological conditions are inconsistent with the proven ones, resulting in construction difficulties; during tunnel operation, the curtain grouting ring gradually fails, and the external water pressure behind the lining structure increases, causing lining crack and water leakage. Therefore, it is necessary to conduct an in-depth study on the life-cycle stability and safety of the tunnel curtain grouting. In view of this, this paper takes on the Baomao Expressway Junchang Tunnel Project as the case to study the stability of lining curtain grouting before, during and after tunnel construction based on comprehensive advanced geological prediction and numerical calculation. This study meets the reasonable value of life-cycle curtain grouting, and is expected to provide reference for the design, construction and operation of tunnels in the water-rich fault fracture zone.

2. Overview of tunnel engineering

The Junchang Tunnel from Cenxi to Shuiwen section of Baotou-Maoming Expressway is located between the Caledonian fold group (consisting of Fuqing syncline, Baishidong syncline and Tangdong syncline) and Shuiwen syncline during Yanshan period. The Rongxian-Cenxi fault and Dalong-Shuiwen fault pass through the northeast and southeast of the tunnel, respectively. The tunnel is a separate small clear-distance tunnel. The left line is 4270m long, while the right line is 4288m long, and the left and right tunnels are 17m apart. The water-rich soft layer of the Junchang tunnel is about 150m long. The survey data reveals that it is located in the tectonic development area with relatively developed rock joints and fractures, relatively broken local rock mass and high groundwater level. In addition, the tunnel passes through the residential area of Shanxin Village, posing great difficulty for construction.

3. Comprehensive advanced geological forecast

The Junchang Tunnel was constructed to the weak water-rich stratum in March 2013, but failed to achieve any footage in the face for two and a half years until August 2015, witnessing the occurrence of three large-scale water and mud disasters with the maximum water inflow of 1200 m^3/h and the maximum mud volume of 2900 m^3. Water and mush gushing caused deformation and cracking of the initial branch and partial cracks of the second lining.

(a) (b)

Figure 1. Water gushing and mud gushing in Junchang Tunnel

Due to the multi-solution of geophysical exploration, a single geological prediction method cannot fully reflect the actual formation. Therefore, after the occurrence of a sudden water disaster in the Junchang Tunnel, the ultra-high-density resistivity method, terrestrial sonar method, transient electromagnetic method and drilling method are used to conduct comprehensive advanced geological prediction analysis of the rich-water weak layer so as to find out the geological and hydrogeological conditions of the unfavorable geological section.

Table 1. Synthetic advance geology forecast results in water-rich fault fracture zone

Forecast method	Function	Conclusion
High density resistivity method	Proving detailed geological structure information and distribution of groundwater resources	DK7+814 and DK7+960 are the lithostratigraphic boundary of left line tunnel. DK7+814~ DK7+960 are the water- bearing fault belt. CK7+950 is the lithostratigraphic boundary of right line tunnel. CK7+838~CK7+950 water- bearing fault belt.
Land-sonar method	Exploration of cavity, Flow channel and fault distribution location in rock Layer	DK7+850~DK8+040 of left line and CK7+840~ CK8+020 of right line are fault fracture zone, the joints and crack development, wide water-rich.
Transient electromagnetic method	Prove hydrogeology such as aquifer geology, Hydraulic passage, Deep irregular water body, et,al	30~50m in front of the working face of the left tunnel, there may be a large crushing zone on the left side, and a large volume of water on the right side of the left tunnel.
Drilling method	Explore the formation lithology, Rock strength and water content in front of tunnel working face	The tunnel working face is mainly medium-weathered migmatite with grey-brown color, developed joints and fissures, fragmentary structure, fragmented rock mass and hard rock.

Table 2. Verification of actual excavation in water-rich fault fracture zone

Mileage	Synthetic advance geology forecast	Verification of actual excavation
DK7+812~ DK7+815	Medium weathered migmatite, Joints and fissures are well-developed, Surrounding rock is relatively fragmented, Presenting massive structure.	The weathered migmatite is mainly characterized by well-developed joints and fissures, mostly in a closed state, presenting massive structure, wet rock mass (local dripping effluent), hard rock and poor self-stabilization ability.
DK7+815~ DK7+828.8	Strongly weathered migmatite, Joints and fissures are well-developed, Surrounding rock is partially fragmented, Presenting fragmentary structure, Large water content.	Strong weathered migmatite is dominant, joints and fissures are developed, yellowish brown sandy clay is filled and cemented, presenting fragmentary structure, hard rock, seepage in many places, poor self-stabilization ability.
DK7+828.8 ~DK7+846	Strongly weathered migmatite, Joints and fissures are well-developed, Presenting fragmentary structure, Large water content. There is a large fault near DK7+839 with an inclination of about 69 degrees and inclination to the direction of import.	Strong weathered migmatite is dominant, joints and fissures are well-developed, presenting fragmentary structure. Grouting veins are found in DK7+839. The predicted results are basically consistent with the fault fracture zone.
DK7+846~ DK7+862	Totally weathered migmatite, Joints and fissures are well-developed, Surrounding rock is extremely fragmented, Presenting loose structure, located in the water-rich permeability zone. There is a large fault near DK7+847 with a dip angle of 90 degrees.	Totally weathered migmatite is dominant, presenting fragmentation-loose structure, seepage in many places, no self-stabilization ability. A large area of grouting veins are found in the arch, agree with the predicted results. There is a large wave of water in the right arch waist, and the waer body is turbid, extremely unfavorable geological conditions and large water inflow.

It can be seen from Table 1 to Table 2 that the water-rich soft layer of the Junchang Tunnel is dominated by fully weathered granite with fractured rock mass, developed fractures, multiple well-connected water guiding channels, large water pressure, complex and varied geological conditions. The comprehensive advanced geological prediction method is used to provide a basis for the full-section curtain grouting method in the water-rich fault fracture zone, and has achieved good results in practice.

4. Fluid-structure coupling calculation during tunnel construction

This section uses midas GTS NX to establish a two-dimensional fluid-structure coupling analysis model to analyze the effect of groundwater in the water-rich stratum on the stability of surrounding rock of the tunnel. The calculation satisfies the following basic assumptions: (1) The surrounding rock of the tunnel is made of a continuous, homogeneous, and isotropic material; (2) The water flow in the surrounding rock is small and can satisfies Darcy's law; (3) The groundwater level at the boundary of the model is constant. The stratum and curtain grouting reinforcement circle adopts the Mohr-Coulomb constitutive model commonly used in geotechnical engineering, and is simulated by plane strain element. Its lining structure adopts elastic constitutive model, the primary lining adopts beam element simulation, and the secondary lining adopts plane element simulation.

The mapping grid is used to divide the model into 4728 units and 4624 nodes. The tunnel calculation model is shown in figure 2 and figure3. The model size is set to be 150m (width) × 108m (height), while the depth of tunnel is 50 m. The horizontal constraint is applied to the left and right sides of the model and the vertical constraint is applied to the bottom, while the top is the free boundary. In order to simulate the seepage field characteristics of the tunnel under water-rich conditions, the hydrostatic pressure boundary is set around the model boundary, with the water pressure increasing linearly with depth.

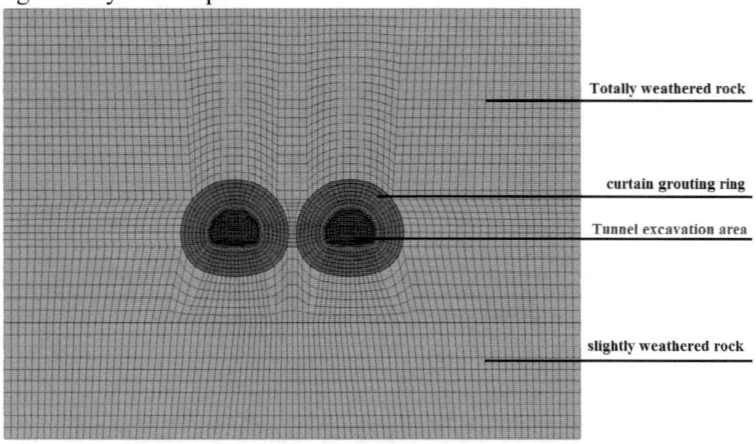

Figure 2. The calculation model of tunnel

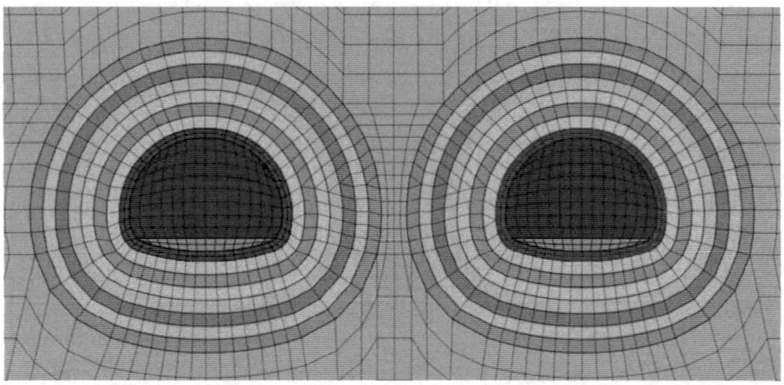

Figure 3. Local enlargement of grids in full-face curtain grouting ring

According to Code for Design of Road Tunnel (JTG D70-2004), geological survey reports and related engineering experience, this paper sets the physical and mechanical parameters of the surrounding rock and supporting structure of the tunnel, as shown in Table 3.

Table 3. Mechanics parameter of model

Material	Unit weight ($kN\cdot m^{-3}$)	Elastic modulus (GPa)	Poisson ratio	Porosity	Cohesive forces(MPa)	The angle of internal friction(°)
Totally weathered rock	19.5	0.02	0.36	0.43	20	20
Slightly weathered rock	21	1.3	0.3	0.3	100	25
Curtain grouting ring	22	1	0.35	0.36	150	27
Shotcrete	25	28	0.2	-	-	-
Secondary lining	25	33.5	0.2	-	-	-

During the analysis process, the curtain grouting reinforcement ring mainly bears the external hydrostatic pressure. According to the actual measurement data of the Junchang Tunnel, the allowable compressive strength of curtain grouting reinforcement ring is 2.5 MPa and the maximum hydrostatic pressure is 1.0 MPa. The equivalent diameter of tunnel excavation is 12.0m. According to the fourth strength theory of thick-walled cylinder formula, the required curtain thickness is calculated to be 4.8m. Based on the experience of tunnel grouting reinforcement and water blocking construction [5], this section sets the reference thickness of curtain grouting to be 5m. Based on this, it is determined that the numerical simulation schemes are all full-section advanced curtain grouting. A total of 8 schemes are used so as to determine the optimal curtain grouting thickness. The calculation results of the full-section curtain grouting reinforcement using the finite element model are shown in Table 4.

Table 4. Comparison of numerical results of full-section curtain grouting ring

Analysis case	Crown settlement (mm)	Ground settlement(mm)	Bottom uplift (mm)	Water inflow of working face($m^3\cdot h$)
1(no grouting)	55.35	26.42	43.89	14.71
2(1m grouting)	42.66	21.09	46.41	9.60
3(2m grouting)	31.68	16.06	40.67	7.46
4(3m grouting)	24.83	12.76	31.43	6.20
5(4m grouting)	20.81	10.65	24.88	5.38
6(5m grouting)	17.95	9.16	20.67	4.78
7(6m grouting)	15.88	8.07	17.97	4.36
8(7m grouting)	14.33	7.25	15.97	3.99

Figure 4 and Figure 5 are the comparation of case1 and case6. The calculation results show that:

(1) The full-weathered granite is treated using full-section advanced curtain grouting method, thus achieving relatively satisfactory water blocking and reinforcement effects. As the curtain thickness increases, the tunnel water inflow gradually decreases. When the curtain grouting reinforcement reaches 5m+ thick, the tunnel water inflow can be controlled within 5m3/h, thereby effectively reducing the risk of water and mud gushing during tunnel construction.

(2) The full-section curtain grouting method can effectively reduce the ground deformation caused by tunnel excavation. When the curtain grouting thickness reaches 5m, the vault settlement and surface subsidence are only 32.4% and 34.7% of that under non-grouted situation, respectively.

(3) From the main stress distribution cloud diagram, it can be seen that the bottom lining may crack when it is not grouted. But when the curtain grouting thickness reaches 5m, the lining tensile stress value reaches 1.39MPa<1.43MPa, at which point the lining structure is basically safe.

Based on guideposts of surrounding rock deformation, water inflow, and lining stress during construction, it can be judged using the thick-walled formula, experience of existing tunnel engineering grouting reinforcement and water blocking construction and finite element comparison that the 5m-thick full-section curtain grouting for reinforcement in the Junchang Tunnel section can meet the structural stability requirements.

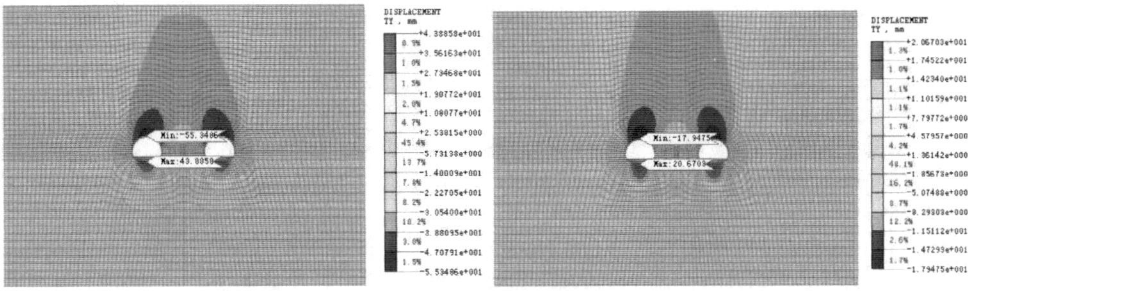

(a) (b)

Figure 4. Vertical deformation contour

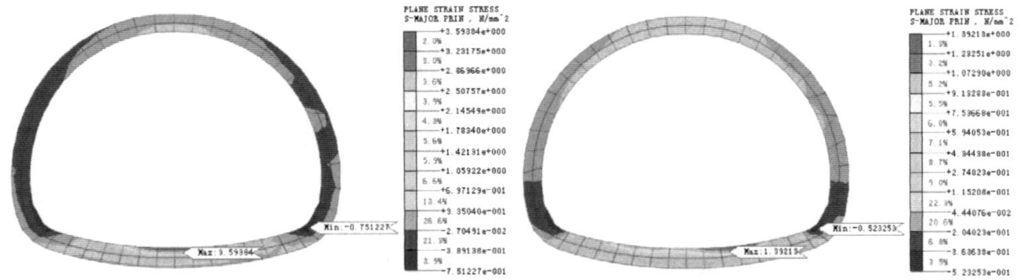

(a) (b)

Figure 5. Principal stress contour

5. Fluid-structure coupling calculation during tunnel operation

Based on the above analysis and actual construction, the 5m-thick full-section curtain grouting can meet the safety requirements during construction. The Junchang Tunnel was successfully completed on October 18, 2016 and was officially opened to traffic on January 5, 2017. However, for more than two years of operation, there have been problems such as lining cracks, water seepage, and water gushing in different sections of the water-rich fault fracture zone. The main reasons include: 1) The lining is thicker than 50cm, and the construction joints and uncompacted vibration occurred during construction lead to water seepage; 2) Under the joint influence of hydrodynamic pressure and overload vehicle disturbance, the water-repellent and bearing capacities of the curtain grouting layer are gradually reduced, leading to greater water pressure on the lining, thereby resulting in the occurrence and development of lining cracks and water seepage.

Therefore, it is necessary to consider the failure possibility of curtain grouting reinforcement ring during operation. In addition, it is worth analyzing the decrease of elastic modulus, cohesion force, internal friction angle of the reinforcement ring layer, the increase of permeability coefficient, the increase of water pressure, the uneven layer settlement, the crack of secondary lining, water leakage and water gushing.

5.1Determination of external water pressure of lining during operation

Before tunnel excavation, the water pressure at each point of the surrounding rock increases linearly with depth. After excavation, the contour is related to the drainage rate of the tunnel lining structure. The extreme working conditions are drainage rate=0% (impermeable, figure6(a)) and drainage rate=100% (water permeable, figure6(b)). Based on the head analysis under the above two working conditions, it can be seen that: (1) 0% drainage rate of the lining structure is equivalent to the adoption of full sealing waterproof measures. After the seepage field of the surrounding rock becomes stable, the pore water pressure at each point of the surrounding rock is basically equal to that of the initial seepage field, and the head height behind the lining structure is almost equal to the initial head height. (2) When the drainage rate of the lining structure is 100%, the pore water pressure is distributed in the shape of the precipitation funnel along the outer contour of the left and right tunnels. This is mainly because the water in the water-rich fault continues to leak into the face after excavation, leading to smaller pore water pressure near the tunnel face. In addition, the head value on the model boundary is not affected because the head boundary is applied around the model. In the fluid-structure coupling analysis model during operation, the head setting is analyzed under the 100% drainage rate condition so as to characterize the safety of the tunnel lining structure under the most unfavorable condition.

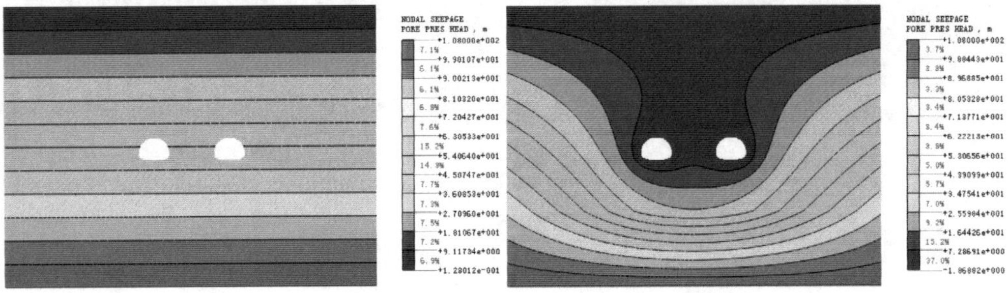

(a) (b)

Figure 6. Pore water pressure contour of surrounding rock under different drainage rate after tunnel excavation

5.2Analysis of fluid-structure coupling calculation results during tunnel operation

Based on the two-dimensional fluid-structure coupling analysis model established in Chapter 4 and the head determination method of the surrounding rock nodes described in Section 4.1, this paper sets the calculation conditions under different failure degrees of reinforcement rings, as shown in Table 5.

Table5. Failure reduction parameter of curtain grouting ring

Analysis case	Parameter reduction	Unit weight ($kN\cdot m^{-3}$)	Elastic modulus(GPa)	Cohesive forces(MPa)	The angle of internal friction(°)
1	90%	21.8	0.92	137	26.3
2	80%	21.6	0.84	124	25.6
3	70%	21.4	0.76	111	24.9
4	60%	21.2	0.68	98	24.2
5	50%	21.0	0.60	85	23.5

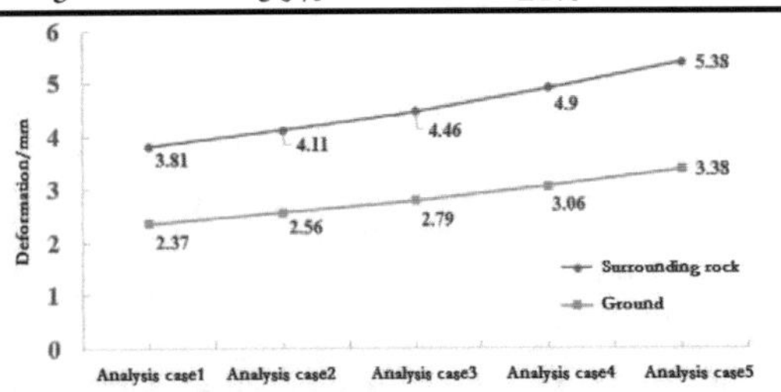

Figure 7. Deformation of 5m curtain grouting ring

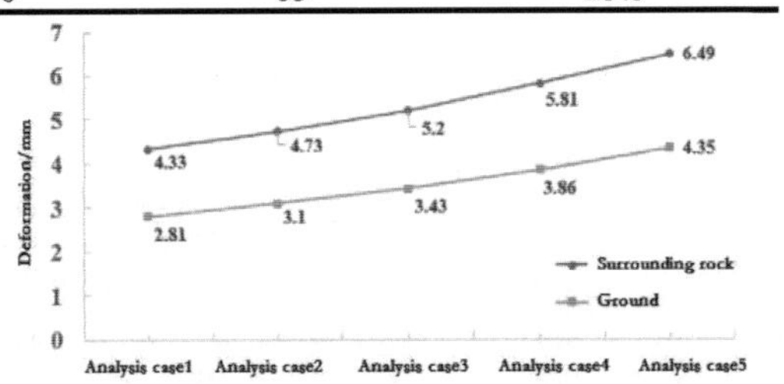

Figure 8. Deformation of 7m curtain grouting ring

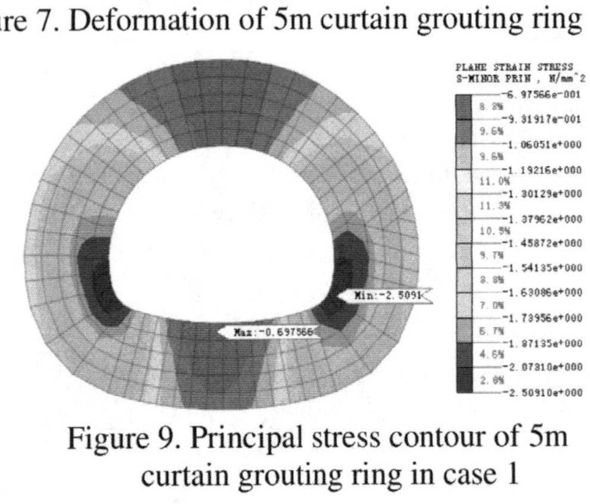

Figure 9. Principal stress contour of 5m curtain grouting ring in case 1

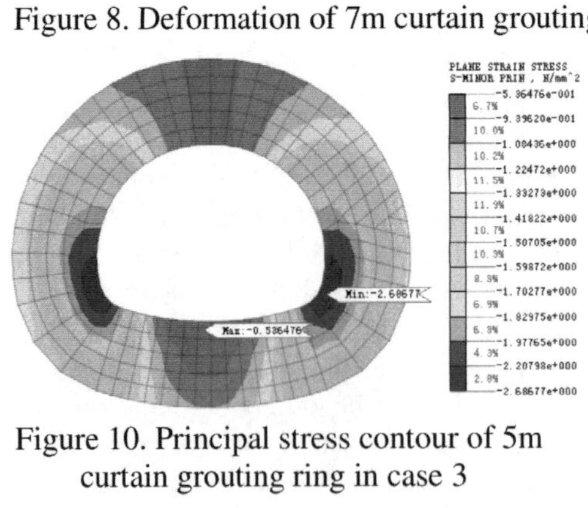

Figure 10. Principal stress contour of 5m curtain grouting ring in case 3

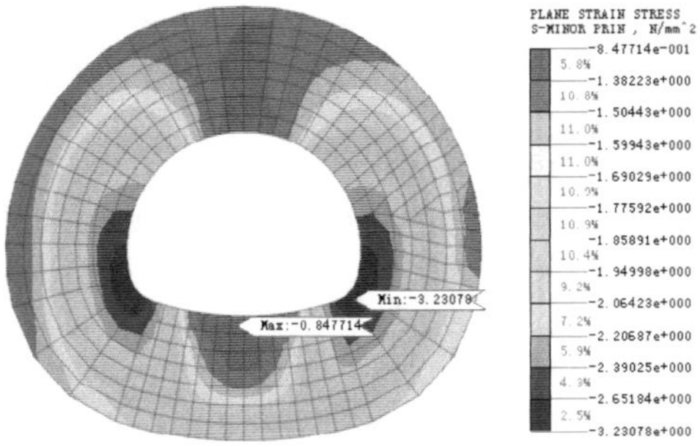

Figure 11. Principal stress contour of 5m curtain grouting ring in case 5

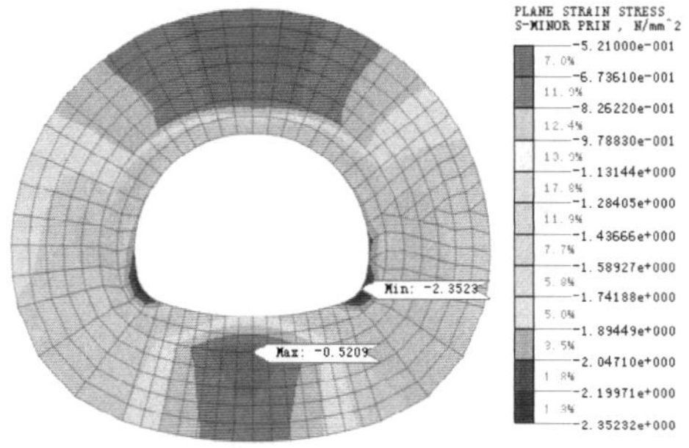

Figure 12. Principal stress contour of 7m curtain grouting ring in case 5

It can be seen from Figure 7 to Figure 12 that:

When the thickness of the full-section grouting reinforcement ring reaches 5m and the parameter is reduced to 90%, the curtain grouting reinforcement ring reaches its critical compressive strength of 2.5MPa; when the parameter is reduced to 50%, the maximum compressive stress value is 3.23 MPa>2.5 MPa, leading to higher failure possibility and further lining problems.

When the thickness of the full-section grouting reinforcement ring is increased to 7m and the parameters are reduced to 50%, the stress value of the corresponding grouting ring is 2.35MPa, still less than the compressive strength value. It can be seen that the increase of the curtain grouting ring thickness can ensure its safety during operation in consideration of the possible material failures.

In summary, 5m-thick full-section curtain grouting can meet the requirements of construction. However, from the perspective of long-term operation, as the grouting effect gradually weakens and the layer penetration increases, there may be confined groundwater after lining, which will have a negative influence on the structure. Therefore, it is feasible to adopt 7m-thick curtain grouting ring, which can effectively ensure the normal working condition of the curtain grouting ring and reduce the possibility of problem occurrence, thereby ensuring the safe operation of the tunnel and greatly reducing the operating expenses.

6. Conclusion and suggestions

Based on the practice of Junchang Tunnel and the fluid-structure coupling calculation model, this paper analyzes the impact of the thickness of the curtain grouting reinforcement ring on the structural safety of the tunnel. The main conclusions are as follows:

(1) The comprehensive advanced geological prediction method combining ultra-high density electric resistance method, terrestrial sonar method, transient electromagnetic method and drilling method can accurately predict the geological conditions of Junchang Tunnel and provide a basis for the construction of curtain grouting construction section. Therefore, it can be applied in similar engineering practices.

(2) Full-section curtain grouting method can effectively reduce the layer deformation caused by tunnel construction. When the thickness of curtain grouting is 5m, the crown settlement and surface subsidence are only 35.5% and 39.5% of that under the non-grouted situation. Based on the thick-walled cylinder formula, the experience of grouting reinforcement and water blocking construction of existing tunnel projects, and finite element comparison, it can be concluded that the 5m-thick full-section curtain grouting reinforcement can meet the structural stability requirements during tunnel construction. This is basically consistent with the existing research results.

(3) During tunnel operation, when the full-section grouting reinforcement ring of the tunnel is 5m thick and the parameters are reduced to 90%, the failure probability of the curtain grouting reinforcement ring significantly increases, leading to shorter service life of the tunnel. When the thickness of the full-section grouting reinforcement ring is increased to 7m and the parameters are reduced to 50%, the safety of tunnel operation can be effectively guaranteed. It can be seen that for the

operation period, the increase of curtain grouting ring thickness can ensure the safety of the tunnel project throughout its life-cycle in consideration of the gradual failure of the grouting curtain.

(4) With the whole life-cycle of the tunnel taken into consideration, the 7m-thick curtain grouting ring is obviously better than the 5m-thick one, as it can more effectively guarantee its normal operation. However, it should be noted that the control from a single index of increasing the curtain grouting thickness has certain limitations. Further research is needed to consider the comprehensive influence of other factors so as to propose corresponding countermeasures.

Acknowledgments
The research work reported herein was made possible by the Guangxi Scientific Research and Technology Development Plan Project (No. Guikegong 1355008-3). The financial support is highly appreciated.

References
[1] ZHANG C.P., ZHANG D.L., WANG M.S., XIANG Y.Y. (2007) Study on appropriate parameters of grouting circle for tunnels with limiting discharge lining in high water pressure and water-enriched region. Chinese Journal of Rock Mechanics and Engineering, 26: 2270-2276.
[2] WANG G. (2014) Study on stability and parameters of curtain grouting reinforcement ring in water-rich strata of tunnel, Jinan: Shandong University.
[3] Liu Q, Li S.C, Li Y.H., ZHANG L.F. (2013) Study on mechanism and treatment of gushing water and burst mudat F2 fault in longtan tunnel. Chinese Journal of Underground Space and Engineering, 9: 1419-1426.
[4] Sorochan E.A., Tolmachev V.V. (2006) A base for building codes and construction methods in karst-prone regions. Soil Mechanics and Foundation Engineering, 43: 216-220.
[5] REN X.H, CHEN X.R., SHAN Z.G. (2004) Water problems and corresponding countermeasures in projects of deep-lying long tunnels located in water-enriched regions. Chinese Journal of Rock Mechanics and Engineering, 23: 1924-1929.
[6] Li L.P. (2009) Study on catastrophe evolution mechanism of karst water inrush and its engineering application of high risk karst tunnel. Jinan: Shandong University.
[7] ZHOU W., LUO G.C., CHEN W.Z. (2008) Key technology study on comprehensive prediction and early-warning of geological hazards during tunnel construction in high-risk karst areas. Chinese Journal of Rock Mechanics and Engineering, 27: 1297-1307.
[8] WANG X.Y., WANG M.S., ZHANG M. (2005) Research on regulating water pressure acting on mountain tunnels by blocking ground water and limiting discharge. Chinese Journal of Geotechnical Engineering, 27:125-127.
[9] YU L., FANG L., DONG Y.C., WANG M.N. (2018) Research on the evaluation method of the hydraulic pressure on tunnel lining according to the range of seepage field. Chinese Journal of Rock Mechanics and Engineering, 37: 2288-2298.
[10] FANG Y., GUO J.N., KANG H.B., XU C. (2016) Influence of voids behind lining on the mechanical behavior of lining structure of highway tunnel in watery strata. Chinese Journal of Rock Mechanics and Engineering, 35: 1648-1658.
[11] ZHANG Z.Q., HE B.G., MA T.F. (2012) Mechanical characteristics of main structure of mining tunnel under water pressure. Journal of Southwest Jiaotong University, 47: 935-941.

Overview of Stability and Failure Characteristics of Expansive Soil Bank Slope

Zhao Xueying[1,2], Zhang Shunfu[1,2]

[1]Research Center on Flood and Drought Disaster Reduction of the MWR, Beijing 100038, China

[2]Research Center on Flood and Drought Disaster Reduction, China Institute of Water Resources and Hydropower Research, Beijing 100038, China

Abstract: Expansive soil is a type of special clay that expands and softens after water absorption and shrinks and cracks after water loss. It tends to cause engineering accidents such as slope deformation and landslide damage. This paper summarizes the interaction between three engineering properties of expansive soil as well as the slope failure characteristics based on case analysis. In this way, this paper can contribute its share to the design, construction and operation management of the expansive soil bank slope.

1. Introduction

At present, the expansive soil is influencing more than 40 countries worldwide, including more than 20 provinces and regions with an area of over 100,000 square kilometers in China [1]. The expansive soil has high particle dispersion and is mainly composed of hydrophilic clay minerals. In the natural state, the expansive soil has high strength. After water absorption, it expands and softens and its strength becomes lower. After water loss, it shrinks and cracks and its strength becomes higher. According to incomplete statistics, the annual economic losses caused by the expansive soil in the development of public facilities, water conservancy projects and houses in the 1970s and 1990s were as high as 10 billion US dollars [2]. For half a century, scholars at home and abroad have made remarkable achievements in the study on physical properties, deformation characteristics and prevention methods of expansive soil. This paper analyzes and summarizes the important engineering properties of the expansive soil and their mutual effect, and explores the failure characteristics of the expansive soil.

2. Engineering properties of expansive soil

The engineering properties of expansive soil mainly include the swelling-shrinkage property, fracture property, over-consolidation property and intensity attenuation property.

Swelling-shrinkage property is an important feature of expansive soil. There are many internal and external factors leading to this feature.

Among the internal factors, the soil microstructure is an important one. Under the same precipitation condition [3], it is found that the plain soil and lime soil that have suffered from long and heavy rainfall show different swelling-shrinkage behaviors: the compaction of the plain soil decreased by about 15% while that of lime soil remains essentially the same. This paper uses the lyophilization method to study the microstructure of the soil [4] and arranges the strength of common soils in China according to their microstructural swelling-shrinkage property: laminar flow structure> turbulent particle structure> matrix particle structure> skeletal particle structure> spongy particle structure> cellular particle structure. This paper also points out that the high Mg^{2+} content exchanged in the expansive soil modified by lime can change the soil microstructure and promotes the formation of cement between the particles, thereby inhibiting the expansion and shrinkage of the soil to a certain extent.

Changes in water content are recognized to be important among all the external factors. A large number of engineering cases show that the bulking failure of the expansive soil bank slope often occurs during and after the rainfall period. The soil swells after absorbing water, which leads to: 1) lower negative pore pressure of the surface, and the effect force decreases with the decrease of the surface suction; 2) softer soil and lower shear strength. Under the above influence, the soil strength is reduced, causing landslide instability.

The fracture is the key factor leading to the instability of the expansive soil. Fractures can not only break the integrity of the soil, but also provide a channel for water infiltration, leading to a rapid increase of water content of the soil near the fracture, a rapid decrease of effective stress, and a great decrease of shear strength. The fracture surface becomes the weak surface of the soil, which further becomes the cause of landslide accidents.

Under the effect of gravity, the expansive soil repeatedly expands, shrinks, and solidifies during the deposition process, leading to over-consolidation property. The over-consolidated expansive soil is featured with low water content, small void ratio, large dry density and high structural strength.

The engineering properties of expansive soil interact with each other. The swelling-shrinkage property causes soil water loss and secondary cracks. The infiltration of water along the fracture accelerates the softening of the internal soil. The over-consolidation property increases the development of the fracture and promotes the softening of the soil. Therefore, swelling-shrinkage property is an internal factor; the fracture property is the key factor affecting the mechanical properties of the soil and the slope stability; the over-consolidation property is the factor stimulating the failure. In this way, the three properties interact with each other.

3. Failure characteristics of expansive soil

Pull-type shallow damage: The front edge of the slope body firstly undergoes sliding deformation, and then gradually develops upward to pull the rear slip mass. According to the statistics of landslide cases, the slope surface shows faulting of steps and sometimes multi-layer sliding surface. This is because secondary cracks begin to develop during the drying and watering cycle [5], gradually destroying the integrity of the soil, leading to lower shear strength of the slope. If the soil strength continues to decrease after the first sliding, subsequent damage will also occur. Sliding will stop only after a new balance is reached.

Generally speaking, the soil 3m beneath the surface layer is significantly affected by the external climate. In a hygrothermal environment, multiple drying and watering cycles destroy the overall soil structure and develop fractures, leading to the shallow damage with the depth of the landslide surface between 2 to 4 meters in the expansive soil slope. Anchor bolt support can be used for reinforcement. The numerical simulation [6] shows that a certain number of anchors arranged at the foot of the slope can deliver a satisfactory supporting effect and reduce the shallow damage.

Seasonal Law: Based on the cases of expansive soil slope landslide, it is not difficult to find that most of these damages occur during the rainfall period. The first heavy rain after drought tends to cause large-scale landslide damage in the expansive soil area, indicating that the soil water content is closely related to the mechanical properties of the expansive soil. Lei et al. [7] found that strong rainfall can cause severe sliding of the bank slope, the rainfall was closely related to the landslide development, and that power function can be used to fit the relationship between the superficial displacement of the monitoring point at the expansive soil bank slope and the amount of rainfall based on the observation of surface displacement of the expansive soil bank slope during the construction period. If the fitting function of the expansive soil in the monitoring section is to be found through experiments, we can set monitoring points at the top, middle and bottom of the bank slope, and predict the amount of landslide displacement of each part based on rainfall monitoring.

Structural characteristics: The fractures of different genesis and occurrence are gradually connected to each other, forming a weak structural layer with a certain thickness inside the soil. Under the external loads (such as excavation unloading) and the change of moist heat conditions, progressive failure occurs, which results in crack penetration and a sliding surface. Under this circumstance, the slope slides under the action of gravity. For example, the excavation unloading changes the stress state of the over-consolidated soil and makes the soil structure more relaxed. The unloading crack is generated on the slope, and the shear strength of the soil is reduced.

4. Conclusion

Deformation and failure of expansive soil pose great difficulties on engineering construction. Since there are many factors affecting soil stability, appropriate preventive measures should be taken in the stages of planning, design, construction, and operation management so as to reduce landslide damage. When the economy and technology permit, soil backfill or improved soil can be adopted in the planning and design stage so as to eliminate or reduce the influence of soil expansion and shrinkage. Anti-seepage measures can be taken at the foundation to prevent external water infiltration, while anchors can be used at the foot of the slope to prevent shallow landslide damage. During the construction, the excavation speed and the foundation and joint treatment should be controlled to ensure that the soil is fully compacted so as to prevent the occurrence of uneven settlement, cracks or seepage channels in the later stage. In the weak structural surface and areas with high landslide possibility, we should use equipment to monitor the seepage pressure and deformation and strengthen daily inspection. Once an abnormal phenomenon is found, we should take effective measures in time to avoid further damage.

Acknowledgments
This paper is supported by National Key R&D Program of China (2017YFC1501202).

Reference

[1] Li Tielin, etc. Engineering geological study of expansive soil in China. Nanjing: Phoenix Science Press, 1992.
[2] Liu Tehong. The problem of expansive soil in engineering construction. Beijing: China Architecture and Building Press. 1997.
[3] Shi Jiangang, Chen Shanxiong, Yu Fei. The effect of rainfall on constructing quality of expansive soil subgrade and its counter-measures. Soil Engineering And Foundation, 2006(01):3-6.
[4] Li Shenglin, etc. Engineering geological study of expansive soil in China. Chinese Journal of Nature, 1997(02):82-86.
[5] Geng Jianbin. The development and its prevention of ground crack in the expansive soil area. Geotechnical engineering technique, 1998(01):21-23.
[6] Hou Dinggui, etc. Research on stability and anchor support parameters of expansive soil slop under the condition of seepage. Construction technology, 2015, 44(24):77-80.
[7] Lei Peiyun, etc. Superficial Displacement Monitoring During the Construction of Expansive Soil Landslide. Chongqing Architecture, 2019,18(02):42-44.

IOP Conf. Series: Materials Science and Engineering **794** (2020) 012020 doi:10.1088/1757-899X/794/1/012020

Experimental study on seismic performance of shear walls with steel fiber recycled concrete

Xiaohu Li[1*], Lei Li[1] and Tianfeng Chang[1]

[1]School of Civil Engineering and Architecture, Zhengzhou University of Aeronautics, Zhengzhou, China

[*]E-mail: xiaohu12066@126.com, lileixiaozhou@163.com, 447200844@qq.com

Abstract. The paper studies the seismic performance of shear walls with steel fiber recycled concrete. Four shear wall specimens were designed and manufactured in this paper and tested under low cyclic loading. The effects of waste fiber and recycled aggregate on the seismic behavior of shear wall were studied. The results show that the incorporation of waste fiber can delay the generation of cracks in concrete. The incorporation of waste fiber can improve the bearing capacity and ductility of shear wall to some extent, while the recycled aggregate has little influence on the bearing capacity and ductility of shear wall.

1. Introduction

With the rapid development of urbanization, a large number of buildings have been demolished and rebuilt, at the same time, a large amount of construction waste such as waste concrete was generated. In order to treat the waste concrete reasonably and achieve the purpose of effective utilization of resources and sustainable development of society, many scholars at home and abroad[1-3] used waste concrete and other wastes to replace natural aggregate of concrete to prepare recycled concrete. However, due to a large number of micro-cracks and more internal defects such as cement slurry on the surface of the recycled aggregate in the crushing process, the strength, toughness and cracking resistance of the recycled aggregate concrete are reduced to different degrees, so its application in practical projects is limited to some extent. According to the research data at home and abroad[4], the waste fiber processed from textile waste could improve the mechanical properties of recycled concrete in a certain range.

At present, most scholars[5-6] at home and abroad focus on the research of recycled aggregate concrete and fiber concrete [7-8]respectively. Most of the research on fiber recycled concrete is about the study of its material properties. There are not many reports on the research of fiber recycled concrete components. This paper has carried out experimental research on the seismic performance of the shear wall with steel fiber recycled concrete, which provides a reference for the research on waste fiber recycled concrete components and lays a theoretical foundation for the application of waste fiber recycled concrete in practical projects in the future.

2. The experiment

2.1Design of mix proportion

The recycled coarse aggregate used in this test is made by artificial breaking of the abandoned concrete used in the laboratory test of the monitoring station. The strength of the original concrete is C40. The waste fiber is made of recycled waste polypropylene carpet, which is cut manually and the length is 19 mm. The natural coarse aggregate is gravel, and the particle size of crushed stone is the same as that of recycled coarse aggregate, which is 5 ~ 20 mm. The cement is 42.5R grade Portland cement. Specific concrete mix ratio is shown in Table 1.

Table 1. Mixture ratio of concrete.

Number	Material usage/kg/m^3					Replacement rate of regenerated coarse aggregate/%	Waste fiber	
	cement	sand	gravel	recycled coarse aggregate	Water		Volume ratio/%	The length of fiber/mm
1	434	706	884	295	206	25	0	19
2	434	706	884	295	206	25	1	19
3	434	706	884	589	206	50	0	19
5	434	706	884	589	206	50	1	19

2.2Design of mix proportion

In order to study the effect of waste fiber incorporation on the seismic performance of shear walls with steel fiber recycled concrete, four shear wall specimens were made.The bottom of the specimen is the foundation beam of C30 ordinary concrete, the middle part is the shear wall with steel fiber recycled concrete, and the upper part is the loading beam made of C30 ordinary concrete. The specimens were numbered SWRC1, SWRC2, SWRC3 and SWRC4. All specimens were of the same size. The size of the specimens was shown in Figure 1. The design scheme of the specimen is shown in Table 2.

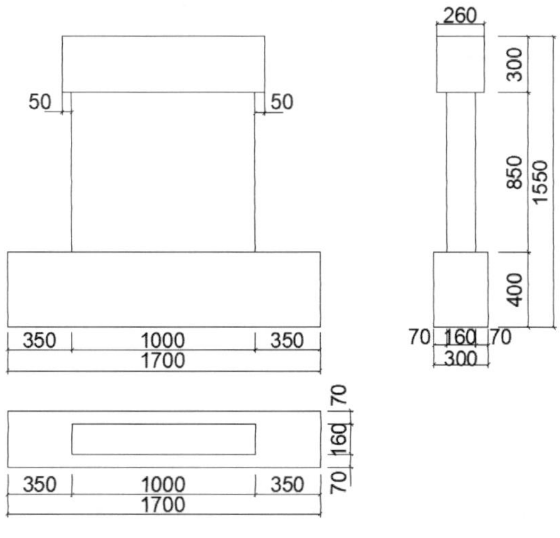

Figure 1. Size of specimen

Table 2. Mixture ratio of concrete.

Number	Volume ratio of waste fiber（%）	Volume ratio of regenerated coarse aggregate（%）
SWRC1	19	25
SWRC2	0	25
SWRC3	0	50
SWRC4	19	50

2.3Test loading device and loading system.

In this test, the hydraulic loading system in the structural laboratory was adopted for the test, and the load-displacement mixed control loading method was adopted. The force is measured by a force transducer during loading. During the test, the hydraulic loader can move horizontally with the deformation of the specimen, so as to ensure that the vertical force test is always in the straight direction, and to minimize the horizontal friction at the support on the top of the hydraulic loader caused by the deformation of the specimen, so as to ensure the correct force boundary conditions of the specimen. The vertical load was calculated as 800kN. Before the repeated horizontal load V loading test, the vertical load was applied to the predetermined value of 800kN and remained unchanged. After the vertical load is stabilized, horizontal load is applied. The horizontal load is controlled by the mixed load and displacement. Each load is 50kN and the cycle is repeated once, increasing step by step until the specimen yields. Displacement loading began after the yielding of the specimen, and record the horizontal displacement as Δy when the specimen was in yield. The step distance of displacement control is 1/8 of Δy. When the horizontal load reached 85% of the peak load of the specimen, the loading stoped and the test ended. The test loading device is shown in Figure 2.

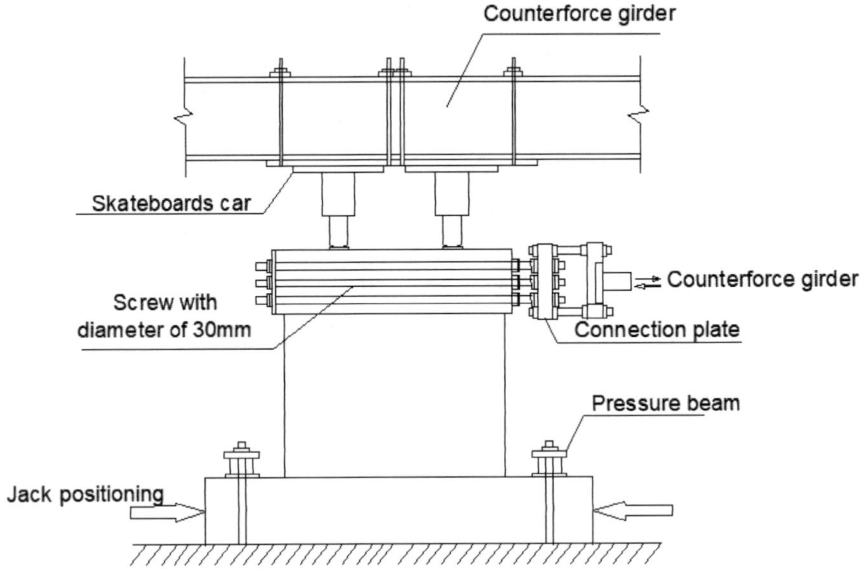

Figure 2. Schematic diagram of loading device

3. Experimental phenomena and results

SWRC2 and SWRC4 have shear failure, which is a relatively normal failure mode of low shear wall. The horizontal cracks are relatively rare and appear at the root of the wall first. When the horizontal load reaches the ultimate load, the original horizontal cracks begin to tilt downward at an angle of 45 degrees. There are inclined cracks in the wall above the wall height. Under the action of repeated load, the "X" shaped shear crack is formed in the core of the wall.

SWRC1 and SWRC3 were damaged by bending shear. The cracks are very thin and appear later, mainly at the root of the wall. There are many horizontal cracks in the wall, which develop rapidly along the root and finally run through. In the case of failure, the corner concrete is crushed, and then the longitudinal reinforcement of the concealed column yields. There are two main reasons for the analysis: first, it is caused by the difference of pouring time. When pouring, the base shall be poured first. Although a rough surface is reserved at the top of the bottom beam, the final part of the wall body and the bottom beam is still insufficient. Second, there is an additional moment effect. During the loading process, the absolute level of the actuator can not be guaranteed, so there will be an additional bending moment. However, this kind of damage can be avoided in the test as long as the construction technology is ensured and the reinforcement of concealed column is increased properly.

From the experimental phenomenon in the loading process, it can be seen that the development form of inclined cracks of recycled concrete shear wall with waste fiber are similar to that of ordinary concrete shear wall. However, due to the constraint of waste fiber on the wall cracks, the wall cracks appear later and develop slowly.

From the final failure mode of the test, it can be seen that the concrete damage in the compression zone of the shear wall of recycled concrete is relatively serious, and the trend becomes more obvious with the increase of the replacement rate of recycled aggregate. This may be due to the damage accumulation in the early stage of recycled concrete, and the strength and stiffness degradation in the later stage is faster. The failure modes of compression zone of some specimens are shown in Figure 3.

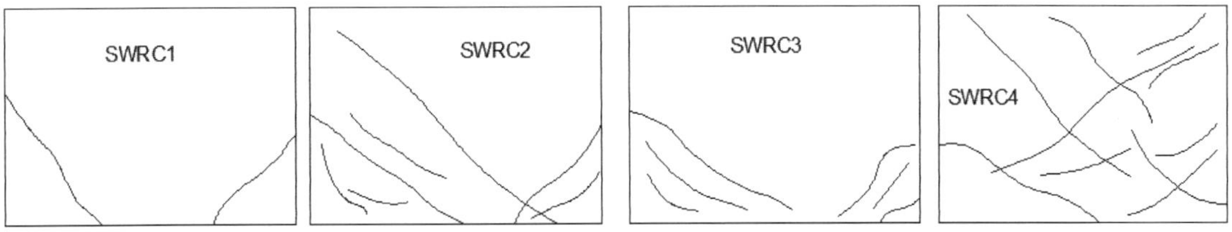

Figure 3. The distribution of cracks

4. Analysis of seismic performance

4.1Analysis of bearing capacity

The measured values of cracking load, yield load and ultimate load of each specimen are listed in Table 3. In Table 3, F_{cr} is the cracking load of the specimen; F_y is the yield load of the specimen; F_u is the ultimate load of the specimen, and $\mu_{yu}=F_y/F_u$ is the ratio of yield load to ultimate load, that is, the flexion strength ratio.

Table 3. Comparison of measured results of load of each specimen.

Number	F_{cr}(kN)	F_y(kN)	F_u(kN)	μ_{yu}
SWRC1	316.07	409.42	461.44	0.68
SWRC2	283.56	389.50	438.17	0.64
SWRC3	300.35	380.59	403.77	0.73
SWRC4	330.86	418.42	472.68	0.66

As shown in Table 3:

(1)The replacement rate of recycled aggregate of SWRC1 is the same as that of SWRC2, the former is added with waste fiber. The yield load and ultimate load of SWRC1 are 5.1% and 5.3% higher than those of SWRC2 respectively. It shows that the waste fiber has a certain influence on the bearing capacity of recycled concrete shear wall.

(2)Compared with SWRC2, the ultimate load of specimen SWRC3 decreased by 4.1%, and its yield load decreased by 2.3%. It can be seen that the content of recycled concrete has little influence on the bearing capacity of recycled concrete shear wall.

4.2Result analysis of ductility performance

The measured displacement and ductility coefficients of each specimen are shown in Table 4. In Table 4, U_{cr} is the horizontal displacement of specimen cracking; U_y is the yield horizontal displacement; U_u is the ultimate displacement; $\mu=U_u/U_y$ is the displacement ductility coefficient.

Table 4. The measured load values of specimen.

Number	U_{cr}(mm)	U_y(mm)	U_u(mm)	μ
SWRC1	1.10	3.18	21.56	6.78
SWRC2	0.94	3.12	20.17	6.46
SWRC3	0.81	3.46	18.97	5.48
SWRC4	1.24	3.72	22.60	6.07

As can be seen from Table 4:

(1) The replacement rate of recycled aggregate of SWRC4 is the same as that of SWRC3 , and its ductility coefficient is increased by 10.9%. It shows that the waste fiber can enhance the ductility performance of recycled concrete shear wall. This is because the waste fiber slows down the development of wall cracks and improves the deformation capacity of wall.

(2) The ductility coefficient of SWRC3 is 18% lower than that of SWRC2. It can be seen that the content of recycled coarse aggregate has a certain influence on the bearing capacity of recycled concrete shear wall.

5. Analysis of seismic performance

This paper mainly introduces the test phenomenon and result analysis of 4 recycled fiber reinforced concrete shear walls, and the bearing capacity and ductility of each specimen are compared and analyzed.

The experimental analysis shows that:

(1) The performance of the recycled concrete shear wall mixed with recycled coarse aggregate is basically similar to that of the ordinary concrete shear wall in bearing capacity and other aspects, and it has good seismic performance. So it is feasible to use recycled coarse aggregate to replace natural aggregate in shear wall structure.

(2) When the replacement rate of recycled coarse aggregate is less than 50%, the failure characteristics and degradation process of recycled concrete shear walls with different replacement rates are basically the same. The replacement rate of recycled coarse aggregate has little impact on the performance of recycled concrete shear walls.

(3) After adding waste fibers, the ductility of recycled concrete shear walls is improved. With the increase of load during the loading process and the development of cracks, the later deformation capacity of walls is improved.

Acknowledgments

The authors gratefully acknowledge funding and support provided by Science and Technology Project of Henan Province (19A560025) and NSFC(U140454).

References

[1] XIAO J, LAN Y.(2006)Experimental study on flexural properties of recycled coarse aggregate concrete beams[J].Special Structures. 23(1): 9-12.

[2] LIU Y S, WANG X J, JIN T et al.(2007)Study on the mechanical properties and constitutive relation of steel fiber reinforced concrete[J]. Journal of university of science and technology of China.37(7) : 717-723.

[3] XU C K, FEI C C.(2016)On anti-crack performance of regenerated concrete beam with various organic fibers[J].Shanxi Architecture.42(2) : 109-111.

[4] ZHOU J H, ZHANG D, YANG Y S.(2013)Test study on flexural properties of waste fiber recycled concrete beams[J]Journal of Shenyang Jianzhu University. 29(2): 290-296.

[5] LIU S, LENG F.(2007)Technology of recycled aggregate concrete[M].China Building Ma- terials Press.Beijing.

[6] Tanaka R, Miura S, Ohaga Y.(2002)Experimental study on the possibility of using permanently recycled concrete for reinforced structures[J].Journal of the Society of Materials Science.51 (8) : 948-954.

[7] LI G W, YANG Y H.(2001)Experimental study on characteristics of polyacrylic fiber concrete[J].Advances in Science and Technology of Water Re- source.(5);15-18.

[8] Andrzej M B.(2008)Fiber reinforced cement-based(FRC) composites after over 40 years of development in building and civil engineering[J].Composite Structure.86;3-9.

IOP Conf. Series: Materials Science and Engineering **794** (2020) 012021 doi:10.1088/1757-899X/794/1/012021

Research on Nanjing Ancient Urban Morphology Evolution Based on Climate Adaptability

Xu Ruofei[1]

[1]Departments of Architecture, University of Southeast University, 210096, China

*Corresponding Author: Xu Ruofei; email: 173065607@qq.com;

Abstract: Evolution mechanisms and rules of urban morphology were explored mostly from cultural, historical and geographical aspects, but rarely from the aspect of climate adaptability. From the aspect of historical geography, the transition of the Nanjing city came out as the layout changes of the natural and social phenomenon, and combined with the gambling and impelling of the relationship between the human and the nature. The research takes the ancient Nanjing city as an example to analysis the climate adaptability from the site selection and change of the urban morphology., the change of site selection and urban river system were caused by the climatic change, which aroused the continuous change of the urban morphology.

1. Introduction

Researchers of Urban morphology mostly had been doing researches based on cultural, historical and geographical sights to explore the evolution of urban morphology, rarely based on the climate adaptability. Schlüter (1899), Sauer (1925) and Conzen (1960) made fundamental contribution to the research of urban morphology.[1] Whitehand treated the research of urban morphology as to explore the relationship between the physical form and the function of urban society.[2][3]Although most area of urban are covered with green space, they obtain less attention from the aspect of urban morphology with the growing acknowledge of their importance. In historical process, climate adaptability existed in the formation and evolution of urban morphology. [4]Researches have explored the spatial relationship between the temperature and morphology of urban green space.[5] This research intends to settle the relationship between the urban morphology and climate adaptability in history. The method and content of urban morphology have not been settled down in academia. [1] The concern of urban fringe belt in China have been increased prominently in recent decades, which proved its weakness in worldwide. [6]According to this research, the change of fringe belt in Nanjing urban morphology-the river bank has certain relevance with the climatic change. The paper attempts to reveal the evolution of urban morphology and the climate adaptability, takes Nanjing as an example and tries to find the relationship between these two items.

2. Material and Methods

2.1 Climate adaptability in site selection

The cite selections of cities in ancient China contained multiple dimensions of wisdoms. The profitable cite selection was "Hidden from the wind and enjoying the water", and was benefit for the eco-system, also was an urban morphology harmonized with topography, climate, ecology and landscape. Wu Qingzhou has concluded the thoughts of cite selection to be: choosing the sky, like the sky and benefit situation. Situation superiority was the consideration of geographical elements, natural environment, and built the city in the cite which was most benefit of defense and production. The city of Nanjing is more suitable for the situation superiority because of the special geographical morphology in the process of being a capital and prefectural city.[7]

The urban planning system of the cities in ancient China was roughly in shape in the early period of Xi Zhou (11th century BC). In the late years of Spring and Autumn period, the official documents of Qi-<*Book of Diverse Crafts*>, one chapter <*Craftsman. Planning the City*> had overall recorded about the construction of city for the first time.[8] "*Craftsman planned the city, nine miles in square, three entrances by the side, nine longitudes and nine latitudes in the middle, nine tracks in the road, ancient temple in the east and altar of land and grain in the west, palace in the front and market in the back.*" These ancient planning systems have not been proved by archaeological evidences, but the cites like Jin Houma, Yan Xiadu and Zhao Handan had palaces as their core and had streets in order which can prove the <chapters of craftsman>'s truth.[9]

2.2 the climate adaptability of Nanjing's Cote selection

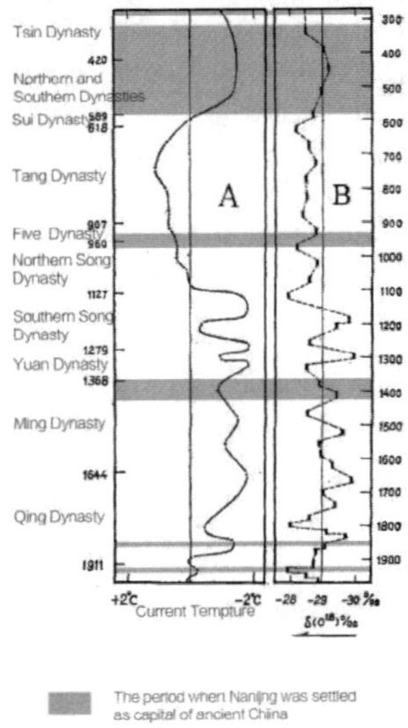

Figure1. The world temperature undulating during the past 1700 years and the period when Nanjing Settled as the capital of China

Figure2. The historical sites of Nanjing and the relationship with the water system

There has been a long time for ancient citizens to utilize the natural resources of the mountains and rivers for the urban planning of the Nanjing city. There were many examples to choose the cities' settlement based on the geographical and topographical advantages, Nanjing was a tropical one. [3] As recorded in <*Jing Ding Jiankang Zhi*>:" The city of Jinling (Nanjing) was located along the Yangze River and took advantage of the terrain. The western part of the city occupied the Stone Mountain and connected to the Changan Mountain." [10]When Zhu Yuanzhang settled Nanjing as the capital of Ming Dynasty, he was trying to utilize the natural resources to construct the city which was advised by Liu Ji. [11](Figure1. The world temperature undulating during the past 1700 years and the period when Nanjing Settled as the capital of China, Figure2. The historical sites of Nanjing and the relationship with the water system)

3. Results

3.1 The evolution of Qinhuai River

3.1.1The change of the Qinhuai River's width

Qinhuai River was quite broad in the ancient times, but was used to be called "small river". In the period of Three kindoms, there were lots of troops stationed in Nanjing because of the broad of Qinhuai River.The high density of population, intensified efforts made the width of Qinhuai River become narrow continously. According to the archaeological discoveries, the distance between the wharfs in two sides was 100 meters approximately in the Six Dynasties, and the width of one side became five meters narrower in Southern Tang Dynasty, the wharf became lower than whadf in the Six Dynasties; the river's width became narrow roughly, about 20 meters by one side, and the river's mud revetment was one meter lower than in Southern Tang Dynasty. Because of the dry season, the width of Qinhuai River became only 50 meters left. The empty revetment was occupied by the citizens and became narrower to be 20 meters. [12]The change of the Qinhuai River's width and the decline of water level were relevant to the chilling in the 12th century in ancient China. [9] (Table1. The width of Qinhuai River in history)

Table1. The width of Qinhuai River in history

Dynasty	Six Dynasty	Southern Tang Dynasty	Song Dynasty	Recent
The distance between the two revetments	100	90	50	20
The height of the ground		Slightly lower than in Six Dynasty	One meter lower than in Southern Tang Dynasty	

Figure3. The Qinhuai river system in 1927

Figure4. The Qinhuai river system in 1967

Figure5. The Qinhuai river system in 2010

3.1.2 The urban morphology changes in both sides of Qinhuai River

Before the beginning of modern times, waterway transportation was in the domain place, the urban morphology was influenced by the waterway mostly, road network was related with the river.There was intense waterway transportation network in Qinhuai River, some of the waterway vanished and some of them became road network. [13]With the development of the transportation, the relevance with the waterway became less until it vanished finally. (Figure6. The urban morphology of the riverside of Qinhuai River in Nanjing in 1927. 2 kilometers X 2 kilometers grid. Figure7. The urban morphology of the riverside of Qinhuai River in Nanjing in 2010. 2 kilometers X 2 kilometers grid.

Figure6. The urban morphology of the riverside of Qinhuai River in Nanjing in 1927. 2 kilometers X 2 kilometers grid.

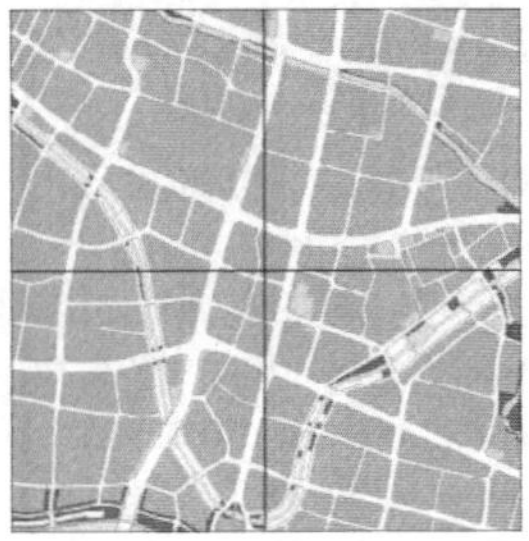

Figure7. The urban morphology of the riverside of Qinhuai River in Nanjing in 2010. 2 kilometers X 2 kilometers grid.

3.2 The transformation from Lake to farmland in Xuanwu Lake

3.2.1 The origin of the Xuanwu Lake

In the end of Quaternary period, the surface of the Nanjing was rolls-and-swells, there were obvious basins. The center of northern basin was in the present position of Xuanwu Lake, and was slightly close. There were two water outlsets in different directions in Xuanwu Lake, which were the reflection of the ancient water relics. The natural drainage of Xuanwu Lake recorded in <*Tong Zhi Liang Jiang Zhi*> and <*The Chorography of the Capital*> were: "*One of the water was along Lulong Mountain to Yangtza River, another one flew to the city through the water gate in Taicheng. There were four waterlocks which were controlled by the water level.*"

3.2.2 The changing of the morphology of Xuanwu Lake

Xuanwu Lake was formed in ancient times, the morphology and function were changed a lot because of the climate. In the cold period of Northern Song, Wang Anshi1thought "*Xuanwu Lake was just a place of interests for the former period, but was empty in this time, and there was no meaning to observe it.*" The magistrate of Shenzhou-Ding Wei had submitted documents about the scale of Xuanwu Lake. According to the record in <*History of the Song Dynasty. The chorography of waterways*>, the water was dried up which was the record of the dry up for the first time. Parts of the lake surface turn into farmland after drought. After the proposal of Ding Wei, Xuanwu Lake was swept again and became free life pond in 1020.[14]

The morphology change of Xuanwu Lake was related to the floods caused by the increased precipitation. According to the record in local chronicles, the first flood happened in 1118, 42 years after the abolish of the Xuanwu Lake in Northern Song Dynasty. The second flood was happened in 1163, about 87 years after the abolish of the Xuanwu Lake. As to the natural disaster like flood, Xuanwu Lake could not protect itself. Xuanwu Lake was dredged only in 1301 and 1342 in Yuan Dynasty (1271-1368), and remained to be as an abolish pool. [13] The record in <*Jing Ding Jiankang Zhi*> in Southern Song Dynasty was not good about Wang Anshi's abolish of Xuanwu Lake: "*The production in land could benefit less people, while the situation in hydraulic engineering benefit most.* "Which meant compared with the flood, the production made by the lake's transition into the farmland was very little. (Figure 8. The section change of Xuanwu Lake in history)

[1]Wang Anshi (1021-1086), a famous ideologist, politician, litterateur in Northern Song Dynasty (960-1127). Wang Anshi had been the magistrate of Jiangning Province (Nanjing) after he retired as a prime minister.

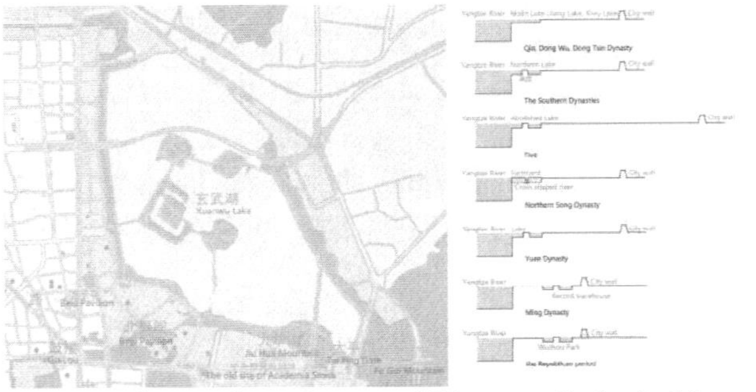

Figure 8. The section change of Xuanwu Lake in history

3.3The morphological change of the river bank

The morphological change of the Yangtze river bank was influenced by the joint effort by the deposit and construction. Because of the deposit, there were several islands formed along the area of river bank. People built several rafts to get into the island in order to take advantage of the islands. The rafts piled up along the river bank which could form the giant island[6] (Figure 9. The morphology comparison of Nanjing river bank according to the Wu region map in <*Jinling Gu Ji Tu Kao*>, <*Jinling Tu Kao*>, Jiangning city in <*Jiang Nan Tong Zhi*>in Qianlong period.)

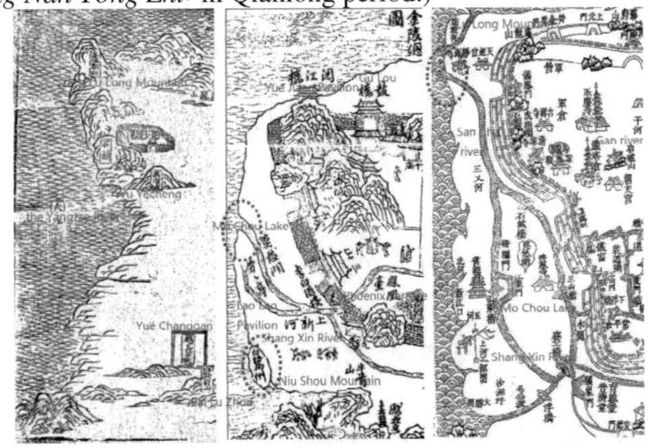

Figure 9. The morphology comparison of Nanjing river bank according to the Wu region map in <*Jinling Gu Ji Tu Kao*>, <*Jinling Tu Kao*>, Jiangning city in <*Jiang Nan Tong Zhi*>in Qianlong period.

4.Discussion

The climatic change in the ancient times was one of the reasons why Nanjing was chosen as the capital of China for several times. Wu (2000) once pointed out that the cite selection was a comprehensive research which contained the knowledge of astronomy, geography, culturology, religious studies, geology, meteorology, hydrology, hydraulics, shipping, biology, ecology, military science, science of disaster. [7]

In the evolution of city, there was natural force like the deposit to change the shape of the river bank in Nanjing. There were ten functions in city water system: military defense, water supply, improving the city environment, transportation, irrigation, preventing the flood, and avoiding the wind and waves, greening and entertainment. [15]This paper takes Nanjing as an example and discusses the climate adaptability in the evolution of the ancient city and the changing rules during the change of the climate.

5. Conclusions

The natural resources influenced the cite selection and construction of the Nanjing city, and formed the unique view of its landscape. During the past thousands years, Nanjing was chosed as the capital of ancient China for ten times because of the climate changes. This paper tries to expolre the climate adaptability from the changes of the width of Qinhuai River, the change of the Xuanwu Lake and the river bank of the Yangtze River in morphological aspect

The paper also tries to analyze the effort between the city and the nature and tries to explore the ancient ecological philosophy inside. In the process of the city's development, the cite selection and the morphological change based on climate adaptability should be considered. Exploring the innovation and rules of the city from the aspect of climate adaptability could also provide advice for the future's urban constructions.

References

[1] Gu Kai. (2001) The theory and method urban morphology—exploring an overall and reasonable research frame[J]. City Planning Review, 25(12):36-41.

[2] Whitehand, J. W. R., & Gu, K. (2006). Research on Chinese urban form: Retrospect and prospect. Progress in Human Geography, 30, 337–355.

[3] Whitehand, J. W. R. (2019) Green space in urban morphology: a historico-geographical approach. Urban Morphology, 23(1),5-17.

[4] Zeng Zhongzhong. (2011) The ancient research on Chinese ancient cities' morphology based on climate adapality. (Doctoral dissertation).

[5] Wang, XJ, Qian, Y, Zou, H. (2015). The Effects of Green Spaces Form on Urban Thermal Field Pattern. 4th International Conference on Energy and Environmental Protection (ICEEP). Shenzhen. 3583-3586.

[6] Whitehand, J. W. R., & Gu, K. (2017) Urban fringe belts: evidence from China. Urban Analytics and City Science, Vol. 44(1) 80-99.

[7] Wu Qingzhou. (2000) The historical experience and usage for reference in the cite selection of ancient Chinese cities (second part) [J]. City Planning Review, (10):39-46.

[8] He Yeju. (1996) The history of urban planning in Chinese ancient cities. Chinese Architectural Industrial Press. Beijing. pp. 204.

[9] Liu Dunzhen. (1984) The History of Chinese Ancient Architecture. Chinese Architectural Industrial Press. Beijing. pp. 36.

[10] Zhou Yinghe. (2009) Jing Ding Jiankangzhi. Nanjing Press. Nanjing. pp. 78.

[11] Xu Ruofei, Wang Xiaojun. (2019) Urban Green Space Research of Southwestern Nanjing based on Historical Documents in Digital Age. Cheng Yuning, Yang Rui. Digital Landscape Architecture: The Fourth Digital Landscape Architecture Forum[M]. Southeast University Press. Nanjing. pp. 304.

[12] Hu Maochuan, Zhang Xingqi, Maxiao. (2012) The conservation and characteristic of the river house along the Qinhuai River in Nanjing. Sichuan Building Science, 38(03):292-295.

[13] Xue Bing. (2015) History of Nanjing City[M]. Nanjing Press. Nanjing. pp. 205.

[14] Liu Zongyi. (2001) Why Wang Anshi abolish the Xuanwu Lake[J]. Jiangsu Chorography, (1):36-37.

[15] Wu Qingzhou. (1991) The urban water system in ancient China. Huazhong Architecture (02), 49+62-68.

Study on pore fabric, dilatancy, dissipation function and yield function for sand

Zhang Weihua[1*], Zhao Chenggang[2], Zhang Yinping[3]

[1]Urban Construction School, Beijing City University, Beijing, 100083, China

[2]School of Civil Engineering, Beijing Jiaotong University, Beijing, 100044, China

[3]Handan Power Supply Company, Handan, Hebei, 056002, China

*Corresponding author's e-mail: 18910610508@163.com

Abstract. The mechanical behavior of sand is affected by the anisotropic fabric. Integated with the micromechanics theory, and linked with void fabric tensor, the relations among macroscopic volumetric strain, shear strain and void fabric are established, and it is further shown that dilatancy and anisotropy under applied shear force of sand are necessarily induced. By means of thermomechanics theory, the inclined elliptical yield function in dissipation space is built, and considering the effects of anisotropic fabric, there exists three important lines, just as the normal consolidation line, phase transformation line and peak value line, and corresponding with three characteristic lines passing through the origin in true stress space which relatively truly reflect the mechanical characteristics for sand, especially dense sand.

1. Preface

The mechanics of granular materials is widely used in engineering, but this theory is still immature. There are still many challenging problems to be studied. There are generally two methods to establish the constitutive model of granular materials, namely the continuum mechanics method and microscopic mechanical methods. At present, most of the existing constitutive models of granular materials are based on continuum mechanics and are used to describe the mechanical properties of soils, such as the use of soil as an isotropic continuum, and to simulate the behavior of soil. Many valuable theories, such as the framework of critical state soil mechanics, are based on this; the second is based on the method based on micromechanics which considers soil as having regular and irregular arrangements. Aggregates of discrete particles. In models, the contact distribution of particles usually uses probability distribution functions to reflect their anisotropy (Horne, 1965)[1]; Oda, 1972a & b[2][3]).

Since the continuous plastic theory can not directly explain the microscopic properties of the irreversible shape of the granular material, the detailed study of the microscopic properties of the granular material is mathematically complicated, which leads to great limitations in the application of this method. This paper starts from the theory of particle mechanics. The pore fabric of sand is discussed, and the relationship between pore structure anisotropy, dissipation function and real stress space yield function is analyzed in detail with thermodynamic theory. The theory of establishing boundary surface constitutive model based on thermodynamic theory is further discussed. frame.

2. Strain tensor expression based on pore fabric

For granular materials, the most interesting study is not the reaction of individual particle but how the particles are arranged in space under external force and their overall reaction. So the mechanical behavior of the granular material is closely related to its microstructure. Here, we assign the microscopic and mesoscopic magnitudes of the particles to the microstructure. In the triaxial compression experiment on sand, Oda (1982)[4] observed that the granular material had different strength due to the different compression direction. Moreover, he observed that the long axis direction of the non-spherical particles tended to rotate perpendicular to the maximum compressive stress, so the mechanical behavior of the granular material is affected by its microstructure. Generally void ratio e or porosity n indicates the degree of compaction of the granular material structure, but these quantities do not indicate the directionality of the granular material structure. The internal structure of the randomly distributed granular material can be described by the geometric statistical properties of the particle and its spatial distribution, The higher-order microstructure variable $\mathbf{d}$ of the "fabric tensor" can usually represent the distribution and orientation of particles and pores (Oda et al., 1982[4]; Mehrabadi, 1982[5]). Using the famous Voronoi-Delaunay checkerboard grid, Bagi(2006[6]) derived the mathematical description of the internal structure of the granular material, and it is pointed out that the spatial distribution of the contact point of the particle is the most basic factor when the dilatancy of the material is considered. Li and Li (2009)[7] proposed to quantify the internal structure of granular materials by contact-based grid system. In three-dimensional case, the continuous pore space is composed of the smallest pore units, and the unit boundaries are statically determined. The microstructure expression of the strain tensor is briefly described below.

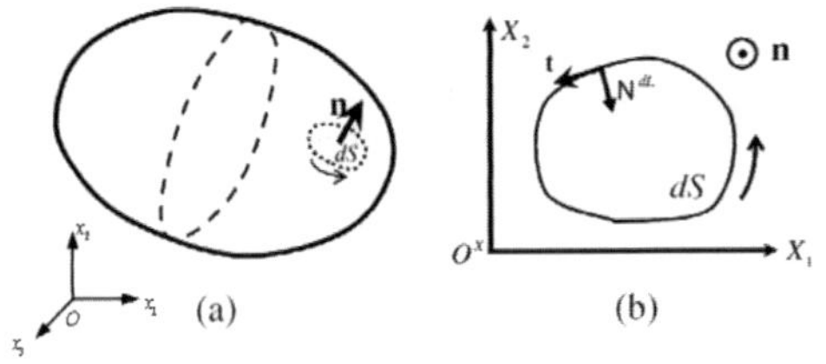

Figure1. The surface increment $\mathrm{d}S$ [7]

Under the condition of uniform strain, an arbitrary representative unit Rev. is set, and the volume is V , and the boundary surface is closed. Then, for the representative unit Rev., the compatibility condition is necessarily satisfied, namely

$$\oint_S \mathrm{d}(u_j S_i)=0 \tag{1}$$

Among them S a continuous boundary surface representing a representative unit (Rev.), as shown in Figure 1, u represents the representative unit displacement. Here the direction is kept consistent with the soil mechanics and the pressure is positive. Under the compressive stress, the average displacement gradient tensor of the representative unit Rev. can be expressed as

$$\bar{e}_{ij} = \frac{1}{V}\oint_v e_{ij}\mathrm{d}V = -\frac{1}{V}\oint_v u_{j,i}\mathrm{d}V \tag{2}$$

According to the scatter theorem, the formula (2) can be re-expressed as

$$\bar{e}_{ij} = \frac{1}{V}\oint_s u_j n_i(\mathbf{x})\mathrm{d}S = \frac{1}{V}\oint_s \mathbf{u}\otimes\mathbf{n}\mathrm{d}S \tag{3}$$

Among them $\mathbf{n}(\mathbf{x})$ represents the normal direction of point $\mathbf{x}$ on the boundary surface $\mathrm{d}S$ that the direction points to the inside. then, By using the statistical probability method to describe the edge $\mathbf{n}$, the displacement gradient of the direction is expressed as (for detailed derivation to see[7])

$$\begin{aligned}\bar{\boldsymbol{e}} &= \frac{1}{V}\sum_{c\in V}\boldsymbol{\mu}^c \otimes \Delta\mathbf{u} = \frac{1}{V}\sum_{\Omega}\bar{\mu}(\mathbf{n})\mathbf{n}\otimes\Delta\bar{u}\mathbf{n}_\Delta\Omega \\ &= \frac{1}{V}E_0\Delta\bar{u}\oint_\Omega \bar{\mu}(\mathbf{n})\mathbf{n}\otimes\mathbf{n}\mathrm{d}\Omega\end{aligned} \tag{4}$$

Here $\boldsymbol{\mu}^c$ represents pore element polygon vector[7] for granular materials which is closely related to the direction distribution of the length of the pore vector. The integral range is the entire space Ω, and E_0 is the standardization factor, and $E_0 = 1/\oint_\Omega d\Omega$ which is respectively equal to 2π and 4π in two and three dimensions, $\Delta\bar{u}$ indicates the average relative displacement at the contact point of the microelement which is set to a constant under uniform strain conditions.

For variables $\boldsymbol{\mu}^c$ is closely related to the direction distribution of the length of the pore vector, so the tensor can be established based on the directional distribution of the pore vector. Considering the pore length $\bar{\mu}(\mathbf{n})$ on the direction of the representative pore unit $\mathbf{n}$ is the direction of the average length of all pore vectors in the sample $\mathbf{n}$. Therefore, the fabric tensor based on the directional distribution of the average pore vector length is defined to describe the shape and orientation of the average pore unit, $\bar{\mu}(\mathbf{n})$ represents the distance from the center of the micro-porous unit to the boundary of the pore unit in space $\Delta\Omega$,and $\mathbf{n}$ is a unit vector representing the direction of the aperture, ignoring the high order, along the direction $\mathbf{n}$, pore vector $\bar{\mu}(\mathbf{n})$ can be approximated as

$$\bar{\mu}(\mathbf{n}) = \bar{\mu}_0(\mathbf{I} + \mathbf{d}) : (\mathbf{n} \otimes \mathbf{n}) = \bar{\mu}_0(1 + \mathbf{n} \cdot \mathbf{d} \cdot \mathbf{n}) \tag{5}$$

Here $\bar{\mu}(\mathbf{n}) = \bar{\mu}(\mathbf{-n})$, $\mathbf{d}$ is a second-order tensor with no trace and represents the anisotropy of the pore structure, $\bar{\mu}_0 = E_0 \oint_\Omega \bar{\mu}(\mathbf{n}) d\Omega$ representing the average length of the pores, Substituting formula (5) into (4), we get

$$\bar{\boldsymbol{e}} = \frac{1}{V} E_0 \Delta\bar{u} \oint_\Omega \bar{\mu}(\mathbf{n}) \mathbf{n} \otimes \mathbf{n} d\Omega = \frac{1}{V} E_0 \Delta\bar{u} \oint_\Omega \bar{\mu}_0 (1 + \mathbf{n} \cdot \mathbf{d} \cdot \mathbf{n}) \mathbf{n} \otimes \mathbf{n} d\Omega \tag{6}$$

The above formula is the expression of the displacement gradient, and the rotation tensor is removed, so the expression of the strain tensor is

$$\bar{\varepsilon}_{ij} = (\bar{e}_{ij} + \bar{e}_{ji}) / 2 \tag{7}$$

Then under the action of compressive stress, the volumetric strain expression in the directional $\mathbf{n}$ is

$$\varepsilon_{tv}(\mathbf{n}) = \bar{\varepsilon}_{ij} \delta_{ij} = \frac{1}{V} E_0 \Delta\bar{u} \oint_\Omega \bar{\mu}_0 (1 + \mathrm{d}_{kl} n_k n_l) n_i n_j \delta_{ij} d\Omega \tag{8}$$

Let's say that $\varepsilon_v = E_0 \Delta\bar{u} \bar{\mu}_0 / V$, then formula (8) can be converted into

$$\varepsilon_{tv}(\mathbf{n}) = \varepsilon_v (1 + \mathbf{d} : \mathbf{n} \otimes \mathbf{n}) \tag{9}$$

Here $\varepsilon_{tv}(\mathbf{n})$ indicates pore volume change rate in the unit vector $\mathbf{n}$ under the compressive stress, and ε_v for the macroscopic volumetric strain of the granular material.

3. Fabric changes due to deformation

The change in particle arrangement and its pore structure due to shear deformation is actually a change in fabric. Since Philofsky and Finn (1967)[8] established a method for measuring strain using the Principle of stereography, early researchers begun to study the relationship between fabric and strain, Kanatani (1984)[9] deeply studied and developed the relationship between strain and different tensor tensor. Satake(1989)[10] proposed that in granular materials, the average strain is a function of the relative displacement between the particles and the branch vector connecting the centers of the particles. For the particle material fabric, the relative position between the particles changes, and accordingly the pore structure also changes. Therefore, for the dual structural system of granular material: pore structure and solid structure[7] are complementary each other, and any structure can describe the internal structure of the granule. Numerical experiments haved proved[7] that the normal anisotropy of particle contact is closely related to the anisotropy of pore structure. The direction distribution of contact normal tends to the direction of pore structure. Therefore, the changes of volumetric strain and pore fabric tensor under compressive stress will be discussed below. We can get the differential expression for formula (9)

$$d\varepsilon_{tv}(\mathbf{n}) = d\varepsilon_v(1 + d_{ij}n_in_j) + \varepsilon_v d(d_{ij}n_in_j) \tag{10}$$

The volumetric strain rate of the pores is summed in all directions:

$$d\varepsilon_{tv} = d\varepsilon_v + \varepsilon_v d(d_{ij}n_in_j) \tag{11}$$

Horne (1965)[1] believed that in the initial state of deformation, the normal direction of particle contact and the direction of its pore structure tended to align along the direction of the maximum principal stress and led to anisotropy. When deformation occured, the sliding between particles was no longer be restricted by a certain direction, the degree of anisotropy was reduced which not only caused the dilatancy to be weakened but also the stress ratio η reduced. Therefore, based on the above description, the tensor of the pore structure $\mathbf{d}$ is a bias and changes with shear strain, using the isotropic tensor function to express the theorem (Boehler, 1987)[11]): $d(d_{ij})=d(d_{ij}(d_{kl,}d\varepsilon_{kl},\varepsilon_v))$, the changes of the fabric with the partial or shear strain ε_{ij} can be connected , in general, this function is very complicated, for the sake of simplicity, it is assumed here dd_{ij} and $d\varepsilon_{ij}$ have the same spindles(Muhunthan et al., 1996[12]), so the relationship can be expressed as

$$\varepsilon_v d(d_{ij}n_in_j) = \alpha d(\varepsilon_{ij}n_in_j) \tag{12}$$

Here α represents the anisotropic parameters of the structure reflecting the basic trend of soil fabric development. For partial strain ε_{ij} ,under triaxial stress $d\varepsilon_{ij}n_in_j$ also represents the shear strain increment.Setting $d(\varepsilon_{ij}n_in_j) = d\varepsilon_q$, and substituting it into formula (11), we can get

$$d\varepsilon_{tv} = d\varepsilon_v + \alpha d\varepsilon_q \tag{13}$$

The above formula shows the rate of change of volumetric strain $d\varepsilon_{tv}$ under compressive stress in granular materials considering anisotropy is coupled to the rate of change with partial strain $d\varepsilon_q$, that is to say, the change of the partial strain will also lead to the change of volumetric strain.

4. Decomposition of plastic strain

In the existing constitutive model of granular materials, it is generally considered that the change in volume of the material is derived from the compression of the pores, and the plastic volumetric strain is only caused by the change of the compressive stress, and has nothing to do with the soil skeleton. In fact, through the above analysis The relationship between volumetric strain and pore structure indicates that for granular materials such as dense sand or medium dense sand, the change of volume under shear stress includes two processes. First, when the axial strain is small, dense sand or medium dense sand produce shear under compressive stress ε_{tv} , and then with the development of anisotropic particle organization, will also produce additional volume changes $\alpha\varepsilon_q$, that is, "Reynolds dilatancy[13]" effects (since the sand particles are significantly larger than the clay, as the shear strain increases, the particles at the shear plane will rise, shift, pull, and pull out, accompanied by an increase in volume - dilatancy). In dense or medium-density sand, another source of plastic volumetric strain is shear-induced volumetric strain-dilatation. Therefore, the macroscopic plastid strain of the granular material can be expressed as

$$d\varepsilon_v^p = d\varepsilon_{tv}^p - \alpha d\varepsilon_q^p \tag{14}$$

As can be seen from the above formula, the macroscopic plastic volumetric strain increment $d\varepsilon_v^p$ includes two components: one component is the change in pore volume caused by the change of compressive stress of the granular material directly $d\varepsilon_{tv}^p$ -shearing, the other component is shear-induced volumetric change $(\alpha d\varepsilon_q^p)$ - dilatancy, since the pore structure of the particles is anisotropic during the loading process, this part of the volumetric deformation has been in a dilatant state and is dominant in the volume change. In the correction of the body strain in the Cambridge model, it usually refers to the first volumetric change-shear, which is completely obtained by compressive stress change, and does not take into account the shear-induced part of the dilatancy.

Therefore, the correction of the stress dilatancy relationship reflected by the Cambridge model has certain limitations. Below, we will further analyze this part of the shear-induced volumetric change based on the thermodynamic framework. $(\alpha d\varepsilon_q^p)$ -the relationship between the dilatancy and the dissipation function and the yield function of the real stress space.

5. Thermodynamic analysis

Today, there are many mechanical models that describe the mechanical behavior of sand and other particulate materials. In order to make these models more realistically reflect the mechanical phenomena observed in the laboratory of particulate materials, these problems are generally solved by increasing the model parameters. However, all These models are inseparable from some basic mechanical concepts. These concepts have been verified by practice and laid a solid theoretical foundation: critical state, stress dilatancy theory and state parameters. Most of the models are based on rate-independent, elastoplastic solids. The classical theoretical framework, therefore, the yield surface, the flow rule, etc. need to be determined, and many models assume the associated flow rule, but in fact the sand should be unrelated flow.

We use thermodynamic theory for it is a more rigorous method and does not require many artificial assumptions in traditional methods. The constitutive equations based on thermodynamics not only satisfy the laws of thermodynamics, but also have a clear physical explanation based on energy storage and dissipation. It is especially important to establish the close relationship between dissipation, yield function and flow law, and plastic dilatancy and induced anisotropy. The yield function , the law of flow, the form of non-correlated flow laws can be inferred from the dissipative part of the energy balance equation. , naturally to get the nature of frictional materials (such as soil, rock, concrete, etc.) in the unstressed flow in the real stress space. In addition, using thermodynamic methods, the micro-macro behavior of simulated particulate materials can be observed and is linked to macroscopic behavior. Therefore, thermodynamic theory further regulates the constitutive equations of a wider range of material types, including soil and granular materials, providing a unified theoretical framework.

6. Sand anisotropy model

6.1Pure friction material shearing model

According to the principle of thermodynamics, it can be known that the two key points to establish the dissipation function $\delta\Phi$ is the choice of stress and strain variables. First, most isotropic hardening comes from compressive strain, and when the pressure is sufficiently large, the particles are rearranged and crushed. Sand or particles are defined herein as pure friction materials (Collins and Muhunthan, 2003[14]), under the action of shearing, this material usually undergoes a large shear deformation before the particle breakage occurs; that is, the normal consolidation stress of the material p_c is very large when the particle breaks. Because this article does not consider particle breakage, the stress variable of the dissipative function $\delta\Phi$ is no longer dependent on the consolidation pressure (Collins and Muhunthan, 2003[14]). Secondly, through the analysis of the mechanism of the plastic volumetric deformation of sand, the macroscopic plastic volumetric strain increment of dense sand or medium dense sand $d\varepsilon_v^p$ includes two parts: one is the volumetric compression strain $d\varepsilon_{tv}^p$ caused by the current compressive stress change. Second, another plastic deformation induced by shear deformation - dilatancy $(\alpha d\varepsilon_q^p)$, which is closely related to the pore fabric anisotropy. In fact, this part of the volumetric increment associated with the pore structure anisotropy $(\alpha d\varepsilon_q^p)$ is the result of particle rigid body motion, which is Reynolds dilatancy, even if the particles are rigid, completely smooth particles will dilate, so Collins (2007)[15] ,Collins and Muhunthan (2003)[14] assume that Reynolds dilatancy only indicates the internal motion constraint of the material, and the pressure of the associated reaction force and the storage energy generated by the shear force component are balanced against each other.

The dissipative energy associated with this part of the dilatancy is zero. Therefore, the dissipative function $\delta\Phi$ should be directly related to stress-related volumetric strain $\mathrm{d}\varepsilon_{tv}^{p}$ which is not related to this part of the dilatation induced by shear (Kanatani, 1982[9]; Goddard, 1990[17]). For convenience of description, here is defined $\alpha=\tan\theta$, based on the analysis of the above two aspects, set the dissipation function $\delta\Phi$ for

$$\delta\Phi = p\sqrt{(\mathrm{d}\varepsilon_v^p + \tan\theta \mathrm{d}\varepsilon_q^p)^2 + M^2 \mathrm{d}\varepsilon_q^{p2}} \tag{15}$$

Thus, the complete plastic work expression is

$$\begin{aligned}\delta W^p &\equiv p\mathrm{d}\varepsilon_v^p + q\mathrm{d}\varepsilon_q^p \\ &= (\rho \mathrm{d}\varepsilon_v^p + \xi \mathrm{d}\varepsilon_q^p) + (\pi \mathrm{d}\varepsilon_v^p + \tau \mathrm{d}\varepsilon_q^p) \\ &= (\rho \mathrm{d}\varepsilon_v^p + \xi \mathrm{d}\varepsilon_q^p) + \\ & p\sqrt{(\mathrm{d}\varepsilon_v^p + \tan\theta \mathrm{d}\varepsilon_q^p)^2 + M^2 \mathrm{d}\varepsilon_q^{p2}}\end{aligned} \tag{16}$$

Dividing both sides of the above equation by $p\mathrm{d}\varepsilon_q^p$, the general dilatancy relationship of pure friction particle materials expressed by angle can be obtained as

$$\begin{aligned}&(\tan\phi_m - \tan\psi) = \\ &\beta(\tan\bar{\theta} - \tan\psi) + \sqrt{(\tan\theta - \tan\phi_m)^2 + \tan\phi^2}\end{aligned} \tag{17}$$

Among them: $\tan\phi_m = q/p$, $\beta = \rho/p$, $\tan\bar{\theta} = \xi/\rho$, ϕ_m、ϕ、ψ represents sliding friction angle, internal friction angle and dilatancy angle, respectively, and satisfy $\mathrm{d}\varepsilon_v^p / \mathrm{d}\varepsilon_q^p = -\tan\psi$.

Based on above formula, corresponding dissipative stress components are

$$\pi \equiv \frac{\partial(\delta\Phi)}{\partial(\mathrm{d}\varepsilon_v^p)} = p^2(\mathrm{d}\varepsilon_v^p + \tan\theta \mathrm{d}\varepsilon_q^p)/\delta\Phi \tag{18}$$

$$\tau \equiv \frac{\partial(\delta\Phi)}{\partial(\mathrm{d}\varepsilon_q^p)} = \frac{p^2((\mathrm{d}\varepsilon_v^p + \tan\theta \mathrm{d}\varepsilon_q^p)\tan\theta + M^2 \mathrm{d}\varepsilon_q^p)}{\delta\Phi} \tag{19}$$

Through the above two formulas can be derived as

$$\tau - \tan\theta\pi = M^2 p^2 \mathrm{d}\varepsilon_q^p / \delta\Phi \tag{20}$$

Then it can get a tilted elliptical yield surface equation in the dissipating space (π, τ)

$$\frac{\pi^2}{p^2} + \frac{(\tau - \tan\theta\pi)^2}{M^2 p^2} = 1 \tag{21}$$

As shown in Figure 2, the pure friction material dissipates the space yield function as a slanted ellipse with a tilt angle of θ.

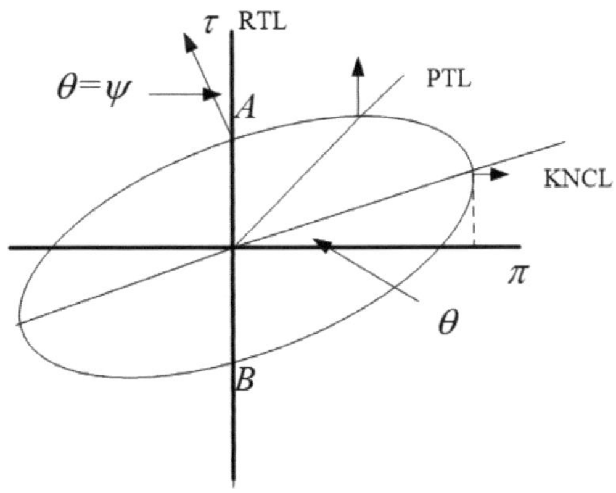

$$\delta\Phi = p\sqrt{(\mathrm{d}\varepsilon_v^p + \alpha \mathrm{d}\varepsilon_q^p)^2 + M^2 \mathrm{d}\varepsilon_q^{p2}}$$

Figure2. Inclined elliptical yield locus in dissipative stress space

Conversion formula (19) and (20), Using formula (21)available, it can get

$$\mathrm{d}\varepsilon_v^p = \delta\Phi\left\{\frac{\pi}{p^2} - \frac{(\tau - \tan\theta\pi)\tan\theta}{M^2 p^2}\right\} \quad \mathrm{d}\varepsilon_q^p = \delta\Phi\frac{(\tau - \tan\theta\pi)}{M^2 p^2} \tag{22}$$

Then dilatation equation D can be derived as

$$D \equiv \frac{\mathrm{d}\varepsilon_v^p}{\mathrm{d}\varepsilon_q^p} = -\tan\psi = M^2\frac{\pi}{\tau - \tan\theta\pi} - \tan\theta \tag{23}$$

There are three important lines in Figure 2, all of which rotate with the development of shear deformation. The first one is the "normal consolidation line of motion" (KNCL), which satisfies $\mathrm{d}\varepsilon_q^p = 0$, the plastic shear strain increment on the line is 0, but the stress state is not isotropic, according to the formula, in the dissipative stress space to meet: $\tau/\pi = \tan\theta$ The second is the "Phase transformation line" (PTL), the total volumetric strain increment on this line is 0, ie $\mathrm{d}\varepsilon_v^p = 0$. At this time, the dilatation D=0, shear-induced dilatancy is offset by stress-induced shearing, in the dissipative stress space: $\tau/\pi = M^2/\tan\theta + \tan\theta$ .The third line is the "Reynolds-Taylor Line" (RTL) which has a dissipative pressure of zero, that is $\pi = 0$, $p = \rho$. According to formula (18), $\mathrm{d}\varepsilon_v^p + \tan\theta \mathrm{d}\varepsilon_q^p = 0$, the only volume change is caused by the Reynolds shear effect and has

$$D = \frac{\mathrm{d}\varepsilon_v^p}{\mathrm{d}\varepsilon_q^p} = D_{max} = -\tan\psi = -\tan\theta \tag{24}$$

6.2Constrained shear model of pure friction particle material

The above analysis is mainly for the general deformation of pure friction materials. In order to study the dilatancy relationship Taylor (1948)[18] proposed, Collins and Muhunthan (2003)[14] defined a constrained shear deformation of a purely frictional particulate material , which is thermodynamically defined as the stored plastic work not considering the pre-consolidation pressure p_c and the pure shear flow, the volume fraction of the dissipation increment is 0, and the plastic free energy is not stored, that is $\mathrm{d}\Psi^p = 0$ .However, the pressure and shear component of the reaction force are not equal to zero. The above viewpoint is consistent with the analysis Arthur et al. (1991)[19]and Dunstan (1998)[1]proposed. They believed that for the shear deformation of granular materials (sand), the energy stored in the force chain is quickly dissipated due to friction shear sliding, so no plastic work is stored and all energy is dissipated. Collins (2003) [14] also stated in the text that for purely frictional particulate materials, as long as the anisotropy is considered in the model, the "quasi-steady-state" shear deformation of the dilatancy must occur, and no plastic properties are stored, but the reaction The force component is not equal to zero.

Based on the above analysis, we can get the correlation of constrained shear deformation of pure friction materials:

$$\pi \mathrm{d}\varepsilon_v^p=0,\quad \tau \mathrm{d}\varepsilon_q^p=\delta\Phi = p\sqrt{(\mathrm{d}\varepsilon_v^p+\tan\theta \mathrm{d}\varepsilon_q^p)^2+M^2\mathrm{d}\varepsilon_q^{p2}} \tag{25}$$

When the material is dilatant, the volumetric strain increment is not zero, so under such conditions, it must be satisfied $\pi=0$. Therefore, according to the formula (21), it can be known that when $\pi=0$, the dissipative stress is at the two vertices of the yield trajectory A、B, as shown in Figure 2, and satisfied $\rho=p$, that is, "Reynolds-Taylor line", according to And, it can get

$$\tau=Mp,\quad \tan\psi=\tan\theta \tag{26}$$

And formula (17) converted into

$$\tan\phi_m=\tan\bar{\theta}+\tan\phi \tag{27}$$

Since the shear deformation of the pure friction material is a pure shear flow, the pre-consolidation pressure p_c is not considered. Storage of plastic free energy, thus plastic free energy increase $\mathrm{d}\Psi^p$ Only related to the Reynolds effect. According to the above description, the plastic part of free energy can be seen as the function for $\mathrm{d}\varepsilon_{tv}^p$

$$\mathrm{d}\Psi^p(\varepsilon_{tv}^p)=\frac{\mathrm{d}\Psi^p(\varepsilon_{tv}^p)}{\mathrm{d}\varepsilon_{tv}^p}\mathrm{d}\varepsilon_{tv}^p=(\mathrm{d}\varepsilon_v^p+\alpha \mathrm{d}\varepsilon_q^p)\frac{\mathrm{d}\Psi^p(\varepsilon_{tv}^p)}{\mathrm{d}\varepsilon_{tv}^p}=(\mathrm{d}\varepsilon_v^p+\alpha \mathrm{d}\varepsilon_q^p)\Psi^{p'} \tag{28}$$

Therefore, the reaction force components are

$$\rho=\mathrm{d}\Psi^p/\mathrm{d}\varepsilon_v^p=\Psi^{p'}\quad \xi=\mathrm{d}\Psi^p/\mathrm{d}\varepsilon_q^p=\alpha\Psi^{p'}=\alpha\rho \tag{29}$$

Then the storage plasticity associated with the Reynolds effect is

$$\rho(-\alpha \mathrm{d}\varepsilon_q^p)+\xi \mathrm{d}\varepsilon_q^p=\rho(-\tan\theta \mathrm{d}\varepsilon_q^p)+(\rho\tan\theta)\mathrm{d}\varepsilon_q^p=0 \tag{30}$$

Formula (30) shows that the storage plastic work related to the Reynolds effect is zero, which satisfies $\xi/\rho=\tan\theta=\tan\bar{\theta}$ From the analysis in the previous section, the normal consolidation line is also satisfied in the dissipation space: $\tau/\pi=\tan\theta$, so the normal consolidation line (kncl) of the dissipative stress space is in the true stress space. p-q a ray of origin further is converted to

$$\tan\phi_m=\tan\theta+\tan\phi \tag{31}$$

The above formula is the classic "Taylor stress dilatancy" relationship[13], that is, the Reynolds-Taylor line, which is also commonly called the line of destruction or peak strength, passing through the real stress space. p-q Origin, which describes the peak state of sand under drainage (the undrained state is the state asymptote).

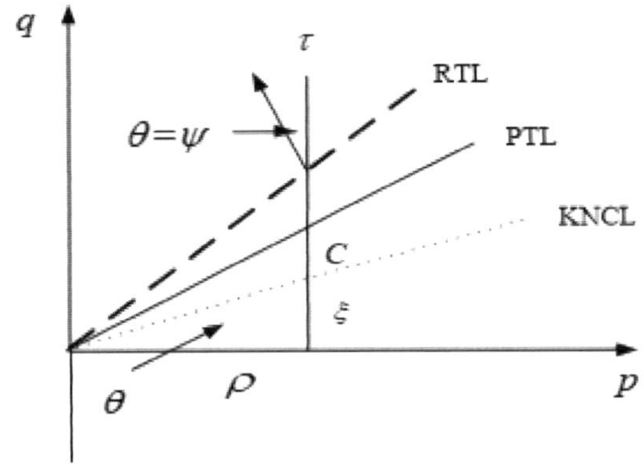

Figure3.Yield locus in true stress space for a constrained shear deformation of a purely frictional material

As shown in Figure 3, point C corresponds to the position of reactive component (ρ,ξ) in the real stress space. Comparing Fig. 2 and Fig. 3, for the shear deformation of the pure friction material, the three characteristic lines in the dissipative stress space respectively correspond to the three characteristic lines of the real space starting from the origin, that is, normal consolidation line (KNCL), phase change line (PTL) and peak line (RTL).

For the sake of simplicity, but highlighting the force characteristics of the granular material, the yield function of the real stress space is set as

$$f = q - \eta p = 0 \tag{32}$$

Here η is the stress ratio, although it does not appear in this yield function $\alpha(\tan\theta)$. But it implied η and anisotropic parameters α ($\tan\theta$) is closely related, α changes as the stress ratio changes, that is, under shear stress, it will inevitably cause dilatancy, and the corresponding stress-induced anisotropy change of - α ,and α also known as the reaction force ratio (in the general stress space, also known as the reaction force than the tensor).

Based on the above analysis, it can be known that the thermodynamic method can properly describe the three characteristic lines of sand as pure friction material: normal consolidation line (KNCL), phase change line PTL and peak line RTL, which better describe the sand. Strain strengthening and softening characteristics, and this is precisely the boundary surface plasticity model of sand (Manzari and Dafalias, 1997)[21].The mechanical characteristics of the sand reflected are well matched.

7. Summary

Firstly, from the perspective of micromechanics, the averaging theory is used to investigate the pore structure of granular materials and the increase of extra volume due to structural changes-dilatancy, which should be reflected in the plastic model. Through thermodynamic analysis, the anisotropic parameters of pore structure are introduced into the dissipative function, and the yield surface function of sand as a pure friction particle material in the dissipative space is established. Due to the influence of anisotropic fabric, it exists in the dissipative space. The three important lines, namely the normal consolidation line, the phase change line and the peak line, are correspondingly extended to the three characteristic lines of the original stress space: the normal consolidation line, the phase change line and the peak line.The ground reflects the mechanical characteristics of sand, especially dense sand. Therefore, the constitutive model of sand based on thermodynamic theory is in good agreement with the plastic model of sand boundary.

IOP Conf. Series: Materials Science and Engineering **794** (2020) 012022 doi:10.1088/1757-899X/794/1/012022

Acknowledgments
Acknowledgments are given to Dr. Zhao chenggang who has given so many ideas during the development of the research project. The research work was greatly supported by my husband Zhang yinping, no words exist to show the appreciation for his help.

References
[1] Horne, M.R. The behaviour of an assembly of rotund, rigid, cohesionless particles II. Proc. Roy. Soc. London, 1965, Series A. 286, No 1404:79-97.
[2] Oda, M. (1972a). Initial fabrics and their relations to mechanical properties of granular material. Soils and Foundations, 1972,12(1): 17-36.
[3] Oda, M. (1972b). The mechanism of fabric changes during compressional deformation of sand. Soils and Foundations, 1972, 12(2) 2: 1-18.
[4] Oda, M., Nemat-Nasser, S. & Mehrabadi, M.M. A Statistical Study of Fabric in a Random Assembly of Spherical Granulates. Int. Jour. Num. Anal. Meth.Geomech, 1982, 6(1): 77-94.
[5] Mehrabadi, M.M., Nemat-Nasser, S. Oda,M. On statistical description os stress and fabric in granular materials. Int.J. Number. Meth. Geomech. 1982, 6, 95-108.
[6] Bagi , K. Analysis of microstructural strain tensor for granular assemblies. Int. J. Solids Struct., 2006, 43(10): 3166–3184.
[7] Li, X.&Li, X.S. Micro–macro quantification of the internal structure of granular materials. Journal of Engineering Mechanics, ASCE , 2009, 135 (7):641-656.
[8] Philofsky, E. M . & Finn, J.E. A New method for measuring strain distribution. In: Stereology, Proc. 2nd Int. Congress for Stereology, Chicago, Springer, Berlin, 1967.
[9] Kanatani, K. Stereological determination of structural anisotropy. International Journal of Engineering Science, 1984, 22(5): 531-546.
[10] Satake, M. Stress and strain in granular materials, Mechanics of granular materials, Japanese Society of Soil Mechanics and Foundation Engineering, 1989,1-15.
[11] Boehler, J.P (Ed.) Application of Tensor Functions in Solid Mechanics, Springer Verlag, 1987.
[12] Muhunthan, B., Chameau, J.L. & Masad, E. Fabric effects on the yield behavior of soils. Soils and Found., 1996, 36(3): 85-97.
[13] Reynolds, O. On the dilatancy of media composed of rigid particles in contact. Phil. Mag., 1885, 5(20): 469.
[14] Collins, I. F. & Muhunthan, B. On the relationship between stress-dilatancy, anisotropy, and plastic dissipation for granular materials. Geotechnique, 2003, 53(7): 611-618.
[15] Collins, I.F. & Muhunthan, B., Tai, A.T. and Pender, M.J. The concept of a "Reynolds Taylor state" and the Mechanics of Sands. Geotechnique, 2007, 57(5):437-447.
[16] Kanatani, K. Mechanical foundation of the plastic deformation of granular materials. Proc. IUTAM Conf. on Deformation and Failure of granular materials, Delft, 1982:119-127.
[17] Goddard, J.D. & Bashir, Y. M. "On Reynolds dilatancy" Recent Developments in Structured Continua, eds. D. DeKee and P.N. Kaloni, New York, Longmans, 1990, 2: 23-35.
[18] Taylor, D.W. Fundamentals of soil mechanics, John Wiley, New York, 1948.
[19] Arthur, J.R.F., Dunstan, T., Dalili, A. & Wong, R.K.S. The initiation of flow in granular materials. Powder Technol., 1991,65: 89-101.
[20] Dunstan, T., Arthur, J.R.F., Dalili, A., Ogunbekun.O.O. & Wong, R.K.S. Limiting mechanisms of slow dilatant plastic shear deformation of granular Media Nature, 1998, 336: 52 – 54.
[21] Manzari, M.T. & Dafalias, Y.F. A critical state two-surface plasticity model for sands Géotechnique, 1997, 47(2), 255-272.

IOP Conf. Series: Materials Science and Engineering **794** (2020) 012023 doi:10.1088/1757-899X/794/1/012023

The Seismic Ductility Capability Influence Factors Analysis of Integral Station-Bridge Frame Structure

Shenxia Wang

China Railway First Survey & Design Institute Group Co., Ltd. Xi'an, ShaanXi. 710043

Abstract: When performance-based structure design is adopted, the structure has a clear structural performance objective under the specified setting criteria, and the performance target is specifically reflected in the damage degree of the structure. The components of the "bridge-building" structure are mainly concentrated on the pier column of the station hall. Therefore, the degree of damage of the pier column of the station hall layer reflects the degree of damage of the structure to a certain extent. This paper describes the earthquake-resistant ductility of the combined bridge structure. The performance indexes were quantitatively analyzed. Then, through the elasto-plastic analysis of the structure, the influence of various factors on the seismic ductility of the structure was obtained. The ductility of the piers under the earthquake was discussed, which provided a reference for structural design.

1. Basic concept of ductile earthquake resistance

The seismic design method based on ductility[1] is the main method of structural design at this stage. The dynamic behavior of the structure exhibiting nonlinear changes is the research basis of the method. Its main idea is to improve the setting of structural plastic hinge and improve the limit of structural members. Deformation ability and hysteresis ability[2]. It is the basic idea of ductility design to use the local deformation of the member to consume energy to ensure the proper ductility and strength of the structure as a whole[3].

2. Quantification of ductility indicators

The curvature ductility ratio is widely used to quantify the ductility index of the structure, which is also commonly referred to as curvature ductility[4].

The inelastic deformation ability[5, 6] of steel malleable members is mostly from the rotation of the section of the plastic hinge. The rotation ability can be reflected by the curvature ratio of the section. The wallpaper of the post-yield curvature and the yield curvature is defined as the curvature ductility ratio. The design is concerned with the maximum curvature ductility ratio u_φ, which is defined as:

$$u_\varphi = \frac{\varphi_u}{\varphi_y}$$

In the formula, φ_y with φ_u The yield curvature and the ultimate curvature of the section of the plastic hinge region are respectively indicated.

3. Model establishment

Guzhenkou South Station is a two-story island station on the road side. The structural form adopts a four-column frame structure of a combined bridge structure system, and the section bridge is overlapped with the station through the ox leg. The total length of the station is 126m, including 85m for the main body of the station and 41m for the auxiliary equipment. The main body of the station is two floors above ground, from the bottom to the top, the exhibition hall floor and the platform floor.

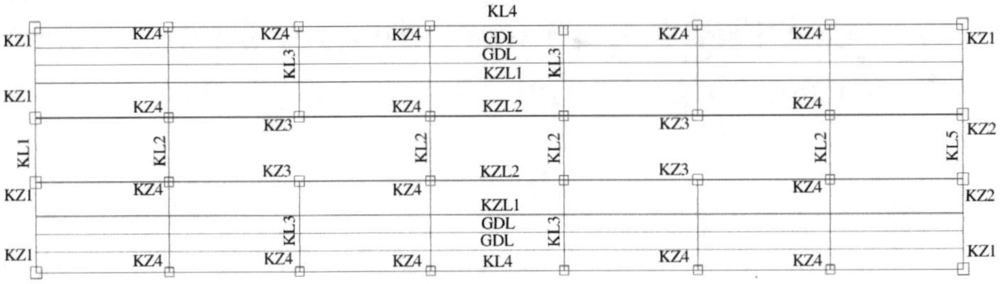

Fig.1 Finite element model of station structure

4. Analysis of ductility performance of pier columns

When the moment curvature analysis of the component is carried out[8, 9], kl1 is selected as the analysis object, and the change rules of the delay characteristics of the other pier columns are the same as those of kl1, and no further analysis is performed. When the whole structure is subjected to linear analysis. Under some factors, the analysis results of other piers are also compared.

4.1 Influence of axial compression ratio

The axial compression ratio is the ratio of the design value of the axial pressure of the pier column to the product of the total cross-sectional area of the column and the design value of the compressive strength of the concrete,

$$u = N / Af_c$$

Where: u is the axial compression ratio, A is the total cross-sectional area of the column, and N is the design value of the column axis pressure. f_c Design values for concrete compressive strength.

In order to avoid the high axial pressure ratio, the axial compression ratio limit selected in this paper is 0.3.

In order to analyze the influence of axial compression ratio on the ductility of the pier column, the pier column kl1 is analyzed in detail, and the response law of the other pier columns is similar to that of kl1. Finally, the ductility of the four piers is compared with the axial pressure.

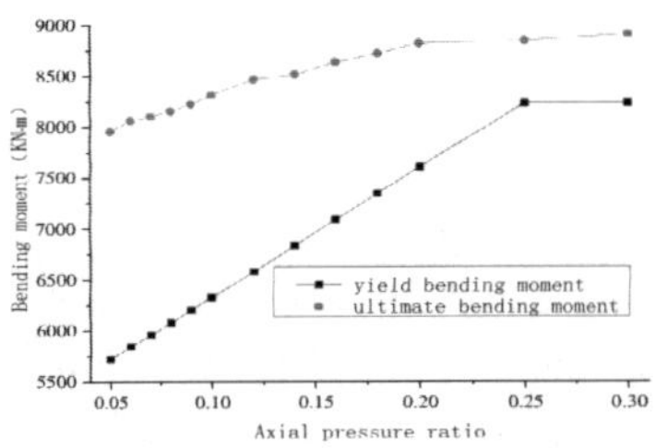

Figure 2 Variations of yield and ultimate moments

Figure 3 shows the change in yield curvature and ultimate curvature as a function of axial compression ratio. As the axial compression ratio increases, the limit curvature of the section decreases, and the yield curvature increases, but the change is not obvious. Therefore, the maximum curvature ductility coefficient of the section is basically controlled only by the section limit curvature.

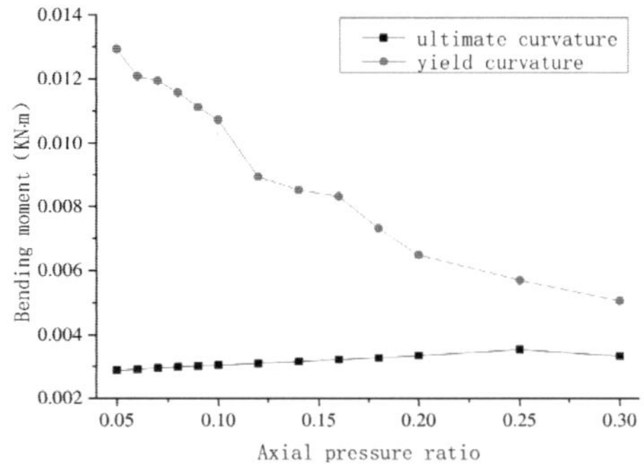

Fig. 3 Yield curvature and ultimate curvature

As shown in Fig. 4, as the axial compression ratio increases, the curvature ductility coefficient decreases. As the axial compression ratio increases, the ultimate curvature decreases and the yield curvature increases, resulting in a decrease in the curvature ductility coefficient, which adversely affects the ductility of the pier. When the axial pressure ratio is increased to 0.25, the value of the curvature ductility coefficient remains substantially unchanged. In order to ensure the ductility of the pier, it is not suitable to use a large axial pressure ratio.

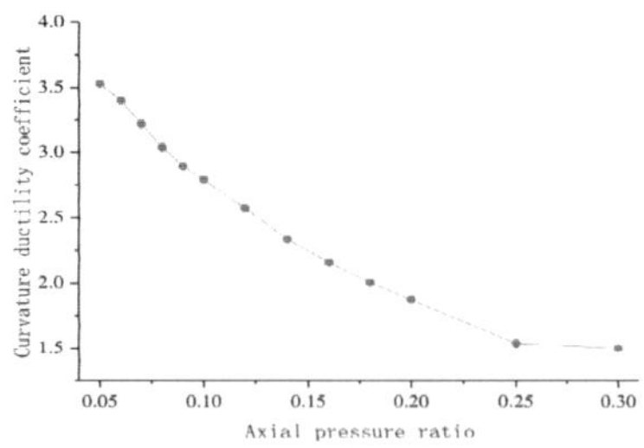

Figure 4 The change of the Curvature ductility coefficient coefficient

Figure 5 shows the variation of the curvature ductility coefficient of four types of pier columns with axial force. It can be seen from the figure that the curvature ductility coefficients of different types of pier columns are different under the same axial compression ratio, but the values are not much different, indicating that the ductility of various types of piers is the same at the same axial compression ratio. almost the same.

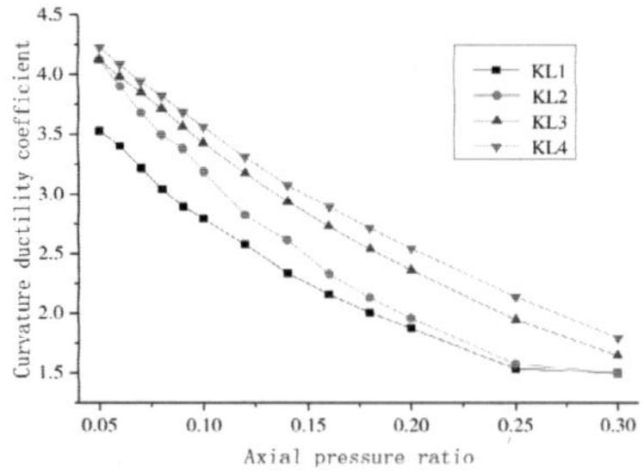

Figure 5 The variation of the ductility coefficient of curvature with axial force

4.2 Influence of longitudinal reinforcement strength

Taking KL1 as the research object, the model with the standard tensile strength of steel bars of 300MPa, 335MPa was established for price analysis.

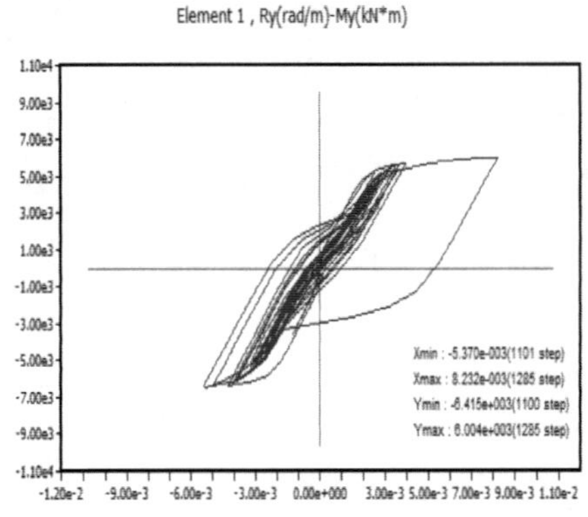

Figure 6 Longitudinal strength 300MPa

Table1 Effect of longitudinal reinforcement strength on ductility

Longitudinal strength	Yield curvature	Equivalent yield curvature	Limit curvature	Curvature ductility coefficient	Yield bending moment	Ultimate bending moment	Response maximum curvature	Pier top displacement
300	0.0022366	0.00760312	0.0126434	5.653	4619.208	6457.6385	0.008232	0.04445
335	0.0024963	0.00312950	0.0118688	4.755	5234.867	6925.9115	0.007421	0.04458
400	0.0029117	0.00372117	0.0101096	3.473	5730.173	7691.3142	0.005711	0.0448

By comparing the calculated data of three strength longitudinal steel bars, it can be seen that increasing the strength of the longitudinal reinforcement reduces the ductility performan-ce by 39%.The response to the maximum curvature is reduced by 31%.Therefore, in order to ensure the ductility of the structure, it is best not to use high-strength steel bars. However, high-strength steel bars can reduce the corner angle of the section, so the design should be considered comprehensively.

4.3 Effect of longitudinal reinforcement ratio

Kl1 is selected as the analysis object, and the influence of the reinforcement ratio on the ductility is studied by changing its reinforcement ratio. In the structural design, if the reinforcement ratio is changed, it is often changed with the same type of pier. In this paper, when changing the reinforcement ratio, not only the one column is changed, but the same type, and the remaining pillar reinforcement ratio remains unchanged.

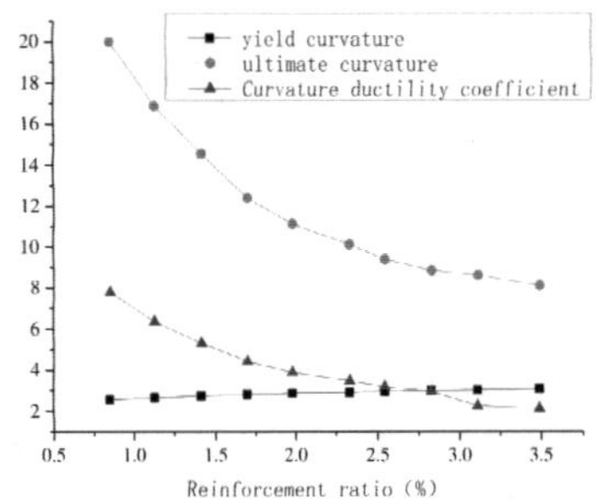

Fig.7 Effect of reinforcement ratio on longitudinal ductility

It can be seen from Fig.7 that increasing the longitudinal reinforcement ratio does not improve the ductility of the pier, but it will reduce it. The maximum curvature and the top displacement will decrease rapidly when the reinforcement ratio is less than 1.5%. There is little change between %~2.8%. Considering the recommendation, the longitudinal reinforcement ratio is about 1.5%.

4.4 Influence of concrete strength

Taking kl1 as the analysis object, in order to compare and analyze the influence of concrete strength on the ductility of the pier, this section selects seven kinds of labeled concrete between c20 and c50 for analysis.

Table 2 Effect of Concrete Strength on Ductility

Concrete strength	Yield curvature	Equivalent yield curvature	Limit curvature	Curvature ductility coefficient	Yield bending moment	Ultimate bending moment
C20	2.8681202	3.95628932	7.8502254	2.737062902	5482.6223	6367.7797
C25	2.9117172	3.883104395	8.0954409	2.780297772	5517.3255	6718.9558
C30	2.9705201	3.818720264	8.3988578	2.827403154	5575.751	7024.9388
C35	3.047452	3.774808452	8.8834437	2.915039721	5635.4936	7280.8953
C40	3.1511692	3.744051198	9.5434138	3.028531032	5686.6181	7519.334
C45	3.2940245	3.721169599	10.109602	3.069073247	5730.1733	7691.3142
C50	3.4701105	3.696502065	10.671084	3.075142488	5777.4579	7843.8554

It can be seen from Table 2 that when other influencing factors remain unchanged, only the concrete strength grade is increased, and the yield curvature, ultimate curvature, yield bending moment and ultimate bending moment of the pier column increase with the increase of concrete strength. Therefore, under the same conditions, the strength of the concrete can be improved, and the ductility of the pier can be improved. However, when the higher concrete grade is used, the improvement of the ductility of the pier is not obvious, so it is not suitable to use high-grade concrete to improve the pier. Ductility.

4.5 Influence of section size

KL1 was used as the analysis object to study the influence of the change of section size on the ductility of the pier column. The change is the KL1 type of pier. The section size varies from 1000mm to 1500mm.

Table3 Effect of section size on ductility performance

section	Yield curvature	Equivalent yield curvature	Limit curvature	Curvature ductility coefficient	Yield bending moment	Ultimate bending moment	Response maximum curvature	Pier top displacement
1000	3.634864	4.50886878	11.6887	3.216	3548.814	4616.94	6.531	4.908
1100	3.256540	4.04453999	10.5860	3.251	4737.302	6190.52	4.294	4.044
1200	2.91171	3.72116959	10.10960	3.472	5730.173	7691.31	5.711	4.448
1300	2.67048	3.34972806	9.56186	3.581	7395.217	9780.03	3.79	3.797
1400	2.37244	2.9950817	8.98849	3.789	9281.67	11847.32	2.955	3.36
1500	2.17320	2.8696752	8.57958	3.954	10785.45	14335.1	2.964	3.447

It can be seen from Table 4 that, under the condition that other influencing factors are unchanged, only the size of the pillar column is changed. The curvature ductility coefficient, the yield bending moment and the ultimate bending moment increase with the increase of the section size, and the section size increases. The axial compression ratio is reduced, and the reduction of the axial pressure can increase the ductility of the pier. Therefore, increasing the size of the pier column has a significant effect on improving the ductility of the pier column.

5. Conclusion

Based on the basic concept of ductile seismic resistance, this paper analyzes some factors affecting the ductility of piers and columns, and obtains the influence of various factors on the ductility of piers.

(1) When the parameters of other influencing factors are constant, as the axial compression ratio increases, the ductility of the pier column decreases, but when the axial compression ratio increases to a certain extent, the axial compression ratio increases to the ductility of the pier column. The impact will be weakened.

(2) As the strength of the longitudinal reinforcement increases, the ductility of the pier column decreases, but at the same time, the curvature of the section is reduced, and the design should be considered comprehensively.

(3) With the increase of the longitudinal reinforcement ratio, the ductility of the pier column is reduced, and the displacement of the pier top and the response curvature are not obvious when the reinforcement ratio is between 1.5% and 2.8%. Therefore, the longitudinal reinforcement ratio is recommended. At around 1.5%.

(4) Under the same conditions, improving the strength grade of concrete can improve the ductility of the pier column, but when the concrete grade is higher, the effect is not obvious;

(5) The increase in section size is advantageous for the ductility of the pier.

References

[1] Jiang Haipeng, Huang Guohui, Yang Qingshan. Study on ductility of large continuous rigid frame bridges under spatial variability earthquakes[J]. Journal of Railway Engineering Society, 2001(2): 37-46.

[2] Liu Peng. Study on seismic behavior of continuous rigid frame bridge with restricted displacement piers[J]. Journal of Earthquake Engineering. 2015(1): 120-125.

[3] Chen Lijun, Hu Ning, Zhang Maohui, et al. Seismic analysis of elevated metro stations [J]. Railway standard design. 2015(12): 88-90.

[4] Zhu Wenzheng. Seismic performance evaluation of multi-span simply supported beam bridge reinforcement measures [J]. Journal of Shenzhen University (Science and Technology Edition). 2015(2): 152-161.

[5] Hu Ning, Chen Lijun, Li Yun, et al. Seismic analysis of elevated metro stations under rare earthquakes [J]. Journal of Wuhan University of Technology (Transportation Science and Engineering Edition). 2015(6): 1229-1232.

[6] Yang Xinran. Research on Key Technology of Rigid Frame Continuous Beam Design of Huzhun Railway Yellow River Bridge [J]. Railway Standard Design. 2017(5): 109-113, 148.

[7] Deng Jihua, Chen Wei. Ductility and Seismic Isolation Design of Long-Span Continuous Beam Bridges [J]. Transportation World (Late). 2017(1): 95-96.

[8] Huang Yue. Preliminary study on seismic design of continuous girder bridge [J]. Shandong Industrial Technology. 2017(10): 116.

[9] Zhang Yongliang, Lu Xiaosu, Chen Xingchong, et al. Seismic design of double thin-walled short pier continuous rigid frame bridge in high-intensity area [J]. Journal of Railway Engineering. 2017(11): 45-50.

The Influence of Knickpoints on the Erosion in Songlin River Basin

Tan Long, Tang Di, Wu Xuefeng and Qi Lijiang*

College of Water Resources and Hydropower, Sichuan Agricultural University, Ya'an 625014

*Qi Lijiang's e-mail: qilij7@163. com

Abstract: Based on fieldwork and the longitudinal section obtained by ArcGIS, seven major knickpoints in Songlin River Basin are located. It is found that the regions with a great average elevation difference between ridge and valley suffer more from disasters such as collapse, landslides, and debris flows. Dimensionless number Vindex is used to quantitatively describe the cross-section of the Songlin River Valley. The Vindex coefficient is below 0 outside the upstream influence of knickpoint and in downstream of knickpoint and it is over 0 in the upstream of knickpoint and at the knickpoint. The overall change trend is "super V shape- U shape- super V shape". The formation and development of knickpoints strongly influence the geomorphological characteristics of the Songlin River Basin. Due to the rising erosion base level of the upstream of knickpoint, sediment is deposited and the incision rate of riverbed decreases or it even stops down cutting. In this way, the average elevation difference between ridge and valley is tend to reduce, which lowers the energy that causes geological hazards and improves the stability of the upstream. The gradient of both riverbanks decreases, forming a flat and wide U-shaped river valley. Although the elevation difference between the river and the downstream watershed of the knickpoint remains constant, the current energy increases and the erosion is still severe, keeping the valley in a V shape.

1. Introduction

Under a balanced condition, the gradient of the river gradually decreases with the increase of the basin area of the downstream, and the longitudinal section of the river appears as a relatively smooth concave. Once this balanced condition is broken, the river system will be in a transient imbalance [1]. The development of knickpoints is one of the most typical characteristics of the transient imbalanced geomorphologic system [2]. Its causes mainly include: different tectonic activities [3], changes in lithology of riverbed [2], river blocking [4], sea-level changes, and the dropping erosion base level caused by river capture [5-7]. As the longitudinal section of the river under a transient imbalance state will continuously adjust itself to a balanced state, the knickpoint can control the erosion process of the entire valley (including the slope and the channel itself) [1].

The knickpoint can significantly change the distribution characteristics of water energy in the river and make the flow energy significantly different upstream and downstream. It can also adjust the erosion capacity of the river, and change the stability of the riverbed and even the entire river basin. Crosby's [5] research on the Waipoua River Basin in New Zealand suggests that the knickpoint is the boundary between the remnant topography and the adjusted topography in the basin, and the erosion effects in the upstream and downstream areas of knickpoint are different. Zhang Kang [8] et al. propose that large knickpoints can change riverbed evolution and river geomorphology. Guo Chenwen

[9] et al. apply the HI value to the study on knickpoint and stability of the bank slope. They confirm the rationality of judging the stability of bank slope based on the HI value, and find that different causes of knickpoints have different effects on the evolution of the river basin. Yu Guoan[10] et al.'s analysis on the geomorphological environmental effects of the collapsed Yansai Dam shows that the long-term existence of the dam generates knickpoints on the riverbed surface. They raise the base level of erosion of the upstream riverbed, increase the flow resistance of the river, control riverbed 's undercutting, and stabilizes the bank slope.

The formation and development of knickpoints are closely related to basin erosion. The Songlin River Basin is rich in water energy resources, but frequently suffers from geological disasters. This paper studies the Songlin River Basin and explores the relationship between knickpoints and basin erosion with quantitative indicators.

2. Introduction to the Songlin River

The Songlin River, formed by the Wanba River (mainstream) and the Hongba River branch, is a first tributary on the right bank of the middle reaches of the Dadu River. The Songlin River basin covers an area of 1445 km2, and the Hongba River basin is 640 km2. The lowest point of the basin is Sonlin River Estuary with an elevation of 882m and the highest point is the top of the watershed of the Songlin River Basin and the Tianwan River Basin, with an altitude of 5946m. The elevation difference of the basin is 5064m. The basin is located in the mountainous area on the western edge of the Sichuan Basin, at the intersection of the north-westward Xianshui River fault zone and the NS-trending Anning River fault zone. The western region belongs to the western Sichuan trough, and the rest belong to the eastern Sichuan platform area. The basin is located between the platform and the trough, where the crustal movement is frequent [11]. The mountain shows a north-south trend, with a terrain of triangular shape. The topography of the mountain is high in the west and low in the east, and there are numerous rigid magmas and metamorphic rocks [12]. The main source of surface runoff is atmospheric precipitation, followed by snow meltwater and glacial meltwater. Changes in flooding and drying flow are large, and the maximum flow is 39 times the minimum flow [11]. The climate in the region is mild. The temperature difference in the four seasons is small, and the dry and wet seasons are clear [12-13].

As the influence of small knickpoints on the geomorphology of the basin is limited, this paper studies the major knickpoints of higher research value. According to the river longitudinal section and field investigation, seven major knickpoints in the Songlin River are identified. The seven knickpoints divide the basin into nine regions, which are named 1-9 (Fig. 1).

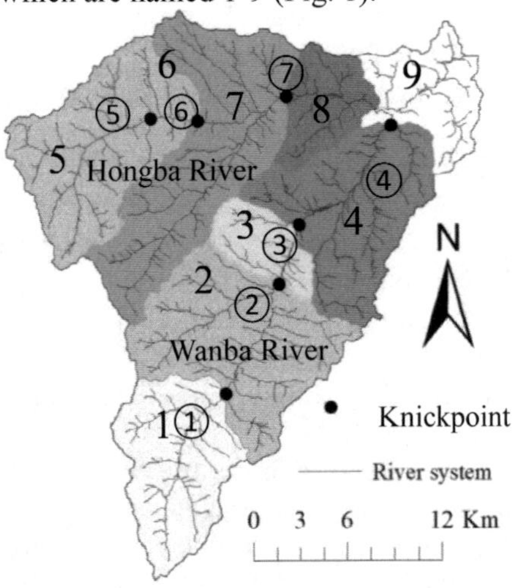

Fig 1. River system and major knickpoints in the Songlin River Basin

3. Research Methods

With ArcGIS, DEM (http://www.gscloud.cn/) with a resolution of 30m was used to extract the longitudinal section of the river. Based on field investigations, the authors identified the knickpoints and used ArcGIS to obtain the river valley cross-section of the upstream and downstream of knickpoint. Then, they calculate the average elevation difference $\overline{H}$ and Vindex coefficients of knickpoints regions, and explore the relationship between the knickpoints and basin erosion.

4. H of Knickpoints Regions

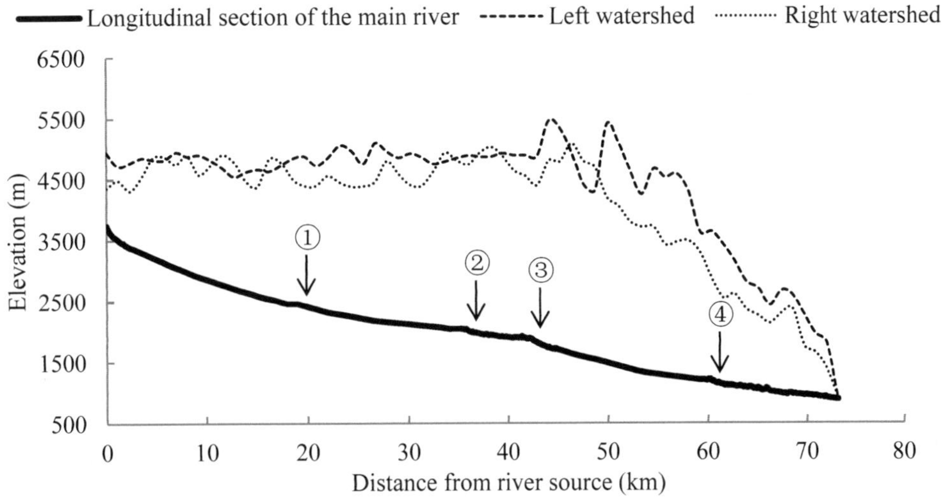

Fig 2. Longitudinal section of the main river and the corresponding watershed and major knickpoints along the way（①-④）

The main river of the Songlin River flows from south to northeast, with a length of 73.2km, a drop height of 2834m, and the average river gradient of 38.7 ‰. Four major cracks were found along the way. The longitudinal section of the main river is as follows (Fig. 2):

The Hongba River is the left tributary of the Songlin River and it flows from southwest to northeast. The Hongba River Basin's area accounts for 44% of the entire Songlin Basin. Therefore, it is studied together with the main river. The Hongba River has a natural drop of 3473m, a length of 49.7km, and the average river gradient of 69.9 ‰. Three main rifts were found along the mainstream. The longitudinal section of the Hongba River is as follows (Fig. 3):

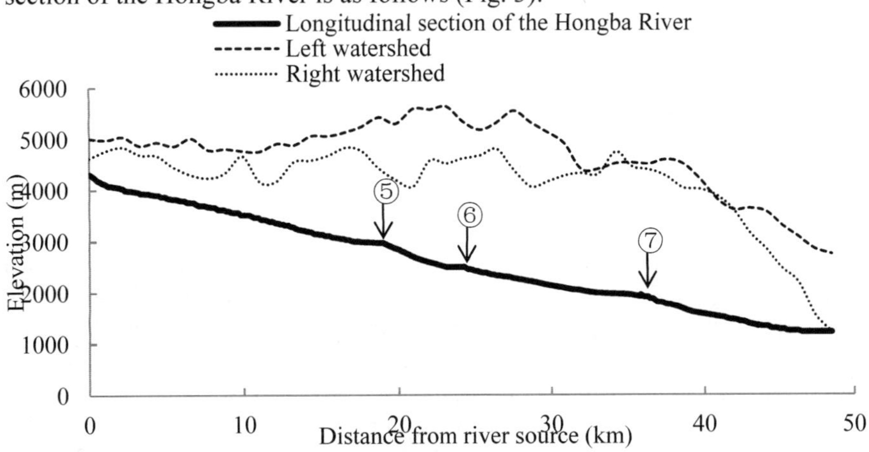

Fig 3. Longitudinal section of the Hongba River and the corresponding watershed and major knickpoints along the way（⑤-⑦）

The basic characteristics of the seven knickpoints are shown in Table 1.

Table 1. Basic characteristics of knickpoints

River	the main river				Hongba Rive		
knickponit	1	2	3	4	5	6	7
Distance from river source (km)	19	35	41	60	19	24	37
Elevation (m)	2460	2040	1880	1210	2945	2440	1970
Upstream specific gradient（‰）	25.8	41.9	16.9	18.2	36.6	22.7	37.0
Downstream specific gradient（‰）	82.1	147.8	93.6	87.2	188.8	96.4	102.8

From Table 1, 6 of the 7 major knickpoints in the basin are above the altitude of 1500m, indicating that there are mainly high-altitude knickpoints in the basin. The difference of gradient between upstream and downstream of the knickpoints is large, and the gradient of downstream is mostly more than three times that of upstream.

Figures 2 and 3 are about the main river, the Hongba River and their corresponding watershed. The steps for drawing the figures are:, make a series of parallel lines at equal intervals (approximately 1km in this paper) and perpendicular to the river from the river source and along the way; then take the elevation of the intersection of each parallel line and the corresponding watershed of the entire river basin as the ordinate. According to the 9 regions, the area enclosed by the longitudinal section and the watershed in each region is divided by the horizontal distance between the knickpoints and the river source or between the knickpoints and the knickpoints to obtain the average elevation difference $\overline{H}$ of each region.

It can be seen from Fig. 4 that the $\overline{H}$ of the main river and the Hongba River along the river from the top to the bottom increase at first and then decrease, while the left and right banks have the same change trend and the $\overline{H}$ of left watershed is greater than that of the right. In the main river basin, the region of maximum elevation difference is region 4, with $\overline{H}$ of 3078m, and the region of minimum elevation difference is region 5, with $\overline{H}$ of 946m. In the Hongba River Basin, the region of maximum elevation difference is region 7, with $\overline{H}$ of 2926m, and t the region of minimum elevation difference is region 6, with $\overline{H}$ of 867m.

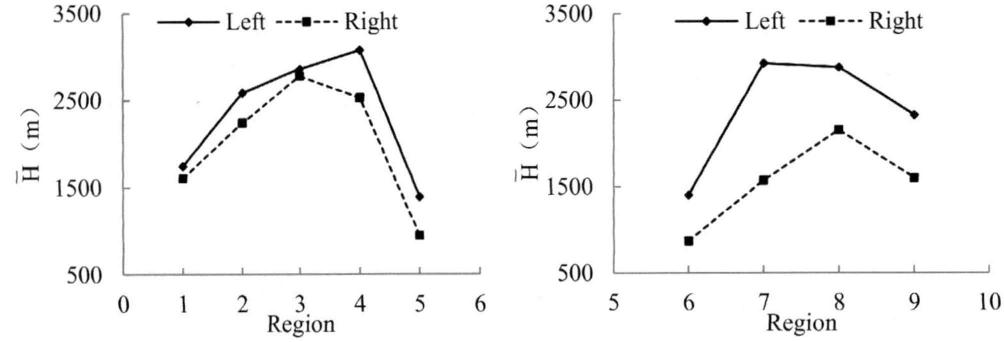

(a) The main river (b) Hongba Rive

Fig 4. $\overline{H}$ of each region

Geological disasters are actually short-distance movements of solid detrital material. The energy to transport detrital material is the potential energy of the elevation difference along the river valley or of hillside [14]. The $\overline{H}$ of the river and the left watershed is greater than that of the right watershed, indicating that at the same time external conditions, the left watershed are easier to cause disasters. In regions with large $\overline{H}$, there are many disasters such as collapse, landslides, and debris flows. According to field investigations, the seven major knickpoints in Songlin River Basin were formed because a large number of rocks and sediment generated by disasters like debris flows in the channel went into the river and they were deposited and blocked the river inflow operation. Five of knickpoints were formed under the influence of disaster in the left branch gully (Fig. 5-6). Only 3 small landslides were observed along with the upstream of the first knickpoint 5 of Hongba River, and more than 30 landslides were observed along the downstream. This is the impact of knickpoints on basin erosion.

The formation and development of knickpoints raise the upstream erosion base level. So, the flow velocity decreases, the sediment is deposited, and the riverbed undercut is controlled, which reduces the average elevation difference of the region. The bank slope tends to be stable. In addition, the erosion in upstream is weakened [10], but the erosion in the downstream is still severe.

The formation of knickpoint 5 decreases the incision rate of upstream, reduces the potential energy provided for the occurrence of geological disasters, and improves the stability of upstream more than downstream. If this knickpoint is destroyed, the incision rate of upstream will increase rapidly, and the average elevation difference of the region will increase. Then the materials on both banks of the slope will have enough potential energy, which will increasingly cause disasters like collapses and landslides.

Fig 5. Debris flow accumulation fan and dammed lake at knickpoint 5

Fig 6. Debris flow accumulation fan at knickpoint 6

5.Vindex Coefficient of River Valley cross-section in Knickpoint Area

The characteristics of the cross-section of the river valley can reflect the erosion of the basin. The evolution of eroded landforms of the incised river in mountainous can be divided into 4 stages [14], and different stages will present different cross-section features. When there is rapid river incision, the valley will form a narrow “V” shape or super “V” shape. When the river incision reaches a certain level, slope instability will cause disasters such as collapse, landslides, and debris flows to widen the valley. Then, it will become a wide “V” shape. When the incision rate slows down and aggradation occurs, the valley develops into a "U" shape. When the supply rate and transport rate of sediment finally match, the flow energy is consumed, the incision stops, and the riverbed becomes stable, forming the valley a wide "U" shape. The slope stability of the U-shaped valley is greater than that of the V-shaped valley. The slope stability of a U-shaped valley is better than that of a V-shaped valley.

In order to quantitatively describe the cross-section shape of the valley in the knickpoint region, this paper uses the Vindex coefficient [15] to describe the deviation of the actual valley cross-section area Ax from the ideal V-shaped valley cross-section area Av, and it can also describe irregular valley cross-section, the expression is:

$$V_{index} = \frac{A_x}{A_v} - 1 \tag{1}$$

For the ideal V-shaped valley, Vindex is equal to 0; for the super V-shaped valley, Vindex is less than 0; for U-shaped valley, Vindex is more than 0. (Fig. 7)

The extraction of valley cross-section must meet three conditions [15-16]: (1) the cross-section is basically perpendicular to the river; (2) the cross-section avoids crossing the branch gully; (3) with the river as the base point, the distance extracted from both banks should be approximately the same and longer than the distance set in the study (1.5km in this paper) so as to ensure that the Vindex coefficient of cross-section is a description of the entire valley, not a local one. To calculate the ideal cross-section area Av of the river valley, the calculation boundary of the cross-section of the ideal V-shaped valley should be determined, the standard is [15-16]: (1) find the first inflection point of the cross-section curve from the river to the banks; if the elevation difference from this inflection point to the next inflection point is more than the set threshold (10m in this paper), take this inflection point as the peak elevation of the bank slope; (2) if no inflection point meets the requirements, then take the elevation of the end position of the cross-section curve as the peak elevation of the bank slope; (3) take the smaller peak of the elevation of the left and right banks as the maximum elevation of the ideal section.

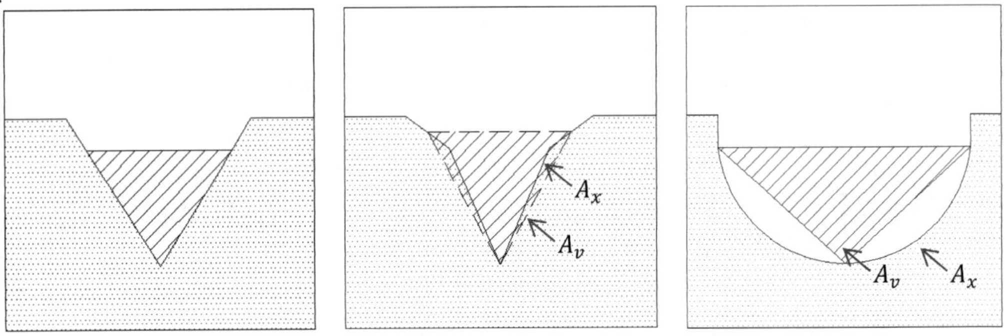

(a) $V_{index}=0$ (b) $V_{index}<0$ (c) $V_{index}=>0$

Fig 7. Schematic diagram of V_{index} calculation[16]

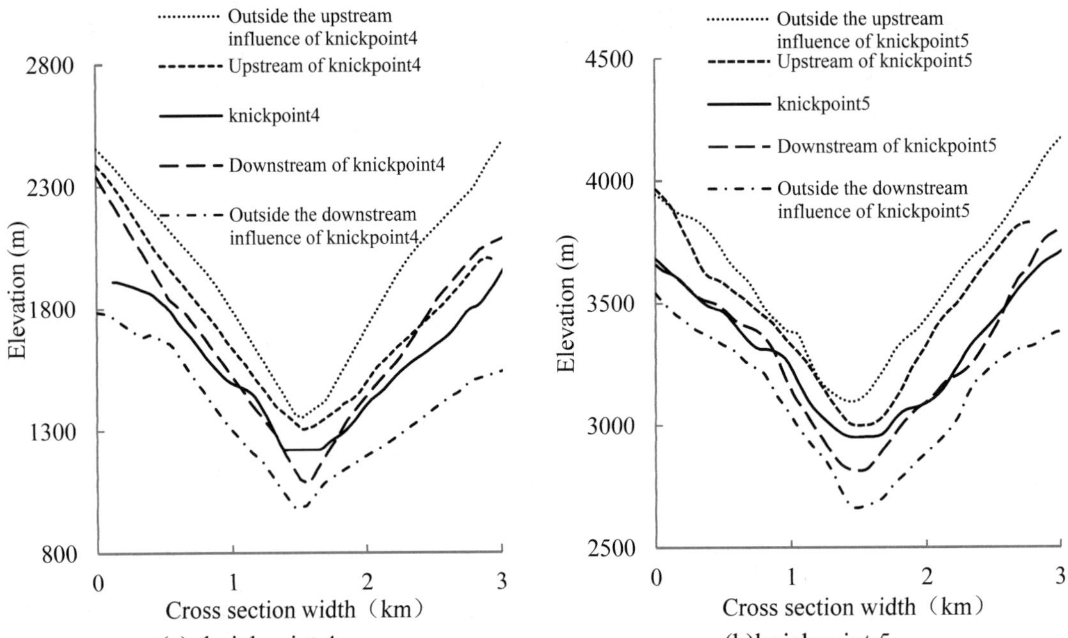

(a) knickpoint 4 (b)knickpoint 5

Fig 8. Schematic diagram of river valley cross-section

According to the above standards, this paper extracts the typical cross-sections of the river valleys in the seven major knickpoints in the Songlin River Basin. The schematic diagram of the typical knickpoints is as follows (Fig. 8):

From Fig. 8 we can see that the bank slope of the river valley cross-section outside the upstream influence of the knickpoint is steeper than that in upstream of the knickpoint and the river valley is narrower. The cross-section at the knickpoint appears to be much wider, and the cross-section of the downstream valley appears to be steep and narrow. The Vindex coefficient of typical river valley cross-section in the Songlin River knickpoint region is calculated in Table 2.

Table 2. V_{index} coefficient of river valley cross section in Songlin River knickponit region

knickponits	Outside the upstream influence of knickpoint	Upstream of knickpoint	Knickpoint	Downstream of knickpoint	Outside the downstream influence of knickpoint
1	-0.05	0.13	0.10	-0.06	-0.08
2	-0.05	0.09	0.05	-0.01	-0.02
3	-0.02	0.02	0.03	-0.03	-0.10
4	-0.05	0.08	0.29	-0.05	-0.06
5	-0.01	0.02	0.29	-0.02	-0.14
6	-0.10	0.04	0.07	-0.11	-0.18
7	-0.06	0.05	0.07	-0.07	-0.19

It is known from Table 2 that the Vindex coefficients of the region outside the upstream influence of the knickpoint and that of downstream are less than 0, and the Vindex coefficients of the typical cross-section of upstream of the knickpoint and at the c knickpoint are more than 0. It indicates that the valleys outside the upstream influence of the knickpoint and downstream are super V-shape, while the cross-section of the valley at the rift and its upstream influence area is U-shape. The change trend of the whole river valley cross-section is "super V-U-super V". And the Vindex coefficient of the typical cross-section outside the downstream influence of knickpoint is less than that within the influence.

The formation and development of knickpoints have a strong influence on the geomorphological characteristics of the Songlin River Basin. The Vindex coefficient of the cross-section of the river valley gradually increases from the position outside the upstream influence of the knickpoint to the position at the knickpoint. The river valley showed a change from a super V shape to a U shape. It shows that the formation and development of the knickpoints can decrease the incision rate of upstream or even stops the incision and turns it into aggradation. It controls the longitudinal section adjustment of the channel, turns it from longitudinal evolution to transverse evolution. In this way, the gradient of the banks becomes small, and a U-shaped river valley is formed [8], enhancing the overall stability. Although the elevation difference between the channel and the downstream watershed of knickpoint remains the same, the flow energy increases, leading to severe erosion, and forming the V-shaped valley. The Vindex coefficient of this section is less than 0, and the stability is poor. The cross-section Vindex coefficient of typical river valley outside the downstream influence of knickpoint is less than that within the influence, which reflects that the formation and development of knickpoint can make the downstream valley narrow and deep.

6.Conclusion

The formation and development of knickpoint significantly influence the geomorphological characteristics of the Songlin River Basin. Its formation causes erosion base level of upstream to rise and reduces the average elevation difference between ridge and valley. Then, the water flow velocity decreases, sediment is deposited, which control river incision, and stabilize the bank slope, forming a "U" river valley, and weakening the erosion of river basins. When the elevation difference between the downstream watershed and the river remains the same, but the flow energy increases to keep severe erosion, a super V-shaped valley with poor slope stability is formed.

References

[1] Whipplc K X. (2004) Bedrock rivers and the geomorphology of active orogens [J]. Annu Rev Earth Planet Sci, 32: 151-185.

[2] Gardner W. (1983) Experimental study of knickpoint and longitudinal profile evolution in cohesive, homogeneous material [J]. Geological Society of America Bulletin, 94(5): 664-672.

[3] Wallace R E. (1980) Degradation of the Hebgen Lake fault scarps of 1959 [J]. Geology, 8(5): 225-229.
[4] Ahmed M F, Ali M Z, Rogers J D, et al. (2019) A study of knickpoint surveys and their likely association with landslides along the Hunza River longitudinal profile [J]. Environmental earth sciences, 78(5): 176.
[5] Crosby B T, Whipple K X. (2006) Knickpoint initiation and distribution within fluvial networks: 236 waterfalls in the Waipaoa River, North Island, New Zealand [J]. Geomorphology, 82(1-2): 16-38.
[6] Bishop P, Hoey T B, Jansen J D, et al. (2005) Knickpoint recession rate and catchment area: the case of uplifted rivers in Eastern Scotland [J]. Earth Surface Processes and Landforms, 30(6): 767-778.
[7] Harkins N, Kirby E, Heimsath A, et al. (2007) Transient fluvial incision in the headwaters of the Yellow River, northeastern Tibet, China [J]. Journal of Geophysical Research: Earth Surface, 2007, 112(F3).
[8] Zhang Kang, Wang Zhaoyin, Liu Huaixiang, et al. (2011) The development of crack points and its influence on the stability of dams [J]. Journal of Mountain Science, 29 (04): 474-482.
[9] Guo Shenwen, Liu Tianxiang, Yao Lingkan. (2019) Study on the relationship between river crack points and valley slope stability based on HI value [J]. Journal of the Yangtze River Scientific Research Institute: 1-8.
[10] Yu Guoan, Wang Zhaoyin, Huang Heqing, et al. (2010) Geomorphic environmental effects of landslide dams (lakes) [J]. Advances in Earth Science, 25 (09): 934-940.
[11] Ya'an Regional Local History Office. (2000) Ya'an Natural Geography [Z].
[12] Chen Weizhong. (1996) Songlin River, a tributary of the Dadu River [J]. Sichuan Water Resources, (04).
[13] Wang Dekang. (1998) Hydrological characteristics of Songlin River Basin [J]. Sichuan Water Resources, (02).
[14] Wang Zhaoyin, Zhang Chendi. (2019) River bed structure and energy dissipation and mitigation mechanism in mountain rivers of Southwest China [J]. Journal of Hydraulic Engineering, 50 (01): 124-134+ 154.
[15] Zimmer P D, Gabet E J. (2018) Assessing glacial modification of bedrock valleys using a novel approach [J]. Geomorphology, 318: 336-347.
[16] Zhang Chendi, Lin Yongpeng, Xu Mengzhen, et al. (2019) Characteristics of the dam on the main stream of the Nu River and its stable riverbed mechanism [J]. Journal of Hydraulic Engineering: 1-12.

IOP Conf. Series: Materials Science and Engineering **794** (2020) 012025 doi:10.1088/1757-899X/794/1/012025

Determination Method for Conversion Sampling Frequency of Structural Deflection Monitoring of Small-and Medium-Span Bridges

Wenlong Hua[1], Runjian Wang[2*], Jianming Hu[3], Haiguang Li[4], Xiaohong Sun[5]

[1]Jinhua Highway Administration, Jinhua, Zhejiang Province,321013, China

[2]Beijing Gongke Bridge Technology Co., Ltd, Beijing, 100088, China

[3]Wuyi Department of Transportation, Zhejiang Province,321013, China

[4]Jinhua Comprehensive transportation hub construction management center, Zhejiang Province,321013, China

[5]Research Institute of Highway Ministry of Transport, Beijing, 100088, China

[*]Corresponding author's e-mail: 591624306@qq.com

Abstract. This paper proposes a method for determining the frequency of interest of static and dynamic deflection responses based on the analysis of components of the deflection deformation of small-and medium-span bridges in service. By analyzing the impact of sampling frequency on the time-domain signal amplitude, this paper determines the maximum sampling frequency limit of undistorted signals, thereby providing reference for the determination of the sampling frequency for the real-time monitoring of the deflection of small-and medium-span bridges.

1.Introduction

Existing bridge health monitoring systems assume that the equifrequent data acquired by the sensor nodes can be used to accurately obtain the true structural deformation status of the bridge. High-frequency sampling can help us to collect the actual deformation of the bridge structure, but it will cause a significant increase in cost and data volume. Low-frequency equifrequent sampling, however, cannot guarantee the retention of key points, resulting in significant errors in the monitoring results and actual data. The structural deformation of the bridge is continuously changing, and equifrequent sampling is just a discrete process of continuous changing deformation. Therefore, there exist problems such as missing key points and curve distortion.

Based on the analysis of the deflection deformation components of the bridge structure in service, this paper divides the frequency period of interest in deflection monitoring, proposes the method for determining the frequency of interest in each period, and analyzes the impact of the sampling frequency on the amplitude. On this basis, this paper puts forward the determination method for the maximum sampling frequency limit of undistorted signals, thereby providing reference for the setting of the conversion sampling frequency for structural deflection monitoring in the bridge health monitoring system.

2. Determination of the limit frequency of interest of dynamic response based on deflection component analysis

The structural deflection of the bridge is an overall response parameter. The performance of the bridge under various loads and environments and the variation of the bridge material can be reflected by deflection. The deflection of the bridge after operation can be categorized into the following types: 1) permanent structural deflection deformation due to internal causes such as concrete shrinkage, creep, and material degradation; 2) deflection deformation caused by component cracking, component spacing connection damage (hinge damage, damage of simple supported-continuous cast-in-situ section; 3) deflection deformation caused by temperature and other effects; 4) deflection deformation caused by vehicle load (including crowd load, etc.). Among the above four aspects of deflection components, the first and second require lower sampling frequency in principle (collecting data in hours or even days can deliver good data results). For temperature-induced deflection deformation, in the general case of non-extreme temperature abrupt changes, the samples are usually taken in the unit between 10 minutes and 1 hour; at the same time, the components of deflection caused by temperature can also be measured. As the vehicle crosses the bridge, its deflection deformation components include the static effect components under the vehicle's own weight load and the dynamic response components under the vehicle vibration. For small-and medium-span bridges, it is necessary to collect data in milliseconds so as to obtain the actual deformation.

The deflection monitoring of the highway bridge structure in service can be divided into two modes according to the characteristics of the monitoring indicators and the types of sensors currently used in this field: ① use dynamic and static displacement sensors (such as magnetostrictive displacement sensor, differential variable-pressure displacement sensor, laser displacement sensor, etc.) to monitor static, quasi-static and dynamic deflection; ② use vibration measurement sensors to test the dynamic deflection of the key parts only.

Generally speaking, the above ① mode can be used for the deflection monitoring of mid- and small-span bridge structures in service; for the long-span bridge structure, the ① mode is often difficult to achieve due to field and economic limitations; therefore, it is feasible to use ② mode supplemented by static test equipment. This research monitors the deflection deformation of small-and medium-span bridge structures. Therefore, the analysis focuses on the frequency limits of static deflection components and dynamic deflection components under the deadweight of vehicles crossing the bridge.

2.1 Division of frequency of interest period for deflection monitoring of small-and medium-span bridges

As shown in Figure 1, the time-history curve of mid-span deflection deformation of a reinforced concrete beam bridge structure under normal operation caused by vehicle load can be divided into three time periods:

(1) The magnitude of the deflection caused by the vehicle is relatively small (which needs to be determined according to the actual structural parameters). At this point, the structure is in a safe state, during which the data collected is generally not used for warning, but often as a record of deflection. Therefore, the upper limit of the static frequency range can be used as the frequency of interest.

(2) When there is no operational load on the bridge, then there is no frequency range of interest. Data collection at this stage can be performed with reference to the sampling frequency of the temperature test. When conditions permit, the frequency of interest can be used as the basis of sampling frequency.

(3) When there exist possibilities that the part with larger deflection deformation exceeds the warning state, as shown in the time period from 17:48:30 to 17:48:42 in Figure 1, the deflection deformation response frequency of interest shall be determined based on the analysis results of the dynamic deflection spectrum.

IOP Conf. Series: Materials Science and Engineering **794** (2020) 012025 doi:10.1088/1757-899X/794/1/012025

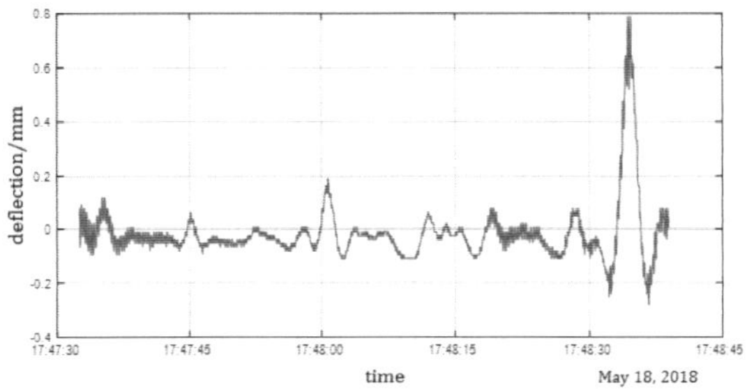

Figure 1 Mid-span deformation due to vehicle load under normal operation

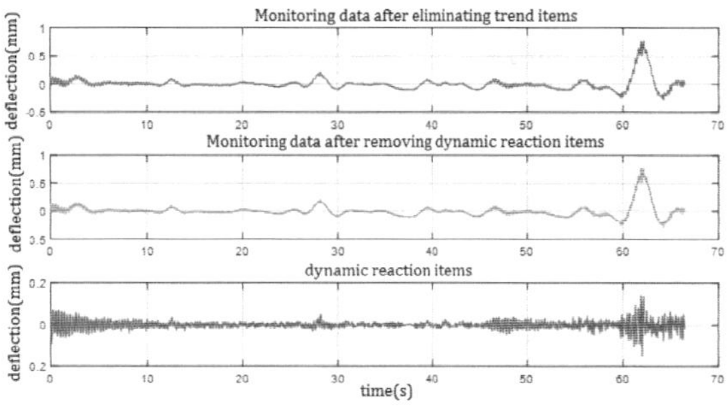

Figure 2 Static and dynamic component decomposition of monitoring data for deflection deformation

2.2 Methods for determining the frequency of interest in each period

Based on the three monitoring frequency periods, it is necessary to decompose the components of the static and dynamic deflections monitored in Figure 1, which can usually be performed using a zero phase difference filtering method.

The low-pass part is the static deflection component under the vehicle's own weight, and the high-pass part is the dynamic deflection component. Then the cut-off frequency of the filter is determined according to the first-order natural vibration frequency of the structure (-3dB is taken for the static deflection separation caused by the vehicle's own weight). Figure 2 shows the results of static and dynamic deflection analysis.

2.3 Methods for determining the frequency of interest of static and dynamic deflection components

The frequency of attention for static and dynamic deflection components can be determined through decomposition of the real-time signal and Fourier transform.

However, it is worth noting that the span of a given bridge structure is fixed and is one of the main factors determining the static component. The other determinant is the speed of the vehicles passing the bridge and the number of vehicles on the bridge. Therefore, it is feasible to select representative traffic conditions during the trial operation of the monitoring system and conduct statistical analysis so as to determine the frequency of interest.

The frequency component of the dynamic deflection is not determined using the upper limit of the natural structural frequency. The frequency component of the dynamic deflection response is often related to the ratio of the excitation frequency to the natural frequency of the vehicles passing the bridge, which is difficult to be made clear using theoretical calculation. However, for most bridge structures under normal circumstances (the vertical natural vibration frequency is not close to the natural frequency of the bridge, and the bridge deck is not seriously uneven), the dynamic deflection response can be basically controlled within the vibration frequency of the first few orders. The specific order can be determined based on the statistical analysis of the typical traffic conditions selected during the trial operation of the monitoring system. Figure 3 is the amplitude spectrum of static and dynamic deflection deformation before separation. Figure 4 is the amplitude spectrum of dynamic deflection deformation.

IOP Conf. Series: Materials Science and Engineering **794** (2020) 012025 doi:10.1088/1757-899X/794/1/012025

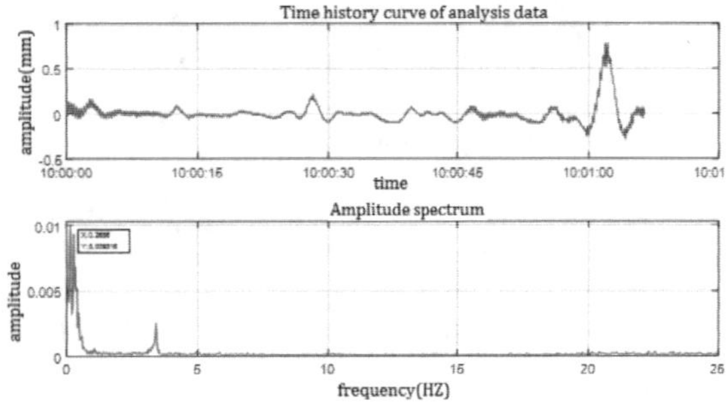

Figure 3 Amplitude spectrum of static and dynamic deflection deformation before separation

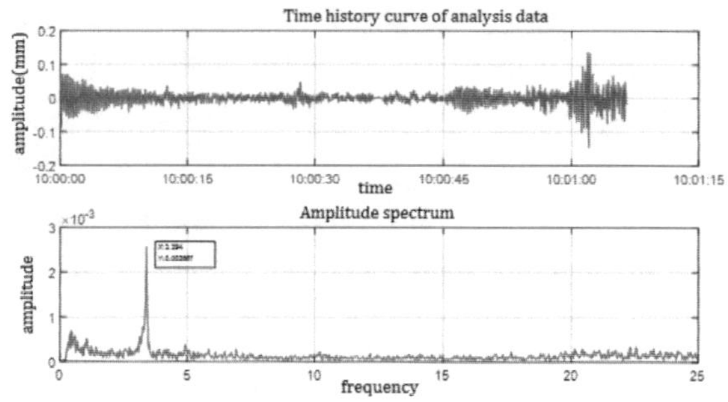

Figure 4 Amplitude spectrum of dynamic deflection deformation

It can be seen from Figure 3 that for the static deflection caused by the vehicle's own weight, the frequency of interest due to the impact of the vehicle speed and bridge span can be controlled at 0.2686 Hz. From Figure 4, it can be seen that for the dynamic deflection, the frequency component of interest should be at least 3.394 Hz.

3. Principles and methods for determining the deflection frequency of bridge structures

3.1 Principles for determining the maximum sampling conversion frequency

The sampling rate (fs) must be greater than twice the highest frequency (fn) component of the tested signal. This frequency is often called the Nyquist frequency.

$$f_s > 2 * f_n \tag{1}$$

If the signal is band-limited and the sampling frequency is greater than twice the signal bandwidth, then the original continuous signal can be completely reconstructed based on the samples. In general, the sampling frequency is guaranteed to be 2.56 to 4 times of the highest signal frequency. The sampling theorem explains the relationship between the sampling frequency and the signal spectrum, which is the basis for the discretization of continuous signals.

The sampling theorem provides the most basic conditions for analog-to-digital conversion. However, under these conditions, the analog signal, even a simple sinusoidal signal waveform, cannot be reconstructed after digitization. Assume there is a certain sinusoidal signal with a frequency of 2.0 Hz and take samples using 3, 5, and 8 times of the signal frequency (6.0 Hz, 10.0 Hz, and 16.0 Hz). The sampling results in Figures 5 to 7 shows that the sampling frequency that only satisfies the sampling theorem is not enough for restoring the waveforms. It is clear that the sampling theorem is the most basic condition to avoid aliasing distortion of the signal in the frequency domain, but not the condition to avoid the distortion of time domain signal.

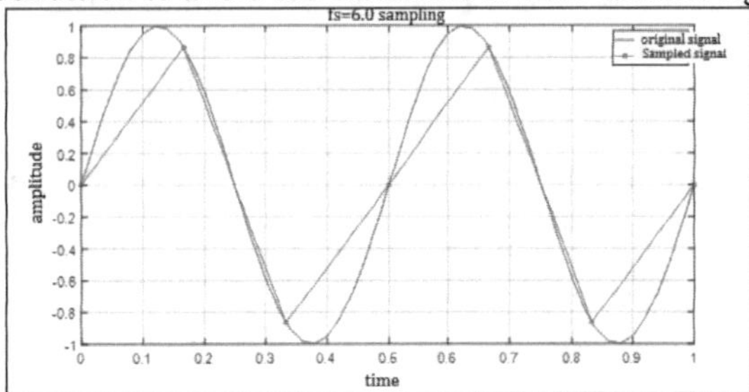

a) Sampling frequency: 6.0 Hz

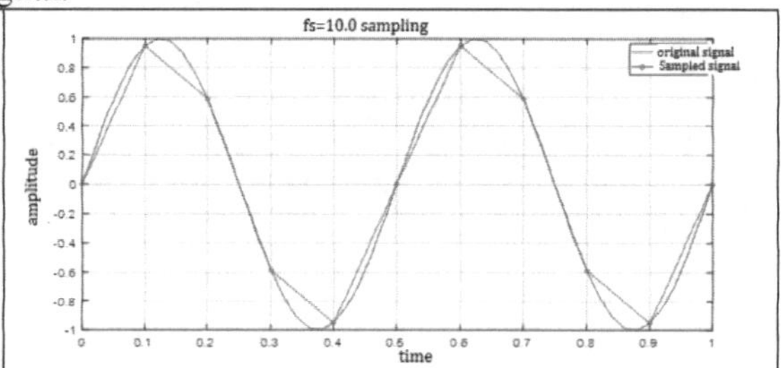

b) Sampling frequency: 10.0 Hz

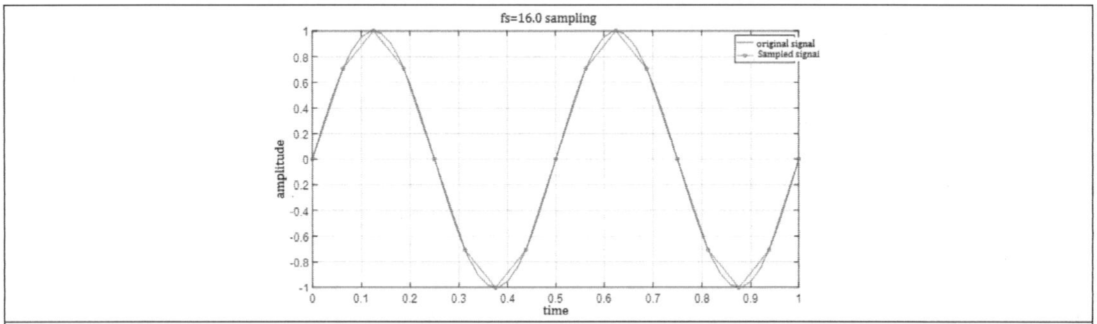

c) Sampling frequency: 16.0 Hz

Figure 5 Comparison between the sampling signal and the original signal of 1.0 Hz sinusoidal signal with different sampling frequencies

3.2 Effect of sampling frequency on time-domain signal amplitude

The data in Table 1 is taken from the article "Effect of Sampling Frequency on Amplitude of Vibration Signal" by Professor Zong from Institute of Structural Engineering and Earthquake Protection, Tongji University.

Table 1 Sampling frequency estimation of signal amplitude distortion

Frequency band	sampling frequency (Hz)	Peak error (%) mean value	Peak error (%) standard deviation	Maximum error
1-10Hz	20	8.3	7.0	37.5
	25	7.0	6.1	28.8
	30	5.3	4.8	21.4
	40	2.3	2.2	11.6
	80	0.99	0.87	3.8
	100	0.62	0.59	2.7

The data in Table 1 shows that when the sampling frequency is 10 times the frequency of interest, the average error of the mean value and standard deviation is less than 1%.

3.3 Determination of the highest sampling frequency for ensuring undistorted signal

It is recommended that the sampling rate shall be at least 5 times the analysis frequency, at which point the average percentage reduction of the maximum peak value of the sampled signal compared to the real peak value can be controlled within 3%. As for the sampling with strict peak requirements, it is still necessary to further increase the sampling rate. When the sampling rate is 8 to 10 times of the analysis frequency, the peak value of the signal is basically close to the real signal peak.

4. Conclusion

Based on the current situation of equifrequent sampling for real-time monitoring of deflection of small- and medium-span bridges in service and the component characteristics of their deflection deformation, this paper puts forward a method for determining the frequency of static and dynamic deflection components. By analyzing the influence of the sampling frequency on the amplitude of time-domain signal, this paper clarifies the range of the maximum sampling frequency limit that ensures undistorted signals, thereby providing reference for the determination of the real-time sampling frequency of the deflection of small-and medium-span bridges.

References

[1] Yulong Ma. (2014) Analysis of vehicle-bridge coupling vibration on highway curved continuous bean bridge[D]. Hunan University.
[2] Gang Wang. (2017) Research on Bridge Frequency Test Method Based on Vehicle-bridge

Coupling Theory[D]. Zhengzhou University.
[3] Yong Tu, Xiaoming Chen, Hui Zhao. (Jun 2019) Continuous T-beam bridge based on the LS-DZNA highway Research on Vehicle-Bridge Coupling Vibration Response Painted. Highway. pp 107-112.
[4] Mige Liu (May 2019) Research on Sampling Theorem. Journal of Xi'an University (Natural Science Edition) pp64-68.
[5] Hao Tang, Jianyun Zhang, Xiaodong Zhang. (May 2019) Application of Multi－Source Data Analysis on Bridge Operation Monitoring. Technology of Highway and Transport. pp88-93.

IOP Conf. Series: Materials Science and Engineering **794** (2020) 012026 doi:10.1088/1757-899X/794/1/012026

Research on Water Allocation Method of Juzhang River in Yichang City

Zhiqiang Yu[1], Jiang Wu[1,2,*], Hao Cao[1], Xiaofeng Hong[1,2], Qingqing Li[1,2] and Ang Li[3]

[1]Changjiang River Scientific Research Institute

[2]Hubei Key Laboratory of River Basin Water Resources and Ecological Environment Science

[3]China Institute of Water Resources and Hydropower Research

[*]Corresponding author's e-mail: cky_wujiang@163.com

Abstract: Water allocation refers to the gradual allocation of the total available water resources or the amount of water that can be allocated to the administrative area, and determines the share of water consumption or the share of water used for living and production in the administrative area. The total amount of available water resources includes the amount of surface water resources and the amount of groundwater resources that can be exploited, minus the overlap between the two. Aiming at the water allocation problem of the Juzhang River, this paper discusses the basic principles of water allocation and the technical route of water allocation plan formulation. A water allocation plan for the Juzhang River Basin was drawn up through steps such as supply and demand forecasting and supply and demand balance analysis. The research can provide a certain reference for the development of water allocation schemes in small and medium-sized river basins.

1. Introduction

A combined system of total amount control and quota management on water use is implemented in China. Regulating runoff and allocating water should be based on river basin planning and long-term water supply and demand planning, and basin-based water allocation plans should be formulated [1-3]. Carrying out water allocation work and formulating river basin water allocation plans are the basic requirements for implementing the Water Law of China and advancing administration according to law [4]. They are the objective need to strengthen macro-control of water resources and achieve sustainable use of water resources [5-6]. They are necessary measures to avoid and eliminate water conflicts, and maintain social harmony and stability [7-8]. It's the foundation for implementing the most stringent water resources management system and establishing a red line for the control of water resources development and utilization.

In order to strengthen the unified management and dispatch of water resources and ensure the safety of water supply of the Juzhang River in Yichang city, it is necessary to develop a water allocation plan for the Juzhang River in Yichang city. It can provide a reference for the compilation of the water allocation plan of the Juzhang River Basin in the later period.

2. Research area and hydrological analysis

2.1 Study area

The Juzhang River is a tributary of the northern bank of the middle reaches of the Yangtze River. It originates from the Helinggou in Baokang County. The river basin spans the four cities of Xiangyang, Yichang, Jingmen, and Jingzhou. The length is 341.5 km and the drainage area is 7284 km². The Juzhang River in Yichang is 221 km long and its drainage area is 3480 km². It involves 17 towns in 3 counties (cities) of Yichang City. Among them, the Ju River is 236 km long, with a catchment area of 3360 km², and 67 tributaries. The Zhang River in the east branch is 196.7 km long, which has a catchment area of 2965 km². A hydrological station is located 81 km upstream of the Juzhang River estuary, with a controlled area of 6325 km². The river above Herong Station is 245 km long, the slope of the river channel is 0.8‰, the density of the river network is 0.3km/km², the length is 157 km, the average width is 40.4%, the average elevation is 451m, and the river bending coefficient is 1.9.

Figure 1. Map of Juzhang River Basin.

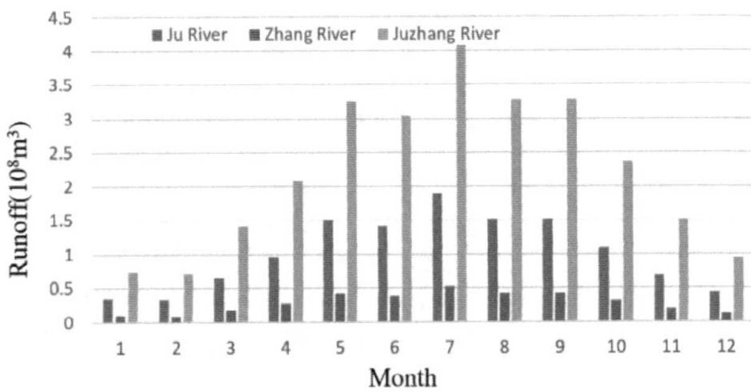

Figure 2. Monthly runoff at multi-year average frequency of the Ju River, Zhang River, and the Juzhang River.

2.2 Surface runoff and available water amount

According to 15 hydrological station data in Juzhang River basin, monthly runoff at multi-year average frequency of the Ju River, Zhang River, and the Juzhang River are obtained as shown in Figure 2.

The flood volume that is difficult to control and use during the flood season refers to the flood volume that cannot be controlled and controlled by engineering measures during the foreseeable period. The amount of flood that is difficult to control during the flood season is calculated based on the analysis of the total storage capacity and the availability of water above the control node at the lowest downstream of the basin. Subtract the natural runoff from the river basin control station during the flood season and subtract the maximum amount of water that can be stored and consumed in the river basin. The maximum amount of water that can be stored and consumed during the flood season is the maximum value of the total amount of water consumption, reservoir storage, and adjustable external basin water in the flood season.

Based on the multi-year average runoff results in the Juzhang River Basin and calculation results that are difficult to control during the flood season in the Juzhang River Basin. The available water volume in the Juzhang River Basin is calculated as shown in Table 1.

Table 1. Available water amount of Juzhang River (10^8m^3).

Station	Frequency	Runoff	Minimum ecological water demand	Water amount that is difficult to control during the flood season	Available water amount

Ju River	multi-year average	11.34	1.134	3.036	8.304
Zhang River		3.327	0.3327	0.9962	2.3308
Main stream		2.4148	0.2415	0.9489	1.4659
Total of the Juzhang River Basin		17.0818	1.7082	4.9811	12.1007

3. Allocation of water resources based on total water use control

3.1 Water demand forecast

The water demand of the Juzhang River in Yichang city is mainly Mingfeng town, Hualinsi town, Jiuxian town, Yangping town, Maopingchang town, Hekou town of Yuan'an county, Yuyang office, Yuquan office, and Baling office, Yuxi town, Miaoqian town, Herong town, Lianghe town, Caobuhu town of Dangyang city, including domestic water, agricultural water, industrial water and ecological water. The water demand results are shown in Table 2.

Table 2. Water demand forecast result.

District	Frequency	Water demand (2025)	District	Frequency	Water demand (2025)
Mingfeng town	multi-year average	1448	Yuquan office	multi-year average	2913
Hualinsi town		924	Baling office		2546
Jiuxian town		1006	Yuxi town		4259
Yangping town		1488	Miaoqian town		3768
Maopingchang town		1586	Herong town		3237
Hekou town		829	Lianghe town		1525
Yuyang office		2937	Caobuhu town		1336
Total			29801		

3.2 Water supply forecast

Combined with the water conservancy census data of Yuan'an county and Dangyang city and statistics of water supply capacity of water conservancy projects in the Juzhang River Basin, the water supply capacity is analyzed and calculated for the surface water storage, diversion, water lifting, water pond and weir projects. The available water supply amount at different frequencies are calculated as shown in Table 3.

Table 3. Water supply forecast amount at different frequencies.

Year	Frequency	Available water supply amount（$10^4 m^3$）
2020	multi-year average	89654
	50%	88495
	75%	76952
	95%	60581
2025	multi-year average	93326
	50%	91569
	75%	80659
	95%	62948

3.3 Water allocation scheme

The total water use control indicators of Yuan'an county in 2020 and 2030 are 165 and 176 million m^3, respectively. The total water use control indicators for Dangyang city in 2020 and 2030 are 330.3 and 350 million m^3, respectively. Taking into account the population, area, socio-economic, and water use levels, the total water use control index of districts is decomposed into the Juzhang River Basin. The water allocation scheme is shown in Table 4.

Table 4. The water allocation scheme of Juzhang River in Yichang City (10^4m^3).

Frequency	District		2030				
			Domestic water	Production		Ecological water	Total
				Industrial water	Agricultural water		
multi-year average	Yuan'an county	Mingfeng town	254.96	1174.80	156.74	11.03	1597.52
		Hualinsi town	77.29	446.41	451.66	4.19	979.55
		Jiuxian town	106.04	632.72	341.53	5.94	1086.23
		Yangping town	158.47	944.20	497.04	8.87	1608.57
		Maopingchang town	105.69	627.12	923.95	5.89	1662.65
		Hekou town	64.53	380.48	427.20	3.57	875.79
		Subtotal	766.99	4205.72	2798.13	39.49	7810.33
	Dangyang city	Yuyang office	458.27	1823.07	1040.81	23.11	3345.26
		Yuquan office	150.77	734.19	2177.25	9.31	3071.52
		Baling office	254.69	1150.11	1381.94	14.58	2801.32
		Yuxi town	237.40	1151.14	3105.67	14.59	4508.80
		Miaoqian town	170.97	862.80	2908.48	10.94	3953.19
		Herong town	242.93	1173.72	2063.87	14.88	3495.40
		Lianghe town	146.28	737.38	796.96	9.35	1689.97
		Caobuhu town	111.04	509.10	821.60	6.45	1448.20
		Subtotal	1772.36	8141.50	14296.58	103.21	24313.65
	Total		2539.35	12347.23	20670.06	142.70	35699.34

4. Conclusions

As the recommended water allocation scheme proposed in this paper is only a water distribution scheme with three frequencies of 50%, 75%, and 95%, and the actual water flow situation in the basin often does not correspond to these three frequencies, Water distribution should be based on the proportion of water distribution. The negotiation of the water allocation plan confirmed that the participation of relevant prefecture-level and county-level governments and their water administrative departments is needed to coordinate water conflicts in the basin. When it is difficult to reach consensus, an authoritative ruling body should be established to determine the plan agreed by the majority, so as to improve the efficiency of the initial water right allocation and prevent long-standing disputes. Water administrative departments at all levels should actively exercise their administrative management and technical guidance functions, give full play to the regulation and storage function of water storage projects, strive to realize flood resource utilization, and try to avoid conflicts in water supply due to uneven spatial and temporal distribution of water resources.

Acknowledgements

This work was funded by the National Key R&D Program of China (2016YFC0402203), the National Natural Science Foundation of China (No. 51779013, 51509009), National Public Research Institutes for Basic R&D Operating Expenses Special Project (CKSF2019209,No.CKSF2019478), Water Conservancy Science and Technology Innovation project of Guang Dong Province(2017-03), Natural Science Foundation of Hubei Province(2017CFB606, 2018CFB655). Special thanks were given to the anonymous reviewers and editors for their constructive comments.

IOP Conf. Series: Materials Science and Engineering **794** (2020) 012026 doi:10.1088/1757-899X/794/1/012026

References

[1] Dinar A, Rosegrant M W, Meinzendick R. Water Allocation Mechanisms: Principles and Examples[J]. 1997, 95(1):30–41.
[2] N. Takama, T. Kuriyama, K. Shiroko,et al. Optimal water allocation in a petroleum refinery[J]. 4(4):251-258.
[3] Charles W. Howe, Dennis R. Schurmeier, W. Douglas Shaw. Innovative Approaches to Water Allocation: The Potential for Water Markets[J]. Water Resources Research, 1986, 22(4).
[4] Olivier Barreteau, Patrice Garin, Alexandre Dumontier, et al. Agent-Based Facilitation of Water Allocation: Case Study in the Drome River Valley[J]. Group Decision & Negotiation, 2003, 12(5):441-461.
[5] Juan Reca, José Roldán, Miguel Alcaide, et al. Optimisation model for water allocation in deficit irrigation systems: I. Description of the model[J]. Agricultural Water Management, 48(2):0-116.
[6] Hamstead M. Water allocation planning in Australia - current practices and lessons learned[J]. 2008.
[7] Speed R, Li Y, Zhou Z, et al. Basin Water Allocation Planning: Principles, Procedures and Approaches for Basin Allocation Planning[J]. 2013.
[8] René Brink, Arantza Estévez-Fernández, Gerard Laan, et al. Independence of downstream and upstream benefits in river water allocation problems[J]. Social Choice & Welfare, 2014, 43.

Manufacturing and Hoisting Technical Measures of A Kind of Glass Fiber Reinforcing Bar

Mingqi Liu [1], Bihui Dai [1], Dewen Liu [1]*

[1]College of Civil Engineering, Southwest Forestry University, Kunming 650224, P.R. China.

*Corresponding author's e-mail: 350376315@qq.com

Abstract. In this paper, the key technologies of manufacturing and hoisting of glass fiber reinforcing bar (GFRB) in a certain project are described. When lowering the reinforcement cage, the two ends are docked to increase its rigidity; in the process of making GFRB cages, it is necessary to take measures to increase the stiffness of the fiberglass cages; the hoisting points need to be placed on the reinforcing bar. And pictures and texts are used to assist in the explanation. This paper has certain reference value for the construction process and related technical engineers.

1. Results and Discussion

For the four walls of the shield hole position of a certain project, due to its special shield needs, the main reinforcement and stirrups of the continuous wall within 1.0 meter around the outer frame of the opening of the tunnel were made of glass fiber reinforcing bars (GFRB) of the same diameter, and the remaining steel bars were still ordinary steel bars[1]. Due to the poor bending and shear resistance of GFRB bars, the self-stability of the framework was poor when the fiber reinforced cage is hoisted[2], and it was easy to loose the cage, therefore the reinforced cage made of GFRB was divided into two sections to lower the reinforced cage. Measures can also be taken to increase the rigidity of the fiberglass cage, such as setting I-beam edging and reinforced trusses that can be removed later to ensure its own stability. When hoisting, but paying attention to cutting off the I-beam and truss ribs during the lowering process. There were four GFRB bars walls. The length of the reinforced cage was 53.75 meters, and it was divided into two sections with an upper length of 27.6 meters and a lower length of 26.15 meters. What is more, 6 points manner was used for hoisting, and U-shaped buckles were used to connect the joints. The connections between the GFRB and the reinforcing bars[1,3], and the reinforcing bars and the reinforcing bars shall all be made of steel U-shaped buckles, the U-shaped buckles shall be affixed to the inside of the U-shaped buckles together with the reinforcing bars with the diameter of the main reinforcing bars, the GFRB shall be affixed to the outside of the U-shaped buckles. The number of U-shaped buckles at the connecting end of each bar should not be less than two, and the overlap length should be not less than 40d.

The connection between the GFRB and the steel bars, and between the GFRB and the GFRB in other parts can be wrapped with iron wire firmly.

Figure 1. The GFRB cage is bandaging by U-shaped buckles

During the production of the GFRB cages, some measures should be taken to increase the rigidity of the GFRB cages (i.e., use I-beam edging on both sides of the cage, the inside of the cage adopts some fiberglass trusses or reinforced trusses that can be removed later) to prevent the large deformation during the hoisting and transportation process from affecting the lowering of the GFRB cage.

Figure 2. Truss bars in glass fiber

Figure 3. Cut off the reinforced trusses in the GFRB during the lowering process

For situations where hoisting is required to place the tendon cage in place, the shield needs to pass through the middle of the tendon cage (along the height), the overlap of the tendon cage generally occurs between the two parts (the GFRB and the reinforcing bar on the stressed main tendon, and between the GFRB and the GFRB)[3,4]. The hoisting method should adopt three lifting points from top to bottom (along the height direction). The first hoisting point is generally located within 1 meter of the uppermost part of the steel cage, and the second hoisting point is generally located within 1 meter above the connection point. The third hoisting point is generally located at a suitable position below the second connection point.

During the rib cage hoisting process, the hoisting points must be placed on the reinforcing bars. It is strictly prohibited to place the hoisting points on the GFRB or the connection position between the GFRB and the reinforcing bars.

Figure 4. The scene when the GFRB cage is lowering

2. Conclusions

Generally speaking, key technologies of producing and hoisting of glass fiber reinforcing bar (GFRB) in a certain project are described. When lowering the reinforcement cage, the two ends are docked to increase its rigidity; in the process of making GFRB cages. It is necessary to take measures to increase the stiffness of the fiberglass cages; the hoisting points need to be placed on the reinforcing bar. We think that this paper will be of great reference value to new building materials, construction process and technology, and related construction personnel.

Acknowledgments
The writers gratefully acknowledge the financial support of National Natural Science Foundation of China (No. 51808467), Yunnan Agriculture Foundation Fund Project (No.2017FG001-097), Yunnan Provincial Department of Education Fund Project (No.2019J0194), Doctor Start-up Fund of Southwest Forestry University and Yunnan Postdoctoral Fund (2018).

References
[1] Cui J.W., Song H., Zhang H.P., Yan X. (2006) Research on mechanical properties of short fiber reinforced PVC. FRP/Composite Materials, 01:31-33.
[2] Li C.J., Yuan X.K., Yu M.H. (2006) Study on preparation of long glass fiber reinforced PVC composites by water suspension-pultrusion, GRP/Composite Materials; 04:41-43.
[3] Mochamed, N., Farghaly, A.S., Benmokrane, B. (2015) Aspects of deformability of concrete shear walls reinforced with glass fiber-reinforced bars. Journal of composites for construction, 19(5):06014001.
[4] Geonwoo, Kim., Mu-Kyung S., Namkyoung C.,Yong-Il K., Ki-Bok K. (2019) Comparison of PZT, PZT based 1-3 composite and PMN-PT acoustic emission sensors for glass fiber reinforced plastics. International Journal of Precision Engineering and Manufacturing, 20(6):1007-1015.

IOP Conf. Series: Materials Science and Engineering **794** (2020) 012028 doi:10.1088/1757-899X/794/1/012028

Research on Hoisting Method of Reinforcement Cage for Underground Continuous Wall

Mingqi Liu [1], Bihui Dai [1], Dewen Liu [1]*

[1]College of Civil Engineering, Southwest Forestry University, Kunming 650224, P. R. China.

*Corresponding author's e-mail:350376315@qq.com

Abstract. The hoisting method of reinforcement cage in a diaphragm wall project is studied. Through certain graphic display, the process of hoisting the reinforced cage was described. In this paper, we propose a specific method of hoisting a 880mm thick of reinforcement cage, and the entire specific process and technology of lifting it is also offered. During the hoisting procedure and the lifting transformation process, reinforcement cage hoisting is an important part of construction, and this article has certain reference value to the construction process.

1. Introduction

In order to save construction time and reduce the adverse effects caused by section production, the reinforcement cage is hoisted into the groove once in a project. When the reinforcement cage is lowered, the double crane is used for lifting and then straightening in the air. With a 260t crawler crane as the main crane, a 150t crawler crane is used as the auxiliary crane. When lifting with crane, make the center of hook coincide with the center of gravity of reinforcement cage, and ensure the lifting balance. And it need 15m (lifting rope) + 14m (connecting rope) long steel wire rope for main crane, 13m + 10m long steel wire rope for auxiliary crane[1,2].

2. Reinforcement cage parameters

The thickness of the reinforced cage of the wall of a project is 880mm (inside and outside concrete protection layer 80mm), separately "-", "L", "T", "Z" four forms. The length of the reinforcement cage of the diaphragm wall on the south side of the standard section (near the second ring ramp bridge) is 55.75m. The length of the reinforced cage of the underground diaphragm wall on the north side of the standard section is 48.95m. The end well is 57.25m. It can be seen from the comparative calculation that the maximum cage weight is the reinforcement cage of the first 5m wide diaphragm wall of the end well. Therefore, the reinforcement cage of the first 4m wide diaphragm wall of the end well is selected for hoisting and checking calculation[3].

Table 1. weight list of end reinforcement cage

Serial number	Name	Specifications	Weight (kg)	Remarks
1	Main tendon	32/28	23578	HRB500
2	Horizontal tendon	20/18	5778	
3	Truss bars	28	2645	
4	Stand up	18	2033	
5	Scissors tendon	28	2601	

6	Rings	36	72	
7	Positioning pad		77	
8	H Section steel		19863	
9	Gross weight		56616	

3.Reinforcement cage lifting process

(1) Direct the two cranes to the lifting position, crane worker install shackles for lifting points respectively.

(2) After checking the installation condition and stress center of gravity of the steel wire rope of the two cranes, start to lift horizontally at the same time.

(3) After hoisting the lower reinforcement cage to 0.3m ~ 0.5m above the ground, check whether the lower reinforcement cage is stable, and then lift the hook with a 260t crane. According to the distance from the bottom of the lower reinforcing cage to the ground and then direct the auxiliary aircraft to cooperate with the hook at any time.

(4) After the steel cage is lifted, the 260t crane rotates to the left (or right) side and 150t crane is turned to the right position, let the cage be perpendicular to the ground.

(5) Instruct the lifting workers to unload the lifting point of the 150t crane on the steel cage. Then keep away from the lifting operation range.

(6) Direct the 260t crane to insert the steel cage into the trough, location, crane travel should be smooth, pull the rope from the steel cage. The reinforcement cage is placed at the mouth of the slot and kept horizontal. When the reinforcement cage is lowering, it must not be forced into the slot[3,4].

(7) After the steel cage is lowered into place as a whole, it is placed flat. After the steel cage is lowered, the next process is performed.

4. Reinforcement cage lifting conversion process

(1) With two machines in place, begin lifting the reinforced cage.

(2) Double machine lifting the steel cage, the main crane lifts the reinforcement cage, the auxiliary crane moves smoothly forward.

(3) Main hoist hook, the auxiliary lifting hook runs slowly, until the main crane lifts the reinforcement cage.

(4) The auxiliary crane unloads the hook, and then the main crane completely lifts the reinforcement cage. The main crane rotates the boom to transfer the reinforcement cage to the lower guide wall. Then, align with the dividing line and start to lower it. In this process, a special person will pull the wire rope of the auxiliary crane. When it is lowered to a node, the main crane will stop lowering and the special person will remove the clasp.

(5) When the auxiliary hoisting ropes are all removed, the main crane continues to be lowered. When the main crane changes the lifting point of the steel wire rope, use the shoulder pole to clamp the place where the reinforcement cage passes through the shoulder pole, and then the main crane puts down the reinforcement cage to make the weight of the reinforcement cage bear on the shoulder pole[3,4].

(6) The lifting rope and connecting rope of the main crane shall be installed, and the hook of the main crane shall be closed to make the steel wire rope of the main crane bear the force, lift the reinforcement cage, and draw out the shoulder pole, and the main crane shall continue to lower the reinforcement cage.

(7) When the reinforced cage is lowered to the first horizontal rib from the top of the cage, use a flat load to clamp the hanging point of the cage head again. Transform the wire rope of the main crane. Install the wire rope of the main crane on the suspension ribs, and the main crane lifts the hook until the steel cage is lifted to the guide wall 10-20cm, then the flat load is pulled out. Continue to lower the reinforcement cage, so that the suspension bars of the reinforcement cage rest on the flat load, and finally remove the shackle of the wire rope, and the entire suspension process of the reinforcement cage is completed.

IOP Conf. Series: Materials Science and Engineering **794** (2020) 012028 doi:10.1088/1757-899X/794/1/012028

5. Conclusion

In general, a 260t crawler crane is used as the main crane, and a 150t crawler crane is used as the auxiliary crane. The entire reinforcement cage is lowered into place and leveled to ensure the safety and stability of the entire suspension process. The waiting lifting method ensures the orderly progress of a project, saves construction time and reduces the adverse effects caused by section production, and completes the requirements for the project to use the steel cage to be hoisted into the groove at one time.

Acknowledgments

The writers gratefully acknowledge the financial support of National Natural Science Foundation of China (No. 51808467), Yunnan Agriculture Foundation Fund Project (No.2017FG001-097), Yunnan Provincial Department of Education Fund Project（No.2019J0194）, Doctor Start-up Fund of Southwest Forestry University and Yunnan Postdoctoral Fund (2018).

References

[1] Zhou, C.H., (2010) Talking about the difficulties and solutions of the underground continuous wall. Heilongjiang Science and Technology Information, 16: 297.
[2] Ge J. (1997) Foundation pit support for high-rise buildings. Journal of Nantong Vocational University, 04: 47-48.
[3] Wu, C.J., Sun, Z.H., Bao, H. (2018) Study of deformation characteristics of diaphragm wall induced by deep large excavation in soft soil region. Yantu Lixue/Rock and Soil Mechanics, 39: 245-253.
[4] Zhou, X., Fang, G., Leng, Y. (2014) Manufacturing and hoisting technology for reinforcement cage of deep foundation excavation bored piles. Advanced Materials Research, 1044-1045: 633-637.

IOP Conf. Series: Materials Science and Engineering **794** (2020) 012029 doi:10.1088/1757-899X/794/1/012029

Constructed wetland technology and its application in water quality control of south lake system

Li Can[1,2,3*], Xu Jing[4]

[1]Zhejiang Guangchuan Engineering Consulting Co., Ltd., Hang Zhou 310020, China

[2]Zhejiang Institute Of Hydraulics & Estuary, Hang Zhou 310020, China

[3]Zhejiang Provincial Key Laboratory Of Hydraulic Disaster Prevention and Mitigation, Hang Zhou 310020, China

[4]Zhejiang University of Water Resources and Electric Power, Hangzhou 310018, China

Abstract. Constructed wetland technology is widely used. It has significant effect in the removal of eutrophication factor N, P. From the wetland development and cleansing mechanism of The paper introduces the constructed wetland technology in the use of water treatment of lake form the development of the constructed wetland and the derergency mechanism. The paper especially introduces its application in water quality control of south lake system. In order to analyze the total phosphorus load, it analyze the annual water diversion by the water balance. It discusses the design points of the constructed wetland used in the south lake system. Finally, the outlook of the research in the constructed wetlands and it's application in eutrophic lake pollution control.

1. Introduction

Constructed wetland is an artificial system that simulates the structure and function of natural wetland. It uses the physical, chemical and biological triple functions provided by matrix, aquatic plants and microorganisms to realize the treatment of sewage. It has many advantages such as: low investment, stable treatment effect, low operating cost, simple operation and so on. Therefore, the constructed wetland technology has broad application prospects in the sewage treatment and water quality control.

2. Development Status of Constructed Wetland and Its Decontamination Mechanism

2.1Development status of constructed wetland

Constructed wetland originated in the 1950s and gradually evolved into the main unit of sewage treatment after the 1970s. At least 6,000 constructed wetland systems in Europe and over 1,000 constructed wetland systems in the United States are currently operating. The number of constructed wetland systems is increasing in South America, Australia, New Zealand, Asia and Africa[1]. However, China did not begin research on constructed wetland until the "seventh five-year" period. According to relevant surveys, before 2006, there were more than 200 constructed wetland systems with an area of 100 m^2 to 8000 m^2 in China[1]. The using of constructed wetlands to control agricultural non-point source pollution can produce better economic and environmental benefits in China.

2.2Decontamination Mechanism of Constructed Wetland

Constructed wetland remove dissolved organic matter by microorganisms living in it. And particulate organic matter is quickly removed by precipitation and filtration. Microorganism removes dissolved organic matter primarily through aerobic and anaerobic degradation. The degrade of heterotrophic bacteria produces carbon dioxide. The degrade of facultative anaerobes and obligate anaerobic bacteria produce methanol, fatty acids, hydrogen, carbon dioxide. Anaerobic degradation of sulfate-reducing bacteria and methanogens produce methane, carbon dioxide, hydrogen sulfide [2].

A part of the inorganic nitrogen in the sewage can be absorbed by the plant and removed by the harvesting of the plant, while the organic nitrogen is mainly removed by ammoniation, nitrification and denitrification. First, the organic nitrogen in the sewage releases ammonia nitrogen through the ammoniation of the wetland microorganisms; secondly, the ammonia nitrogen is converted into nitrite nitrogen by the action of nitrous acid bacteria, and then further oxidized to nitrate nitrogen; finally, under non-oxidative conditions and by the action of denitrification bacteria, the nitrate nitrogen is reduced to nitrogen or nitrogen dioxide by the action of denitrification bacteria, which escapes from the system, thereby achieving the purpose of removing nitrogen.

Plants in constructed wetlands assimilate phosphorus in wastewater, producing organic components such as ATP, DNA, and RNA, and removing phosphorus from the system by harvesting the plants. Studies have shown that matrix adsorption and precipitation play the most important role in the removal or fixation of phosphorus in constructed wetlands[3].

3. Application of Constructed Wetland Technology in Controlling Exogenous Pollution of Lakes

3.1The possibility of applying constructed wetland technology in eutrophication lakes restorations

Agricultural non-point source pollution is one of the main factors for the formation of eutrophic lakes in China. The agricultural non-point source pollution is scattered, the amount of sewage is large, and the content of N and P in the sewage is high. It is easy to cause eutrophication of the lake by directly discharging into the lake. It is difficult to solve this problem in general treatment technology. Constructed wetland has significant effects in removing N and P which are eutrophication factors, and has the advantages of small investment, simple construction, low operation and management costs, and significant ecological benefits. Therefore, it is possible to consider the application of constructed wetland technology to eutrophic lake treatment.

3.2Key points of Constructed Wetland Design in eutrophication lakes restorations

The constructed wetland system can be divided into three main types: surface flow wetland, horizontal subsurface wetland and vertical flow wetland according to the difference of water flow state. Surface flow wetland has some features of low cost, low hydraulic load and great effect by season; horizontal subsurface wetland has better effect on removal of organic matter and heavy metals such as BOD and COD, and is few affected by seasonal influence; vertical flow wetland is better than the former two, but Its construction requirements are high and it has not been widely used [3]. Which type of wetland to use is decided by the local pollution sources, the amount of sewage, and the investment amount of the project in a specific engineering practice.

Aquatic plants are an integral part of the constructed wetland system and are the basis for the survival of many peri-cluster organisms. They can absorb and intercept the organic matter in the sewage and provide a living environment for the microorganisms. In the selection of aquatic plants, it is necessary to investigate the aquatic plant species in the local lakes, and try to use local aquatic plants to avoid species invasion.

In constructed wetland systems, the matrix plays the most important role in the removal of phosphorus and provides a carrier and medium for the survival of aquatic organisms and microorganisms. Yuan Donghai et al. studied the mechanism of seven kinds of matrix materials for purifying phosphorus in sewage. The results show that slag and coal ash have good removal effect on phosphorus, followed by vermiculite, yellow cinnamon soil and lower loess, and the zeolite and sand is the poorest[4]. According to Zhu Xizhen et al., the constructed wetland of coal-ash matrix has better removal effect on organic pollutants, and the removal rates of COD and BOD are 71%~88% and 80%~89%[5]. In constructed wetland designs, the matrix can be configured based on the major contaminants in the sewage and the local material supply.

4. Application of constructed wetland in water quality control of Nanhu Lake system

The Nanhu Water System Project is located in the southeast of Shuangliu County, 1 km away from Chengdu City. The total planned land area is 17 km^2. The Nanhu Water System is the core of the Nanhu area planning. In the Nanhu water system, six artificial lakes are planned, with a total lake area of 376 m^2 and a total water storage capacity of 16.87 million m^3.

4.1Analysis of exogenous pollution in Nanhu Lake system

The annual average water intake of the Nanhu Lake system can be calculated by the water balance calculation. Average annual total phosphorus load in the Nanhu lake system can be obtained by the calculation model of the intake water pollution load index.

4.1.1Analysis of water balance in Nanhu Lake

According to the water circulation movement characteristics and operation scheduling mode of the Nanhu water system, the water balance equation of Nanhu can be established:

$$\Delta W = R_i + P + Q - E - R_o$$

Where: E is the amount of water evaporating from the lake; R_i is the runoff of surface water into the lake; R_o is the runoff flow from the lake; Q is the amount of other forms of lake water inflow; P is the total amount of precipitation on the lake; ΔW is the change of water storage in the lake during a certain period of time .

(1) The amount of water intake into the lake

There is no large-scale development in the area, which is a natural lake basin. Therefore, the scope of the catchment area and the way and direction of runoff and runoff input are basically the same as those of ordinary natural lakes. When the lake is formed, the amount of water shortage has the amount of water replenished by the old south main channel that passes through the area. Based on the above situation, the surface water input items are divided into two parts: one is the runoff input of the actual catchment area outside the lake, and the other is the input of the water intake of the old south main channel.

① Annual average amount of water that surface runoff flows into Nanhu

The rainfall area of the Nanhu Lake system is 15.04km^2. The total lake surface area of the lake system under normal water level is 3.759 million m^2. The rainfall area of each lake and the corresponding lake area under normal water level are shown in Table 1. The area of the runoff yield is 11.281 million m^2, and the average annual runoff depth of the area is 450 mm. Therefore, the annual average amout of water that runoff flows into the Nanhu Lake system is 5,078,500 m^3.

Table 1. Characteristic parameter table of lake system watershed

	Controlling catchment area (km^2)	Lake surface area under normal water storage level (Ten thousand m^2)
Lake A	3.55	100
Lake AT	1.47	51.9
Lake B11	2.54	26.8
Lake B12	3.93	104
Lake B2	2.88	87.6
Lake C	0.67	5.6
total	15.04	275.9

② Average annual rainfall on the lake

IOP Conf. Series: Materials Science and Engineering **794** (2020) 012029 doi:10.1088/1757-899X/794/1/012029

The average annual rainfall of the lake system can be calculated using the average annual rainfall measured by the Shuangliu County Meteorological Bureau and the lake area under the normal water level. The average annual rainfall measured by the Shuangliu County Meteorological Bureau is 902.7mm, and the lake area under the normal water level of the lake system is 3.759 million m^2. The average annual rainfall of the lake system is 3,393,200 m^3.

③ Annual average water diversion of the south main channel

Through years of average adjustment calculations, it can be seen that the annual average water diversion of the Nangan Canal is 819,000 m^3.

Then, the average annual lake intake in Nanhu is the sum of the above three items, namely:

507.85+ 339.32 + 81.90 = 929,070,000m^3

(2) Consumption of water

At this stage, the water consumption of Nanhu Lake only includes the loss of lake surface evaporation and the leakage loss of the lake system.

① Annual water evaporation quantity from the lake surface

The lake surface evaporation quantity can be calculated by the lake surface area and the annual average water surface evaporation quantity. The annual average water surface evaporation can be converted by the average annual evaporation which can use Shuangliu County Meteorological Bureau's measured average annual evaporation of 931.3 mm. The project is located in the western part of the Sichuan Basin and belongs to the subtropical humid monsoon climate. According to the *Hydrological Calculation Code for Water Resources and Hydropower Engineering* (SL278-2002), the water surface evaporation conversion coefficient is a=0.9, and the average annual water surface evaporation is 931.3 × 0.9 = 838.17 mm. Then, the average evaporation quantity of the lake in the Nanhu Lake system is 3,150,700 m^3.

② Annual seepage quantity of Lake system

The hydrogeological conditions of the Nanhu Lake System are good according to the engineering geological section of the Chengdu Nanhu Headquarters Economic and Creative Industry Development Land Consolidation Project - Lake System Engineering Feasibility Study Report.The lake bed is a non-seepage layer, and the groundwater surface is close to the reservoir surface. The seepage quantity of Lake system is approximately 1% of the normal storage volume of the reservoir. The annual seepage quantity of the lake system is 1,975,600 m^3 from the results of years average adjustment calculation.

The change in the annual storage of the lake

According to the principle of water balance, the multi-year average of changes in lake water storage should be zero.

Annual flow of the lake

According to the above data and water balance model, the runoff output from the Nanhu Lake system is:

$$Ro = Ri + P + Q\text{-}E$$

Run the equation: Ro = 4,152,500 m^3

4.1.2 Calculation of total phosphorus load in Nanhu Lake

(1) Pollution load in the water diversion

The point source pollution in Nanhu is mainly from the pollutants carried in the diversion channel of the old South Canal. Field investigations were carried out on the Nanhu Lake system from the old south canal diversion channel section, and the typical channel section was selected to monitor the pollutant content. The calculation model of the diversion pollution load index is:

$$P_i = Q \times C_i$$

Where, P_i is the load of a load indicator; Q is the discharge of sewage; C_i is the concentration of a certain load indicator.

Through the distribution of sample points and the detection concentration analysis, the total phosphorus concentration in the Nangan Canal sample was 0.095 mg/L. The annual average water diversion of the Nanhu Lake system is analyzed by water balance, and the amount is known to be 819,000 m^3. Then:

Annual average total phosphorus load in water = sample total phosphorus concentration × annual average water intake = 0.095 mg / L × 81.90 million m^3 = 77.81 kg

(2) Atmospheric deposition

The atmospheric sediments on the lake surface were measured and collected by an atmospheric sediment trap. Through analysis, the average condition of atmospheric deposition in Nanhu Lake in Chengdu is 246.80 mg/d·m^2 in the nearshore area and 85.12 mg/d·m^2 in the central area. Based on this calculation, the average annual atmospheric deposition of the Nanhu Lake in Chengdu is 17,7048.9 kg, of which the total phosphorus (0.17% of the total sediment particles) is 296.73 kg/a, and the lake area is 3,759,900 m^2.

(3) Input by atmospheric precipitation

Through analysis, the total precipitation phosphorus content in the Nanhu catchment area is 0.0462 mg/L. The annual average receiving rainfall of the Nanhu Lake system is 3,393,200 m^3. Then: annual average atmospheric precipitation input total phosphorus load = total phosphorus concentration in precipitation × annual average atmospheric precipitation = 0.0462 mg / L × 3,393,200 m3 = 156.77 kg / a.

(4) Non-point source pollutant load in the Nanhu Basin

The non-point source pollution in the Nanhu water system is mainly from farmland and its formation of surface runoff.

Referring to related research, the pollutant output load model is:

$$Q = R \times C_i$$

In the formula, Q is the annual load of a certain load index; R is the annual runoff; C_i is the concentration of a certain load indicator.

Through background investigation, the average value of total phosphorus content of surface runoff is 0.37 mg/L; the average annual runoff of Nanhu Lake system is 5.078 million m^3 calculated by water balance, then: annual total phosphorus load = annual average runoff × total Phosphorus concentration = 5,078,500 m^3 × 0.37 mg / L = 1879.03 kg

(5) Total amount of external pollution load in Nanhu Lake

According to the above analysis, the annual average total phosphorus pollution load of Nanhu Lake is 2410.33 kg, as shown in Table 2 below.

Table 2. Annual average load of total phosphorus in Nanhu Lake

Total phosphorus pollution source		Load（kg）
South Main Canal Diversion Load		77.81
Wet and dry sediment load	Atmospheric subsidence load	296.73
	Atmospheric precipitation load	156.77
Non-point runoff load		1879.03
Total		2410.33

4.2Design of constructed wetland system

The hydration treatment is carried out by using a horizontal subsurface flow constructed wetland treatment system in low tail area of the lake. The horizontal submerged artificial wetland treatment system has good moisture retention, and the phosphorus removal rate can reach 60-70%, which is less affected by the season.

4.2.1 Hydraulic retention time

The hydraulic retention time of the constructed wetland is:

$$t = V \cdot \varepsilon / Q_{av}$$

Where t is the hydraulic retention time; V is the wetland volume; is the wetland porosity; is the average flow. In order to ensure the effect of denitrification in the subsurface wetland, the hydraulic retention time is 2~4d.

4.2.2 Water level control

Considering the normal growth of plant roots and ensuring the nitrification effect, the depth of the system is 1~2 m, from the water surface to the depth is about 0.8 m, the sunlight can penetrate, the photosynthesis is strong, the dissolved oxygen is sufficient to show aerobic state, and the organic pollution in the sewages such as COD, NH3-N, TP, etc can be processed. Below this depth, the deposited sludge achieves anaerobic fermentation and decomposition, reducing sludge production. In order to improve the hydraulic retention time, the section is designed to push flow through the reasonable arrangement of the retaining wall, as shown in Figure 1.

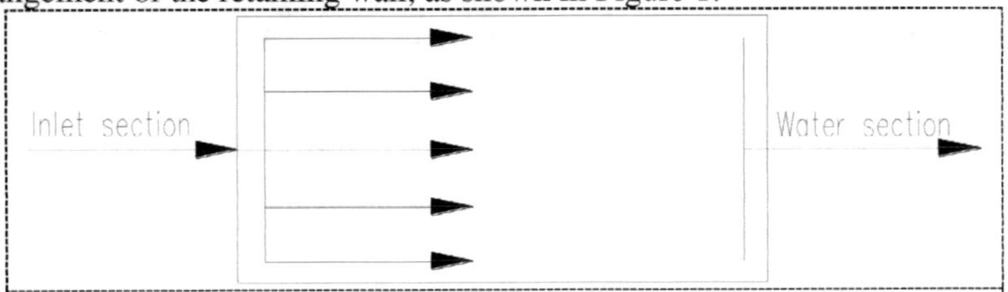

Figure 1. Push flow process

4.2.3 Plants and planting

To form a good biological chain, local aquatic herbs such as reeds and lotus are choosed. Aquatic plants are selected based on conditions such as temperature, light, mixing, and pretreatment. Because the growth of aquatic plants is faster, the initial planting density should not be too large. The planting density of reeds is 5 clusters/$m^{2,}$ and the planting density of lotus is 1 plant/3 m^2. The filter material can be selected from natural river sand, ceramsite, coal gangue, fluorite sand and other filter materials, and placed in appropriate areas for easy maintenance.

5. Conclusion

The constructed wetland technology is in a stage of rapid development. The water quality problem of lakes in China is significant, and the most important one is the eutrophication of lakes. The constructed wetland technology has a good prospect in the treatment of eutrophic lakes. The research on the decontamination mechanism of constructed wetlands is relatively backward, and the kinetic model of the movement and transformation of pollutants in wetlands has not been established. Therefore, in-depth study of the transformation mechanism of pollutants in wetland systems, optimization of hydraulic models and explanation of kinetic models will be beneficial to improve the level of research on constructed wetlands in China, and it should also have far-reaching significance for constructed wetland technology.

References

[1] Dong liu, YingGe, JieChang, ChanghuiPeng, BinheGu, Gilbert YS Chan, and Xiaofu Wu. Constructed wetlands in china: recent developments and future challenges. The Ecological Society of America, 2009, 7(5): 261-268.

[2] Liang Ruirong, on the application development of constructed wetlands [J]. Urban Roads Bridge and Flood Control, 2009, (5): 99-101.

[3] Zhang Yingying, Ding Weimin, Qian Yanyan, Wang Lei, Pang Wei.Process and design of constructed wetland wastewater treatment technology[J].Industrial water and wastewater,2009,(01):5-10.

IOP Conf. Series: Materials Science and Engineering **794** (2020) 012029 doi:10.1088/1757-899X/794/1/012029

[4] Yuan Donghai, Jing Lijie, Gao Shixiang, et al. Analysis of phosphorus pollution performance in concentrated constructed wetland substrate [J]. Environmental Science, 2005, 26(1): 51-55.
[5] Zhu Xizhen, Cui Lihua, Wen Xiaolu, et al. Purification Effect of Different Substrates Vertical Flow Constructed Wetland on Municipal Sewage[J].Journal of Agro-Environment Science,2003,22(4):4542-4571.
[6] Fan, H.L., Wang, F. 2016. Similarities and differences in spatial distribution and its influencing factors of two kinds of national wetland parks in China. Acta Scientiarum Naturalium Universitatis Pekinensis. 52(3), 535-544.

Experimental study on bonding performance of stainless steel bar and concrete

Li Qingfu[1*], Kuang Yihang [1] and Guo Wei [1]

[1]School of Water Conservancy Science and Engineering, Zhengzhou University, Zhengzhou, China

*lqflch@zzu.edu.cn

Abstract. The purpose of this paper is to study the bond behavior between stainless steel bar and concrete. Through the bonding test between reinforcement and concrete, the average bonding force and the distribution of the bonding force between stainless steel bar and concrete are studied. The test results showed that the bond performance between stainless steel bars and concrete was excellent. Within a certain range, the bond performance between stainless steel bars and concrete was proportional to the compressive strength of concrete. The bonding performance of stainless steel bars and concrete was related to the diameter of reinforcement. The larger the diameter of reinforcement, the greater the average bonding stress between reinforcement and concrete. In addition, the area affected by the bonding stress between reinforcement and concrete showed a linear growth trend with the diameter of reinforcement roughly affected as 10 times as the diameter of reinforcements. Stainless bar were similar to ordinary bar, and the peaks of their bond stress with concrete were roughly distributed near the loading end. As the load increased, the effective bond length increased slowly, while the peak stress increased significantly. The peak stress showed an internal shift only when the specimen was near damage.

1. Introduction

In recent years, stainless steel reinforced concrete has been applied in the project of serious corrosion area of the word, especially in the bridge project of coastal area. At present, domestic and foreign scholars have carried out many theoretical and experimental studies on stainless steel reinforced concrete, but most of these studies focus on the corrosion resistance of stainless steel reinforced concrete, and the bond between stainless steel reinforced concrete is less[1]. The bond between stainless steel bars and concrete belongs to the interface problem of stainless steel bars concrete. The interface problem of stainless steel bars concrete is a very important and special problem in the structural analysis and design of stainless steel bars concrete, which is related to the rationality and reliability of structural analysis and design model. At present, there are generally three kinds of testing methods for the adhesion of ordinary reinforcement bars, pull out test, beam test and axial pull test [2]. In order to study the mechanical properties of stainless steel bars, Calderon et al. [3]analyzed the effects of corrosion on the properties of reinforced concrete by simulating the marine environment. The results show that the effect of curing age on the bond strength of duplex stainless steel bars was significant. Higher water-cement ratio and steel corrosion degree would reduce the bond strength of stainless steel bars. BaiZhang et al.[4] investigated the bond performance between corroded steel bars and concrete after high temperature exposure. The results show that the decrease of the bond property, including bond strength, bond stiffness, bond energy dissipation and residual bond stress, were affected by the corrosion and exposure temperature. S. Coccia [5] analysed the corrosion effects at the steel–concrete interface. An analytical model was developed to evaluate the internal pressures caused by corrosion products. Zhao, YX[6] studied the bond between normal aggregate concrete (NAC) or recycled aggregate concrete (RAC) and corroded steel bars. The results show that the corrosion - induced bond degradation characteristics of NAC and RAC specimens were similar. HongweiLina[7]studied the potential relationship between crack width and bond strength. The results show that the width of surface crack is closely related to the degree of corrosion. Congqi Fangevaluate[8] studied the effect of corrosion on bonding and bond-slip behavior, and believed that when the degree of corrosion was very low, the bond strength would increase with the increase of the degree of corrosion.

2. Experimental design and method

Figure 1 shows a schematic of the specimen used in the average bonding test. The specimens tested for average cohesive force were drawn from the cube center without transverse reinforcement, and the size of the concrete test block was 150mm×150mm×150mm. The center of the test block was equipped with steel bar, which could be divided into three types: 12mm, 16mm and 32mm. In order to avoid local damage, the 25mm of each end of the steel in the concrete was covered with a plastic sleeve to form a non-bonded length.

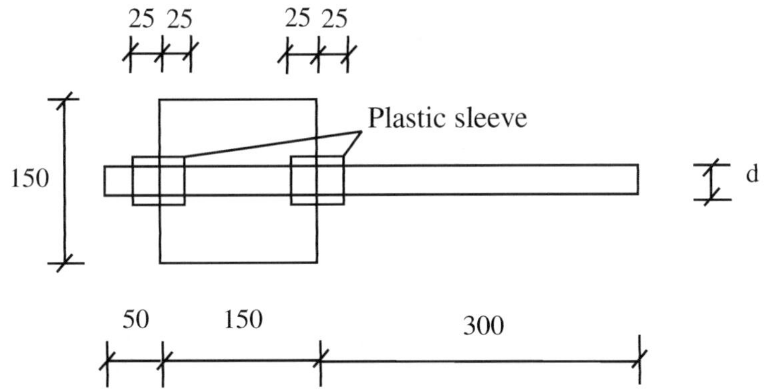

Figure 1. The schematic plot of average bond test block.

In this test, considering that the strength of concrete and the diameter of reinforcement may affect the bonding force between reinforcement and concrete, three kinds of concrete with different strengths C30, C40 and C50, were designed. And the steel bar was also divided into 12mm, 16mm and 32mm according to different diameters. The test was carried out in three batches according to the concrete strength grade. The main instrument used was a 300-ton hydraulic servo test machine. In addition, in order to prevent the use of impact cracks, the puller pull-out device and the 300-ton press were used for slow loading until the tester replacement or twisting curve could not be reloaded instead of zero. Theoretically, it is believed that the end of the test should be caused by the following two conditions: one is the yielding phenomenon of the steel bar or the steel bar is pulled out; the other is the concrete test block splitting or cone crushing [9]. Pulling the test for each batch of samples can obtain the average load stress of the damaged load of the sample and concrete. It can be obtained by using Equation (1), Equation (2) and Equation (3) [10].

$$\tau_F = \alpha . \tau \tag{1}$$

$$\alpha = \frac{f_{cu,k}}{f_{cu}} \tag{2}$$

$$\tau = \frac{1000F}{\pi d I_a} \tag{3}$$

Where τ_F is the average bond stress of steel and concrete, α is the concrete compressive strength correction factor, τ is the average bond stress without considering the concrete strength deviation,

IOP Conf. Series: Materials Science and Engineering **794** (2020) 012030 doi:10.1088/1757-899X/794/1/012030

$f_{cu,k}$ is the concrete strength grade, f_{cu} is the measured compressive strength of concrete, F is the breaking load, d is the diameter of the steel bar, l_a is the bond length of steel and concrete.

The average value of the average bond stress τ and τ_F of a steel bar and concrete is $\overline{\tau}$, $\overline{\tau_F}$, respectively. They determined by referring to Equation (4) and Equation (5):

$$\overline{\tau}=\sum_{i=1}^{n}\tau_i / n \tag{4}$$

$$\overline{\tau_F}=\sum_{i=1}^{n}\tau_{F_i} / n \tag{5}$$

After the end of each batch of drawing test, the compressive test was carried out on the reserved concrete block, and the failure load of the reserved concrete block can be measured, so that the compressive strength f_{cu} of the reserved concrete block can be obtained, and the calculation is performed, as presented in the Equations (6):

$$f_{cu}=\frac{F}{A} \tag{6}$$

Equation 7 represents the compressive strength of a reserved concrete block of a strength grade.

$$\overline{f_{cu}}=\sum_{i-1}^{n}f_{cu_i} / A \tag{7}$$

Table 1 shows the failure load, compressive strength and average value of the obtained reserved concrete block.

Table 1. Compressive test data of reserved concrete blocks.

Classification	C30PYL	C40PYL	C50PYL
$\overline{f_{cu}}$(Mpa)	52.55	61.16	69.91

3. Experimental phenomena

Through observation of the test process, it was found that the test was generally divided into five stages from the start of loading to the end of the test: (1) When the test machine began to load, the load was proportional to the time. At this stage, the bond between the steel and the concrete was small, mainly due to the action of the chemical bond, the sliding at the loading end was small, and the free end did not slip. As the load increased, the free end begins to show a slight slip. At this stage, the amount of slip was small, called the micro-slip phase. (2) The test load increased continuously, and the concrete on the surface of the sample showed tiny cracks. At this stage, the amount of slip increased, the slip accelerated, and rapidly developed to the free end, called the slip stage. is called slip stage. (3) As the test was carried out, the load slowly rose with time, and the load growth rate slowed down. At this stage, slight dislocation occurred between the reinforcement and concrete, and the reinforcement of some test blocks was even pulled out directly. This stage became the pull-out stage. (4) The test load continued to increase until the maximum value was reached. The surface crack of the specimen quickly developed to the free end, and the free end slip amount and the load end slip amount were almost close. This stage is called the splitting stage. (5) After loading to the maximum load, the load time curve dropped rapidly, and the specimen had cracked and destroyed, and this stage became the descending stage.

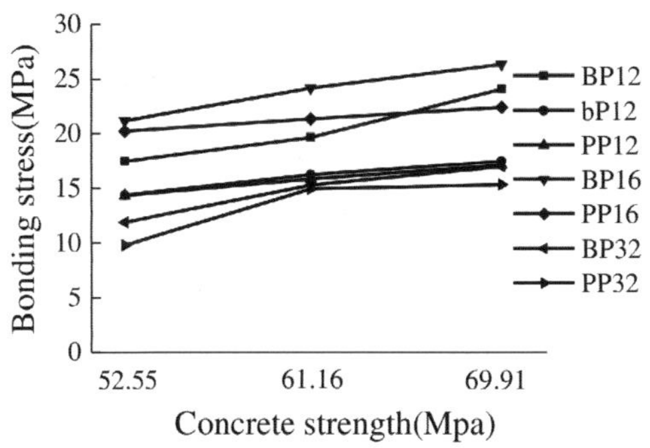

Figure 2. Average bond stress under different concrete strengths.

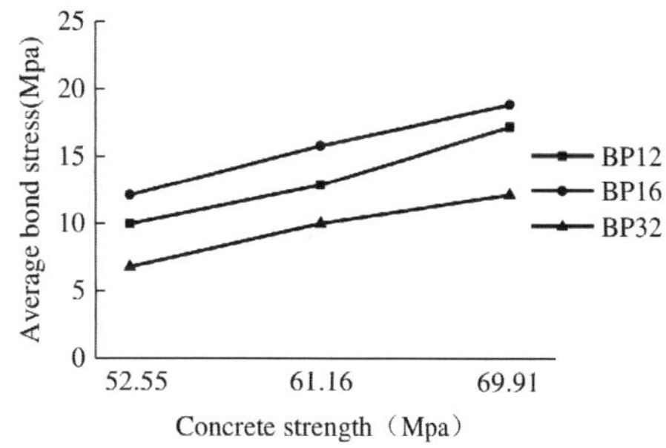

Figure 3. Mean value of 022Cr22Ni5MO3N stainless steel reinforcement's average bond stress.

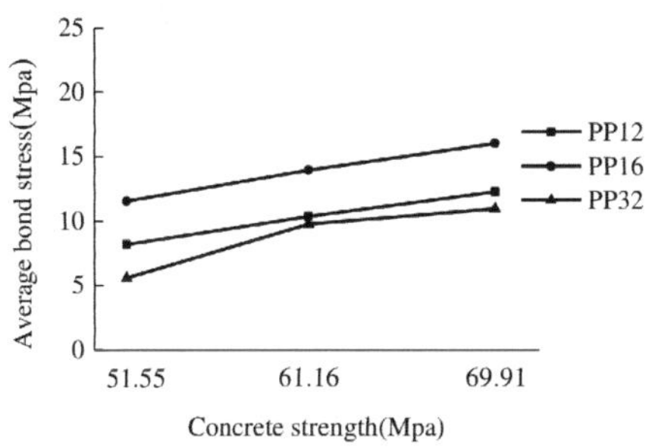

Figure 4. Mean value of HRB335 ordinary steel reinforcement's average bond stress.

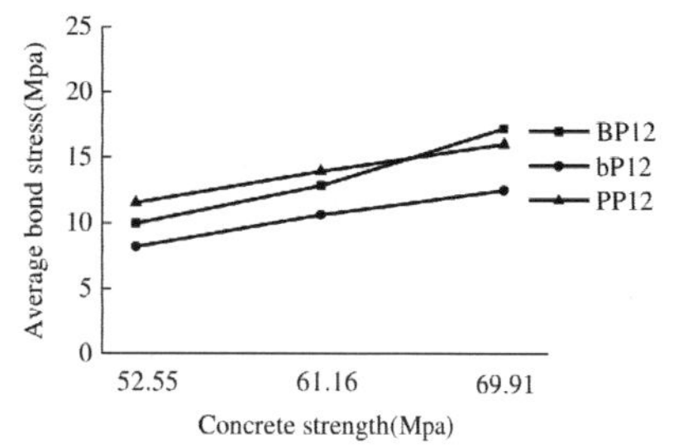

Figure 5. Average bond stress of three kinds of diameter 12mm steel

4. Experiment analysis

Figure 2 shows the average bond stress under different concrete strengths. It can be seen from Figure 2 that the bond stress and concrete strength of stainless steel bar and stainless steel bar and ordinary steel increase linearly.

The average bond stress of stainless steel bar and ordinary steel bar are respectively drawn as graphs, which are represented by Figure 3 and Figure 4. It can be seen from Figure 3 and Figure 4 that under the same concrete strength condition, except for BP32, the concrete bond stress of stainless steel bar and the ordinary steel bar increases with the increase of steel bar diameter. The bonding stress of steel bars with a diameter of 32 mm is obviously smaller than that of steel bars with diameters of 12 mm and 16 mm. This is because the larger the diameter of the steel bar, the greater the influence range of the bond stress between the steel bar and the concrete, which is generally considered to affect the range of 10 times the diameter of the bar. The concrete dimensions of the specimens are all 150mm×150mm×150mm, so for the steel bar test block with a diameter of 32mm, the influence range of bond stress between steel bar and concrete has exceeded the size of the test block concrete.

Figure 5 shows the average bond stress between stainless steel bar with a diameter of 12mm and an ordinary steel bar. It can be seen from Fig. 5 that the three kinds of steel bars of 022Cr22Ni5MO3N stainless steel rib, 06Cr17Ni12MO2 stainless steel rib and HRB335 ordinary steel bar with diameter of 12mm are in the same concrete strength condition, the order of bond stress of concrete is 022Cr22Ni5MO3N stainless steel bar, 06Cr17Ni12MO2 stainless steel bar and HRB335 ordinary steel bar. And the bonding stress of 06Cr17Ni12MO2 stainless steel bar and HRB335 ordinary steel bar is similar. The average bond stress between stainless steel bars and concrete can be calculated according to Equation 8.

$$\tau_b = \frac{f_{cu,k}}{f_{cu}} \cdot \frac{1000F}{\pi d I_b} \quad (8)$$

Where τ_b is the average bond strain between deformed stainless steel bars and concrete, $f_{cu,k}$ is the concrete strength grade, f_{cu} is the measured compressive strength of concrete, F is the measured value of the load when the deformed stainless steel bar is broken, d is the diameter of the deformed stainless steel bar, I_b is the effective bond length between the deformed stainless steel bar and the concrete.

5. Conclusion

The following conclusions can be drawn from the average bonding force test and the bonding force distribution test of reinforcement and concrete:

1) The bond performance between deformed stainless steel bars and concrete was better than that between deformed ordinary steel bars, and the bond performance between deformed stainless steel bars and concrete was the best.

2) In a certain range, the bond performance of deformed stainless steel bars and concrete was directly proportional to the strength of concrete, the greater the strength of concrete, the better the bond performance of stainless steel bars and concrete.

3) In the range of common deformed steel bars, the bond performance between stainless steel bars and concrete was related to the diameter of the steel bars. The larger the steel bar diameter, the greater the average bond stress between stainless steel bars and concrete. Moreover, the bond stress affected zone between steel bar and concrete increased with the increase of steel bar diameter, which was about 10 times of the diameter of steel bar.

4) Whether it was stainless steel or ordinary steel, the peak of the bond stress of concrete was mostly near the loading end. As the load increases, the effective bond length increases slowly, while the stress peak increases significantly, only in the sample. At the end of the damage, the peak of the stress will appear to shift. As the load increased, the effective bond length increased slowly, while the stress peak increased significantly. Only when the specimen was near the damage, the peak stress appeared to shift.

Acknowledgment

The authors would like to acknowledge the financial support provided by the National Natural Science Foundation of China (Grant No. 51679220).

References

[1] Zhang, G., Wu,M. (2006) Application and development of stainless steel rebars for reinforced concrete structures. J.Journal of Foshan University (Natural Science Edition)., 10-13.
[2] Zhao, G. (1999) Reinforced Concrete Theory and Analyse. Beijing: China Electric Power Press.
[3] Calderon, I., Briz, E., Larrinaga, P. (2018) Bonding strength of stainless steel rebars in concretes exposed to marine environments.J. Construction and Building Materials., 172: 125-133.
[4] Zhang, B., Zhu, H., Chen, J. (2019) Evaluation of bond performance of corroded steel bars in concrete after high temperature exposure. J. Engineering Structures., 198: 109479.
[5] Coccia, S., Imperatore, S., Rinaldi,Z. (2016)Influence of corrosion on the bond strength of steel rebars in concrete.J. Materials and Structures/Materiaux et Constructions. , 49(1-2): 537-551.
[6] Zhao,Y., Lin,H., Wu,K. (2013) Bond behaviour of normal/recycled concrete and corroded steel bars.J. Construction and Building Materials., 48: 348-359.
[7] Lin,H., Zhao,Y., Ožbolt,J. (2017) Bond strength evaluation of corroded steel bars via the surface crack width induced by reinforcement corrosion.J. Engineering Structures., 152: 506-522.
[8] Fang,C., Lundgren,K., Chen,L. (2004)Corrosion influence on bond in reinforced concrete.J. Cement and Concrete Research., 34(11): 2159-2167.

[9] Li,Z. (2009)Analysis of steel sheathing force test .J. China's new technology and new products., (17): 90.
[10] GB50152-92. Standard Methods for Testing of Concrete Structures.,S.

IOP Conf. Series: Materials Science and Engineering **794** (2020) 012031 doi:10.1088/1757-899X/794/1/012031

Analysis on cooperative working performance of curtain wall glass and steel hanging columns

Ke Gong, Wei Luan*, Xiwen Yang, Yuyang Zhou

China Construction Eighth Engineering Division Co., Ltd., Shanghai 200135, China

*Corresponding author's e-mail:lwei13@163.com

Abstract. The curtain wall of Xi'an Silk Road International Conference Center is a large-span combined steel column supported glass curtain wall system. The curtain wall glass is directly installed on the steel hanging columns of the external suspension structure, and there are no curtain wall hanging columns that are independent of the main structure. The vertical, tangential and radial deformation of the steel hanging columns under service load may lead to vertical shear, in-plane extrusion, in-plane sliding and out of plane bending of the curtain glass. Therefore, finite element model of curtain wall glass was established using the finite element program ABAQUS. Considering the four possible failure modes of the curtain glass, the collaborative working performance of the curtain wall glass under the most unfavourable deformation conditions of steel hanging columns was analyzed.

1. Introduction of the glass curtain wall system

The external curtain wall of Xi'an Silk Road International Conference Center is a large-span combined steel column supported glass curtain wall system with a special-shaped curved surface. The curtain wall span is 189m and the height is 37.3m to 46.7m. In order to meet the requirements of the architectural design effect, the curtain wall glass is directly installed on the steel hanging columns of the external suspension structure, and there are no curtain wall hanging columns that are independent of the main structure. Compared with the conventional glass curtain wall system, new requirements have been put forward on the installation accuracy of the steel hanging columns and the cooperative working performance between the curtain wall glass and the steel hanging columns. The standard partition size of curtain wall glass is 2400mm × 3500mm, and its main features are as follows: hollow double laminated ultra-white glass was used in the curtain wall system, which integrated safety, lighting, sound insulation and self-explosion protection together. The vertical keel of the curtain wall was replaced by the steel hanging columns, and the glass was supported by cantilevered steel plate welded on the steel hanging columns.

2. Analysis of the most unfavourable deformation of curtain wall hanging columns

Based on the calculation and analysis of the whole structure model under the load combination without earthquake operated by Zhou[1], it is found that the maximum vertical relative deformation of the adjacent curtain wall columns appears in the situation listed in the following table 1 and figure 1.

Table 1. Corresponding situation of coordinated deformation and failure mode of curtain wall glass and steel hanging column.

Failure mode of curtain wall glass	Coordinated deformation mode of	Location of maximum deformation occurs	Maximum coordinated deformation

IOP Conf. Series: Materials Science and Engineering **794** (2020) 012031 doi:10.1088/1757-899X/794/1/012031

	steel hanging column		
vertical shear	Vertical relative deformation	South side 'figure 1a'	25mm
in-plane extrusion	Tangential internal deformation	Southwest side 'figure 1b'	5mm
in-plane sliding	Tangential external deformation	Southwest side 'figure 1c'	20mm
out of plane bending	Relative radial deformation	West side 'figure 1d'	150mm

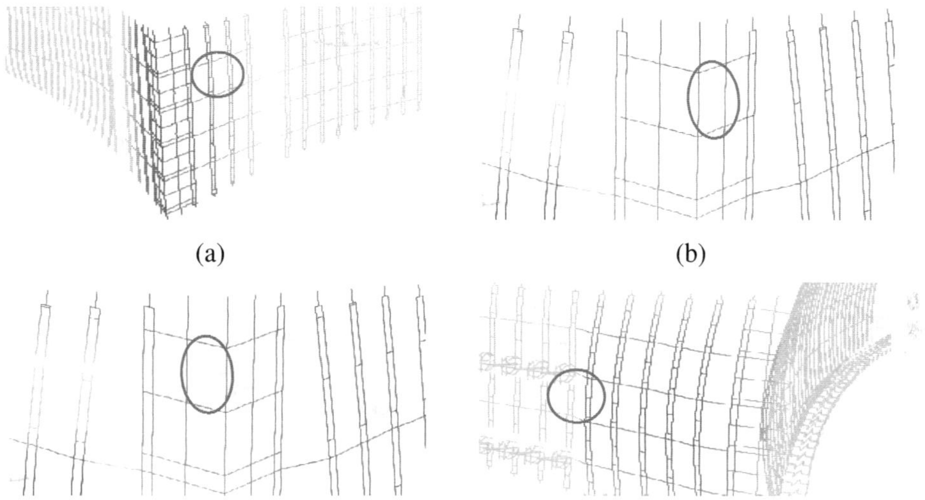
(a) (b)
(c) (d)

Figure 1. Location of maximum deformation occurs

3. Mechanical performance analysis of curtain wall glass

3.1Model of curtain wall glass

Considering the structural characteristics and calculation conditions, numerical simulation on the curtain wall glass of Xi'an International Conference Center was carried out using the finite element program ABAQUS. The size of glass panel is 2400mm × 3500mm, the width of adhesive strip is 12mm, and the thickness is 8mm. Because the upper and lower layers of double-layer glass are not related to each other, only one side need to be calculated, and the 8 (TP) + 1.52pvb + 8 (TP) side was selected for analysis[2]. The finite element model of curtain wall glass is shown in figure 2, the parameters were set as follows:

Glass parameters: elastic modulus is 72000Mpa, Poisson's ratio is 0.2.

Strip parameters: elastic modulus is 7.8MPa, Poisson's ratio is 0.47.

Element and mesh: solid element is used which was divided into 8 layers along the thickness direction.

Contact setting: smooth contact is set between two layers of glass, and the contact surface cannot be separated.

Boundary condition: since the glass panel only adheres to the adhesive strip on both sides, the upper and lower sides are free ends, and the displacement in the plane (U2) and out of the plane (U3) are constrained by the width direction of the adhesive strip on the left and right sides[3-4].

The geometric nonlinearity was also considered. After parameter setting was completed, the model is as shown in figure 3.

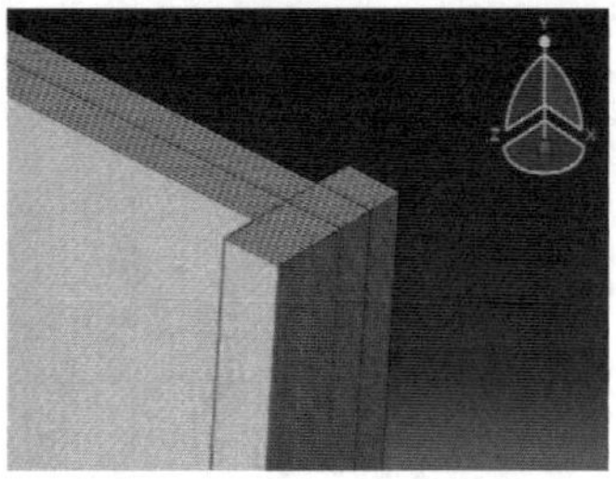

Figure 2. Finite element model of glass Figure 3. Model of curtain wall glass after parameter setting

3.2Vertical shear analysis

According to table 1, the maximum vertical relative deformation of the unit curtain wall glass panel is 25mm. Based on the given boundary conditions, the calculation results are as follows:

The maximum deformation of the whole glass panel is only about 5mm, but the vertical deformation of the end rubber strip is up to about 25mm due to its small elastic modulus.

The maximum Mises stress of the curtain wall glass is 1.0 MPa (figure 4- figure7), which appears at the junction of glass and rubber strip. According to the Technical code for glass curtain wall engineering (JGJ 102-2003) [5], the design value of side strength of glass with thickness of 8mm is 19MPa. Therefore, the overall stress level is small, and the safety of the glass can be guaranteed.

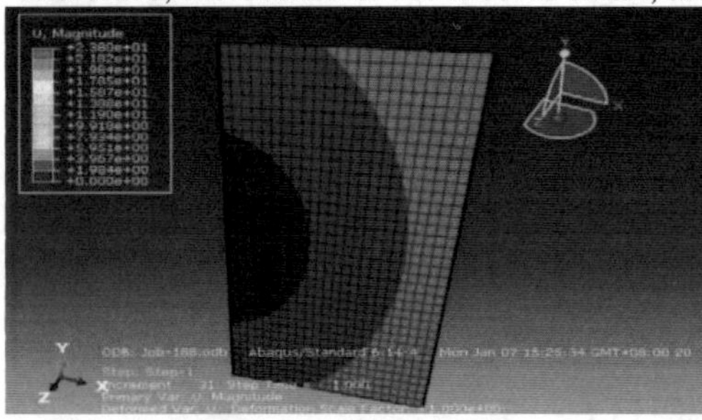
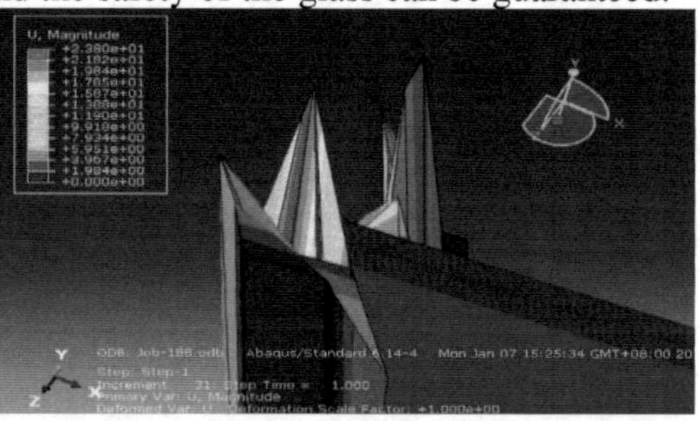

Figure 4. Glass deformation in shear state Figure 5. Local deformation in shear state

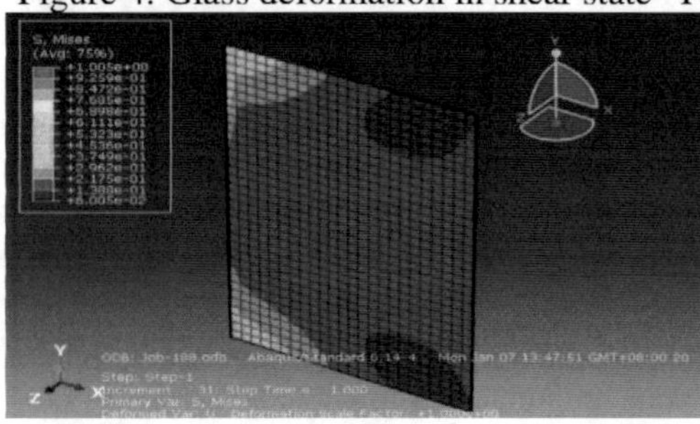
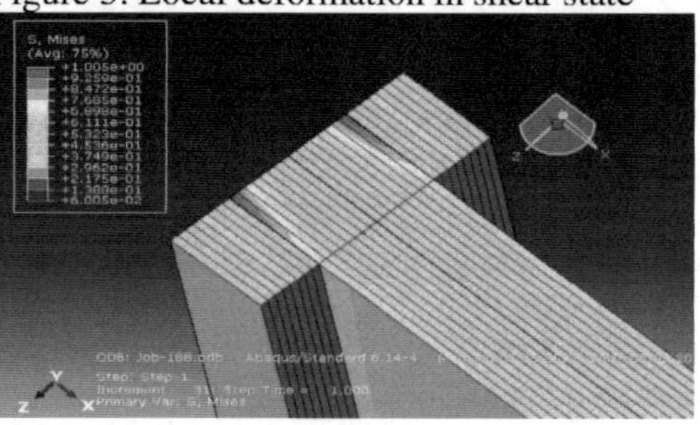

Figure 6. Stress in glass under shear state Figure 7. Local stress in shear state

3.3In plane extrusion analysis

According to the supporting structure of the curtain wall glass, there are still 12mm margins on both sides of the depth of the glass entry groove, that is, 24mm tangential deformation space exists. Therefore, the in-plane squeezing problem does not occur.

3.4In plane slippage analysis

The depth of the glass that has entered the groove is 23mm on both sides, that is, there is a 23mm tangent outward deformation space. Therefore, in-plane slippage will not occur.

3.5Out of plane bending analysis

According to table 1, the maximum radial relative deformation of unit glass panel is 150 mm. The calculation results are as follows:

The maximum value of total deformation is 150.5mm, among which the maximum value of in-plane tangential deformation caused by bending is about 5.4mm, which is less than 23mm into the groove depth, so the glass will not slide due to bending deformation.

The maximum Mises stress of the curtain wall glass is 0.048 MPa, which appears at the junction of glass and rubber strip (figure 8-figure 11). Therefore, the overall stress level is small, and the safety of the glass can be guaranteed.

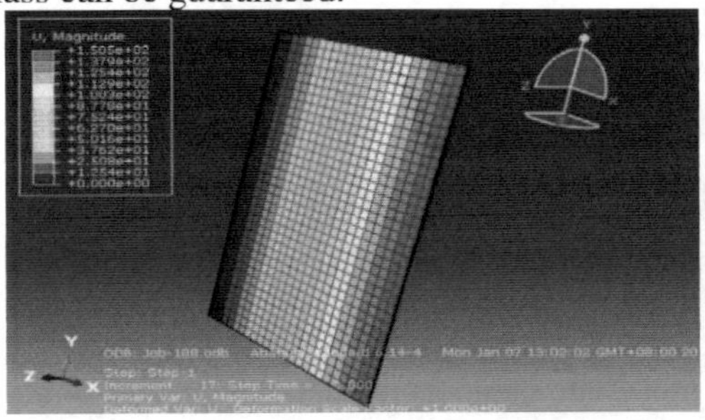

Figure 8. Overall deformation calculation results

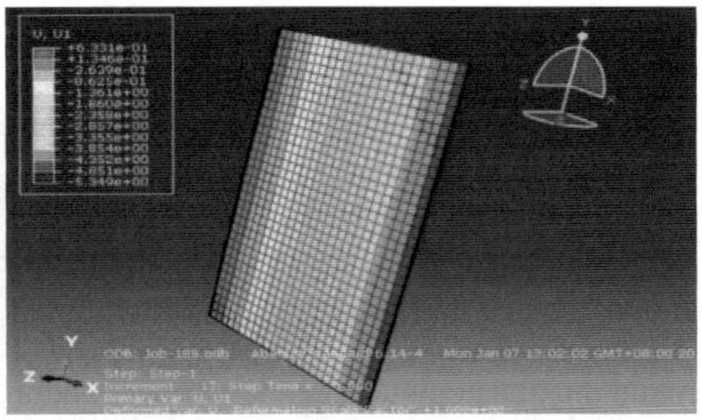

Figure 9. In-plane deformation calculation results

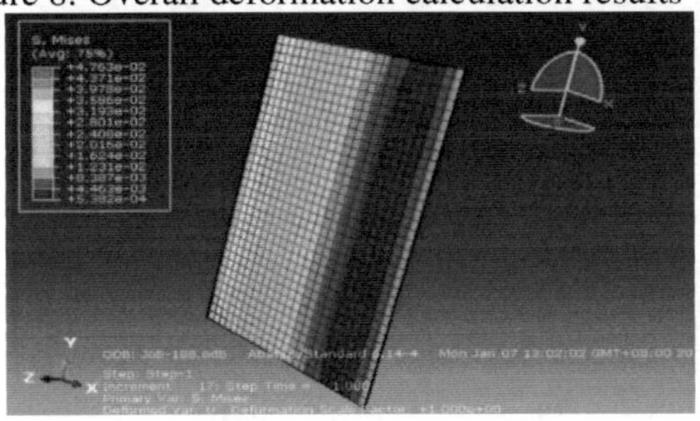

Figure 10. Overall Mises stress calculation results

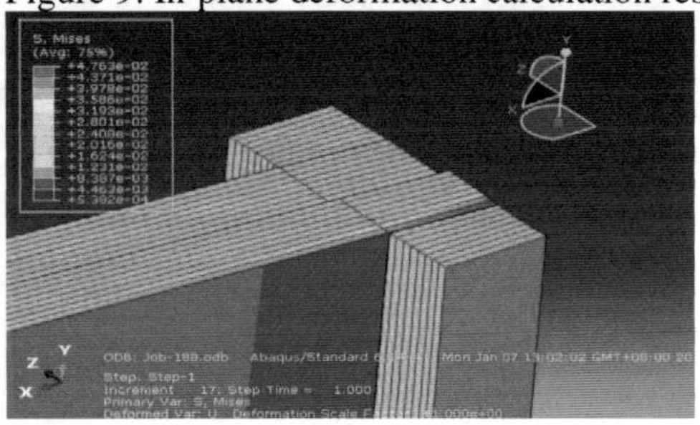

Figure 11. Local Mises stress calculation results

4. Conclusion

Using the finite element program ABAQUS, an accurate finite element model of the curtain wall glass of the Xi'an Silk Road Conference Center was established. Focused on four possible failure modes of curtain wall glass such as shear failure of glass caused by vertical deformation of curtain wall column, extrusion failure of glass caused by tangential internal deformation of curtain wall column, sliding of glass caused by the outward deformation of curtain wall column and out of plane bending of glass caused by radial deformation of curtain wall column, finite element analysis were carried out. The results showed that under the most unfavorable deformation condition of steel hanging column in normal operation stage, the curtain wall glass will not be damaged by in-plane extrusion or in-plane sliding, the vertical shear deformation and out of plane bending deformation are also very small, the wall glass and steel hanging columns work well together, and the structure is safe and reliable.

References

[1] Zhou Y.Y. (2019) Research on Construction Process of the Complex Suspended Steel Structure. Master's thesis of Tongji University, Shanghai.

[2] Zhao X.A. (2007) Curtain wall glass. J. China Construction Metal Structure, (10):24-27.

[3] Zhou X.Q. (2007) Construction quality control of glass curtain wall project. J. Shanxi Architecture, 33(11): 226-227.

[4] Lu, W.S., Huang B.F., Mosalam, Khalid M. (2016) Experimental evaluation of a glass curtain wall of a tall building. J. Earthquake Engineering & Structural Dynamics, 45(7):1185-1205.

IOP Conf. Series: Materials Science and Engineering **794** (2020) 012031 doi:10.1088/1757-899X/794/1/012031

[5] China Academy of Building Sciences (2003) Technical code for glass curtain wall engineering. China Construction Industry Press, Beijing.

IOP Conf. Series: Materials Science and Engineering **794** (2020) 012032 doi:10.1088/1757-899X/794/1/012032

Calculation and test on the construction support deformation for the viaduct of the subway Line 1 of Changzhou

Wang Weixing[1], Liu Teng[2], Pan Darong[3*], Niu Longlong[3], Dai Tianjiao[3]

[1]Jiangsu Yanghu Construction Project Administration Co., Ltd, Changzhou, Jiangsu, 213157, China;

[2]Beijing Municipal Construction Group Co., Ltd, Beijing, 100045, China

[3]School of Architecture Engineering, Nanjing Institute of Technology, Nanjing, Jiangsu, 211167, China

[*]Corresponding author's e-mail: pandarong@msn.com

Abstract: When the high formwork support collapses during construction, the bars present obvious deformations that exceed the normal limit. The deformation characteristics of the high formwork support were calculated and analyzed by nonlinear buckling finite element method, and the deformation test was carried out by using the automatic deformation monitoring system based on machine vision. The results show that the FE results are in good agreement with the tested results, and the FE calculation can predict the deformation behavior of the structure relatively well.

1.Introduction

In recent years, accidents of the bridge structures constructed with full space support have been frequently caused by the collapse of the supports. For example, in the construction of Yanba expressway approach bridge in Shenzhen, there was a large height error in the column, and no cross bracing in the lateral direction of the support, while the number of cross bracings in the longitudinal direction of the support is insufficient. All of these lead to the instability of the support body and the collapse of the beam body, causing that 69 people fell and 19 injured (with 5 seriously injured)[1]. In these accidents, the collapse of formwork support system with the height more than 8m (high formwork support) accounts for a very high proportion[2]. The main reason of the accidents is the overall instability of the supporting structure caused by human negligence, design defects, and construction problems[3-4]. In the construction process of the structure, due to the initial bending of the support members and other reasons, these unsteady support accidents, which are characterized by nonlinear instability, have infallible indications that the deformations of the members exceed the normal limits[5]. Therefore, the deformations of the high formwork support of the viaduct of Changzhou Metro Line 1 during the construction were calculated by nonlinear buckling finite element method, and the deformation characteristics of the whole structure at nonlinear buckling were studied in this paper. At the same time, an automatic deformation monitoring system based on machine vision was used to continuously monitor the deformations of the support at the construction site to verify the calculation and study results.

2.Structure situation

2.1 Main girder

The monitored structure under construction was a viaduct on Changzhou Metro Line 1. Its main girder was a 30 + 45 + 30m single-boxed and single-chambered continuous beam. The pier shaft was of 12.5m height, and the detailed sizes of this structure is shown in Fig. 1.

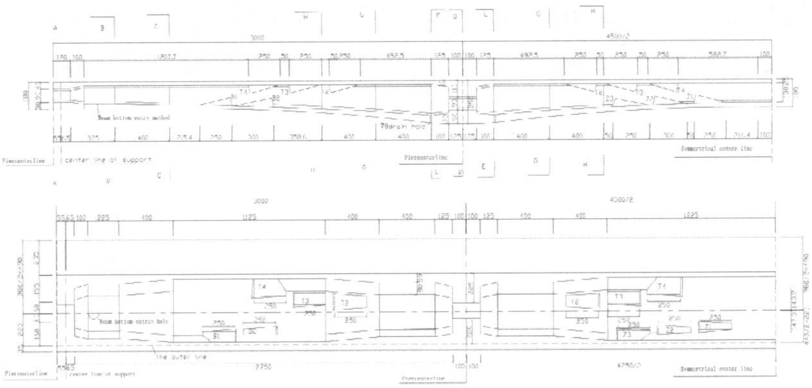

Fig. 1 Main structural drawing of the box girder

2.2 Support

Steel pipes with the diameter of 45mm were used as the support columns. Near the end diaphragm, in the range of 2.4m on the single side, the longitudinal and lateral spacing of the columns was 60cm × 60cm. Near the middle diaphragm, in the range of 2.7m on the single side, the longitudinal and lateral spacing of the columns was 30cm × 60cm, and in the range between 2.7m and 7.2m, that was 30cm × 60cm. Near the web position, the longitudinal and lateral spacing of the columns was set to 90cm × 60cm. At the hollow position, the longitudinal and lateral spacing of the columns was set to 90 cm × 90 cm. Near the flange position, the longitudinal and lateral spacing of the columns was set to 90cm × 90cm. The detailed arrangement of the support was shown in Fig. 2.

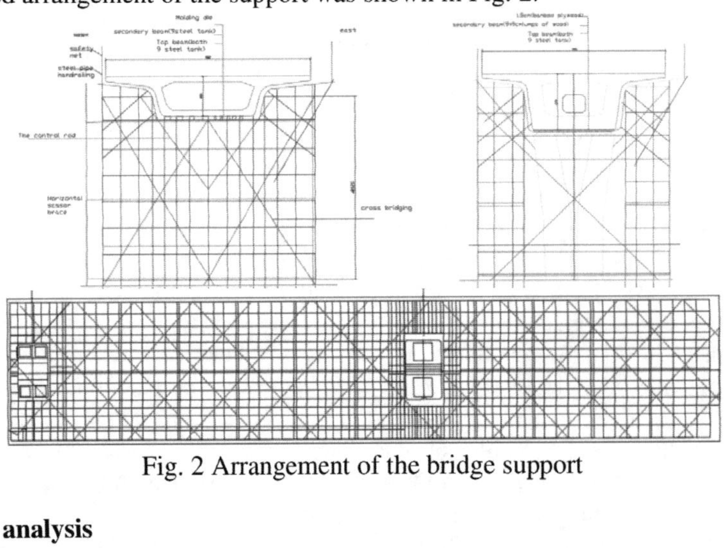

Fig. 2 Arrangement of the bridge support

3.Finite element analysis

3.1 Nonlinear buckling analysis theory

Nonlinear buckling analysis is more accurate than linear buckling analysis, so it can be used in the design or calculation of actual structures. This analysis method is to use a nonlinear static analysis technique which gradually increases the load to obtain the critical load when the structure becomes unstable. With the application of nonlinear technology, the model can contain the characters of initial defects, plasticity, gaps, large deformation responses, and so on. The nonlinear buckling caused by the initial bending defects of the members is mainly considered in this paper.

3.2Value of the initial bending defects
The Value of initial bending defect was taken from the statistical data of reference [6]. The initial bending rate = initial bending deflection / steel pipe length. The mean value of the initial bending rate was 2.4 ‰, and the standard deviation was 1.5 ‰, which obeyed normal distribution. When the finite element model was established, N sample parameters were randomly generated according to the above mean value and standard deviation, Where N was the total number of the members. Then, the N sample parameters were randomly assigned to each member, so that each member had an initial bending rate which obeyed the above mentioned statistical distribution. In addition, considering the randomness of the member bending direction in the actual structure, the bending direction angle was also set for each member randomly when the member model was established.

Finite element model of the structure

The general finite element software ANSYS was used for nonlinear buckling calculation. The Beam188 linear beam element was used to simulate the supporting members, the Solid185 solid element was used to simulate the main girder, and the Shell181 shell element was used to simulate the steel formworks. The arc length method was used to solve the mechanical equations considering geometric nonlinearity[7-10].

The structural loads mainly contained the structure self-weight, personnel and construction equipment load, load from vibrating concrete and horizontal wind load. The structure self-weight was taken as the actual value, personnel and construction equipment load was taken as 1.0kN/m^2, load from vibrating concrete was taken as 2.0kN/m^2, and the standard value of horizontal wind load was taken as 0.308kN/m^2according to the design code[11].

The connections between the main girder formwork and the support members were simulated as hinge joints. The right angle fastener between the horizontal member and the vertical member was the semi-rigid connection between the rigid connection and the hinged connection, and was simulated as torsion spring. According to the literature[12], the influence of the torsional rigidity of qualified right angle fastener on buckling bearing capacity is relatively weak. Taking the intermediate value 22.9kN·m/rad of the torsional stiffness range[12] of the qualified fasteners as the torsional rigidity of the torsion spring. The finite element model of the structure is shown in Fig. 3.

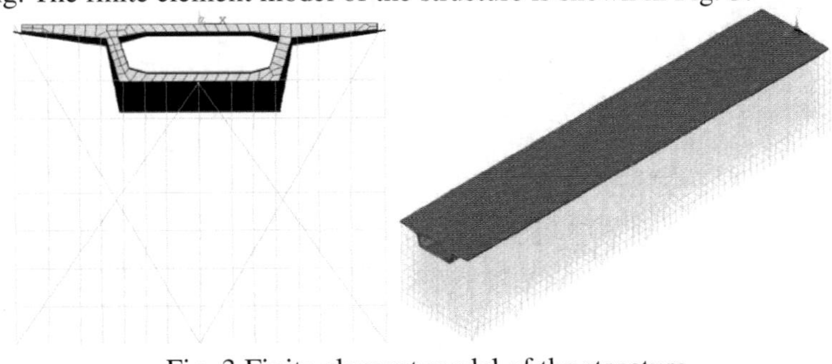

Fig. 3 Finite element model of the structure

3.3 Analysis on the finite element results

3.3.1 Lateral (X direction) deformation
The lateral (x direction) deformation of the support structure is shown in Fig. 4. As can be seen from the figure, the maximum lateral deformation of the structure is about 0.5 mm, approximately in the middle part of the support structure.

3.3.2Vertical (Y direction) deformation
The vertical (Y direction) deformation of the support structure is shown in Figure 5. It can be seen from

the figure that the maximum vertical deformation of the structure is 2.6mm, which is approximately at the upper part of the support structure.

3.3.3Longitudinal (Z direction) deformation

The longitudinal (Z direction) deformation of the support structure is shown in Fig. 6. As can be seen from the figure, the maximum longitudinal deformation of the structure is 9.3 mm, approximately at the upper part of the support structure.

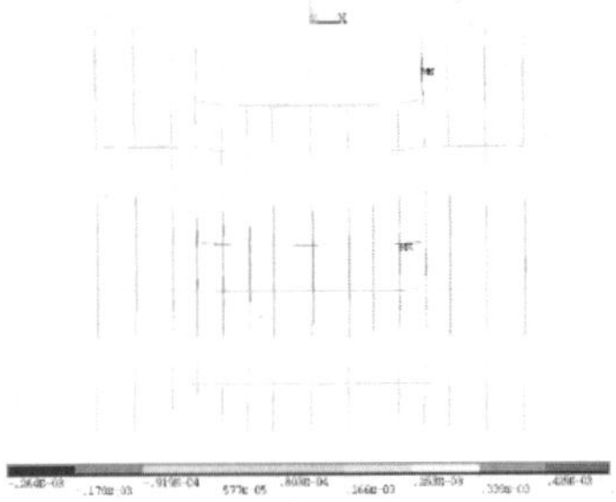

Fig. 4 Lateral deformation

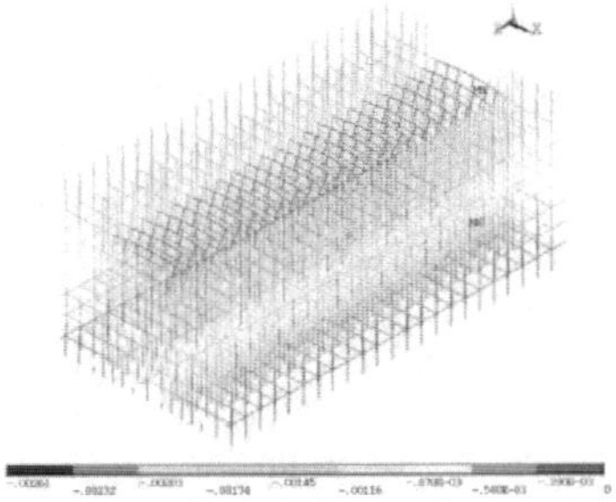

Fig. 5 Vertical deformation

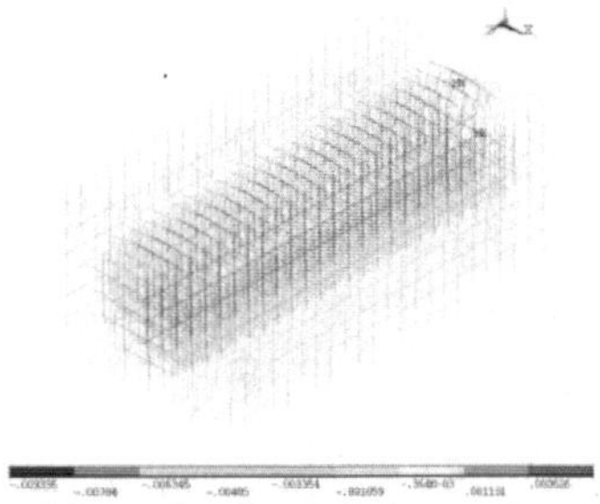

Figure. 6 Longitudinal deformation

4.Deformation monitoring experiment

For the high formwork support structure, it is difficult to work effectively for the traditional deformation monitoring method. If to dynamically monitor the deformations directly through displacement sensor, on the one hand, the installation position of the sensors could not be well solved; on the other hand, it would also cause interference to the construction process. To use total station to measure displacement had less interference to the construction process, but it cannot conduct dynamic monitoring. Indirect measurement of structural deformation information through acceleration sensor required twice integration, which would led to error accumulation[13]. Therefore, an automatic deformation monitoring system based on machine vision developed by the project team was used in this paper to test the real-time deformations of the support structure[14].

4.1 Test device

The hardware of the dynamic monitoring system included an image acquisition module, a data transmission and processing module, an LED light compensation module, a measurement target, and a monitoring system server. During the construction process, the image acquisition module automatically collected the real-time image of the measurement target on the support in the monitoring area. When the support deformed, the position of the observation point on the measurement target changed, the monitoring system automatically recognized the deformation observation point and then sent the data to the monitoring system server. When operating at night, the LED light compensation module was used for optical compensation to improve the recognition effect. The detail of the system is shown in Fig. 7.

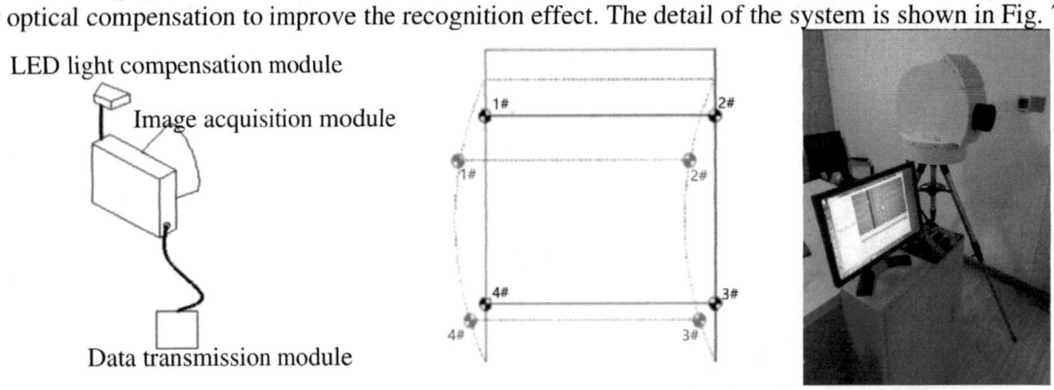

Fig. 7 Dynamic deformation monitoring system used in the high formwork construction

4.2 Layout of the measured points

According to the finite element results and the in-field measurement conditions, the targets were set on the east side of the coping, namely the observation points D1 ~ D5 respectively, shown in Fig. 8.

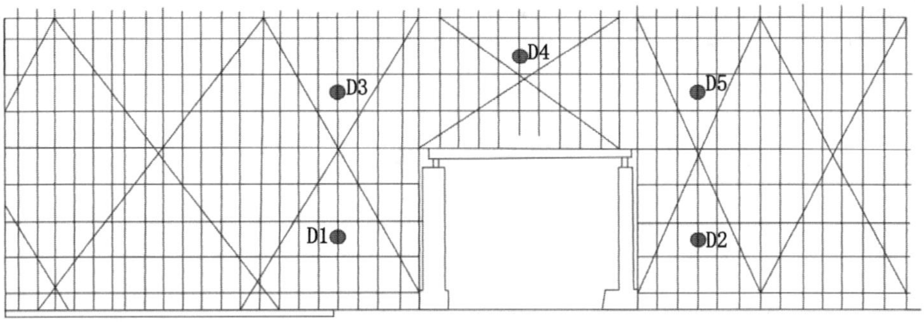

Fig. 8 Measurement points of the bridge framework support

4.3 test result analysis

The measure tracked the whole process of concrete pouring, from 18:00 to 02:10 in the next morning. Photos were taken every 30 minutes. The deformation curves of each observation point are shown in the following figures.

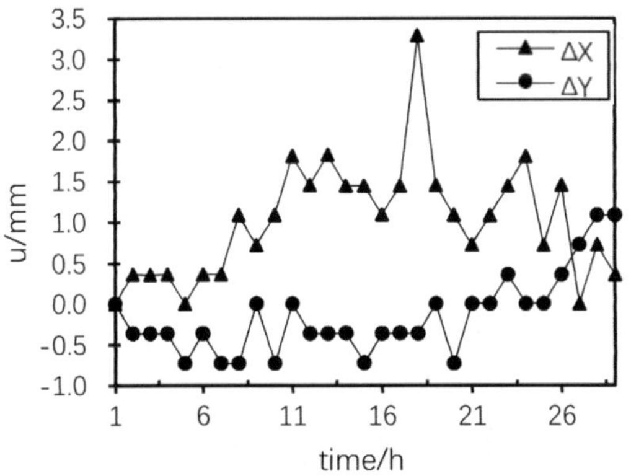

Fig. 9 Deformation curves of point D1

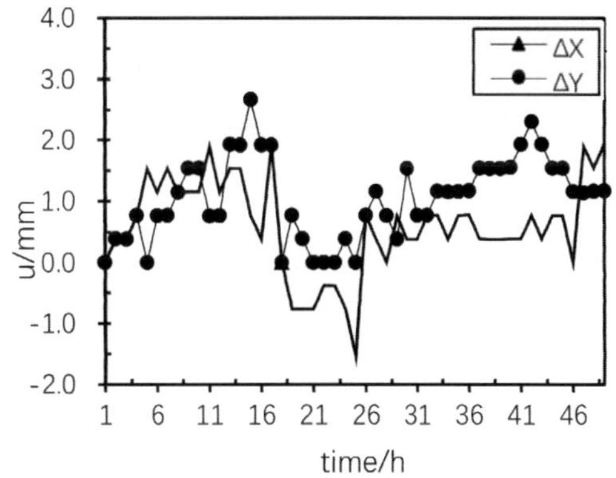

Fig. 10 Deformation curves of point D2

It can be seen from the deformation curves of D1 that the measured maximum vertical deformation is 2mm, and the maximum horizontal deformation is 8mm at the completion of pouring.

It can be seen from the deformation curves of D2 that the measured maximum vertical deformation is 3mm, and the maximum horizontal deformation is 4mm at the completion of pouring.

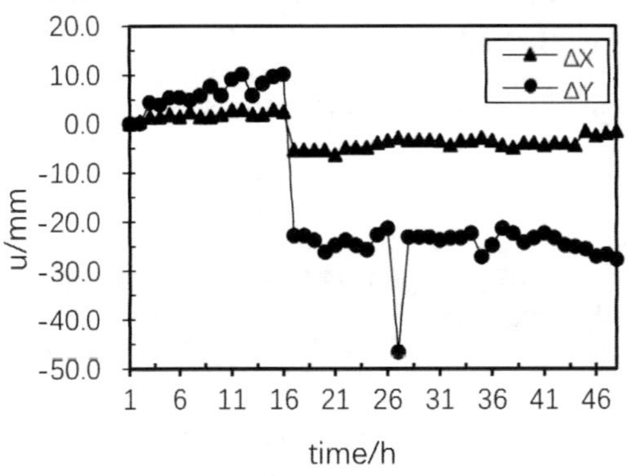

Fig. 11 Deformation curve of point D3

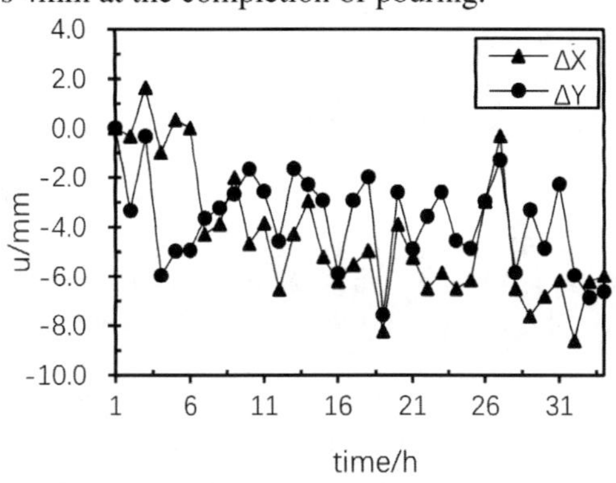

Fig. 12 Deformation curve of point D4

It can be seen from the deformation curves of D3 that the measured maximum vertical deformation is 62mm, and the maximum horizontal deformation is 8mm at the completion of pouring.

It can be seen from the deformation curves of D4 that the measured maximum vertical deformation is 8.6mm, and the maximum horizontal deformation is 7.6mm at the completion of pouring.

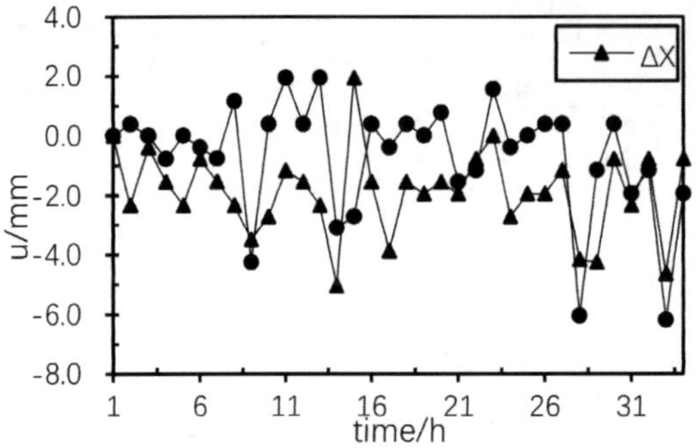

Fig. 13 Deformation curve of the observation point D5

It can be seen from the deformation curves of D5 that the measured maximum vertical deformation is 6.2mm, and the maximum horizontal deformation is 5.0mm at the completion of pouring.

The calculated values of the observation point deformations were extracted and compared with the measured values, as shown in Tab. 1.

Tab. 1 Comparison between the calculated deformations of the observation points and the measured deformations

Deformation		D1(mm)	D2(mm)	D3(mm)	D4(mm)	D5(mm)
Calculated value	X direction (Lateral)	0.2	0.3	0.4	0.5	0.4
	Y direction (Vertical)	0.7	0.8	2.5	2.6	2.5
	Z direction (Longitudinal)	2.4	2.5	6.8	6.9	6.7
Measured value	Y direction (Vertical)	2	3	62	8.6	6.2
	Z direction (Longitudinal)	8	4	8	7.6	5.0

It can be seen from the table above that there are some deviations between the calculated values and the measured values due to the inaccuracy of the initial defects and the joint stiffnesses, and the regardless of the complicated factors in the finite element calculation. However, the calculated deformations of most measurement points are not much different from the measured values, basically in the same order of magnitude; the calculated deformation laws are relatively consistent with those gained by measure. This indicates that the calculation results have certain reference significance, and the finite element method can calculate the buckling possibility of this kind of structure. The calculation values of the observation point D3 are quite different from the measured values, this is because the D3 target was moved from the original position before the 16th measure.From the curves recorded by this system, the change is quite obvious. It indicates from another angle that the deformation monitoring system can respond automatically, and a system alarm event will be generated when the preset alarm value is reached.

5.Conclusion

Through the finite element calculation and monitoring test of the high formwork support deformation, it is found that the vertical deformation and horizontal deformation of the supporting structure of the viaduct of Changzhou Metro Line 1 are small, which do not exceed the design requirements, so the structure is safe. The differences between the calculation results and the test results are small, which verifies the correctness of the calculation method. The calculation method and results can provide reference for other similar bridge support construction, and have certain reference significance for the studies on high formwork support structure safety.

Acknowledgments
This work is Supported by Jiangsu construction system science and technology project (2017ZD131).

Reference

[1] http://www.chinanews.com/2000-11-28/26/58450.html
[2] Nan Xie, (2012)Renzhong Liang, Jingjing Wang, Occurrence of human errors in high falsework and influence on structural safety [J]. Engineering mechanics,29(Supplement I): 63-67.
[3] Tingsheng Zhao, Zehui Wang, Xianglin Gu. (2005)Statistical Parameters of the Formwork Supports [J]. Construction technology, 34(3): 49-52.
[4] Renzhong Liang, Nan Xie, (2010)The safety assessment analysis of supports for formwork in china [J]. Industrial safety and environmental protection, 36(8): 36-37.
[5] Xuexia Yuan, Weiliang Jin, Zheng Lu, Xin Liu , Tianmin Chen.(2006) A study on the stability bearing capacity of fastener – style tubular steel formwork-supports [J]. China civil engineering journal, 39(5):43-50.
[6] Nan Xie, Yong Wang, (2008)Study on load-carrying capacity of super high supports for formwork, 25(Supplement I): 148-153.
[7] Hongbo Liu, Qiuhong Zhao, (2010)Xiaodun Wang, et al. Experimental and analytical studies on the stability of structural steel tube and coupler scaffolds withoutX-bracing [J]. Engineering Structures, 32(4):1003-1015.
[8] NanXie, Xiaohui Fu, Lifeng Wang, Hang Hu, Tong Wu, (2016)Design method of load and resistance factor for high falsework with couplers [J]. Engineering mechanics, 33(10): 68-75.
[9] Tianxia Song, (1996)Finite element calculation of nonlinear structure [M]. Wuhan: Huazhong University of Technology Press.
[10] Ji Chen. (1994)Theory and application of steel structure stability [M].Beijing, Science and Technology Literature Press.
[11] JGJ130-2011, (2011)Safety technical code of construction fastener type steel pipe scaffold [S]. Beijing: China Building Industry Press.
[12] Yong Wang, Nan Xie. (2007)The influence of horizontal strengthened story and vertical diagonal bracing to fastener style bearing capacity of super high supports for formwork [J]. Architecture, (16): 36-37.
[13] Zhicheng Qiu . (2004)A new analysis method of Photogrammetry error [J] . Science of Surveying and Mapping, (3): 10-13 .
[14] Bing Yao, Qilin Zhao, Ting Rui, Yong Ding. (2009)Distance measurement method for non-contact detection of concrete bridge cracks [J]. China Municipal Engineering, (03):38-40.

IOP Conf. Series: Materials Science and Engineering **794** (2020) 012033 doi:10.1088/1757-899X/794/1/012033

Research on coupling vibration of hydropower house

Shuhe Wei[1*], Xin Wang[2] and Liaojun Zhang[3]

[1]School of Architectural Engineering, Kaili University, Kaili, Guizhou, 556011, China

[2]Key Laboratory of Navigation Structure Construction Technology, Nanjing Hydraulic Research Institute, Nanjing, Jiangsu, 210029, China

[3]College of Water Conservancy and Hydropower Engineering, Hohai University, Nanjing, Jiangsu, 210029, China

[*]674665017@qq.com

Abstract. In order to describe and express the hydraulic vibration source as the exciting force in the vibration analysis of hydropower house, a new coupling method for vibration analysis of house structure-flow passage turbulence is proposed in this paper, and then, the finite element coupling model of Xiangjiaba house is developed. Based on the turbulent field and house structure vibration calculated in the time domain, the coupling vibration relationship between hydraulic vibration source and hydropower house is studied. Fluid pressure fluctuations are mainly caused by low frequency vortex in taper pipe of draft tube and Rotor/Stator interaction; both the low frequency vibration displacements on generator floor which are at primary frequency of 0.625 Hz and 1.25 Hz and the low frequency vibration velocities on generator floor which are at primary of 0.626 Hz and 9.375 Hz are caused by low frequency vortex in draft tube. The results show that the coupling method and corresponding finite element model initiated in this paper succeed in solving the description and expression of hydraulic vibration source. Due to short calculation time and inadequate setting of fluid initial condition, the distributions of maximum vibration value need further research.

1. Introduction

The vibration study of hydropower house mainly relies on prototype observation and numerical simulation. Time domain analysis method is used to solve the vibration response of hydropower house structure and unit in time domain by numerical approach[1-6]. Differences between research productions are mainly manifested as different processing method for excitation load. We take hydraulic vibration source as example. Based on pressure fluctuation obtained by prototype observation or model test in limited positions, interpolations are used to get the hydraulic loads on the surface of flow passage. Due to violent turbulence, it is difficult to estimate the error. In view of the development of Computational Fluid Dynamics (CFD), a new coupling method for vibration analysis of house structure-flow passage turbulence is proposed in this paper, then, the finite element coupling model of Xiangjiaba house structure is developed. Thus, structure vibration of Hydropower house is studied in a new way.

2. House-turbulence coupling vibration analysis method

Based on the two-way coupling iterative solution method, the sliding mesh technology and arbitrary Lagrangian-Eulerian (ALE) method are combined to propose the house-turbulence coupling vibration

IOP Conf. Series: Materials Science and Engineering **794** (2020) 012033 doi:10.1088/1757-899X/794/1/012033

analysis method, and the vibration quantity of power plant structure and turbulent flow field in channel are solved.

2.1Fluid-structure iterative computing method of two-way coupling

The incompressible Navier-Stokes equations are considered to be the governing equations and its weak forms are established using the Galerkin method; governing equations for solid are the structural dynamic equilibrium equations and employ the virtual displacement principle. Finite element equations for fluid and structure are established in frame of finite element method, so the finite element equations of the coupled fluid-structure system can be expressed as

$$\mathbf{F}[\mathbf{X}] \equiv \begin{bmatrix} \mathbf{F}_f\left[\mathbf{X}_f, \underline{\mathbf{d}}_s(\mathbf{X}_s)\right] \\ \mathbf{F}_s\left[\mathbf{X}_s, \underline{\boldsymbol{\tau}}_f(\mathbf{X}_f)\right] \end{bmatrix} = 0 \tag{1}$$

where, $\mathbf{F}_f$ and $\mathbf{F}_s$ are finite element equations corresponding to fluid and solid respectively. Iterative computing of two-way coupling method is also called the partitioned method. The fluid and solid solution variables are fully coupled in the solution. The fluid equations and the solid equations are solved individually in succession and the computational procedure can be summarized as follows.

We start with the initial solution guess $\underline{\mathbf{d}}_s^{-1} = \underline{\mathbf{d}}_s^{0} = {}^t\underline{\mathbf{d}}_s$ and $\underline{\boldsymbol{\tau}}_f^{0} = {}^t\underline{\boldsymbol{\tau}}_f$. For iterations $k = 1,2,\ldots$, the following equilibrium iteration is performed to obtain the solution ${}^{t+\Delta t}\mathbf{X}$.

①Solve the fluid solution vector $\mathbf{X}_f^k$ from the fluid equation;

$$\mathbf{F}_f\left[\mathbf{X}_f^k, \lambda_d \underline{\mathbf{d}}_s^{k-1} + (1-\lambda_d)\underline{\mathbf{d}}_s^{k-2}\right] = 0 \tag{2}$$

② if only the stress criterion needs to be satisfied, the stress residual is computed and checked against the tolerance. If the criterion is satisfied, steps ③ to ⑤ are skipped;

③solve the fluid solution vector $\mathbf{X}_f^k$ from the fluid equation

$$\mathbf{F}_s\left[\mathbf{X}_s^k, \lambda_\tau \underline{\boldsymbol{\tau}}_f^k + (1-\lambda_\tau)\underline{\boldsymbol{\tau}}_f^{k-1}\right] = 0 \tag{3}$$

Note that the fluid stresses have also been relaxed using a stress relaxation factor λ_τ ($0 < \lambda_\tau \leq 1$);

④the fluid nodal displacements are computed with the prescribed boundary conditions

$$\underline{\mathbf{d}}_f^k = \lambda_d \underline{\mathbf{d}}_s^k + (1-\lambda_d)\underline{\mathbf{d}}_s^{k-1} \tag{4}$$

⑤if only the displacement criterion needs to be satisfied, the displacement residual is computed and checked against the tolerance. If both the stress and the displacement criteria need to be passed, both convergence conditions are checked. If the iteration has not converged yet, the program goes back to step ① and continues for the next iteration unless a maximum number of FSI iterations have been reached;

⑥save the fluid and solid solutions.

The iterative method requires less memory than the direct method in two-way couplings. It is more effective for large problems.

2.2Sliding mesh

The fluid meshes in turbine runner rotate with runner. Flux transfer and momentum conservation are achieved between rotating mesh in runner and static mesh around runner. The sliding mesh technology is designed to allow meshes in different regions to move relative to each other, while the physical variables across the sliding interface remain continuous. The nodal points M, M1 and M2 are sitting at different sides of the sliding mesh interface. The control volume of M is composed of upper and lower sub-volumes. In order to calculate the flux/force over the control volume represented by the node M, we need to deal with upper and lower sub-volumes respectively. For the lower half, there is no difficulty to get the flux using the information from node M. Meanwhile, the flux for the upper half control volume is calculated using the neighboring nodes M1 and M2, as shown in Figure 1.

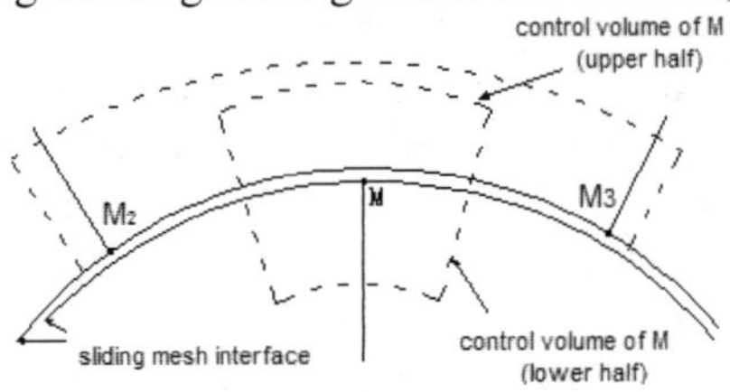

Figure 1. Sliding mesh interface coupling

2.3Arbitrary Lagrangian-Eulerian (ALE) formulation

Boundary change of the fluid domain is caused by guide vane opening of hydraulic turbine and the problem cannot be overcome by using Eulerian formulation. ALE formulation is introduced into coupling method to deal with mesh adjusting caused by change of the fluid-solid interface.

Let $\mathbf{v}$ be the fluid velocity and $\hat{\mathbf{v}}$ be the velocity of a reference domain, then the time derivatives in the continuity and momentum equations are obtained by considering an arbitrary volume of the reference domain. For example, the continuity equation is

$$\frac{\partial}{\partial t}\int_V \rho dV + \int_S \rho(\mathbf{v} - \hat{\mathbf{v}})\cdot \mathbf{n} dS = 0 \tag{5}$$

S is the surface of V and $\mathbf{n}$ is the unit normal vector. In practice, the velocity of the reference domain is the velocity of the finite element mesh. Special algorithms based on the solution of Laplace equations can be used to calculate the new nodal point positions in fluid mesh.

Time integration for both fluid and solid equations must be consistent due to iterative computing of two-way coupling, so the consistent time integration is employed. Shear Stress Transport model is used to take into account the turbulence effect. Integrating sliding mesh and ALE formulation, hydropower house structure-turbulence coupling vibration analysis method is proposed on the basis of iterative computing method of two-way coupling.

2.4Numerical model

According to actual size and material properties of Xiangjiaba hydropower house, the finite element coupling model includes house concrete structures and unit and turbulence in the fluid passage, as shown in Figure 2. Furthermore, the responses of fluid and structure are calculated under normal operating conditions in time domain.

Synchronous rotation speed of Xiangjiaba hydropower unit is 75 r/min, that is, f_n=1.25 Hz. Time step size is set to 0.002 s and total time steps 4000 in the finite element model.

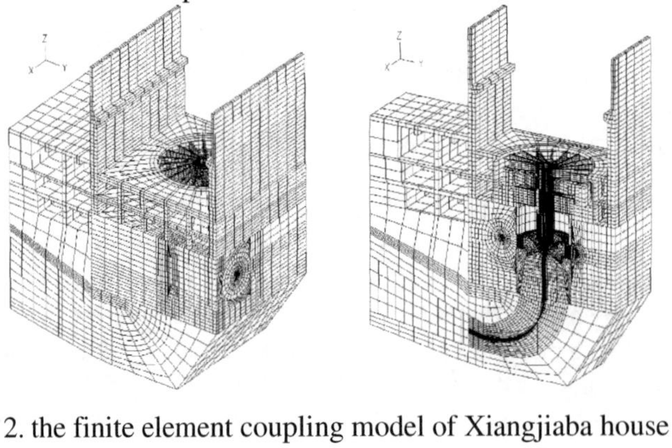

Figure 2. the finite element coupling model of Xiangjiaba house

IOP Conf. Series: Materials Science and Engineering **794** (2020) 012033 doi:10.1088/1757-899X/794/1/012033

3. Analysis of unsteady flow in whole passage

According to calculations, pressure fluctuation characteristics of unsteady flow are studied. Pressure contours in whole passage is shown in Figure 3.

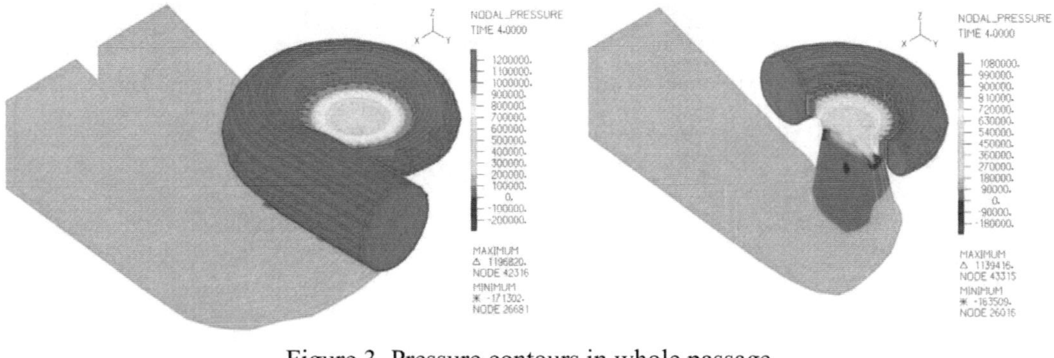

Figure 3. Pressure contours in whole passage.

Table 1. Primary frequency and amplitude of whole passage flow.

Frequency (Hz) / Amplitude (Pa)	0.626	1.25	18.77	56.32	93.87	131.41
Inlet section of volute	2156	515	410	353	268	**898**
Runner inlet	3265	2283	**10668**	**1317**	**343**	**459**
Back of runner blade	3958	4659	604	636	142	358
Taper pipe of draft tube	**10298**	**5228**	782	631	296	404
Elbow pipe of draft tube	3245	1003	225	146	71	74
Divergent pipe of draft tube	215	31	39	12	3	2

The primary frequency and amplitude of whole passage flow are shown in Table 1. Under normal operating conditions, pressure fluctuations in passage can be divided into two classes, low frequency vortex at primary frequency of 0.626 Hz and 1.25 Hz and middle frequency fluctuations at primary frequency of 18.77 Hz and 56.32 Hz and 93.87 Hz and 131.41 Hz.

The amplitude of low frequency vortex reaches the highest level in taper pipe of draft tube and decreases along upstream and downstream, so it is concluded that this is low frequency vortex happened in draft tube. In the same way, the amplitude of middle frequency fluctuations reaches the highest level in front of runner and decreases along downstream. Furthermore, its frequency value divided by the product of blade number and speed frequency are odd numbers. It is clear that the middle frequency fluctuations are caused by rotor/Stator interaction.

4. Vibration analysis of hydropower house

In order to investigate the internal vibration condition, Maximum vibration value contours and profiles of hydropower house are shown in Figure 4. The vibration time history and corresponding frequency spectrum analysis of the generator floor are shown in Figure 5.

IOP Conf. Series: Materials Science and Engineering **794** (2020) 012033 doi:10.1088/1757-899X/794/1/012033

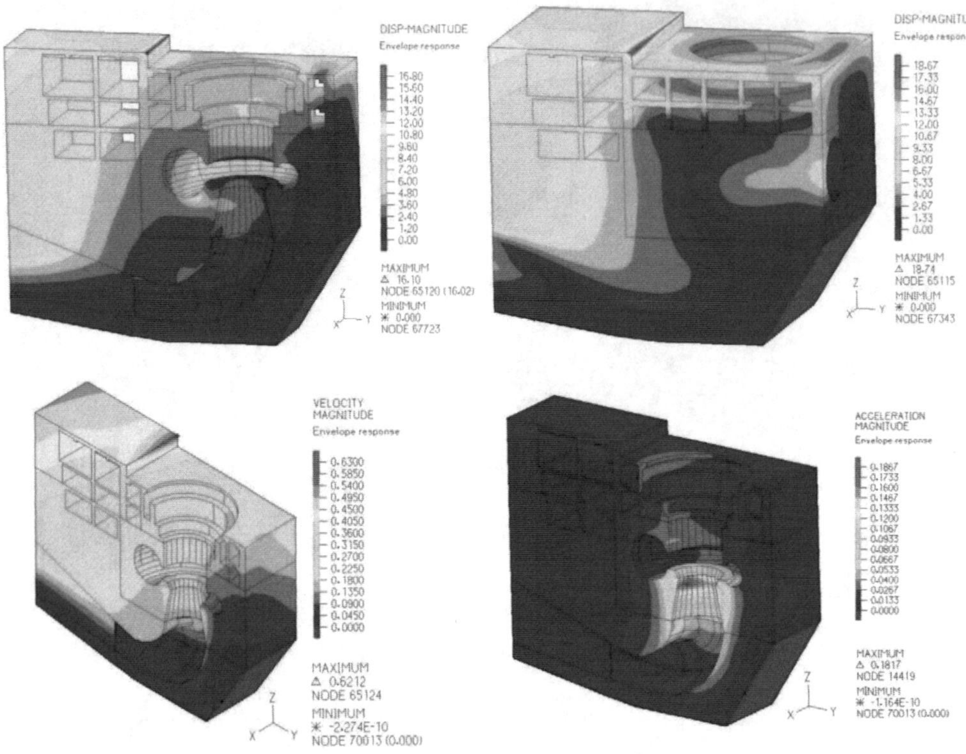

Figure 4. Maximum vibration value contours of hydropower house.

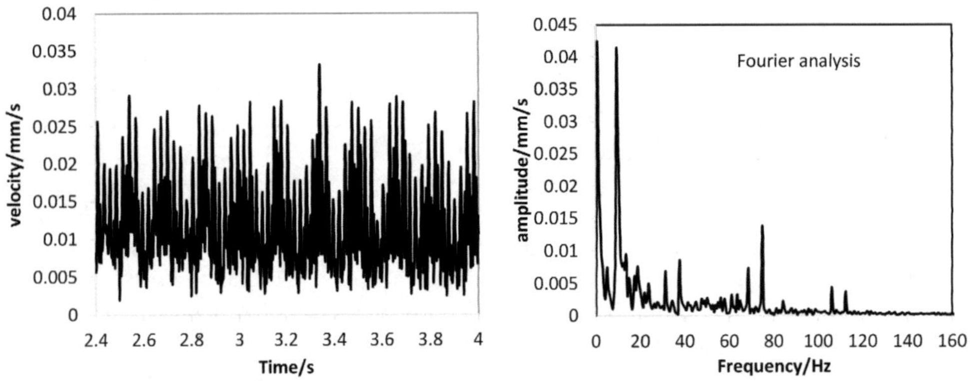

Figure 5. Time history of vibration on the generator floor and spectrum analysis.

Combining with maximum vibration value contours and time history of vibration on the generator floor and spectrum analysis, some conclusions are gotten:

- both the low frequency vibration displacements on generator floor which are at primary frequency of 0.625 Hz and 1.25 Hz and the low frequency vibration velocities on generator floor which are at primary of 0.626 Hz and 9.375 Hz are caused by low frequency vortex in draft tube;
- both the middle frequency vibration displacements which are at primary frequency of 18.77Hz and 56.25 Hz and middle frequency vibration velocities on generator floor which are at primary frequency of 37.5 Hz and 75.0 Hz and 112.5 Hz are caused by Rotor/Stator interaction;
- the degree of vibration displacement and velocity and acceleration is not high, and the calculation results comply with the relevant regulations under normal operating conditions.

IOP Conf. Series: Materials Science and Engineering **794** (2020) 012033 doi:10.1088/1757-899X/794/1/012033

5. Conclusions

Under normal operating condition, based on the finite element coupling model of Xiangjiaba house, unsteady flow field in the whole passage and the time history vibration of house structure are calculated. Furthermore, the coupling vibration relationships between turbulence in passage and house structure are studied. Some conclusions are gotten for normal operating condition:

- fluid pressure fluctuations are mainly caused by low frequency vortex in taper pipe of draft tube and Rotor/Stator interaction;
- the vibrations on the generator floor are mainly caused by low frequency vortex and Rotor/Stator interaction;
- the hydropower house structure-turbulence coupling vibration analysis method proposed in this paper succeeds in solving the description and expression of hydraulic vibration source.

Due to short calculation time and inadequate setting of fluid initial condition, the distributions of maximum vibration value need further research.

Acknowledgments

First, I would like to express my gratitude to my supervisor Professor Zhang in Hohai University for his various help. He has walked me through all the stages of the writing of this thesis. Without his consistent and illuminating instruction, this thesis could not have reached its present form. Second, this work was supported by National Natural Science Foundation of China (51779151). This work was also supported by the Specialized Research Fund for the Doctoral Program of Kaili University (BS201809). Last, I own my sincere gratitude to my fellow classmate Professor Wang in Nanjing Hydraulic Research Institute who offers me his helps and money in helping me deal with various difficulties for the past several years. My wife and my parents gave me a lot of encouragement when I was working at this thesis. They always share my weal and woe. I feel much grateful and heartily owe my achievement to them.

References

[1] Cao, W., Zhang, Y.L., Ma, Z.Y., and Chen, J. (2007). Vibration analysis of roof overflow powerhouse. Journal of Hydraulic Engineering, 38 (9), 1090-1095.

[2] Lian, J.J., Qin, L., Tian, H.J., Zheng, C.L. and He, C.L. (2006). Vibration analysis of hrdropower-house during start-up process. Journal of Hydroelectric Engineering, 25 (2), 1-5.

[3] Ouyang, J.H., Chen, H.Q., and Li, D.K. (2005). Computation of vibration in powerhouse of Three Gorges Project and prototype verification. Journal of Hydraulic Engineering, 36 (4), 484-490.

[4] Shen, K., Zhang, Z.Q., and Liang, Z. (2003). Hydraulic vibration calculation of Yantan hydropower house. Hydroelectric Energy, 21 (1), 73-75.

[5] Song, Z.Q., Ma, Z.Y., Chen, J., and Zhang, Y.L. (2008). Dynamic analysis on powerhouse of Longtoushi hydropower station. Journal of Hydraulic Engineering, 39 (8), 916-921.

[6] Xing, X.L., Chen, J.K., and Liao, C.G. (2013). Vibration response of large underground powerhouse. Journal of Vibration and Shock, 32 (9), 21-27.

IOP Conf. Series: Materials Science and Engineering **794** (2020) 012034 doi:10.1088/1757-899X/794/1/012034

Research on static load test of foundation piles by self-blance method

Yuan Chao [1], **Zhao Xiao Yong**[2], **Zheng Song Wen** [3,*]

[1]Guizhou Construction Science Research and Design Institute of CSCEC, Guiyang 550000, China

[2]Chengdu Municipal Waterworks Co., Ltd., Chengdu Environment Investment Group Co., Ltd., Chengdu 610000, China

[3]School of Civil Engineering, Southwest Jiaotong University, Chengdu 610000, China

*Corresponding author's e-mail: qazzsw1995@163.com

Abstract. The pile foundation static load test is a technique for detecting pile foundation bearing capacity in engineering. In determining the ultimate bearing capacity of a single pile, it is a relatively accurate and reliable test method, which can accurately reflect the stress condition and deformation characteristics of a single pile, and determine the vertical compressive bearing capacity of a single pile. Based on the static test of pile foundation, the process of self-balancing loading method is expounded.

1.Introduction

The static load test of pile adopted in China is the most reliable method to test the bearing capacity of pile. However, for the bearing capacity of pile with large diameter and large tonnage, the equipment required is bulky and expensive, which is difficult to meet the bearing capacity test of pile with special site and large tonnage. The self-balancing test of pile bearing capacity solves the problem of pile bearing capacity test of large tonnage and special site. The self-balancing method of pile bearing capacity has the advantages of large test tonnage, free from the limitation of site conditions, and fast and convenient test. This paper analyzes the bearing capacity and displacement of foundation pile through the test of self-balancing method in a certain area of guizhou province, providing a reference for the future application of self-balancing method of pile bearing capacity.

2.Project Overview

Based on on-site geological survey and drilling data combined with regional geological data, the underlying bedrock of the site is the Triassic Lower Anshun Formation (T_1a) argillaceous dolomite, which is under the red clay layer. Purplish red, gray white, gray yellow, thin - thick layer, fine crystal structure, dissolution ditch, stone bud more developed, the buried depth is generally 0.00~9.40m, the maximum buried depth is about 17.90m, and the bedrock surface is undulating.

3.Test pile determination

Corresponding to the three engineering piles of ZK2-41, ZK2-92 and ZK2-59 (construction numbers C6, C27, C34 respectively), the static load test is carried out, and the hole forming method adopts manual digging, and the test method is self-balance loading method.

Table 1. The parameters of the pile.

No.	Pile No.	Pile diameter /mm	Enlarged head diameter/mm	Pile length /m	concrete grade/KN	Design bearing capacity/KN
1	C6	1000	1800	12.53	C30	5343
2	C27	1000	1600	14.87	C30	5343
3	C34	1000	1600	14.19	C30	5343

4.Test methods and experimental settings

4.1Test principle

4.1.1Self-balance loading method

This test was conducted using the self-balancing method. The detection principle is to put a special loading device-load box, which is embedded in the corresponding position in the pile together with the steel cage before the concrete is poured, and the pressure tube of the loading box and other required test devices are led from the pile to the ground, then infused into piles. The pressure pump is loaded on the ground facing the load box, and the load box generates forces in the upper and lower directions and is transmitted to the pile body. Due to the self-acting reaction force of the pile body and the surrounding rock mass of the pile, we will get the data equivalent to two static load tests: above the load box, we obtain the corresponding reaction series parameters of the upper part of the pile during reverse loading; In the lower part of the load box, we obtain the corresponding reaction parameters of the lower part of the pile during forward loading. This principle test pile is called the self-blance force test pile in the domestic industry. This method can be used to provide data basis for design, and can also be used to test the bearing capacity of engineering piles [1-2].

4.2Experimental device

4.2.1Loading system

The loading system includes a high pressure oil pump, a special jack, and a high pressure oil pipe. The jack is a professional jack with patent technology. It is characterized by: During the design, the parameters such as the shape and layout of the load box fully consider the space reserved for the tasks such as grouting and replenishing. The design of the jack diameter and the loading area fully takes into account the medium and low pressure of the loading hydraulic pressure and the high bearing capacity after the pile test. The jack is designed with built-in special supercharging technology to generate a large loading force with very low oil pressure, which can greatly reduce the failure rate of the loading system[3-5].

4.2.2Installation of load boxes

In order to ensure the quality of the pile foundation and the success of the test pile, the following safety measures will be taken when the load box is buried: In order to avoid damage to the pile due to stress concentration during loading, the reinforcement cage stirrups near the load box are properly encrypted. The connection strength between the load box and the steel cage is appropriate to facilitate opening the load box during the test. Because the load box is placed in the middle of the pile, there is a through hole in the center. In order to ensure the thickness of the sediment at the bottom of the pile, it is recommended that the load box and the steel cage be connected into one body. After the hole is closed, the pile bottom is placed at one time.

4.2.3Determination of vertical ultimate bearing capacity of single pile

Obtained the ultimate bearing capacity of the upper pile of the load box and the ultimate bearing capacity of the lower pile of the load box. Refer to the calculation formula of the bearing capacity in the Technical Code for Self-Balance Measurement Method of pile Bearing Capacity (DBJ/T14-055-2009). Vertical compressive ultimate bearing capacity of piles:

$$Q_U = \frac{Q_{UU} - W}{\gamma} + Q_{UD} \tag{1}$$

Q_U: vertical compressive ultimate bearing capacity (kN) of single pile;

Q_{UU}: measured ultimate bearing capacity (kN) of the upper pile of the load box;

Q_{UD}: the measured ultimate bearing capacity (kN) of the pile in the lower part of the load box;

W: the weight of the upper pile of the load box;

γ: The correction coefficient of the pile side resistance of the upper part of the load box, usually taking the value of 0.75~1.0.

5.Experimental monitoring results and analysis

5.1Test pile C6 test process and data analysis

According to the field measured data, draw the load box bottom downward, top upward displacement curve Q-s, s~lgQ curve. The correlation curves are shown in Figures 1 and 2. Vertical compressive ultimate bearing capacity of single pile = ultimate bearing capacity of upper pile + ultimate bearing capacity of lower pile, calculated according to formula (1) is 11370kN, which is greater than 2 times of vertical bearing capacity of design single pile, ie greater than 10686kN, and the upward and downward displacements of the load box are small, 12.03mm and 14.68mm. Respectively, so the final test bearing capacity of the test pile is 11370kN.

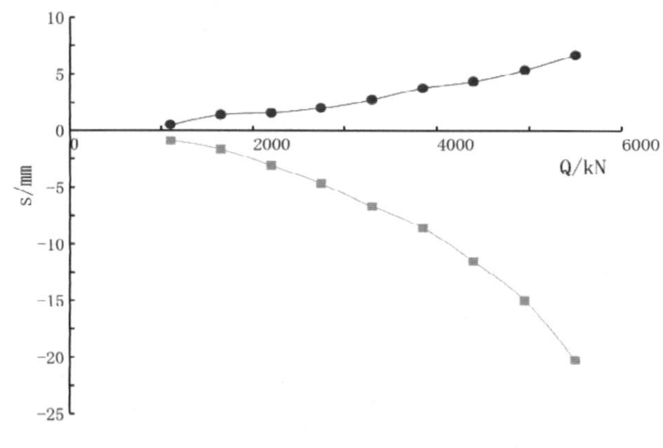

Figure 1. Q-s curve diagram of C6 test pile in area C of museum.

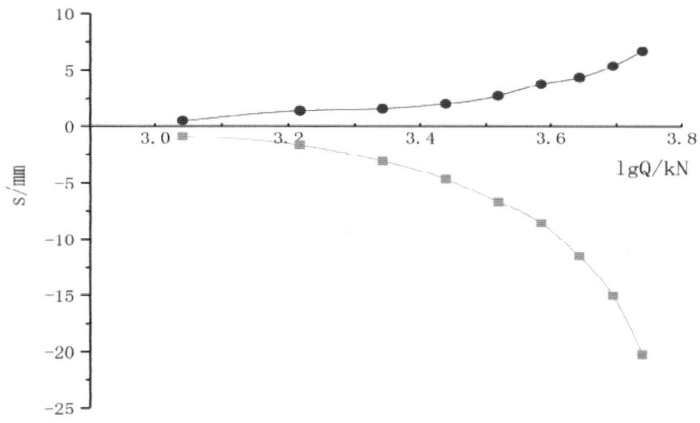

Figure 2. s-lgQ curve diagram of C6 test pile in area C of museum.

5.2Test pile C27 test process and data analysis

According to the field measured data, draw the load box bottom downward, top upward displacement curve Q-s, s~lgQ curve. The correlation curves are shown in Figures 3 and 4. Vertical compressive ultimate bearing capacity of single pile = ultimate bearing capacity of upper pile + ultimate bearing capacity of lower pile, calculated according to formula (1) is 11319kN, which is greater than 2 times of vertical bearing capacity of design single pile, ie greater than 10686kN, and the upward and downward displacements of the load box are small, 13.18mm and 10.00mm. Respectively, so the final test bearing capacity of the test pile is 11319kN.

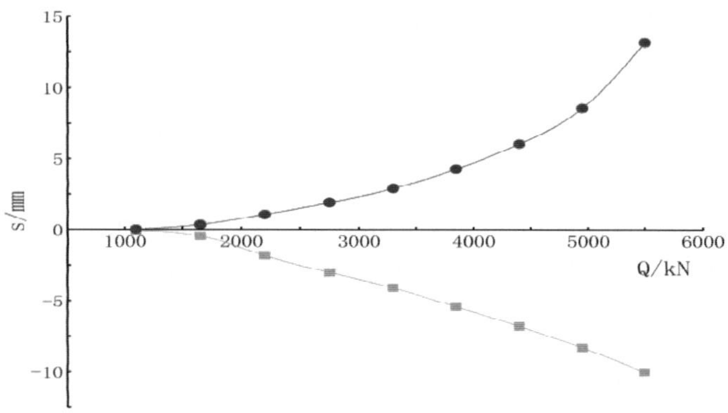

Figure 3. Q-s curve diagram of C27 test pile in area C of museum.

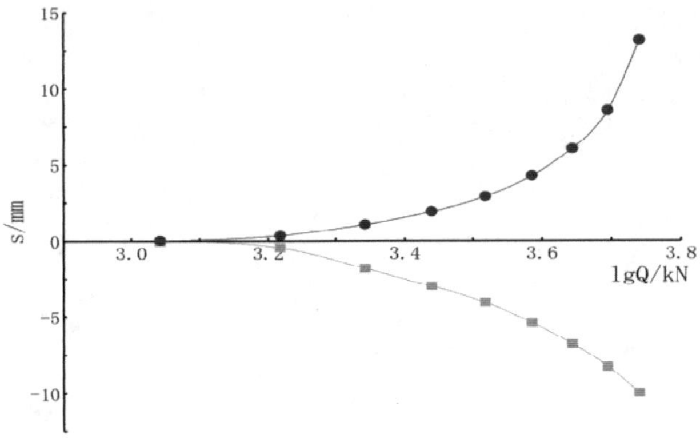

Figure 4. s-lgQ curve diagram of C27 test pile in area C of museum.

5.3Test pile C34 test process and data analysis

According to the field measured data, draw the load box bottom downward, top upward displacement curve Q-s, s~lgQ curve. The correlation curves are shown in Figures 5 and 6. Vertical compressive ultimate bearing capacity of single pile = ultimate bearing capacity of upper pile + ultimate bearing capacity of lower pile, calculated according to formula (1) is 11345kN, which is greater than 2 times of vertical bearing capacity of design single pile, ie greater than 10686kN, and the upward and downward displacements of the load box are small, 16.15mm,13.09mm. Respectively, so the final test bearing capacity of the test pile is 11345kN.

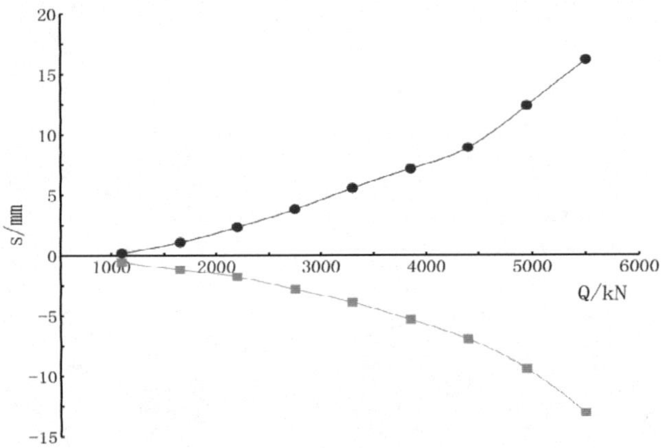

Figure 5. Q-s curve diagram of C34 test pile in area C of museum.

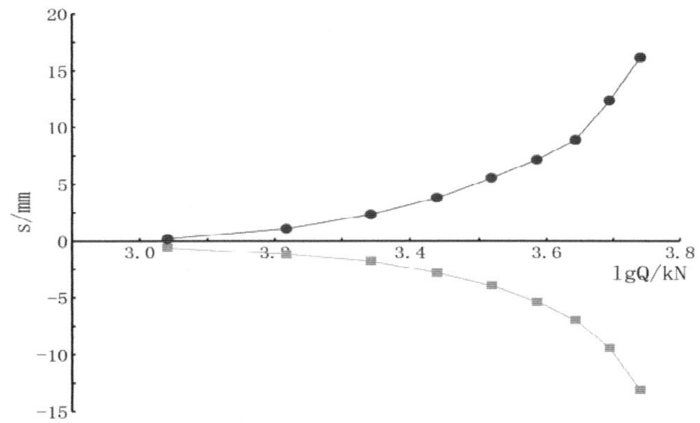

Figure 6. s-lgQ curve diagram of C34 test pile in area C of museum.

6.Conclusion

1.The self-blance method pile foundation static load test is convenient and cost-effective. The loadable range of the loading box can also achieve satisfactory experimental results for large-tonnage test installation, and the cost saving is more obvious.

2.The loading box is a one-time product, and the placement position will also have an impact. It should be considered before construction to avoid waste.

3. According to the formula of vertical ultimate bearing capacity of single pile, this paper calculates that at 5500kN, the upper and lower pile sections of loading box are basically in the elastic stage, and the bearing capacity of single pile is more than twice the vertical bearing capacity of single pile.

Acknowledgments

The authors gratefully acknowledge the support from Guizhou Provincial Science and Technology Department (QKH-ZC [2018]2833).

References

[1] Shi Peidong, Lu Yi. The development of the Osterberg static load test method for 10 years [J]. Industrial Construction, 1999, 29 (13)

[2] Shi Peidong, Huang Qin. New static load test technology for piles [A]. Pile Foundation Engineering Technology [M]. Beijing China Building Industry Press, 1996.

[3] DBJ/T14-055-2009, Technical Code for Self-Balance Measurement Method of pile Bearing Capacity.

[4] Gong Weiming, Dai Guoliang, Jiang Yongsheng, et al. Theory and practice of self-equilibrium test of pile bearing capacity [J]. Journal of Building Structures, 2002, 23(1).

[5] Dai Guoliang. Theory and Practice of Self-balancing Test Method for Pile Bearing Capacity [D]. Doctoral Dissertation, Southeast University, 2003.

IOP Conf. Series: Materials Science and Engineering **794** (2020) 012035 doi:10.1088/1757-899X/794/1/012035

Dynamic Centrifuge Modelling Tests for Asphalt Concrete Core Dam

Guofeng Ren[1,2], Xingwen Gu[1,2*] and Nianxiang Wang[1,2]

[1]Geotechnical Engineering Department, Nanjing Hydraulic Research Institute, Nanjing, Jiangsu, 210029, China

[2]Key Laboratory of Failure Mechanism and Safety Control Techniques of Earth-rock Dam of the Ministry of Water Resources, Nanjing, Jiangsu, 210029, China

[*]Corresponding author's e-mail: xwgu@nhri.cn

Abstract. Dynamic centrifuge modelling tests are carried out to study the seismic behaviour of a sand-gravel dam with asphalt concrete core wall. The acceleration response, deformation, core wall stress and stability are analysed under design and check seismic intensity. The crest acceleration amplification factor is 2.7~3.0. The crest settlement rate is 0.442% and 0.481%, respectively under design and check intensity. Maximum seismic stress is recorded in core wall of 2/3~4/5 dam height. Under design intensity, failure cracks appear in crest and upstream dam revetment at water level. Under check intensity, more cracks are found. Furthermore, landslide occurs in upstream slope above 2/3 dam height under check intensity. Downstream slope is stable. No evident separation is observed between dam shell materials and core wall. The dam is unstable under design and check earthquake. Aseismic measures are necessary according to test results.

1. Introduction

Sand-gravel dam with asphalt concrete core wall is employed in a hydropower project. The maximum dam height is 129 m. The dam zone of the project is with high earthquake intensity. The design aseismic intensity is IX degrees. It is necessary to study the seismic response, deformation, core wall stress and stability by means of dynamic centrifuge modeling tests.

2. Dynamic centrifuge modelling tests

The technique of centrifuge shaking table has been developed since around 1980 for dynamic centrifuge modeling test of earthquake problems[1]. The use of dynamic centrifuge modelling tests is particularly important for hydraulic engineering as it offers high gravity field for scaled models by centrifuge and earthquake loading using a centrifuge shaking table in flight. The technique has been recognized as one of the most effective and advanced research methods[2]. Dynamic centrifuge modeling tests has shown its advantages in research of dam seismic failure mode, aseismic design and numerical model verification, etc.[3-6].

2.1Similarity laws

The similarity laws for dynamic centrifuge modeling are shown in table 1[7], according to the third law of similarity theory and based on the similarity laws for shaking table test. η is defined as scaling factor of model to prototype for parameters.

Table 1. Similarity laws of dynamic centrifuge modelling

Symbol	Parameter	Dimension	Centrifuge model
L	Length	L	η_l
ρ	Density	ML^{-3}	η_ρ
g	Acceleration	LT^{-2}	η_g
C	Modulus		η_C
σ	Stress	$ML^{-1}T^{-2}$	$\eta_\sigma=\eta_l\eta_\rho\eta_g$
E	Elasticity modulus	$ML^{-1}T^{-2}$	$\eta_E=\eta_l^{1/2}\eta_\rho^{1/2}\eta_g^{1/2}\eta_C$
c	Cohesion	$ML^{-1}T^{-2}$	$\eta_{c'}=\eta_l\eta_\rho\eta_g$
ϕ	Friction angle		$\eta_{\phi'}=1$
ε	Strain		$\eta_\varepsilon=\eta_l^{1/2}\eta_\rho^{1/2}\eta_g^{1/2}\eta_C^{-1}$
u	Displacement	L	$\eta_u=\eta_l^{3/2}\eta_\rho^{1/2}\eta_g^{1/2}\eta_C^{-1}$
v	Velocity	LT^{-1}	$\eta_v=\eta_l^{3/4}\eta_\rho^{1/4}\eta_g^{3/4}\eta_C^{-1/2}$
k	Coefficient of permeability	LT^{-1}	$\eta_k=\eta_l^{3/4}\eta_\rho^{-3/4}\eta_g^{-1/4}\eta_C^{-1/2}$
T	Time	T	$\eta_T=\eta_l^{3/4}\eta_\rho^{1/4}\eta_g^{-1/4}\eta_C^{-1/2}$
f	Frequency	T^{-1}	$\eta_f=\eta_l^{-3/4}\eta_\rho^{-1/4}\eta_g^{1/4}\eta_C^{1/2}$
ξ	Soil damping ratio		$\eta_\xi=1$

2.2Facilities

The tests are carried out in Nanjing Hydraulic Research Institute (NHRI). The NHRI-400gt large centrifuge and its shaking table are engaged. The centrifuge has a maximum radius of 5.5 m and can accelerate a maximum payload of 2000 kg up to 200 g. Indexes of the shaking table are as table 2. Inner space of the model container is 700 mm long, 350 mm wide and 650 mm high.

Table 2. Indexes of centrifuge shaking table

Centrifugal acceleration	Seismic acceleration	Payload	Duration	Frequency	Wave mode	Direction
80 g	20 g	500 kg	3 s	20~200 Hz	Sinusoid, Site wave, Random	Horizontal

2.3Modelling tests

In centrifuge modeling, η_l is normally equal to $1/\eta_g$, meaning that the prototype gravity field is fully modelled. Here the prototype dam is too large to be fully scaled in any centrifugal shaking table in the world. Therefore η_l is not equal to $1/\eta_g$ and the prototype gravity field is partially simulated. Also reasonable simplification is made.

2.3.1Mosel setup. The section of maximum dam height is partly simulated. The berms are ignored and dam shell material is treated as uniform. The model is scaled in 1/450 and centrifugal acceleration is 40 g. 2 tests of the same model setup (figure 1) are carried out, under design (P_{50}=2%) and check (P_{100}=2%) seismic intensity respectively. The peak acceleration is 516.5 gal and 643.3 gal for design and check intensity.

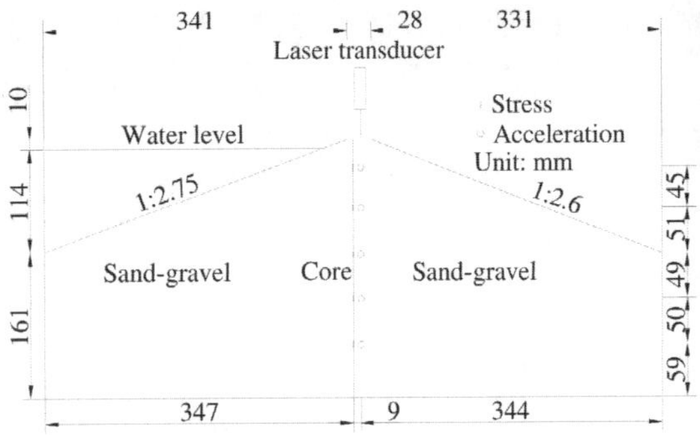

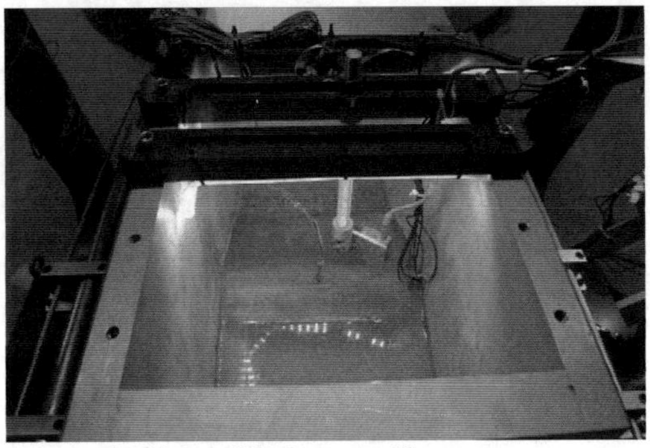

Figure 1. Model setup and the real model on shaking table in centrifuge

2.3.2Dam material simulation. The maximum size is 40 mm for model dam material considering scale effect of footings and boundary effect in centrifuge modelling [8]. Model dam material is simulated according to Specification of soil test: SL237—1999 [9].

2.3.3Core wall and revetment simulation. The thickness is 700~1400 mm and modulus is 500 MPa of prototype asphalt concrete core wall. Model core wall is a perspex sheet of 1.2 mm thick and its modulus is 2.3 GPa. Thus the flexural rigidity ratio meets the similarity laws. The model revetment is a thin layer of plain concrete.

3. Analysis of test results

All given results are converted to prototype according to similarity laws.

3.1The acceleration response

Figure 2 gives recorded and aim waves at basement for design and check earthquake. The peak acceleration is 516.5 gal and 643.3 gal of aim waves. The recorded peak acceleration is 501.9 gal and -523.1 gal for design earthquake. The recorded peak acceleration is 645.0 gal and -645.0 gal for check earthquake. Also the frequency responses are satisfying. The shaking table simulates the earthquakes accurately.

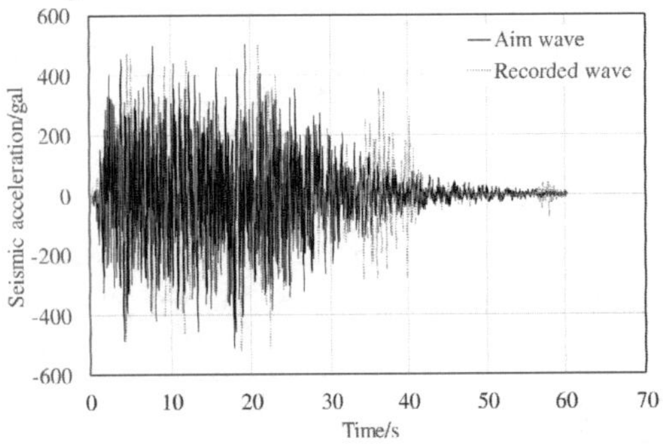

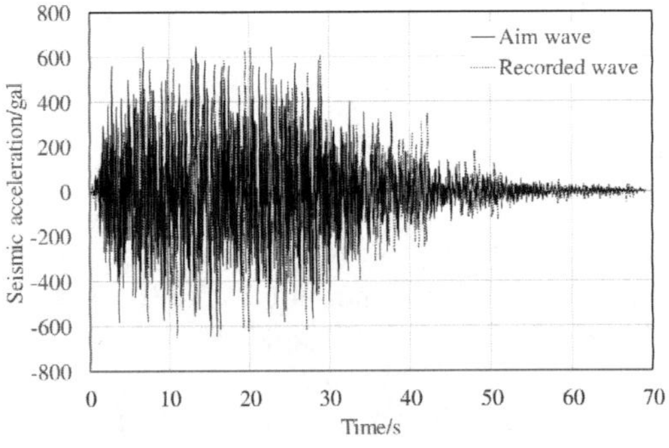

Figure 2. Recorded and aim waves (left for design and right for check earthquake)

Figure 3 shows peak acceleration and amplification factor profiles at dam axis. Table 3 gives detailed data. H is the dam height and h is the height of acceleration transducer to basement.

The acceleration increases with elevation in dam body under input earthquake loading. The acceleration amplification effect is obvious. The acceleration response of the dam can be divided into 2 linear lines at 2/3 dam height. The acceleration amplification effect is stronger for the upper part of dam body. The acceleration amplification factor is lower for the stronger earthquake. The crest acceleration amplification factor is about 2.7~3.0.

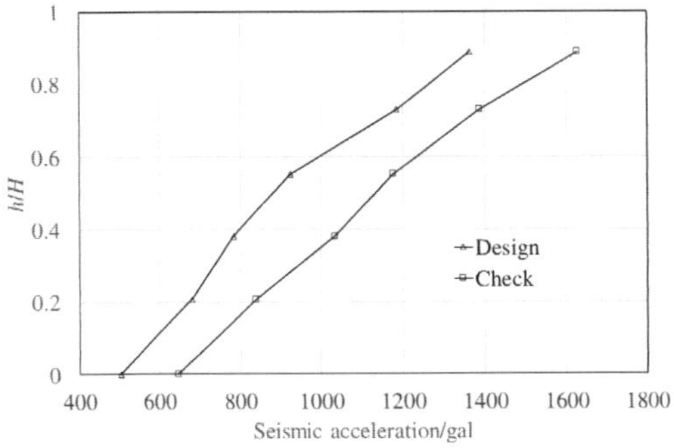

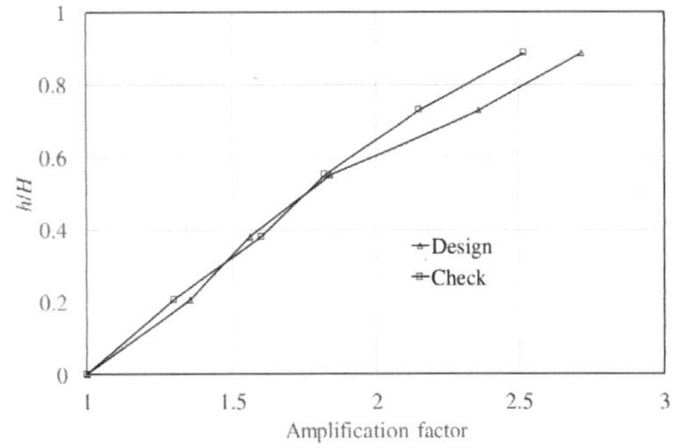

Figure 3. Acceleration amplification effect at dam axis

Table 3. Acceleration response data at dam axis

Elevation/m	*h/H*	Design earthquake		Check earthquake	
		Peak acceleration /gal	Amplification factor	Peak acceleration /gal	Amplification factor
2176	0	501.937	1	646.388	1
2202.4	0.20	680.152	1.355	839.462	1.299
2224.8	0.38	783.289	1.560	1034.831	1.601
2246.9	0.55	924.614	1.842	1178.099	1.822
2269.8	0.73	1185.229	2.361	1387.335	2.146
2290.1	0.89	1362.938	2.715	1628.741	2.520

3.2The crest settlement

Figure 4 shows the crest settlement development under design and check earthquake. It is obvious from the data that the crest settlement waves and increases with earthquake wave and tends to be stable. The stronger the earthquake is, the larger settlement is recorded. The crest settlement is 568 mm and 618 mm, and the settlement rate is 0.442% and 0.481%, respectively under design and check seismic intensity.

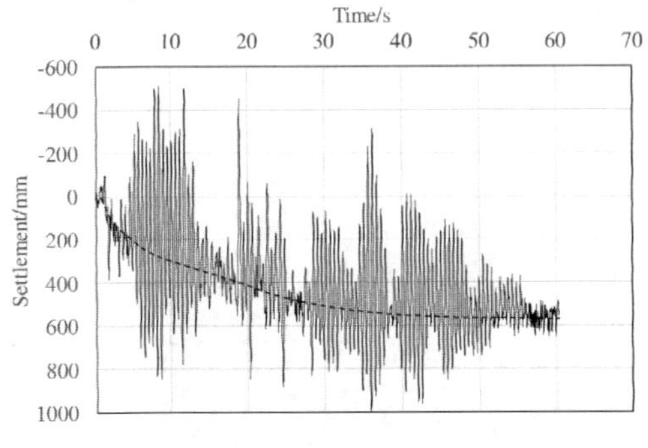

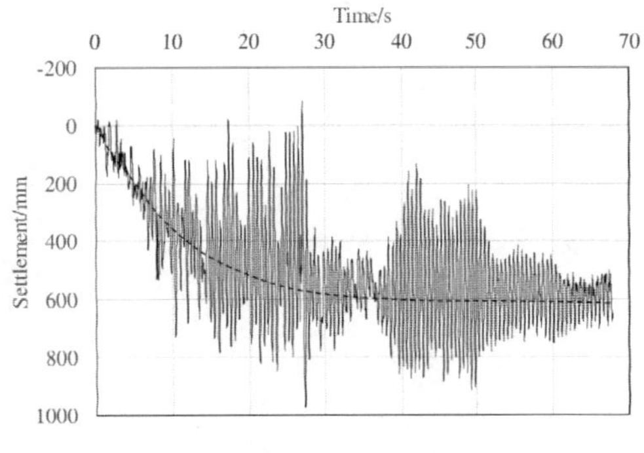

Figure 4. Crest settlement during earthquake (left for design and right for check intensity)

3.3The seismic stress of core wall

Figure 5 is a typical core stress curve. Figure 6 shows the core stress profiles after earthquake. Some text.

Maximum seismic stress is recorded in core wall of 2/3~4/5 dam height. The maximum seismic tensile stress during and after design earthquake is 232 kPa and 192 kPa, respectively. The maximum seismic tensile stress during and after check earthquake is 501 kPa and 474 kPa, respectively.

IOP Conf. Series: Materials Science and Engineering **794** (2020) 012035 doi:10.1088/1757-899X/794/1/012035

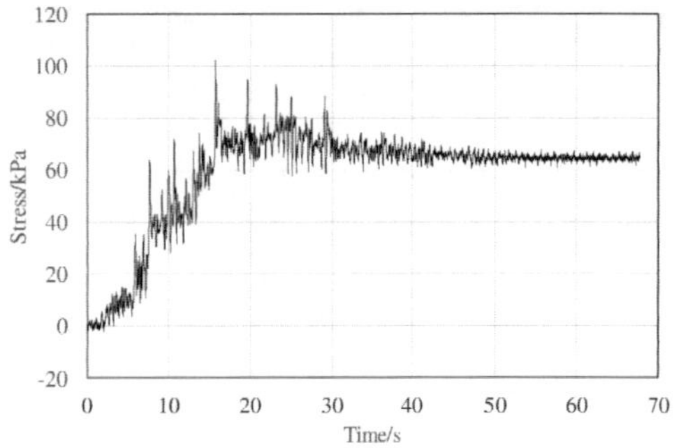

Figure 5. Typical core stress curve

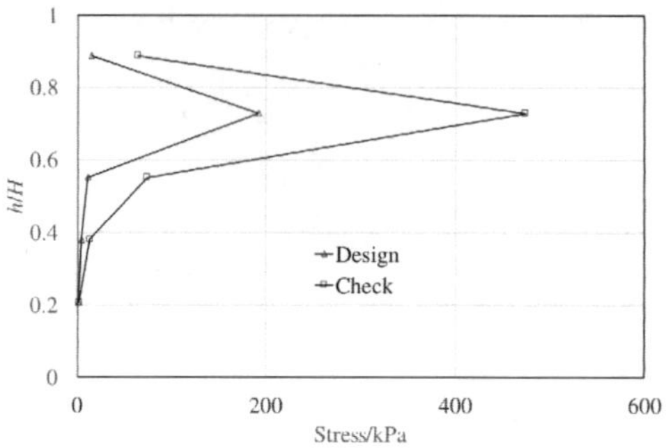

Figure 6. Core stress profiles after earthquake

3.4The seismic failure mode

Figure 7 and figure 8 show model dams after design and check earthquake.

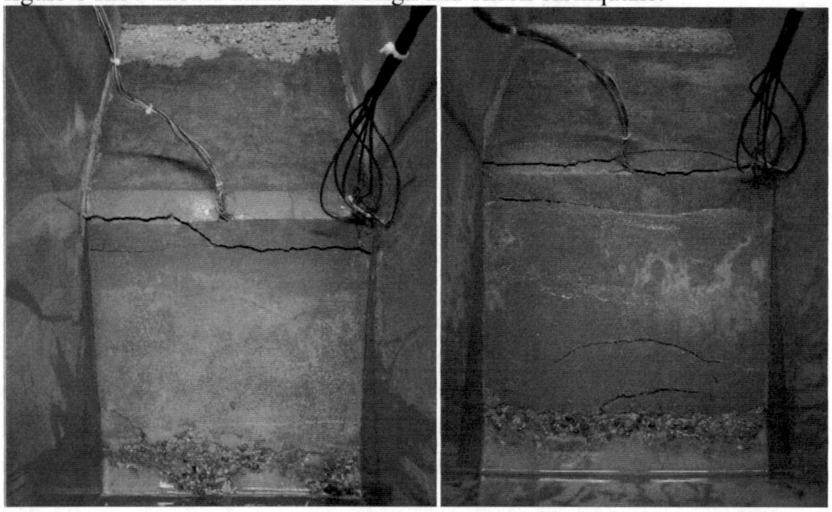

Figure 7. Cracks in upstream revetment and crest (left for design and right for check earthquake)

Figure 8. Material settlement (left for design and right for check earthquake)

Under design intensity, failure cracks appear in crest and upstream dam revetment at water level. Under check intensity, more cracks are found. Furthermore, landslide occurs in upstream slope above 2/3 dam height. Downstream slope is stable. No evident separation is observed between dam shell materials and core wall. The seismic failure mode is mainly the separation between upstream revetment and rockfill material induced by settlement, causing revetment to crack or even collapse. The dam is unstable under design and check earthquake.

4. Conclusions

According to test data analysis, conclusions are as follows.

(1) The acceleration increases with elevation in dam body under input earthquake loading. The acceleration amplification effect is obvious. The acceleration response of the dam can be divided into 2 linear lines at 2/3 dam height. The acceleration amplification effect is stronger for the upper part of dam body. The acceleration amplification factor is lower for the stronger earthquake. The crest acceleration amplification factor is about 2.7~3.0.

(2) The crest settlement waves obviously during earthquake. The settlement increases and tends to be stable. The stronger the earthquake is, the larger settlement is recorded. The crest settlement is 568 mm and 618 mm, and the settlement rate is 0.442% and 0.481%, respectively under design and check seismic intensity.

(3) Maximum seismic stress is recorded in core wall of 2/3~4/5 dam height. The maximum seismic tensile stress during and after design earthquake is 232 kPa and 192 kPa, respectively. The maximum seismic tensile stress during and after check earthquake is 501 kPa and 474 kPa, respectively.

(4) Under design intensity, failure cracks appear in crest and upstream dam revetment at water level. Under check intensity, more cracks are found. Furthermore, landslide occurs in upstream slope above 2/3 dam height. Downstream slope is stable. No evident separation is observed between dam shell materials and core wall.

(5) The seismic failure mode is mainly the separation between upstream revetment and rockfill material induced by settlement, causing revetment to crack or even collapse.

(6) Aseismic measures are necessary under design and check seismic intensity.

Acknowledgments

Authors wishing to acknowledge financial support from National key research and development program 2017YFC0404905, the central level public welfare scientific research institute basic research business special fund Y319001 and Y319002.

References

[1] Taylor R. N. (1995) Geotechnical Centrifuge Technology. Blackie Academic and Professional, Glasgow.

[2] Wang N.X., Zhang W.M. (2015) Geotechnical Centrifuge Technology and Application. China Building Industry Press, Beijing.

[3] Baziar M.H., Salemi S.H., Merrifield C.M. (2009) Dynamic centrifuge model tests on asphalt-concrete core dams. J. Géotechnique, 59(9): 763–771.

[4] Wang N.X., Zhang W.M., Gu X.W., Wu X.Y. (2009) Dynamic Centrifuge Model Test for Changhe Dam. J. Water Power, 35(5): 67–70. (in Chinese)

[5] Wang N.X., Wang Z.J., Jiang Z.X., Gu X.W., Ren G.F. (2018) Centrifuge shaking table test on aseismic measures of concrete face rockfill dam of Altash Water Control Project. J. Water Resources and Hydropower Engineering, 49(S1): 58–63. (in Chinese)

[6] Zhang X.D., Li G., Wei Y.Q., Zhang Z.T., Liang J.H., Hu J. (2018) Examining the performance of geo-grid reinforcement using dynamic centrifuge tests. J. Journal of China Institute of Water Resources and Hydropower Research, 16(6): 591–597. (in Chinese)

[7] Wang N.X., Zhang W.M., Gu X.W., He S.B., Du S.L. (2012) Combined model for seismic responses of high core rock-fill dams. J. Chinese Journal of Geotechnical Engineering, 34(5): 798–804. (in Chinese)

[8] Xu G.M., Zhang W.M. (1996) Study on scale effect of footings and boundary effect in centrifuge modeling. J. Chinese Journal of Geotechnical Engineering, 18(3): 80–86. (in Chinese)

[9] Specification of soil test: SL237—1999. S. China WaterPower Press, Beijing. (in Chinese)

IOP Conf. Series: Materials Science and Engineering **794** (2020) 012036 doi:10.1088/1757-899X/794/1/012036

Parameter calculation and verification of concrete plastic damage model of ABAQUS

Li Qingfu[1*], Guo Wei [1] and Kuang Yihang [1]

[1]School of Water Conservancy Science and Engineering, Zhengzhou University, Zhengzhou, China

[*]lqflch@zzu.edu.cn

Abstract. The purpose of this paper is to study the method for determining the parameters of ABAQUS CDP model based on the concrete stress-strain curve provided by the code for concrete structure design. A finite element model was established to calibrate the value of the stress-strain curve cutoff and the damage factor of the concrete, and a calculation method of damage factor was recommended. The results show that the relative errors between first crack load, ultimate load and deflection at ultimate load with the experimental results were less than 5%. It is accordingly concluded that the finite element analysis results adequately reflected the experimental results.

1. Introduction

ABAQUS software shows good applicability in solving nonlinear analysis. The concrete damage plasticity model (CDP model) provided by ABAQUS was first proposed by Lubliner[1], and the concept of stiffness recovery introduced by Lee[2] improved its model. The CDP model realizes stiffness change through damage factor, and considers crack development, crack closure and partial stiffness recovery of concrete under cyclic loading[3-5]. The basic rules include: yield criterion, flow rule and hysteresis rule. The CDP model has been widely used in seismic analysis. It can be used to simulate the behavior of concrete structure and composite structure under cyclic loading, and can properly predict the punching shear response of the slab-column composite structures[6-8]. Rama et al.[9] used the concrete plastic damage model to evaluate the fracture energy and characteristic length for different grades of concrete using CDP model, and studied the effect on fracture properties of different input parameters like dilatation angle and eccentricity. Wosatko et al.[3] studied the development and closure of cantilever beam cracks in the concrete plastic damage model based on the plasticity theory. The study showed that the model could be applied to isotropic plastic materials, and the reduction and recovery of stiffness were considered in both the tensile stage and the compression stage, which could be used to analyze crack closure. Fang et al.[10] simulated the reinforced concrete under cyclic loading and compared the influence of concrete uniaxial stress-strain curve on the simulation results. It is recommended to use the formula provided in the Chinese code as the tensile curve and other parts of the concrete. Reasonable selection of CDP model parameters is the key to improve the accuracy of numerical simulation. In this paper, based on the constitutive relationship provided in the concrete structure design code[11] (GB 50010-2010), a test beam was selected to establish a finite element model, and to calibrate the CDP model parameters.

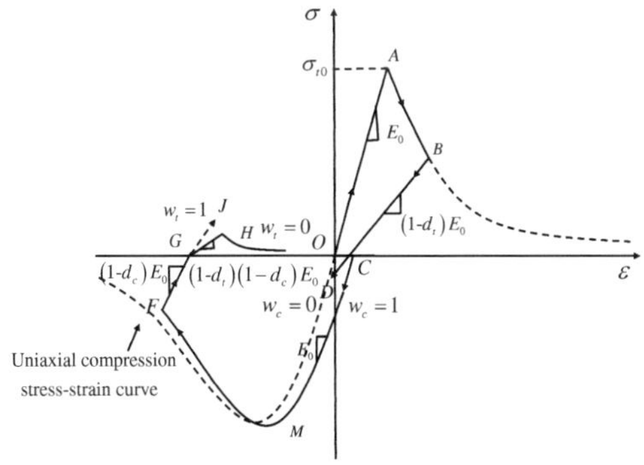

Figure 1. Schematic diagram of elastic modulus recovery of concrete under cyclic loading.

2. Concrete damage plasticity model

The concrete damage plasticity model (CDP model) assumes that the concrete material is continuous, isotropic and uniform, and the elastic modulus of concrete will be lost when the concrete shows inelastic strain. As the stiffness reduction under cyclic loading can be partially restored, the stiffness recovery coefficient w is defined to control this phenomenon. Under uniaxial cyclic loading, stiffness reduction and restoration of concrete under tensile and compressive loading are different. Therefore, the stiffness recovery factor needs to be defined separately in both directions. The damage factor d which is defined as the reduction of secant modulus relative to the initial elastic modulus at a certain point, as presented in the Equations (1) and (2).

$$E=(1-d)E_0 \tag{1}$$

Where the damage factor d is a function of stress state and uniaxial tensile and compressive damage variables d_t and d_c. Under uniaxial cyclic loading, ABAQUS assumes:

$$(1\text{-}d)=(1-s_t d_c)(1-s_c d_t) \tag{2}$$

Where s_t and s_c are respectively functions of the stiffness recovery stress state related to stress inversion, and are determined as:

$$\begin{cases} s_t=1-w_t r^*(\bar{\sigma}_{11}) & 0\le w_t\le 1 \\ s_c=1-w_c\left[1-r^*(\bar{\sigma}_{11})\right] & 0\le w_c\le 1 \end{cases} \tag{3}$$

$$r^*(\bar{\sigma}_{11})=H(\bar{\sigma}_{11})=\begin{cases} 1 & (\bar{\sigma}_{11}>0) \\ 0 & (\bar{\sigma}_{11}<0) \end{cases} \tag{4}$$

Where w_t and w_c are the tension and compression stiffness recovery coefficient. Under the uniaxial cyclic load (tension-compression-tension), the reduction and recovery of concrete stiffness are shown in Figure 1. When the concrete is in an axially stretched state, the tensile stress increases linearly. When the peak tensile stress of concrete (point A) is reached, the concrete cracks and then loads to point B. When the tensile stiffness of concrete decreases, the unloading will be carried out according to the effective stiffness $(1\text{-}d_t)E_0$, just as the path BC. When the axial pressure is applied to the concrete in the reverse direction, if $w_c=0$ (the compression stiffness does not recover after the tensile damage occurs), it is loaded by the path CD, and if $w_c=1$, it is loaded by the path CMF. When point F is reached, it is unloaded and then reverse loaded and stretched. If $w_t=1$, it is loaded according to path GJ. If $w_t=0$, it is loaded by path GH.

3. Model parameters determination method

3.1Concrete model

The concrete constitutive relationship is not given in the CDP model, and the ABAQUS user needs to define the constitutive relationship. The most representative concrete constitutive relations are three kinds: the Chinese code[11] (GB50010-2010), the European code[12] (EN 1992-1-1: 2004) and the full stress-strain curve equation proposed by American scholar Kent-Park[13]. The European code does not give the tensile stress-strain curve of concrete. The stress-strain curve in the Chinese code introduces the damage evolution parameters, which has better convergence in calculation compared with the Kent-Park model. Therefore, the stress-strain curve provided by the Chinese code is used to determine the parameters in the CDP model. The uniaxial stress-strain curve in the code is calculated by the following Equation (5).

$$\sigma=(1\text{-}D)E_0\varepsilon \tag{5}$$

Where D is the damage evolution parameter and E_0 is the initial elastic modulus. $f_{t,r}$ and $f_{c,r}$ use the standard values of axial tensile strength and axial compressive strength.

3.2Damage factor

The CDP model directly uses the damage evolution parameters provided by the Chinese code, which leads to the difficulty of convergence of the calculation. According to the principle of energy equivalence, the calculation method of concrete damage factor proposed by Sidoroff[14] can be applied to the calculation of concrete damage factor under tension and compression. The results are practical and easy for iterative convergence. The damage factors were determined by referring to Equation (6).

$$d=1-\sqrt{\frac{\sigma}{E_0\varepsilon}} \tag{6}$$

Where σ and ε are determined by equation (5), E_0 is the initial modulus of elasticity of the concrete.

3.3CDP model parameters

The CDP model considers the inelastic behaviour of concrete by defining damage factors in both compression and tension. It is assumed that the failure modes of concrete mainly include crack and crush. The evolution of yield surface is mainly controlled by tensile equivalent plastic strain $\tilde{\varepsilon}_t^{pl}$ and compressive equivalent plastic strain $\tilde{\varepsilon}_c^{pl}$. The strain input in the model is inelastic strain, and it is necessary to ensure that the corresponding equivalent plastic strain and $\tilde{\varepsilon}_c^{pl}$ are greater than 0 and increase with the increase of damage factor. Accordingly, the stress-strain curve section of concrete between 0.5 $\varepsilon_{c,r}$ (0.5 times the peak compressive strain of concrete corresponding to uniaxial compressive strength) and 0.6 $\varepsilon_{t,r}$ (0.6 times the peak tensile strain of concrete corresponding to uniaxial tensile strength) is selected as the elastic section.

Figure 2 shows the relationship between the compressive strains of the model. The compressive inelastic strain can be calculated by the following equations:

$$\tilde{\varepsilon}_c^{in}=\varepsilon_c-\varepsilon_{0c}^{el} \tag{7}$$

$$\varepsilon_{0c}^{el}=\frac{\sigma_c}{E_0} \tag{8}$$

$$\tilde{\varepsilon}_c^{pl}=\tilde{\varepsilon}_c^{in}-\frac{d_c}{(1-d_c)}\frac{\sigma_c}{E_0} \tag{9}$$

Where $\tilde{\varepsilon}_c^{in}$ is the compressive inelastic strain, and ε_{0t}^{el} is the elastic compressive strain corresponding to the initial elastic modulus, $\tilde{\varepsilon}_t^{pl}$ is compressive plastic strain, d_t is the compressive damage factor.

Figure 3 shows the relationship between the tensile strains of the model. The tensile cracking strain can be calculated by the following equations:

$$\tilde{\varepsilon}_t^{ck} = \varepsilon_t - \varepsilon_{0t}^{el} \tag{10}$$

$$\varepsilon_{0t}^{el} = \frac{\sigma_t}{E_0} \tag{11}$$

$$\tilde{\varepsilon}_t^{pl} = \tilde{\varepsilon}_t^{ck} - \frac{d_t}{(1-d_t)}\frac{\sigma_t}{E_0} \tag{12}$$

Where $\tilde{\varepsilon}_c^{ck}$ is the tensile cracking strain, and $\tilde{\varepsilon}_c^{ck}$ is the elastic tensile strain corresponding to the initial elastic modulus, $\tilde{\varepsilon}_t^{pl}$ is tensile plastic strain, d_c is the tensile damage factor.

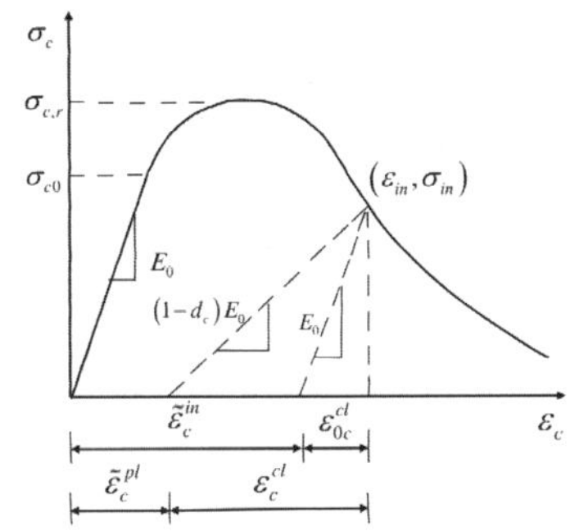

Figure 2. Uniaxial compression stress-strain curve of CDP.

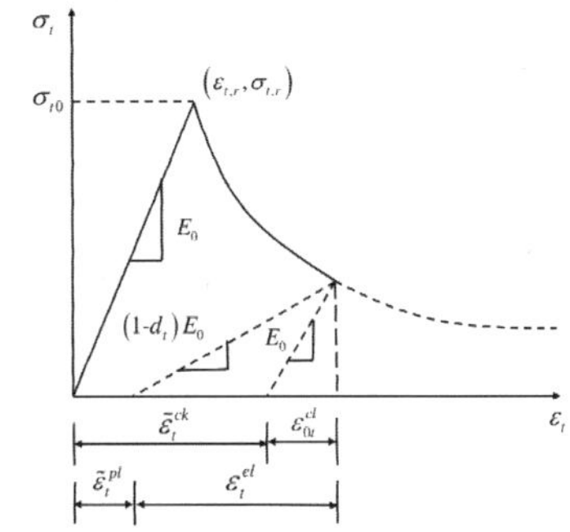

Figure 3. Uniaxial tensile stress-strain curve of CDP.

Table 1. Some parameters of CDP.

ψ(°)	λ	f_{b0}/f_{c0}	K	μ
30	0.1	1.16	0.6667	0.0005

Table 1 shows the recommended values for other parameters of the CDP model for ordinary concrete. The expansion angle ψ and eccentricity λ are parameters related to the yield surface flow rule. K is the parameter that controls the shape of the yield surface. f_{b0} is the concrete biaxial compressive strength, f_{c0} is the uniaxial compressive strength, and the ratio of the two is taken as the recommended value. μ is the viscous parameter defined by the CDP model. The larger μ is, the easier it is to converge. The smaller μ is, the higher the accuracy is. When μ is 0.0005, the accuracy and convergence can be satisfied.

4. Model parameters calibration

The CDP model does not specify the ultimate inelastic strain of concrete. Therefore, it is necessary to calibrate the truncation of tension and compression stress of concrete. Taking the test beam STEEL-40S in the literature [14] as an example, the finite element analysis was carried out in order to test the damage factors calculated by the formula recommended in this paper. At the same time, the cut-off of the stress-strain curve of the concrete was preliminarily calibrated.

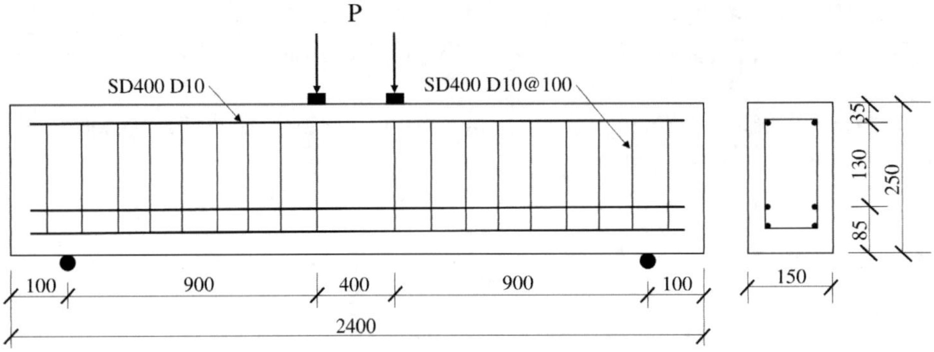

Figure 4. Detailed view of the beam and specimen cross-section.

Table 2. Material properties of reinforcement.[15]

Diameter(mm)	Area(mm^2)	Young's modulus E (GPa)	Yield strength f_y (MPa)	Ultimate strength f_u (MPa)
9.5	71.3	173	446	577

Table 3. Parameters of concrete mechanical behaviour.

x_c [a]	Compressive stress σ_c	Compressive inelastic strain $\tilde{\varepsilon}_c^{in}$	Compressive damage factor d_c	x_t [b]	Tensile stress σ_t	Tensile cracking strain $\tilde{\varepsilon}_t^{ck}$ ($\times 10^{-5}$)	Tensile damage factor d_t
0.5	21.46843	0.000000	0.000000	0.6	1.712712	0	0.000000
0.8	26.56861	0.000460	0.200082	0.8	2.187247	1.6620105	0.104482
1	27.30000	0.000757	0.274751	1	2.410000	3.07461538	0.159226
1.5	22.75000	0.001696	0.459431	2	1.260460	17.1016608	0.570048
2	17.06250	0.002669	0.594573	3	0.809976	28.9777670	0.718586
3	10.50000	0.004468	0.740319	6	0.423590	61.6366447	0.856098
6	4.55000	0.009442	0.879125	10	0.278854	104.041989	0.909560
9	2.86364	0.014285	0.921703	20	0.164949	209.292464	0.950815
15	1.63669	0.023905	0.954150	40	0.099964	419.292418	0.972925
20	1.20477	0.031903	0.965932				

[a] $x_c = \varepsilon_c / \varepsilon_{c,r}$, ε_c is the real compressive strain of concrete, and $\varepsilon_{c,r}$ is the peak compressive strain of concrete corresponding to uniaxial compressive strength $\sigma_{c,r}$.

[b] $x_t = \varepsilon_t / \varepsilon_{t,r}$, ε_t is the real tensile strain of concrete, and $\varepsilon_{t,r}$ is the peak tensile strain of concrete corresponding to uniaxial tensile strength $\sigma_{t,r}$.

4.1Test beam and parameters

Figure 4 shows detailed views of the simply supported beam STEEL-40S. The beam was designed and fabricated with a length of 2400 mm (span 2200 mm), a cross sectional area of 150 × 250 mm, the depth of outer reinforcing bar of 215 mm and the depth of inner reinforcing bar of 165 mm. SD400 D10 reinforcement was used for compression reinforcement, tensile reinforcement and stirrup. Table 2 shows the material properties of SD400 D10. The specified design strength of concrete applied for forming the specimens was 40 MPa. The measurements of compressive strength were 40.91 MPa.

The standard value of the compressive strength after conversion is f_{ck}=27.3MPa .The standard value of concrete compressive strength is used in the calculation of CDP model parameters. Table 3 lists the calculated damage factors and related parameters.

The model was established according to the geometry, loading mode and constraint position of the STEEL-40S test beam. The material properties of concrete adopt the parameters in Table 3. The constitutive model of reinforcement adopted bilinear model. In the model, the concrete elements were represented using C3D8R hexahedral elements, while the reinforcement is simulated using truss elements. The reinforcement was embedded in the concrete and the degrees of freedom are the same as the values of the concrete element's degrees of freedom at the corresponding position. The element size for both concrete and steel was 50mm^3. This model adopted the method of setting steel plate to avoid the phenomenon of stress concentration at the loading point and the supporting point and adopted the displacement method in the loading process.

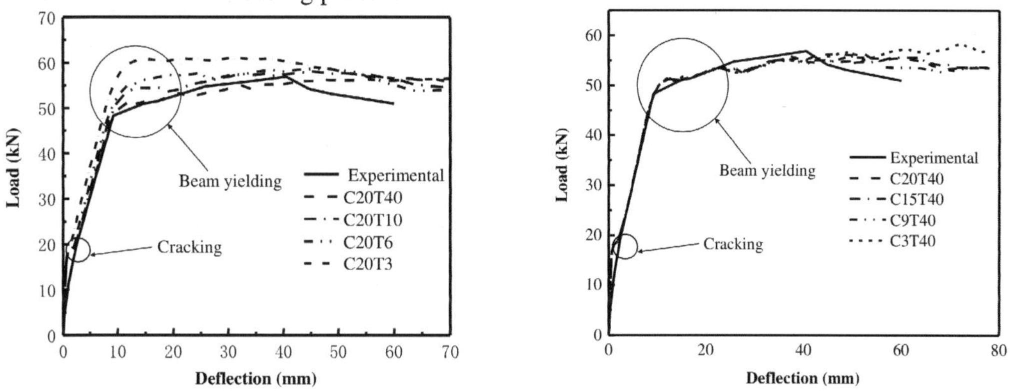

(a) $x_{c\max} = 20$. (b) $x_{t\max} = 40$.

Figure 5. Loading-deflection curves at mid-span of different parameters.

4.2Analytical results and discussion

In Figure 5, the test results were compared with the results of the finite element analysis. In order to study the influence of the truncation of the stress-strain curve and the value of damage factor on the calculated results, different values were taken for A and B in the finite element analysis. When A and B were less than 3, the load-mid-span deflection curve of the finite element model was greatly different from the test value. Therefore, finite element models were established to take 3, 9, 15, and 20 for A, and take 3, 6, 10 and 40 for B. The cracking load of the test beam STEEL-40S is 13.28kN, the ultimate load is 56.91kN, and the deflection at the ultimate load is 40.51mm. Table 4 shows the first crack load and ultimate load in the finite element analysis. When A, the theoretical value of the ultimate load was 60.69kN, and the deflection at the ultimate load was 14.20mm, which did not reflect the actual situation of the test. When B, the theoretical value of the cracking load was 3.26kN from the true value, there was no falling section in the load-span deflection curve within deflection=100mm. When A, the values of tensile damage factor and compression damage factor were both greater than 0.95, and the relative errors between first crack load, ultimate load and deflection at ultimate load with the experimental results were less than 5%. It is accordingly concluded that the finite element analysis results adequately reflected the experimental results.

Table 4. The first crack moment and the experimental and analysis values of the maximum moment.

Specimen	$x_{c\max}$	$x_{t\max}$	First crack load P_{cr} (kN)	Ultimate Failure load P_u (kN)
C20T3	20	3	18.92	60.69

C20T6	20	6	17.96	58.74
C20T10	20	10	17.64	58.05
C20T40	20	40	16.54	56.37
C3T40	3	40	16.54	-
C6T40	6	40	16.54	57.41
C9T40	9	40	16.54	55.74
C15T40	15	40	16.54	55.99

5. Conclusion

This paper investigated the effects of the value of the stress-strain curve cutoff and the damage factor of the concrete on finite element analysis. Through the above discussions, the following conclusions can be drawn:

1) According to the damage factor calculation formula recommended in this paper, the concrete stress-strain curve provided by the code for concrete structure design was applied to the finite element model. The obtained model parameters have been verified, and the finite element model can be used for the crack resistance check and bearing capacity check.

2) When the values of tensile damage factor and compression damage factors were both greater than 0.95, the relative errors between first crack load, ultimate load and deflection at ultimate load with the experimental results were less than 5%. The deformation and damage of the finite element model were consistent with the experimental phenomena. Therefore, the recommended damage factor is greater than 0.95.

Acknowledgment

The authors would like to acknowledge the financial support provided by the National Natural Science Foundation of China (Grant No. 51679220).

References

[1] Lubliner, J., Oliver, J., Oller, S., Oñate, E. (1989) A plastic-damage model for concrete International Journal of Solids and Structures 25:299-326

[2] Lee, J., Fenves, G.L. (1998) Plastic-Damage Model for Cyclic Loading of Concrete Structures. Journal of Waterway, Port, Coastal and Ocean Engineering 124:892-900

[3] Wosatko, A., Genikomsou, A., Pamin, J., Polak, M.A., Winnicki, A. (2018) Examination of two regularized damage-plasticity models for concrete with regard to crack closing. ENG FRACT MECH 194:190-211

[4] Ortiz, M. (1985) A constitutive theory for the inelastic behavior of concrete. MECH MATER 4:67-93

[5] Chi, Y., Yu, M., Huang, L., Xu, L. (2017) Finite element modeling of steel-polypropylene hybrid fiber reinforced concrete using modified concrete damaged plasticity. ENG STRUCT 148:23-35

[6] Sarikaya, A., Erkmen, R.E. (2019) A plastic-damage model for concrete under compression. INT J MECH SCI 150:584-93

[7] Genikomsou, A.S., Polak, M.A. (2015) Finite element analysis of punching shear of concrete slabs using damaged plasticity model in ABAQUS. ENG STRUCT 98:38-48

[8] Alfarah, B., Lopez-Almansa, F., Oller, S. (2017) New methodology for calculating damage variables evolution in Plastic Damage Model for RC. structures ENG STRUCT 132:70-86

[9] Rama, J.R.J.S., Chauhan, D.C.D.R., Sivakumar, M.S.M.V., Vasan, A.V.A., Murthy, A.M.A.R. (2017) Fracture properties of concrete using damaged plasticity model -A parametric study STRUCT ENG MECH 64:59-69

[10] Zihu, F., Haijun, Z., Shaoying, L., Qiang, X. (2013) Choose of ABAQUS concrete stress-strain curve. Building Structure A2:559-61

[11] GB 50010-2010. (2011) Code for concrete structure design.

[12] EN 1992-1-1: 2004. (2004) Eurocode 2: Design of concrete structures.
[13] D.S B (1982) Stress-Strain Behavior of Concrete by Overlapping Hoops at Low and High Strain Rates ACI Journal 13-27
[14] Kim, S., Kim, S. (2019) Flexural behavior of concrete beams with steel bar and FRP reinforcement J ASIAN ARCHIT BUILD 18:94-102

IOP Conf. Series: Materials Science and Engineering **794** (2020) 012037 doi:10.1088/1757-899X/794/1/012037

Experimental Study on the Effects of Fly Ash on Sand Fixation with Microbial Induced Carbonate Precipitation

Zhang Biaozhi[1*], Zhang Jiaming[2], Zhang Guoqing[3], Li Zheng[1], Zhu Jikang[2] and Zhou Yang[2]

[1]Research Institute of Highway, China Ministry of Transport, Beijing, 100088, China

[2]Faculty of Engineering, China University of Geosciences, Wuhan, Hubei, 430074, China

[3]Beijing Xin-Qiao Technology Development Co., LTD, Beijing, 100088, China

[*]Corresponding author's e-mail: bz.zhang@rioh.cn

Abstract. In order to study the sand-fixing effect of microbe induced mineralization after mixing fly ash with eolian sand, different samples of eolian sand were treated. The effects of blending fly ash on microbially induced mineralization and sand fixation were evaluated based on the comparison of permeability, water retention, surface strength, and wind erosion resistance. After three times of microbial induced mineralization treatment with 30% fly ash contained, it showed:(1)the permeability coefficient decreased by 79.4%;(2) the cumulative evaporation decreased by 26.5%;(3)the surface strength increased by 19.9%;(4)and the wind erosion rate decreased by 21.2%.The yield of calcium carbonate was negatively correlated with the wind erosion rate and positively correlated with the surface strength. From the experiment, it can be concluded that mixing fly ash is more practical than simply using microbial induced mineralization to fix sand, which can significantly reduce the permeability, improve the water retention performance, improve the surface strength and wind erosion resistance of the solidified layer of aeolian sand.

1. Introduction

According to statistics, the area of desertified land in China accounts for about 17.87% of the total land area [1], of which the area of desertified land in the west accounts for more than 90% of the total area of desertified land. Land desertification has become a key factor restricting the development of western regions in China. Wind sand control measures such as mechanical sand fixation, chemical sand fixation, and biological sand fixation [2] are often inadequate for long-term sand fixation, high cost, and high added value. Finding a sand-fixing material with good sand-fixing effect, convenient application, and environmental friendliness has become an urgent task in the research of wind-proof and sand-fixing in the western region.

Microbial Induced Calcite Precipitation (MICP) technology can effectively improve the stiffness and strength of soil and reduce the permeability of soil， which has been applied in the fields of soil improvement, leakage and plugging of rock and soil, biological cement, contaminated soil restoration, oil extraction, and protection of stone cultural relics [3].Rehabilitation, oil extraction, protection of stone cultural relics, etc. [3].The application of MICP technology in windbreak and sand fixation has also been studied,and current researches focused more on the direct effects of mineralization on aeolian sand[4][5][6][7].But wind-blown sand particles are small, less powder and clay content, poor water retention performance. If only the microbial induced mineralization treatment technology is used, the amount of mineralization agent is large, the cost is high, and the microbial treatment agent has a good permeability, it will be difficult to control the amount of application alone, and it is easy to cause a large number of treatment agent infiltration into the deep soil, resulting in waste.

As the main solid waste residue of coal industry, fly ash may cause, if improperly handled, regional groundwater pollution ,induce environmental problems such as dust, and occupy a large amount of land[8].But, it was discovered that adding fly ash to soil can significantly improve soil and aeolian sand properties.[9-14]

In order to explore the effect of microbial induced mineralization sand fixation under the participation of fly ash, indoor solidification study was carried out on aeolian sand from Xinjiang region of China. The effect of mixed fly ash on microbial induced mineralization of sand fixation was evaluated, and the mechanism of sand fixation was analyzed based on the yield of calcium carbonate.

2. Materials and methods

2.1Test materials

Aeolian sand is taken from the mobile dune in the southern margin of the Guerbantonggute desert in Xinjiang region. The parent material of the soil is mainly alluvial parent material and aeolian parent material, with no cohesion between particles, natural moisture content of 0.48% and dry density of 1.49g/cm^3.[15] The content of sand, powder and clay in aeolian sand is 90.27%, 7.47% and 2.26% respectively. Before test, the dirt was removed from sand by screening. The fly ash used in the test is of high-calcium with particle size distribution range from 38.5μm to 50μm.The samples used in this test include plain aeolian sand and 30% fly ash content aeolian sand. The corresponding particle grading curve and corresponding soil type are shown in figure 1.

The microbe used was Sporosarcina pasteurii (ACTT 11859), purchased from the national germplasm preservation center of the United States. Bacillus pasteurelis octadecans is a kind of high efficiency urease microorganism, with biosafety level 1 and facultative anaerobic type.

The composition and content of culture medium for expanded culture are shown in table 1. Except urea, all the other components of the medium were sterilized by heat and humidity. According to the inoculation amount of 2% (v/v), the mother microbe solution was added into the culture medium, and the microbe solution was obtained by oscillating culture for 24h at 30°C.

Table 1 composition of medium

composition	Casein peptone	The urea	Sodium chloride	Deionized water	pH
content	15.0 g	20.0 g	5.0 g	1.0 L	8.5

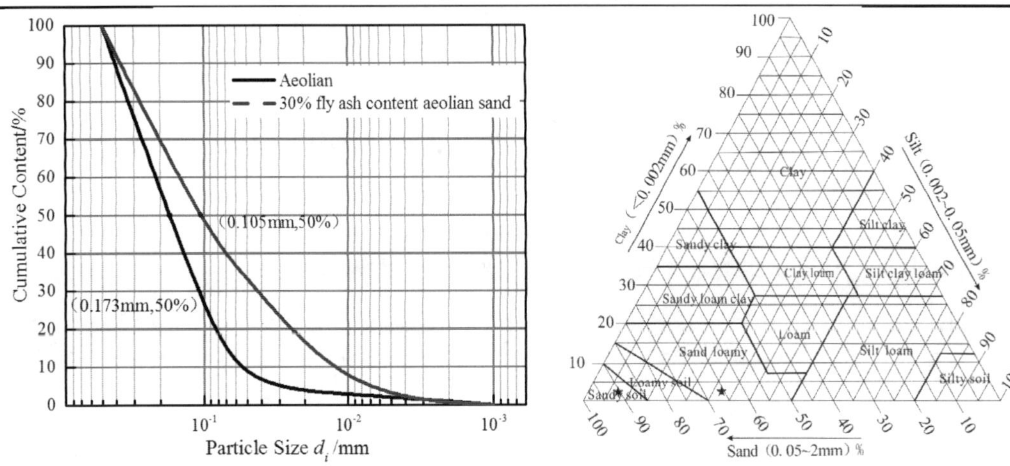

Figure 1. granule grading curve and soil type diagram

2.2Sample preparation

The samples were divided into 8 groups according to the soil used for the test, the type of treatment agent and the number of treatments. For details, please refer to table 2.

Table 2 test scheme

Test number	The experiment with	Treatment agent type	Processing times
S1T	Wind-blown sand	Microbial mineralization treatment agetn	1 time
S2T	Wind-blown sand	Microbial mineralization treatment agent	2 times
S3T	Wind-blown sand	Microbial mineralization treatment agent	3 times
S1W	Wind-blown sand	Distilled water	1 time
SF1T	30% fly ash content aeolian sand	Microbial mineralization treatment agent	1 time
SF2T	30% fly ash content aeolian sand	Microbial mineralization treatment agent	2 times
SF3T	30% fly ash content aeolian sand	Microbial mineralization treatment agent	3 times
SF1W	30% fly ash content aeolian sand	Distilled water	1 time

In the experiment, the microbial induced mineralization treatment agent contained bacteria solution and cementing solution, and the preparation method of the bacteria solution was shown in 1.1. The cementing solution was 0.5mol/L mixed solution of calcium chloride and urea (equal molar concentration).

Sample preparation method: Firstly, 1500 g aeolian sand was loaded into a sand table with the size of 30mm×20mm×3.5mm, and the surface of the sand sample was scraped. Secondly, bacteria solution and cementing solution were sprayed successively on the surface of aeolian sand according to the application amount of $3L/m^2$, with the sand table being naturally dried at room temperature for one week to ensure the continuous internal reaction of the sand sample. Thirdly, electric oven was used to completely dry the sample and terminate the internal reaction of the sand sample. Then the test sample was prepared separately for the permeability test of aeolian sand. The test mold was made of a cylindrical PVC pipe with a diameter of 5cm and a height of 20cm, which was filled with 200g soil samples. A glass petri dish with a diameter of 10cm and a height of 1.5cm was used for the water retention test of aeolian sand, where 80g soil samples were added to the petri dish and equal amount of distilled water was sprayed.

2.3Test methods

The observation items in this study include the wind-erosion resistance, surface strength and calcium carbonate yield of the eolian sand samples. The specific test methods are as follows.

*2.3.1*Permeability: based on the method described in literature [8], the permeability coefficient of the cylindrical sample was measured by the method of variable head.

2.3.2 Water retention: the Petri dishes were kept in a oven at 40°C for continuous drying for 6 hours, the changes of water content before and after drying were recorded, and the evaporation rate was calculated.

*2.3.3*Wind erosion resistance: the self-made wind erosion blowing device is used to test the wind erosion resistance of microbial induced mineralization sand fixation, as shown in figure 2.During the test, the sand table was placed in the middle section of the wind tunnel and was in the same straight line with the wind direction. Under the wind speed of 10m/s, it was continuously blown away for 8min to record the mass loss of the sand sample.

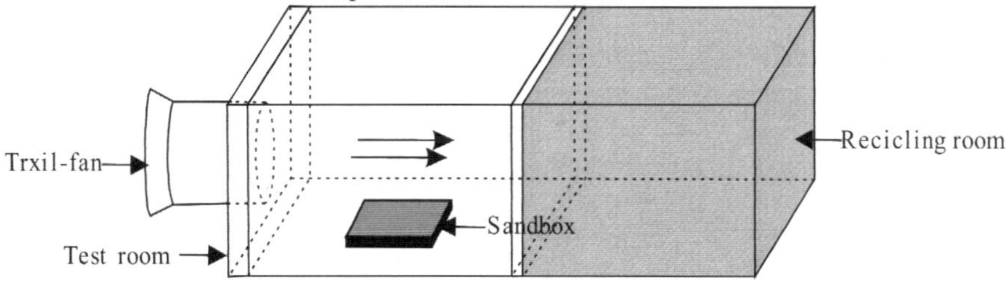

Figure 2. Schematic diagram of wind erosion device

*2.3.4*Surface strength: SF digital dynamometer was used to measure the surface strength of the sample.Each sample was measured at 5 different positions to ensure that the distance between the measuring points was not less than 3cm, the pressure peak of the five measuring points was read, and the ratio of the average maximum stress to the probe area was taken as the surface strength.

*2.3.5*Calcium carbonate yield: the increased weight of the sand table after the test was taken as the mineral yield induced by microbe mineralization. The sand table before and after the treatment was weighed and the calcium carbonate yield during the treatment was calculated.

3. Results and analysis

3.1Permeability test

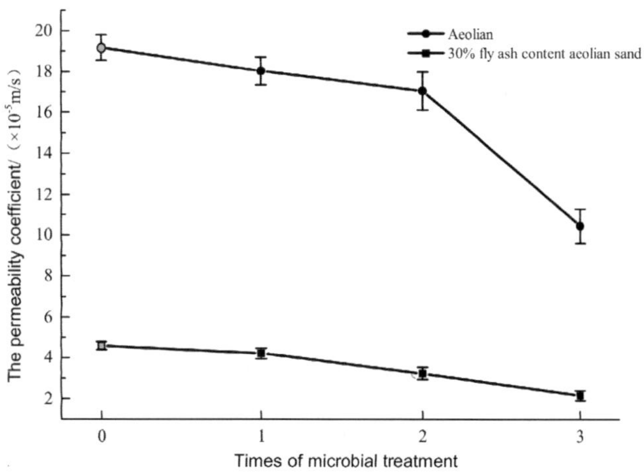

Figure 3. influence of mixed fly ash on the permeability of samples

Figure 3 shows the different effects of microbial induced mineralization and fly ash on the permeability of aeolian sand samples. The permeability coefficient of the samples decreased gradually with the increase of the number of microbial treatments. Among them, the permeability coefficient of S3T sample changed most significantly, and its permeability coefficient (1.04×10^{-4}m/s) was about 54% of that of S1W sample (1.92×10^{-4}m/s).By comparing the permeability coefficient of eolian sand samples before and after mixing fly ash, it can be found that the permeability coefficient changed more significantly after mixing fly ash into eolian sand. Among them, the permeability coefficient of S1W sample (1.92×10^{-4}m/s) is about 4.2 times that of SF1W sample (4.6×10^{-5}m/s), and the permeability coefficient of S3T is about 4.8 times that of SF3T sample (2.15×10^{-5}m/s).For the eolian sand samples mixed with fly ash, the permeability coefficient is generally smaller than that of the plain eolian sand samples, but the overall change is relatively gentle,the reason for which should be related to the fact that the pores between the particles of eolian sand are filled tightly with fly ash, so that the water cannot penetrate rapidly.

3.2Water retention test

The figure 4 shows the influence of mixed fly ash on water retention of samples.The evaporation of water in desertified area is large, which requires sand fixation material not only excellent sand fixation performance, but also good water retention performance, which is beneficial to the growth of sand plants.The evaporation test results show that the untreated samples (S1W) have poor water retention performance and lose more than 90% water after 6h. The cumulative evaporation rate of the samples (SF1W) mixed with fly ash is 70%. With the increase of the number of microbial treatment, the calcium carbonate crystals formed by microbial mineralization gradually increased, filling the pores between particles. Therefore, the cumulative evaporation rate of the two types of aeolian sand samples showed a downward trend. In general, the water-retention performance of the samples mixed with fly ash is better than that of the samples mixed with fly ash. After three times of microbial treatment, the cumulative evaporation rate of the SF3T samples is about 73.5% of that of S3T.

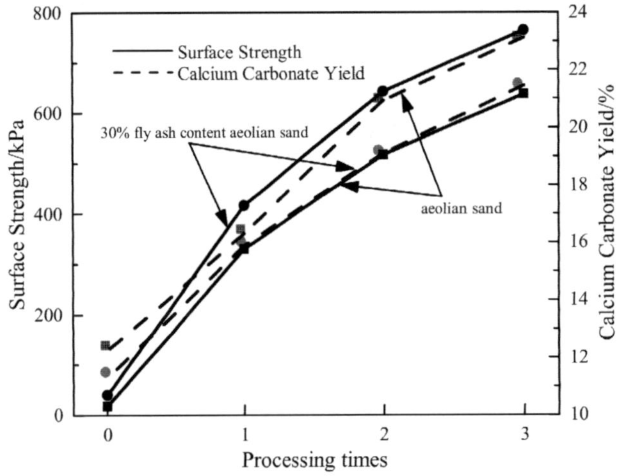

Figure 4. Influence of mixed fly ash on sample surface strength

3.3Wind erosion resistance test

The wind erosion rate is characterized by the mass loss of sand table per unit time.The results showed that the wind-erosion resistance of aeolites was significantly improved after microbial mineralization treatment, and the mass loss rate could be reduced to 0.246g/min after a single microbial treatment, that is, only a small amount of aeolites particles in the surface layer were lost.By comparing S1W and SF1W samples, it is found that mixing fly ash is beneficial to enhancing the wind-erosion resistance of aeolian sand, but there are still obvious wind-erosion phenomena.After single microbial treatment, the wind erosion of both samples was significantly restrained. The results showed that the effect of multiple microbial treatment was not significant compared with that of single microbial treatment. Therefore, single spraying of microbial treatment agent can meet the demand for the key areas to prevent wind erosion.

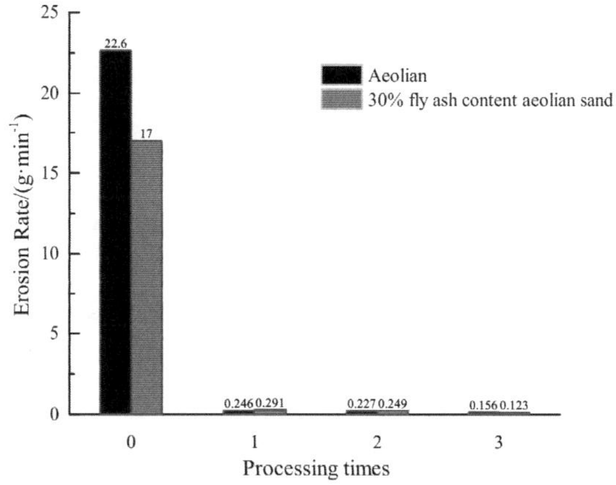

Figure 5. influence of mixed fly ash on wind erosion rate of samples

4. Discussion

MICP technology is realized by injecting a variety of reaction fluids, including bacteria solution, into the soil to be strengthened. Its core principle is urea-hydrolysis reaction, and its main catalytic component is urease. Urea decomposition is slow under normal conditions, and the rate of urea hydrolysis can be greatly improved by introducing urease microorganisms into the reaction system. Another function of urease microorganisms is reflected in their electronegative cell membrane. When the environment contains a large number of calcium ions, the calcium ions tend to be distributed around the microbial body, and the ureolytic products combine with the calcium ions to form calcium carbonate.[16]Taking the solidification of aeolian sand as an example, when the content of calcium carbonate in the reaction system reaches supersaturated state, part of it will be deposited in the vicinity of the pores of aeolian sand particles in the form of precipitation, filling the pores between particles and reducing the porosity. The other part ACTS as a "bridge" connecting particles, which is reflected in the improvement of mechanical strength of aeolian sand on the macro level, as shown in figure 7.The above effects also effectively control the permeability coefficient, cumulative evaporation rate and other indicators of aeolian sand samples.

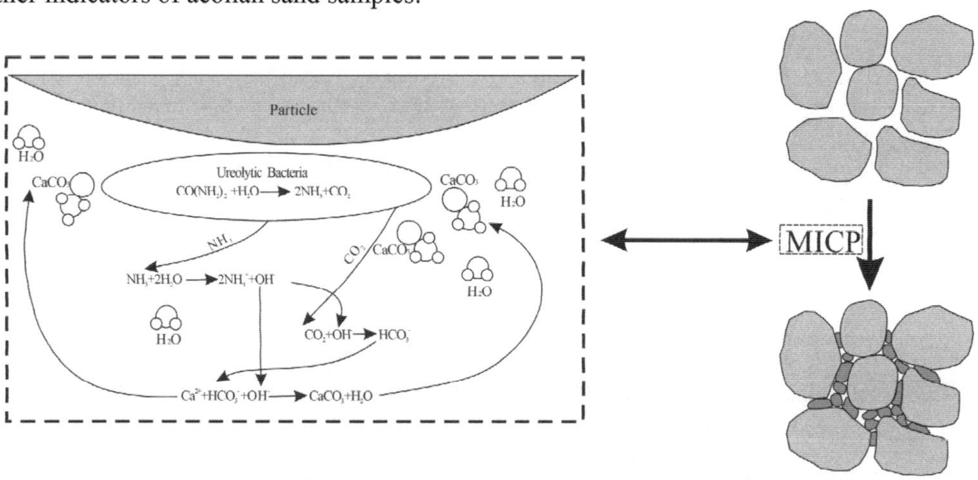

Figure 6. schematic diagram of MICP principle (adapted from literature [17])

Incorporation of fly ash increases the proportion of powder content in aeolian sand and the cohesion between grains. On the other hand, fly ash particles fill the pores of sand particles, increasing the contact area between sand particles and improving the integrity of aeolian sand.[12]In addition, the change of particle size composition will enable more calcium carbonate to be used for cementing aeolian sand particles and fly ash particles during microbial treatment of samples, so the surface strength, wind erosion resistance and other indicators are also improved compared with the unmixed samples.

5. Conclusions

(1) When fly ash is mixed in aeolian sand, the content of powder particles and the cohesion between grains increase, which can significantly reduce the permeability of aeolian sand and improve its water retention performance.

(2) Microbe induced mineralization can be used to improve the surface strength and wind-erosion resistance of aeolian sand, and the effect is significantly improved after mixing fly ash.

(3) The production rate of calcium carbonate is negatively correlated with the wind erosion rate and positively correlated with the surface strength, and the growth rate gradually slows down with the increase of treatment times.

References

[1] Ding xue, lei guoping, xu duanyang, et al. Effects of desertification evolution on regional ecosystem service value in Inner Mongolia from 1981 to 2010 [J]. Soil and water conservation research,2018,25(01):298-303.

[2] Wang lixian. Causes of desertification and its control measures in China [J]. World forestry research,2000,13(6):32-37.

[3] Phillips A J, Gerlach R, Lauchnor E, et al. Engineered applications of ureolytic biomineralization: A review[J]. Biofouling, 2013, 29(6): 715-733.

[4] Maleki M, Ebrahimi S, Asadzadeh F,Et al. Performance of microbial-induced precipitation on wind erosion control of sandy soil[J]. International journal of environmental science and technology, 2016, 13(3): 937-944.

[5] Khaleghi M, Rowshanzamir M a. Biologic improvement of A sandy soil using single and mixed cultures: A comparison study[J]. Soil and Tillage Research, 2019, 186:112-119.

[6] Li chi, wang shuo, wang yanxing, et al. Field test study on microbial induced mineralization and its stability in desert [J]. Rock and soil mechanics,2019,40(04):1291-1298.

[7] Li duo. Study on microbial induced calcium carbonate precipitation solidifying desert aeolites [D]. Northwest a & f university,2018.

[8] Wang jianxin, ulrich jing, zhao shibao, et al. Research progress and prospect of resource utilization of fly ash in China [J]. Chinese silicate bulletin,2018,37(12):3833-3841.

[9] Chang, A, C, Lund L J, Page A L, et al. The Physical Properties of Fly Ash - Amended Soils 1 [J]. Journal of Environmental Quality, 1977, 6 (3) : 267-270.

[10] Ghodrati M, Sims J T, Vasilas B l. Evaluation of fly ash as a soil amendment for the Atlantic Coastal Plain:I. Soil hydraulic properties and elemental leaching [J]. Water, Air, and Soil Pollution, 1995, 81 (3-4) : 349-361.

[11] Pathan S M, Aylmore L A G, et al. Properties of several fly ash materials in relation to use as soil amendments[J]. Journal of environmental quality, 2003, 32(2): 687-693.

[12] Yang kai, tang zejun, zhao zhi, et al. Wind tunnel test of sand fixation effect of fly ash and polyacrylamide [J]. Chinese journal of agricultural engineering,2012,28(04):54-59.

[13] Ye chenlei. Wind tunnel experiment study on fly ash improving desert soil [A]. Beijing dynamics society. Proceedings of the 23rd annual meeting of Beijing dynamics society [C]. Beijing dynamics society: Beijing dynamics society,2017:2.

[14] Cheng gang, zhao liang, sun pengcheng, et al. Effect of fly ash on surface evaporation of sandy soil [J]. Journal of irrigation and drainage,2011,30(06):135-137.

IOP Conf. Series: Materials Science and Engineering **794** (2020) 012037 doi:10.1088/1757-899X/794/1/012037

[15] Xinjiang geographical society. Handbook of xinjiang geography [M]. Urumqi: xinjiang people's publishing house,1993:13550-134.
[16] Ivanov V, Chu j. Applications of microorganisms to geotechnical engineering for bioclogging and biocementation of soil in situ [J]. Journal of Reviews in Environmental Science and Bio/Technology, 2008, 7 (2) :139-153.
[17] DeJong J T, Mortensen B M, Martinez B C, et al. Bio-mediated soil improvement[J]. Ecological Engineering, 2010, 36(2): 197-210.

IOP Conf. Series: Materials Science and Engineering **794** (2020) 012038 doi:10.1088/1757-899X/794/1/012038

Research on prediction performance of multiple monitoring points model based on support vector machine

Kaiyi Min[1], Yanlong Li[1*], Qiaogang Yin[1] and Lifeng Wen[1]

[1]State Key Laboratory of Eco-hydraulics in Northwest Arid Region, Xi'an University of Technology, Xi'an, Shaanxi, 710048, China

[*]Corresponding author's e-mail: liyanlong@xaut.edu.cn

Abstract: The multiple monitoring points model is an important means of dam structure health monitoring. Combined with the strong nonlinear mapping capability of support vector machine, the fitting and prediction accuracy of the model is further improved. According to the point selection of the multiple monitoring points SVM model, five kinds of testing schemes are designed. The influence of points selection on the prediction performance of the model is verified by case analysis. The results show that the predictive ability of multiple monitoring points model based on SVM is greatly affected by the correlation degree of the monitoring points. Therefore, it is very important to reasonably select the point data with high similarity as the training samples for the effective prediction of deformation monitoring model.

1. Introduction

Dam safety monitoring is an important part of dam safety management, and it is also the main means to master the dam's operational behavior. Establishing a reliable dam deformation monitoring model is of great significance for timely detecting safety hazards and ensuring stable operation of dams[1]. In recent years, some experts and scholars have analyzed many monitoring points from the whole, and established a multiple monitoring points model (MMP) which can reflect the displacement relationship between each monitoring point of the dam, so that the analysis and evaluation of the dam working performance move forward in a more in-depth direction[2-4].The MMP model breaks through the limitation of single monitoring point modeling, and transforms the previous "single-point" analysis method into " cross-section " analysis method, so as to better reflect the whole deformation state of the dam. However, there are no uniform rules for determining the sample selection of the monitoring points range, and the selection criteria are relatively vague. In some papers [5-6], all the monitoring points in the dam section are directly selected as the training samples. Such selection method has large randomness and human factors. Due to the diversity of influencing factors such as load method, constraint conditions and material properties, the dam deformation distribution is regional, and the heterogeneity between different parts is large. The unified analysis of the deformation values of all monitoring points is easy to ignore the heterogeneity between the points, leading to large errors in the prediction results.

In view of the advantages of the support vector machine (SVM) in dealing with small samples, nonlinear data classification and regression problems[7], this paper combines SVM with MMP displacement models. Compared with the traditional statistical model, the SVM-based monitoring model can effectively improve the fitting and prediction accuracy[8]. In order to improve the predictive ability of multiple points model based on SVM, this paper designs different test schemes, and analyzes the selection range of suitable sample points by using case analysis. It has certain reference value for constructing MMP model and analyzing the working behavior of dam.

2. Multiple monitoring point statistical model

At present, the single monitoring point model is widely used in the dam displacement prediction mathematical model, that is to establish a model for the certain monitoring point displacement in a certain direction. In the traditional statistical model, the cause of settlement deformation at a single point of the dam body is composed of three parts: water pressure component（δ_H）, temperature component（δ_T）and time effect component（δ_Θ）[9]. The general expression of the model is as follows:

$$\begin{aligned} \delta &= \delta_H + \delta_T + \delta_\theta \\ &= \sum_{i=1}^{3} H^i + \bar{H} + \sum_{i=1}^{2} (\cos\frac{2\pi it}{365} + \sin\frac{2\pi it}{365}) + \theta + \ln\theta \end{aligned} \tag{1}$$

where H is the water level difference between the upstream and the downstream on the date of observation; $\overline{H}$ represents the average air temperatures in the 3 days before the observation date; t is the cumulative number of days from the monitoring date to the starting date; $\theta = t/100$.

For the deformation distribution of MMP model, the position parameters x, y and z of monitoring points are usually introduced to expression, and the control range is expanded from one-dimensional to three-dimensional plane through the coordinate position of the settlement point as the input quantity. The spatial–temporal distribution model of the displacement field is established, so as to show the effect of the spatial distribution on settlement [10]. The MMP model combines each environmental impact factor and pure coordinate factor to form a new independent variable factor set containing the water pressure component, temperature component and time effect component of the statistical model. In summary, the model can be expressed as followed:

$$\hat{\delta}=\delta(H,T,\theta,x,y,z)=\begin{bmatrix} u(H,T,\theta,x,y,z) \\ v(H,T,\theta,x,y,z) \\ w(H,T,\theta,x,y,z) \end{bmatrix} \tag{2}$$

where $\hat{\delta}$ is displacement vector; u, v, w are three coordinate axis directions of space rectangular coordinate system.

The three-dimensional, three-direction MMP model represented by the above formula can degenerate into two-dimensional, one-dimensional and two-direction, single-direction. The specific expression of the two-dimensional MMP settlement model is:

$$\begin{aligned} \bar{\delta} &= f(H,T,\theta,x,y) \\ &= \bar{\delta}_{H,x,y}(t) + \bar{\delta}_{T,x,y}(t) + \bar{\delta}_{\theta,x,y}(t) \end{aligned} \tag{3}$$

where x is the distance from the location of the monitoring point to the center line of cross-section, and y is the distance from the location of the monitoring point to the dam crest.In addition, the one-dimensional MMP model in this study is similar to the two-dimensional model, so it will not be discussed here.

3. Support vector machine for multiple monitoring points model

Support vector machine is a supervised learning model proposed by Vapnik based on statistical learning theory in the mid-1990s. SVM is widely used in pattern classification, optimal control, regression analysis, nonlinear modeling and prediction. For the complex nonlinear problem of dam deformation prediction, the SVM regression prediction model can achieve better results.

In the two-dimensional MMP statistical model，due to the amount of independent variable data is large (43 items), the workload of direct calculation is huge and the uncertainty of factors will lead to low fitting accuracy. In this study, four factor sets are becomed as the preset factor set of regression analysis, and stepwise linear regression analysis is carried out. The impact factors of higher correlation are retained, and the factors with poor correlation are eliminated, so that the model can better fit the measured data. The final filtered impact factor is brought into the model as an input variable. In the one-dimensional MMP statistical model, the monitoring points participating in the modeling are located on the same cross-section and the same straight line of the dam, and the number of independent variable factors is significantly reduced compared with the two-dimensional MMP model.Since the SVM has strong nonlinear mapping ability, the factors which are related to deformation can be directly selected as the input variable in the one-dimensional model.The measured values of the input and output variables are taken as the training samples of the model, the output variables are the dam deformation observation values, and the settlement value of the face rockfill dam is selected as the output value.

All simulations are carried out in the MATALB R2016b environment. Before running the models, all data for SVM algorithm have been normalized to [-1,1]. In the process of model training, the values of penalty function C and kernel function parameter g have a great influence on model performance. In this study, cross validation (CV) method is used to select parameters, which can get the optimal parameters in a certain sense, effectively avoid the occurrence of over learning and under learning States.

4. Case study

The monitoring data of a concrete face rockfill dam of are selected for analysis. The dam started to be filled in October 2008, and the main body of the dam was completed in September 2009. The construction period of face slab concrete is from March 15, 2010 to May 14, 2010, and the water storage began on October 14, 2010. The project reservoir is a daily regulating reservoir with a normal water storage level of 1856 *m*. The total storage capacity of this reservoir is about 294 million m^3. In order to effectively monitor the displacement of face rockfill dam, three straight lines of settlement displacement monitoring points are embedded in the dam body for monitoring. The ES4 and ES1 series are located in the primary rockfill, and the ES2 upper monitoring points are located in the downstream rockfill area, as shown in Figure 1. In this study, the 36 monitoring points on the dam cross-section will be analyzed. The data set is built over a period from December 2010 to June 2015, which is from the completion of the first impoundment to the completion of the third impoundment. Each monitoring point has totally 200 groups of monitoring data, the first 190 groups of data are used for fitting and training, and the last 10 groups of data are used for prediction. The large range of monitoring points can reflect the deformation of the whole rockfill dam under the external load, and its wide range of variation is helpful to study the prediction performance of the model.

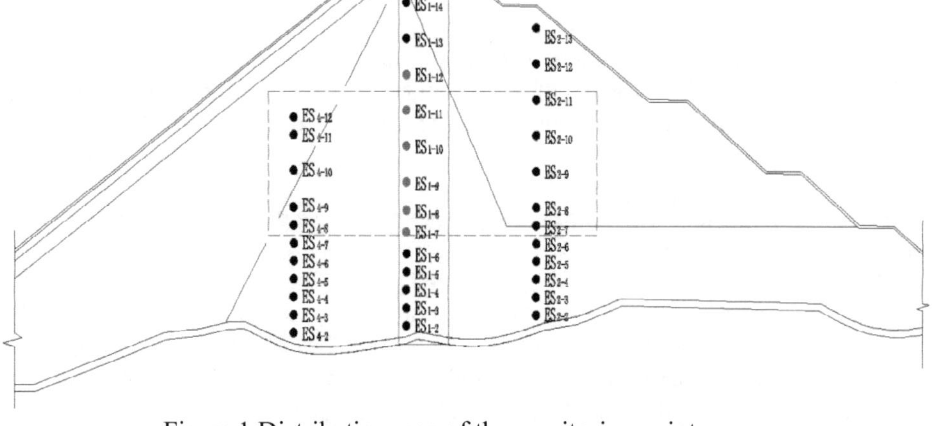

Figure.1 Distribution map of the monitoring points.

In order to accurately explore the influence of the selection of monitoring points on the prediction ability in the multiple monitoring points model, five different schemes are designed for comparison. Scheme 1 is the traditional single monitoring points model, scheme 2-4 is the multiple monitoring points model, the monitoring point selection and independent variable parameters of each scheme are different, and the specific content is shown in Table 1. Among them, the selection range of scheme 3 and 4 is indicated by a dotted line frame and a solid line frame in Figure.1, and the measurement points of scheme 5 are marked with red punctuation.

Hu and Shao[11,12] pointed out that the specific monitoring points in the dam have similar deformation laws. Ji[13] called the monitoring points groups with the same straight line, the same monitoring instrument and the same monitoring direction as the homologous monitoring points series. In the four schemes of MMP model, the monitoring points range is gradually reduced from the whole dam cross section to homologous monitoring points series, and the number of monitoring points is gradually reduced from 36 to 6, but the correlation degree of monitoring points is gradually increasing.

Table.1 The description of schemes.

	Number of points (Remark)	Impact factors
Scheme 1	One (ES1-9)	$\{h, h^2, h^3, \bar{h}, \theta, \ln\theta, \sin\frac{2\pi t}{365}, \sin\frac{4\pi t}{365}, \cos\frac{2\pi t}{365}, \cos\frac{4\pi t}{365}\}$
Scheme 2	Thirty-six (Cross-section of dam)	$\{x, y, \ln\theta, \theta x, \bar{h}, \theta y, \bar{h}x, h^2, \bar{h}y, \ln\theta y, \cos\frac{2\pi t}{365}, \sin\frac{2\pi t}{365}, \sin\frac{4\pi t}{365}\}$
Scheme 3	Fifteen (Maximum settlement area of dam body)	Same as scheme 2
Scheme 4	Thirteen (The series of ES1)	$\{x, h, hx, h^2, h^3, \bar{h}, \bar{h}x, \theta, \theta x, \ln\theta, \sin\frac{2\pi t}{365}, \sin\frac{4\pi t}{365}, \cos\frac{2\pi t}{365}, \cos\frac{4\pi t}{365}\}$
Scheme 5	Six (ES1-7~ES1-12)	Same as scheme 4

As the ES1-9 monitoring point is close to the actual maximum settlement point of the dam, the prediction result of this point has important analytical value. The performance of each scheme will be compared by their prediction results of ES1-9 monitoring point. In this study, correlation coefficient (*R*), mean absolute error (*MAE*) and mean absolute percentage error (*MAPE*) are used as evaluation criteria for model performance. Their expressions are as follows:

$$R = \frac{\sum_{i=1}^{n}(y_D(i)-\bar{y}_D)(y(i)-\bar{y})}{\sqrt{\sum_{i=1}^{n}(y_D(i)-\bar{y}_D)^2\sum_{i=1}^{n}(y(i)-\bar{y})^2}} \tag{4}$$

$$MAE = \frac{1}{n}\sum_{1}^{n}\left|y_D(i)-y(i)\right| \tag{5}$$

$$MAPE = \frac{1}{n}\sum_{1}^{n}\left|\frac{y(i)-y_D(i)}{y(i)}\right| \tag{6}$$

where y_D and $\overline{y}_D$ represent the simulated values and average value. y and $\overline{y}$ denote the measured values and the average value. and n is the number of observation.

Table.2 Prediction accuracy for four models.

	MAE	*MAPE*	*R*
Scheme 1	10.31	2.42%	0.816
Scheme 2	9.97	2.28%	0.829
Scheme 3	8.70	1.81%	0.831
Scheme 4	6.39	1.12%	0.907
Scheme 5	5.78	1.00%	0.913

The prediction performance of each scheme is shown in Table 2, and the specific prediction values are shown in Figure 2. The settlement prediction curves of scheme 2 and scheme 3 fluctuate have large fluctuations up and down, which are not well consistent with the measured curves. The guess reason is that the selected monitoring points are distributed in a large range. The heterogeneity among the monitoring points leads to the low fitting degree of deformation law. In general, the predicted displacement of scheme 4 and scheme 5 is in good agreement with the change trend of measured displacement, and the residual value of scheme 5 is the smallest, indicating that the prediction accuracy of scheme 5 model is the highest. It can be seen that the selection of monitoring point samples has a great impact on the prediction accuracy.

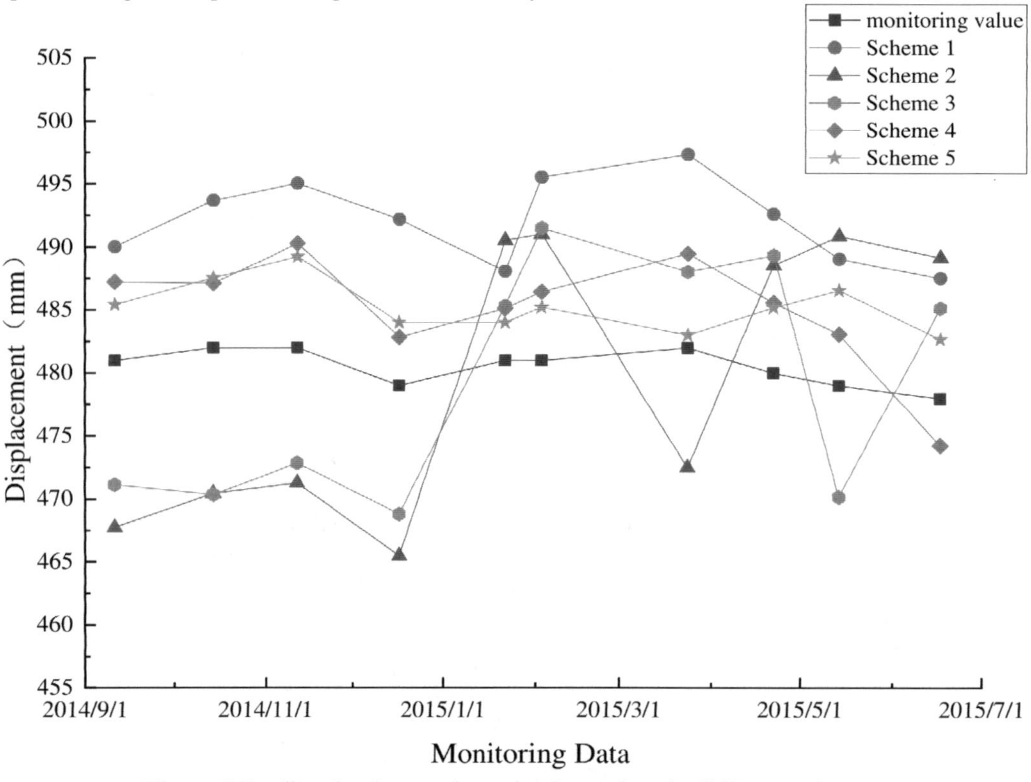

Figure.2 Predicted values and monitoring values in different schemes.

As shown in Table 2, we find that the prediction performances of the multiple monitoring points models are better than that of the traditional single point model, which shows that the MMP model can better reflect the dam deformation law. In the four schemes of MMP model, the three evaluation indexes of MAE, MAPE and R are basically inversely proportional to the number of monitoring points. Comparing scheme 2 with scheme 5, we can see that MAE is reduced from 9.97 to 5.78, MAPE is reduced from 2.28% to 1.00%, and correlation coefficient R is increased from 0.829 to 0.913, which shows that the prediction performance is significantly improved with the reduction of the point selection range. From the range of three evaluation indexes, when the selection range transforms from the maximum settlement area to the ES1 series straight line, the prediction accuracy improves the most, but it should be noted that the number of monitoring points does not change much at this time (The number of points changed from 15 to 13). In the process of scheme 2-3 and scheme 4-5, the number of selected monitoring points is almost reduced by half, but the prediction accuracy of the model is only slightly improved, and the change range is small, indicating that the correlation between the number of monitoring points and the prediction ability is low. However, in the above several cases, the model prediction performance shows a relatively obvious correspondence with the selection of the monitoring points in the scheme. As the correlation degree of the selected points is gradually improved, the prediction accuracy is also significantly improved. It can be seen that the simple increasing or decreasing number of monitoring points has little effect on the prediction performance of the model. The essential reason for the improvement of the model performance is to eliminate the weakly correlated monitoring points data of "misleading" model, because the retained monitoring point samples have higher similarity, so it can better fit the actual deformation law.

5.Conclusion

This paper selected different points to establish MMP deformation prediction models based on SVM. The prediction ability is studied by analyzing the error between the prediction value and the monitoring value under different schemes. The case analysis shows that:

(1) The MMP model has good predictive performance. Compared with the single point model, it can better reflect the overall deformation law, and has higher application value for dam structure safety evaluation and prediction.

(2) When too many points are directly selected and brought into the training model, the prediction performance is poor. The reason is that there are some weakly correlated and uncorrelated monitoring points, which will disturb the fitting of multiple monitoring points model to the deformation law of the target monitoring point, resulting in the decline of the fitting and prediction accuracy of the model.

(3) It is the number of high correlation points that affects the prediction performance of the multiple monitoring points model. Since the points in the homologous series have high correlation, in the case of a certain monitoring point deformation prediction in the actual engineering, points on the same straight line of the target point can be directly selected to construction a one-dimensional MMP model, so as to obtain a stable and reasonable prediction model.

(4) In the design of model schemes, there is a certain degree of subjective selectivity and randomness in the points selection. Therefore, how to objectively select suitable and highly correlated monitoring points to construction model needs further research.

Acknowledgments

The research described in this paper was funded by the National Natural Science Foundation for Excellent Young Scientists of China (Grant No. 51722907); the National Natural Science Foundation of China (Grant No. 51979224) and the Project 2019JLZ-14 supported by Joint Foundation of Shaanxi.

References

[1] Salazar, F., Morán, Rafael, Toledo, Miguel á., Oñate, E. (2017) Data-based models for the prediction of dam behaviour: a review and some methodological considerations. Archives of Computational Methods in Engineering, 24(1), 1-21.
[2] He, J.P., Li, Z.Z.(1994) Research on Mathematical Model of Multi-measurement Point of Dam

Structural Behavior. Journal of Wuhan University, (02), 134-136+1378+138-142. (in Chinese)

[3] He, J.P., Tu, Y.Y.,Shi, Y.Q. (2011) Fusion model of multi monitoring points on dam based on bayes theory. Procedia Engineering, 15(none), 2133-2138.

[4] Stoian, I., Kovacs, I., Ignat, S., Ghiran, O., Oltean, G., Gordan, M. (2018) Multisensorial Data Fusion and Correlation for Dams Structural Health Monitoring. In: IEEE International Conference on Automation, Quality and Testing, Robotics (AQTR). Cluj-Napoca, Romania.

[5] Chai, L.S., Qi, D., Wu, H. (2014) Application of multi-point multi-direction BP network model in dam deformation monitoring. Water Resources and Power, (3), 94-97. (in Chinese)

[6] Huang, D.W., Liu, B., Dai, W.J. (2017) Multi-measurement displacement model of Wuqiangxi dam based on neural network model. Geotechnical Geotech Invest Survey, (5). (in Chinese)

[7] Su, H., Chen, Z., Wen, Z. (2016) Performance improvement method of support vector machine-based model monitoring dam safety. Structural Control and Health Monitoring, 23(2), 252-266.

[8] Rezapour, T.M.M., Zarif, S.H.R. (2018) Prediction of the intermediate block displacement of the dam crest using artificial neural network and support vector regression models. Soft Computing.

[9] Shi, Y., Yang, J., Wu, J., He, J. (2018) A statistical model of deformation during the construction of a concrete face rockfill dam. Structural Control & Health Monitoring, 25(2), e2074.

[10] He, J.P. (2010) Dam Safety Monitoring Theory and Application. China Water & Power Press, Beijing. (in Chinese)

[11] Shao, C., Gu, C., Yang, M., Xu, Y., Su, H. (2017) A novel model of dam displacement based on panel data. Structural Control and Health Monitoring, e2037.

[12] Hu, Y.T., Shao, C.F., Gu, C.S., Meng, Z.Z. (2019) Concrete Dam Displacement Prediction Based on an ISODATA-GMM Clustering and Random Coefficient Model.11, 714.

[13] Ji, W., Liu, X.Q., Lin, C.N. (2019) New multi-measurement displacement model of dam based on BP neural network. Water Resources and Hydropower Engineering, 50(02), 115-120. (in Chinese)

General Analytical Solution to Consolidation of Unsaturated Soil with Vertical Drain considering Drain Resistance

Aifang Qin*, Weifang Xu, Tianyi Li and Lianghua Jiang

Department of Civil Engineering, Shanghai University, Shanghai, 200072, China

E-mail: qinaifang@shu.edu.cn

Abstract. This paper presented a general analytical solution to consolidation of unsaturated soil with vertical drain considering drain resistance. Based on Fredlund's one-dimensional consolidation theory and the equal-strain assumption, the coupled governing equations of excess pore-water and pore-air pressures are first decoupled by introducing two new variables. the method of eigenfunction and undetermined coefficients are then used to obtain the general solution under the drain resistance condition. Moreover, the good agreement between the equal-strain results and the typical free-strain results confirms the validity of the proposed solution. The consolidation behaviours, concluding average excess pore-air and pore-water pressures, are investigated against the drain resistance and the ratio of influence radius to vertical drain radius.

1. Introduction

In dry climate areas, it is possible that vertical drains installed to a great depth of soil may encounter some unsaturated layers in the proximity of the ground surface. The consolidation behaviours of these unsaturated layers, which have considerable differences with saturated soils, should be analysed.

On the Fredlund's consolidation theory of unsaturated soil[1], Qin et al.[2-3] obtained semi-analytical solution to consolidation of unsaturated soils with the free drainage well by using Laplace transform. Wang et al.[4] proposed a semi-analytical solution with symmetric semi-permeable drainage boundary by applying Laplace transform. Ho et al.[5] presented an analytical solution for the axisymmetric consolidation with linearly depth-dependent initial conditions. Shan and Zhou[6] predicted the axisymmetric consolidation behaviour by introducing the differential quadrature method (DQM). All abovementioned studies are based on the free-strain consolidation theory, Ho et al.[7] introduced the equal-strain theory and derived the analytical solution to unsaturated soil stratum considering the smear effects. Zhou et al.[8] derived an equal-strain analytical solution by using the method of eigenfunction. The equal-strain theory simplified the calculation process and captured more realistic behaviours of vertical drain foundation.

This paper obtains a general analytical solution to predict the unsaturated consolidation in vertical drain foundation considering drain resistance. Firstly, the governing equations considering the drain resistance condition are decoupled into equivalent homogeneous PDEs. Then, the method of eigenfunction and undetermined coefficients are used to derive the final solution. Additionally, effects of the drain resistance factor and the ratio of influence radius to vertical drain radius on the dissipation of average excess pore pressures are highlighted.

2. Mathematical model

Figure 1 illustrates a typical foundation of unsaturated soil with vertical drain, which includes a finite depth H, the radius of influence zone r_e, the radius of vertical drain r_d. Respectively, the coefficients k_w,

k_a, k_{w_d} and k_{a_d} are used to represent the differences in permeability coefficients of the influence zone and vertical drain. Only radial flow of influence zone is considered and the top and bottom surfaces and the lateral surface are assumed to be impervious to water and air in this paper. And the top surface of vertical drain foundation subjected to the arbitrary load $q(t)$.

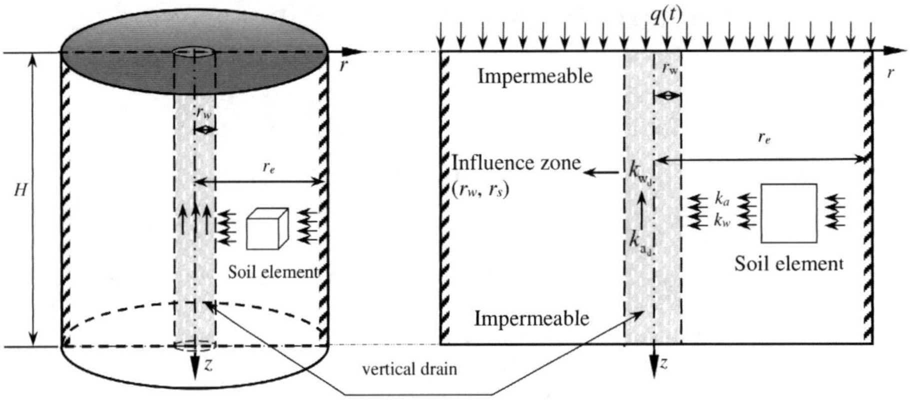

Figure 1. Consolidation model of unsaturated soils with vertical drain considering drain resistance

2.1 Basic assumptions

The basic assumptions for the consolidation model of unsaturated soil are the same as in Ref.[2], and the other assumptions are made as follows:

- The change of radial excess pore pressures in vertical drain can be neglected.
- The vertical drain foundation has no lateral strain, and the vertical strains at the same depth are equal (equal-strain assumption).
- The coefficients of permeability related to the air and water phases and the volume changes for the soil are constant.

The permeability change is nonlinear properties in the consolidation of unsaturated soil. However, it is acceptable that these coefficients remain to be constant at a very small loading increment[5].

2.2 Governing equations

Governing equations of air and water phases for consolidation in unsaturated soil with vertical drains under equal-strain conditions are as follows[2]:

$$\frac{\partial \bar{u}_a}{\partial t} = -C_a \frac{\partial \bar{u}_w}{\partial t} - C_v^a \left(\frac{\partial^2 u_a}{\partial r^2} + \frac{1}{r}\frac{\partial u_a}{\partial r} \right) + C_q^a \frac{\partial q}{\partial t} \tag{1a}$$

$$\frac{\partial \bar{u}_w}{\partial t} = -C_w \frac{\partial \bar{u}_a}{\partial t} - C_v^w \left(\frac{\partial^2 u_w}{\partial r^2} + \frac{1}{r}\frac{\partial u_w}{\partial r} \right) + C_q^w \frac{\partial q}{\partial t} \tag{1b}$$

where, $\bar{u}_a = \int_{r_w}^{r_e} u_a 2\pi r dr \Big/ \left[\pi\left(r_e^2 - r_w^2\right)\right]$, $\bar{u}_w = \int_{r_w}^{r_e} u_w 2\pi r dr \Big/ \left[\pi\left(r_e^2 - r_w^2\right)\right]$ (2)

$C_a = m_2^a \left[m_{1k}^a - m_2^a - u_{atm} n_0 (1 - S_{r0}) / (\tilde{u}_a^0)^2 \right]^{-1}$; $C_v^a = k_a RT \left\{ g \bar{u}_a M \left[m_{1k}^a - m_2^a - u_{atm} (1 - S_{r0}) n_0 \Big/ \left(\tilde{u}_a^0\right)^2 \right] \right\}^{-1}$;

$C_q^a = m_{1k}^a \left(m_{1k}^a - m_2^a - u_{atm} n_0 (1 - S_{r0}) / \tilde{u}_a^0 \right)^{-1}$; $C_w = m_{1k}^w - m_2^w / m_2^w$; $C_v^w = k_w / \gamma_w m_2^w$; $C_q^w = m_{1k}^w / m_2^w$.

where u_a, u_w, u_{a_d}, u_{w_d} are the excess pore-air and pore-water pressures in the influence zone and in vertical drain (kPa), respectively, $\bar{u}_a$ and $\bar{u}_w$ are the average excess pore-air and excess pore-water pressure in the entire influence zone (kPa). R is the universal air constant (~8.314 J(mol·K)$^{-1}$); T (ie, $t°$ + 273) is the absolute temperature (K); u_a^0 and u_w^0 are the initial excess pore-air and pore-water pressure (kPa); $\tilde{u}_a^0$ (ie, $u_a^0 + u_{atm}$) is the absolute initial excess pore-air pressure(kPa); u_{atm} is the atmospheric pressure (~100kPa); M is the air mass molecular(~0.029 kg/mol); n_0 and Sr_0 are the initial porosity and saturation; m_{1k}^a, m_2^a, m_{1k}^w and m_2^w, respectively, are the coefficients related to the variations of net normal stress and suction for the air and water phases(kPa^{-1}); γ_w is the unit weight of water (~9.8 kN/m^3).

2.3 Initial and Boundary Conditions

The initial condition shows in figure 1:

$$u_a(r,z,0)=u_a^0,\ u_w(r,z,0)=u_w^0 \tag{3}$$

The boundary conditions are:

$$u_a(r_w,z,t)=u_{a_d}(z,t),\ u_w(r_w,z,t)=u_{w_d}(z,t) \tag{4}$$

$$\frac{\partial u_a(r_e,z,t)}{\partial r}=\frac{\partial u_w(r_e,z,t)}{\partial r}=0 \tag{5}$$

$$u_{w_d}(0,t)=u_{a_d}(0,t)=\frac{\partial u_{w_d}(H,t)}{\partial z}=\frac{\partial u_{a_d}(H,t)}{\partial z}=0 \tag{6}$$

The drain resistance conditions are expressed as at $r=r_w$:

$$\frac{\partial u_a(r_w,z,t)}{\partial r}=-\frac{r_w}{2}\cdot\frac{k_{a_d}}{k_a}\cdot\frac{\partial^2 u_{a_d}}{\partial z^2},\quad \frac{\partial u_w(r_w,z,t)}{\partial r}=-\frac{r_w}{2}\cdot\frac{k_{w_d}}{k_w}\cdot\frac{\partial^2 u_{w_d}}{\partial z^2} \tag{7}$$

3. Analytical Solution

Eqs. (1a) and (1b) can also be transformed as follows:

$$\frac{\partial \bar{u}_a}{\partial t}=A_a\left(\frac{\partial^2 u_a}{\partial r^2}+\frac{1}{r}\frac{\partial u_a}{\partial r}\right)+A_w\left(\frac{\partial^2 u_w}{\partial r^2}+\frac{1}{r}\frac{\partial u_w}{\partial r}\right)+A_q\frac{\mathrm{d}q}{\mathrm{d}t} \tag{8a}$$

$$\frac{\partial \bar{u}_w}{\partial t}=W_a\left(\frac{\partial^2 u_a}{\partial r^2}+\frac{1}{r}\frac{\partial u_a}{\partial r}\right)+W_w\left(\frac{\partial^2 u_w}{\partial r^2}+\frac{1}{r}\frac{\partial u_w}{\partial r}\right)+W_q\frac{\mathrm{d}q}{\mathrm{d}t} \tag{8b}$$

where, $A_a=-C_v^a/(1-C_aC_w)$; $A_w=C_aC_v^w/(1-C_aC_w)$; $A_q=C_q^a-C_aC_q^w/(1-C_aC_w)$;
$W_a=C_wC_v^a/(1-C_aC_w)$; $W_w=-C_v^w/(1-C_aC_w)$; $W_q=C_q^w-C_wC_q^a/(1-C_aC_w)$.

Eqs. (8a) and (8b) can be converted into equivalent PDEs by introducing variables of Φ_1 and Φ_2:

$$\frac{\partial}{\partial r}\left(r\frac{\partial \Phi_1}{\partial r}\right)=\frac{r}{Q_1}\left(\frac{\partial \bar{\Phi}_1}{\partial t}-\beta_1\frac{\mathrm{d}q}{\mathrm{d}t}\right) \tag{9a}$$

$$\frac{\partial}{\partial r}\left(r\frac{\partial \Phi_2}{\partial r}\right)=\frac{r}{Q_2}\left(\frac{\partial \bar{\Phi}_2}{\partial t}-\beta_2\frac{\mathrm{d}q}{\mathrm{d}t}\right) \tag{9b}$$

where, $\bar{\Phi}_1=\bar{u}_a+q_{21}\bar{u}_w$, $\bar{\Phi}_2=q_{12}\bar{u}_a+\bar{u}_w$ (10)

β_1 (ie, $A_q+q_{21}W_q$) and β_2 (ie, $A_qq_{12}+W_q$), the derivation process and relevant parameters are presented in Appendix A. Based on the boundary conditions provided in Eq. (4) and (5), integrating Eqs. (9a) and (9b) against the radius yields:

$$\Phi_1=-\frac{1}{2Q_1}\left[r_e^2\ln\left(\frac{r}{r_w}\right)-\frac{r^2-r_w^2}{2}\right]\left(\frac{\partial \bar{\Phi}_1}{\partial t}-\beta_1\frac{\mathrm{d}q}{\mathrm{d}t}\right)+\Phi_{1d} \tag{11a}$$

$$\Phi_2=-\frac{1}{2Q_2}\left[r_e^2\ln\left(\frac{r}{r_w}\right)-\frac{r^2-r_w^2}{2}\right]\left(\frac{\partial \bar{\Phi}_2}{\partial t}-\beta_2\frac{\mathrm{d}q}{\mathrm{d}t}\right)+\Phi_{2d} \tag{11b}$$

where, $\Phi_{1d}=u_{a_d}+u_{w_d}q_{21}$, $\Phi_{2d}=u_{a_d}q_{12}+u_{w_d}$.

Substituting Eqs. (11a) and (11b) into Eq. (10) gives:

$$\bar{\Phi}_1=\eta_1\left(\frac{\partial\bar{\Phi}_1}{\partial t}-\beta_1\frac{\mathrm{d}q}{\mathrm{d}t}\right)+\Phi_{1d} \tag{12a}$$

$$\bar{\Phi}_2=\eta_2\left(\frac{\partial\bar{\Phi}_2}{\partial t}-\beta_2\frac{\mathrm{d}q}{\mathrm{d}t}\right)+\Phi_{2d} \tag{12b}$$

where, $\eta_1=\left[r_e^2/8Q_1N^2\left(N^2-1\right)\right]\left(-4N^4\ln N+3N^4-4N^2+1\right)$; $\eta_2=\left[r_e^2/8Q_2N^2\left(N^2-1\right)\right]\left(-4N^4\ln N+3N^4-4N^2+1\right)$;

N (ie, r_e/r_w) is the ratio of influence radius to vertical drain radius.

Then substituting Eq. (7) into the first integration of (9a) and (9b) at domain $r\in(r_w, r)$ yields:

$$\frac{\partial^2\Phi_{1d}}{\partial z^2}=\rho_1\left(\frac{\partial\bar{\Phi}_1}{\partial t}-\beta_1\frac{\mathrm{d}q}{\mathrm{d}t}\right) \tag{13a}$$

$$\frac{\partial^2\Phi_{2d}}{\partial z^2}=\rho_2\left(\frac{\partial\bar{\Phi}_2}{\partial t}-\beta_2\frac{\mathrm{d}q}{\mathrm{d}t}\right) \tag{13b}$$

where, $\alpha_d=k_a/k_{a_d}=k_w/k_{w_d}$; $\rho_1=\alpha_d\left(N^2-1\right)/Q_1$; $\rho_2=\alpha_d\left(N^2-1\right)/Q_2$.

Combining Equation (12a), (12b), (13a), and (13b), as follows:

$$\frac{\partial^3\Phi_{1d}}{\partial z^2\partial t}-\frac{1}{\eta_1}\frac{\partial^2\Phi_{1d}}{\partial z^2}+\frac{\rho_1}{\eta_1}\frac{\partial\Phi_{1d}}{\partial t}=\frac{\rho_1\beta_1}{\eta_1}\frac{\mathrm{d}q}{\mathrm{d}t} \tag{14a}$$

$$\frac{\partial^3\Phi_{2d}}{\partial z^2\partial t}-\frac{1}{\eta_2}\frac{\partial^2\Phi_{2d}}{\partial z^2}+\frac{\rho_2}{\eta_2}\frac{\partial\Phi_{2d}}{\partial t}=\frac{\rho_2\beta_2}{\eta_2}\frac{\mathrm{d}q}{\mathrm{d}t} \tag{14b}$$

Based on the homogeneous boundary conditions given in Eqs. (5) and (6) and the eigenfunction method, Φ_{1d} and Φ_{2d} can be expanded that:

$$\Phi_{1d}(z,t)=\sum_{m=0}^{\infty}\phi_1(t)\sin\left(\frac{M}{H}z\right),\quad \Phi_{2d}(z,t)=\sum_{m=0}^{\infty}\phi_2(t)\sin\left(\frac{M}{H}z\right) \tag{15}$$

where $\phi(t)$ is the undetermined function with respect to t., and $M=(2m+1)\pi/2$, ($m=0,1,2,\cdots$).

Based on the orthogonality of the sine function (ie, sin (Mz/H)), substituted Eq (15) into Eqs. (14a) and (14b) and the differential equations with respect to t will be converted:

$$\frac{\mathrm{d}\phi_1(t)}{\mathrm{d}t}+\delta_1\phi_1(t)=\zeta_1\frac{\mathrm{d}q(t)}{\mathrm{d}t} \tag{16a}$$

$$\frac{\mathrm{d}\phi_2(t)}{\mathrm{d}t}+\delta_2\phi_2(t)=\zeta_2\frac{\mathrm{d}q(t)}{\mathrm{d}t} \tag{16b}$$

where, $\delta_1=\dfrac{M^2}{\rho_1H^2-M^2\eta_1}=\dfrac{Q_1}{d_e^2}\left(\dfrac{1}{M^2}\dfrac{N^2-1}{N^2}G-\dfrac{Q_1\eta_1}{d_e^2}\right)^{-1}$; $\zeta_1=\dfrac{2\rho_1\beta_1\delta_1H^2}{M^3}$;

$\delta_2=\dfrac{M^2}{\rho_2H^2-M^2\eta_2}=\dfrac{Q_2}{d_e^2}\left(\dfrac{1}{M^2}\dfrac{N^2-1}{N^2}G-\dfrac{Q_2\eta_2}{d_e^2}\right)^{-1}$; $\zeta_2=\dfrac{2\rho_2\beta_2\delta_2H^2}{M^3}$;

G (ie, $\alpha_d(H/d_w)^2$) is the drain resistance factor.

Hence,

$$\phi_1(t)=\mathrm{e}^{-\delta_1t}\left(C_1+\int_0^t\zeta_1\frac{\mathrm{d}q(\tau)}{\mathrm{d}\tau}\mathrm{e}^{\delta_1\tau}\mathrm{d}\tau\right) \tag{17a}$$

$$\phi_2(t)=\mathrm{e}^{-\delta_2 t}\left(C_2+\int_0^t \zeta_2 \frac{\mathrm{d}q(\tau)}{\mathrm{d}\tau}\mathrm{e}^{\delta_2\tau}\mathrm{d}\tau\right) \tag{17b}$$

Substituting Equation (17a) and (17b) into Equation (15):

$$\Phi_{1\mathrm{d}}(z,t)=\sum_{m=0}^{\infty}\mathrm{e}^{-\delta_1 t}\left(C_1+\int_0^t \zeta_1 \frac{\mathrm{d}q(\tau)}{\mathrm{d}\tau}\mathrm{e}^{\delta_1\tau}\mathrm{d}\tau\right)\sin\left(\frac{M}{H}z\right) \tag{18a}$$

$$\Phi_{2\mathrm{d}}(z,t)=\sum_{m=0}^{\infty}\mathrm{e}^{-\delta_2 t}\left(C_2+\int_0^t \zeta_2 \frac{\mathrm{d}q(\tau)}{\mathrm{d}\tau}\mathrm{e}^{\delta_2\tau}\mathrm{d}\tau\right)\sin\left(\frac{M}{H}z\right) \tag{18b}$$

Eqs (18a) and (18b) can now be substituted back into Eqs. (13) and (14), and then the final solution obtains by combining Eq. (10):

$$\bar{u}_{\mathrm{a}}=\frac{1}{q_{12}q_{21}-1}\left[q_{21}\bar{\Phi}_2-\bar{\Phi}_1\right] \tag{19a}$$

$$\bar{u}_{\mathrm{w}}=\frac{1}{q_{12}q_{21}-1}\left[q_{12}\bar{\Phi}_1-\bar{\Phi}_2\right] \tag{19b}$$

where, $\bar{\Phi}_1=\sum_{m=0}^{\infty}\frac{2}{M}\left[\left(u_{\mathrm{a}}^0+q_{21}u_{\mathrm{w}}^0\right)\mathrm{e}^{-\delta_1 t}+\beta_1\int_0^t \mathrm{e}^{\delta_1(\tau-t)}\frac{\mathrm{d}q(\tau)}{\mathrm{d}\tau}\mathrm{d}\tau\right]\sin\left(\frac{M}{H}z\right)$

$$\bar{\Phi}_2=\sum_{m=0}^{\infty}\frac{2}{M}\left[\left(q_{12}u_{\mathrm{a}}^0+u_{\mathrm{w}}^0\right)\mathrm{e}^{-\delta_2 t}+\beta_2\int_0^t \mathrm{e}^{\delta_2(\tau-t)}\frac{\mathrm{d}q(\tau)}{\mathrm{d}\tau}\mathrm{d}\tau\right]\sin\left(\frac{M}{H}z\right)$$

4. Verification and Examples

A typical example is used to validate the general analytical solution and to illustrate the consolidation behaviours of vertical drain foundation, as shown in figure 1. This study mainly examines the effects of drain resistance factor G (ie, $\alpha_{\mathrm{d}}(H/d_{\mathrm{w}})^2$) and ratio of influence radius to drain radius N (ie, $N=r_{\mathrm{e}}/r_{\mathrm{w}}$) on the consolidation behaviours of vertical drain foundation subjected to arbitrary loads. Following the Qin et al.[2], the example in this section adopts the material parameters as follows:

q_0 = 100 kPa (u_a^0 = 20 kPa, u_w^0 = 40 kPa), r_{w} = 0.2 m, r_{e} = 1.8 m (N=9), H = 10 m, $k_{\mathrm{a}} = 1\times10^{-8}$ m/s, $k_{\mathrm{w}} = 1\times10^{-10}$ m/s, S_{r} = 80%, n_0 = 50%, $m_{1\mathrm{k}}^{\mathrm{w}} = -5\times10^{-5}$ kPa^{-1}, $m_{1\mathrm{k}}^{\mathrm{a}} = -2\times10^{-4}$ kPa^{-1}, $m_2^{\mathrm{w}} = -2\times10^{-4}$ kPa^{-1}, $m_2^{\mathrm{a}} = 1\times10^{-4}$ kPa^{-1}, $C_{\mathrm{v}} = 2.4\times10^{-9}$ kPa^{-1}.

4.1 Verification.

Based on the special consolidation example without considering drain resistance (ie, G=0) and subjected to instantaneous load (ie, $q(t)=q_0$), the validity of the equal-strain model is verified by comparing with the free-strain model proposed by Qin et al.[2].Figures 2 demonstrates the dissipation curve of the average excess pore-air and pore-water pressures with different $k_{\mathrm{a}}/k_{\mathrm{w}}$ values under the equal-strain and free-strain conditions. It can be observed that the solution obtained from these two different models are close to each other. The slight difference mainly appears at the earlier stage of the consolidation process, but the difference has negligible influence on the average excess pore pressures, which was also mentioned by Zhou et al.[7]. It shows that the equal-strain model is simple and efficient in solving the unsaturated soil consolidation problem. Moreover, there are two stages (ie, before and after the "plateau period") of dissipation curve of average excess pore-water pressures in figure 2B. It is predicted that the single curve converged at second stage are mainly due to the average excess pore-air pressure almost completely dissipates. With an increase of $k_{\mathrm{a}}/k_{\mathrm{w}}$, the dissipation rate of average excess pore-air and pore water pressures increase gradually.

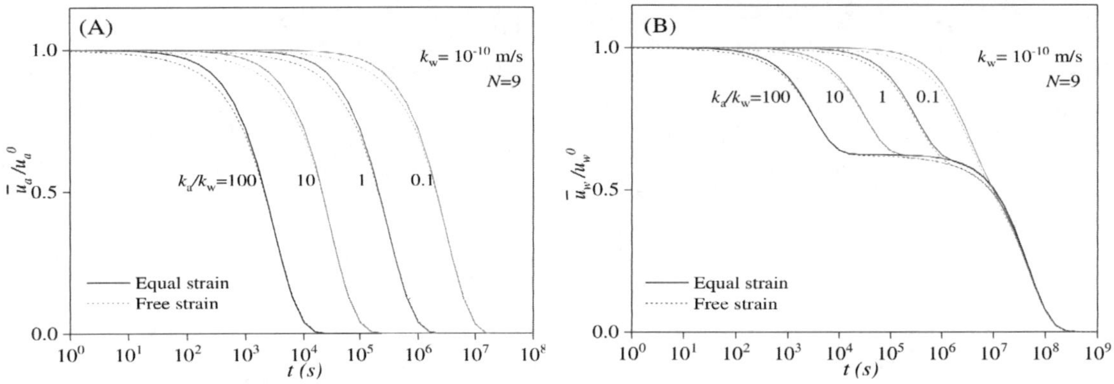

Figure 2. Dissipation of (A) $\bar{u}_a$ and (B) $\bar{u}_w$ with different ratios of k_a/k_w

4.2 Influence of Drain Resistance.

Figure 3 illustrates the dissipation curve under the exponential load varying with the drain resistance factor G (~0 (no drain resistance), 0.1, 0.5, 1 and 2). The exponential loads as follow:

$$q(t) = q_0 + Aq_0\left(1 - e^{-bt}\right)$$

where A is load constant, and A = 1; b is the load parameter, and $b = 5\times10^{-4}\ s^{-1}$.

Obviously, the variation of drain resistance factor G has a significant impact on the dissipation of the average excess pore-air and pore-water pressures, the smaller the drain resistance factor G is, the faster the dissipation is. It is predicted that the drainage capacity of vertical drain will be increases as G decreases, which accelerates the process of consolidation, and G has a similar influence on the dissipation of the two stages (ie, before and after the "plateau period") in figure 3B. However, when G is less than 0.1, the drain resistance has little effect on the dissipation, and it is suggested that drain resistance should not be considered in practical engineering.

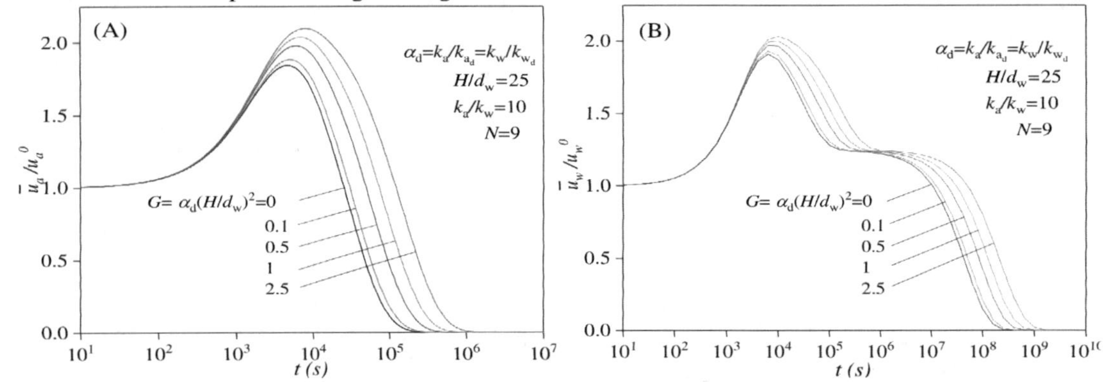

Figure 3. Dissipation of (A) $\bar{u}_a$ and (B) $\bar{u}_w$ under different G

Figure 4 illustrates the comparison curves to dissipation of $\bar{u}_a$ and $\bar{u}_w$ under exponential load with different N (ie, $N = r_e/r_d$), which is used to describing the influence of influence radius on the consolidation process. In this part, G= 0.5 and r_d = 0.2 are applied, and the variation of N value depends on r_e. The smaller N is, the faster the dissipation is in figure 4. The drainage path reduces with the decrease of r_e, and that accelerates the consolidation of vertical drain foundation as shown in figure 4. Otherwise, the average excess pore pressures firstly increase gradually with the external load increasing, and then reach to the peak at about t = 10^4 in figure 3. The reason for this phenomenon is that the increasing rate of the exponential load is faster than the dissipation rate of average excess pore pressures during the early stage.

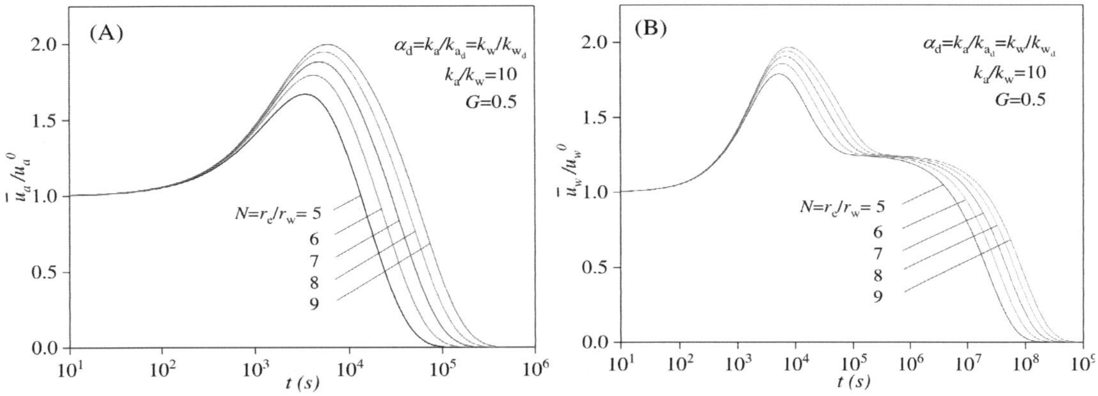

Figure 4. Dissipation of (A) $\bar{u}_a$ and (B) $\bar{u}_w$ under different N

5. Conclusion

Based on the equal-strain hypothesis, the general analytical solution to consolidation of vertical drain foundation considering drain resistance were proposed. The solution of the equal-strain model established in this study has slight difference with the free-strain model at the earlier stage of the consolidation process, but it has a negligible influence on the consolidation behaviours. Specially, the calculation process was simplified by using the equal-strain model. The influence of drain resistance (ie, G) and ratio of influence radius to vertical drain radius (ie, N) on the unsaturated consolidation of vertical drain foundation are took into account. When N decreases, the drainage path of vertical drain foundation reduces, and then the average excess pore pressures would dissipate more quickly. Moreover, the dissipation rate of average excess pore pressures will be increases as drain resistance decreases, and when G is less than 0.1, it is suggested that drain resistance should not be considered in practical engineering.

6. Appendix A.

By adopting the two arbitrary constants q_1 and q_2, Eqs. (8a) and (8b) can be summarized as follows:

$$\frac{\partial(\bar{u}_a q_1+\bar{u}_w q_2)}{\partial t}-(A_a q_1+W_a q_2)\left(\frac{\partial^2 u_a}{\partial r^2}+\frac{1}{r}\frac{\partial u_a}{\partial r}\right)-(A_w q_1+W_w q_2)\left(\frac{\partial^2 u_w}{\partial r^2}+\frac{1}{r}\frac{\partial u_w}{\partial r}\right)=(A_q q_2+W_q q_2)\frac{\mathrm{d}q}{\mathrm{d}t} \quad \text{(A1)}$$

In order to give a true statement to the Eq. (A1), the Q must satisfy the following condition:

$$Qq_1=A_a q_1+W_a q_2,\ Qq_2=A_w q_1+W_w q_2 \quad \text{(A2)}$$

The roots Q_1 and Q_2 of the quadratic equation Eqs. (A2) can be expressed as:

$$Q_{1,2}=\frac{1}{2}\left[A_a+W_w\pm\sqrt{(A_a-W_w)^2+4A_w W_a}\right] \quad \text{(A3)}$$

When $Q = Q1$, the solution of $q1$ and $q2$ in Eq. (A2) are q_{11} and q_{21}, respectively. Similarly, $q1$ and $q2$ in Eq. (A3) are q_{12} and q_{22} when $Q = Q_2$, respectively.
where,

$$\frac{\partial\bar{\Phi}_1}{\partial t}-Q_1\left(\frac{\partial^2\Phi_1}{\partial r^2}+\frac{1}{r}\frac{\partial\Phi_1}{\partial r}\right)=\beta_1\frac{\mathrm{d}q}{\mathrm{d}t} \quad \text{(A4)}$$

$$\frac{\partial\bar{\Phi}_2}{\partial t}-Q_2\left(\frac{\partial^2\Phi_2}{\partial r^2}+\frac{1}{r}\frac{\partial\Phi_2}{\partial r}\right)=\beta_2\frac{\mathrm{d}q}{\mathrm{d}t} \quad \text{(A5)}$$

$$\bar{\Phi}_1=\bar{u}_a+q_{21}\bar{u}_w,\ \bar{\Phi}_2=q_{12}\bar{u}_a+\bar{u}_w \quad \text{(A6)}$$

Where, $\beta_1=A_q+q_{21}W_q$; $\beta_2=A_q q_{12}+W_q$; $q_{12}=W_a/(Q_2-A_a)=(Q_2-W_w)/A_w$; $q_{21}=A_w/(Q_1-W_w)=(Q_1-A_a)/W_a$.

Generally, it is reasonably assumed that $q_{11} = q_{22} = 1$, so Eqs. (7) and (8) can be obtained by substituting

Eqs. (5) and (6) into Eqs. (A1) and (A3), respectively.

Acknowledgments
We sincerely acknowledge the National Natural Science Foundation of China (Grant No. 41372279 and Grant No. 11672172) for financial support.

References

[1] Fredlund D. G., Hasan J U. (1979) One-dimensional consolidation theory: unsaturated soils. J. Can Geotech, 17:521 -31.
[2] Qin A. F., Sun D. A., Yang L P. (2010) A semi-analytical solution to consolidation of unsaturated soils with the free drainage well. J. Computers and Geotechnics, 37(7): 867-875.
[3] Qin A. F., Sun D. A., Tan Y. W. (2010) Semi-analytical solution to one-dimensional consolidation in unsaturated soils. J. Appl Math Mech (English Edition), 31(2):215–26.
[4] Wang L., Sun D. A., Qin A. F. (2017) Semi-analytical solutions to one-dimensional consolidation for unsaturated soils with symmetric semi-permeable drainage boundary. J. Computers and Geotechnics, 38(6): 831-850.
[5] HO L., Fatahi B. (2016) Axisymmetric consolidation in unsaturated soil deposit subjected to time-dependent loadings. J. International Journal of Geomechanics, 17(2):04016046.
[6] Zhou W. H. (2013) Axisymmetric consolidation of unsaturated soils by differential quadrature method. J. Mathematical Problems in Engineering, 497161.
[7] Zhou W. H. (2014) A simple analytical solution to one-dimensional consolidation for unsaturated soils. J. International Journal for Numerical and Analytical Methods in Geomechanics, 38.
[8] HO L., Fatahi B. (2018) Analytical solution to axisymmetric consolidation of unsaturated soil stratum under equal-strain condition incorporating smear effects. J. International Journal for Numerical and Analytical Methods in Geomechics, 42(15):1890-1913.
[9] Zhou F., Chen Z., Wang X. D. (2018) An equal-strain analytical solution for the radial consolidation of unsaturated soils by vertical drains considering drain resistance. J. Advances in Civil Engineering, 5069159

Experimental study on energy dissipation of annular overflow shaft

GU Jinde, Li Yuan, Zhao Jianjun

Nanjing hydraulic research institute, Nanjing 210029

Corresponding author's e-mail:*Jdgu@nhri.cn

Abstract: The annular shaft, which is widely used in tailings and some large and medium water conservancy project, has the characteristics of simple structure and strong discharge capacity. With the increase of working head, the energy dissipation problem of shaft is becoming much important. Model test results shows that the peak of the hydrodynamic pressure at the bottom of the shaft happens when the water jet merge together, for the water depth H in the shaft is low then, the pre-excavated hydrostatic depth is the main factor of energy dissipation. The fitting formula of the maximum impact pressure is obtained and it is suggested that the energy dissipation depth h should be 0.8D.

1. Introduction

With the characteristics of simple structure, strong cavitation resistance and large discharge ability, annular shaft is a common drainage facility in tailings pond and fly ash pond. In recent years, vertical shaft drainage was always adopted by some large and medium-sized water conservancy projects [1-4]. With the development of high dam engineering technology, the overflow head of vertical shaft has reached more than 100m. At the same time, due to the direct drop of water flow, energy dissipation of high speed flow is more serious.

The most common energy dissipation method of annular shaft is to make pre-excavate depth in the bottom of the drainage tunnel to form a water cushion energy dissipation area, so that the energy of falling water tongue can be buffered and the direct impact pressure on the bottom can be reduced, which can ensure the stability of the shaft structure. The reasonable energy dissipation depth should not only ensure the safety of shaft structure, but also take economy into consideration. The energy dissipation depth can be determined in many ways. For example, some engineers suggest it to be determined by conjugdined depth of intake flow, and some suggest it to be determined by energy dissipation rate per unit water body[5-6]. Nevertheless, model study, by which dynamic pressure can be get, is thought to be the most effective way. Many scholars [7][8] have put forward Suggestions on energy dissipation depth based on model test results and different conclusions have been drawn due to different test conditions. Therefore, the problem of energy dissipation depth in circular shaft still needs to be further discussed.

Dynamic water impact pressure at the bottom of a shaft is the most direct indicator of energy dissipation effect, and the energy dissipation depth should be determined by combining dynamic water pressure with the impact resistance of the shaft floor. Therefore, the calculation method of fall-flow water pressure is the key to determine the energy dissipation depth. Based on the research background of a high tailings pond drainage shaft, dynamic pressure at the bottom of the shaft under different energy dissipation depth is analyzed, the empirical formula impact pressure is summarized based on the test data, and energy dissipation water depth in the high drainage shaft is discussed, which provides a basis for related projects.

2. Model construction

The model is built according to a drainage system of overflow tower-drainage tunnel of a tailings pond project. The diameter of the drainage shaft was 5.5m, and the bottom of the shaft was connected to the drainage tunnel. The section of the drainage tunnel was 3.3m×3.3m, which is straight wall semicircular arch. The model is designed according to the gravity similarity criterion by scale of 1:25. The shaft and the drainage tunnel are all made of plexiglass, which not only satisfy the similar roughness, but also easy for observation of the flow pattern. The overflow area at the top of the shaft is enclosed in the steel water tank, while the shaft and downstream drainage tunnel are outside, enhanced by steel structure. Pressure sensors are placed at different place of the shaft, the DASP system of Oriental noise research institute is used for pressure collection and processing. The model test system is shown in Figure1.

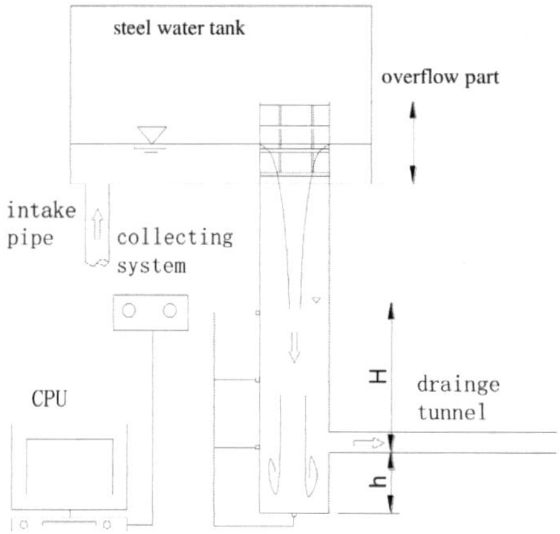

Figure 1 Model diagram

When the drainage system is running, the water level in the shaft rises with the increase of discharge. In this paper, the height from the bottom of the drainage tunnel to the water surface in the shaft is denoted as H, which depends on the discharge ability of the drainage tunnel. The distance from the bottom of the shaft to the bottom of the drainage tunnel is the initial energy dissipation depth, which is denoted as h. The shaft diameter is D. During the test, it can be observed that, with the increase of the discharge flow of the overflow weir, the wall sticking flow, concentrated flow and collision merge flow occur successively in the shaft. In order to compare the variation of the pressure at the bottom of the shaft with different energy dissipation water depths, different initial energy dissipation depth schemes with h= 0.2d, h= 0.5d and h=1D is designed, and the dynamic pressure is tested.

3. Typical flow pattern in shaft

It can be observed that there are two kinds of falling flow patterns in the shaft during the experiment:1.Uncollision flow, when discharge is very small, falling jet almost no contact with each other ,drop points are near the central of shaft, boundary of flow is rather clear, this flow pattern is defined as flow patternI,as shown in Figure 2(a).2. Collision flow. With the increase of discharge, the falling jet becomes thicker and the throw distance increases. The flow of different directions crash in the air, The collision effect of jet is more obvious while discharge increase. The air jet collides, merges and disperses, flow boundary blurs under intensive aeration during the drop, the water roll shaft, this flow pattern is defined as flow patternII, as shown in Figure 2(b).

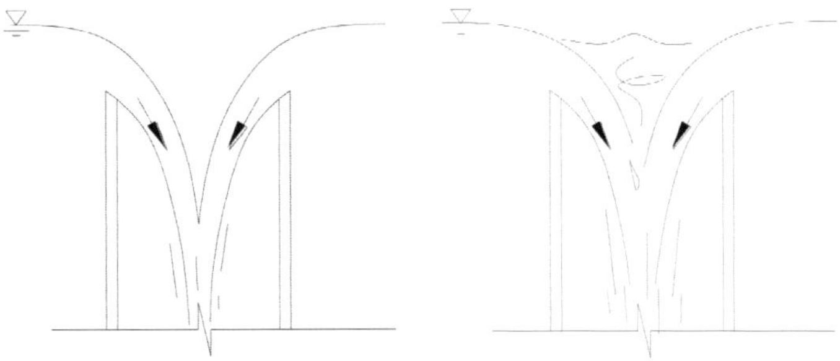

(a) (b)
Figure 2 Shaft overflow type

4. Hydraulic characteristics of energy dissipation zone at shaft bottom

Pressure sensors were installed at different elevations, both in the centre of shaft floor and the side wall, to measure the pressure under different flow rates. Test results are shown in Figure 3. As can be seen, the average pressure in the bottom increases rapidly with the increase of the flow rate under low discharge rate, pressure reach to the peak at flow rate of 15m³/s. Then it gradually decreases with the increase of the flow and reaches to the valley when the flow is about 40m³/s, then the average pressure gradually rises again with the increase of discharge. As can be seen from the perspective of pressure growth trend, when the flow rate exceeds 60m³/s, the pressure growth trend at all measurement points is basically the same. The pressure difference between different measuring points is equivalent to the hydrostatic pressure corresponding to the elevation difference. Peak pressure forms due to insufficient water cushion, and it decreases while the water depth increase. Shaft peak during low flow is insufficient, because the shaft in the inland waters cushion depth drainage of water impact plate, with the increase of discharge shaft after the internal pad increase, direct impact pressure eliminations in the water cushion while discharge is higher than 60m³/s.

The shaft discharge is dimensionless by shaft diameter D, denoted as dimensionless discharge $Q^* = Q/\sqrt{gD^5}$, The impact pressure at the bottom is defined as the difference between the maximum time-averaged pressure and the hydrostatic pressure, the expression is $\Delta P_{max} / \gamma = P_{max} / \gamma - (H + h)$, in which $(H + h)$ is the actual energy dissipation water depth inside the shaft. The impact pressure under each flow rate is shown in Figure 4. As can be seen from the Figure, the impact pressure increases rapidly with the increase of the flow. It peaks around the dimensionless discharge of 0.06~0.07, and then decreases rapidly with the increase of the discharge. When the flow increases above 0.3, the impact pressure is very small.

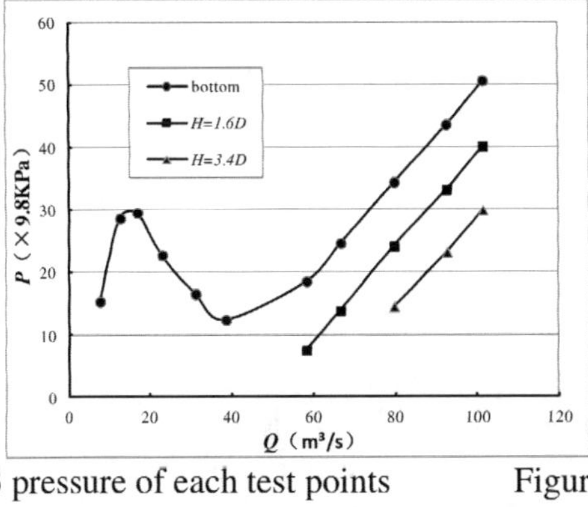

Figure3 pressure of each test points

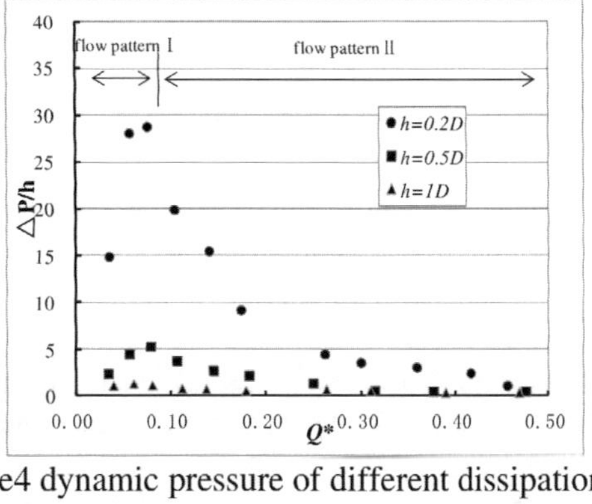

Figure4 dynamic pressure of different dissipation depth

Flow pattern in the shaft is observed during the model test, when the dimensionless flow rate is less than 0.02, overflow concentration increased gradually to the shaft center, the drainage tunnel inlet are in a state of free flow (flowI); When the dimensionless flow rate is higher than 0.02, the flow collision in the air, water tongue collision point rise gradually with further increases in discharge, water level in shaft rise at the same time (flow II).

5. Discussion on initial energy dissipation depth at shaft bottom

The cause of impact pressure is that the high velocity head acts directly on the bottom , and the velocity head is converted into kinetic energy, that is $\Delta P/\gamma = v_m^2/2g$, in which v_m is the bottom velocity of discharge flow. The initial flow velocity drop into the water surface of the shaft is denoted as v_m, it can be inferred that $\frac{v_m}{v_0} = k(\frac{x}{d_0})^n$ According to the diffusion law of free jet in water cushion. Where, k and n are constants, d_0 is the thickness of the flow, and can be evaluated by $d_0 = q/v_0$, where q is the flow rate of single width. Take the common logarithm of both sides of the formula:

$$\lg\frac{v_m}{v_0} = \lg k + n\lg(\frac{x}{d_0})$$

It can be obtained that $k = 1.79$, $n = 0.7663$ by binary linear regression analysis of model test data, than the formula is $\frac{v_m}{v_0} = 1.79(\frac{x}{d_0})^{-0.7663}$,as shown in Figure 5,which can be used for dynamic pressure evaluation in approximate engineering.

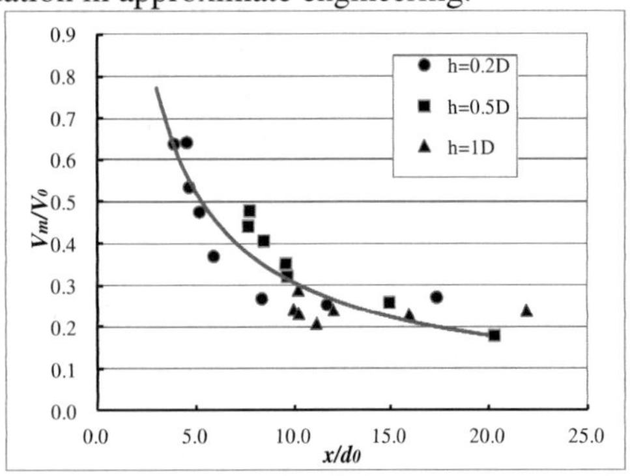

Figure 5 bottom velocity fitting curve

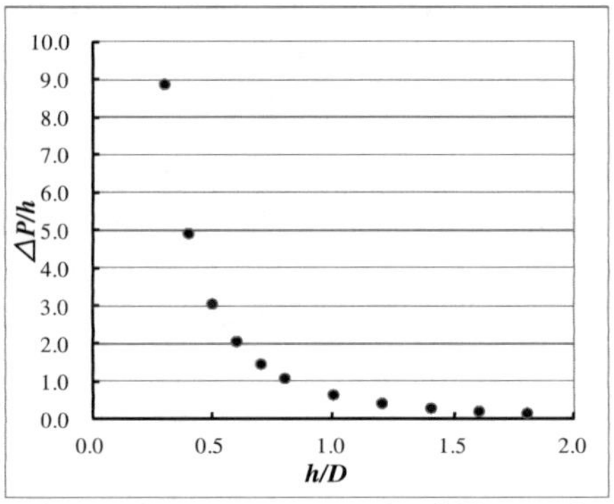

Figure6 Maximum impact pressure in different depth

From the test results, peak dynamic load is found in the region between the flow pattern I and flow patternII, where dimensionless flow rate is about 0.06. dynamic water impact pressure at the bottom of the shaft under different initial dissipation depth can be calculated according to the formula, and the calculated results are shown in Figure 6. As can be seen, when the initial energy dissipation depth h increases from 0 to 0.8d, the impact pressure decreases rapidly. When h>0.8d, the dynamic water impact pressure decreases slowly while it is already very small, and the change trend tends become flat. According to references [5][6], initial energy dissipation depth are suggested to be $\Delta P/\gamma h = 1$, the corresponding h/D is about 0.8 in Figure6. It is shown that 0.8D is an appropriate initial energy dissipation depth of this project according to the calculation results,

6. Conclusions

The following conclusions can be drawn from the 1:25 hydraulic model test of a discharge shaft:1. The dynamic pressure at the bottom of shaft increases first and then decreases with the increase of the discharge. When the annular falling flow begins to merge together, the dynamic water load at the bottom of the shaft come to a peak value. Then, under the influence of the rapidly increasing water depth, the dynamic water load gradually decreases.2. Initial energy dissipation depth has a direct effect on the peak pressure, the empirical formula, which is fitted with the test data, can be used to obtain the impact pressure at the bottom of the shaft under different initial energy dissipation depths. From the calculation results, initial energy dissipation water depth of is 0.8D more appropriate for this project.

References

[1] Ni hangen, Zhou jing, A tentative idea of dams with glory hole spillway[J]. Journal of Da lian university of technology, 1999, 39(2):318-325. (in chinese)
[2] Dong xinglin, Guo jun, Xiao baiyun, et al. Design principle of high head and large discharge vortex drop spillway[J],Journal of hydraulic engineering, 2000(11):27- 33. (in chinese)
[3] Dong xinglin, Gao jizhang, Zhong yongjiang, et al. Study on transferring diversion tunnels into vortex type of vertical shaft spillways [J], Water power, 1995, (3): 32- 37. (in chinese)
[4] Wang tingzheng, Experimental Studies on Flow Characteristics through Shaft and Tunel outlet of a Mine Tailing Reservoir [J]. Journal of Fuzhou University (Natural Science Edition), 1991(04):98-104. (in chinese)
[5] Han mei et al. Summary of Hydraulic Calculation Study of Draft Water-Drop Energy Reduction [J], Journal of Engineering of Heilongjiang University, 1996(3):39-45. (in chinese)
[6] Tan zhiwei, Discussion on hydraulic calculation of falling water in vertical shaft[J]. China rural water and hydropower, 2008(02):97-99+102. (in chinese)
[7] Lv mingcong, Zhu li, Chen ruihua et al. Pressure Characteristics of Vortex Flow Intake Shaft at the Bottom of Energy Dissipation Zone[J], Yellow river, 2015(05):105-108. (in chinese)
[8] Dang ning, Tailings drainage well experimental study on shape optimization of flood discharge system[D]. Xi 'an university of technology, 2010. (in chinese)

IOP Conf. Series: Materials Science and Engineering **794** (2020) 012041 doi:10.1088/1757-899X/794/1/012041

Analysis of Scouring and Silting Changes in the Middle and Lower Reaches of the Yangtze River after the Three Gorges Project Operation

Xiao Zhang[1] and Guangdong Wu[2*]

[1]Changjiang Survey Planning Design and Research Co., Ltd, Wuhan, Hubei, 430010, China

[2]Yangtze River Scientific Research Institute, Wuhan, Hubei, 430010, China

[*]598105302@qq.com

Abstract. The annual runoff of the middle and lower reaches of the Yangtze River has little change before and after the Three Gorges Project is put into operation, but the annual distribution of runoff has changed, and the sediment discharge of the reach has fallen in a cliff-like manner. The application of the Three Gorges Project has changed the scouring and silting situation of the main stream and the two lakes, making them change from silting to scouring direction, causing the water level of the rivers and lakes to drop. The changes of scouring and silting conditions, flow velocity, water depth and water level have a certain impact on the survival and reproduction of the four major domestic fish, porpoise and waterfowl. In order to protect the fragile aquatic ecosystem from the adverse effects of the operation of the Three Gorges Project, this paper puts forward three suggestions, namely, the delimitation of protected areas of top priority, the spatial management and control of river and lake shorelines, and the ecological scheduling, which is of great significance to the scientific decision-making of the water administrative department in the process of protection of the Yangtze River.

1. Introduction

There are large areas of wetlands in the middle and lower reaches of the Yangtze River, mainly including rivers, lakes and riverbank wetlands. As an important part of the water ecosystem in the middle and lower reaches of the Yangtze River, wetland play an important role in maintaining ecological balance, providing habitats for animals and plants, protecting biodiversity, and providing necessary animal and plant resources for human survival and reproduction [1].

Wetland is an important part of the earth's surface ecosystem. It is a natural synthesis with special functions formed by the interaction of land and water, and its response to hydrological process is very sensitive [2,3]. Since the completion of the Three Gorges Dam, due to the regulation and storage of the Three Gorges reservoir, the conditions of water and sediment from the middle and lower reaches of the Yangtze River have changed. The impoundment operation of the Three Gorges reservoir has changed the hydrological situation of the main stream, tributaries and lakes and wetlands of the Yangtze River. At the same time, due to the intercepting effect of the reservoir on the upstream sediment, the clear water has been discharged for quite a long time, which has caused the changes of the downstream channel sediment erosion and deposition, changed the river situation and the relationship between rivers and the lakes [4], which may have a certain impact on the structure and function of the water ecosystem in the middle and lower reaches of the Yangtze River [5,6]. Therefore, this paper briefly introduces the impact of the Three Gorges Project on the hydrology, water and sediment processes in the middle and lower reaches of the Yangtze River, and selects key river sections and two lakes to identify the degree of changes in river and lake sediment erosion and deposition before and after the operation of the Three Gorges Project. Finally, based on the reduction of important biological resources such as finless porpoise and waterfowl after the operation of the Three Gorges project, the paper puts forward countermeasures for the protection of the water ecosystem in the middle and lower reaches of the Yangtze River.

2. The influence of the Three Gorges Project on the hydrologic and sediment processes in the middle and lower reaches

2.1Operation mode of the Three Gorges Project

The Three Gorges project is located in the middle section of Xiling Gorge of the Yangtze River, the control basin area is 1 million square kilometer, annual average runoff is 451 billion cubic meter and annual average sediment discharge is 530 million ton. The designed normal water level is 175 m, the total storage capacity is 39.3 billion cubic meter and the flood control capacity is 22.15 billion cubic meter.

The operation of the Three Gorges project is a key measure to deal with the flood control problems in the middle and lower reaches of the Yangtze River, and it has brought huge comprehensive economic benefits in power generation, shipping, water supply and the development of the reservoir economy. According to the regulation for cascade operation of the Three Gorges (normal operation period) - Gezhouba Water Control Project, the water level of the Three Gorges water control project in flood season is controlled by 145.0m of flood control limit water level. During the real-time operation, the water level of the reservoir can change within a certain range of the flood control limit water level. The impoundment time of the Three Gorges project should not be earlier than September 10. When the water level of Shashi station and Chenglingji station is lower than the warning water level (43.0m and 32.5m respectively), and the forecast does not exceed the warning water level in the short term, the impoundment plan can be implemented. Generally, on September 10, the water level of the Three Gorges reservoir is not higher than 150.0m, and the control water level at the end of September is 162.0m. With the consent of the national defense general, the water level at the end of September can be adjusted to 165.0 m according to the water situation, and 175.0 m at the end of October. From 175.0m to the end of the year, the Three Gorges reservoir should maintain high water level operation as much as possible. In the real-time operation, the need of weekly regulation and daily peak regulation can be taken into account, and appropriate amplitude should be reserved below 175.0m. From January to May of the following year, the water level of the Three Gorges Reservoir gradually subsided under the condition of comprehensive consideration of shipping, power generation, water resources and water ecological demand. Generally, the water level of the reservoir at the end of April shall not be lower than 155.0m, and on May 25 shall not be higher than 155.0m. In case of extremely low water year, the emergency operation of water resources may not be limited by the above water level and flow.

2.2Influence of Three Gorges project operation on hydrological process

The trend of annual runoff in the middle and lower reaches of the Yangtze River is not significant before and after the operation of the Three Gorges project. Compared with the period before the Three Gorges project operation from 1960 to 2002, the annual runoff of typical representative stations in the middle and lower reaches of the Yangtze River after the operation of the Three Gorges project was reduced by 5% - 8%, as shown in Table 1, and the annual runoff of the upper reaches of the Yangtze River was also reduced by about 8% compared with the annual runoff between 1950-2002. On the premise that the impact of water vapor cycle change in the Three Gorges Reservoir area is ignored, it can be considered that the operation of the Three Gorges Project on the annual runoff of the middle and lower reaches of the Yangtze River is not significant. In recent years, the decrease of annual runoff in the middle and lower reaches of the Yangtze River is mainly caused by the decrease of inflow in the upper reaches of the Yangtze River.

Table 1. Average annual runoff of stations in the middle and lower reaches of the Yangtze River before and after the operation of the Three Gorges Project

Period	Runoff/100 million m³			
	Yichang	Shashi	Hankou	Datong
1960-2002	4369	3942	7112	9052
2003-2015	4005	3760	6711	8438
Compared with the runoff between 1960-2002 /%	-8%	-5%	-6%	-7%

However, the operation of the Three Gorges project has a greater impact on the annual distribution of runoff in the middle and lower reaches of the Yangtze River. The operation of the Three Gorges project has changed the annual inner diameter flow process in the downstream of the dam. Through the comparison of the annual average daily flow process of Hankou station in 1952-2002 and 2003-2015, as shown in Figure 1, it is found that after the impoundment of the Three Gorges reservoir, the runoff process of Hankou station is more uniform, the proportion of runoff increases in the dry season and decreases in the flood season, but the period of flood recession becomes shorter and the water process becomes steeper.

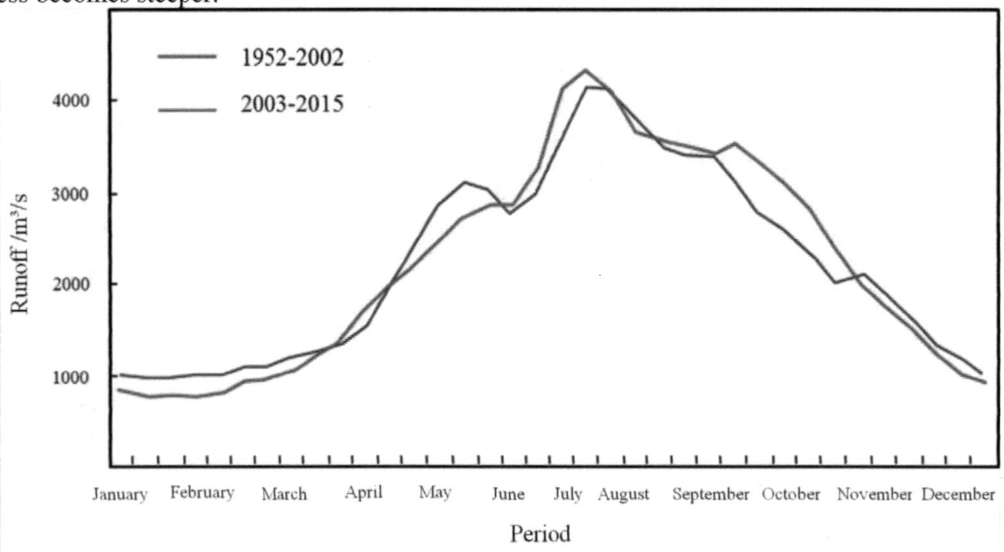

Figure 1. Annual average daily flow process of Hankou Station before and after the operation of the Three Gorges Project

2.3Influence of Three Gorges project operation on hydrological process

Compared with that before the operation of the Three Gorges Project from 1950 to 2002, the annual sediment transport observed by the main hydrological stations in the middle and lower reaches of the Yangtze River decreased significantly from 2003 to 2015. It decreased by 67% - 91% along the way, as shown in Figure 2.

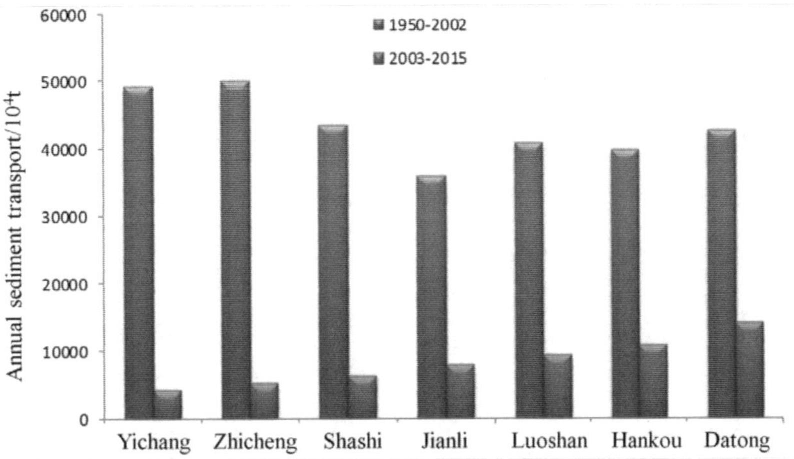

Figure 2. Variation of annual average sediment transport at stations in the middle and lower reaches of the Yangtze River before and after the operation of the Three Gorges Project

Sediment transport in the lower reaches of the Three Gorges project changes significantly before and after the operation of the Three Gorges Project. Before the operation of the Three Gorges reservoir, the trend of sediment transport along the course decreases gradually from the upstream to the downstream, but increases gradually after the operation. This shows that the riverbed scours along the course after the operation of the Three Gorges project, and the sediment concentration of the lower reaches of the river has recovered.

3. Changes of erosion and deposition in the main stream of the Yangtze River

3.1Changes of typical sections

Since the operation of the Three Gorges reservoir, the main channel has been cut down. The section form of Yichang station is irregular "U" shape. The section changes of erosion and deposition and riverbed are still in constant adjustment. Compared with 2002, the section changes little in 2010. The cross section of Zhicheng station is irregular "V" shape. Compared with 2002, the low flow channel was scoured seriously in 2010, and the lower part of the channel is nearly 13 M. The section form of Shashi station is irregular "W" shape. Compared with 2002, the low flow channel and flat beach were scoured by current at the same time in 2009, especially the channel, whose elevation decreased nearly 15 m. Like Shashi station, the section form of Luoshan station is also irregular "W" shape, but the section is scoured by beach silting channel, and the amount of scour is mainly concentrated in the low flow channel, and the riverbed is silted between the low flow and the flat water level, which leads to higher beach with large height and lower channel with small elevation. The sections of Hankou station and Nanjing station are all irregular "U" shape, and the river channel is also scoured to different degrees, but the scour degree of the river channel is smaller than that of Zhicheng station and Shashi station, as shown in Figure 3.

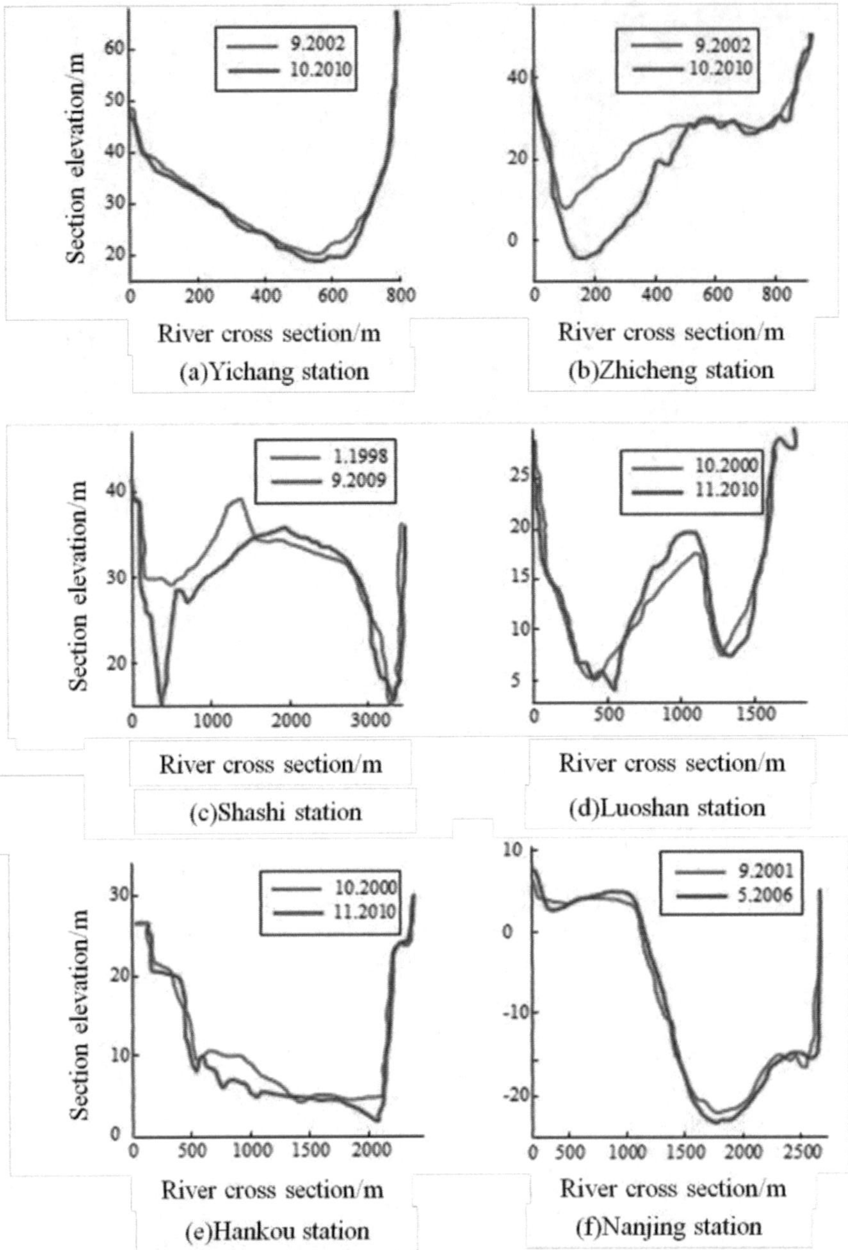

Figure 3. Changes of cross sections before and after the operation of the Three Gorges Project

3.2Changes of deep channel profile

After the operation of the Three Gorges project, the plane form of the middle reaches of the Yangtze River is still generally stable, but the evolution of the local river regime caused by the development of the deep channel of the riverbed and the swing of the main stream line is still in adjustment. The riverbed in the middle reaches of the Yangtze River is dominated by the longitudinal scouring depth. From October 2002 to October 2011, the scouring depth from Yichang reach to Zhicheng reach was cut down 3.6 m on average and the maximum scouring depth was 18.0 m.

The reach from Zhijiang to Chenglingji is the closest sandy reach to the Three Gorges reservoir, which is affected by the discharge of "clear water" from the Three Gorges reservoir. It scoured violently in this reach, but the overall river regime is basically stable. Since the Three Gorges reservoir operation, the vertical profile of Jingjiang river channel has been scoured and silted alternately, which is mainly manifested in the deep scour depth of bend, branch or transition section at the upstream of bend and branch, and the change of the elevation of the straight section is relatively small, as shown in Figure 4. In October 2015, the vertical elevation of the upper Jingjiang deep elevation decreased significantly compared with that in 2002, and the maximum deep elevation of the reach decreased by 13 m, while that of the Lower Jingjiang deep elevation decreased and increased compared with that in 2002. The average scour depth of Jingjiang river reach is 1.4m, the maximum scour depth is 13.8m, and the cross section of some riverbed changes violently.

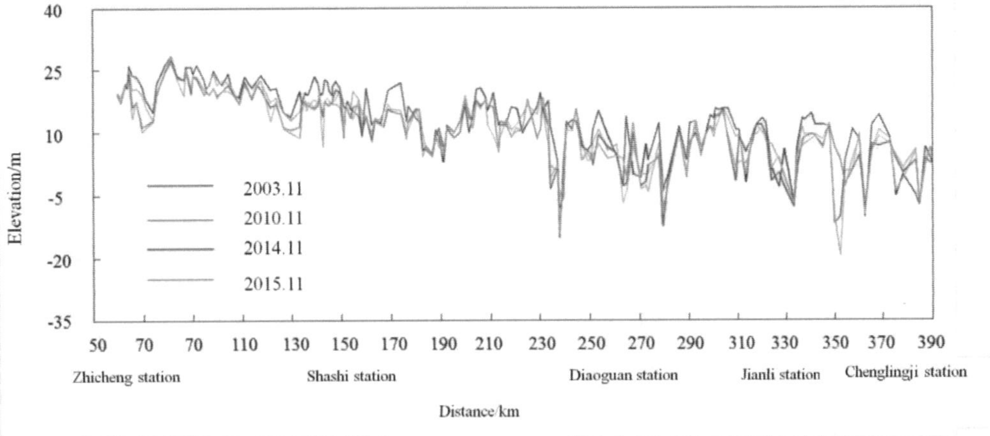

Figure 4. Changes of deep channel profile in the Jingjiang channel

Since the operation of the Three Gorges reservoir, the reach from Chenglingji to Hankou has been scoured, as shown in Figure 5. From November 2003 to November 2015, the longitudinal section of deep elevation in the reach from Chenglingji to Hankou was scoured and silted alternately. Generally, it was mainly scoured and undercut, with an average of 0.75m. Among them, the scour depth of the section, transition and near the bridge was relatively large, such as the maximum scour depth in nanmenzhoutou and Wuhan Yangtze River Bridge, which were 15.1m and 7.2m respectively. There was siltation near the curve section, such as Chibi. The maximum siltation height near the bay is 1.5m and 1.2m respectively, and the other siltation amplitude is relatively small.

In addition, due to the comprehensive influence of the Three Gorges Reservoir and other factors, Chengtong reach is also dominated by scouring undercutting, with an average scouring undercutting of 0.75 m and a maximum scouring depth of 15.4 m, as shown in Figure 6.

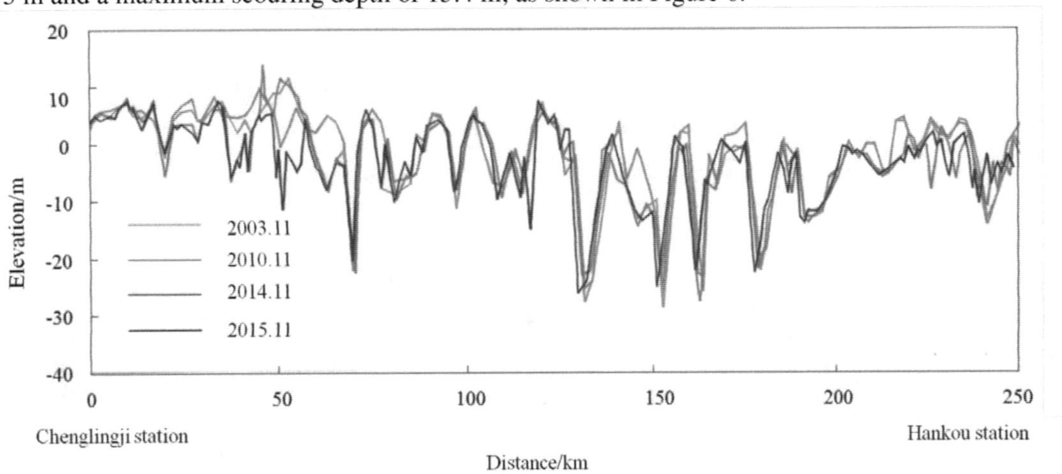

Figure 5. Changes of deep channel profile in the channel from Chenglingji to Hankou

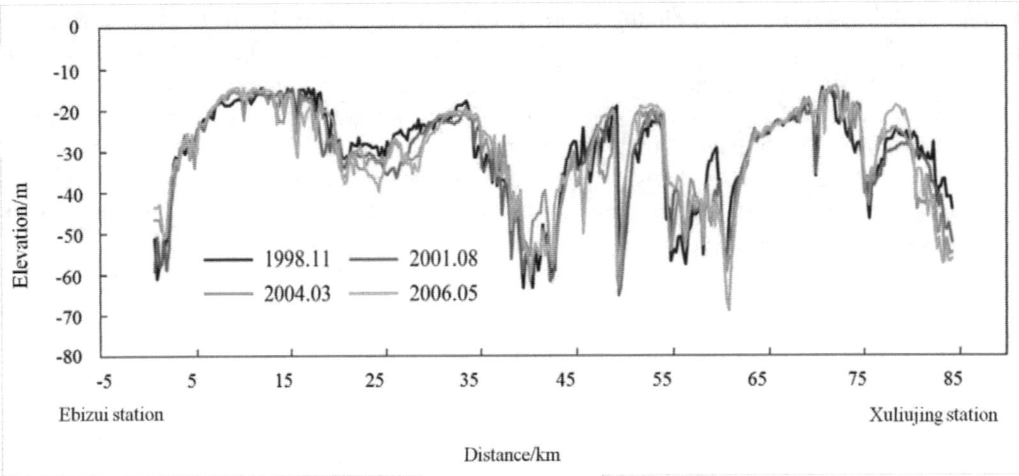

Figure 6. Changes of deep channel profile in the Chengtong channel

4. Changes of erosion and deposition in Dongting Lake and Poyang Lake

4.1Changes of erosion and deposition in Dongting Lake

From 1956 to 2015, the annual sediment inflow and outflow of Dongting Lake showed a downward trend, especially after the application of the Three Gorges Project. Before the operation of the Three Gorges project, the annual average sediment inflow of Dongting Lake was 123.5 million ton per year, and after the operation of the Three Gorges Project from 2003 to 2015, it decreased to 09.6 million ton per year, decreased by 92.2%; before the operation of the Three Gorges project, the annual average sediment inflow of the four waters of Dongting Lake was 28.2 million ton per year, and after the operation of the Three Gorges Project from 2003 to 2015, it decreased to 8 million ton per year, decreased by 71.6%. Therefore, after the Three Gorges project is put into operation, the amount of sediment from three outlets and four waters drops precipitously. Compared with the four waters, the amount of sediment from three outlets has a larger drop. The proportion of sediment from three outlets into the lake has also dropped from 81.4% before the Three Gorges Project was put into operation in 1956-2002 to 54.0% after the Three Gorges project was put into operation in 2003-2015. In general, the annual sediment inflow of Dongting Lake before the operation of the Three Gorges Project from 1956 to 2002 was 151.7 million ton per year, which decreased to 176.6 million ton per year from 2003 to 2015, decreased by 88.4%, and the sediment inflow into the lake decreased significantly, as shown in Table 2.

Table 2. Analysis of sediment inflow and outflow of Dongting Lake in different periods

Period	sediment entering the lake/100 million $t \cdot a^{-1}$				sediment leaving the lake/100 million $t \cdot a^{-1}$	erosion and deposition/100 million $t \cdot a^{-1}$
	Sankou	Sishui	total	proportion/%		
1956—2002	1.235	0.282	1.517	81.4	0.393	1.124
2003—2015	0.096	0.08	0.176	54.0	0.192	-0.016

From 1956 to 2002, the average annual sediment yield of Dongting Lake was 39.3 million ton per year. before the operation of the Three Gorges Project. From 2003 to 2015, the annual sediment yield of Dongting Lake decreased to 19.2 million ton per year, decreased by 51.1%, far less than that of Dongting Lake. The dynamic change of sediment inflow and outflow of Dongting Lake has caused the change of the situation of erosion and deposition of Dongting Lake. Before 2002, the area of Dongting Lake was dominated by sedimentation, with an average annual sedimentation of 112.4 million ton per year. from 2003 to 2015, the area of Dongting Lake showed a trend of erosion, with an average annual erosion of 16 million ton per year.

4.2Changes of erosion and deposition in Poyang Lake

From 1956 to 2002, the average annual inflow and outflow sediment of Poyang Lake were 14.25 million ton per year and 9.91 million ton per year respectively, the inflow sediment was larger than the outflow sediment, and the average annual deposition sediment in the lake area was 4.34 million ton per year; after the operation of the Three Gorges Project from 2003 to 2015, the average annual inflow and outflow sediment of Poyang Lake were 4.97 million ton per year and 12.1 million ton per year respectively, the inflow sediment was smaller than the outflow sediment, and the average annual erosion sediment in the lake area was 7.13 million ton per year. As shown in table 3, which also includes the sand mining volume of river channel.

Table 3. Analysis of sediment inflow and outflow of Poyang Lake in different periods

Period	sediment entering the lake /100 million t•a^{-1}	sediment leaving the lake/100 million t•a^{-1}	erosion and deposition/100 million t•a^{-1}
1956—2002	1425	991	434
2003—2015	497	1210	-713

5. Analysis of the impact of the Three Gorges Project on the water ecosystem in the middle and lower reaches of the Yangtze River

After the operation of the Three Gorges Project from 2003 to 2015, during September to November, the reservoir impoundment, the reservoir discharge decreased, the water level of the main stream and the outlet control stations of the two lakes in the middle and lower reaches decreased, and the low water appeared in advance; during December to may, the Three Gorges project played the role of storing and supplementing the low water, the low water flow of each station in the middle and lower reaches increased, and the water level rose accordingly, and the operation of the Three Gorges project changed the hydrological situation of the middle and lower reaches of the Yangtze River. In addition, the clear water discharge of the Three Gorges reservoir will cause the scour of the river and lake bed, change the water depth and water level of the main stream reach, Dongting Lake and Poyang Lake in the middle and lower reaches of the Yangtze River, and have a certain impact on the habitat and reproduction of aquatic organisms and water birds in the middle and lower reaches of the Yangtze River [7].

The four great fishes in the middle reaches of the Yangtze River have valuable germplasm resources, but since the application of the Three Gorges project, their early resources have declined year by year. Suitable water rising conditions and water temperature are important conditions for the natural reproduction of tetradactylus. Serious gas supersaturation may threaten the survival of tetradactylus eggs and juveniles [8]. The operation of the Three Gorges project makes the flow process in the middle and lower reaches of the Yangtze River tend to be gentle, which changes the water rising conditions on which the four great fishes depend for reproduction. At the same time, the clear water discharged from the Three Gorges Reservoir scours the river and lake bed, reduces the water level, increases the water depth, changes the physical characteristics of the four great fishes spawning ground [9], and has a certain impact on the population activities of the four great fishes.

The Yangtze River finless porpoises generally live in groups near the Yangtze River beach. The riverway and the side beaches of the two lakes in the middle and lower reaches of the Yangtze River have typical habitat characteristics of finless porpoises [10]. However, after the application of the Three Gorges project, the scour balance of the river / lake bed in the middle and lower reaches of the Yangtze River has changed, and the scour effect of the river / lake bed has become prominent, which has changed the habitat environment of finless porpoises to a certain extent, and has a negative impact on the habitat and reproduction of finless porpoises. Influence.

In addition, the survival of waterfowl is also affected by the operation of the Three Gorges Project. There are two main aspects. One is that the operation of the Three Gorges Project causes the migration of rare birds to miss the opportunity of food exposure; the other is that the water surface of the main habitat depression of rare birds is reduced. These two aspects lead to the reduction of food in the habitat, which seriously threatens the food supply of waterfowl and has an important impact on the survival and reproduction of waterfowl. Ring [11].

6. Protection Countermeasures of water ecosystem

After the application of the Three Gorges project, the habitats of aquatic organisms and waterfowls in the middle and lower reaches of the Yangtze River are facing serious threats, so it is urgent to take relevant measures to protect the water ecosystem on which human beings and aquatic organisms depend. This paper mainly puts forward three suggestions, namely, the delimitation of priority protection areas, the spatial management and control of river and lake shorelines, and the ecological regulation.

6.1Delimitation of priority reserves

In view of the habitats of endangered species in the middle and lower reaches of the Yangtze River and the areas where important habitats such as aquatic organisms and waterfowls have changed significantly before and after the Three Gorges project is put into operation, priority protection zones are designated to protect the habitat environment from human activities.

6.2Strengthen the spatial control of river and lake shoreline

Strictly control the water ecological space such as water area and coastline, delimit the management scope of rivers and lakes, water project management scope and protection scope according to laws and regulations, and clarify the management responsibilities. We will implement zoning management and use control of the water coastline, strengthen the protection, conservation and intensive use of the coastline, and scientifically divide the water coastline protection areas, reserved areas, restricted development areas and development and utilization areas. In view of the shoreline of the waters around the important habitat of aquatic organisms and waterfowls, it is included in the scope of water shoreline protection zone, the construction project permission is strictly controlled, and it is strictly prohibited to set up sewage outlets, and to some extent, the restoration of the aquatic habitat under the influence of the Three Gorges project operation is carried out to ensure the effective protection of the important water ecosystem in the middle and lower reaches of the Yangtze River.

6.3Implement ecological scheduling

The change of hydrological process in the middle and lower reaches of the Yangtze River under the operation of the Three Gorges project has an important impact on the spawning and reproduction of the four great fishes. Therefore, in order to simulate the natural flow in the river, it is necessary to put forward the ecological regulation scheme of the three Gorges reservoir which is conducive to the spawning and reproduction of the four great fishes, so as to ensure that the water discharge conditions of the reservoir meet the spawning needs of the four great fishes.

7. Conclusion

Before and after the operation of the Three Gorges project, the trend of annual runoff in the middle and lower reaches of the Yangtze River is not significant. However, the operation of the Three Gorges project has a greater impact on the annual distribution of runoff in the middle and lower reaches of the Yangtze River. In addition, after the operation of the Three Gorges project, the sediment discharge of the river section decreased in a cliff like manner. The clear water discharge of the Three Gorges project has changed the scour and deposition situation of the main stream and the two lakes, making them change from siltation to scour direction, causing the water level of the rivers and lakes to drop. Among them, the scour and deposition situation of different river sections varies greatly due to the influence of local river situation, distance from the Three Gorges project, human activities and other factors.

The changes of scour and deposition conditions, flow velocity, water depth and water level caused by the operation of the Three Gorges project have a certain impact on the survival and reproduction of the four great fishes, finless porpoises and waterfowls. In order to protect the fragile water ecosystem from the adverse effects of the operation of the Three Gorges project, it is urgent to define priority protection areas, implement spatial control and ecological scheduling of river and lake shorelines.

References

[1] Chen Y., Liu T., Huang Y., et al. (2014) Remote sensing research of wetland current status and change in the Yangtze River basin. Resources and Environment in the Yangtze Basin, 23(6):801-808.
[2] Yu W., Zhou G., Chi D., et al. (2007) Advance in Wetland Ecohydrological Process Research. Water Saving Irrigation, (1)19-23.
[3] Zhang Li., Yin J., Jiang Y., et al. (2012) Relationship between hydrological conditions and vegetation communities in Poyang Lake national nature reserve of China. Advances in Water Science, 23(6):768-775.
[4] Zhu L., Chen J., Yuan J., et al. (2014) Sediment erosion and deposition in two lakes connected with the middle Yangtze River and impact of Three Gorges Reservoir. Advances in Water Science, 25(3):348-357.
[5] Xie P. (2017) Biodiversity crisis in the Yangtze River: The culprit was dams, followed by overfishing. Journal of Lake Sciences, 29(6):1279-1299.
[6] 6 Tullos D. (2009) Assessing the influence of environmental impact assessments on science and policy An analysis of the Three Gorges Project. Journal of Environmental Management, 90:208-223.
[7] Zhou J., Zhao Y., Song L., et al. (2014) Assessing the effect of the Three Gorges reservoir impoundment on spawning habitat suitability of Chinese sturgeon (Acipenser sinensis) in Yangtze River, China. Ecological Informatics, 20(3):33-46.
[8] Peng Q., Liao W., Li C., et al. (2012) Impacts of four major Chinese Carps' natural reproduction in the middle reaches of Changjiang River by Three Gorges Project since the impoundment. Advanced Engineering Sciences, (s2):228-232.
[9] Huang Y., Fan B. (2008) Impact of TGP on spawning environment of 4 major Chinese fishs in middle and downstream of the Yangtze River. Yangtze River, 39(19):38-41.
[10] Wei Z., Wang D., Zhang X., et al. (2002) Population size, behavior, movement pattern and protection of Yangtze finless porpoise at Balijiang section of the Yangtze River. Resources and Environment in the Yangtze Basin, 11(5):427-432.
[11] Xu J., Chen J. (2013) Study on the impact of Three Gorges reservoir on Poyang Lake and some proposals. Journal of Hydraulic Engineering, 44(7):757-763.

IOP Conf. Series: Materials Science and Engineering **794** (2020) 012042 doi:10.1088/1757-899X/794/1/012042

Research on key technologies for construction of super-deep railways stations near existing lines in the flood land area of the Yangtze River

Yang Qiongpeng[1], Ma Bin[2], Li Wei[3], Sun Bin[4*]

China Railway Tunnel Stock Co. LTD, Zhengzhou, Henan, 451470, China

China Railway Tunnel Stock Co. LTD, Zhengzhou, Henan, 451470, China

China Railway Tunnel Stock Co. LTD, Zhengzhou, Henan, 451470, China

Geotechnical Engineering Institute, Hohai University, Nanjing, Jinagsu, 210024, China

377749875@qq.com, 190958908@qq.com, 377749875@qq.com, *191304010069@hhu.edu.cn,

Abstract. In this paper, taking Nanjing Mochou Lake Metro Station as an example, the key technologies for the construction of super-deep stations near existing lines in the flood land area of the Yangtze River are studied. Mochou Lake Station is located in Mochou Lake Park on the Qinhuai River. It is a four-level underground island-type platform station, which is interchanged with the existing Line 2 Mochou Lake Station. The station used the "two-light and two-dark" construction method. The first and the second underground floor of the station used an open excavation order method, and the third and the fourth underground floor used to cover and cut-bottom up method. According to the monitoring results and settlement analysis results of the main structure, the construction method of "two- light and two-dark " can help control the deformation of the construction area and ensure the safety of the surrounding structures and existing stations.

1. Introduction

Mochou Lake Station is located in Mochou Lake Park, Jianye District, Nanjing. The excavation depth of the foundation pit is 35.26m. This station is a four-level underground island-type platform station, which is interchanged with the existing Line 2 Mochou Lake Station. The site liquefaction index is 0.45 to 1.12, which is a slightly liquefied site. The buried depth of the soft cohesive soil layer is shallower than that of the base of the main structure of the station. During the construction, the soil stress will be unbalanced, which will adversely affect the construction of the project. The trough wall of the station envelope is reinforced by TRD (trench cutting re-mixing deep wall) construction method of cement-soil mixing wall, 45m deep and 600mm wide [1]. In the range of 3m below the second floor of the underground floor and 5m below the foundation pit bottom, the RJP (rodin jet pile) construction method is used to spin the pile skirt drawbar reinforcement [2]. The ground connection wall has a depth of 64m (3-6m into the weathered rock) and the width of the wall is 1.2m, with a milling joint used. A blocking wall is set in the middle of the pit to divide the foundation pit into two small foundation pits with a length of 100m and 60m respectively [3]; the standard section support form is the first, fifth, and sixth concrete supports and the second, third, and fourth steel supports[4].

Wells precipitation pipe is used in the foundation pit, and observation wells are located outside the foundation pit.

2. Construction of the main structure of the station "two-bright and two-dark" [5]

- Excavate the soil to the first supporting bottom, and pour the first concrete support.
- Excavate to 0.5m below the steel support, install the steel enclosing purlin and set up the steel support.
- Open 0.5m below the 4th support, install steel enclosing purlin and set up the 4th steel support.
- Excavate the soil to 1.8m below the negative second-floor slab, construct the negative second-floor cushion, erection of short scaffolding, pouring the floor and 1.8m side wall under the pouring floor.
- After the bottom plate of the negative second floor reaches the design strength, remove the 4th steel support, and make the lower half of the negative second floor sidewall in the open excavation.
- After the lower sidewall of the negative second layer reaches the design strength, remove the third steel support and steel enclosing purlin, and make the upper part of the negative second layer sidewall, column, and plate in clear cut.
- After the superstructure of the negative second layer reaches the design strength, remove the 2nd steel support, enclosing purlin, and construct the side wall and column of the negative first layer. Set up the scaffold system, tie up the roof reinforcement, set up the formwork, pouring concrete.
- After the roof reaches the designed strength, cut the first concrete support, construct the excavated retaining wall, and partially backfill the roof.
- After the roof is backfilled, continue to lower the excavator from the bottom plate of the negative second floor to excavate the soil of the sixth floor and construct the fifth concrete support.
- After the 5th concrete support reaches the designed strength, the 7th layer of soil shall be excavated to 1.8m below the negative three-layer bottom plate, and the construction cushion shall be constructed, and the plate-hook scaffold shall be constructed, and the negative three-layer bottom plate and 1.8m side wall shall be constructed.
- After the bottom plate of the negative three layers reaches the designed strength, the eighth layer of soil is excavated and the sixth concrete support is constructed.
- After the 6th concrete support reaches the designed strength, continue to dig the 9th-floor earthwork, and construct the grounding net, plain concrete cushion, waterproof layer, and negative four-layer bottom plate.
- After the negative fourth-floor reaches the designed strength, remove the sixth concrete support and construct the negative fourth floor sidewall.
- After the fifth support is removed, the negative three-storey and the lower half of the side wall shall be constructed.

3. Monitor measurement analysis

3.1 monitoring scope and level

In view of the geological conditions of the newly built Mochou Lake Station and the surrounding complicated structures, the excavation depth of 3 times the foundation pit is taken as the scope of influence of this station. According to the relevant specifications, the risk level of the foundation pit project itself is level one, the surrounding environment risk level is level two, and the project monitoring level is comprehensively determined as level one.

3.2 Analysis of settlement data during construction

(1) Analysis of vertical settlement and horizontal displacement of the pile top of the retaining structure

A total of 20 settlement points and 20 horizontal displacement points were arranged for the top of the wall. During the whole construction period, the settlement change, horizontal displacement and changing trend of the wall roof are relatively small.

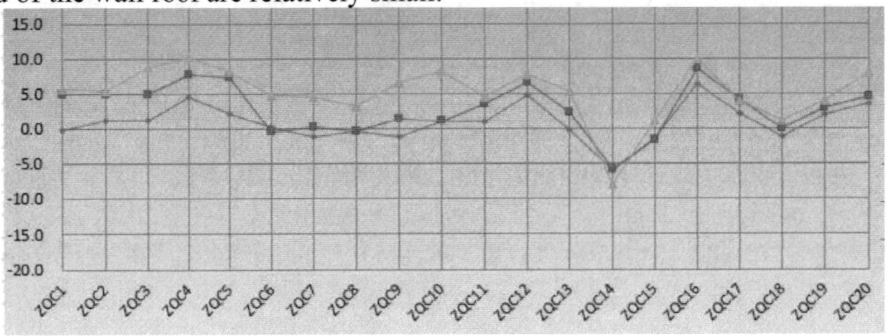

Figure 1. Wall top settlement curve

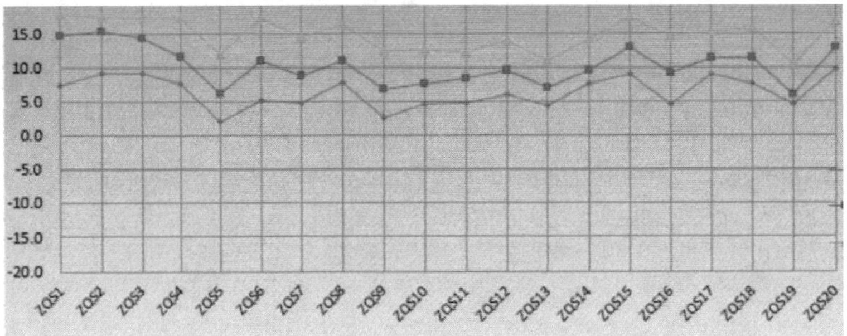

Figure 2. Curve of horizontal displacement of wall top

(2) Analysis of vertical displacement of column pile

A total of 8 column settlement points were established. During the entire construction period, the uplift of the column is relatively small and does not exceed 14mm.

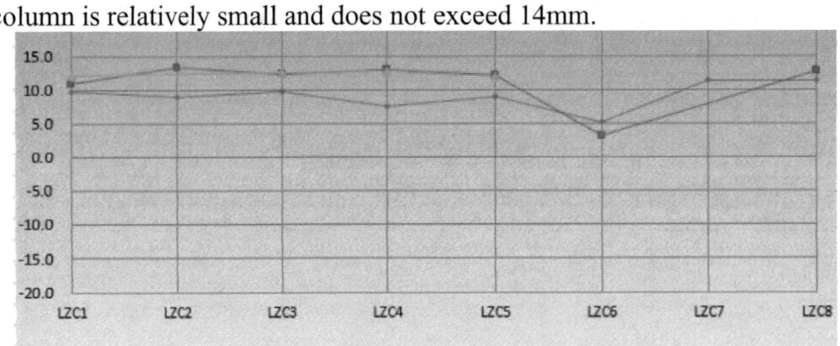

Figure 3. Change curve of column settlement

(3) Analysis of ground settlement around foundation pit

The soil at Mochou Lake Station is relatively loose. Affected by construction during the construction process, especially the construction of heavy-duty machinery and equipment, the surface settlement of the site originally belonged to the bush greening area was relatively large, and the surface settlement of the existing Line 2 farther away was relatively small. (from August 9 to December 4)

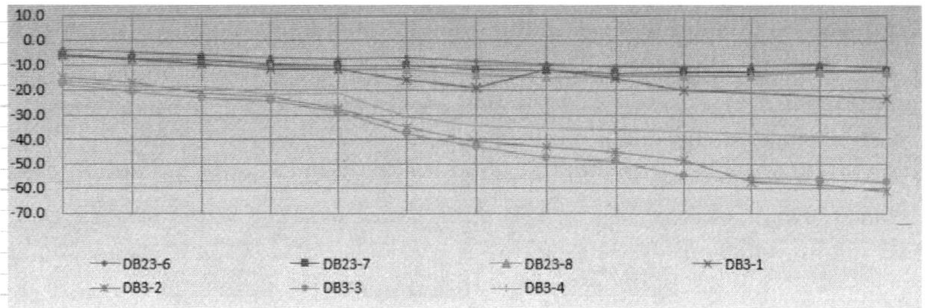

Figure 4. Change curve of surface subsidence

(4) Settlement analysis of surrounding buildings (structures)

The existing buildings (structures) within 3 times the excavation depth of the foundation pit are monitored, and the settlement speed rate of surrounding buildings are small. (from July 7 to December 4)

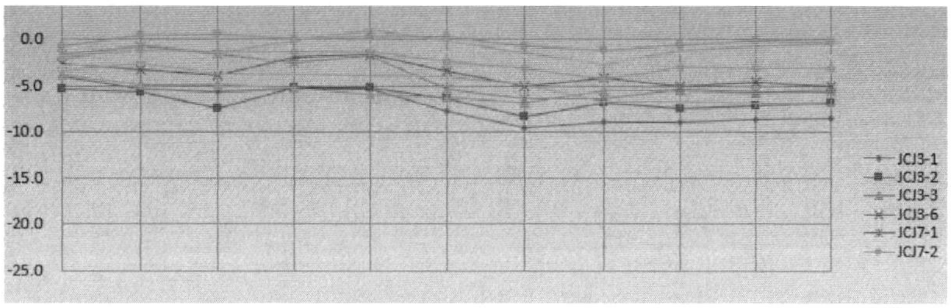

Figure 5.House settlement curve

3.3 Settlement analysis of existing stations

The affected area (length 370m) was protected and monitored. The monitoring data show that the tunnel settlement is -14.2mm. The maximum monthly change rate of the tunnel settlement is Y27: -0.119mm. After on-site inspections, no new cracking or collapse has been found. The existing structure of Line 2 has a small impact.

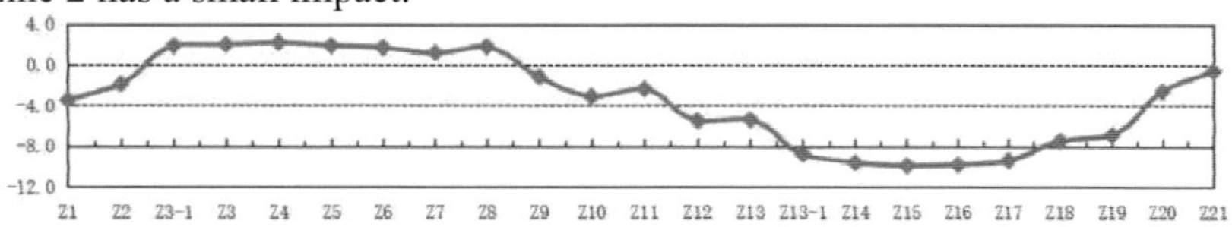

Figure 6. Vertical displacement change map of the downward line automatic track bed

4. Conclusions

The research work presented in this paper analysed the interaction between soil and the new composite retaining structure and recommendations are obtained.

- Abundant groundwater, weak strata and prone to subsidence in the flood land areas. The "two-bright and two-dark" construction method helps control the deformation of the foundation pit, effectively controls the settlement of surrounding structures and ensures the safety of the foundation pit.
- The application of the "two-bright and two-dark" construction method saves materials and construction time, reduces construction safety risks, and has achieved good technical results and economic benefits.
- During the construction of the envelope structure, the construction quality of the diaphragm wall is strictly controlled to ensure the construction quality of the envelope structure. The continuity and uniformity of the TRD cement-soil mixing wall reduce the risk of water leakage from the wall joints.

- The underground continuous wall into the rock effectively cuts off the connection between the inside and outside of the foundation pit. The foundation pit uses well pipe precipitation to carry out precipitation according to construction requirements, and each time the water level is lowered to 1-1.5 meters below the excavation surface.
- During the excavation of the foundation pit, the erection of the support in time, and the tunnelling will follow the erection. If the erection of the support is not timely or over-excavation, it may easily cause the foundation pit to deform and settle, and cause leakage in the wall joints.

References

[1] Wang Weidong, Di Guoen. Technology and engineering practice of cement-soil mixing wall with TRD method and other thicknesses [J]. Chinese Journal of Geotechnical Engineering, 2012, 34 (S1): 628-633.

[2] Hu Xiaohu, Chuan Tianchong, Zhongxi Kangqing, Li Jinjun. RJP high-pressure rotary spraying method and its application in Japan [J]. Chinese Journal of Geotechnical Engineering, 2010, 32 (S2): 410-413.

[3] Lin Feng. Optimization of deep foundation pit design scheme under complicated environmental conditions [J]. Chinese Journal of Geotechnical Engineering, 2014, 36 (S1): 39-45.

[4] Wang Wang. Research on stability evaluation and support technology of foundation pit at Yuanjialing Station of Changsha Metro Line 2 [D]. Central South University, 2012.

[5] Yu Fucai, Zhang Dingli, Mu Weigang, Wang Jianchen. Research on optimization of excavation scheme at the intersection of open excavation and underground excavation in subway station [J]. Journal of Beijing Jiaotong University, 2015, 39 (06): 69-74.

Seismic liquefaction evaluation of deposit soils under engineering loading or unloading conditions

YushengYang[1,2*], Xiaosheng Liu[1,2], Jianming Zhao[1,2], Jie Zhou[1], Bing He[1,2], LongWang[1,2]

[1]State Key Laboratory of Simulation and Regulation of Water Cycle in River Basin, China Institute of Water Resources and Hydropower Research, Beijing, 100048, China

[2]Key Laboratory on construction and security of water projects of Ministry of Water Resources, Beijing, 100048, China

*Corresponding author's e-mail: yangysh@iwhr.com

Abstract. Engineering activities such as water conservancy and hydropower, water transportation and harbor and navigation often lead to loading or unloading conditions, which makes the stress state of overlay different before and after construction. Relevant codes in the field of water conservancy and hydropower projects recommend the linearization correction method based on stress equivalence expressed by buried depth, considering the influence of stress state of overlying soil on its liquefaction resistance under loading and unloading conditions. On the basis of demonstration and analysis, this paper points out that the linearization correction method based on stress equivalence expressed by buried depth will overestimate the anti-liquefaction ability of soil under engineering loading conditions, and the liquefied layer may be evaluated as non-liquefaction, which brings potential risks to the project. Under unloading conditions, the anti-liquefaction capacity of soil will be underestimated, and the non-liquefied layer may be evaluated as liquefaction, which will lead to the increase of foundation treatment cost. Aiming at the defect of linearization correction method based on stress equivalence expressed by buried depth, a new non-linearization correction method is proposed, which takes into account the influence of the stress state of overburden soil caused by loading and unloading of dams and other projects on its liquefaction resistance. This method can effectively avoid the potential danger or waste of investment caused by the miscalculation of the original formula.

1. Introduction

In the construction of water conservancy and hydropower projects, water transport projects, port and harbor projects, and civil and industrial construction projects (such as airports) on river alluvial plains, deep cover problems are often encountered. For the overburden foundation, water conservancy and hydropower engineering, water transport engineering and port and harbor engineering construction activities often lead to the emergence of loading or unloading conditions, making the stress conditions of the overburden layer before and after the construction different, that is, the stress conditions of the overburden foundation in the survey and design stage of the project are different from the stage under the application conditions after the completion of the project. For example, in the engineering construction of water circulation control and hydropower resources development upper damming basin of the overburden, it is often encountered overburden soil loading caused by high filling or overburden soil unloading caused by deep excavation. For the water-retaining construction, the groundwater level near the project site will also increase after the construction of the project. Before and after the construction of the project, the ground elevation and groundwater level change greatly, resulting in changes in the stress conditions of the corresponding parts of the overburden. Similarly, in the construction of airports on river alluvial plains, engineering loading and unloading conditions of the large-scale excavation and filling often take place. When the basic seismic intensity on the project site is high and the soil in the overburden is sensitive to seismic loads, the seismic liquefaction stability of the overburden is a concern of the project construction. This requires evaluation on the seismic liquefaction stability of the overburden foundation in the normal operation phase after the construction during the survey and design stage of the project. That is, the seismic liquefaction evaluation is carried out under the condition that the stress conditions of the soil in the overburden layer are changed due to engineering loading and unloading of the damming and other engineering.

Aiming at the situation that the stress conditions of the overburden foundation are often changed due to loading or unloading caused by engineering activities in the construction of water conservancy and hydropower projects, Mr. Wang Wenshao recommended the formula (1) to collect the actually measured standard penetration test (SPT) blow count in the Anti-seismic Design Code of Hydraulic Structures (SDJ10-78). The corrected SPT blow count was compared with the liquefaction critical SPT blow count under the operating conditions after the completion of the project, to determine the possibility of liquefaction of the original standard penetration test point under the operating conditions after the completion of the project. Therefore, whether the correction of the measured SPT blow count could better reflect the influence of the stress condition on the SPT blow count is directly related to engineering safety.

In the case of setting a standardized overburden homogeneous sand layer model (the groundwater level h_w was 2m, the natural bulk density of the sand layer was 18kN/m^3, and the saturated bulk density was 19kN/m^3.), formula (1) can be transformed into the form shown as formula (2). The current code (GB 50487-2008) stipulates that when the depth of the penetration point in the standard penetration test and the groundwater level are below depth of the test ground, different from the operation after the completion of the project, the actually measured SPT blow count shall be corrected in accordance with formula (3), and the corrected SPT blow N should be used as the basis of reconsideration.

$$N = N'\left(\frac{p/p_a + 0.7}{p'/p_a + 0.7}\right) \tag{1}$$

$$N = N'\left(\frac{d_s + d_w + 7.8}{d_s' + d_w' + 7.8}\right) \tag{2}$$

$$N = N'\left(\frac{d_s + 0.9d_w + 0.7}{d_s' + 0.9d_w' + 0.7}\right) \tag{3}$$

where: N - the corrected SPT blow count, which represents SPT blow count of the original test point when the groundwater level and ground elevation were changed after the completion of the project; N' - the SPT blow count of the standard penetration test point before the construction; p - the effective overlying stress of the original standard penetration test point under the change of groundwater level and ground elevation after the completion of the project, kPa; p' - the effective overlying stress of the standard penetration test point before the construction of the project, kPa; d_s, d_s' - the depths (m) of the standard penetration point below the ground at the time of operation and standard penetration test after the completion of the project, respectively ; d_w, d_w' - the depth (m) of the groundwater level below the ground at the time of operation and standard penetration test after the completion of the project (m), respectively, and when the ground is submerged below the water surface, the value is 0 .

Formula (1) reflects the effect of stress conditions on the SPT blow count. From the evolution of the standard formula (1) to formula (3), it can be seen that the formula (3) still substantially reflects the influence of the stress condition on the SPT blow count. In practical engineering applications, it is found that when engineering loading taking place in engineering activities leads to the increase of the overburden effective stress on the overburden soil, the SPT blow count obtained by the formula (3) seems to be too large. The liquefaction occurs when the liquefaction possibility is judged according to the standard penetration test point stress condition in the standard penetration test, and the liquefaction no longer takes place after the engineering loading happens in the engineering activities, which leads to the increase of the overburden effective stress on the overburden layer. When engineering loading happened in engineering activities leads to the decrease of the overburden effective stress on the overburden soil, the SPT blow count obtained by the formula (3) seems to be too small. The liquefaction does not take place when the liquefaction possibility is judged according to the standard penetration test point stress condition in the standard penetration test, and the liquefaction occurs after the engineering loading happens in the engineering activities, which leads to the decrease of the overburden effective stress on the overburden layer. Therefore, it is necessary to examine the applicability of formula (3) under the conditions of engineering loading and unloading.

This paper systematically summarizes the domestic and overseas research results of the influence of stress conditions and compact state on SPT blow count, and based on the SPT in the seismic liquefaction field and the test results of the indoor full-scale standard penetration model, the effects of stress conditions and compact state on the SPT blow count are investigated. Besides, aiming at the loading and unloading conditions (working conditions) encountered in engineering, the linearization correction method based on stress equivalence expressed by buried depth in the original specification is demonstrated. Under the engineering loading conditions, the soil liquefaction resistance is overestimated, and the liquefaction layer is likely to be judged as non-liquified, which will bring potential risks to the project. Under the condition of engineering unloading, the anti-liquefaction ability of the soil will be underestimated, and the non-liquefaction layer may be judged as liquefied, resulting in an increase in the cost of foundation treatment. In view of the defects of the method in the original specification, on the basis of the argumentation analysis, a new nonlinear correction method is proposed to consider the influence of the change of the overburden solid stress conditions on the liquefaction resistance of the soil due to the loading and unloading of the dam.

2. Influence of overburden effective stress and relative density on SPT blow count

In practical engineering applications, the measured SPT blow count under different overburden effective stress in the overburden foundation soil is normalized to the SPT blow count under the effective stress of 100 kPa, which is denoted by the formula (4).

$$(N_1)_{60} = C_N N' \tag{4}$$

where: $(N_1)_{60}$ - SPT blow count at 100 kPa; C_N - normalization coefficient of overburden effective stress.

On the basis of the indoor full-scale standard penetration model or based on the SPT on the overburden site, different researchers proposed normalized coefficient correction formulas of different overburden effective stress. The article [1] first established the relationship between the mechanical index of overburden sand soil, SPT blow count, and the sand compaction state index, the relative density. Since then, the article [2] and [3] carried out indoor full-scale penetration tests on different sands with different stress conditions and at different density states. The article [4] performed on-field penetration tests on the fine sand packing layer on the island, and the article [5] carried out on-field penetration tests on the seismic sand foundation of Niigata in Japan in 1964. The results of the indoor full-scale penetration model test and the field penetration tests further proved that the penetration blow count is closely related to the relative density and stress conditions: When the stress conditions of the overburden sand are the same, the penetration blow count and the relative density are positively correlated; When the physical state (relative density) of the overburden sand is the same, the penetration blow count and the overburden effective stress are positively correlated. According to the results of overburden sand penetration tests under different relative density and different overburden effective stress, the article [6] firstly proposed the relationship between the penetration blow count, the relative density and the stress condition as the formula (5):

On the basis of the indoor full-scale standard penetration model test or the SPT in the overburden site, different researchers proposed different normalization coefficient correction formulas for overburden effective stress.

$$\frac{N^{'}}{D_{\mathrm{r}}^{2}}=a+b\frac{\sigma_{\mathrm{v0}}^{'}}{p_{\mathrm{a}}} \tag{5}$$

where: N_{m}—measured penetration blow count; D_{r}—relative density; $\sigma_{\mathrm{v0}}^{'}$—the overburden effective stress of the penetration point when the penetration test is performed, kPa; p_{a}—atmospheric pressure; a and b—test constants, both increase with increasing particle size, deposition age, and over consolidation ratio. Wherein, the relative density D_{r} is between 0.35 and 0.85 and the overburden effective stress $\sigma_{\mathrm{v0}}^{'}$ is between 50 kPa and 250 kPa.

2.1Influence of overburden effective stress

For the sand layer of the same homogeneous overburden foundation, the relative density is the same. From formula (5), the expression of the SPT blow count after correction in the case of engineering loading or unloading can be derived.

$$N=\frac{a/b+\sigma_{\mathrm{v1}}^{'}/p_{\mathrm{a}}}{a/b+\sigma_{\mathrm{v0}}^{'}/p_{\mathrm{a}}}N^{'} \tag{6}$$

where: $\sigma_{\mathrm{v1}}^{'}$ - the effective overlying stress of the original standard penetration test point under the conditions of engineering loading or engineering unloading, kPa.

Let $\sigma_{\mathrm{v1}}^{'}$=100 kPa in the formula (6), then the formula (6) is converted into the same form as the formula (4).

$$\left(N_{1}\right)_{60}=\frac{a/b+1}{a/b+\sigma_{\mathrm{v0}}^{'}/p_{\mathrm{a}}}N^{'} \tag{7}$$

That is

$$C_{N}=\frac{a/b+1}{a/b+\sigma_{\mathrm{v0}}^{'}/p_{\mathrm{a}}} \tag{8}$$

Meyerh of (1957) [6] suggested a/b≈1.71 based on test data, Tokimastsu et al. (1983) [7] suggested a/b=0.7, which is in the range of a/b=0.6-0.8 for over consolidated sand. Skempton (1986) [9] suggested that a/b = 1.0 for medium-density fine sand in normal consolidation state, a/b=2.0 for compact coarse sand in normal consolidation state, and a/b = 0.6~0.8 for over consolidated soil. Kayen et al. (1992) [12] suggested a/b=1.2 for medium-density sand, which was recommended to be used in the NCEER method (improved Seed method) by Youd et al. (2001) [11] to determine the seismic liquefaction of sand and calculate C_{N}.

Peck et al. (1974) [7], Seed (1976), and Seed (1983) [8] respectively recommended to utilize the formula (9), formula (10), and formula (11), to calculate the overburden effective stress normalization coefficient C_{N}.

$$C_N = 0.77\lg\left(20 \cdot \frac{p_a}{\sigma'_{v0}}\right) \tag{9}$$

$$C_N = 1 - 1.25\lg\frac{\sigma'_{v0}}{p_a} \tag{10}$$

$$C_N = \left(\frac{p_a}{\sigma'_{v0}}\right)^{1/2} \tag{11}$$

Under the condition that the overburden soil is homogeneous sand, combined with the formula (4), the formula (12) to (14) can be derived from the formula (9) to (11).

$$N = \frac{\lg\left(20 \cdot p_a / \sigma'_{v0}\right)}{\lg\left(20 \cdot p_a / \sigma'_{v1}\right)} \cdot N' \tag{12}$$

$$N = \frac{1 - 1.25\lg\frac{\sigma'_{v0}}{p_a}}{1 - 1.25\lg\frac{\sigma'_{v1}}{p_a}} N' \tag{13}$$

$$N = \left(\frac{\sigma'_{v1}}{\sigma'_{v0}}\right)^{1/2} \cdot N' \tag{14}$$

2.2Influence of sand relative density

Seed (1983) [8] proposed to use the overburden effective stress correction as shown in Table 1, on the basis of the data analysis by Bieganousky et al. (1976) [3] of sand penetration test under different stress conditions and different compact states.

Table 1.Cn of normally consolidated sand recommended by seed et al.

σ'_{v0} /kPa	50	75	100	150	200	250	300
Dr=40%~60%	1.36	1.14	1	0.8	0.69	0.6	0.54
Dr=60%~80%	1.36	1.14	1	0.84	0.74	0.67	0.61

Skempton (1987) reanalyzed the experimental data on the basis of formula (9) and proposed a formula (15) that is of more general meaning than formula (11).

$$C_N = \left(p_a / \sigma'_{v0}\right)^m \tag{15}$$

where m is a function of relative density. Boulanger (2003) [13] analyzed the SPT data of Marcuson et al. (1977) [14, 15] and found that m has a linear relationship with relative density D_r.

$$m = 0.784 - 0.521D_r \tag{16}$$

Combined with formula (4), formula (16) can be derived.

$$N = \left(\frac{\sigma'_{v1}}{\sigma'_{v0}}\right)^m \cdot N' \tag{17}$$

When the relative density $D_r = 54\%$, $m = 0.5$, and then formula (17) is transformed to formula (14).

According to the analysis, under the conditions of engineering loading and unloading, the corrections of the original SPT blow count are linear by formula (1), (2), (3), (6), (14) and (15).

2.3Comparison of the effects of overburden effective stress and relative density on SPT blow count
In order to compare the different formulas describing the relationship between the overburden effective stress and the relative density, a homogeneous sand layer model for standardized overburden foundation was set, with a thickness of 40 m and a groundwater level of 2.0 m. The natural bulk density and saturated bulk density of the sand layer above the groundwater level and below the groundwater level were 18kN/m^3 and 19kN/m^3, respectively, while the SPT blow count at 100kPa was 24.5 hits. According to the formulas recommended by different scholars, the SPT blow count at 100 kPa can be converted into the corresponding SPT blow count under the conditions of different depth and overburden effective stress, as shown in Table 2, and N~σ_{v0}' curve as shown in Fig. 1 can be obtained.

Table 2.Blow counts of SPT under different effective stress conditions calculated by different formulas.

σ_{10}'	Formula (6)				Formula(12)	Formula(13)	Seed(1983)		Formula (17)					
	a/b						Dr		Dr					
	0.7	1	1.2	2			40-60%	60-80%	30%	45%	55%	65%	75%	85%
25	13.7	15.3	16.1	18.4	16.7	14.0			10.2	11.4	12.3	13.2	14.2	15.3
50	17.3	18.4	18.9	20.4	19.9	17.8	18	18	15.8	16.7	17.3	18	18.7	19.4
75	20.9	21.4	21.7	22.5	22.3	21.2	21.5	21.5	20.4	20.9	21.2	21.6	21.9	22.2
100	24.5	24.5	24.5	24.5	24.5	24.5	24.5	24.5	24.5	24.5	24.5	24.5	24.5	24.5
125	28.1	27.6	27.3	26.5	26.4	27.9			28.2	27.7	27.4	27.1	26.7	26.4
150	31.7	30.6	30.1	28.6	28.3	31.4	30.6	29.2	31.6	30.6	30	29.3	28.7	28.1
175	35.3	33.7	32.9	30.6	30.1	35.2			34.9	33.3	32.4	31.4	30.5	29.6
200	38.9	36.8	35.6	32.7	31.8	39.3	35.5	33.1	37.9	35.9	34.6	33.4	32.2	31
225	42.5	39.8	38.4	34.7	33.5	43.8			40.8	38.3	36.8	35.2	33.7	32.3
250	46.1	42.9	41.2	36.8	35.2	48.7	40.8	36.6	43.6	40.5	38.7	36.8	35.1	33.5
275	49.7	45.9	44.0	38.8	36.9	54.3			46.3	42.7	40.6	38.4	36.5	34.6
300	53.3	49.0	46.8	40.8	38.6	60.7	45.4	40.2	48.9	44.8	42.4	40.0	37.7	35.6

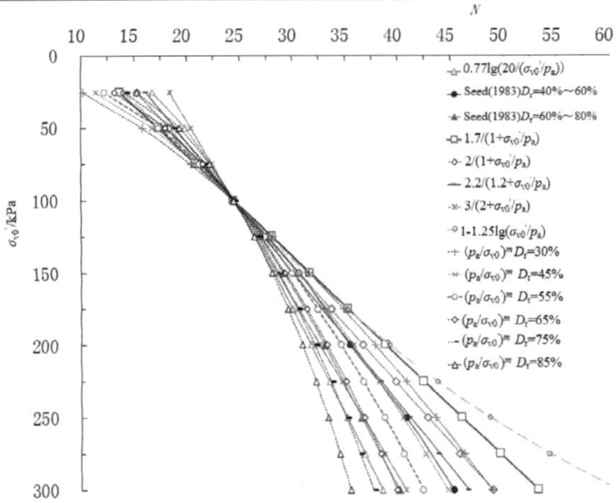

Figure 1. Standard penetration number under different overlying effective stress conditions.

It can be seen from the analysis of Table 2 and Fig. 1 that, in the common overburden effective stress range (50~300kPa) in the seismic liquefaction case of sand layer: The formulas that do not consider the influence of relative density are only applicable to sands at specific states such as loose, medium-density or compact. At a certain relative density of the sand, the N~$\sigma_{v0}^{'}$ curve calculated by the formula (17) is close to the relationship curve between the SPT blow count and the overburden effective stress calculated by the formulas recommended by different scholars. If the curve is calculated by the formula (12) and the relative density in the formula (17) is 75%, the calculated curves are close when the overburden effective stress is greater than 100 kPa. In formula (6), if the value of a/b is 0.7, 1.0, 1.2, and 2.0, respectively, the calculated curves are close to those when the formula (17) take relative density of 30%, 35~40%, and 45%,65%, respectively. The calculated curves when the correction results of the relative density recommended by Seed are 40%-60% and 60%-80%, are close to those when the relative density in formula (17) is 45% and 65%. Therefore, the formula (17) established on the basis of different overburden effective stress and the standard penetration test data of the sands at different relative density, can better describe the influences of relative density and stress conditions on the SPT blow count, and is suitable for sands at different density states.

It can be seen from Table 2 and Fig. 1 that, in general, when the overburden effective stress of the overburden soil caused by the engineering loading happened in the engineering activities increases, the denser the sands, the bigger the SPT blow count obtained by different formulas corrected to the same overburden effective stress. When the overburden effective stress of the overburden soil caused by the engineering loading happened in the engineering activities decreases, the denser the sand, the smaller the SPT blow count obtained by different formulas corrected to the same overburden effective stress. In same kind of formulas, the change trends of correction to the two directions are reversed. However, in practical engineering, it is difficult for engineering design and geological survey workers to choose different formulas to correct according to the actual state of the sand layer. In order to facilitate the application, it is necessary to combine the actual conditions of the project construction to investigate the rationality of the proposed project and propose reasonable correction formulas based on the argument.

3. Comparison of seismic liquefaction evaluation methods for overburden under the conditions of engineering loading and unloading

The ratio of the corrected SPT blow count N corresponding to the overburden effective stress on the standard penetration test point in the engineering application to the actually measured SPT blow count $N^{'}$ at corresponding test point, is defined as the correction coefficient $C_N^{'}$ of the overburden effective stress. The corrected SPT blow count N corresponding to the overburden effective stress in the operation conditions after the completion of the project can be expressed as:

$$N = C_N^{'} N^{'} \tag{18}$$

As can be seen in the formula (18), N is proportional to $C_N^{'}$, and the larger $C_N^{'}$ is, the larger N is. When the corrected N is used to evaluate the liquefaction of the overburden sand foundation, the liquefaction evaluation result tends to be insecure. On the contrary, it tends to be secure.

According to the standard formula (3), it can be obtained:

$$C_N^{'} = \left(d_s + 0.9d_w + 0.7\right) / \left(d_s^{'} + 0.9d_w^{'} + 0.7\right) \tag{19}$$

It can be obtained from formula (17)

$$C_N^{'} = \left(\sigma_{v1}^{'} / \sigma_{v0}^{'}\right)^m \tag{20}$$

It can be seen from the foregoing discussion that the formula (17) can better describe the relationship between the SPT blow count and the relative density and the overburden effective stress, with a wide application scope. Therefore, this section compares formula (20) with formula (19) to demonstrate the applicability of the standard formula.

The above-mentioned assumed homogeneous sand layer is still used as the analysis object, and the variation of the SPT blow count with the overburden effective stress after the correction of the dam under the conditions of engineering loading and unloading is investigated. The investigation includes two working conditions: 1) engineering loading, that is, engineering loading happened in the engineering activities leads to an increase of the overburden effective stress on the overburden soil. For example, in order to enhance the stability of the dam slope after the earth dam is built or to improve the seismic stability of the overburden near the foot of the dam, the measures such as weighting the slope toe are adopted. Or the reduction of the groundwater level after the completion of the project leads to an increase in the overburden effective stress of the original standard penetration test point, and or the overburden effective stress enlarges due to the loading of the filling project during the construction; 2) Engineering unloading, that is, engineering unloading happened in the engineering activities leads to an decrease of the overburden effective stress on the overburden soil. For example, after the completion of the project, the increase of the groundwater level will result in the reduction of the overburden effective stress, and the excavation unloading may happen. For ease of understanding, the increase or decrease of the overburden effective stress is equivalent to the buried depth of the standard penetration test point.

3.1Engineering loading

Here two situations are provided: 1) After the completion of the project, the engineering loading is caused by the increase of the overburden effective stress on the original test point due to the drop of the groundwater level; 2) During the construction of the project, the engineering loading is caused by the increase of the overburden effective stress on the original standard penetration test point due to the filling.

As a typical example, two cases are given. Assuming that before the construction of the project, the buried depth of the test point in the standard penetration test is 10 m and the groundwater level is 2 m. When the drop of the groundwater level after the completion of the project gives rise to engineering loading, the variation of overburden effective stress correction coefficient is shown in Fig. 2(a). Before the construction of the project, the buried depth of the test point is 3m, and the groundwater level is 2m. The variation of the overburden effective stress correction coefficient under engineering loading caused by the increase of the overburden effective stress due to the filling in the project construction is shown in Fig. 2(b).

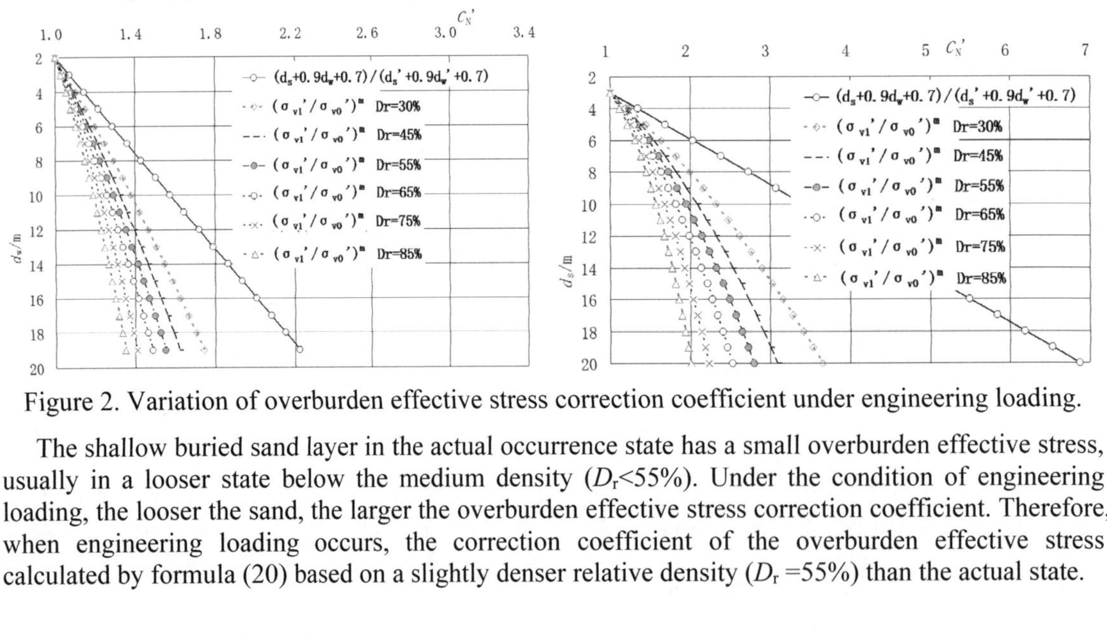

Figure 2. Variation of overburden effective stress correction coefficient under engineering loading.

The shallow buried sand layer in the actual occurrence state has a small overburden effective stress, usually in a looser state below the medium density (D_r<55%). Under the condition of engineering loading, the looser the sand, the larger the overburden effective stress correction coefficient. Therefore, when engineering loading occurs, the correction coefficient of the overburden effective stress calculated by formula (20) based on a slightly denser relative density (D_r =55%) than the actual state.

3.2Engineering unloading

Two situations are provided here: 1) After the completion of the project, the engineering unloading is caused by the increase of the overburden effective stress on the original test point due to the rising of the groundwater level; 2) During the construction of the project, the engineering loading results from the decrease of the overburden effective stress on the original standard penetration test point due to excavation.

As a typical example, two cases are given. Assuming that before the construction of the project, the buried depth of the test point in the standard penetration test is 20 m and the groundwater level is 20 m. When the rising of the groundwater level after the completion of the project gives rise to engineering loading, the variation of overburden effective stress correction coefficient is shown in Fig. 3(a). Before the construction of the project, the buried depth of the test point is 20m, and the groundwater level is 2m. The variation of the overburden effective stress correction coefficient under engineering unloading caused by the decrease of the overburden effective stress due to the excavation in the project construction is shown in Fig. 3(b).

It can be seen from Fig. 3 that, in the case of engineering unloading, when they are corrected to the same overburden effective stress, the value of $C_N^{'}$ calculated by the formula (20) increases with the increasing relative density, and the value of $C_N^{'}$ calculated by the correction formula (19) is significantly smaller than that calculated by formula (20) (D_r=30%). As can be seen, under the condition of engineering unloading, if the actually measured SPT blow count is corrected by the formula (3) in the liquefaction evaluation, the result tends to be more insecure than the corrected result calculated by the formula (16) at the loose state (D_r=30%). Therefore, under the condition of engineering unloading, the standard formulas tend to be conservative.

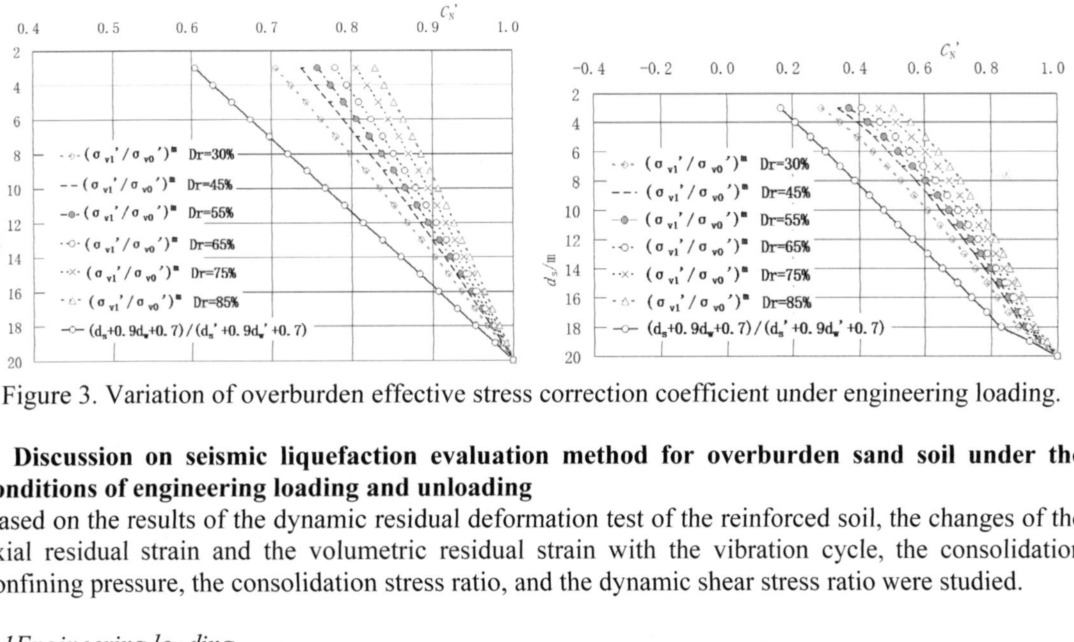

Figure 3. Variation of overburden effective stress correction coefficient under engineering loading.

4. Discussion on seismic liquefaction evaluation method for overburden sand soil under the conditions of engineering loading and unloading

Based on the results of the dynamic residual deformation test of the reinforced soil, the changes of the axial residual strain and the volumetric residual strain with the vibration cycle, the consolidation confining pressure, the consolidation stress ratio, and the dynamic shear stress ratio were studied.

4.1Engineering loading

In the practical engineering, for the shallow buried sand layer, during the construction of the project, it is either excavated or filled caused by engineering loading (such as the adjacent sand layer caused by weighting the slope toe). At this time, the value of $C_N^{'}$ calculated by the standard correction formula (19) is significantly larger than that calculated by the formula (20) in the loose state (D_r=30%). Meanwhile, when engineering loading happens, the value of $C_N^{'}$ calculated by formula (20) decreases with the increasing relative density. Therefore, the linearization correction method based on stress equivalence expressed by buried depth in the code during engineering loading, tends to be insecure to correct the actually measured SPT blow count of the overburden soil considering the influence of the overburden effective stress.

The shallow buried sand layer in the actual occurrence state has a small overburden effective stress, usually in a looser state below the medium density (D_r<55%). Under the condition of engineering loading, the looser the sand, the larger the overburden effective stress correction coefficient. Therefore, when engineering loading occurs, the correction coefficient of the overburden effective stress calculated by formula (20) based on a slightly denser relative density (D_r =55%) than the actual state.

4.2Engineering unloading

In the practical engineering, for the deep buried sand layer, during the construction of the project, engineering unloading caused by excavation is often encountered. At this time, the value of $C_N^{'}$ calculated by the correction formula (19) in the code is significantly smaller than that calculated by the formula (20) in the loose state (D_r=30%). In the meantime, when engineering unloading occurs, the value of $C_N^{'}$ calculated by formula (20) increases with the increasing relative density. Therefore, the linearization correction method based on stress equivalence expressed by buried depth in the code during engineering loading, tends to be conservative to correct the actually measured SPT blow count considering the influence of the overburden effective stress.

The deep buried sand layer in the actual occurrence state has a quite large overburden effective stress, usually in a compact state above the medium density (D_r>55%). Under the condition of engineering unloading, the looser the sand, the smaller the overburden effective stress correction coefficient. Therefore, when engineering unloading occurs, the correction coefficient of the overburden effective stress calculated by formula (20) based on a slightly looser relative density (D_r =55%) than the actual state.

5. Conclusion

This paper systematically summarizes the relevant research results of the overburden effective stress normalization coefficient determined in the on-site standard penetration test and the indoor full-scale standard penetration model test, deduces the correction formulas of the SPT blow count under the conditions of loading and unloading in the engineering activities such as dam construction, and combining with the correction method for the SPT blow count under the conditions of dam loading and unloading in relevant codes in the field of water conservancy and hydropower engineering, demonstrates the relationship between the SPT blow count and the overburden effective stress and relative density. On this basis, it is pointed out that the correction method for the SPT blow count under dam loading and unloading in relevant codes probably leads to misjudgment of the sand layer liquefaction or not. On the basis of argumentation, the new correction method instead of the original method is suggested to be utilized to evaluate the seismic liquefaction of the overburden sand soil under the condition of engineering loading and unloading.

(1) Under the condition of homogeneous sand layer, the relationship between the SPT blow count and the overburden effective stress is nonlinear. The linearization correction method based on stress equivalent expressed by the buried depth cannot truly reflect the influence of engineering loading and unloading on SPT blow count. Under the condition of engineering loading, the liquefaction resistance of the soil will be overestimated, and the liquefaction layer may be judged as non-liquefied, which brings potential risks to the project. Under the condition of engineering unloading, the liquefaction resistance of the soil will be underestimated, and the non-liquefaction layer may be judged as liquefied, resulting in an increase in the cost of foundation treatment.

(2) Through the investigation of the homogeneous sand layer, it is shown that in the nonlinear method, aiming at the defect of linearization correction method based on stress equivalence expressed by buried depth, a new non-linearization correction method is proposed, which takes into account the influence of the stress variation of overburden soil caused by loading and unloading of dams and other

projects on its liquefaction resistance. This method can effectively avoid the potential danger or waste of investment caused by the miscalculation of the original formula.

(3) In the practical engineering construction, the shallow buried sand layer in the medium-density and loose state usually undergoes engineering loading, and the deep buried sand layer in the medium-density and dense state usually undergoes engineering unloading. The demonstration shows that the results that the actually measured SPT blow count considering the influence of engineering loading and unloading is corrected by the formula $C_N^{'} = \left(\sigma_{v1}^{'} / \sigma_{v0}^{'}\right)^{1/2}$ (corresponding to Dr=55%) and the seismic liquefaction of the overburden sand soil during the operation of the project is evaluated, trend to be secure.

(4) It is recommended to revise the seismic liquefaction evaluation method of the overburden under the condition of overburden effective stress change in the current codes of the water conservancy industry, so as to more reasonably consider the influence of engineering and unloading conditions on the seismic liquefaction of the overburden, and avoid potential risks or unnecessary investment waste caused by misjudgment.

Acknowledgments
This research was supported by the National Key R&D Program of China [Grant No.2016YFC0401601]; National Fundamental Science Foundation of China [Grant No. 51679264]; and IWHR Research and Development Support Program (GE0145B562017).

References

[1] Terzaghi, K, and Peck, R B. Soil Mechanics in Engineering Practice [M]. New York: Wiley, 1948

[2] Gibbs, H J & Holtz, W G. Research on Determining the Density of Sands by Spoon Penetration Testing [C]/Proceedings, Fourth International Conference on Soil Mechanics and Foundation Engineering, London, 1957, 1: 35-39.

[3] Bieganousky, W A & Marcuson, W F. Laboratory standard penetration tests on Reid Bedford model and Ottawa sands.Research [R] Report S-76-2no.1,Waterways Experiment Station,Vichsburg,1976.

[4] Saito, A. Characteristics of penetration resistance of a reclaimed sandy deposit and their change through vibratory compaction [J]. Soils and Foundations,1977,17(4):31-43

[5] Ishihara, K & Koga, Y. Case studies of liquefaction in the 1964 Niigata earthquake [J]. Soils Found., 1981,21(3), 35–52.

[6] Meyerhof, G G. Discussion on research on determining the density of sands by spoon penetration testing [C]./Proc.4th Int.Conf.Soil Mech. Fdn Engng, London. 1957,3:110.

[7] Peck, R. B.,Hanson,W.E. & Thornburn,T.H. Foundation engineering [M].2nd edn.New York:Wiley,1974

[8] Seed, H B, Idriss, I M. & Arango, I. Evaluation of liquefaction potentiall using field performance data [J].J.Geotech.Engng Div.Am.Soc.Civ.Engrs,1983, 109,458-482.

[9] SKEMPTON A W. Standard penetration test procedures and the effects in sands of overburden pressure, relative density, particle size, aging and overconsolidation [J]. Geotechnique, 1986,37(3):425-447

[10] Liao, S & Whitman, R. V. Overburden correction factors for SPT in sand [J]. J. Geotech. Engrg., ASCE, 1986,112(3):373–377.

[11] Youd T L, Idriss I M, Andrus, et al. Liquefaction resistance of soils: summary repor [J].Journal of Geotechnical and Geoenvironmental Engineering, ASCE, 2001, 127(4): 297-313.

[12] Kayen, R E, Mitchell, J K, Seed, R B, Lodge, A, Nishio, S, and Coutinho, R. Evaluation of SPT-, CPT-, and shear wave-based methods for liquefaction potential assessment using Loma Prieta data [C]./ Proc., 4th Japan-U.S. Workshop on Earthquake-Resistant Des. of Lifeline Fac. and Countermeasures for Soil Liquefaction, 1992, 1:177–204.

IOP Conf. Series: Materials Science and Engineering **794** (2020) 012043 doi:10.1088/1757-899X/794/1/012043

[13] Boulanger, R W. High overburden stress effects in liquefaction analyses [J]. J. Geotechnical and Geoenvironmental Engineering, ASCE, 2003,129(12):1071-1082.
[14] Marcuson, W F, III and Bieganousky, W A. Laboratory standard penetration tests on fine sands [J]. J. Geotech. Engrg. Div., ASCE,1977,103(6): 565–588.
[15] Marcuson, W. F., III and Bieganousky, W A. SPT and relative density in coarse sands [J]. J. Geotech. Engrg. Div., ASCE, 1977, 103(11), 1295–1309.

Particle breakage of coral sand in direct shear test

WeiguangWang[1], WanLi[2]

[1, 2]Teaching and research office of airport construction engineering, Air force Engineering University, Xi 'an, Shaanxi province, 710038, China.

[1]18392371637@163.com.

[2a]17778249496@163.com.

Abstract. Due to its unique development environment and historical origin, coral sand often breaks up in the process of loading, which is quite different from ordinary siliceous land-based sand in mechanical properties. Therefore, it is of great significance to study coral sand. A particle analysis testwas conducted on the coral sand from a reef in the South China Sea to study its particle size distribution, and then direct shear tests were conducted to study the influence of two factors, namely relative compactness and vertical normal stress, on the crushing of coral sand particles. The results show that when the vertical normal stress is constant, the particle breakage rate increases with the increase of the relative compactness, and when the relative compactness is the same, the increase of the vertical normal stress level can further increase the crushing of coral sand particles.

1. Introduction

Coral sandis an important material for marine engineering construction, and is widely used in port wharves, reclamation of islands, foundations reclamation, buildings and other infrastructures. The particle crushing of soil in the process of loading will seriously affect the mechanical deformation performance of soil.Particle breakage has an important impact on the compression deformation, shear characteristics and permeability characteristics of carbonate sediments.It has great influence on the actual project. Therefore, it has become the research focus of scholars[1-4].

Shahnazari and Rezvani carried out triaxial compression tests on calcareous sand at different locations in the Persian gulf (Bushehr Port and Hormuz Island) to study the influence of confining pressure, relative compactness, grain size distribution and other parameters on soil particle breakage.[1]. Through a series of ring shear tests, Coop et al. studied the development process of carbonate sand particle crushing. It was believed that the final gradation of particle crushing was related to the normalstress applied on the samples and the initial gradation of samples[5]. Miao G and Airey, Dcarried out ring shear test and one-dimensional compression test on carbonate sand, and the research results showed that the particle crushing law was consistent, but the limit stability state was significantly different[6].

Chen HD, Wei HZ, et al conducted triaxial consolidation drainage shear test on calcareous sand, analyzed the effect of relative compactness and confining pressure on particle crushing and studied the effect of particle crushing on shear performance of calcareous sand. The research results showed that under the condition of low confining pressure, the particle breakage of soil samples is less. With the increase of confining pressure, the particle breakage becomes more serious. However, the increment of particle breakage decreases until the particle breakage rate approaches a stable limit [7]. Cai ZY, Hou HYet al carried out triaxial consolidation drained shear test and uniaxial compression test under different stress level and different compactness, research suggests that the soil particle breakage increases with the increase of relative compactness, and linear increase with increasing of confining pressure or vertical load[8].

As the mechanical properties of coral sand are greatly affected by the development environment, the crushing and mechanical properties of coral sand particles in different areas are different. For example, the yield strength of Bushehr Port sand is lower than that of Hormuz Island sand[1]. In this paper, the coral sand taken from a reef in the South China Sea was studied. Two variables, vertical normal stress and relative compactness, were set in the test to study the particle crushing condition under different test conditions.The research results of this paper are of great help to deeply understand the influencing factors of particle breakage of coral sand and explore the degree of particle breakage in the shearing process, which can provide useful help for practical engineering construction.

2. Materials and methods

2.1 Test material and sample preparation method

The coral sand comes from aisland in the South China Sea, and the partial particle groupis shown in the Fig1. Self-made sieves were used for screening test, and specific mesh diameter was shown in the Tab1. Since coral sand is a loose soil granule material, self-made sampler is used for sample pressing. The size of ring cutter is 61.8×20mm, and the sample preparation tool is shown in the Fig2.

Tab 1. Particle composition

Diameter range	1.43-2.0mm	1.0-1.43mm	0.85-1.0mm	0.5-0.85mm	0.3-0.5mm	0.2-0.3mm	0.1-0.2mm
Proportion	0.0563	0.0768	0.0918	0.1352	0.3592	0.1495	0.1311

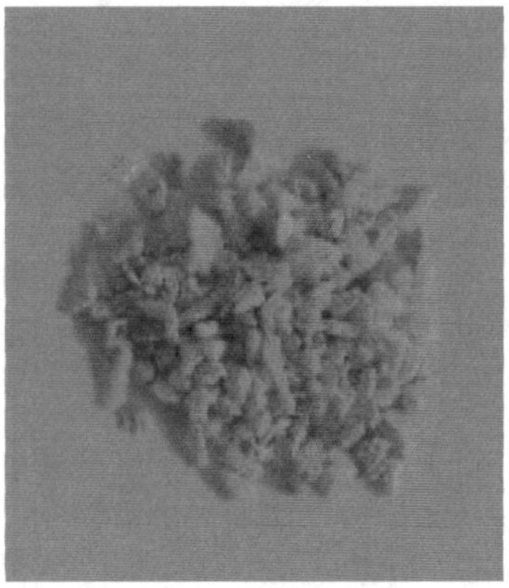

Fig 1. Coral sand

Fig 2. Sample preparationtool

2.2Testscheme

In order to study the influence of relative compactness and vertical normal stress on the crushing of coral sand particles, the traditional direct shear instrument was used for the test. The specific experimental scheme is shown in the Tab 2. Before and after the test, the particle analysis test was carried out. The water content of the samples was 25%.

Tab 2. Test scheme and sample labels

Vertical normal stress /kPa	Relative compactness D_r			
	0.4	05	0.6	0.7
100	P100-0.4	P100-0.5	P100-0.6	P100-0.7
200	P200-0.4	P200-0.5	P200-0.6	P200-0.7
300	P300-0.4	P300-0.5	P300-0.6	P300-0.7
400	P400-0.4	P400-0.5	P400-0.6	P400-0.7

3. Test results and discussion

3.1Particle breakage index

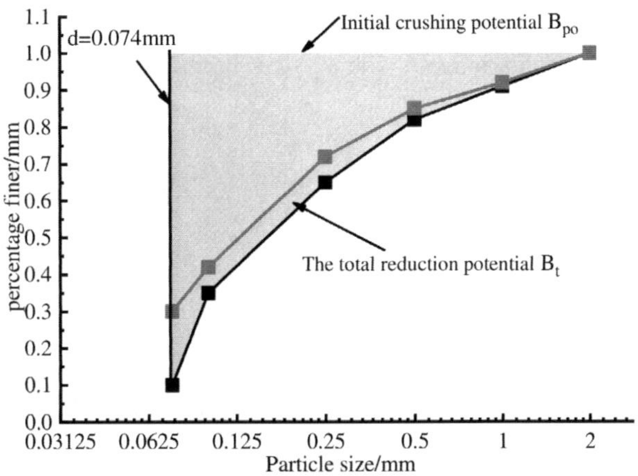

Fig 3. Relative particle breakage rate index

The determination of particle crushing index is of great significance to measure the influence level of various factors on samples.For the definition of particle crushing index, many scholars have put forward their own determination methods, such as B_{15} proposed by Lee and Farhoomand [9], B_{rE} proposed by Einav [10]. In this paper, the crushing index B_r proposed by Hardin is adopted [11], and its formula is defined as follows:

$$B_r = \frac{B_{PO}}{B_t} \tag{1}$$

The specific area range diagramof$B_{PO}$and $B_t$is shown in the Fig 3.

3.2Particle breakage rate

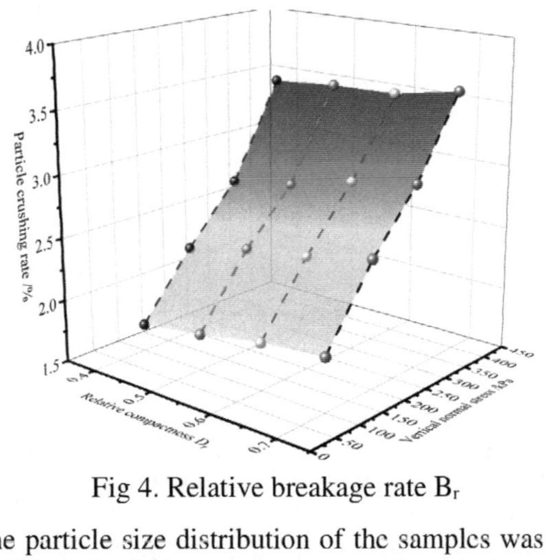

Fig 4. Relative breakage rate B_r

Before and after the test, the particle size distribution of the samples was confirmed, and the particle breakage rate under the condition of relative compactness and vertical normal stress was acquired.The relationship between the dual influence indexes of relative compactness and vertical normal stress and the particle breakage rate was plotted in a three-dimensional figure, as shown in the Fig 4. The influence of the dual influence indexes on the test results could be directly displayed.

It can be seen from the comparative analysis of the test results that the particle breakage of coral sand gradually increases with the increase of relative compactness. The results of the experiment under a certain vertical normal stress condition can be obtained arbitrarily. For example, when the vertical normal stress is 100kPa. The vertical normal stress also has an important impact on the particle breakage of coral sand. Under the condition of the same relative compactness, the particle breakage rate increases with the increase of the vertical normal stress, such as the particle breakage rate under various stress conditions when Dr=0.5.

4. Conclusion

Coral sand has become an important engineering material for the construction of ocean and port infrastructure due to its rich reserves and easy access. However, because of its high calcium carbonate content, it is prone to particle breakage under stress, and its mechanical properties are different from those of land-based sand. Therefore, it is of great significance to conduct research on the particle breakage of coral sand.

In this paper, the particle analysis of coral sand from a reef in the South China Sea was carried out, and the direct shear test was carried out to study the crushing condition of coral sand particles under different influencing factors. The results show that the relative compactness and vertical normal stress have significant effects on the particle breakage of coral sand. At the same vertical normal stress, particle breakage increases with the increase of relative compactness. According to the comparison and analysis of the same relative compactness, the increase of vertical normal stress will increase the particle breakage of coral sand. The research results of this paper are of great significance to further understand the particle breakage of coral sand.

Reference

[1] Habib Shahnazari, Reza Rezvani. Effective parameters for the particle breakage of calcareous sands: Anexperimental study [J].Engineering Geology,2013,159(9):98-105.

[2] Wang X Z , Jiao Y Y , Wang R , et al. Engineering characteristics of the calcareous sand in Nansha Islands, South China Sea[J]. Engineering Geology, 2011, 120(1-4):40-47.

[3] Ma L, Li Z , Wang M , et al. Effects of size and loading rate on the mechanical properties of single coral particles[J]. Powder Technology, 2018.

[4] LvY, Liu J, Xiong Z. One-dimensional dynamic compressive behavior of dry calcareous sand at high strain rates[J]. Journal of Rock Mechanics and Geotechnical Engineering, 2018.

[5] Coop M R,Sorensen K K,Bodas Freitas T, et al. Particle breakage during shearing of a carbonate sand[J]. Géotechnique, 2004, 54(3):157-163.

[6] Miao G, Airey, D. Breakage and ultimate states for a carbonate sand [J]. Geotechnique, 2013,63(14):1221-1229.

[7] Chen HD, Wei HZ, Meng QS, et al. The study on stress-strain-strength behaviour of calcareous sand with particle breakage [J]. Journal of Engineering Geology, 2018, 26(06):85-93. (In Chinese)

[8] Cai ZY, Hou HY, Zhang JX, et al. Experimental study on the influence of density and stress level on particle breakage of coral sand [J]. Journal of Hydraulic Engineering,2019,50(2):184-192. (In Chinese)

[9] Lee K L, Farhoomand I. Compressibility And Crushing Of Granular Soil In Anisotropic Triaxial Compression[J]. Canadian Geotechnical Journal, 1967, 4(1):68-86.

[10] EinavI. Breakage mechanics—Part I: Theory[J]. Journal of the Mechanics and Physics of Solids, 2007, 55(6):1274-1297.

[11] Hardin, Bobby O. Crushing of Soil Particles[J]. Journal of Geotechnical Engineering, 1985, 111(10):1177-1192.

IOP Conf. Series: Materials Science and Engineering **794** (2020) 012045 doi:10.1088/1757-899X/794/1/012045

Application of Three-Dimensional Design in Substation Design

Hailiang Wu[1*], Lin Xing[1], Zhong Liu[1], Wuchen Zhang[1], Nan Cheng[1], Teng Li[1], Yamin Wang[1] and Wen Zhang[1]

[1]State Grid Hebei Economic Research Institute, Shijiazhuang, Hebei, 050000, China

[*]Corresponding author's e-mail: 463289609@qq.com

Abstract: With the improvement of design requirements and computer technology, the three-dimensional design with multi-professional collaborative design, refined design and engineering digitization as the core, is being popularized and used in the substation engineering design of State Grid Corporation. This paper analyzes concrete measures, conducts benefit analysis of three-dimensional design in engineering design, and discusses the application of three-dimensional design in substation design combining with practical application cases.

1. Introduction

Design, the key in engineering construction, can provide drawings required for engineering construction, strong underlying data information for procurement, operation, maintenance and repair[1], possibility for the extension of value-added design services[2],and also data for creating the full life cycle of projects.

As engineering construction puts higher requirements on design and technology advances, design in the power industry has changed greatly from the traditional two-dimensional (2D) design to three-dimensional (3D) design. 3D design facilitates the collaborative work among electricity, architecture, structure, water-heating and other professions and significantly improve its quality and efficiency through meticulous design[3].

The application of 3D design in substations is now a research focus in design institutes[4] [5].

2. Advantages of 3D Design in Engineering

2.1 3D Design Platform

3D design platforms for substations are built with the database of model library and standardization schemes at the core. Through data correlation and sharing and integrated 3D design, collaborative designs in electricity, structure, construction, water-heating and other professions of the same substation can be realized. The model library is set up to store the equipment models and the civil engineering models of substations. These models are constructed according to the State Grid Corporation of China's Technical Guide for 3D Design in Power Transmission and Transformation Engineering Section 1: Substation (Conversion Station) and Modeling Specifications for 3D Design of Substation (Conversion Station). By doing so, all models are unified and standardized and the quality and efficiency of design are improved.

IOP Conf. Series: Materials Science and Engineering **794** (2020) 012045 doi:10.1088/1757-899X/794/1/012045

2.2 Collaborative 3D Design

The collaborative 3D design [6] system based on the same 3D design platform and substation engineering model not only improves the efficiency of data exchange among various professions, but also reduces errors and workloads of data exchange and duplication of work to the utmost extent. Specialists in primary, secondary substations and civil engineering of substations are able to design in the same substation engineering model and update the data and model in real time. Any collision and conflict among these professions are presented timely and vividly to avoid errors and omissions in data exchange and collisions of models, improving the efficiency and quality of design effectively.

2.3 Detailed design

The meticulous 3D design makes it possible to carry out collision detection and safety clearance verification in the foundation of buildings and underground pipelines, and to generate construction drawings and automatically keep statistics of engineering workload based on the 3D model. As a result, we can reduce design errors and omissions and the impact of modifying designs on project duration and cost.

2.3.1 Collision check. Carrying out the 3D design with the same model of transformer substation will enable specialized persons to check the collision between foundations of substation, foundations and underground utilities, and underground utilities according to specification requirements and exchange of data. The 3D model can also display the collision to specialized persons so that they can fix the problem quickly and effectively prevent designing errors from happening.

2.3.2 Safety clearance check. We can verify safety clearance by setting parameters to automatically check whether the safety distance between equipment at different voltage levels and the distance between equipment and ground meet the specification requirements and real engineering conditions. When wires in substation are complex and are crowded in a small space, energized equipment and wire will be intersected. Therefore, verification of safety clearance merely based on design paper will be time-consuming and cause high error rates. 3D design is visualized and accurate and can verify safety clearance, which greatly improves the efficiency of spatial design.

2.3.3 Extract 2D drawing. Using 3D designing collaboration technology, designing data and information can be shared and updated in parallel between different drawings. Based on the 3D model, designing files such as 2D plane, graphic and profile drawing, equipment axonometrical drawing, and material reports can be extracted, which not only makes the design file accurate and efficient, but greatly saves the design workload.

2.3.4 Quantity Statistics. 3D model of the substation makes statistics of engineering quantity accurate and easy to be calculated. By setting the range, engineering quantity statistics of the entire substation, areas of power distribution units at different voltage levels or quantity of different specialized lines can be calculated to meet requirements of engineering accuracy and depth at different designing stages.

3. 3D design benefit analysis

3.1Improving the quality of engineering design

First, we should use 3D design to find the weak points so that we can target at them to improve the quality; second, we should use 3D collaborative design to improve the efficiency of different specialized lines; third, we should use 3D visual design to lift refined design to a higher level.

3.2Realizing targeted investment of engineering

First, with 3D design, the engineering quantity can be calculated accurately. Information of main

IOP Conf. Series: Materials Science and Engineering **794** (2020) 012045 doi:10.1088/1757-899X/794/1/012045

equipment, designing data and material data based on the 3D model of the substation can be automatically counted to make targeted budget estimation; secondly, meticulous and qualified design will save engineering investment; thirdly, calculation and analysis based on 3D model will be more accurate.

3.3Strengthening engineering safety

Soft and hard collision check and safety clearance verification of 3D design allow designers to avoid errors and omissions in the early stage of engineering design, which can reduce the impact of design changes on construction days and cost and can help engineering management. For example, with accurate and visualized calibration of distance between electrical and gas equipment, designers will produce reasonable designing scheme. With stereoscopic and visualized 3D cable laying, designers will make less construction errors.

3.4 Improve the Quality of Project Construction Management

First, the participating parties can directly grasp the design plan, and improve the construction management efficiency with the assistance of 3D design. Second, the material procurement can automatically extract the material list according to the equipment coding and characteristic parameters of the equipment model. The third is to connect the factors and data produced during the construction process by using the 3D model, achieve 4D dynamic visualization management of the project, and ultimately enables more meticulous management of the construction process.

3.5 To Achieve Whole-Process Application of the Engineering Data

The 3D design platform is equipped with a strong database. First, the business system of the infrastructure and the operation inspection department can extract relevant structured data from the 3D model data. Second, the 3D model can correlate the information and data generated from each stage and each business system, which is quick and convenient. Third, after the 3D design platform is transferred to the operation inspection department, it can support the visual management of operation and maintenance, and satisfy the need for functional development of the 3D model during the operation inspection.

4. The Application of 3D Design in Substation Engineering

Longquan 220kV substation is the first batch of 220kV modular pilot project of China's state grid. Taking the new project as an example, the specific application of 3D design in substation engineering design process will be discussed.

Firstly, the substation's unified column grid is established by the civil engineering profession. Next, the substation will perform one conversion. Then, the civil engineering will fixate the reference point on the design platform, and carry out collaborative design work on the substation's column grid.

The electrical profession establishes the product model according to the equipment drawings provided by the manufacturer shown in figure 1, and assigns values to the equipment parameters of the adopted product model, in order to facilitate further operation and maintenance of the substation. Through the use of 3D design platform with safety clearance verification (show in figure 2), secondary cable automatic laying (show in figure 3), wire tension verification (show in figure 4), lightning protection design and other functions, the platform helps to not only achieve design visualization, but also optimize the overall station layout, and reduce the occurrence of design errors to a large extent.

IOP Conf. Series: Materials Science and Engineering **794** (2020) 012045 doi:10.1088/1757-899X/794/1/012045

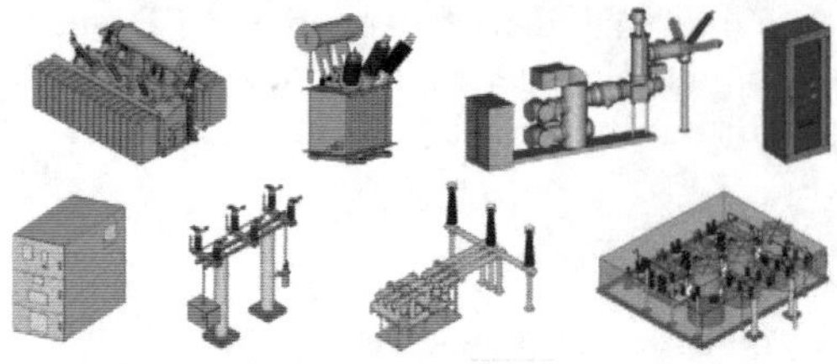
Figure 1. Electrical equipment model

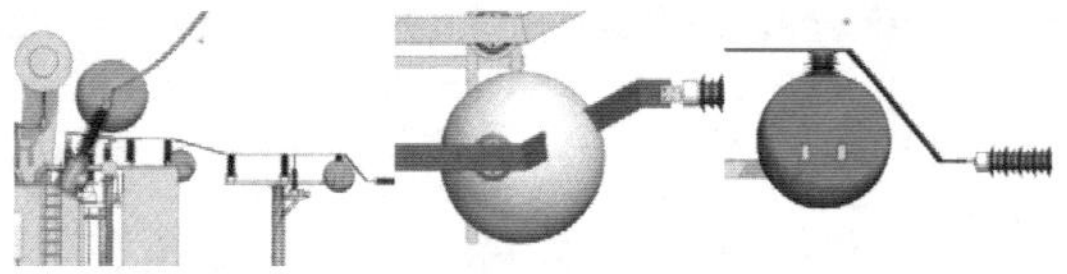
Figure 2. Main transformer zone safety clearance check

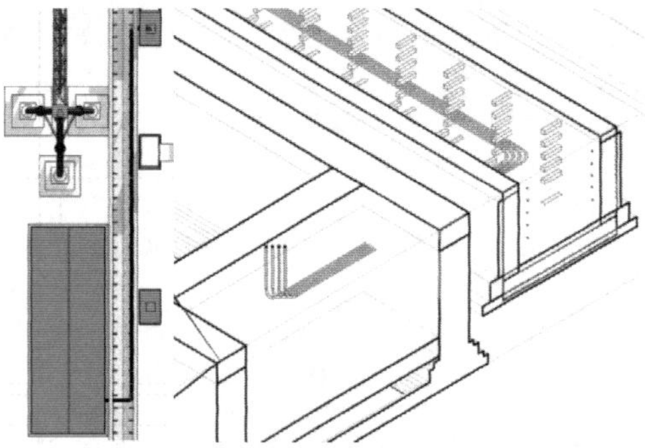
Figure 3. Automatic cable laying

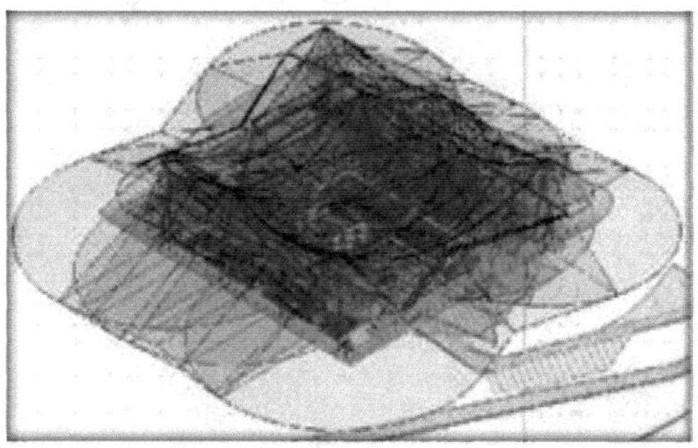
Figure 4. Automatic cable laying

The civil engineering profession uses 3D design software to generate 3D model of the site through 2D topographic map. The model can accurately simulate the natural terrain and landform where the site is located, besides, the coordinates and the elevation of each point in the model are consistent with the actual natural terrain (show in figure 5). By accurately simulating the natural topography of the site, and considering the location, drainage method, road allowable slope and earthwork balance, etc., the vertical layout of the general plane of the site and the amount of earthwork will then be determined.

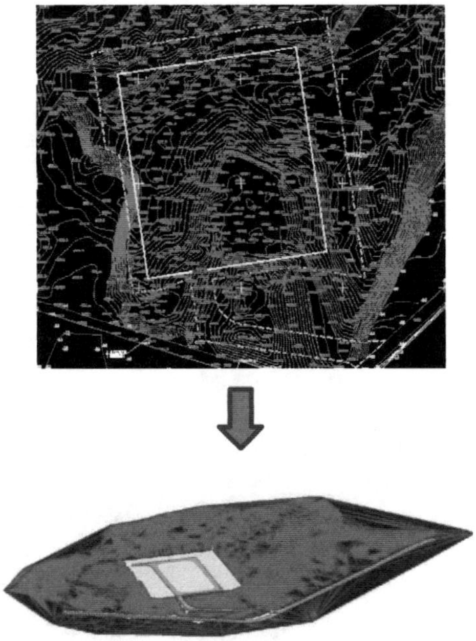

Figure 5.Two - dimensional topographic map to establish a three - dimensional site model

Longquan 220kV substation uses 220-A1-1 plan. The 3D design of civil engineering includes: wall and gate, buildings, capacitor foundation, GIS foundation, supporting structure foundation, cable trench, fire pipe network, water supply and drainage pipe network, etc.

Collision detection among all professions is carried out to ensure the quality of design. Collisions among underground pipelines and between pipelines and foundations are adjusted and revised timely to avoid modifying designs in the later part of construction (show in figure 6) , saving both cost and time for project construction.

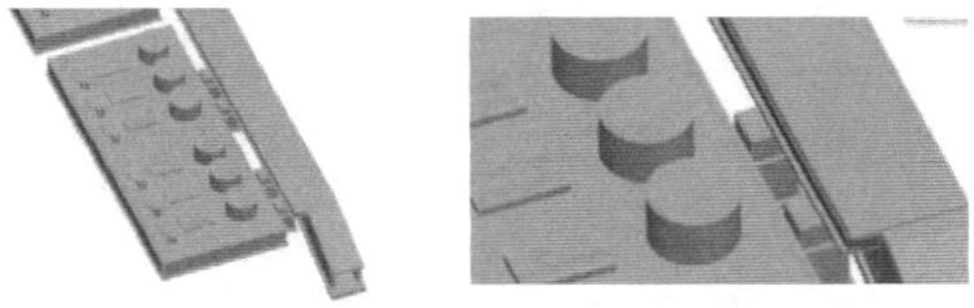

Figure 6. The capacitor disconnecting switch base collids with the cable trench

After the 3D substation design is completed, we slice the overall 3D substation model to generate the plan, elevation and section of each building and also the electrical power distribution unit diagrams and panel sections (show in figure 7), which automatically produces equipment parameters and other information . If the master plan is adjusted, accordingly power distribution unit diagrams and panel sections are automatically adjusted (show in figure 8), significantly reducing design errors and omissions.

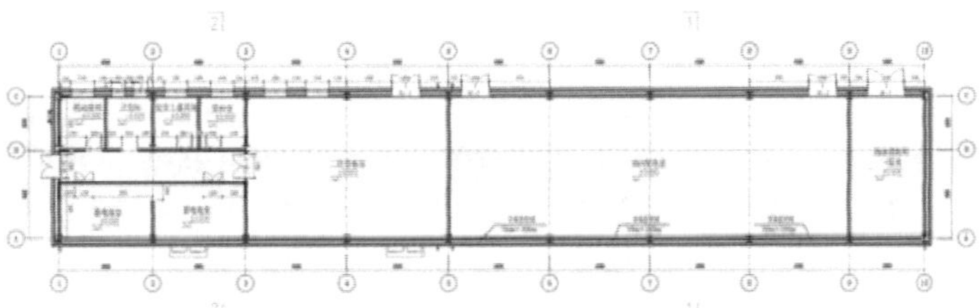

Figure 7. Distribution unit room plan

IOP Conf. Series: Materials Science and Engineering **794** (2020) 012045 doi:10.1088/1757-899X/794/1/012045

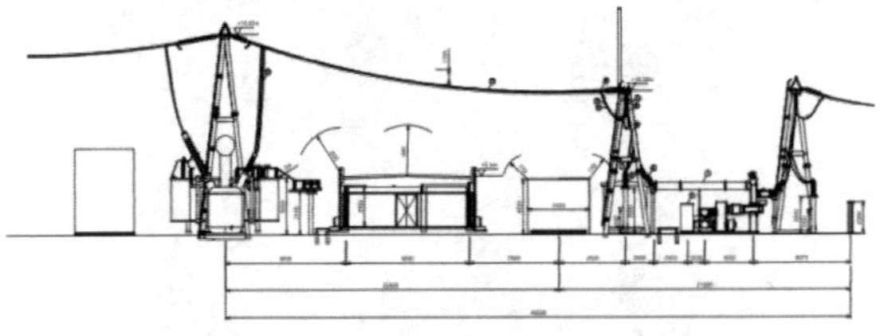

Figure 8. 110kV entry profile

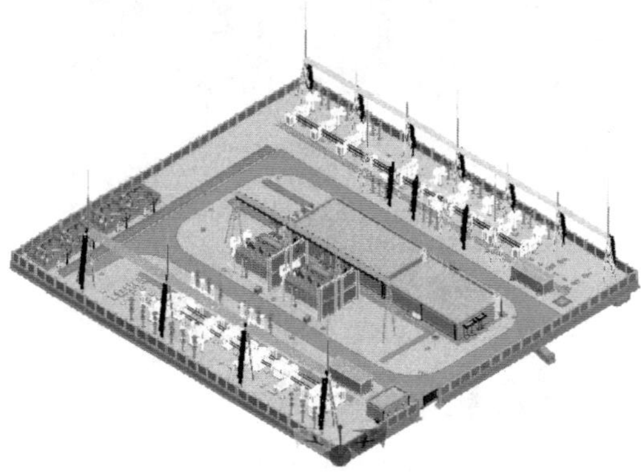

Figure 9. Substation three-dimensional model

The Longquan 220kV substation project adopts the 3D design method,which the 3D model of the whole substation is shown in figure 9. Through 3D collaborative design and meticulous design, professions can achieve in-time communication and close cooperation, which will minimize design errors and risks, thus significantly improving design quality and efficiency. Under the 3D collaborative design mode, the design files and 3D models are updated in real time, and the 3D design results can be used for collision check and space clearance verification at any time. Through program optimization and design optimization, the project can solve the design defects that occur in time, and achieve cost reduction, construction period reduction and sustainable development.

The Longquan new project has multiple advantages by employing 3D optimization design. Compared with the 2D design, the total area covered is reduced by 60 m2, and the cost by 18,000 yuan; the road area in the station by 21 m2, and the cost by 4,200 yuan; the cable trench length by 6 m2, and the cost by 3,500 yuan; the grounding materials by 762 m, the cost by 24,000 yuan; the steel used in the construction of building by 5.5t, and the cost by 42,000 yuan. The investment saved totaled 92,000 yuan, accounting for 0.26% of the construction and installation costs of the project.

5. Conclusions

The application of 3D design is a revolution in the field of substation engineering design. Under the 3D collaborative design mode, the design information and 3D model can be updated and interacted in real time, which is beneficial to the optimization and adjustment of the design scheme, and improves the efficiency and quality of the engineering design. Through the visualized 3D model and the meticulous design results provided by the 3D design, the data integration application is formed, and the multi-link sharing of power grid engineering management is achieved, which helps to improve engineering construction and production operation and maintenance management. This has laid a solid foundation for building a digital power grid and serving full life cycle management of the project.

References

[1] Chen jing. (2013) Application Research of 3D Collaborative Design in Intelligent Substation Design. Electrotechnical Application, S1:539-544.

[2] Ran Ruijiang. (2012) Discussion on the Application of Bentley Software in Substation 3D Digitalization Design. China Electric Power(Technology Edition):83-86.
[3] Qie Xin, Qi LiZhong,Hu Junhui. (2012) Application of 3D digital technology in power transmission and transformation engineering.Power Grids and Clean Energy,28(11):23-26.
[4] Li Zhihai. (2015) Research on digital 3D substation design technology. Electrical Engineering,16(11):83-86.
[5] Zhao yayuan,Xu Wanjun,Han Yongxing,etc. (2014) Development and application of intelligent substation 3d design in shaanxi institute. Electric Power Survey & Design,4:8-9.
[6] Xiang Ling,Shao Junwei,Tan Hailan,etc. (2012) 3d collaborative design of smart substation. China Electric Power,11:454-459.

Tempo-spatial Evolution Law of Valley Width Deformation of High Arch Dams and Analysis of the Causes

Cheng Heng[1], Zhang Guoxin*[1,2], Liao Jianxin[3], Liu Yi[1,2], Zhou Qiujing[1], Jiang Chenfang[1]

[1]China Institute of Water Resources and Hydropower Research, Beijing, 100038

[2]State Key Laboratory of Simulation and Regulation of Water Cycles in River Basin, Beijing, 100038

[3] China Three Gorges Corporation, Beijing, 100038

Abstract: In China, some high arch dams have undergone valley width contraction after water storage, which affects the behavior and long-term safety of the dams.This paper reviews typical cases of high arch dam valley width deformation in the world and analyzes its characteristics and causes.From this it continues to address the deformation of the Xiluodu arch dam valley and studies its width deformation monitoring data, revealing the temporal evolution and spatial distribution characteristics of the deformation. The mechanism of the deformation is then studied in the context of the engineering geology and hydrogeological conditions of the dam site.The results suggest that with the seepage force acting on $P_2\beta_n$ after water storage, the effective stress in this rock layer fell and the rock mass might undergo displacement and creep deformation along $P_2\beta_n$, causing the valley to narrow.Therefore, the contraction of the Xiluodu arch dam valley was caused by the reduction of the effective stress in the rock mass during the evolution of the seepage field following water storage under the specific hydrogeological and geostress field distribution conditions.

1. Introduction

A high dam typically involves a high water head, high and steep bank slopes, and complex geological conditions. The original equilibrium state of the slopes is disrupted by excavation or water storage, inducing deformation in the slopes.There are a large number of bank slope deformation cases, both in China and abroad. The Beauregard arch dam (Italy), for instance, experienced gravity sliding deformation in the left bank slope in the early stage of water storage, with the sliding rate in proportion to the rise of the reservoir water level[1, 2].ZeuzierArch Dam, in Switzerland, after 21 years of normal operation, was diagnosed with seeping saturated rock mass, caused by the excavation of the Gotthard tunnel 400m beneath the dam base, which changed the base rock mass from a saturated into an unsaturated state, resulting in uneven settlement of the foundation and valley contraction[3-5].The Maoping landslide body of the Geheyan Reservoir, on the Qingjiang River, began to deform after 1993, when the reservoir started to store water. By 2004, the cumulative horizontal displacement amounted to 2747.1mm and the vertical displacement to 548.4mm[6,7].The Laxiwa Hydropower Station, in China, developed toppling deformation on its bank following water storage. At present, the reservoir water has stabilized to the normal level. The maximum displacement of the front edge of the bank slope is about 40m, but the deformation is still growing, not fully checked[8].

China's high-dam reservoirs have one after another entered their storage and operation stage. Some high arch dams, such as Jinping I, Xiluodu, and Lijiaxia, have developed valley shrinkageafter storage--deformation towards the center of the valley. By April 2019, Xiluodu, in particular, has a maximum valley contraction, since storage, of about 93.47mm, and the figure was 34.4mm in the case of the Lijiaxia arch dam[9]. With the Jinping I, the maximum shrinkage was about 36 mm by March 2018. This figure is about 6.59mm in the case of Ertan[9].

In view of the contraction cases of high arch dam valleys, domestic and foreign scholars, after analyzing its factors and causes, have suggested some explanations.Yang Jie et al[10] attribute the Lijiaxia arch dam valley deformation to the shrinkage, the increased seepage pressure, and the decreased rock mass strength of the fracture zones under higher water pressure after water storage.The latest development of the valley deformation of the Xiluodu arch dam and the Jinping I arch dam has drawn intense attention among the dam community.Yang Qiang et al[11-13]analyze the deformation of the Jinping I arch dam valley from the perspective of effective stress change, pointing out that the fissure water pressure in the initial stage of water storage altered the equilibrium state of the rock mass and caused plastic deformation in the reservoir rock mass, which is the main driving force for the valley width deformation during water storage. Zhang Guoxin et al[14,15] assert that the deformation of the Xiluodu arch dam valley is mostly time-dependent, caused by altered hydrogeological conditions in the near-dam zone, which deformation is dominated by creep deformation. Zhou Zhifang et al[16] believe that under certain geological and hydrogeological conditions of the reservoir area, changed water pressure in the Yangxin limestone aquifer after water storage is the main cause of the deformation of the Xiluodu reservoir basin and its valley width.He Zhu et al.[17] argue that the additional seepage pressure and creep deformation of interlayer and inter-layer disturbed zones following water storage are the main causes of the deformation of the Xiluodu arch dam valley. Liang Guohe et al.[18] analyze the width deformation of Xiluodu arch dam valley and find that the correlation between valley contraction and reservoir water level and temperature is small, suggesting that there might be other factors.On the whole, existent studies focus mostly on qualitative analysis of the deformation mechanism, which is unable to give a full explanation for the characteristics and law of the deformation observed in the works.

These arch dam works have their own topography, stratum lithology, geological structure and hydrogeology. The law of their post-storage valley deformation varies from one another; so are the inducing factors of the deformation.The research on the origin and development of valley contraction is insufficient. Therefore, to reveal the temporal and spatial evolution law and the intrinsic dynamic mechanism of valley deformation it is necessary to review and analyze the deformation law and influencing factors of valley deformation of arch dams in different parts of the world. In this paper, typical cases in the world of high arch dam valley width deformation are reviewed, some valley deformation monitoring data are gathered and consolidated to develop and analyze the tempo-spatial distribution characteristics of the deformation in each case.Then, the monitoring data of the Xiluodu arch dam valley width are analyzed, to reveal the tempo-spatial evolution law of the deformation. The mechanism of the deformation of the valley is also discussed.

2. Review of typical high arch dam valley width deformation cases

2.1Beauregard Arch Dam, Italy

Beauregard, a concrete gravity arch dam and 132m high, is located in the Aosta valley, northwestern Italy.In the left-bank slope there is a deep shear zone, composed of fragmented gneiss and fault mud.During the first years of water storage (1958--1968), gravity-sliding deformation occurred in the left bank slope, and over the 60 years after water storage the valley has shrunk 100mm to 200mm.Due to the large deformation in the left-bank slope, the sliding of the left-bank foundation caused the arch dam to be squeezed, making the dam to deform upstream (see Fig. 1) and inducing horizontal cracks on the downstream bottom (red lines in Fig. 2). The upstream horizontal joints then opened.

Having analyzing the long-term monitoring data, Barla[2] concludes that after the water level rose the shear strength of the deep shear zone fell, inducing the deformation. The fall of the shear strength of the deep shear zone, which happened after coming into contact with water, does not happen all at once. It is time-dependent.

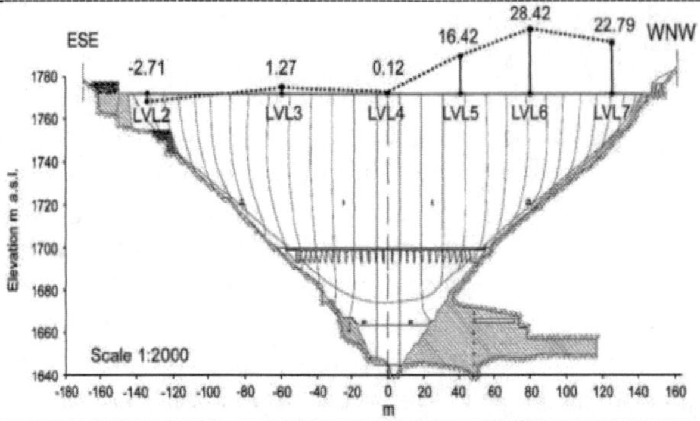

Figure 1. Deformation at the dam crest

Figure 2. Cracking of the downstream face of Beauregard Arch Dam

2.2Zeuzier Arch Dam, Switzerland

Zeuzier Arch Dam is located in Sion, southwestern Switzerland. It is a concrete hyperbolic thin arch dam with a height of 156m. Its construction was completed in 1957 and the reservoir stored water in the same year.In 1976, a highway tunnel was built 400m beneath the dam base. Subsequent observation revealed noticeable foundation settlement and valley contraction, the latter reaching 60 mm. The contraction squeezed the dam causing its crest to deform upward by 110 mm. A number of cracks resulted, as shown in Fig. 3.

The settlement and valley deformation seen in the Zeuzier Arch Dam case are believed to be connected to the construction of the highway tunnel. When the work started on excavation of the tunnel, the seepage exceeded 1000L/s, a high seepage that drained the saturated rock mass and caused it to transit into an unsaturated state, ultimately leading to the dam foundation settlement and the valley contraction[3-5].

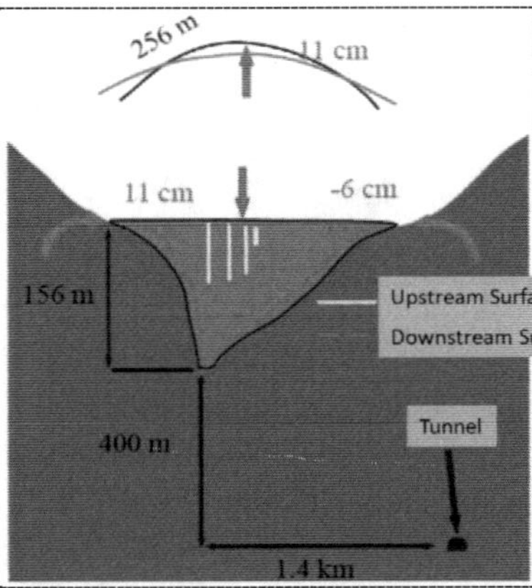

Figure 3. Schematic diagram of the overall deformation of the Zeuzier Arch Dam

2.3Jinping Stage I Arch Dam

Jinping I Arch Dam sits on the main stream of the Yalong River, in Sichuan Province. The dam stands 305m high, with a crest elevation of 1885m. The normal water storage level is 1880m.From the lithological perspective, the dam site is made mainly of marble and sand slate. The foundation is characterized by 'up-soft and down-hard' on the left bank and 'up-hard and down-soft' on the right bank.Fig. 6 and 7 show the valley width deformation process line of the typical monitoring lines at some upstream, abutment and downstream points. The valley width monitoring lines were set up as shown in Fig. 4 and 5. It can be seen that at the initial stage of water storage there exists a good correlation between the upstream valley width deformation and the water level. A rising water level corresponds to apparent deformation. When the water level falls or stabilizes, the deformation slows down; the downstream valley width deformation is less affected by the water storage process and the deformation value is smaller than the upstream valley; In October 2016, when the water level reached the normal water level for the third time, the deformation rate observed along each monitoring line dropped evidently, with the deformation tending to converge. By March 2018, the upstream valley contraction reached a maximum of about 36mm, while the maximum downstream contraction was about 18mm.

Yang Qiang et al[11~13]analyze the valley width deformation mechanism of Jinping I Arch Dam from the perspective of effective stress change, pointing out that the fissure water pressure in the initial stage of water storage altered the equilibrium state of the rock mass and caused plastic deformation in the reservoir rock mass, which is the main driving force of the valley width deformation during water storage.

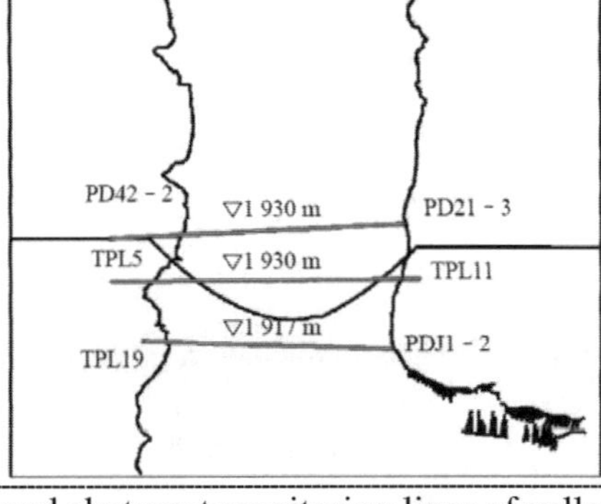

Figure 4. Upstream and abutment monitoring lines of valley width deformation

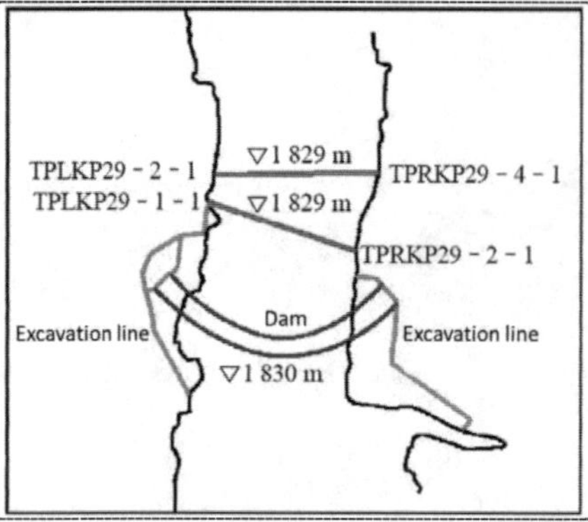

Figure 5. Downstream monitoring lines of valley width deformation

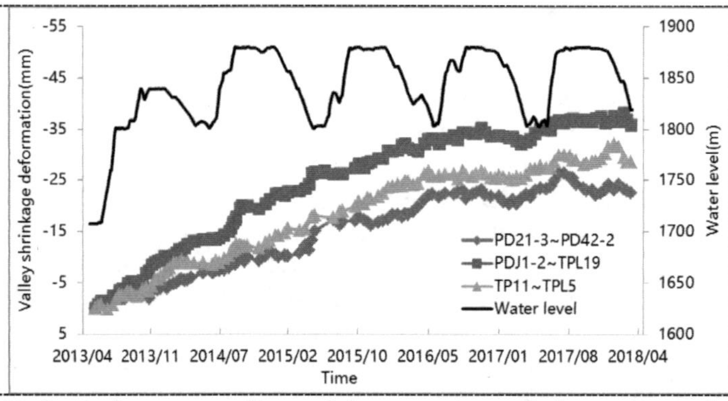

Figure 6. Deformation process lines of the upstream and abutment monitoring lines

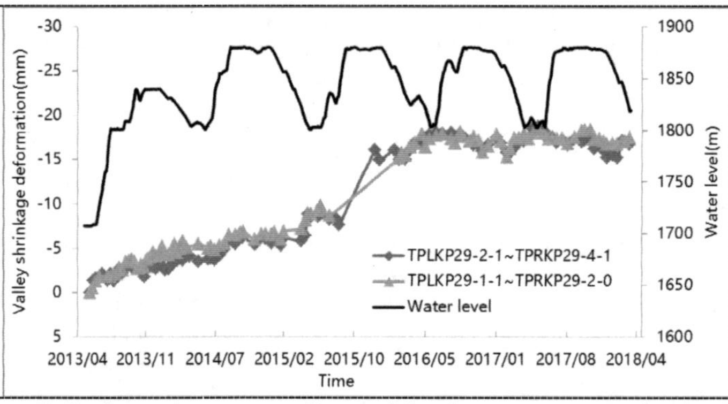

Figure 7. Deformation process lines of the downstream monitoring lines

2.4Lijiaxia Arch Dam

This is a three-center double-curved arch dam, with a dam crest elevation of 2185m and a maximum height of 155m. The geological conditions of its reservoir and the dam site are complex, with developed geological structure and a number of weak foundation locations. There are some slope stability problems with the Lijiaxia reservoir area and the near-dam banks. The most prominent is the deformation in the left abutment and the downstream high slopes. There are established five valley width monitoring lines downstream of the arch dam, at an elevation of 2130~2185m, as shown in Table 1.

Table 1.Arrangement of valley width monitoring points

Description	No. of monitoring points	Elevation (m)
Valley width monitoring	TP17~LJ03	2185
	LJ05~LJ07	2155
	TP15~LJ07	2160
	TP16~TP06	2155
	TP13~TP02	2130

Water storage began in December 1996 and continued until the beginning of 2002. The deformation process lines, developed from the five valley width monitoring lines, are given in Fig. 8. It can be seen that the five lines basically suggest the same deformation law--the valley contraction increases as the water level rises and, when the water level basically stabilizes at the normal water level, the deformation shows signs of convergence and stability. From the perspective of spatial distribution, the higher the elevation of the monitoring line, the greater the value of the observed deformation.

Yang Jie et al[10]analyze the displacement and safety status of the left-bank rock mass of Lijiaxia Arch Dam.It is suggested that there are obvious geological defects in the left reservoir bank near the dam. The increased seepage pressure and the decreased shear strength following water storage induced the gravity creep deformation of the bank slopes toward the center of the valley, causing the valley to contract.

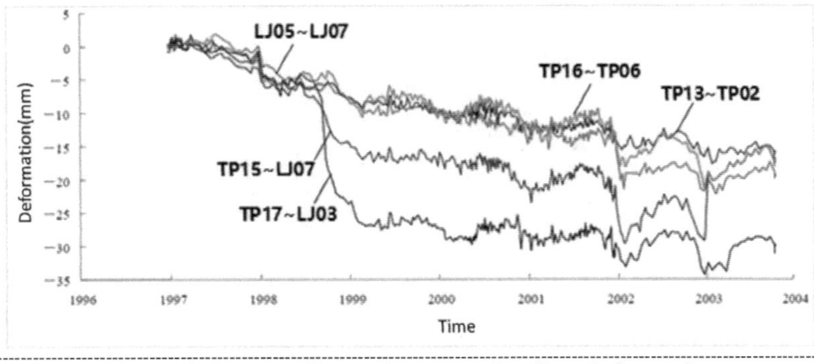

Figure 8. Process lines of measured valley deformation

The above analysis shows that, with a high water head, high and steep slopes, and complex geological conditions associated with a high dam, excavation or water storage alters the seepage field in the dam site, changing the hydrogeological conditions. In the course of seepage field evolution, the change of effective stress on the rock mass structural planes and the softening of the rock mass after soaking are the main inducing factors of the contraction.The tempo-spatial distribution and evolution of the bank slope deformation are to a large extent affected by the geomorphology, stratum lithology, geological structure, and hydrogeology at the dam site. With the Xiluodu arch dam, obvious contraction deformation of the valley has been observed after water storage and, compared with those already observed in other arch dam cases, the deformation is greater, more extensive, and more complex in mechanism.Next, based on the monitoring data of the contraction deformation of the Xiluodu Arch Dam valley, we will analyze the tempo-spatial distribution of the deformation and, by factoring into the post-storage evolution of hydrogeological conditions of the dam site, explore the causes of the deformation.

3. Analysis of the tempo-spatial law of the width deformation of the Xiluoduarch dam valley

Four upstream valley width monitoring lines (Line 01~04) and three downstream monitoring lines (Line 05~07) were set up, and an additional Line 08 in July 2014 and Line 09 in January 2016 were set up.The detailed layout of the monitoring lines is illustrated in Fig. 9.By April 2019, the upstream valley has narrowed by 69.42~93.47mm, the maximum value being observed at Line VDL03-VDR03, at an elevation of 722m. The downstream valley has narrowed by 78.83~80.51mm, and the maximum narrowing was at an elevation of 610m, observed along Line VDL06-VDR06.As far as spatial distribution goes, the contraction magnitudes of the upstream and downstream valleys are essentially the same, and the deformation varies very slightly along the elevation, with the deformation at a higher elevation being a little greater.

The elev. 559m monitoring line extends to the ADR20 and ADL20 drainage tunnels.The monitoring lines inside the two bank tunnels are both about 140m long, as shown in Fig. 10.By June 8, 2018, the valley had contracted by 37.73mm as measured at the tunnel openings. The tunnel opening monitoring point on the left bank shortened by 0.05mm relative to the monitoring point inside this tunnel. In the case of the right-bank tunnel, the narrowing was 0.89mm.All of this indicates that the relative displacements in both bank horizontal drainage tunnels are relatively small, suggesting that the contraction of the valley is not a superficial deformation but one present in a larger extent.

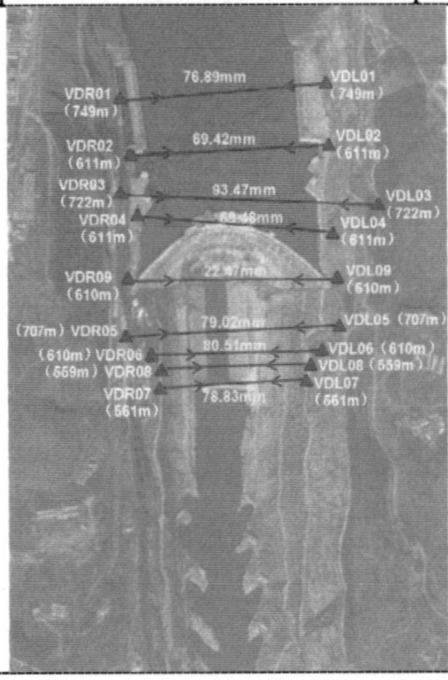

Figure 9. Schematic of the spatial distribution of the valley deformation

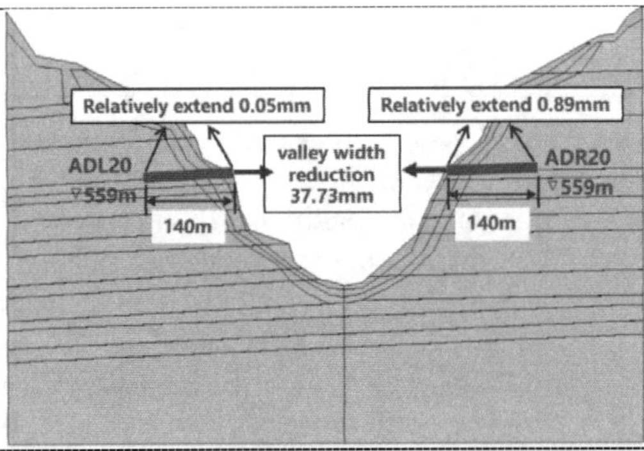

Figure 10. Valley width deformation observed in the ADL20-ADR20 drainage tunnels

Fig. 11 and 12 show the process lines of the deformations measured along the upstream and downstream monitoring lines. As can be seen, the valley contraction is not much connected with the water level, for the contraction does not decrease with the fall of the water level. The contraction continues after the reservoir is filled up with water. In May 2015, when the water level fell to its lowest level (545m) for the first time, the contraction trend slowed down clearly. At this moment, though the deformation has not converged, the rate is falling. A deformation process comparison between the upstream and downstream valleys suggests that contractions occurred synchronously upstream and downstream of the dam and were comparable in magnitude, implying that the effects of reservoir storage on the upstream and downstream bank slopes and dam foundation are synchronous.

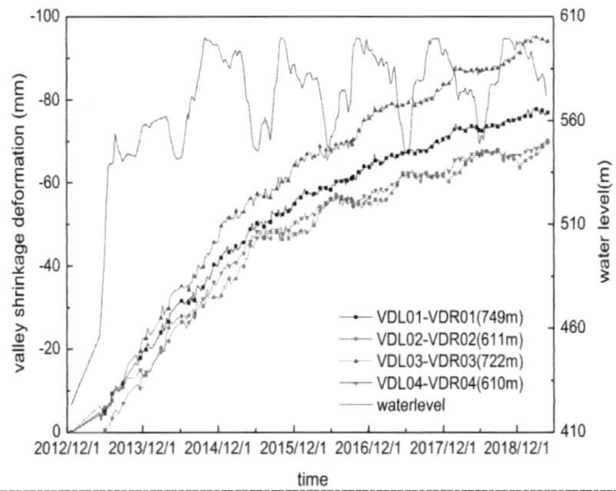

Figure 11. Process line of upstream valley contraction after water storage

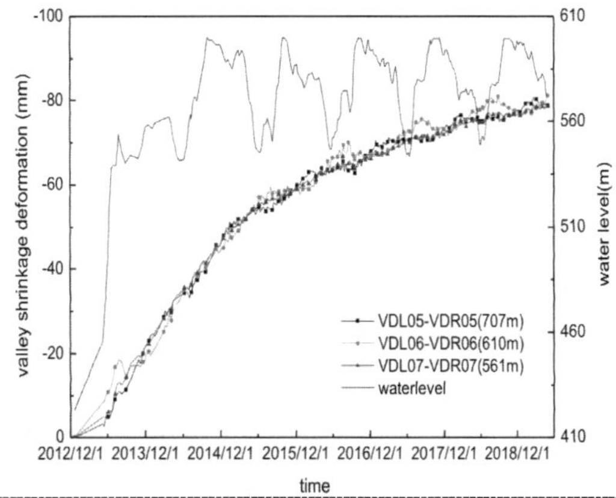

Figure 12. Process line of downstream valley contraction after water storage

4. Analysis of the deformation causes of the Xiluodu Arch Dam valley

4.1Hydrogeological conditions

Xiluodu Arch Dam is located in the Xiluodu gorge on Jinsha River, somewhere between Leibo County of Sichuan Province and Yongshan County of Yunnan Province. It is a concrete double-curvature arch dam with a height of 285.5m, a dam crest level of 610m and a normal water storage level of 600m. The dam bedrock and the slopes on both banks are mainly composed of basalt ($P_2\beta$) and limestone (P_1m), as shown in Fig. 13. The total thickness of the basalt is 490-520m. The limestone outcrops at the valley bottom bedrock, near the mouth of the Doushaxiriver in the dam area, tilting downstream while plunging beneath the basalt. The limestone, as revealed by hole boring, is 260-280m thick. Between limestone and basalt is disposed a layer of shale deposits ($P_2\beta_n$), of 2~3m thick and with a maximum thickness of 5.1m. No large fractured fault is found near the dam site. Inter-layer and intra-layer fault zones are the prominent geological structures in the dam site area, which macroscopically dictate the layer-like structure and geotechnical behavior of the rock mass.

IOP Conf. Series: Materials Science and Engineering **794** (2020) 012046 doi:10.1088/1757-899X/794/1/012046

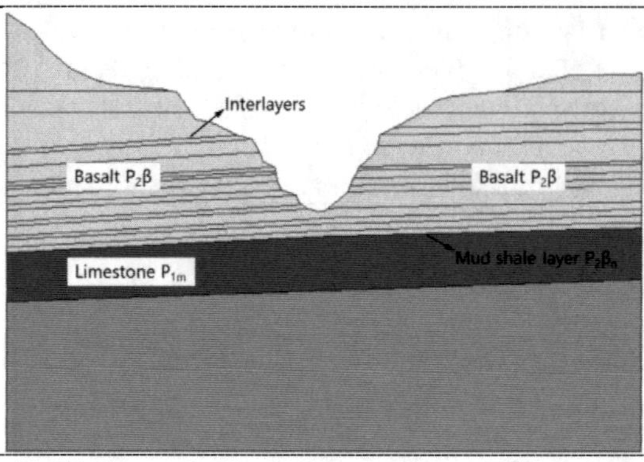

Figure 13. Hydrogeological cross section of Xiluodu dam site

Due to the deep-cut bottom fault around, the dam site area is in a regional geothermal anomaly zone.According to the riverbed borehole temperature data, the ground temperature increases with the depth. Generally, the influence of the river water is apparent within 10~20m below the bedrock surface and the temperature is low and varies appreciably, ranging between 18~25°C. As the depth grows, the temperature rises gradually, at a rate of about 3~5°C/100m owing to the geothermal warming effect.The limestone beneath the dam site contains groundwater with a higher temperature, which reaches 37~42°C at elevation 150~200m.

Fig. 14 shows a profile of the rock mass permeability at the Xiluodu dam site.$P_2\beta$ basalt mostly has a weak to faint permeability (k<3Lu), with moderately permeable rock mass (k=10~100Lu) and weak to moderately permeable rock mass (k=3~10Lu) only appearing in unloaded zone and in interlayer and intra-layer disturbed zones.The P^{1m} limestone has more developed karst in the upstream outcrop area, while the limestone below the dam base is weakly developed in karst. In the riverbed, at the dam base, there is at the top of the limestone (205~250m in elevation and 30~50m in thickness) a downstream, sloping, tapering wedge-shaped rock mass (with a moderate permeability, k=10~20Lu), under which is a faintly permeable rock mass (k=1~3Lu).The $P_2\beta_n$ mud shale layer has a low permeability, thus a relatively water-tight formation.

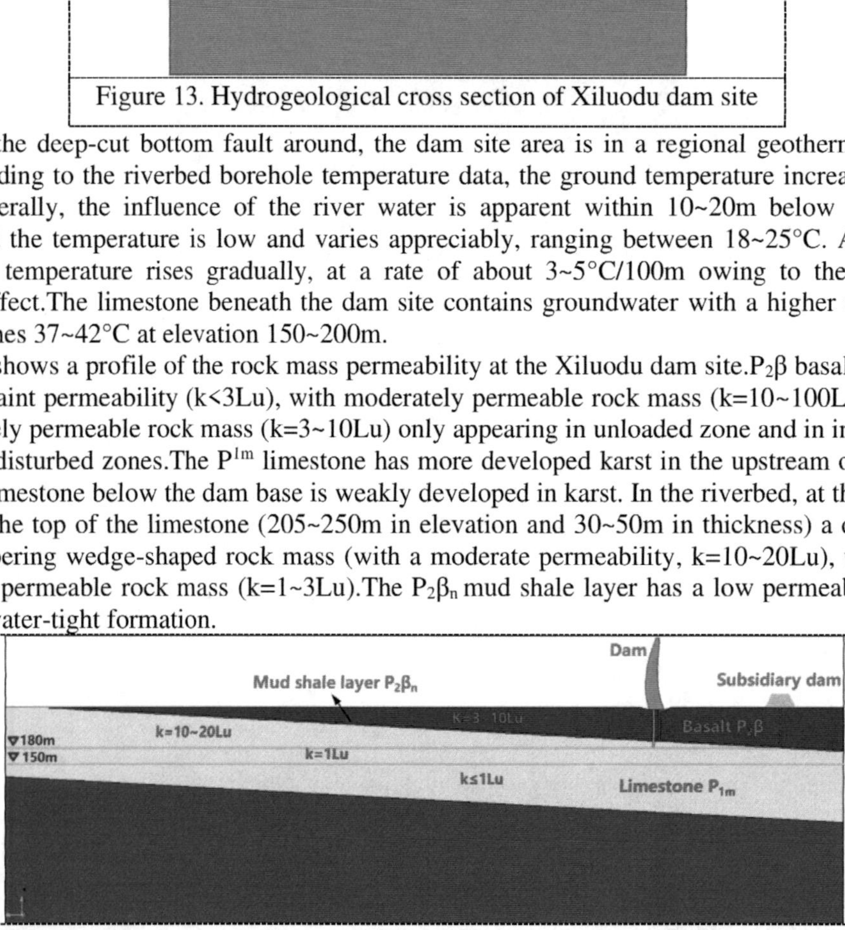

Figure 14. Zoning map of permeability profile of rock mass at Xiluodu dam site

4.2Cause analysis

Following water storage, the seepage field in the dam site was altered, bringing about changes in the hydrogeological conditions.Factors inducing valley contraction may include:

(1) Changes in material properties. The groundwater level at the dam site was low before the commencement of construction, and bank slope rock mass was first in a dry or unsaturated state. Water infiltration following water storage drove up the water content of the rock mass, ultimately softening it. However, there are no large faults and weak rock belts in the Xiluodu dam site. Prominent structural planes are disturbed belts developed inside or between rock layers, and those disturbed belts are mostly filled with basalt breccia and fragments, with small inclusions of rock debris and virtually no mud.Moreover, generally the basalt in the dam site area is monolithic, with shallow weathering

unloading, and thus has a high strength.The softening of the rock mass after water storage is of a weak degree, and therefore the softening of the rock mass is not the main cause of the valley contraction.

(2) Temperature change. The dam site being in a regional geothermal anomaly zone, the geothermal temperature of the underlying limestone is higher. Nevertheless, as the geothermal heat comes from deeper dam foundation, the reservoir water infiltration after storage merely cools down the superficial bedrock, without substantially altering the distribution of the geothermal field in the area.Therefore, the change in bedrock temperature is not the cause of the valley contraction.

(3) Stress changes. Some researchers measured the in-situ stress of the arch dam site before the dam construction[19, 20](see Fig. 15 and Fig. 16).The geostress values suggest that the dam is at a medium geostress area, dominated by near-horizontal tectonic stress. The geostress field is of a generally balanced distribution, with the stress increasing with the depth. The measured maximum horizontal principal stress range between 11 and 16MPa (with a maximum of 20MPa), essentially parallel to the river stream.The measured minimum horizontal principal stress is 7-11 MPa, basically perpendicular to the stream.From about 40m below the riverbed center to $P_2\beta_n$, the stress increases rapidly, and stress concentration occurs at about an elevation of 270m. The minimum horizontal principal stress is 8MPa, while the maximum is 14MPa.Cake cores are developed in the limestone formation below $P_2\beta_n$, indicating a higher geostress, a larger difference between the major and minor principal stresses, and a greater shear stress.

Therefore, before water storage, the shear stress was great near the aquiclude $P_2\beta_n$below the dam base, probably in a critical equilibrium state, predisposing the rock mass to deform along the $P_2\beta_n$ under extrusion.After water storage, with the seepage force acting on $P_2\beta_n$, the effective stress fell and the rock mass underwent disturbed deformation along $P_2\beta_n$, causing the valley to contract.

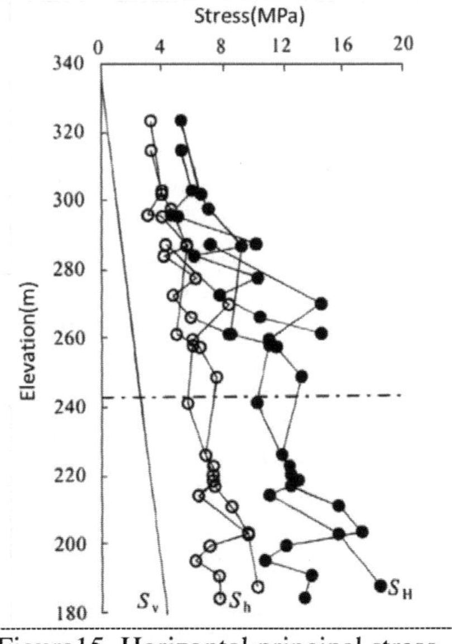

Figure15. Horizontal principal stress vs. depth in 3 river center boreholes[19]

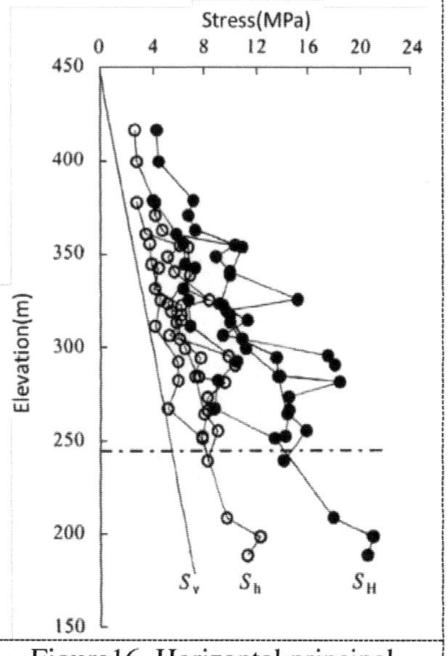

Figure16. Horizontal principal stress vs. depth in 4 left-bank boreholes[19]

5. Conclusions

This paper reviews typical cases of high arch dam valley width contraction in the world and analyzes the tempo-spatial distribution characteristics and causes of such deformation, to find that the main inducing factors of valley width contraction are the change of effective stress on the rock mass structural planes during the evolution of the dam site seepage field after excavation or water storage and the softening of the rock mass.From this it continues to examine the valley width monitoring data of Xiluodu Arch Dam, shedding light on the temporal evolution and spatial distribution characteristics of the deformation. The mechanism of the deformation causes is then studied in the context of the engineering geology and hydrogeological conditions of the dam site.It is found that the contraction of the Xiluodu arch dam valley is great, and the contraction magnitudes of the upstream and downstream valleys are essentially the same, the deformation varying very slightly along the elevation, with the deformation at a higher elevation being a little greater.The valley contraction is not much related to the water level, for the contraction continued after the reservoir was filled up with water. At this point, though the deformation has not converged, the rate is falling.The contraction of the Xiluodu arch dam valley was caused by the reduction of the effective stress in the rock mass during the evolution of the seepage field following water storage under the specific hydrogeological and geostress field distribution conditions.

Acknowledgement
This work was supported by the National Key Research and Development Project of China(2018YFC0406700); China Three Gorges Corporation Research Project (XLD/2114); National Natural Science Foundation of China (51779277).

References

[1] Frigerio A, Mazza G. The rehabilitation of Beauregard dam: the contribution of the numerical modeling. 12th International Benchmark Workshop on Numerical Analysis of Dams, Graz, Austria. 2013, 343-352.
[2] Barla G, Antolini F, Barla M, et al. Monitoring of the Beauregard landslide(Aosta Valley, Italy) using advanced and conventional techniques[J]. Engineering Geology, 2010, 116(3): 218-235.
[3] Lombardi E G. Ground-water induced settlements in rock masses and consequences for dams[C]//IALAD-Integrity Assessment of Large Concrete Dams Conference, Zurich. 2004, 24.
[4] Zangerl C, Evans K F, Eberhardt E, et al. Consolidation settlements above deep tunnels in fractured crystalline rock: Part 1—Investigations above the Gotthard highway tunnel[J]. International Journal of Rock Mechanics and Mining Sciences, 2008, 45(8): 1195-1210.
[5] Ehrbar H, Bremen R, Otto B. Gotthard Base tunnel-Tunnelling in the influence zone of two concrete arch dams[J]. Geomechanics and Tunnelling, 2010, 3(5): 428-441.
[6] Yan Fuzhang, Wang Sijing, XuRuichun. Deformation mechanism and development tendency of Maoping Landslide after impounding of Geheyanreservior on Qingjiangriver[J]. Journal of Engineering Geology, 2003(01): 15-24.
[7] Li Shouding, Li Xiao, Liu Yanhui. Study on generation and evolution of Maoping landslide on Qingjiangriver[J]. Chinese journal of rock mechanics and engineering, 2006, 25(2): 377-384.
[8] Zhang Haiping. Stability prediction and instability forecasting of the Guobo bank slope of the Laxiwa Hydropower Station on the Yellow River [D]. Chengdu: Chengdu University of Technology, 2011.
[9] Wu Mingxin, Jiang Hui, Zhang Chuhan. General rules of dam-valley deformation due to reservoirimpoundment [J]. Journal of hydroelectric engineering, 2019, 38(8): 1-14.
[10] Yang Jie, Hu Dexiu, Guan Wenhai. Analysis of High Slope Rock Deformation and Safety Performance for Left Bank of Lijiaxia Arch Dam[J]. Chinese Journal of Rock Mechanics and Engineering, 2005, 24(19): 3551-3560.
[11] Yang Qiang, Pan Yuanwei, Cheng Li, et al. Mechanism of Valley Deformation of High Arch Dam and Effective Stress Principle for Unsaturated Fractured Rock Mass[J].Chinese Journal of Rock Mechanics and Engineering ,2015,34(11):2258-2269.
[12] Yang Qiang, Pan Yuanwei, Cheng Li, Liu Yaoru. Impounding influence of slope and foundation deformation on high arch dam[J]. Chinese journal of rock mechanics and engineering, 2015, 34(suppl.-2): 3979-3986.

[13] L. Cheng, Y.R. Liu, Q. Yang, Y.W. Pan, Z. Lv. Mechanism and numerical simulation of reservoir slope deformation during impounding of high arch dams based on nonlinear FEM[J]. Computers and Geotechnics, 2017, 81: 143-154.

[14] Zhang Guoxin, Cheng Heng, Zhou Qiujing, Liu Youzhi. Analysis of mechanism of valley creep deformation of high arch dam during impoundment [J]. China sciencepaper, 2019, 14(1): 77-84.

[15] LIU Youzhi, ZHANG Guoxin, CHENG Heng, et al. High arch dam of the narrow valley causes and the influence on dam deformation and stress analysis[C]. 2014 annual conference of Chinese national committee on large dams. Zhengzhou: The yellow river conservancy press, 2014:51-60.

[16] Zhou Zhifang, Li Mingwei, Zhuang Chao, GuoQiaona. Impact factors and forming conditions of valley deformation of Xiluodu hydropower station[J]. Journal of Hohai University, 2018, 46(6): 497-505.

[17] He Zhu, Liu Yaoru, Yang Qiang, Cheng Li, XueLijun. Mechanism of valley deformation of Xiluodu arch dam and back analysis and long-term stability analysis[J]. Chinese Journal of Rock Mechanics and Engineering, 2018, 37(supple.-2): 4198-4206.

[18] LIANG Guohe, HU Yu, FAN Qixiang, et al. Analysis on valley deformation of Xiluodu high arch dam during impoundment and its influencing factors[J].Journal of Hydroelectric Engineering, 2016, 35(9): 101-110.

[19] Ding Lifeng, AnQimei, Wang Haizhong, Zhao Shiguang. Research on crustal stress measurement with hydrofracturing in the Xiluodu Hydropower Station on the Jinshariver[J]. Earthquake Research in China, 2004, 20(1): 95-100.

[20] Ding Lifeng, An Qimei, Wang Haizhong, Zhao Shiguang, WANG Haizhong[J]. Hydrogeology and Engineering Geology, 2004, (6): 56-59.

Experimental study on autogenous volume deformation of RCC mixed with MgO

Van-Chinh Nguyen[1,4], Xiu-Song Zhang[2,*], Van-Nghia Nguyen[3], Dinh-Tuan Phan[4], and Gang Liu[1]

[1]College of Hydraulic and Environmental Engineering, China Three Gorges University, Yichang, Hubei, 443002, China

[2]China Water Resources BeiFang Investigation, Design and Research CO.LTD, Tianjin,300000, China

[3]Division of Hydropower and Renewable Energy, Thuyloi University, Hanoi, Vietnam

[4]Hydraulic Construction Institute, Vietnam Academy for Water Resources, Hanoi, Vietnam

[*]Corresponding author's e-mail: zxs5718@126.com (Xiu-Song Zhang)

Author's e-mail: nguyenvanchinhcivil@gmail.com (Van-Chinh Nguyen),

nghia_nvn@tlu.edu.vn(Van-Nghia Nguyen), dtuan99@gmail.com (Dinh-Tuan Phan), and hkliugang@126.com (Gang Liu)

Abstract. The application of MgO to control thermal stress for concrete structures has achieved many accomplishments over the years. MgO is an important type of materials for the shrinkage cracking control of mass concrete in general, and Roller Compacted Concrete (RCC) dams in particular.Previous studies have investigated the application of MgO concrete and have proven the cracking prevention of MgO concrete.The purpose of this paper is to measure the Autogenous Volume Deformation (AVD) of RCC mixed with the MgO-expansive agent. And these experiments were conducted under different MgO content, temperature and curing time conditions. The experimental results of the AVD process show that the AVD process increased with the MgO content, curing time and temperature condition. The reseach outcomes are evidence that the hydration of expanded materials in concrete can compensate for shrinkage.

1. Introduction

There are many reasons for concrete cracking, including thermal stress and shrinkage during concrete construction period[1]. Massive concrete dams suffer from the limitations of foundation and the increase of thermal hydration of cement, and due to the large concrete masses, poor internal heat transfer performance. The cracks occur when the tensile stress exceeds the maximum allowable tensile stress of the concrete. The occurrence and development of these cracks not only affect the aesthetics of the project, but also destroy the integrity of the concrete dam, change the stress state of the dam, and even result in the concentrated leakage passage of dam, which causes hidden dangers for the dam. Therefore, the control of stress and temperature in concrete is a crucial task in the construction of concrete, and gravity concrete dams. In RCC dam construction, temperature control is even more critical[2].

In order to control the temperature in concrete, some typical methods are usedincluding using the cooling water pipe, precooling aggregate, increasing construction time, adding fly ash, and other traditional temperature control methods [3]. Although these methods contribute effectively to the control of concrete temperature in the dam, they increase construction time and cost money. Therefore, scientists have been continuously looking for new methods to prevent concrete cracks, including the addition of MgO-type expansion agents that can compensate for shrinkage of concrete and relieve the cracking risk [4]. Fact shows that MgO, after being heated at high temperatures, will naturally inflate during hydration to transform to MgOH., which has good latency, and a unique developmental delay by using this characteristic of MgO, the construction process of mass concrete can be effectively simplified, the concrete cracking of the dam body can be avoided, the construction progress can be accelerated, and the engineering investment can be saved.

History shows the changes in the proportion of MgO in concrete content. In the 1880s, cement with 16% to 30% MgO content was used in building bridges in France[5]. In the 1970s, research results at the Liujiaxia and Hengren dams in Jilin province, China demonstrated the reason for fewness cracks, whichthe AVD of MgO concrete made compensation for shrinkage at a lower temperature. The autogenous deformation of the Liujiaxia dam was $60{\cdot}10^{-6}$ and that of the Hengren dam was $100{\cdot}10^{-6}$ when concrete mixed with 4.5% to 6% MgO content. Baishan Dam used 5% MgO mixed into the concrete without concrete cracks. This result is the cause of the hydration of MgO cement [6].Therefore it can be asserted that the use of MgO in concrete is an important scientific foundation to improve the autogenous crack prevention of concrete. Previous studies as Lei Guo et al.[7], S. H.Liu et al. [8], P.W. Gao et al. [9] made predictions about the AVD process of MgO concrete. These studies and applications have shown that the expansion properties of MgO materials will determine the expandability of MgO cement. However, the quality of MgO expansion materials is closely related to production conditions, calcination systems, grinding fineness and particle size distribution, product uniformity and stability. The expansion behavior depends on many parameters including its reaction activity, dosage, mix proportion of concrete, curing temperature and age, etc.

For the reasons mentioned above, this paper studies the Autogenous volume deformation of Roller compacted concrete mixed with MgO. The experiment process under laboratory conditions took into account factors such as the content of MgO, the temperature, and the curing age. In the next sections, the experimental methods and results are first introduced, followed by the formula for calculating the AVD coefficient of RCC mixed with MgO. Finally, the visual results of the AVD process of MgO concrete will be presented with discussions.

2. The applications ofMgO

According to the previous statement, traditional techniques for controlling the temperature rise of concrete have been developed relatively. Nowadays, the technology of adding magnesia concrete to construct dams and RCC arch dams which have gradually become a research hotspot. So for thermal stress and temperature control, it is necessary to study the AVD properties of RCC mixed with MgO.As of 1995, the application of MgO concrete on the whole dam had been adopting the first way, for example, the dams in Liujiaxia, Hengren, Panjiakou; all of them adopted cement which contained 2-5% MgO. Since then, the method of adding MgO into the cement in the factory or adding MgO into the concrete while mixing was adopted in the dam, such as Shitang, Tongjiezi, Shuikou, Doufeng, Yanjinqiao, Lijiaxia dam, Qingxi, and Feilaixia dams in China [6].

3. Experimental study on the expansion of RCC mixed with MgO

3.1The experimental methods

As mentioned above, applications of using MgO to concrete have been widely used to increase the construction speed and prevent cracks of concrete dams. Besides the expansibility of MgO, the quality and duration of storage of cement, the type and content of mixtures, the manner, and content of MgO, aggregate gradation and the number of adhesives are the factors that affecting the expansibility of

MgO concrete. A large number of experimental studies and engineering applications show that the expansibility of MgO cement mainly depends on the expansibility of MgO material. However, the quality of MgO expansive material is closely related to the production conditions, calcination system, grinding fineness and particle size distribution, product uniformity and stability. In this paper, the authors focus on the experiment of the AVD process of MgO concrete under different MgO content, temperature and curing time conditions. There are three duplicate concrete specimens of size 150x150x500(mm)in each experimental group,in which the expansive material mixed with 3%, 4%, and 5% of MgO contents, respectively.And all experimental groups mixed with a fixed content of 60% Fly ash (FA) content. Although the effective content of MgO and FA will definitely affect the AVD process, however, in this article, the authors only emphasize the expansive deformation and development of the AVD process.The AVD of MgO concrete depends on the MgO content, the curing time and temperature. Therefore, the experiment of RCCmixed with MgO which were carried out by China National Standard DL/T5433-2009[10]. The material mixing proportions of experiment specimens as shown in Table 1 following.In particular, the type of MgO used has a reaction time of 120 (s) and an effective MgO content greater than 90%.

Table 1.The mix proportions of concrete specimens (kg/m^3)

Strength grade	MgO content %	Fly ash %	W/C	The mix proportions of concrete (kg/m^3)							
				Cement	Fly ash	Water	Sand	Coarse aggregate size			MgO
								40-60 mm	20-40 mm	5-20 mm	
	3	60	0.55	66	99	91	729	425	566	425	4.8
$C_{90}15$	4	60	0.55	66	99	91	729	425	566	425	6.4
	5	60	0.55	66	99	91	729	425	566	425	7.9

3.2 The experimental conditions

In the experiment on the AVD process of MgO concrete, the concrete specimens were stored at 20°C for the first 2 days after casting. And specimens were stored in HBY-28B Concrete Standard Curing (CSC) box which was cured at 20°C and relative humidity above 95% for 2 days before removing the mold. This equipment was manufactured according to the China National StandardGB/T50081-2002 Standard for Test Method of Mechanical Properties on Ordinary Concrete[11], and JG238-2008 Standard Curing Chamber for Concrete[12]. This experimentaltool has intelligent digital to control the temperature and humidity under standard conditions. During this period, the concrete specimens may shrink, however concrete is still relatively soft and adjustable, so the amount of shrinkage deformation will be very low. After removing the mold, the specimens will befurther experimented for 207 days with HBY-28B, and under laboratory conditions. Each test group was tested at 20°C, 30°C, 40°C, and 50°C, respectively. During the 207 days of this experiment, the length of the MgO concrete specimens after expansion will be accurately measured through two Linear Variable Differential Transformers (LVDTs). The experimental setting is shown in Fig. 1 following.

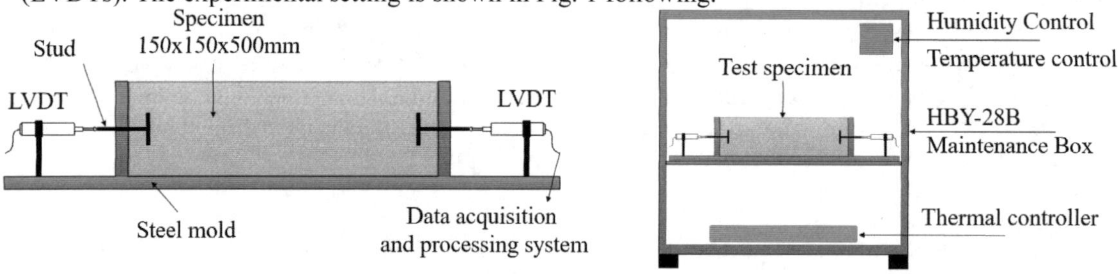

Fig. 1.Deformation measurement set

According to the China National Standard DL/T5296-2013[13], for small linear expansion deformation in three-dimensional space of an isotropic material, the AVD of the MgO concrete specimen will be measured as Eq. (1)following:

$$\varepsilon_\tau = 3 \cdot \frac{L_\tau - L_0}{L_0 - 2 \cdot \Delta} \tag{1}$$

Where: $\varepsilon_\tau$is thecoefficient of volume deformation of the MgO concrete sample at τ-days of age. L is the measurement length of MgO concrete which was carefully measured at each time throughout the 207-day experimental period where the time of measurements is shown in Figs. 2,3,4 ,*mm*. $L_0$is the reference length of the test specimen which was accurately measured immediately after removing the mold, mm.Δis the length of the stud which was partially positioned into the specimen and be connected with LVDTs, *mm*.Based on the experimental procedure described above, the experimental results of the AVD process of concrete are calculated by Eq. (1) and shown in Figs. 2,3,4 following.

4. Results and discussions

According to Eq. (1) theexpansion calculation results of RCC mixed with 5%, 4%, and 3% content of MgO which are calculated and shown in Figs. 2,3,4 following. As can be seen, the expansion appeared as a result of the expansion of MgO concrete.

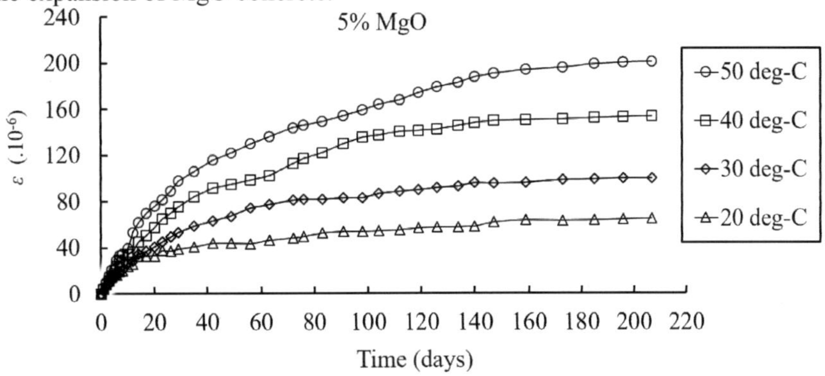

Fig. 2.The AVD process of the specimen mixed with 5% MgO content

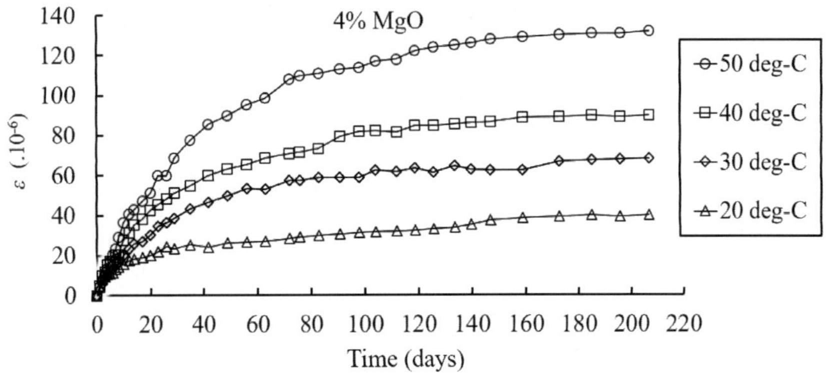

Fig. 3.The AVD process of the specimen mixed with 4% MgO content

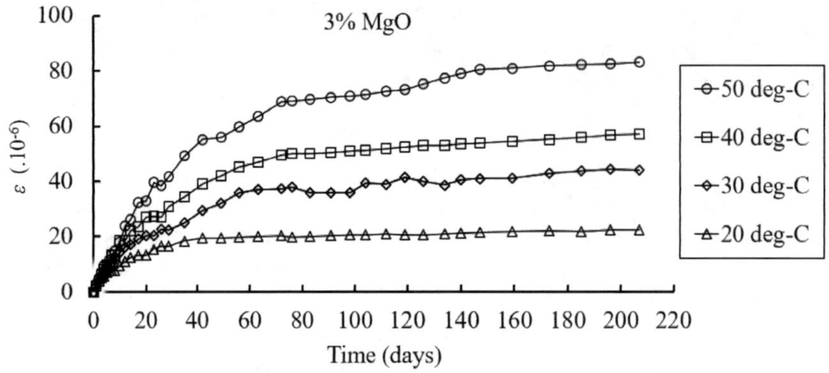

Fig. 4.The AVD process of the specimen mixed with 3% MgO content

The maximum AVD value ε_{max} of the experimental results corresponding to each case are calculated and listed in Table 2. In this experimental, $\varepsilon_{\tau=207\text{-}day}$ is the AVD value at the 200th experiment day.

Table 2.The values of $\varepsilon_{\tau=207\text{-}day}$ for three experimental specimens

MgO (%)	Specimen mixed with 5% MgO				Specimen mixed with 4% MgO				Specimen mixed with 3% MgO			
Temperature (T/°C)	20	30	40	50	20	30	40	50	20	30	40	50
$\varepsilon_{\tau=207\text{-}day}$ ($\cdot10^{-6}$)	65	95	158	209	40	68	90	127	23	44	57	81

As mentioned above, during 207 days of experiments, all concrete specimens were tested with HBY-28B (CSC) box in different curing temperature 20°C, 30°C, 40°C, and 50 °C, respectively. These figures show the AVD process of test samples was mixed with 5%, 4% and 3% of MgO content, respectively. The experimental results in this paper show that the AVD process of MgO concrete increased the MgO contents, the curing time and temperature increased. It proves that the shrinkage of concrete has been compensated by the hydration of the expanded materials mixed in concrete [14]. Figures 2,3,4 show that MgO concrete at early age tends to increase rapidly, while at later ages it tends to be constant. Specifically, the AVD process of MgO concrete increased rapidly in the first 200 days, and slower after that. These expansive deformation process can be surmised by mathematical models, from which it will be applied in the calculation of concrete dam temperature control, contributing to the crack prevention.

5. Conclusions

The expansive experimental model of RCC mixed with MgO is proposed to consider the AVD process under different MgO content, curing temperature and time conditions. This paper used three duplicate concrete specimens in each experimental group mixed with 3%, 4%, and 5% MgO, respectively. The specimens were cured at 20°C, 30°C, 40°C, and 50°C, and relative humidity above 95% under standard experimental conditions. Experimental results have demonstrated the expansion effectiveness of MgO concrete, and with the increasing of MgO-expansive agent content, curing temperature and time, the AVD of RCC increased. The results of this study will contribute to the scientific foundations in the use of MgO in general and its application in concrete crack control.

Acknowledgments

This research was supported by the National KeyR & D Program of China (Grant No.2017YFC1501100), the National Natural Science Foundation of China (Grant No. 51279090, 51679129) and sponsored by Research Fund for Excellent Dissertation of China Three Gorges University. The authors thank the reviewers for useful comments and suggestions that helped to improve the paper.

IOP Conf. Series: Materials Science and Engineering **794** (2020) 012047 doi:10.1088/1757-899X/794/1/012047

References

[1] H. Su, J. Li, and Z. Wen, "Evaluation of Various Temperature Control Schemes for Crack Prevention in RCC Arch Dams During Construction," *Arabian Journal for Science and Engineering*, vol. 39, no. 5, pp. 3559–3569, 2014.
[2] J. Noorzaei, K. H. Bayagoob, W. A. Thanoon, and M. S. Jaafar, "Thermal and stress analysis of Kinta RCC dam," *Engineering Structures*, vol. 28, no. 13, pp. 1795–1802, 2006.
[3] B. F. Zhu, "On the feasibility of building high quality arch dams without cracking and the relevant techniques," *Journal of Hydraulic Engineering*, vol. 30, no. 10, pp. 1155–1162, 2006.
[4] ChongJiang. Du, "A Review of Magnesium Oxide in Concrete," *Concrete International-Detroit*, vol. 27, pp. 45–50, 2005.
[5] Lea. F. M, "The chenistry of cement and concrete," *Chemical Publishing Company*, 1971.
[6] Y. Zhu, Z. Xu, and L. Zhang, "Comment on dam construction technology with MgO concrete," *Hongshui River*, vol. 21, no. 3. pp. 45–49, 2002.
[7] L. Guo and H. Y. Cai, "Experimental Study on RCC Character of Additive MgO and Simulation of Crack Prevention in High Temperature Construction," *Advanced Materials Research*, vol. 368–373, pp. 651–654, 2011.
[8] S. H. Liu and K. K. Fang, "Study on autogenous deformation of concrete incorporating MgO as expansive agent," *Environmental Ecology And Technology Of Concrete*, vol. 302–303, pp. 155–161, 2006.
[9] P. W. Gao, S. Y. Xu, X. Chen, J. Li, and X. L. Lu, "Research on autogenous volume deformation of concrete with MgO," *Construction and Building Materials*, vol. 40, pp. 998–1001, 2013.
[10] China National Standard DL/T5433-2009, *Test Code for Hydraulic Roller Compacted Concrete*. Changjiang Water Resources Commission Yangtze River Scientific Research, 2009.
[11] China National Standard: GB/T50081-2002, *Standard for Test Method of Mechanical Properties on Ordinary Concrete*. Ministry of Construction, 2002.
[12] China National Standard JG238-2008, *Standard Curing Chamber for Concrete*. Ministry of Construction, 2008.
[13] China National Standard DL/T5296-2013, *Technical Specification of MgO for use in Hydraulic Concrete*. Ministry of Construction, 2013.
[14] P. W. Gao, S. Y. Xu, X. Chen, J. Li, and X. Lu, "Research on autogenous volume deformation of concrete with MgO," *Construction and Building Materials*, vol. 40, pp. 998–1001, 2013.

IOP Conf. Series: Materials Science and Engineering **794** (2020) 012048 doi:10.1088/1757-899X/794/1/012048

Modalidentification of a building structure by frequency domain decomposition

Jun Ma[1*],Wei Luan[2]

[1]Engineering research institute, China construction eighth engineering division, Shanghai, 200122, China

[2]Northwest branch company, China construction eighth engineering division, Xi'an, Shaanxi, 710000, China

[*]Corresponding author's e-mail:mjbppc@163.com

Abstract. Frequency domain decomposition (FDD) and Enhanced frequency domain decomposition(EFDD) methods were adopted to estimate modal parameters of an18-story reinforced concrete building. Ambient vibration test was conducted on the 18-story reinforced concrete building and the two modal identification methods above were applied. The results show that FDD and EFDD methods can estimate modal parameters accurately and effectively from ambient vibration data.

1. Introduction

Modal identification techniques can identify modal parameters from dynamic measurements and provide important information reflecting the dynamic characteristics of structures. Traditionally, input-output modal identification methods are used to estimate modal parameters. However, input-output modal identification methods require artificialexcitations to produce input signals. As structures are becoming higher and larger, it is difficult or even impossible to excite a large structure artificially nowadays.

Output-only modal identification techniques become popular as the difficulties obtaining input signals are avoided. Based on the white noise assumption, various output-only modal identification methods have been developed. In the frequency domain, Peak-picking (PP) method has been applied for decades. Frequency domain decomposition (FDD) method was later developed by Brincker et al [1,2] and it overcame disadvantages of the PP method. Enhanced frequency domain decomposition(EFDD) method was subsequently developed to estimate modal damping [3].

In this paper, FDD and EFDD methods were adopted to estimate modal parameters of an 18-story reinforced concrete building. Ambient vibration test was conducted on thebuilding, and the two modal identification methods were employed. The application procedure was presented. The results demonstrate that FDD and EFDD methods can effectively and accurately estimate modal parameters from ambient vibration data.

2. Frequency domain decomposition method

Frequency domain decomposition method removes limitations of the classical approaches. By taking singular valuedecomposition (SVD) of the spectral matrix, the spectralmatrix is decomposed into a set of auto spectral densityfunctions, and each function corresponds to a single degree of freedom (SDOF) system. Modal parameters can then be obtained.

Considering an n degree of freedomstructure subjected to an input $x(t)$. The relationship between the measured output response y(t) and the unknown input can be expressed as

$$G_{yy}(j\omega)=\bar{H}(j\omega)G_{xx}(j\omega)H(j\omega)^T \tag{1}$$

where$G_{xx}(j\omega)$ is the power spectral density matrix of the input, $G_{yy}(j\omega)$ is the power spectral density matrix of the output, H(jω) is the frequency response function (FRF), the overbar and the superscript T denote the complex conjugate and transpose, respectively.

The frequency response function can be written as

$$H(j\omega)=\sum_{k=1}^{n}\frac{R_k}{j\omega-\lambda_k}+\frac{\bar{R}_k}{j\omega-\bar{\lambda}_k} \tag{2}$$

with

$$R_k=\phi_k\gamma_k^T \tag{3}$$

where$n$ is the DOF number and the mode number, R_k is the residue,λ_k is the pole, ϕ_k is the mode shape vector, γ_k is the modal participation vector.

If the input $x(t)$ is ambient excitation, with the white noise assumption, $G_{xx}(j\omega)$ is reduced to a constant matrix C. Based on the Heaviside partial fraction theorem, Eq. (1) is reduced to a form as follows

$$G_{yy}(j\omega)=\sum_{k=1}^{n}\frac{A_k}{j\omega-\lambda_k}+\frac{\bar{A}_k}{j\omega-\bar{\lambda}_k}+\frac{R_k}{-j\omega-\lambda_k}+\frac{\bar{R}_k}{-j\omega-\bar{\lambda}_k} \tag{4}$$

whereA_k is the $k$thresidue matrix of $G_{yy}(j\omega)$, and can be written as

$$A_k=R_kC\left(\sum_{s=1}^{n}\frac{\bar{R}_s^T}{-\lambda_k-\bar{\lambda}_s}+\frac{R_s^T}{-\lambda_k-\lambda_s}\right) \tag{5}$$

In case that the structural damping is small and there are no close modes, Eq. (5) can be simplified as follows

$$A_k=d_k\phi_k\phi_k^T \tag{6}$$

where$d_k$is scalar constant.

Taking Eq. (6) into Eq. (4), $G_{yy}(j\omega)$ can be simplified as follows

$$G_{yy}(j\omega)=\sum_{k=1}^{n}\left(\frac{d_k\phi_k\phi_k^T}{j\omega-\lambda_k}+\frac{\bar{d}_k\bar{\phi}_k\bar{\phi}_k^T}{j\omega-\bar{\lambda}_k}\right)=\Phi\cdot diag\left(2\mathrm{Re}\left(\frac{d_k}{j\omega-\lambda_k}\right)\right)\cdot\bar{\Phi}^T \tag{7}$$

The power spectral density matrix of the kth order can be represented as follows

$$\underset{\omega\to\omega_k}{G_{yy}}(j\omega)=\phi_k\cdot diag\left(\frac{2d_k\zeta_k\omega_k}{(\zeta_k\omega_k)^2+(\omega-\omega_k)^2}\right)\cdot\bar{\phi}_k^T \tag{8}$$

From Eq. (8), the singularvalue decomposition form can be noticed, and it is noticed that only one mode can contribute to the maximum of power spectral energy. Therefore, with the measured response data induced by ambient excitation, the modal parameters can be estimated by the FDD method.

In order to estimate structural damping as well, Enhanced frequency domain decomposition method was developed. In EFDD method,thepiece of the SDOF density function obtained aroundthe peak of the PSD is taken back to the time domain by the Inverse discrete Fourier transform

IOP Conf. Series: Materials Science and Engineering **794** (2020) 012048 doi:10.1088/1757-899X/794/1/012048

(IDFT)method, and the damping ratio can simply be obtained through inspection of the decay of autocorrelation functions.

3. FDD and EFDD methods applied to a building

FDD and EFDD methods were applied to ambient vibration data of an 18-story building. The test building extends from 5.3m underground to 53.1m above the basement level. The plan of a standard story is 22.2m long by 11.4m wide, and the story height is 2.95m.The lateral and gravity load-resisting system consists of reinforced concrete shear walls. Layout of the structural system is shown in Figure 1. Ambient vibration test was conducted on the building. Two accelerometers were set horizontally, one along the length direction and the other along the width direction, and the remaining accelerometer was set vertically.The response acceleration data along the width directionand the associated PSD function are shown in Figure 2. Due to the ambient excitation, the response amplitude is relatively small, and it can be observed that the modes are scatteredly distributed.

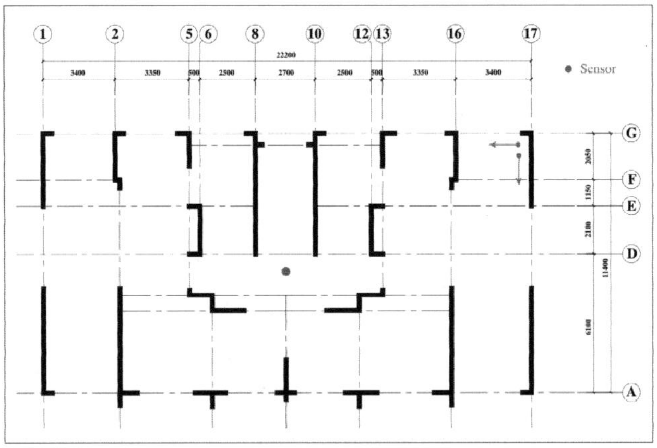

Figure 1.Layout of the structural system

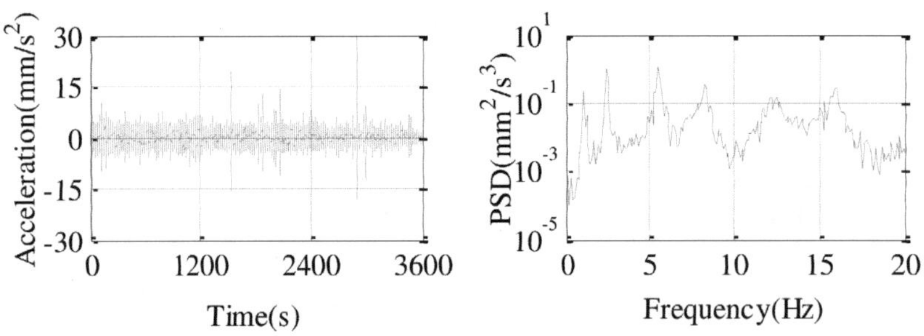

(a) Acceleration along width direction (b) Power spectral density

Figure 2.Acceleration record along width direction

Singular values of the PSD matrix of the output are obtained, as shown in Figure 3. The close modes can be clearly indicated in the plot. The 1st mode and the 2ndmode are coupled, and the 4th mode and 5th mode are coupled. The first six modes are indicated in the figure and the natural frequencies are listed in Table 1.

Thesingular values corresponding to the first mode are obtained, as shown in Figure 4. A reasonable part of the PSD functions is identified, and the IDFTtechnique is applied on the identified part to obtain the normalized autocorrelation function. Finally, the damping ratio is obtained by the logarithmic decrement technique. The estimated damping ratios of the first six modes are listed in Table 1.

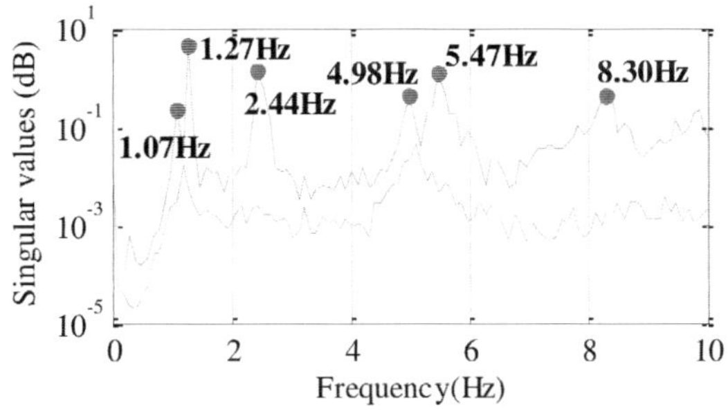

Figure 3.Singular values of the output response PSD matrix

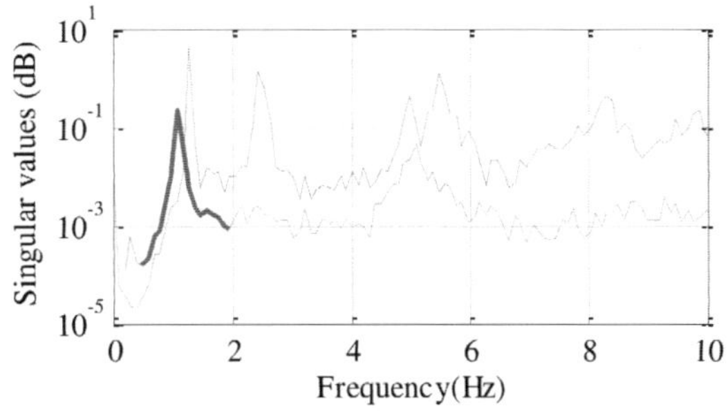

Figure 4.Identification of the first SDOF system

Table 1.Natural frequencies and damping ratios of the building.

Mode	Natural frequency (Hz)	Damping ratio (%)
1st	1.07	1.11
2nd	1.27	0.81
3rd	2.44	1.36
4th	4.98	1.35
5th	5.47	1.31
6th	8.30	2.29

4. Conclusions

In this paper, FDD and EFDD methods are introduced and are applied to ambient vibration data of a building. The results clearly indicate that FDD technique is able to extract modal information from ambient vibration data, and can estimate close modes with high accuracy. The EFDD method extends the application of FDD method and leads to an estimation of structural damping. By combined utilization of FDD method, EFDD method, and the ambient excitation technique, modal parameters can be identified, and the testing cost is significantly reduced.

References

[1] Brincker, R., Zhang, L.M.,Andersen, P. (2001) Modal identification of output-onlysystems using frequency domaindecomposition. Smart Mater.Struct., 10: 441-445.
[2] Brincker, R., Ventura, C.E., Andersen, P. (2001)Modal identification from ambient responses using frequency domain decomposition. In: 18th International Modal analysis Conference.
[3] Jacobsen, N.J.,Andersen, P., Brincker, R.(2006) Using enhanced frequency domain decomposition as a robust technique to harmonic excitation in operational modal analysis. In:International Conference on Noise and Vibration Engineering.

Feasibility Study of Applying BIM Technology in Shantytowns Renocation Project

Shaochun Ma*, Hao Wang, Peng Bao, Hao Dong and Sen Wang

School of Civil Engineering and Architecture, Henan University, Kaifeng, Henan, 475004, China

*Corresponding author's e-mail: mashaochun2627@126.com

Abstract. The transformation of shantytowns is a very important livelihood project in our country. After more than 10 years' efforts and vigorous efforts of the Chinese government, shantytowns have been improved and optimized to some extent. Based on BIM technology, this paper applies it to the actual shantytown reconstruction project. By using the powerful functions of Building Information Modeling (Hereinafter referred to as BIM) technology, the shantytowns renovation project is discussed from aspects of schedule control, construction simulation, collision inspection, rapid measurement and valuation, and process monitoring. Compared with the traditional methods, BIM technology has shown a good result, which has played a better positive role in the fine management level of shantytowns renovation projects, and promoted the healthy and orderly development of the construction industry.

1. Introduction

BIM is a tool that uses all kinds of information data collected from construction projects to build a 3D building model [1], and expresses the physical and functional characteristics of construction projects. BIM Technology can not only simulate the real environment of the building, but also directly reflect the design results [2]. BIM technology develops rapidly in China, not only obtains the support of national policy, but also receives high attention in theoretical research, and is often widely used in engineering projects [3,4]. The application of BIM Technology in projects can change the traditional management mode, so that the enterprise can realize fine management and enhance its core competitiveness.

Shantytown is a specific historical product in the process of urban construction and development [5]. Generally, it refers to the places where houses are concentrated in the scope of urban reconstruction and expansion, such as old cities, old areas and old buildings,which lack of planning, and have the disadvantages of poor performance of earthquake resistance, insufficient fire prevention and flood control, as well as inconvenient traffic. The existence of shantytowns in every city has affected the overall image of the city and restricted the speed of healthy development of the city. In the context of comprehensively promoting the transformation of shantytowns nationwide, there will be more and more shantytowns transformation projects. The government has worked out a popular support project for the reconstruction of old dilapidated houses, the improvement of housing quality and the improvement of living environment of poor families [6]. The shantytown reconstruction project has the characteristics of large scale and large volume, so it puts forward higher requirements for the construction unit in the management of talents and funds and the management of the collaborative work of all participants. BIM Technology has the advantages of platform effect, engineering data sharing, collision inspection, pipeline comprehensive optimization, etc., which can better improve the quality of architectural design and meet the construction requirements of building function, social and environmental benefits, cost, construction quality, time, etc. BIM Technology is applied to the design stage of the project. By collecting information of each stage of the project, BIM information platform is built to enhance the cooperation ability among all parties, realize information integration and management, prevent project risks, improve project management level [7], and promote and ensure the successful completion of the project. It is of great significance to promote people's livelihood projects, solve the problem of shantytown transformation, and improve people's living environment and quality of life.

2. Advantages and value of BIM Technology in project engineering

2.1engineering measurement

The calculation of engineering quantity is the most basic task in project engineering. Only when the quantities are accurately counted can we continue to work out the project budget and project cost. Only by completing the work quantity statistics accurately, can the project successfully and efficiently complete the cost management work in the project settlement, project bidding, design change and other stages.

BIM software is used to build a three-dimensional model of BIM building. By means of parametric modelling, the components are parameterized, the construction process is standardized, and the materials have time information and price information. The two-dimensional of the project is converted into three-dimensional, which is more convenient for the analysis of the construction structure of the project and improves the efficiency of project management. BIM measurement software is used to calculate the quantities based on BIM model, and there is no need to model again. As long as the quantities in the project cost in Revit model file are exported, the quantities of each discipline can be accurately obtained in the project.

2.2In terms of project pricing

The most important core of project management is to maximize project benefit and control cost effectively. Cost management is the basic project whether the project can achieve the expected effect, so it is necessary to control the cost effectively in the whole process. In the construction project management, BIM Technology can improve the level and efficiency of cost management. In the specific operation, it can be realized in the following aspects:

2.2.1Model optimization

Based on the quota design and effective investment control, combined with the actual requirements of the owner and the outstanding problems of each specialty, the BIM model is optimized and an effective control scheme is formulated to avoid the impact in the construction process.

2.2.2Monitor and adjust the data information existing in the model

Based on BIM Technology, fine management of engineering project can be realized. During the design period, the intelligent design of building structure can be directly carried out by adjusting parameters to ensure the overall rationality of building design. At the same time, during the construction period, it is necessary to obtain the basic data of the project in time, so that the construction party and the project department can realize the timely synchronization of information. With a detailed data base, it can better realize the dynamic cost control of the whole process of the project, and effectively control the project risk.

2.2.3Build database and share data and information

BIM Technology is applied to upload the real-time data information of the project to the platform. Through the adjustment, addition and modification of these data information, the overall situation of the project can be reflected in a timely and accurate manner. Then through the correlation with the data, cost risk management and decision quality can be realized, so as to improve the project quality, reduce the unprovoked loss of the project cost and increase the project profit.

2.3In terms of project progress

Based on the BIM related software timeliner, the project schedule is controlled. Based on the BIM three-dimensional building model, the 3D model and schedule are combined to form the 4D of BIM. The actual progress and BIM schedule model are compared to monitor the construction progress in real time. This method is used for visual dynamic control of the construction progress, so as to ensure the realization of the construction progress management.

2.4In terms of project quality

Based on the BIM project software naviswork, the quality control of the project is carried out. Based on the construction of BIM three-dimensional building model, using collision inspection software, the existing collision problems can be detected in time, so as to effectively prevent problems in the construction. Through the application of BIM Technology, it can realize the information sharing of all participants, improve the collaborative ability of managers, and play a very important role in improving the level of project quality management.

3. The application of BIM in the process management of the shed renovation project

3.1Project overview

This project is No. 3 building of a resettlement community project of shantytown reconstruction in Zhengzhou City, Henan Province, with 8 floors above ground and a building area of 12699.89 m^2. This project has a large number of subprojects, a large amount of funds and a long project cycle. BIM Technology has a great advantage in the management of the whole process of the project. This paper takes building 3 as an example.

3.2BIM modeling

Using the method of turnover, using the existing civil engineering CAD construction drawings, using Revit software to create the three-dimensional model of BIM. Import the construction plan of the project into Revit, and build a 3D model according to the 2D construction drawing of CAD, as shown in Figure 1 and figure 2.

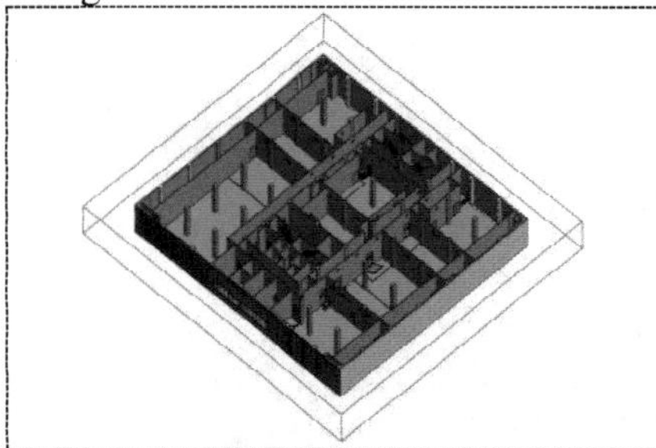

Figure 1. BIM model establishment

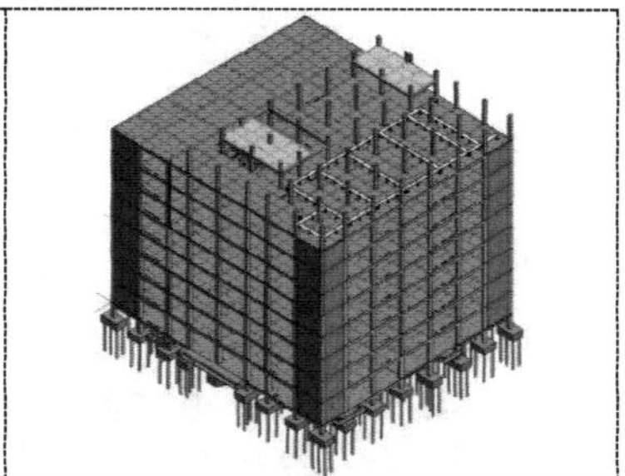

Figure 2. BIM structure model

3.3Application of BIM in rapid measurement and pricing of projects

In the bidding process of a project, it takes a long time and low efficiency to use traditional methods to prepare bidding control price [8]. The project company can use BIM software to build BIM building model, and through the use of Guanglianda pricing software, it can quickly and accurately calculate the project quantity and prepare the bidding control price. At the same time, the project bill of quantities can be compared with the relevant engineering data in BIM database to check the rationality of the project bill of quantities and provide basis for the formulation of project cost.

3.3.1Preparation of bidding control price

On the basis of completing BIM building model, use GFC plug-in to import BIM model in GFC format into Guanglianda BIM civil engineering calculation software GCL, and use automatic statistical summary to get the bill of quantities. Then import the GFC file into the software GGJ of BIM reinforcement calculation of Guanglianda, and get the following model as shown in Figure 3. If different modules are selected, various bill of quantities can also be obtained, and the bill of quantities for reinforcement is shown in Figure 4.

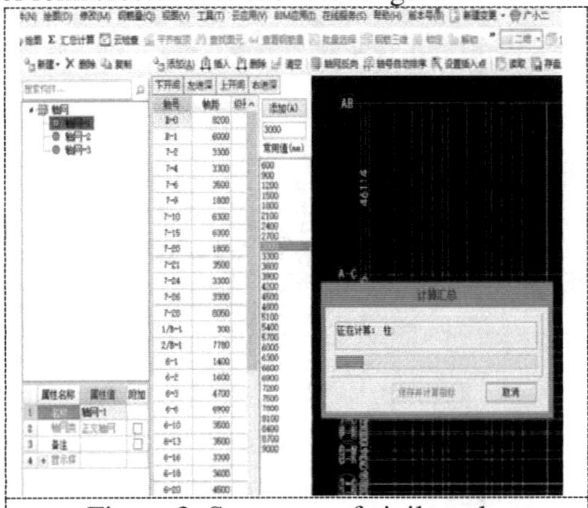

Figure 3. Summary of civil works

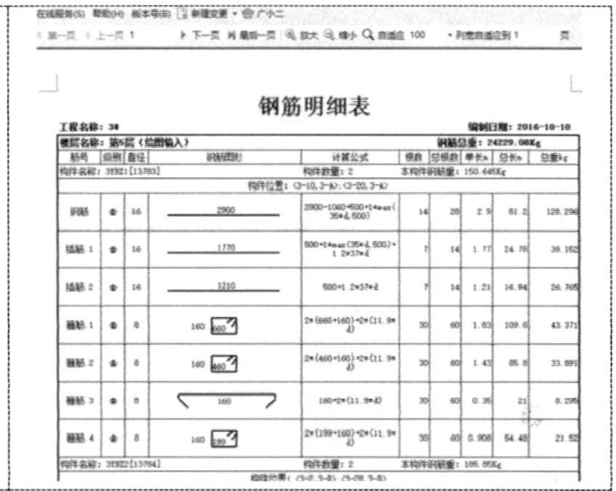

Figure 4. Bill of quantities of reinforcement

Then import the bill of quantities automatically summarized into the pricing software of Guanglianda, automatically apply the quota sub items, and finally get the bidding control price of the project. The specific steps are as follows: the establishment of BIM model - the use of Guanglianda BIM civil engineering calculation volume - obtaining a bill of quantities - the use of Guanglianda pricing software - the use of Guanglianda to prepare the bidding documents.

Through the application of BIM Technology, different effects can be achieved in different project types, especially in the common concrete and reinforcement projects and the final bidding control price, and the error is relatively small, as shown in Table 1. It can be seen that the project company's operation of this method to prepare the bill of quantities and determine the bidding control price can quickly complete the task and generate time and cost-effectiveness.

Table 1. Benefit analysis based on BIM Technology.

Type	**Number of personnel**	**Measurement and pricing**	**Concrete**	**rebar**	**Bidding control price**
Traditional measurement	10	60(h)	88.9 (10k RMB)	98.5 (10k RMB)	668.8 (10k RMB)
BIM technical measurement	5	60(h)	83.5 (10k RMB)	89.7 (10k RMB)	638.8 (10k RMB)
Benefit percentage	50.00%	66.70%	6.10%	8.90%	4.50%

3.3.2Quick measurement and pricing

When the designer uses the traditional method to arrange all professional pipelines, he can only coordinate according to the two-dimensional plan drawings, and lacks the elevation of the pipelines. Therefore, when facing the complex problem of staggered pipelines, he can only rely on the designer's experience and imagination of the three-dimensional space position to judge the rationality of pipeline layout, which is not only time-consuming and labor-consuming, but also inefficient. The use of BIM Technology (Navisworks) can solve the problem of collision between different disciplines that may occur in the construction process in advance [9]. Generally, after the BIM building model is completed in the design stage, collision detection is carried out to reduce the occurrence of changes in the construction process, so as to improve the construction efficiency and reduce the project loss.

IOP Conf. Series: Materials Science and Engineering **794** (2020) 012049 doi:10.1088/1757-899X/794/1/012049

(1) simulated collision inspection

First, transfer the model to naviswork for collision detection as NWF file from Revit software, as shown in Figure 5 and Figure 6. Select the required collision tolerance, run the detection, and finally get the number and location of collision points, and generate the three-dimensional solid diagram of collision points.

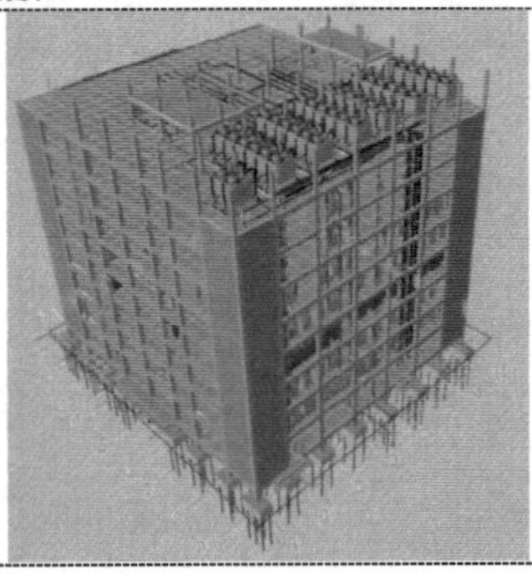

Figure 5. Building and structure combination model

Figure 6. Model visualization structure

(2) quick measurement and valuation of the changed quantities

In case of design change, the designer can modify the BIM model directly according to the requirements of the owner, and automatically calculate the quantities after modification. Through data association and remote update of the design model file, the construction enterprise can directly see the model after design change and the changes of the quantities after design change through updating the data on the platform.

In engineering, we often encounter the situation of change and calculation of quantity and price. BIM Technology can make the tedious work more accurate and fast. It can realize the effective management of design change and dynamic control of cost, reduce the labor volume of budget personnel, and improve the efficiency of cost personnel. The rapid measurement and pricing with BIM Technology enables the project company to quickly complete the visa work after the change, as shown in Table 2.

Table 2. Rapid measurement and benefit analysis based on BIM Technology.

Type	Number of personnel	Measurement and pricing	Rebar engineering	Change and increase
Traditional metering	3	8(h)	20.5 (10k RMB)	15.9 (10k RMB)
BIM Technology measurement	1	3(h)	19.5 (10k RMB)	15.2 (10k RMB)
Benefit percentage	66.70%	62.50%	4.90%	4.40%

3.4Application of BIM in project schedule quality management

In the process of construction, the government issued the key action of comprehensive air pollution control, which is known as the "stop work order". The stop work lasts for up to four months, even longer if the situation is special. This shutdown has caused great impact on cost control and schedule control of the project. BIM Technology can be used to reasonably arrange the construction schedule after the shutdown in advance to improve the efficiency of schedule management.

Firstly, the NWC file is exported by BIM 3D building model, and then the NWC file is imported into naviswork. As shown in Figure 7. Set various task progress information, related costs and other expense conditions in timliner. The actual start time and end time of the project, the planned start time and end time of the project, and various expenses are shown in Figure 8.

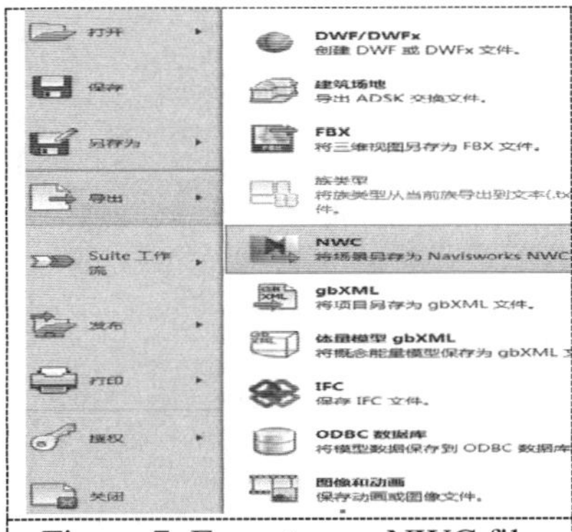

Figure 7. Export save NWC file

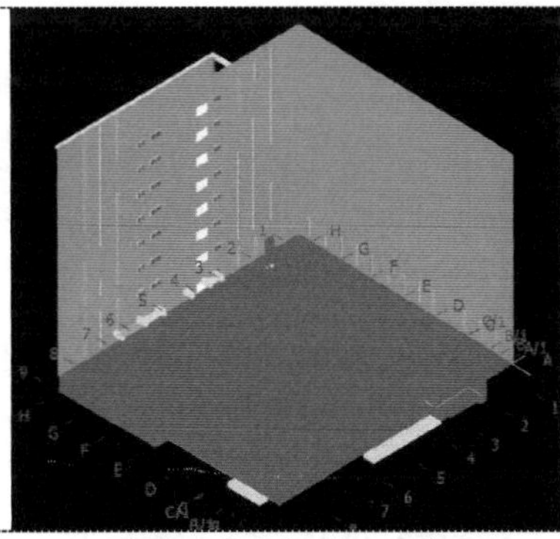

Figure 8. Timliner plan

Compared with the traditional method of progress bar chart, it is not easy to adjust when the project is complex and the design changes occur. Using the BIM software (naviswork) to reasonably plan the construction progress after the stop work order is completed, and using the dynamic construction process simulation can make the construction party more intuitive and vivid to show the progress. BIM Technology is used for project schedule quality management, and the data information related to schedule in BIM model is associated with the model itself, so that the construction schedule can be observed and managed dynamically in real time. The project company uses BIM Technology to control the progress, which can realize the fine management of the project and improve the management level of the project, and produce the time benefit and cost benefit.

4. Conclusion

BIM Technology, as the main development direction of the construction industry, does not conflict with the traditional construction technology. The shantytown reconstruction project has the characteristics of large volume, long cycle, large capital demand and large risk, which is representative. The application of BIM Technology in the renovation project will highlight the advantages of BIM Technology throughout the whole project construction cycle and collaborative management. In this paper, through the feasibility study of the application of BIM Technology in the shantytown reconstruction project, the main conclusions are as follows:

(1) BIM Technology can effectively improve the efficiency of cost control. It has a good advantage in determining the bidding control price and quick measurement and pricing after design change. In the process of engineering construction, it can solve the problem of lag in process measurement.

(2) BIM Technology is used to effectively collect design information of all disciplines, so that designers can grasp the project as a whole, and better complete the project in progress quality control, collision inspection and other work.

(3) The time span of shantytown reconstruction project is long and the work quantity is complex. There are too many participants in the project and the information is not transmitted in time, which will lead to the difficulty of collaborative management of project managers and the inability to achieve refined management. By using BIM Technology, we can accurately control the cost, accurately complete the work of splitting and statistics, and improve the efficiency of construction and capital use. BIM Technology provides a new method for project cost management.

Acknowledgments

The authors would like to express heartfelt gratitude to the financial supported by the Science Technology of the Ministry of Housing and Urban–Rural Development of the People's Republic of China (2018-K9-065), China Postdoctoral Science Foundation funded project (2018M632805), Henan provincial key scientific and technological project (182102210234), Science and Technology Plan Project of Henan Provincial Department of Housing and Urban-Rural Development (K-1823), and supported by Foundation of Henan Educational Committee (18A560007).

IOP Conf. Series: Materials Science and Engineering **794** (2020) 012049 doi:10.1088/1757-899X/794/1/012049

References

[1] Zhou Z.S.(2018) Research on index system of efficiency evaluation of assembly building industrialization based on BIM. Beijing Jiaotong University for the Degree of Master.
[2] Chan D.W.M., Olawumi T.O., Ho A.M.L. (2019) Perceived benefits of and barriers to Building Information Modelling (BIM) implementation in construction: The case of Hong Kong. J. BUILD. ENG., 25: 1–10. (DOI:10.1016/j.jobe.2019.100764)
[3] WangX. Q., Chen H., Qi L., Qiu X. (2019) Study on investment efficiency of BIM hierarchical application. CONSTR. ECON. 40(07):95-99.
[4] Chen Z.Y.(2018) Research on application of BIM technology in evaluation of building industrialization level. Xi'an University of Technology for the Degree of Master.
[5] Meng Y.C, Zheng X.Y, Gu H. (2018) The Analysis of shantytowns reform policies in China from 2007 to 2017 from the incrementalism perspective. J. TSINGHUA UNIV.: PHIL.SOC. SCI. ED.33(03):184-189+194
[6] Miao P.(2018) A study on the construction of the resettlement house in the shantytown of Zhengzhou. Zhengzhou University for the Degree of Master.
[7] Xue R., Wang X.Y., Shi K. (2019) Constructionproblems and countermeasure analysis of prefabricated building based on BIM. IND CONSTRUCT. 48(11):207-210.
[8] Xue J.Y., Tan P., Meng F.M. (2019) Research on the Application of BIM and Earned Value Method in Construction Progress and Cost Control. CONSTRUCT ECON. 40(06):115-119.
[9] Zhang H.D., Xu N.,Fang K. (2019)Structural design and BIM application of a fabricated residential project. BUILD. STRUCT. 49(11):62-66.

Statistical Analysis of Temporal Characteristics of Natural Runoff in Golmud River Basin

Yiwei Zhen[1], Shuguang Liu[1*], Zichen Hu[1] and Jiyu Liang[1]

[1]Department of Hydraulic Engineering, College of Civil Engineering, Tongji University, Shanghai, 200092, China

[*]liusgliu@tongji.edu.cn

Abstract: Based on the 45-year natural runoff time series from 1956 to 2000 of Golmud hydrological station, the control station of Golmud River basin, the inter-annual variation characteristics involving tendency, periodicity and abrupt change and intra-annual distribution of runoff were researched by using the linear tendency estimate method, cumulative departure method, Mann-Kendall method, Morlet wavelet analysis method, Yamamoto method and Pettitt method, combined with the analysis of annual average distribution process and complete adjustment coefficient. The results indicate that the overall inter-annual fluctuations of the runoff in the Golmud Basin are relatively small, and the overall annual runoff shows a significant increasing trend. After the abrupt change occurred in 1966, the runoff increased significantly. The variation of runoff has 3-year, 6-year and 18-year periods, and the 6-year period is especially obvious. Furthermore, there is a change rule of wet and dry with an 8-year alternation in the 18 years periodic variation. The intra-annual distribution of runoff is uneven, showing a "single peak" distribution, and the degree of unevenness tends to more concentrated.

1. Introduction

Golmud River is the second largest inland river in the Qaidam Basin and the primary water source for the largest city in the basin, Golmud City. In recent years, Golmud has developed rapidly economically and socially and has become an important base for the development of salinization and petrochemical resources in western China, which is of great strategic significance for the development of Qaidam Basin. Meanwhile, the urban comprehensive water use is gradually increasing, and the demand for the development and utilization of water resources in Golmud River basin is continuously increasing. At present, Chinese scholars have made some achievements in the study of water resources in theGolmud River basin[1~2], but there are still relatively few studies on the change of natural runoff in Golmud River basin. Given this, this paper adopts a variety of hydrological analysis methods to comprehensively analyze the trend, periodicity, mutability and annual distribution characteristics of natural runoff of Golmud River from 1956 to 2000 and studies the change law of runoff, to provide scientific basis for the development and utilization, optimal allocation, disaster prevention and mitigation and sustainable development of water resources in the basin.

2. Study area and data

Golmud River originates from AketanQiqin Mountain, a branch of Kunlun Range. Its upper reaches are divided into three branches: east branch of Xueshui River, west branch of Najin River and south branch of South River. The three branches then converge to one river, which is called Golmud River. From south to north, it flows through Golmud City and finally into Dabusun Lake, with a total length

IOP Conf. Series: Materials Science and Engineering **794** (2020) 012050 doi:10.1088/1757-899X/794/1/012050

of 378.5 km and a total river basin area of 18648 km^2. Due to various reasons, the control stations of Golmud River- Golmud Station has experienced several changes of location. The runoff generation and confluence conditions of the first, second and third stations have not changed obviously, with no large tributaries in the section, thus the data of them can be combined to use. While the data of the fourth Golmud Station has been affected by upstream water conservancy projects development since the 1990s and needs to be restored for use. In this study, the restored natural runoff data from 1956 to 2000 for a total of 45 years was used to analyze the runoff law.

3. Analysis of inter-annual variation of natural runoff

3.1 Interannual and chronological variation characteristics

The annual runoff sequence of Golmud was analyzed by plotting the interannual variation process line (Figure 1) and calculating the characteristic value of annual runoff variation in each age (Table 1). By analysis, it is known that: (1) In terms of annual change, the average annual natural runoff of Golmud Station from 1956 to 2000 was 24.8m^3/s, the maximum average annual runoff 52 m^3/s (the Year 1989), the minimum average annual runoff 17.4 m^3/s (the Year 1963), the annual extremum ratio 2.99, the variation coefficient 0.23. The extremum ratio and variation coefficient were relatively small, which means that the interannual fluctuations of the annual runoff of the Golmud River are relatively small. (2) In terms of age change, the average annual runoff of Golmud Station in the 1950s, 1960s and 1990s was less than the average annual runoff, while the average annual runoff in the 1970s and 1980s was higher than the average annual runoff, among which the average annual runoff in the 1980s was the largest. The extremum ratio of natural runoff in each age was between 1.21-2.48, and the coefficient of variation was between 0.09-0.32. Both the maximum values of these two characteristic values appeared in the 1980s, which was mainly influenced by the annual runoff maximum value in 1989. (3) From the 5-year moving average process, it can be seen that the natural runoff inGolmud Station showed a less-more-less change process., which is about once every 20 years.

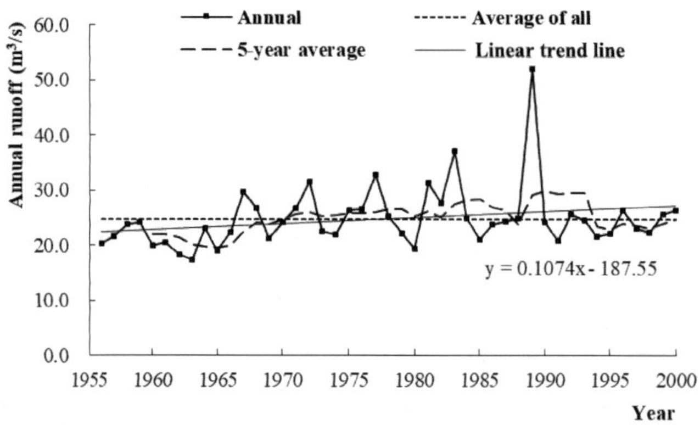

Figure 1. Interannual variation of annual natural runoff in Golmud Station.

Table1 Characteristic values of natural runoff in Golmud Station.

Year	Average (m^3/s)	Maximum (m^3/s)	Minimum (m^3/s)	Extremum ratio	Standard deviation	Coefficient of variation
1956-1959	22.0	24.1	19.9	1.21	1.97	0.09
1960-1961	22.2	29.7	17.4	1.71	3.84	0.17
1970-1979	25.6	32.9	19.5	1.69	4.25	0.17
1980-1989	29.2	52.0	21.0	2.48	9.27	0.32
1990-1999	23.9	26.4	20.9	1.26	2.07	0.09
1956-2000	24.8	52.0	17.4	2.99	5.71	0.23

3.2 Trend analysis of runoff change

The linear trend estimation method and cumulative anomaly method[3] were used to analyze the annual runoff trend change of Golmud Station. The linear trend rate of the natural runoff process of Golmud Station is positive (Figure 2), indicating that the annual natural runoff was increasing in statistical years. The cumulative anomaly process line shows that the runoff of Golmud Station was decreasing from the 1950s to the mid-1960s, and there was a rising trend from the mid-1960s to the early 1990s, then a slow decline after the early 1990s (Figure 2), which is consistent with the results of the 5-year moving average process.

At the same time, the Mann-Kendall test method[4] was used for further quantitative analysis and research on the annual runoff evolution trend of Golmud Station. The results show that the Mann-Kendall (M-K) test values of the annual and monthly runoff are all positive (Figure 3), indicating that the annual runoff of the station was increasing, which is consistent with the previous qualitative analysis results. And the annual runoff series passed the test of the confidence level of 90%, indicating a significant increasing trend. In the runoff series of each month, the runoff in February, March, May and July passed the 90% confidence test, and the runoff in March passed the 99% confidence test. Thus, the runoff variation trend of Golmud Station in February, March, May and July has significantly increased, among which the increasing trend was the most obvious in March, while the increasing trend of other months was not significant.

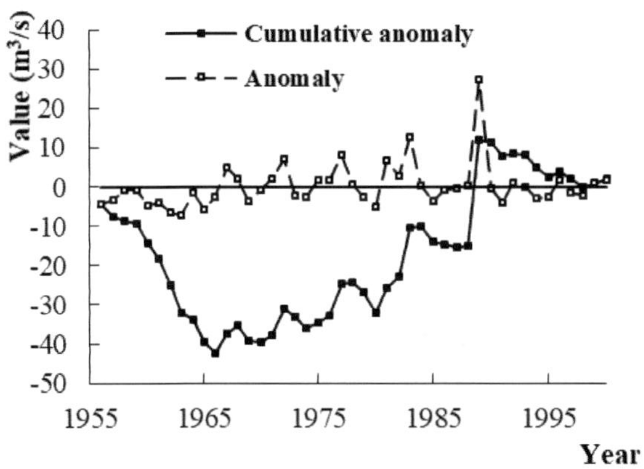

Figure 2.Cumulative anomaly curve of annual runoff in Golmud Station.

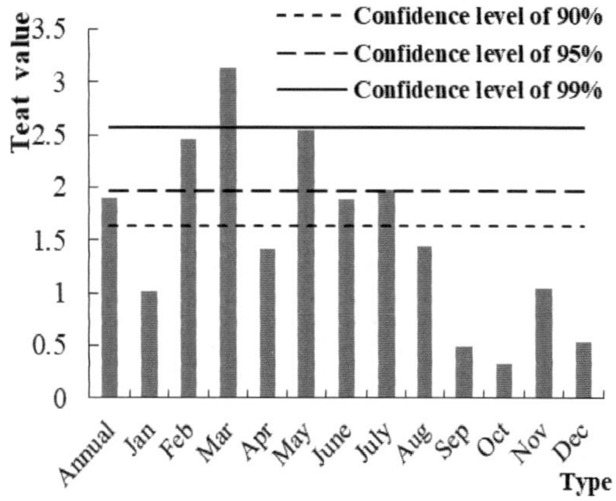

Figure 3. M-K statistic values of annual and monthly runoff inGolmud Station.

3.3 Periodic analysis of runoff changes

In order to identify the periodicity of the natural runoff series of Golmud Station, the Morlet wavelet method[5] was used to analyze the annual natural runoff of Golmud Station on multiple time scales and reveal the regularity of its periodic change. The time-frequency of wavelet coefficients real component of the wavelet transform was analysis to identify the runoff period. In the contour map of the real part of the wavelet transform (Figure 4), the solid line indicates that the real component of the corresponding wavelet coefficients is positive, that is, the runoff increases, while the dotted line indicates that the real component of the corresponding wavelet coefficients is negative, that is, the runoff reduces. In the wavelet variance diagram (Figure 5), the periodic variation law can be found according to the distribution relation of the periodic oscillation intensity of the runoff series over time.

The natural runoff of Golmud Station has obvious features of time-cycle variation, with periodic fluctuations at different scales of 2-year ~ 3-year, 5-year ~ 7-years and 17-year ~ 19-year(Figure 4), and the periodic oscillations show complex nested structures from large to small. On the small-scale periodic 2-year ~ 3-year and 5-year ~ 7-year, wet and dry changed violently alternately, and the fluctuation intensity generally increased first and then decreased. On the large-scale periodic 17-year ~ 19-year, the change of wet and dry is relatively stable, and it changes alternately once every eight years. There are three low-water periods(1958-1966, 1976-1983 and 1993-2000) and two high-water periods (1967-1975 and 1984-1992).

It can also be further identified that the annual natural runoff inGolmud Station has three maximum values at 3-year, 6-year, and 18-year, with the maximum value at 6-year (Figure 5). The three peaks correspond to three periods of annual natural runoff variation, of which 6-year is the main period.

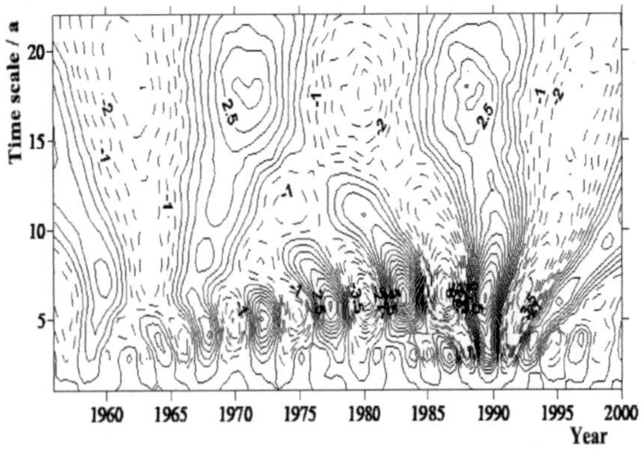

Figure 4. The contour map of the real component of the wavelet transform.

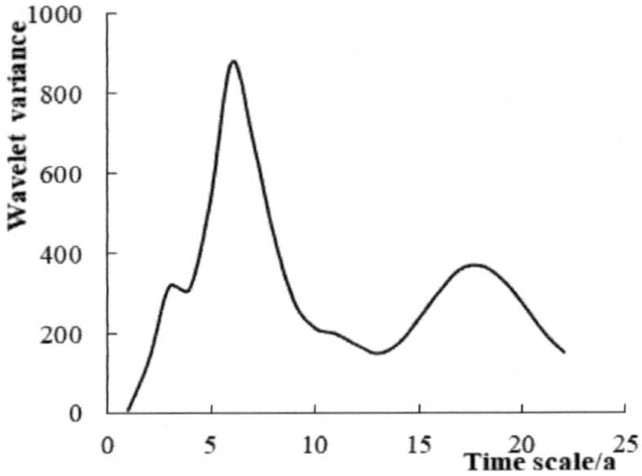

Figure 5. Wavelet variance of theannual runoff inGolmud Station.

3.4 Shift analysis of runoff changes

Influenced by various factors such as natural changes and human activities, it is often difficult to ensure the accuracy of results by adopting a single method. In this paper, the Mann-Kendall test, Yamamoto method and Pettitt method were used to identify the abrupt change points of annual natural runoff in Golmud Station.

The abrupt change analysis of the natural runoff series of Golmud Station was carried out by the Mann-Kendall method[6], the results of which presented in figure 6. Except 1962-1966, the UF curve was all greater than 0, indicating that the annual natural runoff of the Golmud river was basically in an increasing trend. The significance of the UF value of 1975 ~ 1994 was beyond 0.05, indicating a significant increase, which was consistent with the previous trend analysis results. Within the 0.05 confidence level limit, the UF and UB curves intersect at three points in 1958, 1960, and 1966, respectively, indicating that the annual runoff series of Golmud Station may have abrupt change at these three points. Combined with the case that the UF value exceeds the significant level, it can be determined that the annual natural runoff shift in 1966 and has increased significantly since the 1970s.

The Yamamoto method[7] was also used to diagnose and verify the abrupt change point of annual natural runoff change in Golmud Station. According to relevant literature and multiple experiments, the optimal sample length of the two sub-sequences divided by the Yamamoto method was 5, and then the signal to noise ratio (SNR) of the Yamamoto method was calculated (See figure 7). According to the calculation results, the maximum SNR of the runoff series appeared in 1966, which was 1.23, greater than 1.0, which proved that 1966 was the abrupt change point of the natural runoff of the station.

Besides, the Pettit method[8] was adopted to further analyze and verify the abrupt change point of the annual runoff series of Golmud Station. As thecalculation results of the Pettit method show (Figure 8), the maximum test statistic 275 appeared in 1966, that is, 1966 was the possible abrupt change point of the series. The P value of this point was calculated to be 0.015, less than the critical value of 0.5, confirming that the change point of annual natural runoff series appeared in 1966.

Through comprehensive diagnosis and analysis, it was determined that there was a mutation in the annual natural runoff sequence of Golmud Station in 1966, and the runoff increased significantly since 1966.

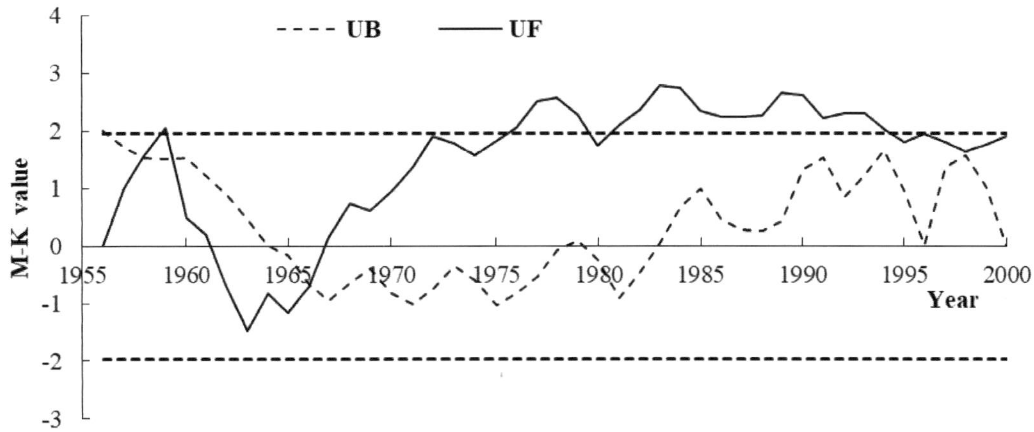

Figure 6.Abrupt change analysis curve of natural runoff by the Mann-Kendall method in Golmud Station.

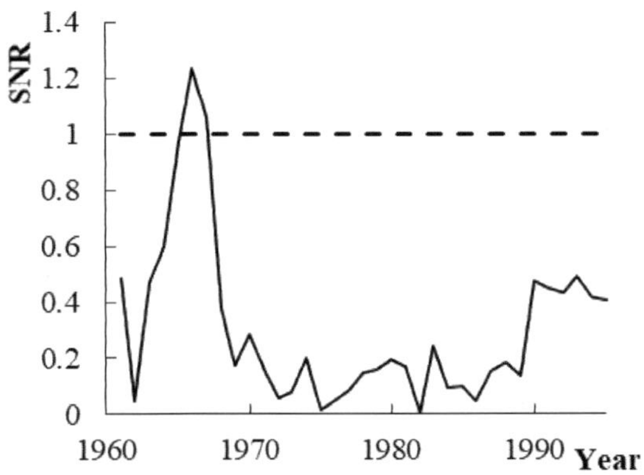

Figure 7. Test results of the Yamamoto method for natural runoff in GolmudStation.

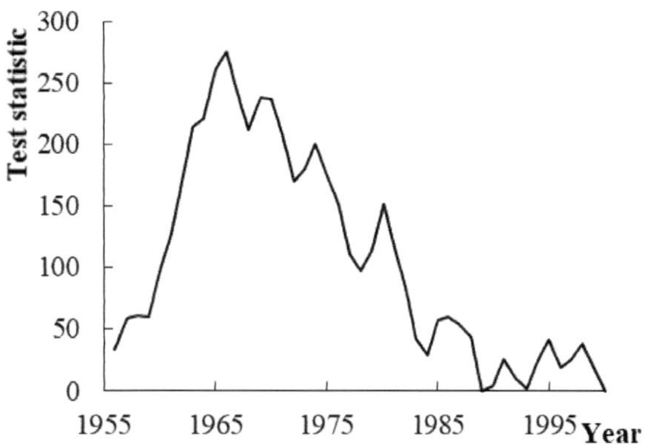

Figure 8.Test results of the Pettit method for natural runoff in Golmud Station.

4. Analysis of annual runoff distribution

4.1 analysis of distribution process within the year

In order to grasp the occurrence law of flood, drought and other disasters and provide the basis for water allocation and water supply for industry and agriculture, this article analyzed the annual runoff distribution characteristics of Golmud Station by drawing the annual runoff distribution process of Golmud Station (Figure 9). As this figure shows, the annual runoff distribution of Golmud Station is uneven, showing a "single peak" distribution. The runoff during the flood season (June to October) accounts for 55.2% of the annual total, mainly concentrated in July and August, accounting for 27.4% of the annual runoff.

4.2 Analysis of distribution characteristics within a year

The annual distribution unevenness of runoff in Golmud Station was quantitativelyanalyzed by using the completed adjustment coefficient method[9]. The completed adjustment coefficient (Cr) of the annual runoff is between 0.078 and 0.348(See figure 10), with a relatively large amplitude of variation. The maximum value appeared in 1989 and the inhomogeneity within the year is relatively large. Combined with the linear trend line, the completed adjustment coefficient shows an upward trend, and the overall runoff distribution tends to be more concentrated.

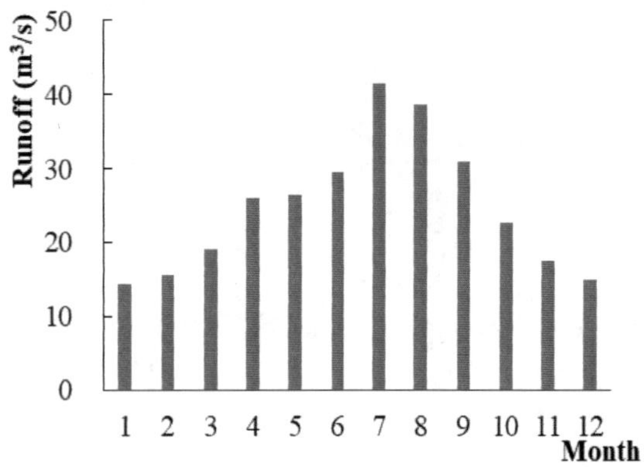

Figure 9.Annual average runoff distribution process of Golmud station.

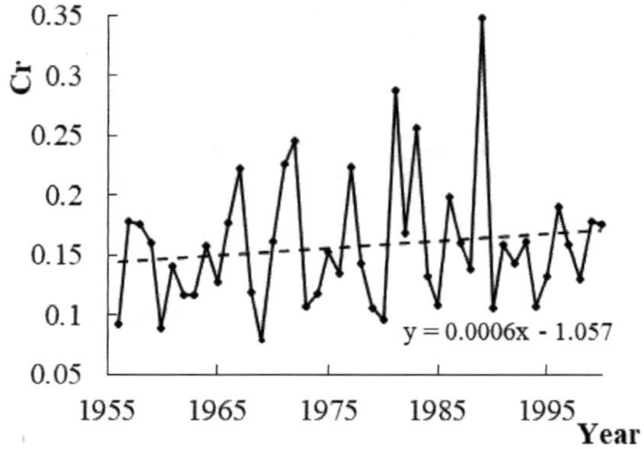

Figure 10.Trend change of completed adjustment coefficient.

5. Conclusions

With the various time series analysis methods, based on the 45-year natural runoff data of Golmud Station, the representative station of the Golmud River in Qaidam Basin, the inter-annual variation of the natural runoff of the river and its distribution characteristics during the year were analyzed. The main conclusions are as follows:

(1) According to the analysis of extreme ratio and coefficient of variation, the overall fluctuations of the inter-annual runoff variation in the Golmud Basin was relatively small. Influenced by the annual runoff maximum value in 1989, the fluctuation range of runoff in the 1980s was relatively large, while that in other ages were relatively small, and the intergenerational variation of runoff was generally stable. The result of the 5-year moving average analysis shows that the annual runoff of Golmud Station presented a change process between high and low flow once every 20 years in the statistical period.

(2)The natural runoff of the Golmud River exhibits an increasing trend as a whole. The results of theMann Kendall test show upwards in the natural runoff trend of Golmud station on the scale of month and year, among which the annual runoff and the runoff in February, March, May and July increase is significant.

(3) The natural runoff of Golmud Station has the characteristics of multi-time scale evolution. According to the wavelet analysis, the annual runoff has 3-year, 6-year and 18-year periods. The periodic change is a complex nesting structure from large to small. On the small-scale period, the runoff changed drastically between abundance and drought. On the large-scale period, the runoff changed relatively lightly between abundance and drought, with three partial dry periods and two partial abundant periods.

(4) Based on the comprehensive judgment of the Mann-Kendall test, Yamamoto method, and Pettitt method, the annual natural runoff series inGolmud Station abrupt changed significantly in 1966, and the runoff increased significantly in the 1970s.

(5) The distribution of the Golmud River runoff is relatively uneven throughout the year, showing a "single peak" distribution, and the flood season accounts for a large proportion, mainly concentrated in July and August.

Acknowledgements

The authors are thankful to the National Key Research and Development Plan during the 13th Five-Year plan period (Grant No. 2018YFD1100401) which supported the study.

References

[1] Li, Y.Z., Wu, Y.Q., Zhang, S.Q., Liu, Z.Y., &Guo, Y.W.(2011) Study on Optimized Allocation of Water Resources in Golmud City.China Rural Water and Hydropower., 02: 46-49.

IOP Conf. Series: Materials Science and Engineering **794** (2020) 012050 doi:10.1088/1757-899X/794/1/012050

[2] An, F.Y., Liu, X.J., Zhang, Q.X., Wang, Y.X., Chen, T.Y., Yu, L.P., Lu, B.L.,& Chang, Q.F. (2018). Drainage geomorphic evolution in response to paleoclimatic changes since 12.8 ka in the eastern Kunlun Mountains, NE Qinghai-Tibetan Plateau. Geomorphology., 319: 117-132.

[3] Lu, J.Y., Zhou, Z.H., Zhu, X.Y., Lu, B.H., Wang, X.B., &Ding, W. (2015) Characteristics of Precipitation Sequence Variation in Yishu River Basin from 1954 to 2010.South-to-North Water Transfers and Water Science & Technology., 06: 1060-1064.

[4] Yang, Y.H., &Tian,F. (2009)Abrupt change of runoff and its major driving factors in HaiheRiverCatchment, China. Journal of Hydrology.,374: 373~383.

[5] Shao, W., Chen, X., Zhou, Z., Liu, J., Yan, Z., Chen, S., & Wang, J. (2017). Analysis of river runoff in the Poyang Lake Basin of China: long-term changes and influencing factors. Hydrological sciences journal, 62: 575-587.

[6] Barbieri, L.F.P., de Fatima Correia, M., da Silva Aragão, M.R., Vilar, R.D.A.A., & de Moura, M.S.B. (2017). Impact of Climate Variations and Land Use Change: a Mann-Kendall Application. RevistaGeama., 3: 127-135.

[7] Liu, Q., & McVicar, T. R. (2012). Assessing climate change induced modification of Penman potential evaporation and runoff sensitivity in a large water-limited basin. Journal of Hydrology., 464: 352-362.

[8] Belihu, M., Abate, B., Tekleab, S., &Bewket, W. (2018). Hydro-meteorological trends in the Gidabo catchment of the Rift Valley Lakes Basin of Ethiopia. Physics and Chemistry of the Earth., 104: 84-101.

[9] Lu, Z.H., Xia, Z.Q., Yu L.L., &Wang Z.J. (2012).Changes of annual runoff distribution in middle reaches of Songhua River. Journal of Hohai University(Natural Sciences)., 40: 63-69.

Study on seismic residual deformation test of reinforced soil

Bing He[1,2], Jianming Zhao[1,2*], Qiwang Liu[1,2],Xiaosheng Liu[1,2], Kaibin Zhu[1,2],LongWang[1,2], YushengYang[1,2],ZhengquanYang[1,2]

[1]Institute of Water Resources and Hydropower Research, Beijing, 100048, China

[2]State Key Laboratory of Simulation and Regulation of Water Cycle in River Basin, Beijing, 100048, China

*Corresponding author's e-mail: zhaojm@iwhr.com

Abstract. The reinforced material can increase the stability of soil material, improve its shear strength and reduce the residual deformation. In this paper, the vibration triaxial tests of coarse-grained soil under different confining pressure, consolidation ratio and dynamic stress are carried out by using the 100t electro-hydraulic servo large-scale static and dynamic triaxial testing machine of Institute of Water Resources and Hydropower Research. The results show that the average reduction ratio of the axial residual deformation of reinforced soil to that of unreinforced soil is 40%-50%, and the average reduction of the volumetric residual deformation is 30-40% It has a good inhibitory effect. With the increase of confining pressure, consolidation ratio and dynamic shear stress ratio, the axial residual deformation and volumetric residual deformation of reinforced soil increase in varying degrees. The reinforcement effect of geogrid material on soil sample under different conditions is analyzed. The lower the dynamic shear stress ratio and the higher the consolidation ratio, the better the reinforcement effect. The lower the confining pressure is, the lower the reduction ratio of the reinforcement to the volumetric strain is.

1. Introduction

Earth-rock dam, with its advantages of convenient local materials, simple construction process, strong adaptability to foundation deformation, and low cost, has become the most widely used type of dam, accounting for more than 80% of all types of dams built in China [1]. China is a country with frequent earthquakes, and the earthquake source is relatively shallow and the earthquake damage is relatively large [2]. Most of the earth-rock dams in China are built in the western regions with complicated geological conditions, seismic frequency and high intensity. Therefore, the safety of earth-rock dams is particularly important in these regions. In order to ensure the safe operation of earth-rock dam in seismic region, it is necessary to calculate the residual deformation of earth-rock dam under seismic conditions and evaluate the seismic stability of earth-rock dam. Therefore, it is of great significance to study the residual deformation of soil under vibration load [3].

Geotechnical reinforcement, with simple construction and low cost, can increase the stability of the structure, improve the shear strength and reduce the deformation characteristics of the structure by the interaction between reinforcement and the soil and the special meshing effect of the grid. Since the 1980s, geotechnical reinforcement technology has been applied in China widely and developed rapidly. At present, many scholars at home and abroad have done a lot of experimental model research on reinforced soil. Bao Huafu studied the stress-strain characteristics of reinforced soil with different reinforcement layers, and proposed that the reinforcement effect of gravel soil will weaken with the increase of confining pressure [4]. Through triaxial test, Wu Jinghai studied the stress-strain characteristics and reinforcement mechanism of geogrid reinforced soil [5]. Bi Jing studied the dynamic characteristics of reinforced coarse-grained soil. The results show that when the dynamic shear strain is small, the reinforcement has little effect on the dynamic shear modulus and damping ratio of coarse-grained soil, and the residual volumetric deformation and residual shear deformation decrease significantly [6]. Yang Shuaidong studied the dynamic elastic modulus of reinforced soil and unreinforced soil under different reinforcement layers, confining pressures and consolidation ratios. The results show that the dynamic elastic modulus of reinforced soil increases with the increase of confining pressure and consolidation ratio [7].

In this paper, the vibration triaxial test is carried out with a 100t electro-hydraulic servo large-scale static and dynamic triaxial test machine of Institute of Water Resources and Hydropower Research. By changing different confining pressure, consolidation ratio and dynamic stress conditions, the residual deformation characteristics of reinforced soil and prototype soil are studied. The reinforcement effect of reinforced soil under different conditions is compared to provide solutions for engineering application.

2. Test content

In this paper, the dynamic triaxial test of saturated consolidated drainage was carried out to study the dynamic axial residual deformation and dynamic volumetric residual deformation characteristics of reinforced soil, and the axial and volumetric residual deformation values of reinforced soil under different confining pressures, consolidation ratios and dynamic stresses were obtained.

According to Seed research, when the initial stress state and the dynamic stress condition are the same, the dynamic residual axial strain and the residual volumetric strain are related to the number. When analyzing the seismic dynamic residual strain, the random seismic load needs to be converted into an equivalent cyclic number under the cyclic load, the equivalent magnitude of the seismic load depends on the magnitude of the earthquake [8]. The results of the 12th, 20th, and 30th magnitudes of the triaxial vibration test correspond to the magnitudes of 7, 7.5, and 8.

2.1Test equipment

In this paper, the dynamic residual deformation of reinforced soil was tested by a 100-ton electro-hydraulic servo triaxial testing machine. The large triaxial testing machine has the characteristic of vertical excitation load, and can conduct static and dynamic tests through the triaxial pressure chamber to study the static and dynamic properties of soil and rock samples. The instrument is sensitive, accurate and has a clear display. Static and dynamic tests are controlled by two independent hydraulic units, which make the operation very convenient.

Representative soil samples were selected for the test. The dry density was controlled to be ρ_d= 2.2g/cm^3, and the relative density was 80%. The grain size accumulation curve of the soil is shown in Figure 1.

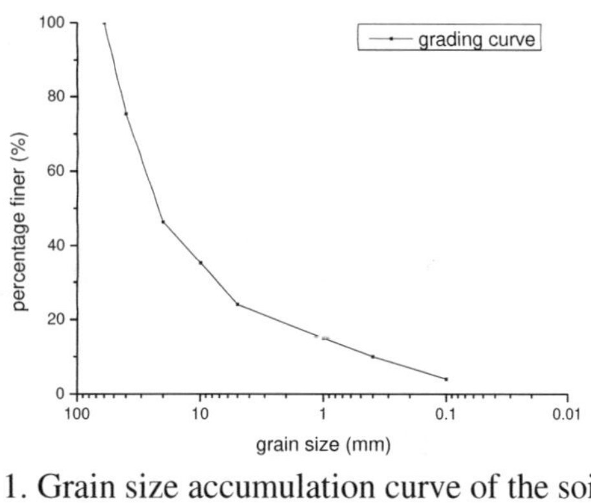

Figure 1. Grain size accumulation curve of the soil.

2.2Test plan
The test uses three groups of control tests to explore the residual deformation characteristics of reinforced soil with seismic strengthening measures under different confining pressures, consolidation ratios, and dynamic shear stress ratios under vibration loading. The control group is Series 1-2. The vibration triaxial tests of reinforced soil and unreinforced soil under the conditions of a confining pressure of 500 kPa, a consolidation ratio of 2.0, a dynamic stress of 50 kN, and a dynamic shear stress ratio of 0.4716 were performed.

2.2.1Dynamic shear stress ratio. The dynamic shear stress ratio test group is series 3-4, and the test chooses 60kN dynamic stress, maintains 2.0 consolidation ratio, and 500kPa confining pressure and explores the residual deformation characteristics of soil samples with and without geogrid material.

2.2.2Consolidation ratio. The consolidation ratio test group is series 5-8. The test uses 2.5 consolidation ratio to investigate the residual deformation characteristics of soil with and without geogrid material at two dynamic shear stress ratios at a confining pressure of 500 kPa.

2.2.3Confining pressure. The confining pressure test group was series 9-13. The confining pressure condition of 1000kPa was selected for the test to explore the residual deformation characteristics of reinforced and unreinforced soil with two dynamic shear stress ratios of consolidation ratio 2.0.

Table 1. Series of large-scale dynamic triaxial test.

Series	confining pressure (kPa)	consolidation ratio	dynamic stress (kN)	reinforced	Dynamic shear stress ratio	Test name
1	500	2.0	50	No	0.4716	unreinforced 500 2.0 50kN
2	500	2.0	60	No	0.5659	unreinforced 500 2.0 60kN
3	500	2.0	50	Yes	0.4716	reinforced 500 2.0 50kN
4	500	2.0	60	Yes	0.5659	reinforced 500 2.0 60kN
5	500	2.5	45	No	0.3638	unreinforced 500 2.5 45kN
6	500	2.5	65	No	0.4850	unreinforced 500 2.5 65kN
7	500	2.5	45	Yes	0.3638	reinforced 500 2.5 45kN
8	500	2.5	65	Yes	0.4850	reinforced 500 2.5 65kN
9	1000	2.0	75	No	0.3537	unreinforced 1000 2.0 75kN
10	1000	2.0	95	No	0.4480	unreinforced 1000 2.0 95kN
11	1000	2.0	75	Yes	0.3537	reinforced 1000 2.0 75kN
12	1000	2.0	95	Yes	0.4480	reinforced 1000 2.0 95kN
13	1000	2.0	105	Yes	0.4951	reinforced 1000 2.0 105kN

2.3Test operation
The large-scale dynamic triaxial test project is carried out in accordance with the relevant requirements of the "vibration triaxial test" in the geotechnical test regulations. The specific operation process of the test is as follows.

2.3.1Sample preparation. Prepare the sample by dry packing. Place a piece of filter paper on the sample base, cover the rubber film on the sample base, and fasten the rubber film on the sample base with a rubber rope to avoid the existence between the base and the rubber film. Gap. Install the split mold, seal the contact seam with rubber clay, turn on the vacuum pump to extract the gas between the split mold and the rubber film to make the two fit together. Load the sample according to the control dry density, and load it in five layers(see Figure 2). A geogrid (see Figure 3) is installed between each two layers, and the sample cylinder is sequentially filled and compacted using the layered vibration method. Pay attention to prevent coarse and fine particles from separating during the filling of the sample, and keep the coarse and the fine soil is mixed evenly. After the sample is installed, put a piece of filter paper on the top of the sample, install the sample cap, and fasten the rubber film and the sample cap with a rubber rope.

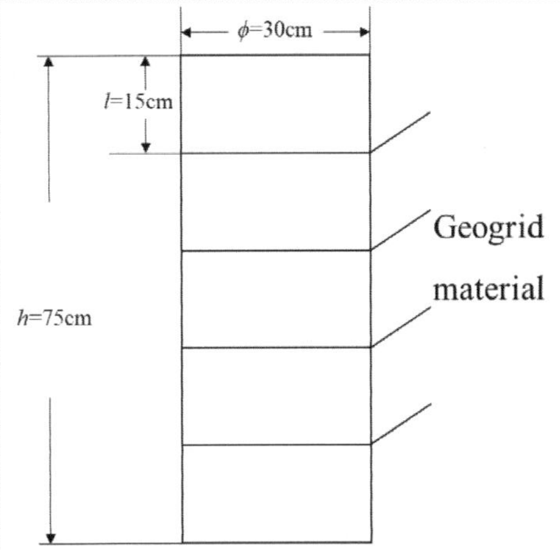

Figure 2. Dimensions of the test material and the geogrid material.

Figure 3. Geogrid material in soil sample.

2.3.2Sample saturation. After the sample is installed, the air in the sample is evacuated under vacuum to apply negative pressure to the sample. Check the air tightness of the sample and remove the split mold. Lift the pressure chamber to the base of the triaxial instrument and tighten the bottom of the pressure chamber fixing screw. Pass in degassing water from the bottom of the sample, and saturate the sample from bottom to top. If necessary, combine the method of applying reverse pressure to saturate the sample. The saturated pore water pressure coefficient of the sample is required to be above 0.97.

2.3.3Sample consolidation. After the sample is saturated, drain the consolidation under the corresponding consolidation stress. Adjust the confining pressure to 500kPa and 1000kPa, remove the top support screw. Connect the consolidation circuit and start the consolidation load. Use grading loading method, each level 100kPa, 10 minutes a level.

2.3.4Confining pressure. According to the requirements of the geotechnical test specification, set the cyclic vibration cycle and cyclic load amplitude on the computer control interface, and then apply dynamic stress to the sample for testing. Computer sensors are used to collect test data during the test.

3. Analysis of test results

Based on the results of the dynamic residual deformation test of the reinforced soil, the changes of the axial residual strain and the volumetric residual strain with the vibration cycle, the consolidation confining pressure, the consolidation stress ratio, and the dynamic shear stress ratio were studied.

3.1Influence of seismic reinforcement measures on residual deformation

IOP Conf. Series: Materials Science and Engineering **794** (2020) 012051 doi:10.1088/1757-899X/794/1/012051

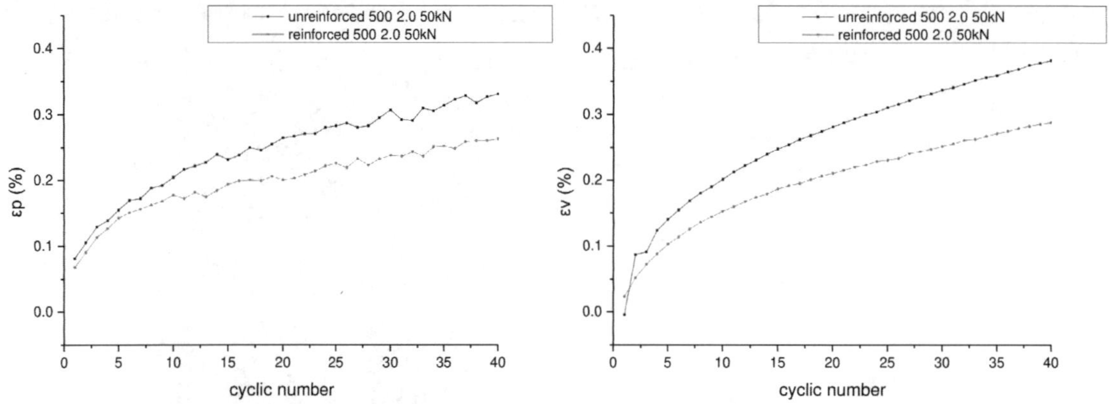

Figure 4. Influence of seismic reinforcement measures on residual strain.

Comparing the test results of series 1 and series 3, the two series maintain the same confining pressure, consolidation ratio, and dynamic stress, and the test variable is whether or not a geogrid material is applied. The test results are shown in Figure 4. From the axial strain to cyclic number It can be seen from the curve that the axial deformation of the reinforced soil is not reduced significantly during the first 5 weeks, and the effect of reducing the axial strain of the reinforced soil after 5 weeks is increased, with the highest reduction ratio reaching 32%. It is stable, and the volumetric strain increases steadily with the increase of vibration frequency, which conforms to the law of volumetric strain. The effect of reducing the volumetric strain of reinforced soil is obvious, and the reduction ratio is about 25%.

To verify the effect law of the reinforced soil on the residual deformation under 2.5 consolidation ratio, a comparative test was carried out in series 6 and 8. The test results are shown in Figure 5.

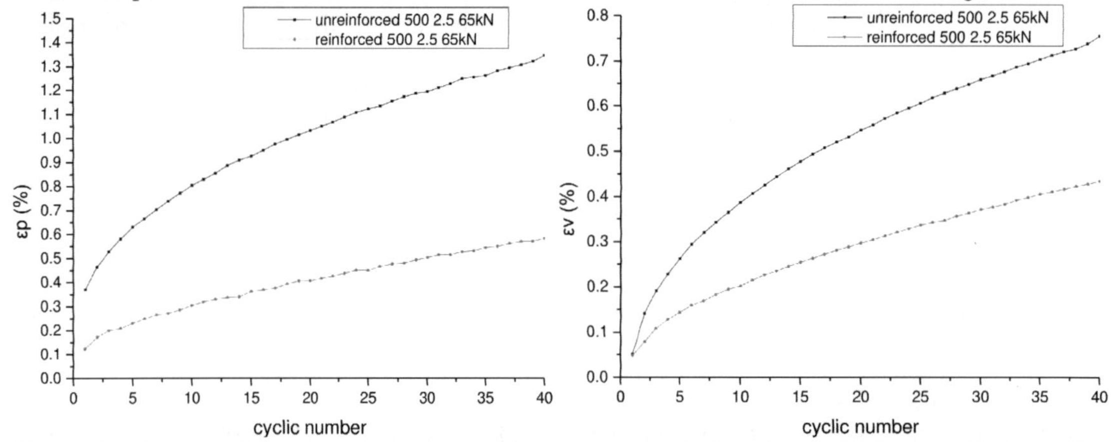

Figure 5. Influence of seismic reinforcement measures on residual strain——high consolidation ratio.

Comparing the test results of working conditions 6 and 8, the two conditions maintain the same confining pressure, consolidation ratio, and dynamic stress, and the test variable is whether or not a geogrid material is applied. The test results are shown in Figure 5. Similar to the result of consolidation ratio of 2.0 also showed that the reinforced soil has the characteristic of inhibiting residual deformation under the condition of a consolidation ratio of 2.5. It can be seen from the axial strain—cyclic number curve that as the vibration order increases, the axial residual deformation inhibits Obviously, the reduction rate of axial residual strain reaches 60%. The volumetric strain—cyclic number curve also shows a good law, and the reduction rate of volumetric residual strain reaches 45%.

To verify the effect law of the reinforced soil on the residual deformation under confining pressure of 1000kPa, a comparative test was carried out in series 9 and 11. The test results are shown in Figure 6.

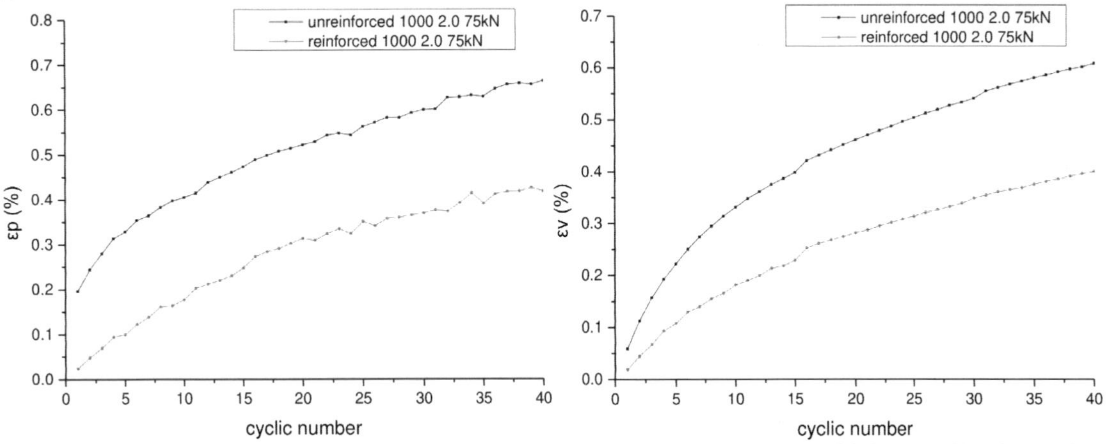

Figure 6. Influence of seismic reinforcement measures on residual strain——high consolidation ratio.

Comparing the test results of condition 9 and condition 11, the two conditions maintain the same confining pressure, consolidation ratio, and dynamic stress. The test variable is whether or not a geogrid material is applied. The test results are shown in Figure 6. Compared to the 500kPa confining pressure test results, when the confining pressure is 1000kPa, the reinforced soil also plays a role in reducing the residual deformation. From the axial strain-vibration number curve, it can be seen that as the vibration number increases, the difference between the axial residual deformation of the unreinforced soil and the reinforced soil is relatively stable, and the average reduction of the axial residual strain under the entire working condition is 50% on average. The volumetric strain-vibration curve in the first 5 weeks of the reinforced soil to suppress the volumetric residual deformation is not obvious. After 5 weeks, the effect of reducing the residual volumetric of the reinforced soil increased, and the reduction of the residual volumtric strain reached 40%.

3.2Influence of dynamic shear stress ratio on residual deformation in reinforced soil

The influence of dynamic shear stress ratio on residual deformation in reinforced soil was investigated. The test results are shown in Figure 7.

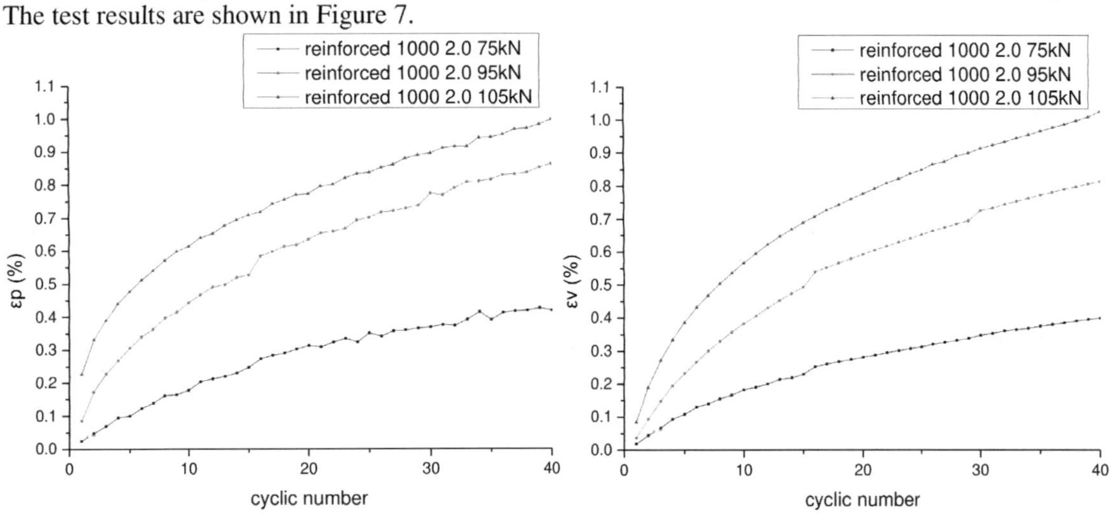

Figure 7. Influence of seismic reinforcement measures on residual strain——high consolidation ratio.

Comparing the test results of series 11, 12, and 13, the three series maintain the same confining pressure and consolidation ratio, and they are all reinforced with geogrid materials. The test variables are different dynamic stresses of 75kN, 95kN, and 105kN. The test results are following Figure 7. In reinforced soil, when the confining pressure, consolidation ratio and geogrid material are the same, the axial strain increases significantly with the increase of dynamic stress. The volumetric residual strain also shows similar conclusions and the increase is more Stable, showing the same research conclusions as the prototype soil sample.

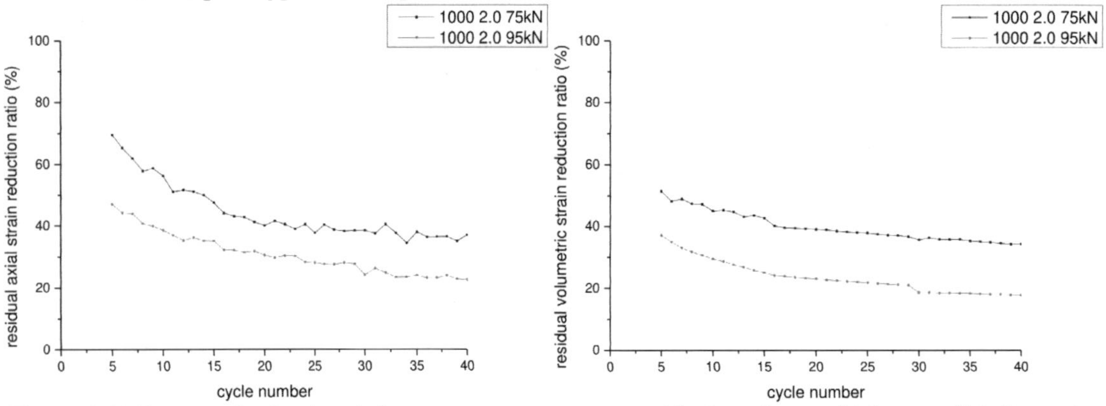

Figure 8. Influence of seismic reinforcement measures on residual strain——high consolidation ratio.

The test results of series 9-12, calculated under the same confining pressure, consolidation ratio, and dynamic stress conditions, the reduction ratio of residual deformation compared to reinforced soil and unreinforced soil, the processing results are shown in Figure 8. Deformation mutations at 4 weeks have a greater impact, so the processing data is discarded from the first 4 weeks of strain data. As the cyclic number increases, the reduction ratio of the axial strain and the volumetric strain decreases. Under the condition of low dynamic shear stress ratio, the reduction ratio of axial strain and volumetric strain is higher, which is stable at more than 40%, and the effect of geogrid material is better than that of high dynamic shear stress ratio.

3.3Influence of consolidation ratio on residual deformation in reinforced soil

The influence of consolidation ratio on residual deformation in reinforced soil was investigated by comparing test series 4 and series 8. The test results are shown in Figure 9.

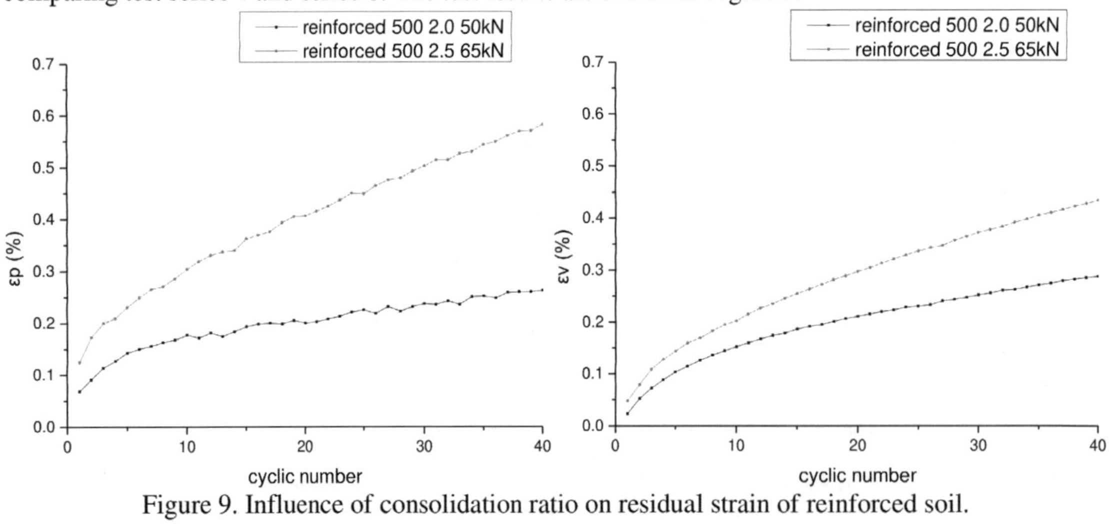

Figure 9. Influence of consolidation ratio on residual strain of reinforced soil.

Comparing the test results of series 4 and series 8, the two conditions maintain the same confining pressure and the dynamic shear stress ratio are similar, and both are applied with geogrid materials for reinforcement. The test variables are different consolidation ratios of 2.0 and 2.5. As shown in Figure 9. In the reinforced soil, as the consolidation ratio increases, the axial residual deformation increases significantly, and the volume residual strain also increases. This shows that for the reinforced soil, the axial residual deformation is affected by the consolidation ratio larger, the residual volume strain is less affected by the consolidation stress ratio.

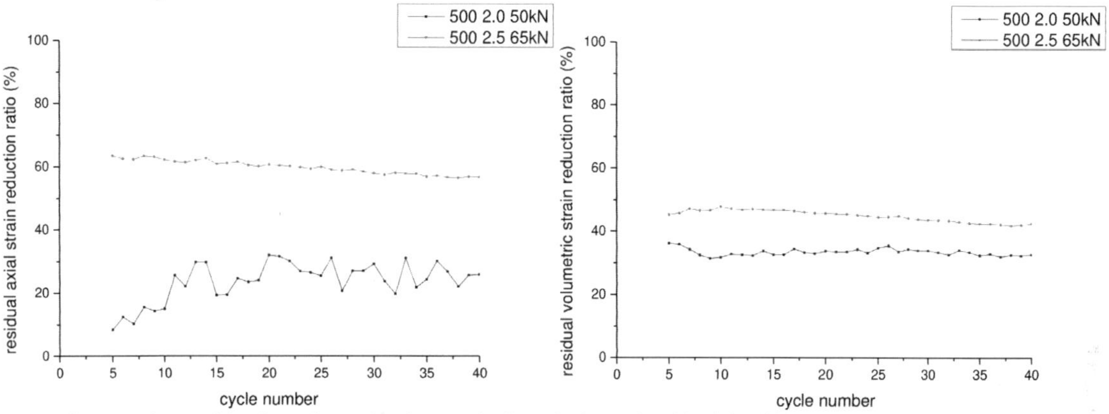

Figure 10. Residual strain reduction ratio in reinforced soil with different consolidation ratio.

The test results of series 1, 3, 6, and 8 are calculated, and the reduction ratio of residual deformation compared with reinforced soil and unreinforced soil under the same confining pressure, consolidation ratio, and dynamic stress is calculated. The data from the previous 4 weeks is also discarded, on the condition of high consolidation ratio, and the reduction ratio of axial strain is higher than that of low consolidation ratio, indicating that the higher the consolidation ratio, the better the reinforcement effect. Under the condition of consolidation ratio 2.0 The reduction ratio of the axial strain fluctuates around 25%, and the reduction ratio of the volumetric strain is about 30%. At the consolidation ratio of 2.5, the reduction ratio of the axial strain is 60%, and the reduction ratio of the volumetric strain is about 40%.

3.4Influence of confining pressure on residual deformation in reinforced soil

The influence of confining pressure on residual deformation in reinforced soil was investigated by comparing test series 3 and series 12. The test results are shown in Figure 11.

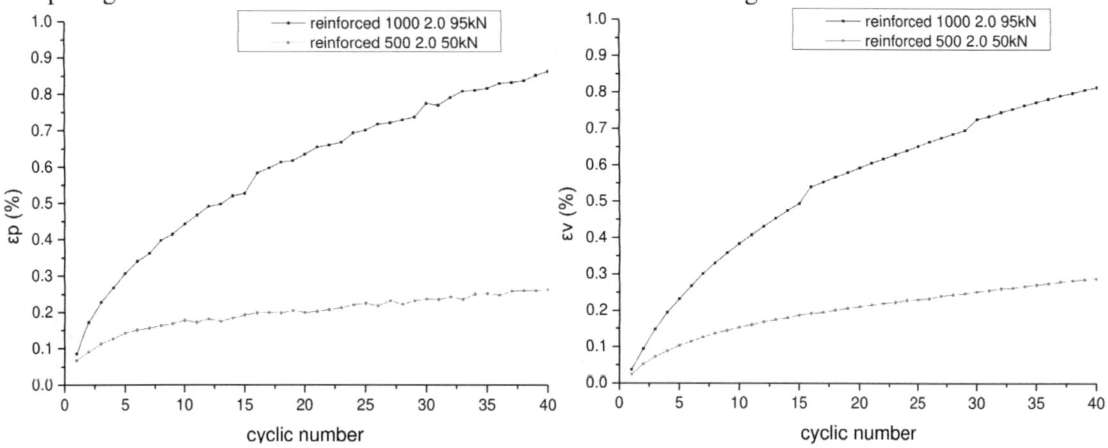

Figure 11. Influence of confining pressure on residual strain of reinforced soil.

Comparing the test results of series 3 and series 12, the two conditions maintain the same consolidation ratio, the dynamic shear stress ratio is approximately equal, and both are applied with geogrid materials. The test variables are 500kPa and 1000kPa with different confining pressures. The test results are shown in Figure 11. In the reinforced soil, when the dynamic shear stress ratio, consolidation ratio and geogrid material are the same, with the increase of the confining pressure, the axial residual deformation and volumetric residual deformation all increase significantly, showing the research conclusions consistent with the unreinforced soil.

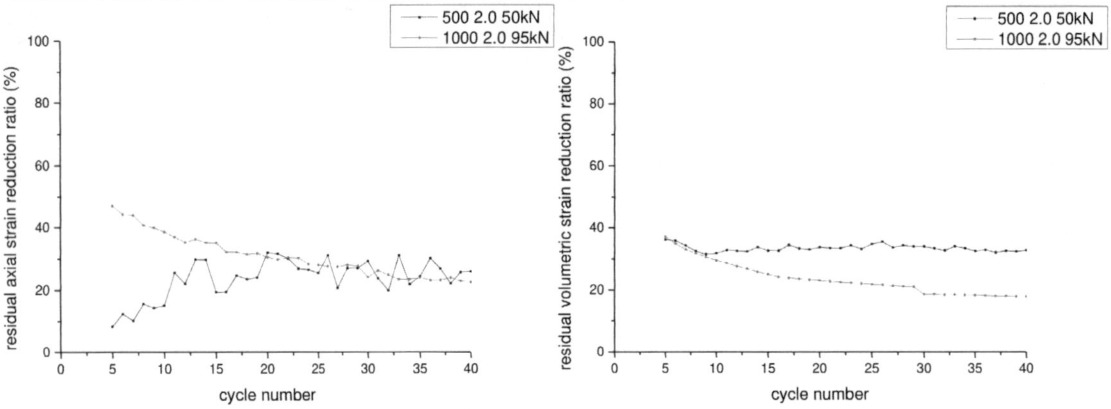

Figure 12. Residual strain reduction ratio in reinforced soil with different confining pressure.

The test results of series 1, 3, 10, and 12 calculate the reduction ratio of residual deformation of reinforced soils with different confining pressures and unreinforced soils under the same consolidation ratio and similar dynamic shear stress ratio. The processing results are shown in Figure 12. After discarding the data for the first 4 weeks, it is concluded that the reduction ratio of volumetric strain is higher in low confining pressure conditions than in high-confining pressure conditions, indicating that the smaller the confining pressure, the better the reinforcement effect. The reduction ratio of the axial strain fluctuates around 25%, and the reduction ratio of the volumetric strain is about 30%. Under the confining pressure of 1000 kPa, the reduction ratio of the axial strain is about 25%, and the reduction ratio of the volumetric strain is about 20%. Due to the weak shear expansion potential of the sample under high confining pressure, the lateral movement of the soil material between the grids is restricted, so the reinforcement effect is reduced.

4. Conclusion

As a seismic reinforcement measure, geogrid material is important to study the residual deformation characteristics of reinforced soil for the seismic safety performance of earth-rock dams. This paper reviews the results of large-scale dynamic triaxial tests on reinforced and unreinforced soil. The main conclusions obtained from the analysis are as follows.

(1) Under the same consolidation ratio, confining pressure and dynamic shear stress ratio, the axial residual deformation and volumetric residual deformation of the soil samples with geogrid in the test materials are reduced in varying degrees. The average reduction of axial residual deformation is 40% - 50% of that of unreinforced soil, the maximum is 60%, and the average reduction of volume residual deformation is 30-40%. The geogrid material has a good inhibition effect on the residual deformation of coarse-grained soil.

(2) Under the same confining pressure and consolidation ratio, the seismic residual deformation of reinforced soil increases with the increase of dynamic shear stress ratio. Under the same confining pressure and dynamic shear stress ratio, the axial residual deformation and volume residual strain of reinforced soil increase with the increase of consolidation ratio. Under the same consolidation ratio and dynamic shear stress ratio, with the increase of confining pressure, the axial residual deformation and volume residual deformation of reinforced soil increase significantly. All the above results are consistent with the unreinforced soil.

(3) The reinforcement effect of geogrid material on soil sample under different conditions is analyzed. The results show that the lower the dynamic shear stress ratio and the higher the consolidation ratio, the better the reinforcement effect. The lower the confining pressure, the lower the volumetric strain reduction ratio of the reinforced soil, and the axial strain reduction ratio is about 25%.

Acknowledgments
This research was supported by the National Key R&D Program of China [Grant No.2017YFC0404905]; National Fundamental Science Foundation of China [Grant No. 51679264]; and IWHR Research and Development Support Program (GE0145B562017).

References
[1] Zhang R. Analysisand Researchof Seismic Effectsand Slope Sseismic Stabilityof High Earthand Rockfill Dams[D]. Dalian UniversityofTechnology,2008.
[2] Gu Y, Shen C, Cen W, et al. Earthquake Dam Earthquake Engineering [M] .Beijing: China Water Resources and Hydropower Press, 2009.
[3] Zhai Y, Liu B, Zhu K, Liu X. Research Progress on Dynamic Residual Deformation Characteristics of Coarse-Grained Soil [J] . Journal of China Institute of Water Resources and Hydropower Research, 2018,6 (06): 555-562.
[4] Bao H, Zhou Y, Zhao C, Huang Y. Study on Georid Reinforced Stone [J] .Chinese Journal of Geotechnical Engineering, 1999 (02): 82-86.
[5] Wu J, Wang D, Wang L, Chen H. Experimental Studyon Geosyntheticreinforcement [J] .Chinese Civil EngineeringJournal, 2002 (06): 93-99.
[6] BiJ. Studyonthe Deformationand Strength Behaviorof Coarse GrainedSoilReinforcedwith Geogrid [D]. Dalian University of Technology, 2009.
[7] Yang S, Yu X, Li Y, Wang Z. Triaxial Test Study on Dynamic Elastic Modulus of Reinforced Soil [J] .World Earthquake Engineering, 2010,26 (S1): 65-69.
[8] Seed, H.B., Idriss, I.M., Makdisi, F.I. and Banerjee, N. Representation of Irregular Stress Time Histories by Equivalent Uniform Stress Series in Liquefaction Analyses, Report No. EERC 75-29, EERC Univ. Of California, Berkeley, California. Oct., 1975.

IOP Conf. Series: Materials Science and Engineering **794** (2020) 012052 doi:10.1088/1757-899X/794/1/012052

The Post-occupancy Evaluation Over the Landscape of the Campus of Zhejiang Agriculture and Forestry University

Shen JiaFeng[1], Huang Wei[1]*, Wu Yun[1]

[1]Landscape Architecture School of Tourism and Health, Zhejiang A&F University, Hangzhou, Zhejiang, 313000, China

*E-mail: 295663188@qq.com

Abstract: This paper obtains the post-occupancy evaluation over the landscape of Zhejiang Agriculture and Forestry University through a questionnaire survey. Then, this paper carries out relevant investigation over the East Lake Teahouse, which is the most important part of the campus environment, and its surrounding environment, as well as its flow of people. After that, this paper discusses and analyses the effect imposed by the landscape environment on teachers and students on campus and tourists off campus from the perspective of environmental psychology, and draws corresponding conclusions.

1. Introduction

Post-Occupancy Evaluation (POE for short) refers to a research method which systematically evaluates the facilities being used that have been designed from the users' perspective [1]. Actually, it mainly applies the methods of interdisciplinary research including sociology, behaviouristics, psychology, statistics, architecture and urban planning for the aim of conducting systematic evaluation over the landscape environment and buildings in the environment. Consequently, it is able to identify comprehensively the extent to which the landscape design is able to satisfy the demands of users, as well as the supervision over design and feedback mechanism brought by POE, which can improve the comprehensive efficiency and quality of landscape and construction projects [2].

2. The Purpose and Significance of the Study

The overall prominent features of the campus of Zhejiang Agriculture and Forestry University are known as large size, green, regularity, seriousness, square buildings, grid roads, and eyeful green, and the lush, clean and quiet atmosphere provides all the students with an ideal learning environment. Promoted by the construction of new campuses in many colleges and universities, the landscape planning of university campus is becoming increasingly significant than ever before. In particular, the university campus spaces in modern times have undergone great changes not only in the design of specific space such as campus teaching, communication and life, but also in the concept and mode about the overall planning of the campus in comparison with the old campus. Then, how to establish a good learning and living environment for both teachers and students in the new campus environment has turned out to be a problem that must be confronted by every campus designer.

Study on the rationality of the public environment on campus plays a significant role for an open university seeking for further development, which consequently not only points out the right direction for the planning of university, but also lays a strong basis for students' excellent learning environment. Therefore, a university is required to conduct a comprehensive investigation and study on the public environment as a top-class university. It is acknowledged that POE study can enable us to understand the situation about the use of school public environment more clearly, so as to enhance the construction on campus to move forward towards the right direction.

Known as a special artificial landscape, campuses of universities not only serve as a simple place for teaching, but also have their own characteristics—the original ground mode and natural landscape would change because of the regularity for the spatial layout of landscape and the distribution of buildings[3]; the landscape system would be relatively open due to the high-intensity flow of people through the landscape space. The East Lake Campus of Zhejiang Agriculture and Forestry University is regarded as an open-ended ecological campus, and the spatial layout of its landscape adapts well to the local conditions. In the environment surrounded by mountains on three sides, the plant landscape on the campus extends based on the original undulating terrain so as to integrate well with the environment. This paper mainly carries out related studies on the Middle East Lake Teahouse on campus with the aim of providing better suggestions for the transformation of landscape on campus in the future, enhancing the scientific rationalization of landscape structure on campus, and offering a beautiful environment to teachers and students for their learning, leisure, entertainment and exercise. Moreover, this paper analyses the regional characteristics of the East Lake Teahouse from the angle of the landscape pattern of plant, summarizes the characteristics and styles possessed by planning and layout, and creates excellent campus plant landscape through the principle of sustainable development so as to from a beautiful campus environment.

3. Analysis on Survey Results

3.1 Investigation and analysis on the flow of people around the area

The flow of people in the East Lake Teahouse is affected by different seasons and different time periods of a day. Through asking the staff working in the teahouse, it is obtained that the flow of people is the most during the spring, for the reason that it is warm in spring when a hundred flowers would be in full bloom and there is a pleasant scenery by the lake. As a consequence, it would attract many people inside and outside the university to drink tea, enjoy the scenery and talk freely here. The second frequently visited season is autumn when the weather is getting cooler, the leaves are withering, and the fallen leaves are colourful, which offers us a totally different scenery. In summer when there is usually hot and students and teachers also have summer vacation, fewer visitors would come to the university, and the flow of people is less. However, some people would come here occasionally to enjoy the cool generated by the lush trees. In winter, the flow of people is also very small due to the cold weather. Besides, the daily flow of people varies because of different reasons, such as different seasons, holidays, and weekends and so on.

During this survey, a total of 150 questionnaires were distributed and 131 valid questionnaires were collected. Table 1 displays the proportion of different users and respective ratios occupied by different genders. Among them, the students in the university account for 55.7%, more than half, and the proportion of women are higher than of men.

Table 1. The proportion of users with different identities and genders

Proportion of Users	Proportion of People/%	Male/%	Female/%
Students in the University	55.7	48.7	51.3
Staff in the University	30.5	32.8	67.2
Off-campus Tourists	13.8	39.7	60.3

3.2 Survey on utilization frequency of environmental public facilities

Whether the design of environmental landscape is excellent or not is usually expressed by the utilization frequency of the landscape[4]. Located around Wuzhou Square, the East Lake Teahouse benefits much from it because the square is an urban public space which attracts people together to carry out leisure activities, and it is also an outdoor public space that vehicles are not allowed to enter. Beside Wuzhou Square, the East Lake Teahouse is able to provide people with convenience to rest.

Table 2 shows the frequency of different users' coming here, and the frequency of students' coming is higher, which shows that up to 70.2% of students would come once a week. Most of the off-campus tourists come here occasionally to enjoy the scenery, and only 16.6% of them come here once a week. It is clear that most of them are residents near the campus, and they'd come here for a walk during their spare time.

Table 2. Utilization frequency of users

Proportion of Users	Once a Week/%	Once a Mouth/%	Above Half a Year/%
Students in the University	70.2	28.3	1.5
Staff in the University	32.6	60.9	6.5
Off-campus Tourists	16.6	51.2	32.2

3.3 Survey on the satisfaction degree over environmental public facilities

In the designing process of landscape, the designer should take into account many environmental and psychological factors, whether it is about the arrangement of a rockery or the layout of a plant space. It is of great necessity to consider not only the spatial positions between them but also the relationship between them and people[5]. It is suggested that designers should fully display the most attractive features of objects through a series of designs for the purpose of controlling people's perception of objects. Therefore, it is obtained that the layout of public facilities in the landscape environment is very important.

It is shown by the results obtained from survey that most visitors are satisfied with the public facilities in the environment, especially the teachers and students in the university whose dissatisfaction rates are only 0.9% and 2.2% respectively. However, some people still hold the view that proper improvement should be conducted for the teahouse. For example, some users propose that railings should be set up near the river so as to prevent children from running and accidentally falling, while others state that railings should not be set up because it would block the view of the sitting viewers.

Table 3. Users' satisfaction degree over environmental public facilities

Proportion of Users	Very Satisfaction/%	Just so so9/%	Dissatisfaction /%
Students in the University	75.6	23.5	0.9
Staff in the University	68.7	29.1	2.2
Off-campus Tourists	65.3	26.7	8.0

3.4 Survey on environmental function

Based on the basic theory of environmental psychology, the guiding ideology of "people-oriented" design can only be fundamentally reflected when the designer understand people's behaviour and psychological laws in the landscape environment, and explore the rationality and scientificity of garden space design[6].Therefore, it is acknowledged that the function of the environment is one of the important factors which can attract users to come here. This study carries out the survey on why users come here, and the results show that on-campus students mainly come here to take photos and relieve boredom, which many students think can relieve the pressure of learning. On the other hand, the staff and off-campus tourists mostly drink tea, chat and enjoy the scenery here.

Table 4. Statistics about the purpose of users

Proportion of Users	Drink tea and Chat/%	Take photos /%	Wandering and relax/%
Students in the University	10.2	52.8	37.0
Staff in the University	42.6	45.9	11.5
Off-campus Tourists	48.4	32.5	19.1

4. Summary

Through the analysis and survey on post-occupancy of the landscape at the East Lake Teahouse and its surrounding environment on the campus of Zhejiang Agriculture and Forestry University, it is found that a good landscape on campus plays a significant role for university students, and a comfortable environment can relieve stress, which is beneficial to the mental health of university students. The environmental landscape on campus is a complex whole, which includes all sorts of plants, public facilities and buildings, and the ingenious combination of all elements is a requisite for an excellent landscape environment. As a special public space, the environment of university campus is able to impose certain effect on shaping and changing the personality of university students, and thus the significance of the public environment on the university campus can be shown clearly.

Human beings are known as the main body of space, and therefore any landscape should take people's demands as the starting point so as to reflect the concern of people, satisfy people's physical and psychological demands, and create a beautiful environment. The application of environmental psychology to landscape and architectural design actually leads to the new research direction of design, promotes the designer's concern over people, and realizes the optimization of landscape environment, which consequently provides many detailed design with theory and basis.

Environmental psychology is actually everywhere, whether in our lives or in the process of designing. Sometimes designers would only carry out design from their own perspective, which would result in some dehumanized designs, while environmental psychology teaches us to better analyse problems by putting ourselves into others' shoes, so that we can really understand what kind of environment the broad masses really need. Otherwise, it may design the ideal environment in the designer's own mind, instead of the ones expected by the masses Therefore, designers are required to make rational use of environmental psychology so as to design a more reasonable and convenient environment for people to use and conduct activities. Only when we sufficiently understand this subject, can we design a human-oriented landscape environment.

References

[1] Marcus, G.C, Francis, C. (2001) People Places--Design Guidelines for Urban Open Space[M]. Bei Jing: China Architecture & Building Press.
[2] Yin, J, Shen, X. (2009) Significance and Methods of Introducing Post Occupancy Evaluation in China's Landscape Architecure Design[J]. Journal of Zhejiang Forestry College, 26(03): 417-420.
[3] Han, C.N, Wang, W.Z, Duan Y. G. (2009) Expression of the Cultural Intension in the Landscape Design for a University—A Case Study of South Campus Square of Northwest A&F University[J]. Journal of Northwest forestry University, 24(06): 169-172.
[4] Huang, Y, Li,.B.R, Weng, S.F. (2017)Study on Landscape Image of Campus in Guangzhou[J]. Chinese Landscape Architecture, 33(01): 88-94.
[5] Han, J, Hu, S.X. (2006) Consider the past, and you shall know the future—The Introduction of the way for Post-Occupancy Evaluation[J]. Architucture Journal, 01: 80-82.

[6] Luo, L.L, Ren, Q.H. (2009) International Progress and Methodological Analysis of Theoretical Breakthroughs for Environmental Psychology Research[J]. Architucture Journal, 07: 13-16.

IOP Conf. Series: Materials Science and Engineering **794** (2020) 012053 doi:10.1088/1757-899X/794/1/012053

System identification from ambient vibration measurements using output-only modal identification techniques

Jun Ma[1*], Wei Luan[2]

[1]Engineering research institute, China construction eighth engineering division, Shanghai, 200122, China

[2]Northwest branch company, China construction eighth engineering division, Xi'an, Shaanxi, 710000, China

[*]Corresponding author's e-mail:mjbppc@163.com

Abstract. Frequency domain decomposition (FDD) method and Random decrement technique are proposed to analyze ambient vibration data. An ambient vibration test was conducted on an 18-story reinforced concrete building, and the proposed methods were applied to estimate modal parameters. The estimated natural frequencies by the FDD method and the RD technique are in good agreement, and comparisons for damping values estimated by the EFDD method and the RD technique show a 72-246 percent difference. The results imply that the proposed methods can accurately evaluate structural dynamic properties from ambient vibration data in a non-destructive manner.

1. Introduction

Output-only modal identification techniques have been popular due to some advantages. In out-put only modal testing, only output response data is measured, and the input signals are ignored. This means the output-only modal testing can be carried out under natural conditions, and the test structure can be in normal operation during the test.

Based on the white noise assumption, some output-only modal identification methods have been developed. The output-only modal identification techniques can mainly be classified into two groups, including the frequency domain methods and the time domain methods. In the frequency domain, the Peak-picking method has been used for decades, and the Frequency domain decomposition (FDD) method improves Peak-picking method by taking singular value decomposition of the output spectral matrix[1,2], later the Enhanced frequency domain decomposition (EFDD) method was developed to estimate modal damping [3].In the time domain, the Ibrahim time domain method (ITD), Eigensystem realization algorithm (ERA), Stochastic Subspace Identification (SSI) and Random decrement (RD)technique have been widely used.

In this paper, the Frequency domain decomposition method, Enhanced frequency domain decomposition method, and Random decrement technique were adopted to estimate modal parameters from ambient vibration response of an 18-story reinforced concrete building. These two techniques represent two different classes of identification. The results clearly illustrate that the output-only modal identification techniques in both frequency and time domain perform excellently in dealing with ambient vibration response data.

IOP Conf. Series: Materials Science and Engineering **794** (2020) 012053 doi:10.1088/1757-899X/794/1/012053

2. Output-only modal identification techniques

2.1Frequency domain decomposition method

The Frequency domain decomposition method removes limitations of the PP method. The response power spectral density matric is decomposed into a set of auto spectral density functions, and each function corresponds to a single degree of freedom (SDOF) system, then modal parameters are obtained.

Considering an n degree of freedom structure. The relationship between the measured output response $y(t)$ and the unknown input $x(t)$ can be expressed as

$$G_{yy}(j\omega) = \bar{H}(j\omega)G_{xx}(j\omega)H(j\omega)^T \tag{1}$$

where $G_{xx}(j\omega)$ is the power spectral density matrix of the input, $G_{yy}(j\omega)$ is the power spectral density matrix of the output, $H(j\omega)$ is the frequency response function (FRF), the overbar and the superscript T denote the complex conjugate and transpose, respectively.$H(j\omega)$ can be written as

$$H(j\omega) = \sum_{k=1}^{n} \frac{R_k}{j\omega - \lambda_k} + \frac{\bar{R}_k}{j\omega - \bar{\lambda}_k} \tag{2}$$

where n is the DOF number and the mode number, R_k is the residue,λ_k is the pole, ϕ_k is the mode shape vector, γ_k is the modal participation vector.

With the white noise assumption, the input PSD matrix $G_{xx}(j\omega)$ is reduced to a constant matrix C. Assume that the structure is lightly damped and there is no closely spaced mode, the Eq.(1) can be simplified as follows

$$G_{yy}(j\omega) = \sum_{k=1}^{n} \left(\frac{d_k \phi_k \phi_k^T}{j\omega - \lambda_k} + \frac{\bar{d}_k \bar{\phi}_k \bar{\phi}_k^T}{j\omega - \bar{\lambda}_k} \right) = \Phi \cdot diag\left(2\mathrm{Re}\left(\frac{d_k}{j\omega - \lambda_k} \right) \right) \cdot \bar{\Phi}^T \tag{3}$$

The power spectral density matrix of the kth order can be represented as follows

$$\underset{\omega \to \omega_k}{G_{yy}}(j\omega) = \phi_k \cdot diag\left(\frac{2d_k \zeta_k \omega_k}{(\zeta_k \omega_k)^2 + (\omega - \omega_k)^2} \right) \cdot \bar{\phi}_k^T \tag{4}$$

From Eq.(4) it is noticed that only one mode can contribute to the maximum of power spectral energy, so that modal parameters can be estimated by the FDD method.

The structural damping can be estimated by the Enhanced frequency domain decomposition method. The piece of the SDOF density function obtained around the peak of the PSD is first isolated and then is taken back to the time domain using IFFT. The damping ratio can then be obtained from the decay of autocorrelation functions.

2.2Random decrement technique

Random decrement technique was first put forward to estimate damping of airplane wings [4]. It assumes the output response data to be white noise. Base on the white noise assumption, the measured random response is divided into a series of short segments, and the free vibration response is extracted by averaging the short segments. The obtained free vibration signal can be written as

$$z(\tau,c) = \frac{1}{2N} \sum_{i=1}^{2N} \mathrm{sgn}[y(t_i + \tau, c)] \cdot y(t_i + \tau, c) \tag{5}$$

where $z(\tau,c)$ is the extracted free vibration signal, $y(t,c)$ is the displacement, velocity or acceleration response at time t with respect to the given trigger value of c, $\mathrm{sgn}[y(t,c)]$ is the signum function of $y(t,c)$, t_i is the i-th time instant corresponding to the trigger value c, $2N$ is the number of time segments.

3. Applications

Frequency domain decomposition method, Enhanced frequency domain decomposition method, and Random decrement technique were applied to evaluate modal parameters from ambient vibration response of an 18-story reinforced concrete building, as shown in Figure 1. Layout of the test building is shown in Figure 2. Ambient vibration test was conducted. The response acceleration data along the width direction and the associated PSD function are shown in Figure 3.

Figure 1.Test building

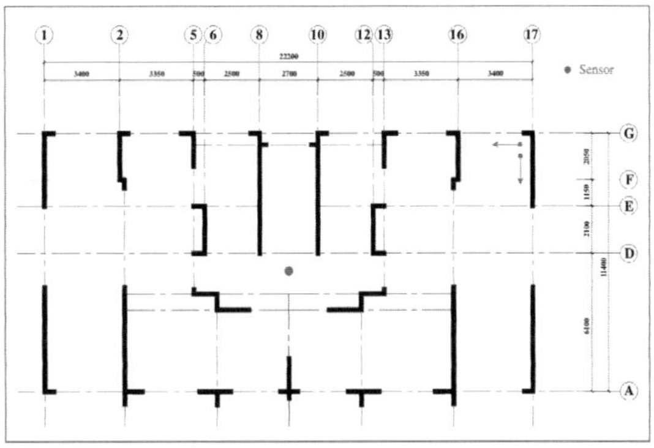

Figure 2.Layout of the structural system

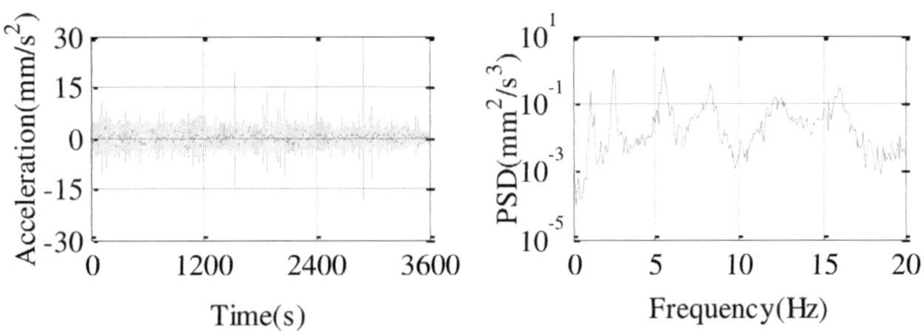

(a) Acceleration along width direction (b) Power spectral density

Figure 3.Acceleration record along width direction

Figure 3 shows singular values of the output response PSD matrix. It can be noticed that the 1st mode and the 2nd mode are coupled, and the 4th mode and the 5th mode are coupled. However, the closely spaced modes are clearly illustrated in the figure.

Figure 4 shows curve fitting process of the RD technique. The free vibration response is extracted from the measured random response by averaging short segments, then the curve fitting technique is applied to the free vibration signal to estimate modal parameters. It can be seen that the curve fitting result fits well with the free vibration data.

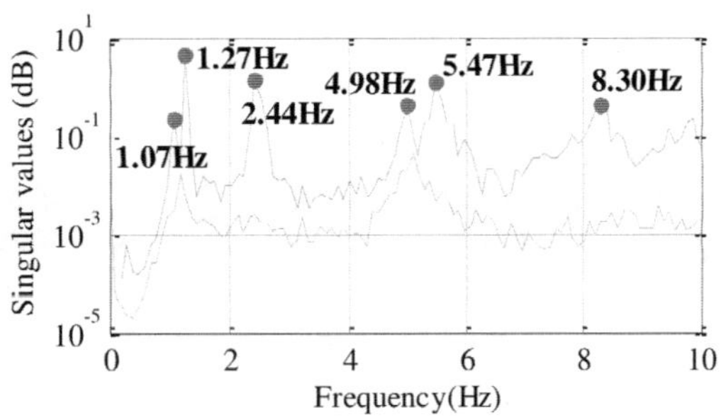

Figure 4.Singular values of the output response PSD matrix

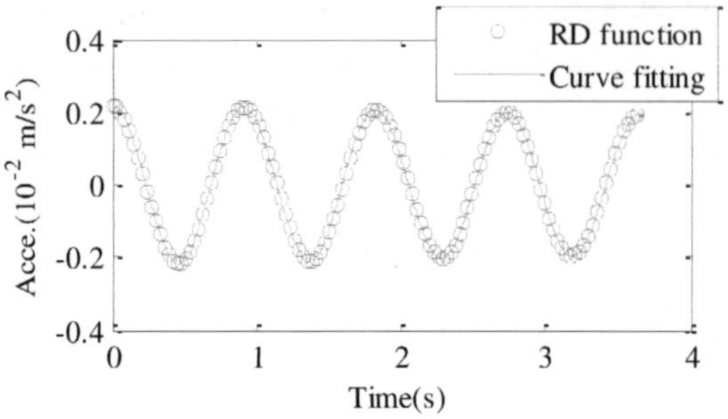

Figure 5.Curve fitting result

The estimated natural frequencies and damping ratios by the FDD method, EFDD method, and RD technique are shown in Table 1. The first six modes are identified by the FDD method, but the RD technique fails in identifying the 4th mode. The estimated natural frequencies by the FDD method and the RD technique are in good agreement. However, an obvious distinction between the EFDD method and the RD technique can be noticed. The damping ratios estimated by the EFDD method are generally larger than those estimated by the RD technique, except the 2nd mode. Comparisons for damping values estimated by the two methods show a 72-246 percent difference.

Table 1.Natural frequencies and damping ratios of the building.

Mode	FDD and EFDD		RD technique	
	Natural frequency (Hz)	Damping ratio (%)	Natural frequency (Hz)	Damping ratio (%)
1st	1.07	1.11	1.07	0.53
2nd	1.27	0.81	1.27	2.54
3rd	2.44	1.36	2.44	0.52
4th	4.98	1.35	-	-
5th	5.47	1.31	5.47	0.76
6th	8.30	2.29	8.30	0.66

4. Conclusions

In this paper, the frequency domain and time domain output-only modal identification methods are introduced, including the FDD method, EFDD method, and the RD technique. The proposed methods are applied to ambient vibration response of a concrete building. The results show that natural frequencies estimated by the frequency and time domain methods are in good agreement, but damping ratios show significant difference. The results imply that both frequency and time domain methods can evaluate modal parameters from ambient vibration data in a non-destructive manner.

References

[1] Brincker, R., Zhang, L.M., Andersen, P. (2001) Modal identification of output-only systems using frequency domain decomposition. Smart Mater.Struct., 10: 441-445.

[2] Brincker, R., Ventura, C.E., Andersen, P. (2001) Modal identification from ambient responses using frequency domain decomposition. In: 18th International Modal analysis Conference.

[3] Jacobsen, N.J., Andersen, P., Brincker, R.(2006) Using enhanced frequency domain decomposition as a robust technique to harmonic excitation in operational modal analysis. In: International Conference on Noise and Vibration Engineering.

[4] Cole, H.A. (1971) Methods and Apparatus for Measuring the Damping Characteristics of a Structure. U.S. Patent 3, 620,069[P].

Study on the Chemical Characteristics of the Water Environment in the Middle and Lower Reaches of the Tao River

ZhaoxiaZhu*, **Jinqi Liu, Chungang Jing**

Water Environment Monitoring Center of Gansu Province, Lanzhou, Gansu Province, 730000, China

*Corresponding author's e-mail:50562668@qq.com

Abstract. In this paper, the water in the middle and lower reaches of the Tao River was selected as the research object, the chemical characteristics of water environment in the middle and lower reaches of Tao River were studied and the water quality was evaluated based on the comparative analysis of the data of three monitoring stations for one year. The results show that the contents of HCO^{3-},SO_4^{2-} and $Ca^{2+}$are high in the middle and lower reaches of Tao River, and the water quality belongs to alkaline medium salinity water, and the salinity increases with the increase of HCO_3^-. The pH value, total hardness, salinity and main ion content in the middle and lower reaches of the river reach the requirements of the GB5749-2006 Sanitary Standard for Drinking Water or the Class III water quality standard of the surface water environment of GB3838-2002, and which proves the water quality is good.

1. Introduction

As an important part of terrestrial freshwater resources, the river is the most active part of the earth's water circle, its physical and chemical properties not only characterize the environmental quality of water body, the chemical characteristics of regional environment, the distribution and migration and transformation of water elements, but also have an important response relationship with climate change, topography and the formation of rock and mineral characteristics[1,2]. In recent years, due to the increase of human activities, human fact has become an important factor affecting the chemical composition of river water. Therefore, through the analysis of the natural conditions and chemical composition of ions in the river basin, the main sources and influencing factors of ions in the river can be distinguished [3]. At present, the hydrochemistry of river water in China is mostly concentrated in Yangtze River, Yellow River, Pearl River and other large water systems [4-8], but there is a lack of systematic research data for some small watersheds. The research of Tao River mainly focuses on the development of water resources and ecological protection. The basic chemical composition and environmental background value of the river have been discussed by hydrochemistry research.

In order to improve the drought and water shortage in the central part of Gansu, the water supply project of the main body of the Jiudianxia Water Control Project has been started, and the development and utilization of water resources in the Tao River watershed will usher in a new challenge, and the water environment quality and the water ecological system safety problems will become a major problem. Therefore, it is important to study the hydrochemistry of the river water in Tao River watershed, and analyze the physical and chemical properties of the river, the composition of the main ions and its temporal and spatial changes, which is also of great significance in the evaluation of water resources and environmental protection.

2. Overview of the study area

The Tao River watershed is located in the south of Gansu Province, between 101°52'N-104°19'N and 34°03'E-35°55'E. The four sides of the river watershed are adjacent to the river watershed and the river watershed of the Yangtze River, and the Longling Mountain and the Daxia River watershed are the boundary of the river watershed, and the north and south are adjacent to the main stream of the Yellow River. Tao River, as the largest tributary of the upper reaches of the Yellow River, has a total watershed area of 25527km^2 and an average runoff of 4.92 billion m^3, which carries the important function of water resources allocation in western China, especially in central Gansu, and directly affects the economic and social development of central Gansu and its surrounding areas.

In this study, the sampling data of three monitoring stations in the middle and lower reaches of Tao River are Minxian Water quality Monitoring Station, Lijiacun Hydrological Monitoring Station and Hongqi Hydrological Monitoring Station (Fig. 1). The sampling time is from February 2018 to December 2018 and sampling every two months, and the sample analysis method is the Water and Wastewater Monitoring and Analysis method of the State Environmental Protection Bureau.

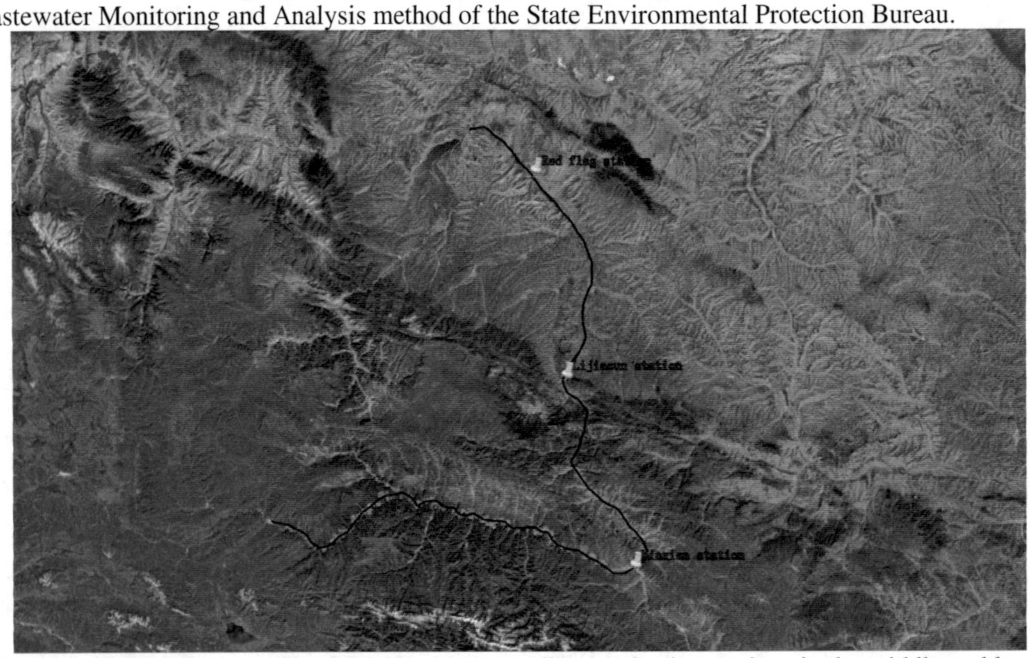

Figure 1. Geographic coordinates of three water quality monitoring stations in the middle and lower reaches of Tao River.

3. Research results and analysis

3.1 Main ions in the water of the middle and lower reaches of Tao River

According to the monitoring data of the three monitoring stations in the middle and lower reaches of the river, the main water chemical composition in the middle and lower reaches of the river is HCO_3^- >Ca^{2+}>SO_4^{2-}>Na^+>Mg^{2+}>Cl^->CO_3^{2-}>K^+.

3.1.1 HCO_3^-. HCO_3^- is the main anion in the middle and lower reaches of Tao River. The annual HCO_3^- of Minxian Water quality Monitoring Station in the middle reaches of Tao River was 186~247mg/L, the mean value was 218.5mg/L, the average annual HCO_3^- of Lijiacun Water quality Monitoring Station in the lower reaches of Tao River was 205mg / L, and the mean annual HCO_3^- of Red Banner Water quality Monitoring Station was 181~239mg/L (Fig.2(a)). $H_2CO_3^-$-HCO_3^--CO_2 in natural water is generally in equilibrium, but when pH value is about 8.5, carbonate ions in river water mainly exist in the form of HCO_3^-, and its content will directly affect the total salt content, and then affect the chemical stability of water.

3.1.2 Ca^{2+}.Ca^{2+} is one of the main cations in the middle and lower reaches of the river. The water quality monitoring station of Minxian County in the middle reaches of the Tao River is monitoring the water quality of the whole year, the average value of Ca^{2+} is 43.4~61.8mg/L, the mean value is 52.1mg/L, and the water quality monitoring station of Lijia Village in the lower reaches of the river is about 41.0~62.0mg/L, the mean value is 52.3mg/L. The monitoring water quality of the water quality monitoring station of the red flag is 31.2~79mg/L, and the mean value is 55.2 mg/L (Fig.2(b)).

HCO_3^- and Ca^{2+} mainly come from the dissolution of carbonate rocks such as marl, limestone, dolomite and so on. The dissolution process is related to the content of CO_2 in water. The anions in the middle and lower reaches of Tao River are mainly HCO_3^-, while the cations are mainly Ca^{2+}, $Mg^{2+}$and K^+. The reason is that the rocks with more carbonate minerals are distributed in the upper reaches of Tao River, and the hydrothermal conditions are sufficient, which creates good conditions for carbonate dissolution. It also leads to the accumulation of enough HCO_3^- and Ca^{2+} in the water, which occupies an absolute advantage in hydrochemistry, and the proportion of HCO_3^- and Ca^{2+} in the river increases gradually with the passage of time.

3.1.3 Mg^{2+}. The annual Mg^{2+} of Minxian Water quality Monitoring Station in the middle reaches of Tao River was 11.3~16.6mg/L, the mean value was 14.13mg/L, the annual Mg^{2+} of the lower reaches of Tao River was 7.2~20.2mg/L, the average value was 15.5mg/L, and the annual Mg^{2+} of Red Banner Water quality Monitoring Station was 6.37~26.00mg/L, the mean value was 14.92mg/L (Fig. 2(c)).

3.1.4 Cl^-.The annual Cl^- of Minxian Water quality Monitoring Station in the middle reaches of Tao River was 1.49~4.93mg/L, the mean value was 2.89mg/L, the annual Cl^- of the lower reaches of Tao River was 2.41~5.46mg/L, the average value was 4.20 mg/L, and the annual Cl^- of Red Banner Water quality Monitoring Station was 8.68~15.8mg/L, the mean value was 11.44mg/L (Fig. 2(d)).

3.1.5SO_4^{2-}. The annual $SO_4^{2-}$of Minxian Water quality Monitoring Station in the middle reaches of Tao River was 15~29.2mg/L, the mean value was 21.2mg/L, the annual SO_4^{2-} of the lower reaches of Tao River was 13~24.7mg/L, the average value was 20.0 mg/L, and the annual SO_4^{2-} of Red Banner Water quality Monitoring Station was 24.2~92.9mg/L, the mean value was 45.0mg/L (Fig. 2(e)).

3.1.6Na^+. The annual $Na^+$of Minxian Water quality Monitoring Station in the middle reaches of Tao River was 3.22~12mg/L, the mean value was 7.82mg/L, the annual Na^+ of the lower reaches of Tao River was 3.3~13.3mg/L, the average value was 8.27mg/L, and the annual Na^+ of Red Banner Water quality Monitoring Station was 5.8~61.5mg/L, the mean value was 23.7mg/L (Fig. 2(f)).

3.1.7 K^+, MnO_4^- and CO_3^{2-}.Through the monitoring and analysis, the proportion of these ions in the river is low, and the variation in the whole year is small, but the overall trend is that the middle reaches of the river is higher than that of the downstream.

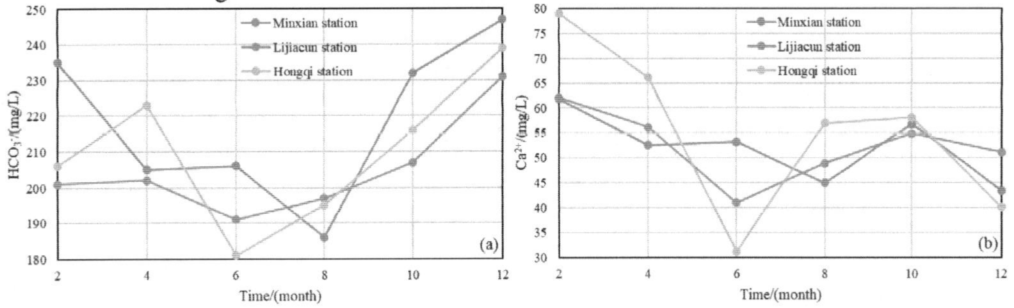

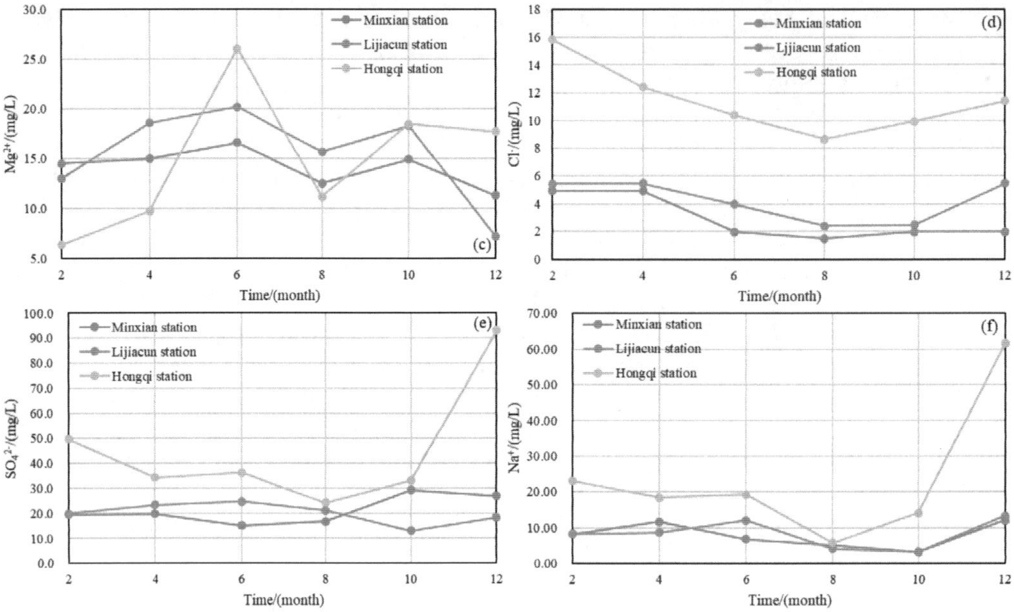

Figure 2. Main ion content of three stations in the middle and lower reaches of Tai River: (a) HCO_3^-; (b) Ca^{2+}; (c) Mg^{2+}; (d) Cl^-; (e) SO_4^{2-}; (f) Na^+.

3.2 Physicochemical properties of river water

*3.2.1 pH value.*The annual pH value of monitoring water quality at Minxian Water quality Monitoring Station in the middle reaches of Tao River was 8.2~8.7, with an average value of 8.37; the annual pH value of Lijiacun Water quality Monitoring Station in the lower reaches of Tao River was 8.3~8.5, with an average value of 8.38; the annual pH value of Red Flag water quality monitoring station was 8.3~8.5, with an average value of 8.37; The pH values of the three monitoring points in dry season are higher than those in high water season (Fig.3(a)). The water quality of the middle and lower reaches of Tao River is alkaline, which is mainly caused by the dissolution of carbonate rocks widely distributed in the upper reaches of Tao River, which leads to the formation of bicarbonate buffer system in water.

3.2.2 Salinity. The annual salinity of Minxian water quality monitoring station in the middle reaches of Tao River is 294~364mg/L, the mean value is 323mg/L, the annual salinity of Lijiacun water quality monitoring station in the lower reaches of Tao River is 216~313mg/L, the mean value is 276.17mg/L, and that of Red Flag water quality monitoring station is 272~446mg/L, the mean value is 346mg/L(Fig.3(b)). Compared with the middle reaches of Tao River watershed, the salinity of the lower reaches is higher.

According to the salinity classification standard of SL395-2007, the water with salinity less than 300mg/L belongs to low salinity water, but 12 of the 18 monitoring data have mineralization value higher than 300mg/L, which indicates that this section of river water belongs to medium salinity water and moderate hard water, which is consistent with the high HCO_3^- content in this section. The salinity shows that the total salt level in the water directly affects the water migration coefficient of the elements, and it is an important index to reflect the physical and chemical properties of water quality. Similarly, according to the salinity greater than 500mg/L, the natural background water quality of surface water has become worse, while the water quality mineralization value of 18 monitoring stations in the middle and lower reaches of Tao River is lower than 500mg/L, which indicates that the natural background water quality in the middle and lower reaches of Tao River is less polluted.

3.2.3 Electrical conductivity. The annual conductivity of Minxian water quality monitoring station in the middle reaches of Tao River is 316~512μs/cm, the average value is 385.67 μs/cm; The annual conductivity of Lijiacun water quality monitoring station in the lower reaches of Tao River is 367~454μs/cm, The average value is 402.33μs/cm; The annual conductivity of Red Flag water quality monitoring station is 403~608μs/cm, the average value is 488μs/cm. It is found that the conductivity of the middle and lower reaches of Tao River in the period of February to April and October to December is relatively larger than that of June to August. At the same time, compared with the middle reaches of Tao River watershed, the electrical conductivity of the lower reaches is higher, which indicates that the water quality in the middle reaches is better than that in the lower reaches (Fig.2(c)).

3.2.4 Dissolved oxygen. The annual dissolved oxygen of Minxian water quality monitoring station in the middle reaches of Tao River is 8.17~10.5mg/L, the mean value is 8.90mg/L, the annual dissolved oxygen of Lijiacun water quality monitoring station in the lower reaches of Tao River is 7.1~10.2mg/L, the average value is 8.8mg/L, and that of Red Flag water quality monitoring station is 7.0~10.3mg/L, the mean value is 9.0mg/L(Fig.3(d)). Compared with the middle reaches of Tao River watershed, the dissolved oxygen of the lower reaches is higher.

*3.2.5 Total hardness (in $CaCO_3$).*The annual total hardness of Minxian water quality monitoring station in the middle reaches of Tao River is 155~214mg/L, the mean value is188.3mg/L, the annual total hardness of Lijiacun water quality monitoring station in the lower reaches of Tao River is 157.1~217mg/L, the average value is 194.36mg/L, and that of Red Flag water quality monitoring station is 173.16~224mg/L, the mean value is 199.53mg/L(Fig.3(e)). Compared with the middle reaches of Tao River watershed, the total hardness of the lower reaches is higher.

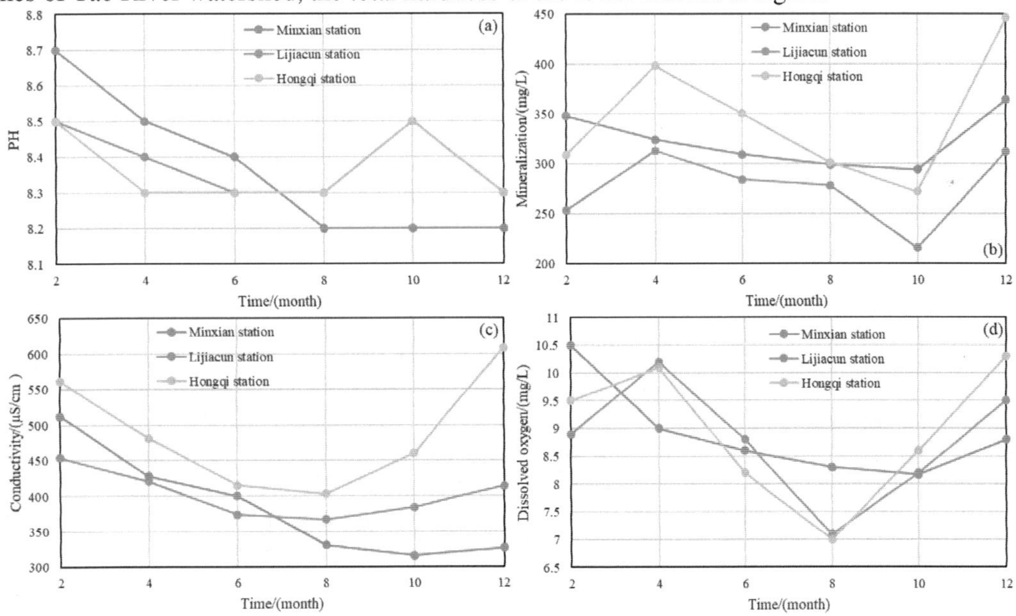

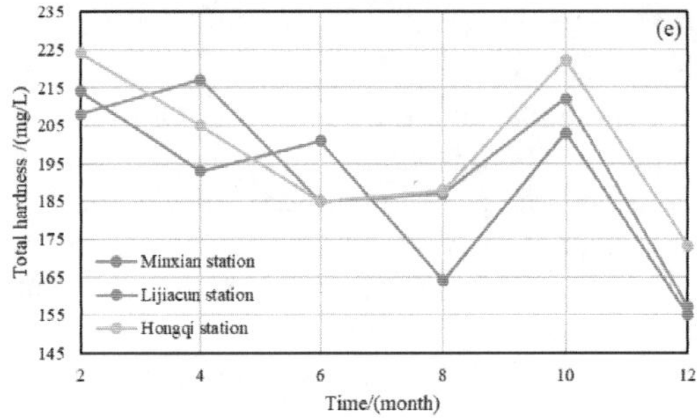

Figure 3. Physical and Chemical Properties of three stations in the Middle and Lower reaches of Tao River:(a) pH value; (b) Salinity; (c) Electrical conductivity; (d) Dissolved oxygen; (e) Total hardness.

4. Conclusions

The main results are as follows:

(1) the water quality of the middle and lower reaches of Tao River belongs to alkaline and medium salinity water, and its salinity increases with the increase of HCO_3^-, and the main hydrochemical composition of the river is $HCO_3^->Ca^{2+}>SO_4^{2-}>Na^+> Mg^{2+}>Cl^-> CO_3^{2-}>K^+$.

(2) In the middle and lower reaches of Tao River, the contents of HCO_3^-, SO_4^{2-} and $Ca^{2+}$are high, and the chemical type is calcium and sodium bicarbonate type water.

(3) In the middle and lower reaches of Tao River, HCO_3^-、K^+、$Cl^-$and Ca^{2+} mainly come from the dissolution of carbonate rocks such as marl, limestone and dolomite, and their contents increase gradually with the passage of time.

(4) The pH value, total hardness, salinity and the content of main ions and trace elements in the middle and lower reaches of Tao River meet the requirements of GB5749-2006 hygienic Standard for drinking Water or GB3838-2002 Surface Water Environmental quality Standard III, so the water quality is good.

Acknowledgments

Financial supports for this project are from the Gansu Water Conservancy Science Experimental Research Project (2018-35) and the Special Project of Basic Scientific Research Fee of Institute of Earthquake Prediction, China Earthquake Administration (Grant No. 2014IESLZ01).

References

[1] Pu Junbing, Yuan Daoxian, Jiang Yongjun, et al..(2010) Hydrogeochemistry and environmental meaning of Chongqing subterranean karst streams in China [J]. Advances in water science, 21(5): 628-634. (in Chinese)

[2] Wang Zhe, Lu Li, et al..(2013) Numerical simulation of the evolution of karst groundwater system [J]. Journal of Yangtze River scientific research institute, 30(7):22-28. (in Chinese)

[3] Bao Li-ran, Li Xiao-dong, Liu Xiaolong. (2010) Space-time variation of chemical composition of major ions in Jialing River [J]. Advances in science and technology of water resources, 30(4): 36-40. (in Chinese)

[4] ELDERFIELD H, UPSTILL -GODDARD R, SHOLKOVITZ E R. (1990) The rare earth elements in rivers, estuaries, and coastal seas and their significance to the composition of ocean waters[J]. Geochimica et CosmochimicaActa, 54(4): 971-991.

[5] REEDER S W, HITCHON B, LEVINSON A A. (1972)Hydrogeochemistry of the surface waters of the Mackenzie river drainage basin, Canada-I. Factors controlling inorganic composition[J]. Geochimica et CosmochimicaActa, 36(8): 825-865.

[6] Chen Jingsheng, Guan Wenrong, et al..(1998) Evolution in water quality and its relation with environmental acidification in the upper and middlereaches of the Yangtze River [J]. Acta scientiae circumstantiae, 18(3): 265-270. (in Chinese)

IOP Conf. Series: Materials Science and Engineering **794** (2020) 012054 doi:10.1088/1757-899X/794/1/012054

[7] Chen Jingsheng, Xia Xinghui. (1999) Progress in research on river hydrochemistry in China [J]. Scientiageographicasinica, 19(4): 290-294. (in Chinese)
[8] Zhang Litian, Chen Jingsheng. (2000) The relationship between the composition of the major ion of river of China and regional natural factors[J]. Scientiageographicasinica, 20 (3): 236-240. (in Chinese)

IOP Conf. Series: Materials Science and Engineering **794** (2020) 012055 doi:10.1088/1757-899X/794/1/012055

Three-dimensional geonet ecological slope protection technology and its engineering application

WANG De-yong[1,2], Zun Qing-jun[1,2], Wang Jing[1,2] and Huang Wei-yuan[3]

[1]CCCC Fourth Harbor Engineering Institute Co., Ltd., Guangzhou 510230

[2]Key Laboratory of Environmental Protection & Safety of Communication Foundation Engineering, CCCC, Guangzhou 510230

[3]Hohai university, Nanjing, 210098

Abstract: A protective structure system for slope with combination of vegetation and 3D geonet network was established. First the mechanism and ecological effects of slope protection with vegetation technology in slope engineering were discussed. Then specific implementation steps were proposed. Taking a slope ecological restoration project as an example, herbaceous plants suitable for slope protection were compared and tested, long term monitoring shows that the slope plants grow well, the protection effect of the new branches and roots were obvious, the stability of the slope was enhanced under comprehensive protection measures.

1. Introduction

More than 80% of the ground surface belongs to sloping land. In recent years, various projects in China have emerged to meet the needs of economic development. A large number of exposed slopes formed by engineering excavation, which damaged existing vegetation, caused a series of ecological and environmental problems such as soil erosion, landslides, and slope instability [1]. The traditional slope protection technology requires excavation and filling, which breaks the local ecological balance, and lacks landscaping and greening effects.

Slope ecological protection refers to measures to stabilize the slope using plants or plants combined civil engineering [2]. The slope ecological protection technology can not only meet the effects of reinforcement and treatment, but also can prevent soil erosion, purify the atmosphere, green and beautify the landscape, reduce noise, and absorb heavy metal pollution, etc. [3-4]. Three-dimensional geonet ecological slope protection is one of the slope ecological protection technologies. This technology has low cost, can effectively protect the slope from wind, rain, and flood erosion, and reinforce the slope. It is mostly used in actual engineering [5-6]. This paper introduces the three-dimensional geonet ecological slope protection technology and its engineering application, and provides theoretical and design basis for slope protection measures with vegetation.

2. Ecological slope protection theory

2.1 Basic idea

Three-dimensional vegetation geonet is a kind of yped seed planting material with three-dimensional structure, which can protect the ground surface being corroded before the turf is growing up, consolidate soil and fix the seeds .The compound protection layer formed after the grass grows up can prevent slope erosion.

IOP Conf. Series: Materials Science and Engineering **794** (2020) 012055 doi:10.1088/1757-899X/794/1/012055

2.2Composite slope protection mechanism

The influence factors of vegetation on the slope system can be divided into three categories: mechanical effects, hydrological effects and biological effects [7-10]. The 3D ecological geotechnical net combined vegetation can resist the tensile and shear stresses on the soil of slope, constrain the deformation of the soil, increase the cohesive force of the soil, and increase the lateral stress of the soil due to the restraint effect of the root. what's more, plant growth activities can achieve the effect of root reinforcement, which can effectively improve the cohesion of nutrient soil and improve the integrity of the fragmented rock blocks on the slope, thereby increasing Anti-sliding force [11] of rock or soil slope.

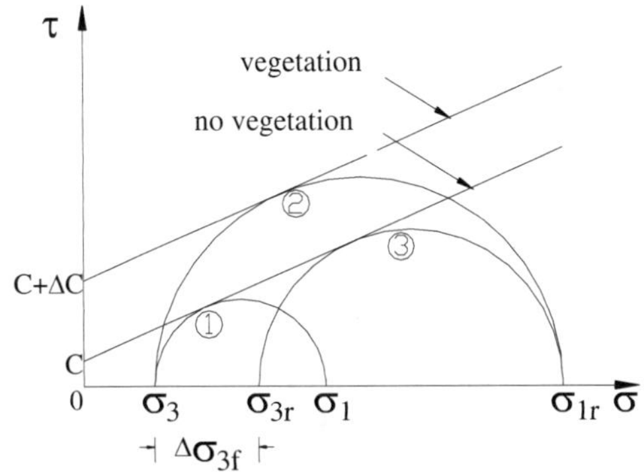

Figure 1Ecological slope protection stress circle

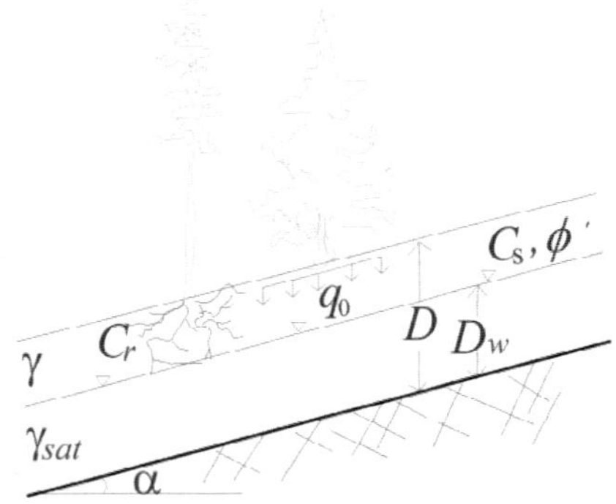

Figure 2 Vegetation infinite slope model

2.3Computing theory

The infinite slope model [12] assumes that: ① the failure surface and the groundwater level surface are parallel to the slope surface; ② the failure surface extends indefinitely; ③ only one layer of soil is considered. When it is a multilayer, the shear strength and internal friction angle of the soil weighted average along the entire sliding surface. When considering the effect of vegetation reinforcement, the expression of safety factor is as follows:

$$FS = \frac{C_r + C_s + cos^2\alpha[q_0 + \gamma(D - D_w) + (\gamma_{sat} - \gamma_w)D_w]tan\phi}{sin\alpha cos\alpha[q_0 + \gamma(D - D_w) + \gamma_{sat}D_w]} \tag{1}$$

Where: FS -safety factor; α -slope angle; D- thickness of total soil; D_w - thickness of saturated soil; q_0 -vegetation load; C_s -soil cohesion; φ -effective internal friction angle; γ - natural soil weight; γ_{sat} - saturated soil weight; γ_w - water weight.

For selecting vegetation, the expressions of soil thickness and critical height of groundwater are as follows:

$$d = \frac{\frac{C}{\cos\alpha} + [q_0 + (\gamma_{sat} - \gamma_w - \gamma)D_w]\cos\alpha\tan\phi - [q_0 + (\gamma_{sat} - \gamma)D_w]\sin\alpha FS}{\gamma\sin\alpha FS - \gamma\cos\alpha\tan\phi} \tag{2}$$

$$D_w = \frac{(q_0 + \gamma D)(\sin\alpha FS - \cos\alpha\tan\phi) - \frac{C}{\cos\alpha}}{(\gamma_{sat} - \gamma_w - \gamma)\cos\alpha\tan\phi - (\gamma_{sat} - \gamma)\sin\alpha FS} \tag{3}$$

The infinite slope model is a two-dimensional model. If the failure surface is very narrow, the load-limit slope model is conservative.

3. implementation of geonet combined vegatation Slope Protection

(1) Slope Stability Analysis

The premise of the design of ecological slope protection is to meet the deep stability requirements.

(2) Slope cleaning and shaping

Trim the slope according to the designed shape, The landslide section is cut first, the slope is leveled and then filled with clay and tamped, and then reinforced with a bamboo grid in the shape of a square grid with a pitch of 1m.

(3) Design of slope interception and drainage system

Sidewalks are set up at a certain interval on the slope surface, with flow channels left on both sides, and stone drainage ditches at the bottom of the slope surface: excavation of interception ditch on slope sections with a slope height greater than 10m (about 100m), and the interception ditch is set inside the top edge At 5m, the intercepting ditch adopts geomembrane composite mortar masonry structure. The construction of intercepting ditch follows the principle of vertical and horizontal, and it is completed continuously from top to bottom. Stepped energy dissipation sills are installed in the longitudinal drainage ditch.

(4) Excavation and backfilling of planting troughs

Excavate the planting troughs in a rectangular grid shape with a spacing of 1m in the horizontal direction and 2m in the vertical direction. Keep the grooves 20cm deep and 20cm wide. After slotting, backfill and compact the soil.

(5) Covered with nutrient soil

Prepare nutrient soil on site and cover the slope with 15cm thickness as base fertilizer.

(6) Laying three-dimensional geonet mat

A layer of geonet mat is laid in order from top to bottom and left to right on the slope covered with nutrient soil, and the net pad is closely attached to the slope. When fixing, the net pad must be smoothly combined with the slope and tightened. When laying the mesh pad on the top of the slope, it must extend 30 ~ 60cm. The geonet is laid in the fixed groove, and U-shaped nails are fixed at a distance of 1m, and the soil is backfilled and compacted.

(7) Cover soil

After laying the three-dimensional geonet, the cohesive soil, fertilizer, cement, etc. are mixed in a certain proportion and stirred uniformly, and then manually sprayed on the surface of the slope in 2 or 3 times.

(8) Yarrow grass seed spraying

According to the local natural, site, soil, climate and other conditions, select the appropriate vegetation type, and spray the mixed solution containing different herbs and shrub seeds, compound fertilizers, water retaining agents, adhesives, anti-transpiration agents, etc.

(9) Covered non-woven fabric

Sprinkle water evenly, then cover the mulch, make sure that the two mulches overlap 20cm, and fix it with a packing rope.

(10) conservation

Regularly check the growth of plants and the humidity of the soil profile, and water, fertilize and spray as needed, and thinning, increasing seedlings, pruning or changing plant combinations according to the density of plant distribution.

(11) Construction is over

After passing the inspection, clear the field to ensure the surrounding environment is clean.

4.engineering practice

4.1 Project Overview

A river slope in the Yonghu area of Huizhou, Guangdong, as shown in Figure 3. Where belongs to the subtropical marine monsoon climate in South China. This slope was a cut slope of weathered rock residual soil. It was partially micro to moderate weathered rock and the slope was acidic. These conditions were not conducive to the growth of plants. Without any vegetation cover, the soil erosion on the slope was serious.

Figure3Pre-construction state of slope

4.2 Vegetation matching principle

The slope of this project has poor soil, high compactness, and poor management conditions. When selecting vegetation types, follow the corresponding principles [13-14]. Vegetation planting experiments were conducted for this project. As can be seen in the figure 4, ryegrass and trifolium were pioneer plants and were suitable for growth in winter. If the ratio of ryegrass and white trifolium was greater than 30%, large-scale wilting and green-yellow disconnection will occur in summer. If it was less than 20%, effective green vegetation cover would be difficult to form in winter, which easily cause soil erosion. The ratio of white trifolium, white happy grass, Bermuda and thrush is 1: 1: 1: 4: 2: 2.

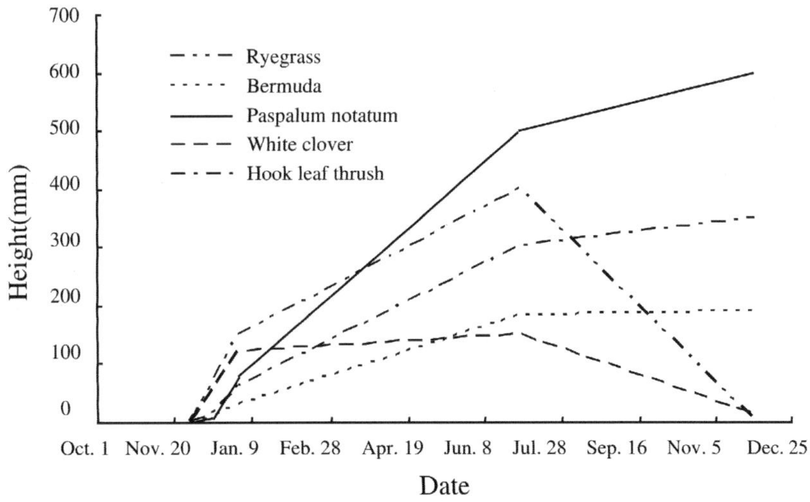

Figure4Vegetation growth status chart.

4.3 Slope prevention and treatment effect

Because the slope was steep and there were too many shallow landslides, by using the 3D vegetation geonet, the sliding force of the local dangerous section can be spread to the stable section next to it, which can significantly improve the stability of the slope, and at the same time the role of soil retention and water retention wa helpful to the formation of vegetation ecological environment. The slope reinforcement design was shown in Figure 5, and the construction was shown in Figure 6.

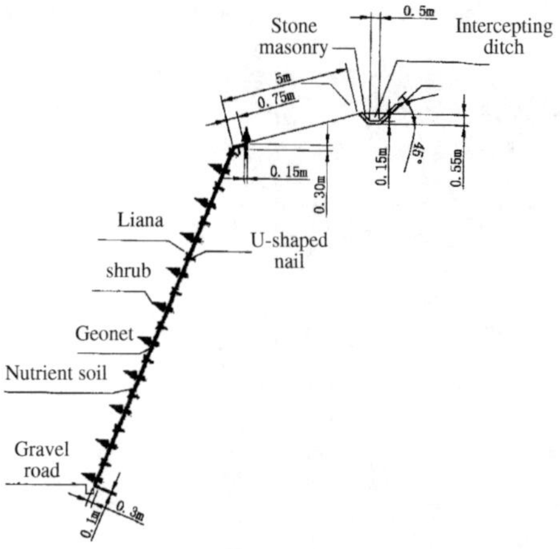

Figure5ecological slope overall map.

Figure6anchoring 3D geonet construction.

In order to quantitatively analyze the function of the anchored geonet composite vegetation system, the bare slope at the same slope location and the slope after the ecological protection were selected for simulated rainfall erosion tests. The slope of the test section was 12m high and the slope angle was 65 °. Transverse width 3m. The rainfall intensity is 2.5mm / min, the rainfall duration is 30min, and the speed of the raindrops landing on the ground is 9m/s. The erosion test results were shown in Fig. 7. The soil erosion on the bare slope without vegetation cover was more serious. The simulated soil erosion rate gradually increased before 19 minutes, and the soil erosion rate began to slow down after 19 minutes. Calculated up to 17kg, and soil erosion can not be seen on the slope protected by the anchored geonet composite vegetation. The long-term observations showed that the ecological slope protection effect is better, which can be seen from Figures 8 and 9.

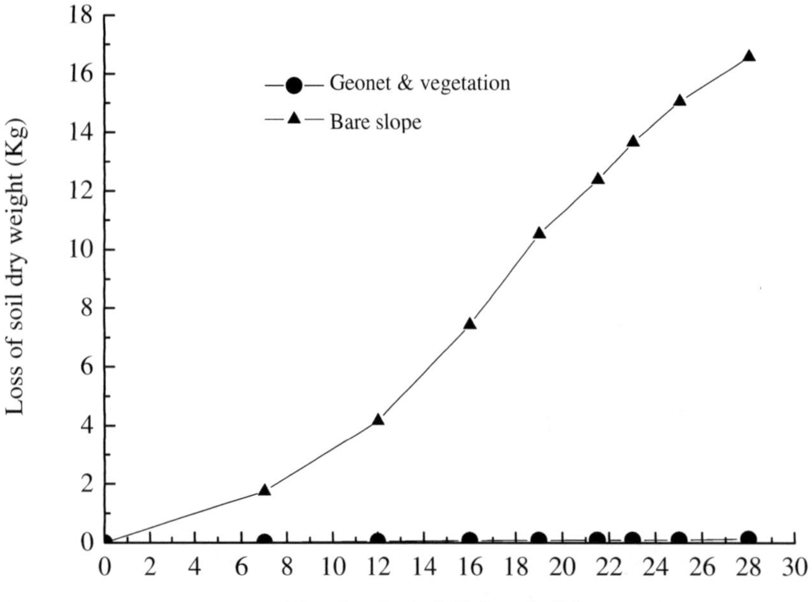

Figure7relationship between soil loss and rainfall intensity.

Figure8After 90 days of planting.

Figure 9After 480 days of planting.

5. conlusion

The three-dimensional vegetation compound geonetcan effectively protect the slope surface from rain erosion, slow down the impact of temperature changes, prevent and delay further weathering, fragmentation and erosion of the rock surface, and protect the overall stability of the slope; Survival and growth create a good water and fertilizer supply system to ensure the long-term growth of vegetation and achieve the purpose of long-term slope protection. Among them, the problems of artificial conservation, persistence of protection effects, and selection and configuration of plants need to be further improved in practical engineering applications.

Acknowledgements

National key research and development plan key project "Transportation Infrastructure Construction Safety Key Technology and Equipment Research" (2017YFC0805300) and CCCC 2018 Applied Basic Research Project " Study on Stability of Residual Soil Slope and New Anchorage Protection Technology under the Seepage and Pipe Flow".

Reference

[1] Galvão, Terezinha Cássia de Brito, Pereira A.R., Parizzi M.G., et al. Bioengineering Techniques Associated with Soil Nailing Applied to Slope Stabilization and Erosion Control. Natural Hazards Review, 2010, 11(2):43-48.

[2] Zhang J Y, Zhou D P,Li S C.Study on greening method for rock slope in highway project. Chinese Journal of Rock Mechanics and Engineering, 2002(09): 1400-1403.

[3] WANG Rong-hua, ZHAO Bao-wei. Research Progress and Development Trend of Highway Slope Ecological Protection in China. Highway, 2007(11): 199-202.

[4] Yang Dong, Wan Fu-xu. Research Development on Bioengineering Technique for Highway Slope Restoration in China. World Forestry Research, 2011, 24(06): 34-38.

[5] ZHANG J R. Application of three-dimensional net grass planting slope protection in water saving reconstruction of Beigaogan Canal. Water Resources Science and Technology, 2014(03): 37-39.

[6] Chen Xiao-juan. Application of 3D mesh spray planting technology for skeleton lattice beam. Urban Road Bridge & Flood Control, 2013(07): 335-336.

[7] Zhou De-pei, Zhang Jun-yun. Vegetation Slope Protection Engineering Technology. Beijing: China Communications Press, 2003.

[8] Khalilnejad A, Ali F H, Osman N. Contribution of the Root to Slope Stability. Geotechnical and Geological Engineering, 2012, 30(2):277-288.

[9] Leung A K, Boldrin D, Liang T, et al. Plant age effects on soil infiltration rate during early plant establishment. Géotechnique, 2017:1-22.

[10] Mulyono A, Subardjia A, Ekasari I, et al. The Hydromechanics of Vegetation for Slope Stabilization//IOP Conference Series Earth and Environmental Science,2018,118(1):012038.

[11] Li Guang-xin, Chen Lun, Cai Fei. A new approach to analysis of stress and deformation of reinforced earth. Chinese Journal of Geotechnical Engineering, 1994, 16(5): 46-53.
[12] Greenwood J R. SLIP4EX -A program for routine slope stability analysis to include the effects of vegetation , reinforcement and hydrological changes.Geotechnical and Geological Engineering, 2006,24(3) : 449-465.
[13] TAN Shao-hua, WANG Yi-min. Research Progress and Thinking of Bioengineering Techniques for Slope Protection in Expressway. Soil and Water Conservation Research, 2004(03): 81-84.
[14] LI Hai-fen, LU Xin-shi, JIANG Yu-lin. Progress in Ecological Restoration Technology of Expressway Slope. Sichuan Grassland, 2006(02): 34-38.

IOP Conf. Series: Materials Science and Engineering **794** (2020) 012056 doi:10.1088/1757-899X/794/1/012056

Groundwater Hydrochemical Characteristics Analysis and Water Quality Evaluation in the Jiaozuo Area

Sheng-yuan Qian[1*], Cong-jian Chen[1], Xiao-dong Zhang[1], Guo-ping Yang[2]and Sheng-feng Jiang[1]

[1]School of Civil Engineering and Architecture, Jiangsu Open University, Nanjing, JiangSu, 210019, China

[2]School of information Science and Engineering, Nanjing City Vocational College, Nanjing, JiangSu,211200, China

*Sheng-yuan Qian e-mail: feixiang_99@sina.com

Abstract. In2017, 42 groundwater samples in the Jiaozuo area were collected and tested. The groundwater hydro-chemical characteristics and reasoning of the area were analyzed by using mathematical statistics, Piper three-line diagram and correlation of components methods. The groundwater quality in the study area were evaluated by using the method of fuzzy comprehensive evaluation method based on principal component analysis. The results show that the domination of groundwater is $Ca^{2+,}$ Mg^{2+}, $SO4^{2-}$ and HCO_3^-. And the shallow groundwater hydrochemical type evolves from HCO_3-Ca to HCO_3·SO_4-Ca·Mg, HCO_3·SO_4·Cl-Ca·Mg, according to the direction of groundwater runoff. The chemical type of the deep groundwater is mainly HCO_3·SO_4-Ca·Mg type, local type is HCO_3·SO_4-Ca·Mg. The change of hydrochemical composition is mainly affected by weathering and leaching of carbonate, sulfate and rock salt. The fuzzy comprehensive evaluation results show that 61% and 75% of the shallow and deep groundwater quality in the study area meet the III water standards and above.

1. Introduction

Groundwater is an important water supply resource in China and an important strategic resource to support China's economic and social development [1]. In recent years, due to economic and social development and changes in the geochemical environment, the groundwater environment has been polluted to varying degrees, especially urban domestic sewage and industrial and mining. The wastewater discharged by enterprises penetrates into the groundwater system and poses a serious quality threat to groundwater[2,3]. Through the research on groundwater chemical characteristics and water quality evaluation, it is possible to grasp the groundwater environment problem, which is of great significance for promoting regional economic development, water resources management and ecological environment protection [4].

Due to the development and utilization of groundwater and the drainage of coal mine pits in the Jiaozuo area, geological environmental problems such as groundwater falling funnels and partial aquifer dewatering, water quality deterioration and ground collapse have formed. At present, the research on the water environment in the focus area is mainly focused on the pollution mechanism, migration prediction and pollution characteristics of shallow groundwater pollution sources, and there are few studies on its water chemistry characteristics, genesis and water quality evaluation. LuoYunyun et al[5,6]used multivariate statistical analysis methods to conduct qualitative and quantitative analysis of shallow groundwater pollution sources in the area, and evaluated the pollution indicators through fuzzy comprehensive evaluation method, and obtained the level of shallow groundwater quality in the study area. HouYusong, Jiao Fangfang et al[7,8] studied the correlation between shallow groundwater hardness, chromium pollutants and other chemical properties of groundwater in the area of focus, and predicted the migration of pollutants. Huang Pinghua et al[9,10] obtained the groundwater recharge characteristics and evolutionary causes of groundwater in the mining area by focusing on the groundwater hydrochemical characteristics of the mining area, but did not evaluate the groundwater quality.

The fuzzy comprehensive evaluation method is a commonly used water quality evaluation method. It considers the fuzzy problem existing in the classification of groundwater quality, and uses membership degree to characterize the degree of subordination of groundwater to various types of water, which is consistent with the ambiguity in the classification of water pollution levels. Objectively reflect the true state of groundwater quality[11,12], however, there are still some shortcomings in the selection of evaluation indicators and the determination of indicator weights. Principal component analysis method is to reduce the dimensionality of most of the original information, and transform the data into several comprehensive evaluation indicators to reflect the water quality. Comprehensive judgment[13]. In view of the large number of groundwater quality evaluation indicators, the influencing factors are complex, and the comprehensive evaluation is difficult, this paper uses the principal component analysis method to carry out principal component analysis, simplifies the number of evaluation factors, and uses the compound operation of membership degree and weight set to obtain water quality and water quality standards. The degree of subordination at each level determines the water quality level of groundwater samples at the maximum degree of membership.

In this study, 42 groundwater samples from the focus area of 2017 were collected and tested. The mathematical statistics of the groundwater were analyzed by using mathematical statistics, Piper three-line diagram and correlation analysis of each component, revealing the groundwater chemistry of the study area. The distribution characteristics and causes, and the fuzzy comprehensive evaluation method based on principal component analysis were used to systematically analyze the groundwater quality in the study area. The research results can provide a basis for water quality assessment and rational development and utilization of water resources in the region.

2. Overview of the study area

The study area ranges from 113°05′ to 113°19′ east longitude and 35°02′～35°19′ north latitude. It is mainly located in the urban area and surrounding area of Jiaozuo, with an area of about 150km2. It belongs to the warm temperate continental climate. Due to the topography and climate, the atmospheric precipitation in the northern mountainous area to the southern plain gradually decreases. The average annual rainfall in the northern mountainous area is 695.7mm; the annual average rainfall in the southern plain area is 594.4mm; the annual average temperature is 14.1. Between -14.4 ° C, the annual evaporation is in the range of 1721-2048 mm. In the north is the Taihang Mountains, the terrain is controlled by the fault structure, and it descends roughly in the south to the north of Jiaozuo City. The central part is the slope plain of the front slope of the mountain, inclined to the south and south to the east, with a slope of 10-17‰, and the south is the alluvial plain of the Huanghua River. The terrain is flat and the slope is less than 3‰. The terrain is generally high in the north, followed by the south and low in the middle. According to the genetic type, material composition and distribution pattern of the geomorphology, the geomorphology of the study area is divided into the slope alluvial slope, the alluvial fan-shaped land, the alluvial fan-slanted land, the alluvial fan front sag and the Huangpi River floodplain. unit.

3. Water sample collection and analysis methods

3.1Water sample collection

According to the groundwater outcrop distribution, hydrogeological conditions, land use characteristics and potential pollution source distribution in the study area, 42 representative sampling points were arranged, including 26 shallow water samples and 16 deep water samples. The sample test is completed by the Penny Test Group Co., Ltd., which mainly tests the total hardness TH, the total dissolved solids value TDS, K^+, Na^+, Ca^{2+}, Mg^{2+}, NH_4^+, HCO_3^-, CO_3^{2-}, Cl^-, SO_4^{2-} and other indicators. Temperature, EC, pH, TDS, dissolved oxygen, turbidity, oxidation-reduction potential and other indicators are detected on site. The collection and analysis of all chemical components are carried out in accordance with the Technical Specifications for Groundwater Environmental Monitoring (HJ/T164-2004), in which pH and dissolved oxygen are measured on-site using a portable water quality analyzer; the main cations such as Na^+, Ca^{2+} and Mg^{2+} are plasma. The mass spectrometry (ICP-MS) was used for the determination; the anions such as SO_4^{2-}, Cl^-, NO_3^-, and F^- were determined by ion chromatography (DX-120); HCO_3^- was determined by acid-base neutralization titration.

3.2Analysis method

After the test is completed, all test results are analyzed by SPSS20 software for statistical calculation and correlation analysis. The Piper three-line diagram is drawn by Aquachem 4.0 software, and the groundwater chemistry type of the area is analyzed, and the fuzzy comprehensive evaluation method based on principal component analysis is applied. Groundwater quality in the study area was evaluated.

4. Results analysis

4.1Groundwater chemical characteristics

The results of water chemical parameters such as pH value, dissolved total solids (TDS), F- and conventional ions in groundwater samples in the study area are shown in Table 1. The analysis found that the main ions in shallow groundwater are SO_4^{2-}, HCO_3^- and Ca^{2+}, with an average content of 477.2 mg/L, 428.15 mg/L and 160.58 mg/L, followed by Cl^-, Na+ and Mg^{2+} with an average content of 186.1 mg. /L, 155.01 mg/L and 102.69 mg/L; the pH value varies from 6.92 to 7.73; the TDS varies from 385 to 7100 mg/L. Shallow groundwater is generally neutral to weakly alkaline, with high TDS and brackish water. Compared with shallow groundwater, deep groundwater is deeply buried, and is less affected by surface environmental factors. The chemical composition of each water decreases with the increase of burial depth of groundwater, and TDS shows a decreasing trend. SO_4^{2-} is still the most important. The anion has an average content of 531.43 mg/L, and its content has a significant increase, while the contents of HCO_3^- and Cl- have decreased to some extent, indicating that there is a certain hydraulic relationship between deep groundwater and shallow groundwater. The cation is mainly Na^+, followed by Ca^{2+} and Mg^{2+}. From the coefficient of variation, the coefficient of variation of SO_4^{2-}, Na^+ and Cl^- exceeds 1, reflecting that it varies greatly in water, and it is easy to change with environmental factors.

Tab.1 Statistical eigenvalues of hydrochemical parameters in Jiaozuo area

project		PH	TDS (mg/L)	Cation (mg/L)			Anion (mg/L)				
				Na^+	Ca^{2+}	Mg^{2+}	Cl^-	SO_4^{2-}	HCO_3^-	NO_3^-	F^-
shallow groundwater	minimal value	6.92	385	13.4	44.3	25.1	15.5	57.7	268	0	0.19
	Maximum value	7.73	7100	877	477	555	1080	3610	604	24.8	1.98
	average value	7.34	1431.46	155.01	160.58	102.69	186.10	477.20	428.15	9.56	0.86
	standard deviation	0.21	1313.44	186.34	96.80	101.55	214.10	731.00	87.56	8.01	0.58
	variation coefficient	0.03	0.92	1.20	0.60	0.99	1.15	1.53	0.20	0.84	0.67
deep groundwater	minimal value	7.11	372	9.62	53.7	20.7	12.9	51.2	224	0.09	0.19
	Maximum value	7.76	7020	1060	432	580	600	4030	455	15.6	1.71

average value	7.46	1251.13	149.38	134.14	91.79	104.32	531.43	342.00	4.54	0.70
standard deviation	0.17	1638.03	257.47	100.76	136.89	141.67	992.72	73.21	4.21	0.56
variation coefficient	0.02	1.31	1.72	0.75	1.49	1.36	1.87	0.21	0.93	0.80

4.2Water Chemistry Type

Using the Piper three-line diagram, and according to the Shukalev classification, the groundwater hydrochemical classification of the study area is shown in Figure 1. There are 6 types of shallow groundwater chemical types in the study area and 4 types of deep groundwater chemical types. The shallow groundwater chemistry types are mainly HCO_3·SO_4-Ca·Mg, HCO_3·SO_4·Cl-Ca·Mg·Na and HCO_3-Mg·Ca, accounting for 73% of the total samples. In the alluvial fan geomorphology, HCO_3·SO_4-Ca·Mg type water with a salinity of less than 0.5g/L, and HCO_3·SO_4·Cl-Ca·Mg with a salinity of not more than 1g/L to the front margin Mg type water, HCO_3·SO_4·Cl-Ca·Mg and HCO_3·SO_4 -Ca·Mg type water with salinity greater than 1.5g/L. The deep groundwater water chemistry types are mainly HCO_3·SO_4-Ca·Mg and HCO_3-Ca·Mg type, accounting for 81% of the total samples. The main cations in shallow groundwater are Ca^{2+} and Mg^{2+}, with milliequivalent percentages of 36% and 35%. SO_4^{2-} and HCO_3^- are the main anions with 45% and 31% equivalents. The main cations in deep groundwater are Ca^{2+} and Mg^{2+}, with a milliequivalent percentage of 37% and 32%, and SO_4^{2-} is still the main anion with a milliequivalent percentage of 56%. It can be seen that the content of Ca^{2+}, Mg^{2+} and SO_4^{2-} and HCO_3^- in the groundwater of the study area is high.

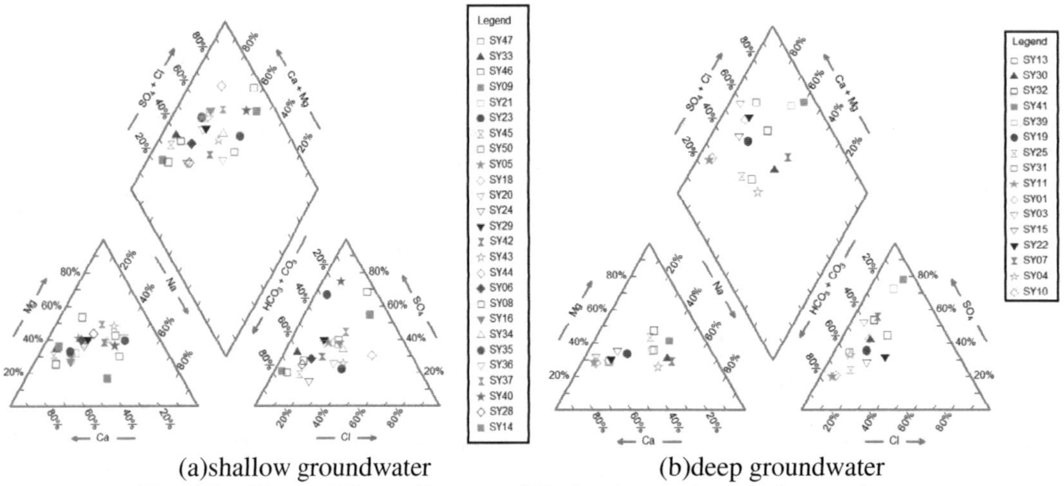

(a)shallow groundwater (b)deep groundwater

Fig.1 The Piper-tri-linear diagram of Hydrochemical type of groundwater

4.3 Correlation analysis

Substances from different sources and hydrochemical processes show different interrelationships between groundwater components, so correlation analysis can reveal similarities and differences in groundwater hydrochemical parameters and consistency and variability in groundwater sources [14,15]. It can be seen from Calculation results that except for HCO_3^- in shallow groundwater, other components are positively correlated with TDS. SO_4^{2-}, Cl^-, Mg^{2+}, Ca^{2+} and Na^+ are highly correlated with TDS, and the correlation coefficient is 0.833-0.986, indicating that the salinity of both shallow and deep groundwater is equal to the concentration of SO_4^{2-}, Cl^-, Mg^{2+}, Ca^{2+} and Na^+. closely related. SO_4^{2-} has a high correlation with Mg^{2+}, Na^+ and Ca^{2+}. The study area may produce the dissolution of sulfate such as gypsum; HCO_3^- has a weak correlation with Ca^{2+}, Na^+ and Mg^{2+}, which may cause carbonates such as calcite and dolomite. Weathering. Cl^- is highly correlated with Na+, and the correlation coefficient reaches 0.966, which may cause the dissolution of rock salt. At the same time, Cl^- is also highly correlated with Mg^{2+}, indicating that rainwater and river water in the study area carry a large amount of soil salt into the groundwater, and Cl^- is the acidification and water quality change of the water body. It has an impact [16]. It can be seen that the main minerals of the water-bearing strata in the study area are calcite, dolomite, gypsum and rock salt. It is speculated that the hydrochemical process of groundwater in this area is mainly based on the weathering and leaching of carbonate, sulfate and rock salt.

4.4Saturation index analysis

Through the calculation of the saturation index of mineral phase such as rock salt, gypsum, fluorite, dolomite, calcite and CO_2, it can be further determined whether the leaching effect occurs in the groundwater, thus truly reflecting the actual form of the groundwater component. The tendency of mineral dissolution or precipitation during groundwater runoff can be calculated using the saturation index SI:

$$\mathrm{SI} = \mathrm{Lg}\frac{\mathrm{IAP}}{\mathrm{K_{sp}}}$$

In the formula, IAP is the activity product, K_{sp} is the concentration product.

When SI>0, the mineral will precipitate in the groundwater; when SI<0, the mineral will dissolve in the groundwater;

When SI=0, the mineral is in a dynamic equilibrium state when its dissolved amount and sediment amount are the same in groundwater.

Using the hydrogeochemical simulation software PHREEQC, the saturation index SI of each mineral phase at the starting point and the end point of the groundwater in the SY18-SY42 path was simulated. It can be seen from Table 2 that the rock salt, gypsum and fluorite were not in the course of runoff. When it reaches saturation, there is a tendency to continue to dissolve, while both calcite and dolomite are saturated, accompanied by dissolution of CO_2.

Tab.2Mineral phase saturation indices on groundwater flow path

water sample	Mineral phase					
	rock salt	gypsum	fluorite	calcite	dolomite	P(CO_2)
SY18	-6.33	-0.8	-2.15	0.43	0.69	-1.22
SY42	-6.27	-1.22	-0.65	0.54	1.31	-1.59

5. Water quality evaluation

5.1Fuzzy comprehensive evaluation

5.1.1Calculate the variance contribution rate and determine the principal component

Calculating the variance contribution rate of principal component F_i according to the correlation coefficient matrix eigenvalue λ_i.

$$E_i=\lambda_i\Big/\sum_{i=1}^{n}\lambda_i \tag{1}$$

Select the first n principal components of $\sum_{i=1}^{n} E_i \geq 85\%$ for comprehensive analysis.

5.1.2Principal component load calculation

The relationship between the principal component load l_{ij} and the correlation coefficient matrix eigenvector u_{ij} is:

$$l_{ij}=u_{ij}/\sqrt{\lambda_i} \tag{2}$$

5.1.3Principal component factor analysis

According to the principal component load size, that is, the degree of correlation with the principal component, the main control factor of each principal component is selected as the evaluation factor of fuzzy comprehensive evaluation.

5.1.4Establishment of fuzzy relation matrix

According to the membership degree of the evaluation factor to the evaluation level, a fuzzy relation matrix is obtained. Let S_{ij} be the membership degree of the i-thevaluation factors to j-th evaluation level indicators, then the membership function is:

$$S_{ij}=\begin{cases} 1 & x_{ij}\leq C_{i(j-1)},\ x_{ij}\geq C_{i(j+1)} \\ \frac{x_{ij}-C_{i(j-1)}}{C_{ij}-C_{i(j-1)}} & C_{i(j-1)}\leq x_{ij}\leq C_{ij} \\ \frac{C_{i(j+1)}-x_{ij}}{C_{ij}-C_{i(j-1)}} & C_{ij}\leq x_{ij}\leq C_{i(j+1)} \\ 0 & x_{ij}=C_{ij} \end{cases} \tag{3}$$

In the formula, $x_{ij}$is the measured value of the i-th evaluation factor of the j-th sample, and C_{ij} is the evaluation standard value corresponding to the i-th evaluation factor.

The fuzzy relation matrix$\boldsymbol{R}$ is calculated by the membership function, namely:

$$\boldsymbol{R}=\begin{bmatrix} r_{11} & r_{12} & \cdots & r_{1n} \\ r_{21} & r_{22} & \cdots & r_{2n} \\ \cdots & \cdots & \cdots & \cdots \\ r_{i1} & r_{i1} & \cdots & r_{mn} \end{bmatrix} \tag{4}$$

$i=1,2,\cdots m;j=1,2,\cdots n$

5.1.5Establishment of fuzzy weight matrix

The weight $\boldsymbol{W}_{ki}$of the evaluation factor in the groundwater pollution weight is determined by the normalized weight calculation formula, namely:

$$\boldsymbol{W}_{ki}=(x_{ki}/S_i)/\textstyle\sum_{i=1}^{n}(x_{ki}/S_i) \tag{5}$$

In the formula, $S_i=(C_{i1}+C_{i2}+\cdots+C_{ij})/j$

Combining the weight matrix $\boldsymbol{W}$ with the fuzzy relation matrix$\boldsymbol{R}$ to obtain a comprehensive evaluation model:

$$B_i=\boldsymbol{A}\cdot\boldsymbol{R}=(W_1,W_2,\cdots W_i)\begin{bmatrix} r_{11} & r_{12} & \cdots & r_{1n} \\ r_{21} & r_{22} & \cdots & r_{2n} \\ \cdots & \cdots & \cdots & \cdots \\ r_{i1} & r_{i1} & \cdots & r_{mn} \end{bmatrix}=(b_1,b_2,\cdots b_i) \tag{6}$$

In the formula, b_i is the membership value of each indicator.

5.1.6Fuzzy synthesis operation and results

According to the above method, the fuzzy comprehensive evaluation of the water quality of 26 groups of shallow groundwater samples and 16 groups of deep groundwater samples was carried out. The subordination degree and evaluation results of water quality of each water sample to each level of water are shown in Table 3. It can be seen that in the water samples of the shallow water 26 groups, the water quality class I water samples 5 groups, accounting for 19% of the total water samples, the class III water samples 11 groups, accounting for 42% of the total water samples; the V water samples 10 Group, accounting for 39% of the total number of water samples, no II, IV water, 61% of the drinking water standards. Among the 16 groups of deep water samples, the number of groundwater samples in the water quality class IV were 9, 0, 3, 0, and 4, respectively, accounting for 56%, 0, 19%, 0, 25% of the total deep water samples. 75% of the drinking water standard.

Tab.3The membership degree of water sample at the sampling site

Sample serial number		I	II	III	IV	V	Comprehensive evaluation
shallow groundwater	SY05	0.22	0.12	0.26	0.20	0.21	III
	SY06	0.31	0.10	0.37	0.03	0.20	III
	SY08	0.35	0.17	0.18	0.02	0.28	I
	SY09	0.23	0.10	0.29	0.07	0.32	V
	SY14	0.07	0.04	0.10	0.02	0.56	V

	SY16	0.17	0.21	0.29	0.16	0.20	III
	SY18	0.15	0.13	0.28	0.08	0.45	V
	SY20	0.23	0.17	0.31	0.12	0.19	III
	SY21	0.24	0.18	0.37	0.04	0.20	III
	SY23	0.28	0.22	0.32	0.18	0.00	III
	SY24	0.64	0.20	0.27	0.00	0.00	I
	SY28	0.47	0.20	0.36	0.02	0.00	I
	SY29	0.27	0.06	0.33	0.21	0.19	III
	SY33	0.32	0.19	0.40	0.14	0.00	III
	SY34	0.17	0.10	0.45	0.18	0.16	III
	SY35	0.14	0.01	0.13	0.18	0.59	V
	SY36	0.14	0.14	0.35	0.23	0.13	III
	SY37	0.08	0.08	0.31	0.20	0.39	V
	SY40	0.06	0.06	0.04	0.04	0.82	V
	SY42	0.17	0.17	0.27	0.07	0.33	V
	SY43	0.13	0.08	0.47	0.17	0.14	III
	SY44	0.14	0.10	0.22	0.20	0.44	V
	SY45	0.59	0.38	0.03	0.00	0.00	I
	SY46	0.46	0.26	0.30	0.04	0.00	I
	SY47	0.04	0.01	0.04	0.01	0.92	V
	SY50	0.14	0.10	0.33	0.08	0.39	V
deep groundwater	SY01	0.34	0.25	0.34	0.07	0.00	I
	SY03	0.32	0.24	0.43	0.01	0.00	III
	SY04	0.62	0.36	0.02	0.00	0.00	I
	SY07	0.39	0.23	0.17	0.00	0.24	I
	SY10	0.67	0.33	0.01	0.00	0.00	I
	SY11	0.64	0.35	0.00	0.00	0.00	I
	SY13	0.22	0.14	0.19	0.04	0.48	V
	SY15	0.34	0.27	0.51	0.04	0.00	III
	SY19	0.46	0.30	0.32	0.00	0.00	I
	SY22	0.57	0.29	0.29	0.07	0.00	I
	SY25	0.45	0.21	0.34	0.01	0.00	I
	SY30	0.21	0.06	0.43	0.18	0.17	III
	SY31	0.44	0.30	0.23	0.16	0.00	I
	SY32	0.20	0.04	0.19	0.06	0.51	V
	SY39	0.06	0.05	0.09	0.20	0.60	V
	SY41	0.04	0.01	0.04	0.00	0.92	V

According to the sampling point location and evaluation results, the shallow groundwater V-type water samples are mainly distributed in the top area of the Dasha River alluvial fan, the middle-eastern junction of the study area and the southern Huangpi River alluvial plain. The over-standard points are mostly in the piedmont slope plain and The intersection of the Huangpi River and the alluvial plains is relatively flat and has frequent industrial and agricultural activities. In particular, there are a large number of toxic and harmful chemical wastes discharged from the upstream of the Dasha River, and the unreasonable use of large-scale pesticides and fertilizers. At the same time, the groundwater level in these areas is relatively shallow, and the degree of impact on human activities is relatively high, and the quality of groundwater is relatively poor. The I and III water samples of deep groundwater are distributed in the study area, indicating that the overall quality of deep groundwater is better in the study area, and the V type water samples are mainly distributed in the southeast of the front of the alluvial fan. According to the site investigation. According to the information, the poor groundwater quality in this area may be caused by the over-flowing of mixed wells.

6. Conclusions and discussion

(1) The groundwater bodies in the study area are dominated by Ca^{2+}, Mg^{2+}, SO_4^{2-} and HCO_3^-.The water chemistry type in the shallow water area is mainly from the HCO_3^-Ca type water in the northern recharge area, along the direction of groundwater runoff, from north to south. From northwest to southeast, it gradually evolved into $HCO_3\cdot SO_4$-Ca·Mg and $HCO_3\cdot SO_4$·Cl-Ca·Mg type water; while the water chemistry type of deep groundwater is mainly $HCO_3\cdot SO_4$-Ca·Mgtype water, and HCO_3-Ca·Mg type water appear locally.

(2) Correlation analysis shows that the change of ion composition in water is mainly affected by the weathering and leaching of carbonate, sulfate and rock salt. In the process of groundwater runoff, dissolution of rock salt, gypsum and fluorite occurs and calcite occurs. A small amount of precipitation with dolomite is accompanied by the dissolution of carbon dioxide.

(3) Using the principal component analysis method to screen the evaluation indicators, not only reduces the problem of participating in the fuzzy comprehensive evaluation factor, but also ensures the reliability of the evaluation results. At the same time, the fuzzy comprehensive evaluation considers the influence of different evaluation indicators on water quality. The weight makes the evaluation result more reasonable and credible. The results show that the shallow groundwater reaches the sampling point that meets the water quality standards of Class III and above, accounting for 61% of the total sampling points, and the IV and V waters account for 39% of the total sampling points. The deep groundwater meets the sampling points of Class III and above water quality standards, accounting for 75% of the sampling points, and the IV and V waters account for 25% of the total sampling points. The poor water quality of shallow groundwater is mainly distributed in the intersection of the piedmont slope plain and the Huanghua River alluvial plain. The deep groundwater quality is mainly distributed in the southeast of the foreland.

References

[1] Y.H.Hu, F.H.Zhang, Z.Y.Niu, etc. (2014) Hydro-chemical characteristics of groundwater in centralized drinking water sources and its quality assessment in northern Anhui Province. Journal of University of Science and Technology of China, 44(11):913-920,925.

[2] Y.S.Lin, J.G.Pei, Y.C.Du, etc. (2016)Temporal and spatial distribution of the Hydro-chemical characteristics of Yaocun underground river in Guangxi province. Journal of Yangtze River Scientific Research Institute, 33(12):6-9,16.

[3] S.B.Yu, Z.M.Ma, H.S.Zhang. (2011) Pollutant characteristics of shallow groundwater in Jiaozuo site of the middle South-to-North water diversion project. Journal of University of Jinan(Sci. and Tech.),26(1):91-95.

[4] L.Lu, J.Y.Yang. (2017) Effect of over exploitation of groundwater on the hydro-chemical characteristics of groundwater. Journal of Yangtze River Scientific Research Institute, 34(9):14-18,23.

[5] Y.Y.Luo, Z.M.Ma, Y.S.Hou, etc. (2013)Apportionment of pollution source of shallow groundwater system in Jiaozuo. Nonferrous Metals(Extractive Metallurgy), 2013(4):58-61.

[6] Y.Y.Luo, Z.M.Ma, Y.S.Hou. (2013)An evaluation of the quality of shallow groundwater in Jiaozuo area. China Rural Water and Hydropower, 2013(8):17-20.

[7] Y.S.Hou, Z.M.Ma, Y.Y.Luo, etc. (2014)Pollution mechanism and migration prediction of hardness of shallow groundwater in Jiaozuo area. Journal of University of Jinan(Sci. and Tech.),28(2):151-156.

[8] F.F.Jiao, Z.M.Ma, Y.S.Hou. (2014) pollution mechanism and migration prediction of Cr(VI) in shallow groundwater in Jiaozuo area. Coal Geology & Exploration, 42(6):82-86,96.

[9] P.H.Huang, J.S.Chen. (2011) The chemical features of ground water and FDA model used to distinguish source of water burst in Jiaozuo mine area. Coal Geology & Exploration, 39(2):42-46,51.

[10] P.H.Huang, J.S.Chen, C.Ning,etc. (2010)Hydro-chemical characteristics and hydro-geochemical modeling of groundwater in the Jiaozuo mining district. Geoscience, 24(2):369-376.

IOP Conf. Series: Materials Science and Engineering **794** (2020) 012056 doi:10.1088/1757-899X/794/1/012056

[11] S.X.Zhang, Z.M.Wang, T.H.Dai. (2019) Chemical characteristics and water quality assessment of karst groundwater in Bijie.Journal of Yangtze River Scientific Research Institute, 36(5):28-33,41.
[12] C.Qian, W.P.Mu, K.Wang, etc. (2016)Fuzzy comprehensive evaluation of groundwater quality based on principal component analysis. Water Resources and Power, 34(11):31-35.
[13] J.K.Du, Y.Fu, X.X.Li. (2015)Application of fuzzy comprehensive evaluation method-principal component analysis to groundwater quality evaluation. Environmental Monitoring in China, 31(4):75-81.
[14] M.Zhang, Q.M.Liu, K.X.Liu. (2018) Hydrochemical Characteristics and cause analysis of groundwater in Gubei mining district. Safety in Coal Mines, 49(5):187-190.
[15] Z.X.Zhang, M.Xu, Q.Zhang, etc. (2018)Hydrogeochemistry, genesis and waterquality analysis of shallow groundwater in red-bed area of the Mianyang city. Science Technology and Engineering, 18(3):168-173.
[16] F.Zhang, F.D.Li, J.Li, etc. (2013)Hydrochemical characteristics of surface water in main rivers of the irrigation districts in the downstream of Yellow river. Chinese Journal of Eco-Agriculture, 21(4):487-493.

IOP Conf. Series: Materials Science and Engineering **794** (2020) 012057 doi:10.1088/1757-899X/794/1/012057

Semi-analytical solutions to consolidation for unsaturated soils by vertical drains under arbitrary time-dependent loading

Aifang Qin*, **Lianghua Jiang, Tianyi Liand Weifang Xu**

Department of Civil Engineering, Shanghai University, Shanghai 200444, China

E-mail: qinaifang@shu.edu.cn

Abstract. This paper presents general semi-analytical solutions to equal strain consolidation for unsaturated soil layer by vertical drains, and the well resistance is considered. Firstly, the coupled governing equations are transformed into ordinary differential equations (ODEs) by Laplace transform and introducing two new variables, and combined with boundary conditions to obtain solutions. On this basis, the semi-analytical solutions in the time domain are achieved by performing inverse Laplace transform with Crump's method. Secondly, the correctness of the semi-analytical solutions are verified by comparison with the existing analytical solutions of radial consolidation for unsaturated soils under constant loading. Finally, the effect of well resistance factor, permeability coefficient ratio k_a/k_w and load parameter on the average degree of consolidation was analyzed under exponential loading.

1. Introduction

At present, the traditional calculation method of saturated soils is usually used to predict the consolidation settlement of unsaturated soils subgrade. But in fact, the existence range of real saturated soil is limited. In the subgrade near the surface, due to the evaporation and transpiration of water, the subgrade is not fully saturated, which is especially common in arid and semi-arid areas. Therefore, there has been an essential need to study the consolidation characteristics of unsaturated soils near the ground.

In recent years, research on the problems related to the consolidation for unsaturated soils has made some progress. The typical one is the solution of the one-dimensional unsaturated soil consolidation problem proposed with the double-stress variable theory by Fredlund and Hanson [1]. On the basis of this theory, for the consolidation of one dimensional (1D) unsaturated soil, Qin et al. [2] obtained semi analytical solutions by Laplace transform and Cayley Hamilton method. Shan et al. [3] studied the analytical solution for different boundary conditions by separating variables, while Zhou et al. [4] introduced the analytical solution under different initial conditions and boundary conditions by difference integral method (DQM). In addition, Wang et al. [5, 6] made further improvement on the consolidation problem of unsaturated soil, and presented the semi analytical solutions of one-dimensional (1D) and two-dimensional (2D) consolidation problem of unsaturated soil. For the application of the axisymmetric consolidation model in unsaturated soil, Qin et al.[7] first proposed a semi analytical solution for the drainage consolidation of unsaturated soil through an ideal sand well. Then, Zhou et al. [8] solved this problem by differential quadrature method (DQM). In addition, Ho et al. [9] and Zhou et al. [10] used equal strain assumption to get the solution of consolidation for unsaturated soil by vertical drains. In practical engineering, the load usually changes with time, but there is no research on the consolidation of unsaturated soil layer by vertical drain under arbitrary time-dependent loading.

In this paper, on the basis of Fredlund's one-dimensional consolidation theory of unsaturated soils, the semi-analytical solutions to equal strain consolidation for unsaturated soils by vertical drains considering well resistance under arbitrary time-dependent loading are derived. The Laplace transform technique and the method of introducing two new variables are used to solve the proposed solution. Then, the dissipation of the average excess pore pressure is compared with the existing analytical solution [10]. And the exponential loading is selected to study the influence of well resistance factor, permeability coefficient ratio k_a/k_w and load parameter on the average degree of consolidation.

2. Mathematical model

2.1Governing Equations

Figure 1 shows the details of consolidation modeling to unsaturated soil by a vertical drain considering well resistance. Dimensions of the unsaturated soil subgrade system include the thickness of unsaturated soil layer H (m), the radius of vertical drain r_w (m) and the radius of influence zone r_e (m). The air and water permeability coefficients of unsaturated soil in the radial direction are k_a and k_w (m/s), respectively. k_{aw} and k_{ww} are the air and water permeability coefficients of the vertical drain in the vertical direction (m/s), respectively. In addition, the top and bottom boundaries (i.e. $z = 0$, $z = H$), and the outer boundary of influence zone (i.e. $r = r_e$) are both impermeable to air and water. $q(t)$ is a loading that varies with time in the vertical direction on the top surface of the foundation (kPa).

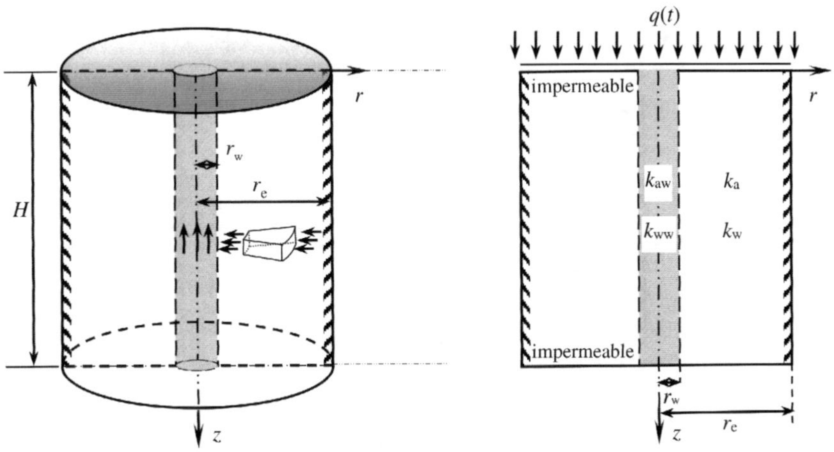

Figure 1.Consolidation modeling of unsaturated soils by vertical drains considering well resistance

The research hypothesis are similar to those proposed by Zhou et al.[10] They are listed as follows:

(1) The unsaturated soil is homogeneous.

(2) The air flux conforms Fick's law, while the water flux refers to Darcy's law. In addition, both flow of those are continuous and independent.

(3) The soil grain and water are incompressible.

(4) The coefficients of permeability withregard to air and water, and volume change for the soil remain constant throughout the consolidation process.

(5) The strain occurring during consolidation is small strain.

Note: the parameters in assumption (4) throughout the consolidation process of unsaturated soils are functions of stress state. However, if these parameters are nonlinear changed during consolidation, It is impossible to get the semi-analytical solution. According to the relevant studies on unsaturated soils[1-10], it is assumed that these parameters remain constant during consolidation process.

Referring to the studies given by Zhou et al. [10] the governing equations are expressed as follows:

$$\frac{\partial \bar{u}_a}{\partial t} = -C_a \frac{\partial \bar{u}_w}{\partial t} - C_v^a \left(\frac{\partial^2 u_a}{\partial r^2} + \frac{1}{r} \frac{\partial u_a}{\partial r} \right) + C_q^a \frac{\partial q}{\partial t} \tag{1a}$$

$$\frac{\partial \overline{u}_{\mathrm{w}}}{\partial t} = -C_{\mathrm{w}}\frac{\partial \overline{u}_{\mathrm{a}}}{\partial t} - C_{\mathrm{v}}^{\mathrm{w}}\left(\frac{\partial^2 u_{\mathrm{w}}}{\partial r^2} + \frac{1}{r}\frac{\partial u_{\mathrm{w}}}{\partial r}\right) + C_{\mathrm{q}}^{\mathrm{w}}\frac{\partial q}{\partial t} \tag{1b}$$

where $\overline{u}_{\mathrm{a}}$ and $\overline{u}_{\mathrm{w}}$ are the average excess pore-air and pore-water pressures (kPa), respectively. They can be expressed as follows:

$$\overline{u}_{\mathrm{a}} = \frac{1}{\pi\left(r_{\mathrm{e}}^2 - r_{\mathrm{w}}^2\right)}\int_{r_{\mathrm{w}}}^{r_{\mathrm{e}}} u_{\mathrm{a}}\, 2\pi r dr \tag{2a}$$

$$\overline{u}_{\mathrm{w}} = \frac{1}{\pi\left(r_{\mathrm{e}}^2 - r_{\mathrm{w}}^2\right)}\int_{r_{\mathrm{w}}}^{r_{\mathrm{e}}} u_{\mathrm{w}}\, 2\pi r dr \tag{2b}$$

where

$$C_{\mathrm{a}} = \frac{m_2^{\mathrm{a}}}{m_{1\mathrm{k}}^{\mathrm{a}} - m_2^{\mathrm{a}} - u_{\mathrm{atm}} n_0 (1-s_{\mathrm{r0}})(\overline{\overline{u}}_{\mathrm{a}}^0)^{-2}},\quad C_{\mathrm{v}}^{\mathrm{a}} = \frac{k_{\mathrm{a}} RT}{g\overline{\overline{u}}_{\mathrm{a}}^0 M\left[m_{1\mathrm{k}}^{\mathrm{a}} - m_2^{\mathrm{a}} - u_{\mathrm{atm}} n_0 (1-s_{\mathrm{r0}})(\overline{\overline{u}}_{\mathrm{a}}^0)^{-2}\right]},$$

$$C_{\mathrm{q}}^{\mathrm{a}} = \frac{m_{1\mathrm{k}}^{\mathrm{a}}}{m_{1\mathrm{k}}^{\mathrm{a}} - m_2^{\mathrm{a}} - u_{\mathrm{atm}} n_0 (1-s_{\mathrm{r0}})(\overline{\overline{u}}_{\mathrm{a}}^0)^{-2}},\quad C_{\mathrm{w}} = \frac{m_{1\mathrm{k}}^{\mathrm{w}} - m_2^{\mathrm{w}}}{m_2^{\mathrm{w}}},\quad C_{\mathrm{v}}^{\mathrm{w}} = \frac{k_{\mathrm{w}}}{\gamma_{\mathrm{w}} m_2^{\mathrm{w}}},\quad C_{\mathrm{q}}^{\mathrm{w}} = \frac{m_{1\mathrm{k}}^{\mathrm{w}}}{m_2^{\mathrm{w}}}.$$

$\overline{\overline{u}}_{\mathrm{a}}$ ($\overline{u}_{\mathrm{a}} + u_{\mathrm{atm}}$) is the absolute average excess pore-air pressure (kPa), and u_{atm} is the atmospheric pressure. $\overline{\overline{u}}_{\mathrm{a}}^0$ is the absolute initial average excess pore-air pressure (kPa). u_{a} and u_{w} mean excess pore-air and pore-water pressures (kPa) in the influence zone, respectively. $m_{1\mathrm{k}}^{\mathrm{a}}$ and $m_{1\mathrm{k}}^{\mathrm{w}}$are the coefficients of air and water volume change with regard to a change in the net normal stress ($q - \overline{u}_{\mathrm{a}}$) (kPa^{-1}), respectively. m_2^{a}andm_2^{w} are the coefficients of air and water volume change with regard to a change in the matric suction ($\overline{u}_{\mathrm{a}} - \overline{u}_{\mathrm{w}}$) (kPa^{-1}), respectively. S_{r0} and n_0 are the initial degree of saturation and initial porosity, respectively. M is the molecular mass of air (0.02895 kg•mol^{-1}). R is the molar gas constant (8.314 J•mol^{-1}•K^{-1}), and T (t_0+273) is the absolute temperature. $t_0$is the ambient temperature (20 °C). g is the gravitational acceleration (9.81 m/s^{-2}), and γ_{w} is the water unit weight (9.81 kN•m^{-3}).

2.2Initial and boundary conditions

In this paper, it is assumed that the initial excess pore-air and pore-water pressures are equally scattered with the depth z. Thus, they are given as follows:

$$u_{\mathrm{a}}\left(r, z, t=0\right) = u_{\mathrm{a}}^0 \tag{3a}$$

$$u_{\mathrm{w}}\left(r, z, t=0\right) = u_{\mathrm{w}}^0 \tag{3b}$$

As shown in figure 1, the boundary conditions are expressed as followed:

$$u_{\mathrm{a,r}}\left(r=r_{\mathrm{e}}, z, t\right) = u_{\mathrm{w,r}}\left(r=r_{\mathrm{e}}, z, t\right) = 0 \tag{4}$$

$$u_{\mathrm{a}}(r=r_{\mathrm{w}}, z, t) = u_{\mathrm{aw}}(z,t) \tag{5a}$$

$$u_{\mathrm{w}}(r=r_{\mathrm{w}}, z, t) = u_{\mathrm{ww}}(z,t) \tag{5b}$$

$$u_{\mathrm{a,r}}(r=r_{\mathrm{w}}, z, t) + \frac{r_{\mathrm{w}} k_{\mathrm{aw}}}{2k_{\mathrm{a}}} u_{\mathrm{aw,z}}(z,t) = 0 \tag{6a}$$

$$u_{\mathrm{w,r}}(r=r_{\mathrm{w}}, z, t) + \frac{r_{\mathrm{w}} k_{\mathrm{ww}}}{2k_{\mathrm{w}}} u_{\mathrm{ww,z}}(z,t) = 0 \tag{6b}$$

$$u_{\mathrm{aw}}\left(z=0, t\right) = u_{\mathrm{ww}}\left(z=0, t\right) = 0 \tag{7}$$

$$u_{\mathrm{aw,z}}\left(z=H, t\right) = u_{\mathrm{ww,z}}\left(z=H, t\right) = 0 \tag{8}$$

3. Derivation of semi-analytical solutions

3.1General semi-analytical solutions

The governing Equations (1) can be integrated with regard to r and the boundary condition presented in Equation (4) is considered. Thus, the following Equations can be acquired:

$$\frac{\partial u_{\mathrm{a}}}{\partial r}=-\frac{1}{2}\left(\frac{(r_{\mathrm{e}})^{2}-r^{2}}{r}\right)\frac{1}{C_{\mathrm{v}}^{\mathrm{a}}}\left(-C_{\mathrm{a}}\frac{\partial\bar{u}_{\mathrm{w}}}{\partial t}-\frac{\partial\bar{u}_{\mathrm{a}}}{\partial t}+C_{\mathrm{q}}^{\mathrm{a}}\frac{\mathrm{d}q}{\mathrm{d}t}\right) \tag{9a}$$

$$\frac{\partial u_{\mathrm{w}}}{\partial r}=-\frac{1}{2}\left(\frac{(r_{\mathrm{e}})^{2}-r^{2}}{r}\right)\frac{1}{C_{\mathrm{v}}^{\mathrm{w}}}\left(-C_{\mathrm{w}}\frac{\partial\bar{u}_{\mathrm{a}}}{\partial t}-\frac{\partial\bar{u}_{\mathrm{w}}}{\partial t}+C_{\mathrm{q}}^{\mathrm{w}}\frac{\mathrm{d}q}{\mathrm{d}t}\right) \tag{9b}$$

Then, an integration of Equations (9) with regard to r while considering the boundary conditions presented in Equations (5) gives:

$$u_{\mathrm{a}}=-\frac{1}{2}\left[(r_{\mathrm{e}})^{2}\ln\left(\frac{r}{r_{\mathrm{w}}}\right)-\frac{r^{2}-(r_{\mathrm{w}})^{2}}{2}\right]\frac{1}{C_{\mathrm{v}}^{\mathrm{a}}}\left(-C_{\mathrm{a}}\frac{\partial\bar{u}_{\mathrm{w}}}{\partial t}-\frac{\partial\bar{u}_{\mathrm{a}}}{\partial t}+C_{\mathrm{q}}^{\mathrm{a}}\frac{\mathrm{d}q}{\mathrm{d}t}\right)+u_{\mathrm{aw}} \tag{10a}$$

$$u_{\mathrm{w}}=-\frac{1}{2}\left[(r_{\mathrm{e}})^{2}\ln\left(\frac{r}{r_{\mathrm{w}}}\right)-\frac{r^{2}-(r_{\mathrm{w}})^{2}}{2}\right]\frac{1}{C_{\mathrm{v}}^{\mathrm{w}}}\left(-C_{\mathrm{w}}\frac{\partial\bar{u}_{\mathrm{a}}}{\partial t}-\frac{\partial\bar{u}_{\mathrm{w}}}{\partial t}+C_{\mathrm{q}}^{\mathrm{w}}\frac{\mathrm{d}q}{\mathrm{d}t}\right)+u_{\mathrm{ww}} \tag{10b}$$

Thereafter, substituting Equations (10) into Equations (2) results in:

$$\bar{u}_{\mathrm{a}}=A_{\mathrm{a}}\frac{\partial\bar{u}_{\mathrm{a}}}{\partial t}+C_{\mathrm{a}}A_{\mathrm{a}}\frac{\partial\bar{u}_{\mathrm{w}}}{\partial t}-C_{\mathrm{q}}^{\mathrm{a}}A_{\mathrm{a}}\frac{\mathrm{d}q}{\mathrm{d}t}+u_{\mathrm{aw}} \tag{11a}$$

$$\bar{u}_{\mathrm{w}}=W_{\mathrm{w}}\frac{\partial\bar{u}_{\mathrm{w}}}{\partial t}+C_{\mathrm{w}}W_{\mathrm{w}}\frac{\partial\bar{u}_{\mathrm{a}}}{\partial t}-C_{\mathrm{q}}^{\mathrm{w}}W_{\mathrm{w}}\frac{\mathrm{d}q}{\mathrm{d}t}+u_{\mathrm{ww}} \tag{11b}$$

where A_{a}, W_{w} and F are the constants of integration.

$$A_{\mathrm{a}}=\frac{(r_{\mathrm{e}})^{2}F}{2C_{\mathrm{v}}^{\mathrm{a}}},\ W_{\mathrm{w}}=\frac{(r_{\mathrm{e}})^{2}F}{2C_{\mathrm{v}}^{\mathrm{w}}},\ F=\frac{N^{2}}{N^{2}-1}\left(\ln N+\frac{1}{N^{2}}-\frac{1}{4N^{4}}-\frac{3}{4}\right).$$

N is the ratio of influence radius to vertical drain radius, $N=r_{\mathrm{e}}/r_{\mathrm{w}}$.

The Laplace transform is used to decouple Equations (11). Accordingly, the average excess pore-air and pore-water pressures in the Laplace transform domain are obtained as follows:

$$\tilde{\bar{u}}_{\mathrm{a}}=\frac{1}{D}\left[(sW_{\mathrm{w}}-1)\tilde{u}_{\mathrm{aw}}-sC_{\mathrm{a}}A_{\mathrm{a}}\tilde{u}_{\mathrm{ww}}+B_{\mathrm{a}}\right] \tag{12a}$$

$$\tilde{\bar{u}}_{\mathrm{w}}=\frac{1}{D}\left[(sA_{\mathrm{a}}-1)\tilde{u}_{\mathrm{ww}}-sC_{\mathrm{w}}W_{\mathrm{w}}\tilde{u}_{\mathrm{aw}}+B_{\mathrm{w}}\right] \tag{12b}$$

where $B_{\mathrm{a}}=(sC_{\mathrm{a}}A_{\mathrm{a}}C_{\mathrm{w}}W_{\mathrm{w}}-sA_{\mathrm{a}}W_{\mathrm{w}}+A_{\mathrm{a}})\bar{u}_{\mathrm{a}}^{0}+C_{\mathrm{a}}A_{\mathrm{a}}\bar{u}_{\mathrm{w}}^{0}+(sC_{\mathrm{a}}A_{\mathrm{a}}C_{\mathrm{q}}^{\mathrm{w}}W_{\mathrm{w}}-sC_{\mathrm{q}}^{\mathrm{a}}A_{\mathrm{a}}W_{\mathrm{w}}+C_{\mathrm{q}}^{\mathrm{a}}A_{\mathrm{a}})(s\tilde{q}-q_{0})$,

$B_{\mathrm{w}}=(sC_{\mathrm{a}}A_{\mathrm{a}}C_{\mathrm{w}}W_{\mathrm{w}}-sA_{\mathrm{a}}W_{\mathrm{w}}+W_{\mathrm{w}})\bar{u}_{\mathrm{w}}^{0}+C_{\mathrm{w}}W_{\mathrm{w}}\bar{u}_{\mathrm{a}}^{0}+(sC_{\mathrm{q}}^{\mathrm{a}}A_{\mathrm{a}}C_{\mathrm{w}}W_{\mathrm{w}}-sA_{\mathrm{a}}C_{\mathrm{q}}^{\mathrm{w}}W_{\mathrm{w}}+C_{\mathrm{q}}^{\mathrm{w}}W_{\mathrm{w}})(s\tilde{q}-q_{0})$,

$D=s^{2}C_{\mathrm{a}}A_{\mathrm{a}}C_{\mathrm{w}}W_{\mathrm{w}}-s^{2}A_{\mathrm{a}}W_{\mathrm{w}}+sA_{\mathrm{a}}+sW_{\mathrm{w}}-1$. $\tilde{\bar{u}}_{\mathrm{a}}$, $\tilde{\bar{u}}_{\mathrm{w}}$, $\tilde{q}$ are the result of the Laplace transform of $\bar{u}_{\mathrm{a}}$, $\bar{u}_{\mathrm{w}}$, $q(t)$, respectively. And $\bar{u}_{\mathrm{a}}^{0}$, $\bar{u}_{\mathrm{w}}^{0}$, q_{0} are initial value of $\bar{u}_{\mathrm{a}}$, $\bar{u}_{\mathrm{w}}$, $q(t)$, respectively.

The boundary condition Equations (6) is substituted into Equations (9), then substituted into Equations (10), and the following Equations is obtained through Laplace transform:

$$\frac{\mathrm{d}^{2}\tilde{u}_{\mathrm{aw}}}{\mathrm{d}z^{2}}=\rho_{\mathrm{a}}\left(\tilde{\bar{u}}_{\mathrm{a}}-\tilde{u}_{\mathrm{aw}}\right) \tag{13a}$$

$$\frac{\mathrm{d}^{2}\tilde{u}_{\mathrm{ww}}}{\mathrm{d}z^{2}}=\rho_{\mathrm{w}}\left(\tilde{\bar{u}}_{\mathrm{w}}-\tilde{u}_{\mathrm{ww}}\right) \tag{13b}$$

where G_{a}, G_{w} are the well resistance factor with regard to air and water, respectively. ρ_{a}, ρ_{w} are two constant parameters, they are defined as:

$$\rho_{\mathrm{a}}=\frac{-8(N^{2}-1)G_{\mathrm{a}}}{H^{2}F},\ \rho_{\mathrm{w}}=\frac{-8(N^{2}-1)G_{\mathrm{w}}}{H^{2}F},\ G_{\mathrm{a}}=\frac{k_{\mathrm{a}}}{k_{\mathrm{aw}}}\left(\frac{H}{d_{\mathrm{w}}}\right)^{2},\ G_{\mathrm{w}}=\frac{k_{\mathrm{w}}}{k_{\mathrm{ww}}}\left(\frac{H}{d_{\mathrm{w}}}\right)^{2}.$$

d_w is the diameter of the vertical drain, $d_w = 2r_w$.

Equation (12) is combined with Equation (13) to obtain the ordinary differential equations (ODEs) of variable z as follows:

$$\frac{\mathrm{d}^2\tilde{u}_{aw}}{\mathrm{d}z^2} = \Omega_a\tilde{u}_{aw} + \Omega_w\tilde{u}_{ww} + \Omega_q \tag{14a}$$

$$\frac{\mathrm{d}^2\tilde{u}_{ww}}{\mathrm{d}z^2} = \psi_w\tilde{u}_{ww} + \psi_a\tilde{u}_{aw} + \psi_q \tag{14b}$$

where

$$\Omega_a = \rho_a\left(sW_w - 1 - D\right)D^{-1},\ \Omega_w = -sC_aA_a\rho_aD^{-1},\ \Omega_q = \rho_aB_aD^{-1},$$

$$\psi_w = \rho_w\left(sA_a - 1 - D\right)D^{-1},\ \psi_a = -sC_wW_w\rho_wD^{-1},\ \psi_q = \rho_wB_wD^{-1}.$$

The above equation (14) is decoupled by introducing two new variablesproposed by Qin et al. [7]

$$\frac{\mathrm{d}^2\tilde{\Phi}_1}{\mathrm{d}z^2} - Q_1\tilde{\Phi}_1 - \left(\Omega_q + \psi_q q_{21}\right) = 0 \tag{15a}$$

$$\frac{\mathrm{d}^2\tilde{\Phi}_2}{\mathrm{d}z^2} - Q_2\tilde{\Phi}_2 - \left(\Omega_q q_{12} + \psi_q\right) = 0 \tag{15b}$$

where

$$q_{21} = \frac{\Omega_w}{Q_1 - \psi_w},\ q_{12} = \frac{\psi_a}{Q_2 - \Omega_a},\ Q_{1,2} = \frac{1}{2}\left\{\Omega_a + \psi_w \mp \left[\left(\Omega_a - \psi_w\right)^2 + 4\Omega_w\psi_a\right]^{\frac{1}{2}}\right\}.$$

$\tilde{\Phi}_1$ and $\tilde{\Phi}_2$ are intermediate variables, $\tilde{\Phi}_1 = \tilde{u}_{aw} + \tilde{u}_{ww}q_{21}$, $\tilde{\Phi}_2 = \tilde{u}_{aw}q_{12} + \tilde{u}_{ww}$.

By combining the boundary conditions (7), (8) after Laplace transform with the Equations (15), the solution can be obtained as follows:

$$\tilde{\Phi}_1 = \xi_1\left(\lambda_1 - 1\right) \tag{16a}$$

$$\tilde{\Phi}_2 = \xi_2\left(\lambda_2 - 1\right) \tag{16b}$$

where

$$\xi_1 = \frac{\Omega_q + \psi_q q_{21}}{Q_1},\ \lambda_1 = \cosh\left[H\left(Q_1\right)^{\frac{1}{2}} - z\left(Q_1\right)^{\frac{1}{2}}\right]\mathrm{sech}\left[H\left(Q_1\right)^{\frac{1}{2}}\right],$$

$$\xi_2 = \frac{\Omega_q q_{12} + \psi_q}{Q_2},\ \lambda_2 = \cosh\left[H\left(Q_2\right)^{\frac{1}{2}} - z\left(Q_2\right)^{\frac{1}{2}}\right]\mathrm{sech}\left[H\left(Q_2\right)^{\frac{1}{2}}\right].$$

Therefore, the average excess pore-air and pore-water pressure can be expressed as:

$$\tilde{\bar{u}}_a = \frac{\left(\lambda_1\xi_1Q_1 - q_{21}\lambda_2\xi_2Q_2\right) + \rho_a\left[\xi_1\left(\lambda_1 - 1\right) - q_{21}\xi_2\left(\lambda_2 - 1\right)\right]}{\rho_a\left(1 - q_{12}q_{21}\right)} \tag{17a}$$

$$\tilde{\bar{u}}_w = \frac{\left(\xi_2Q_2\lambda_2 - q_{12}\xi_1Q_1\lambda_1\right) + \rho_w\left[\xi_2\left(\lambda_2 - 1\right) - q_{12}\xi_1\left(\lambda_1 - 1\right)\right]}{\rho_w\left(1 - q_{21}q_{12}\right)} \tag{17b}$$

Referring to Fredlund and Hasan's [1] method of using stress state variables to express the volumetric strain of soil unit, the average degree of consolidation in the Laplace transform domain can be expressed as:

$$\tilde{\bar{U}} = \left|\frac{\tilde{\varepsilon}_v(s)}{\tilde{\varepsilon}_v(\infty)}\right| = \frac{\left(m_2^s - m_{1k}^s\right)\left(\tilde{\bar{u}}_a - \bar{u}_a^0/s\right) - m_2^s\left(\tilde{\bar{u}}_w - \bar{u}_w^0/s\right) + m_{1k}^s\left(\tilde{q} - q_0/s\right)}{\left(m_{1k}^s - m_2^s\right)\bar{u}_a^0 + m_2^s\bar{u}_w^0 + m_{1k}^s\left(q_\infty - q_0\right)} \tag{18}$$

where $\tilde{\bar{U}}$, $\tilde{\varepsilon}_v(s)$, $\tilde{\varepsilon}_v(\infty)$ are the average degree of consolidation degree, the volumetric strain and the final volumetric strain in the Laplace transform domain, respectively. m_{1k}^s ($m_{1k}^s = m_{1k}^a + m_{1k}^w$) and m_2^s ($m_2^s = m_2^a + m_2^w$) are the coefficients of soil element volume change with regard to the net normal stress ($q - \bar{u}_a$) (kPa^{-1}) and the matric suction ($\bar{u}_a - \bar{u}_w$) (kPa^{-1}), respectively. q_∞ is the final value of $q(t)$.

The general semi-analytical solutions can be obtained by the inverse Laplace transform of Equations (17) and (18) with Crump's method. Details of the Crump's method can refer to Wang et al. [5] The common time-dependent loading after Laplace transform $\tilde{q}$ can be referred to Xu et al. [11]

3.2Verification

In order to verify the reliability of the proposed semi-analytical solution, a special consolidation case under constant loading (q= 100 kPa) is considered, and compared with this of the analytical solution obtained by Zhou et al. [10] The following unsaturated soil examples are used for the verification and the analysis of consolidation behavior and influence factors later. It is worth noting that the calculation method of initial value of excess pore pressure under constant loading can refer to Ho et al. [9]

$r_w = 0.2$ m, $r_e = 1.8$ m (i.e. $N = 9$), $H = 10$ m, $S_{r0} = 80\%$, $n_0 = 50\%$,
$k_w = 1\times10^{-10}$ m/s, $m_{1k}^{a} = -2\times10^{-4}$ kPa^{-1}, $m_{1k}^{w} = -5\times10^{-5}$ kPa^{-1}, $m_{1k}^{s} = -2.5\times10^{-4}$ kPa^{-1},
$m_2^{a} = 1\times10^{-4}$ kPa^{-1}, $m_2^{w} = -2\times10^{-4}$ kPa^{-1}, $m_2^{s} = 1\times10^{-4}$ kPa^{-1}, q_0=100 kPa (u_a^0=20 kPa,u_w^0=40 kPa).

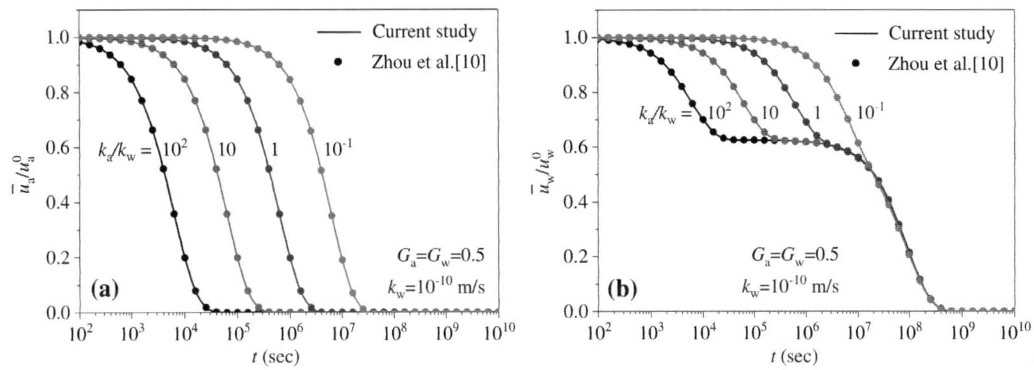

Figure 2. Variations of (a) average excess pore-air pressure and (b) average excess pore-waterpressure with different ratio of k_a/k_w

Figure 2(a) and (b) are respectively the dissipation curves of average excess pore-air pressure and pore-water pressure with different permeability coefficient ratio of k_a/k_w on the equal strain assumption. During this analysis, it is assumed that the permeability of water is constant (i.e. $k_w = 10^{-10}$ m/s), and the well resistance factor with respect to air and water are equal (i.e. $G_a = G_w$). As shown in figure 2, the dissipation curve of the average excess pore pressure obtained from the semi analytical solution under different permeability coefficient ratio of k_a/k_w in this paper has an almost-identical dissipation trend with this achieved from the analytical solution by Zhou et al. [10] Thus, it can be concluded that the current semi analytical solution is reliable.

4. Case study and discussions

In the case study, it mainly discusses the influence of permeability coefficient ratio k_a/k_w, well resistance factor with regard to air and water (i.e. G_a, G_w) and load parameter λ on the consolidation of unsaturated soils by vertical drains under exponential loading. The exponential loading is a simplified mathematical expression of construction loading, and it can be expressed as:

$$q(t) = q_0 + bq_0\left(1 - e^{-\lambda t}\right) \tag{19}$$

where q_0 is the initial surcharge (100 kPa), b is the load constant and b = 1 in this paper. λ is the loading parameter, which controls the loading rate of exponential loading. Hence, $\tilde{q} = 2q_0 s^{-1} - q_0\left(\lambda + s\right)^{-1}$.

Figure 3(a), (b) shows the effects of well resistance factor with regard to air and water (i.e. G_a, G_w) on the average degree of consolidation under exponential loading. The adopted parameters for the analysis include $r_w = 0.2$ m, $N = 9$, k_a/k_w =10, $k_w = 10^{-10}$ m/s and $\lambda = 5\times10^{-4}$ s^{-1}. Figure 3(a) presents the

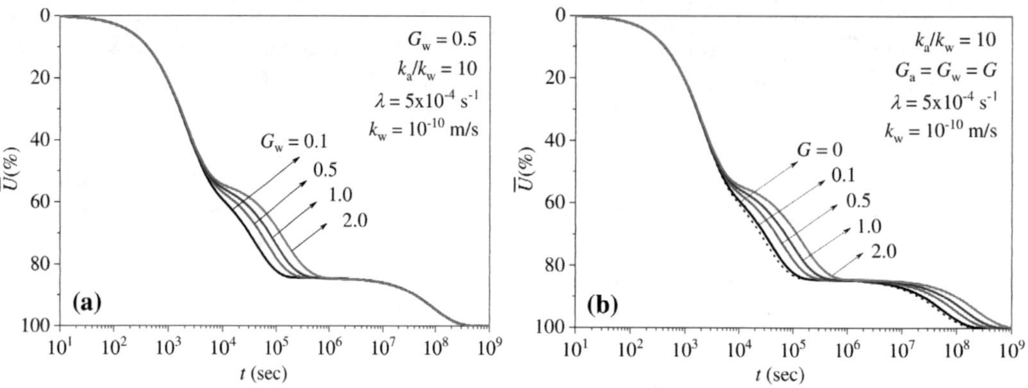

Figure 3. Variations of average degree of consolidation with time under exponential loading for (a) different G_a with $G_w = 0.5$and (b) different well resistance factor G with $G_a = G_w = G$

average degree of consolidation varying with time for different G_a with $G_w = 0.5$. As shown in the figure, the change curve of average degree of consolidation is obviously divided into two stages, and when G_w is constant, the change of G_a only affects the first stage. When other conditions are the same, the time needed to reach the same degree of consolidation increases with the increase of G_a. After entering the second stage, the curve tends to be consistent. This can be explained by the fact that after entering the second stage, the excess pore-air pressure has completely dissipated. This phenomenon is consistent with Zhou et al. [10] in the case of constant loading.

Figure 3(b), on the other hand, illustrates the average degree of consolidation varying with time for different well resistance factor G with $G_a = G_w = G$. Where $G = 0$ means to ignore the influence of well resistance. When comparing it with other curves, it can be known that when the well resistance factor reaches a certain value, its influence on the average degree of consolidation is obvious. Additionally, the time needed to reach the same degree of consolidation decreases with the decrease of the well resistance factor. Moreover, when the well resistance factor G is less than 0.1, the influence of the well resistance can be neglected.

Figure 4(a), (b) depicts the average degree of consolidation varying with permeability coefficient ratio of k_a/k_w and loading parameter λ, respectively. The adopted parameters are r_w =0.2m, $N = 9$, $G_a = G_w = G = 0.5$and $k_w = 10^{-10}$ m/s. The variation of the average degree of consolidation with time in figure 4(a) is similar to that of figure 3(a). That is because the k_w value is assumed to be constant of consolidation only occurs in the first stage, and the consolidation becomes faster as k_a increases. The effect of loading rate on average degree of consolidation is shown in figure 4(b). The consolidation process is accelerated with the increase of load parameter λ (i.e. the faster the loading rate). It is intere-

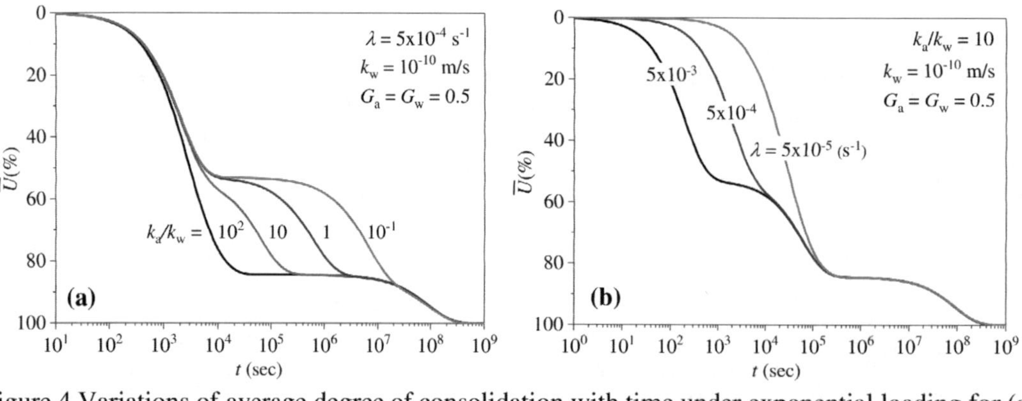

Figure 4.Variations of average degree of consolidation with time under exponential loading for (a) different k_a/k_w with λ =5x10^{-4} s^{-1} and (b) different values of λ with k_a/k_w =10

sting to observe that when the load parameter λ is greater than $5x10^{-4}$ s^{-1}, a new stage appears in the first stage when the excess pore pressure is not completely dissipated. The reason for this phenomenon may be that when the time reaches 10^4 s, the load of $\lambda = 5x10^{-4}$ s^{-1} has reached the maximum value and remains constant. The load of $\lambda = 5x10^{-5}$ s^{-1} remains at the stage of load increase. The increase in load accelerates the rate of consolidation.

5. Conclusions

Based on the equal strain assumption, well resistance and arbitrary time-dependent loading are considered in this paper to get the semi-analytical solutions of consolidation for unsaturated soils by vertical drains. The reliability of the obtained solution is verified by comparing with the analytical solution of radial consolidation of unsaturated soil considering well resistance under instantaneous loading conditions. Then, the influence of well resistance factors, permeability coefficient ratio k_a/k_w, and load parameter on the average degree of consolidation are analyzed under exponential loading. In addition, the general semi-analytical solutions can be used to analyze the consolidation problem with more complex loads in practical engineering.

Acknowledgement

We sincerely acknowledge the National Natural Science Foundation of China (Grant No. 41372279) for financial support.

References

[1] Fredlund, D.G., Hasan, J.U. (1979) One-dimensional consolidation theory: unsaturated soils. J. Can Geotech J., 17: 521-31.
[2] Qin, A.F., Sun, D.A, Tan Y.W. (2010) Semi analytical solution for one-dimensional consolidation of unsaturated soils. J. Applied mathematics and mechanics. 31(2): 199-209. (in Chinese)
[3] Shan, Z.D., Ling, D.S., Ding, H.J. (2012) Exact solutions for one-dimensional consolidation of single-layer unsaturated soil. J. International Journal for Numerical and Analytical Methods in Geomechanics, 36 (6):708-722.
[4] Zhou, W.H., Zhao, L.S. (2013) Influence of initial and boundary conditions on one-dimensional consolidation of unsaturated soil. J. Chinese Journal of geotechnical engineering, 35(s1): 305-311. (in Chinese)
[5] Wang, L., Sun, D.A., Qin, A.F. (2017) General semi-analytical solutions to one-dimensional consolidation for unsaturated soils. J. Applied Mathematics and Mechanics (English Edition), 38 (6): 831-850.
[6] Wang, L., Xu, Y.F., Xia, X.H. et al. (2019) Semi-analytical solutions of two dimensional plane strain consolidation in unsaturated soils subjected to the lateral semipermeable drainage boundary. J. International Journal for Numerical and Analytical Methods in Geomechanics, 2019(2).
[7] Qin, A.F., Sun, D.A., Yang L.P., et al. (2010) A semi-analytical solution to consolidation of unsaturated soils with the free drainage well. J. Computers and Geotechnics, 37(7): 867-875.
[8] Zhou, W.H., Tu S. (2012) Unsaturated consolidation in a sand drain foundation by differential quadrature method. J. Procedia Earth & Planetary Science, 5(8): 52-57.
[9] Ho, L., Fatahi, B. (2018) Analytical solution to axisymmetric consolidation of unsaturated soil stratum under equal strain condition incorporating smear effects. J. Numerical and Analytical Methods in Geomechanics. 42(15): 1890-1913.
[10] Zhou, F., Chen, Z., & Wang, X. (2018). An equal-strain analytical solution for the radial consolidation of unsaturated soils by vertical drains considering drain resistance. J. Advances in Civil Engineering, 2018.
[11] Xu, C. J., Cai Y. Q., Wu S. M. (1999) One dimensional consolidation of layered elastic foundation under arbitrary load. J. Journal of civil engineering, 1999(04): 57-63. (in Chinese)

IOP Conf. Series: Materials Science and Engineering **794** (2020) 012058 doi:10.1088/1757-899X/794/1/012058

Flow Patterns and Rectification Measures in the Forebay of Pumping Station

Yuan LI [1], Jinde GU [1], Chao GUO [1], Chenlu ZHOU[2]

[1]Nanjing Hydraulic Research Institute, Nanjing, Jiangsu, 210029, China

[2]The Eastern Route of South-to-North Water Diversion Project Jiangsu Water Source Co., Ltd, Nanjing, Jiangsu, 210029, China

liyuan@nhri.cn

Abstract. Due to land limitation, the forebay in the pumping station is generally not long enough, some disadvantageous flow patterns are easy to appear in the forebay, which is not beneficial to the safety and economic operation of the pump. In this paper, the flow pattern in the forebay of pumping station is taken as the research object, and we focus on the physical model test method, the analysis method of the flow pattern and the rectification measures such as beam, breast wall and column. At the same time, the research results show that the above-mentioned rectification measures are helpful to improve the pumping station inlet conditions and operation reliability.

1.Introduction

The forebay is an important part of the pumping station. Its function is to make the stream spread evenly and smoothly when the stream flows from the approach channel to the inlet conduit, thus providing a better flow pattern for the pumping station[1]. In order to improve the unfavorable flow patterns, such as flow deviation, reverse current and vortex caused by the bend flow in the forebay of pumping station, certain measures should be used to meet the requirements of engineering operation.

At present the domestic and foreign scholars have done a lot of research on the hydraulic characteristics of forebay[2-4]. Based on the Fluent software, Jiang wen et al. simulated the flow field of side-inlet pumping station with the Navier-Stockes equations and the SIMPLEC algorithm method, the complex flow pattern in the forebay of the lateral intake pump station is revealed[5]. Feng jiangang et al. carried out numerical simulation of the flow patterns based on RNG k-ε turbulence model in the forebay of the pumping station, the computed results are consistent with those from physical experiments[6]. Zhang xue et al. analyzed the inlet condition of forebay, aiming at proposing rectifying measures to improve the flow pattern[7]. We do the further research to study on flow patterns and rectification measures by physical experiment.

This pumping station is built as sea water cooling system for the GNPower Dinginin coal fired power station. In the 3 circulating water pumps, these are 2 in operation and 1 standby in all the operation conditions.

2.Model Details

According to the requirement of the research, the simulation range includes the inlet pipe, the inlet forebay of the pump suction chamber, the suction bell mouth. Some part would be simplified simulated.

2.1Model scale

Research of pump house is always carried out based on big scale model, in which vortex can be studied and accurate result can be gain. In accordance with the relevant model test procedures and taking into account the accuracy of model test results, model scale is no less than 1:10, diameter size of bell mouth throat should be not less than 100mm in physical model, Reynolds number in bell mouth should be more than 105, Weber number in bell mouth should be more than 240, width of suction room in physical model should be not less than 450mm (300mm in Hydraulic Institute Standards), the minimum water depth is more than 150mm（Hydraulic Institute Standards）.

In this project, the results are reasonable when the model scale is 10, by comparing the model parameters at different model scales. The model parameters include model discharge, diameter of bell mouth in model, width of suction room in model, area of bell mouth in model, velocity on bell mouth section in model, Froude number of bell mouth in model, Reynolds number of bell mouth in model, Weber number of bell mouth in model, etc.

2.2Similarity law

Research on the velocity distribution and vortex in pump house, the Fr criterion and Re criterion should all be considered according to the similarity law. But in the model and the prototype, it cannot satisfy both criteria. Our research model is designed by the Fr criterion, in order to revise the influence of dissimilar of Re number, vortex is checked by increasing velocity in model. One way is to build model in a big model scale; another way is to increase flow velocity in the model. In this test, we choose the second plan, vortex is observed under 1.5 Fr number, according to the Hydraulic Institute Standards.

2.3Model materials

Model intake tank is made of grey plastic and reinforced with steel. The water diversion pipe, circulating pump house (including forebay, filter room and suction room), bell mouth, are made up of Plexiglas for all-round observation. The outlet pipe of the model is made of PVC pipe. Several water-measuring weirs are installed at the bottom of the model to measure the flow rate. As it shows in Figure 1.

Figure 1. Photo of model

2.4Velocity measurement

Velocity measuring sections are set as Figure 2. Coordinate system is established in each section to describe the velocity distribution. As shown in Figure 3, central line of each section is set to be zero, right is positive and left is negative. Velocity in different height is measured in each section.

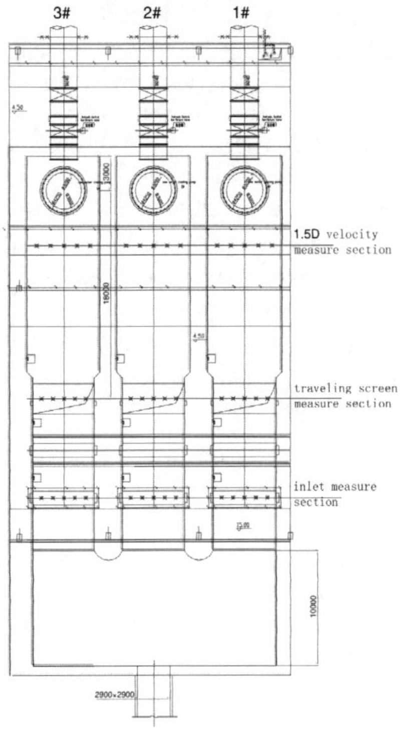

Figure 2. Measuring section of circulating water pump house

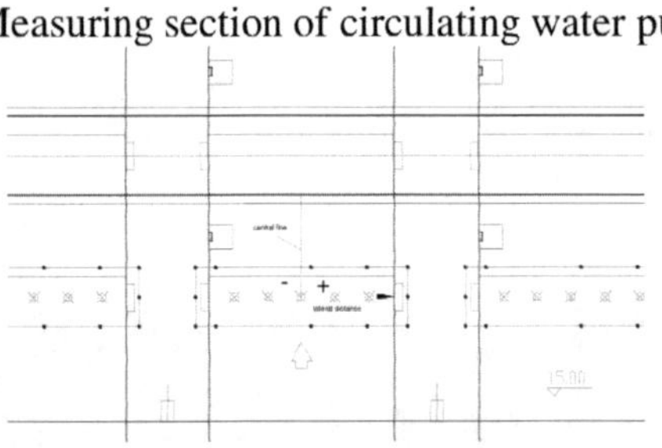

Figure 3. Coordinate system in each test section

3.Model test

In this project, operation combination can be classified to two kinds: symmetrical and unsymmetrical operation. Both are analyzed in preliminary test. The result in unsymmetrical operation is worse than that in symmetrical operation. Undesirable flow patterns always happen when the water level is low. For this reason, preliminary model test is carried out under working condition of low water level. This article takes unsymmetrical operation and low water level as example: the water level is -3.0m, discharge is 11.6m³/s, 1# and 2# are working.

3.1Preliminary model test and analysis

In the preliminary test results, velocity distribution is even in the 1.5D section in the suction room, which means flow pattern become better through the flow channel. But as it shows in Figure 4, velocity is higher than 1m/s in some points in the 2# entrance section. At the same time, it can be observed that velocity distribution in 2# channel is highly impacted by the inlet flow, water level fluctuates violently.

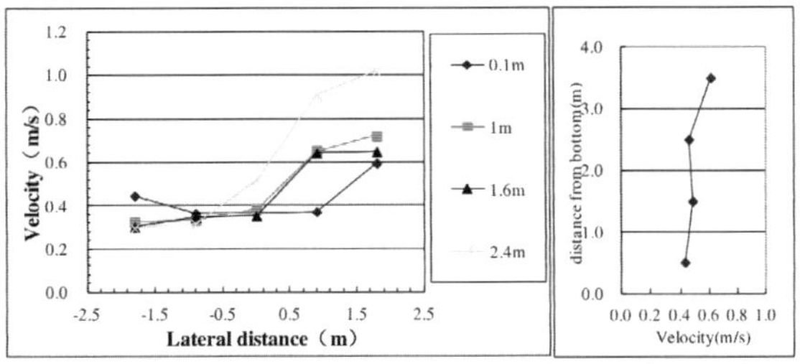

Figure 4. Velocity distribution in 2# channel

3.2Model test with rectification measures in forebay

In order to acquire better flow pattern, these are three options chosen to help flow diffusion, further observation is carried out after the phenomenon of channel drift disappears in all sections.

3.2.1Model test of cross beam scheme. Cross beam of 2m height is set in the middle of the forebay, facing to the middle of the inlet tunnel, model test is carried out to evaluate the effect of the cross beam. Flow pattern in 2# channel entrance is refined according to the Figure 5, which indicated high efficiency of cross beam. In the result, velocity distribution in 1.5D section is even, the water level is more stable in this scheme. But when the discharge enlarges to 1.5 times, vortex can be observed in the suction room, which can induce vibration in the pump and reduce the pump efficiency.

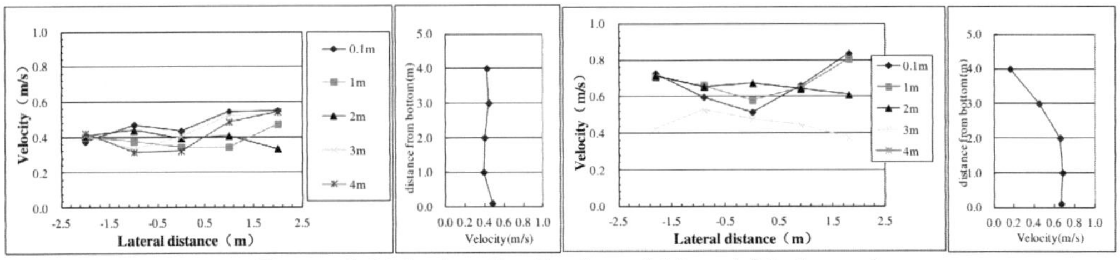

Figure 5. Velocity distribution of 1# and 2# channel entrance

3.2.2Model Test of Cross Beam and breast wall Scheme. The inlet flow diffuses well with the help of cross beam, and the vortex in the suction chamber disappears after the breast wall is set. The Hydraulic Institute Standards states that "Average swirl angle indicated by the swirl meter must be less than 5 degree, maximum short-term swirl angle can be up to 7 degree". According to the test results, velocity in the flow channel seems reasonable and swirl angle is less than 3 degree. Unfavorable flow patterns, such as back flow and large bias flow, are not found in all operation modes. It is considered that the cross beam and breast wall scheme can ensure a smooth flow pattern in the channel.

3.2.3Model Test of Columns Scheme. There are 2 rows of columns with 0.5m×0.5m size, column space of the first row is 0.6m and 0.7m in the second row. The first row of column is 3m far from the entrance, the second row is 2m far from first row. As it shows in Figure 6 and Figure 7, column arrangement of scheme is considered helpful in flow diffusion for this project. Swirl angle is far less than 3 degree in all operation modes according to the test results.

IOP Conf. Series: Materials Science and Engineering **794** (2020) 012058 doi:10.1088/1757-899X/794/1/012058

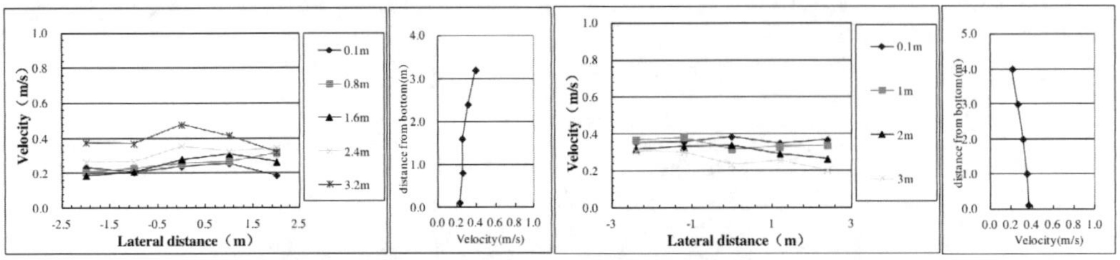

Figure 6. Velocity distribution of 1# and 2# channel entrance

Figure 7. Flow pattern observation by dye method

4Conclusion

This study provides a reference for the physical model research of forebay in pumping station, including the design and manufacture of the model, the measurement method of the flow pattern and so on. As it shows that the preliminary test results are not satisfactory, which will cause vibration and reduce the efficiency of the pump due to the influence of the inlet pipe velocity, the water level fluctuation and the vortex in the suction room. Cross beam, breast wall and columns are useful to obtain better flow pattern, these rectification measures can help to diffuse the flow in the forebay and prevent the vortex.

Acknowledgments

This research is financially supported by the National Key R&D Program of China (Grant No. 2018YFB1600400).

References

[1] Blazek J. Computational Fluid Dynamics: Principles and Applications [M]. Netherlands: Elsevier Ltd, 2005: 227-270.
[2] Tian Y.,Yang P., Lu J. Numerical simulation of flow patterns in forebays of the urban drainage pumping stations[J]. Pump Technology,2008(4):24-26.
[3] Xu H., Zhang L. Review of flow pattern modifying technology for pump sump with side-inlet[J]. Advances in Science and Technology of Water Resources,2008,28(6):84-88.
[4] Cheng L., Liu C., Zhou J.R.. Numerical simulation of sill flows in the forebay of pumping station[J]. Journal of Hohai University (Natural Sciences),2001,29(3):42-45.
[5] Jian W., Yu J.Z., Fu Z.F., Tan Q. Flow Patterns and Rectification Measures in a Forebay of Pumping Station Designed at Bend River [J]. YELLOW RIVER, 2019,41(4),83-88
[6] Feng J.G., Qian X.D., Zhang R. Flow patterns and rectification measures in forebays of urban water pumping station[J]. South-to-North Water Transfers and Water Science &Technology, 2018,38(2),77-83
[7] Zhang X., Zhou J.R., Liang J.D. Experimental research on flow pattern of forebay of side-inlet pumping station[J]. South-to-North Water Transfers and Water Science &Technology, 2016,14(3),101-105

IOP Conf. Series: Materials Science and Engineering **794** (2020) 012059 doi:10.1088/1757-899X/794/1/012059

A light piston well-flushing technique for clogged relief wells along Yangxin Dyke of Yangtze River in China

Mi Tian[1], Xiaotao Sheng[2*]

[1]School of Civil Engineering, Architecture and Environment, Hubei University of Technology, Wuhan, Hubei 430068, China

[2]Key Laboratory of Geotechnical Mechanics and Engineering of Ministry of Water Resources, Yangtze River Scientific Research Institute, Wuhan, Hubei 430010, China

[*]Corresponding author's e-mail: shengxiaotao@163.com

Abstract. Plastic relief wells along Yangxin Dyke of Yangtze River in China have been seriously clogged in recent years. It is necessary to dredge the wells to restore its disaster prevention and mitigation function. However, traditional piston well-flushing technique is suitable to restore steel and iron wells, and it can cause damage to plastic relief wells. A new light piston well-flushing technique is proposed to overcome the limitation of traditional piston well-flushing technique. The proposed light piston has the advantages of simple structure, convenient use and low cost. By the light piston moving down and up quickly the relief well, the nylon brush can clean the side wall and filters of relief well, and the negative pressure can make the deposited materials in the aquifer and inverted filter sucked out. The proposed well-flushing technique is applied to restore the plastic wells along Yangxin Dyke of Yangtze River in China. Its efficiency is illustrated using field pumping test data. It is shown that the drainage capacity and permeability of clogged relief wells can be greatly improved after using the proposed technique. The light piston well-flushing technique can provide an effective way to restore clogged relief wells along Yangtze River levees.

1. Introduction

Relief wells can provide pressure relief and controlled seepage outlets that offer little resistance to flow and, at the same time, prevent sand boils or piping[1][2]. From 1995 to 2001, 164 relief wells were constructed along Yangxin Dyke of Yangtze River in China for controlling seepage and piping along the lower Yangtze River levees[3][4]. The structure of constructed relief wells is shown in Figure 1. The diameter of the wells is 30 cm and the designed well depth is 18.0~30.0m. However, these relief wells have been seriously clogged in recent years. A large amount of debris such as silt and sand were deposited in the wells. The deposits block the filters of relief wells and drainage channel of aquifer, and affect the wells' seepage control efficiency[5][6], which brings a tremendous threat for the safety of Yangtze River levees. Therefore, it is, necessary to dredge the clogged relief wells to restore its disaster prevention and mitigation function.

IOP Conf. Series: Materials Science and Engineering **794** (2020) 012059 doi:10.1088/1757-899X/794/1/012059

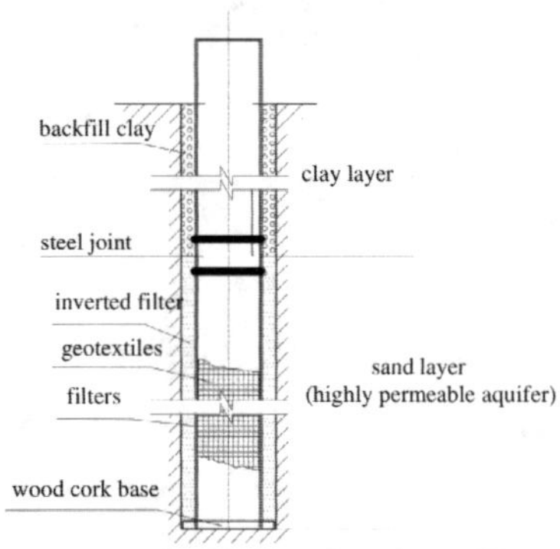

Figure 1. Structure of relief wells along Yangxin Dyke

Generally, the traditional piston well-flushing technique has been widely used in the maintenance and restoration of relief wells[7][8]. By moving the piston up and down quickly, instantaneous negative pressure and hydraulic impact are formed at the bottom of relief wells to bring out the deposits[7][8]. However, the traditional piston well-flushing technique is applicable for the restoration of steel and iron wells. The 164 relief wells along Yangxin Dyke are made of plastic[3]. When the piston moves up and down quickly, the huge instantaneous negative pressure will cause damage to the plastic wall of relief wells and a large number of inverted filter sand outflow. Eventually, the filters (as shown in Figure 1) are destroyed, affecting the performance of relief wells.

This paper explores a new well-flushing technique for mechanically clogged relief wells. Firstly, the light piston well-flushing technique that is applicable for the restoration of plastic wells is proposed, and its dredging mechanism is introduced. Then, the proposed well-flushing technique is applied to restore the plastic relief wells along Yangxin Dyke of Yangtze River in China. Finally, the efficiency of the proposed technique is illustrated using field pumping test data.

2. Light piston well-flushing technique

Figure 2(a) shows the schematic diagram of the light piston. The length of piston is 1.5 m. The light piston mainly consists of four parts: pump pipe, internal plastic pipe, epoxy external pipe and nylon brush. The diameter of pump pipe is 50mm. The inner and external diameter of internal plastic pipe are 50 mm and 100 mm, respectively. The inner and external diameter of epoxy external pipe are 100 mm and 150 mm, respectively. The nylon brush is fixed on the epoxy external pipe. The nylon brush consists of two parts: the long part and the short part. The long part is 270 mm in length and the short part is 255 mm in length.

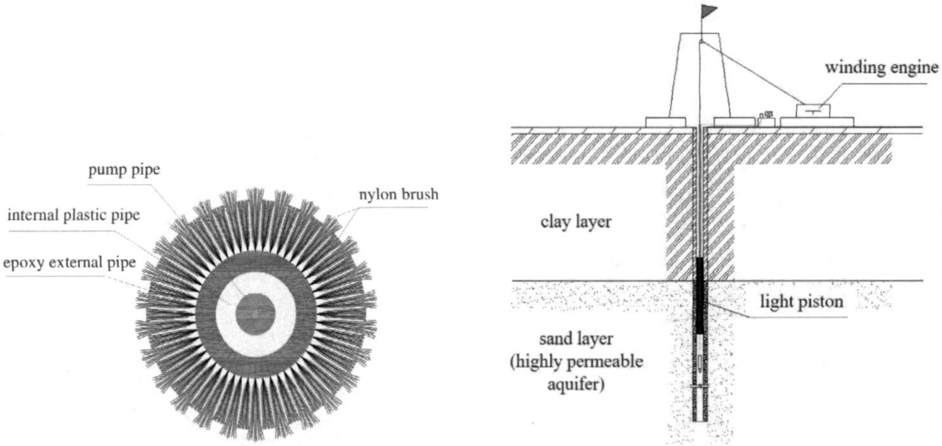

(a) cross-section view of light piston (b) working schematic

Figure 2. Schematic diagram of light piston well-flushing technique

Figure 2(b) shows the working schematic of light piston well-flushing technique. The light piston is lowered down the well through a tripod and lifted by the winding engine. When the light piston moves down the relief well, the nylon brush keeps rubbing against the side wall of the well and cleaning the side wall and filters of relief wells (as shown in Figure 1). When the light piston moves up, a great negative pressure forms at the bottom of the piston. Under the negative pressure, the deposited materials in the aquifer and inverted filter are sucked out. Then, the drainage channel of relief wells is dredged and the permeability of aquifer and filters are improved. The light piston is simple in structure and economical in use and maintenance. The light piston is also easy and flexible to operate.

3. Application and efficiency of light piston well-flushing technique

The proposed well-flushing technique for clogged relief well is applied to the 4 plastic relief wells (i.e., 64#~67# wells) along Yangxin Dyke of Yangtze River in China. The diameter of 64#~67# wells is 30 cm and the designed well depth is 30.0 m. The filters and drainage channel of aquifer of these 4 relief wells are silted up with silt and sand seriously.

In order to illustrate the efficiency of light piston well-flushing technique, single-hole steady flow pumping test and non-steady flow water level recovery test are conducted[9-11]. The discharge of unit drawdown is obtained from steady flow pumping test to evaluate the drainage capacity of these four relief wells. The transmissibility coefficient of aquifer is evaluated by the non-steady flow level recovery test to reflect the permeability of aquifer.

The discharge of unit drawdown q (m^3/h. m) is calculated as

$$q = \frac{Q}{s} \tag{1}$$

in which Q (m^3/h) represents the correlation coefficient; s (m) is the drawdown of relief wells.

The transmissibility coefficient T (m^2/d) of aquifer is obtained as follows:

$$T = \frac{0.183Q}{[i]} \tag{2}$$

where $[i]$ is the slope of the straight line section of the curve of residual drawdown and lg$t$ data in the recovery stage; t is the stop time for water level recovery test.

Steady flow pumping test and non-steady flow water level recovery test are conducted to obtain q and T of these 4 clogged wells before and after applying the proposed well-flushing technique, respectively. Table 1 shows the results of pumping test for the 4 relief wells. It can be seen that the values of q of these 4 wells increase obviously after using the proposed light piston well-flushing technique. The q values are 1.3 to 3.8 times that before dredging. It is indicated that the drainage capacity of relief wells has been improved after the application of the proposed well-flushing technique.

Table 1. Pumping test results of 4 clogged relief wells along Yangxin Dyke

Relief well number	Before well flushing		After well flushing	
	q (m^3/h. m)	T (m^2/ d)	q (m^3/h. m)	T (m^2/ d)
64#	6.22	27.65	8.51	45.79
65#	3.42	25.92	8.39	45.79
66#	1.23	13.82	4.71	44.06
67#	3.53	21.60	6.20	33.70

Table 1 also shows the values of T of these 4 clogged wells. There are great difference between before and after the application of the proposed technique. Before well flushing, T values range from 13.82 m^2/ d to 27.65 m^2/ d. After dredging the wells by light piston, T values range from 33.70 m^2/ d to 45.79 m^2/d. It can be clearly seen that the T values increase significantly after using the proposed well-flushing technique. It is suggested that the permeability coefficient of aquifer increases. It also indicates that the drainage channel of aquifer has been dredged after using the proposed well-flushing technique.

4. Summary and conclusions

This paper analyzed the clogging mechanism of relief wells along Yangxin Dyke of Yangtze River in China. Because the traditional piston well-flushing technique is applicable for the restoration of steel and iron wells and it can cause damage to plastic relief wells. A new light piston well-flushing technique is proposed to overcome the limitation of traditional piston well-flushing technique.

The proposed light piston consists of four parts: pump pipe, internal plastic pipe, epoxy external pipe and nylon brush. It has the advantages of simple structure, convenient use and low cost. By the light piston moving down and up quickly, the nylon brush can clean the side wall and filters of relief wells and the negative pressure formed at the bottom of the piston can make the deposited materials in the aquifer and inverted filter sucked out. The proposed well-flushing technique is applicable for the restoration of plastic wells.

The proposed light piston well-flushing technique was applied to dredge relief wells along Yangxin Dyke of Yangtze River in China. Its efficiency was illustrated by field pumping test data. It was shown that the drainage capacity and permeability of relief wells can be improved by using the proposed well-flushing technique. The light piston well-flushing technique can effectively clean the deposits in the filters and aquifer of clogged relief wells. The proposed light piston well-flushing technique provides an effective way to restore clogged relief wells along Yangtze River levees.

References

[1] Hadj-Hamou, T., Tavassoli, M.R., Sherman, W.C. (1990) Laboratory testing of filters and slot sizes for relief wells. Journal of Geotechnical and Geoenvironmental Engineering, 116(9): 1325-1346

[2] Mansur, C. I., Postol, G., Salley, J.R. (2000) Performance of relief well systems along Mississippi River levees. Journal of Geotechnical and Geoenvironmental Engineering, 126:727-738.

[3] Yangxin County Water Conservancy Records Editorial Committee. (2010) Yangxin county water conservancy records (1986~2005). China Water Power Press, Beijing, China.

[4] Huang, Y.Y., Peng, L.Y., Luo R.H. (2002) Engineering geology analysis and evaluation of dangerous section of Yangxin Yangtze River Dike. Yangtze River, 33(7): 12-22.

[5] Wu, C.Y., Zhang, W., Sun, H.C. (2005) Present Situation on Relief-well Mechanism Research. Journal of Yangtze River Scientific Research Institute, 22(2): 59-62.

[6] Wu, C.Y., Zhang, W., Li, S.S., et al. (2009) Research on mechanical clogging mechanism of releaf well and its control method. Rock and Soil Mechanics, 30(10):3180-3187.

[7] Cao, G. (1992) Discussion on the causes and treatment measures of silting wells on the Yangtze River Levees in Anqing city. Yangtze River, 7: 28-32.

[8] Duan, X.B., Yang, C., Xie, L.F. (2011) Analysis on mechanism of relief-well clogging and experiment on a new technique for its restoration. Water Resources and Hydropower Engineering, 42(3): 45-48.

[9] Water conservancy standards of the People's Republic of China (SL320-2005). (2005) Test rules for borehole pumping in water conservancy and hydropower projects. China WaterPower Press, Beijign, China.

[10] China Geological Survey. (2012) Hydrogeology manual (second edition). Geological Publishing Press, Beijing, China.

[11] Zhang, Y., Dang, C.H., Dong, D., et al. (2019) Intelligent optimization calculation of hydrogeological parameters. Science Press, Beijing, China.

IOP Conf. Series: Materials Science and Engineering **794** (2020) 012060 doi:10.1088/1757-899X/794/1/012060

Water Resources Allocation of the Changjiang To Huai River Water Diversion

Zhijun Wu[1,2], Bo Yan[1,2*], Hao Cao[1,2], YongqiangWang[1,2], Feng Huang[3]

[1]Water Resources Department (International River Management Department), Changjiang River Scientific Research Institute, Wuhan 430010, China

[2]Hubei Provincial Key Laboratory of Basin Water Resources and Ecological Environmental Sciences, Changjiang River Scientific Research Institute, Wuhan 430010, China

[3]College of Hydrology and Water Resources, Hohai University, Nanjing 210098, China

*Corresponding author's e-mail: byanhhu@163.com (B Yan)

ABSTRACT. The Changjiangriver to Huai river water diversion project plans to divert water from the lower reaches of the Changjiangriver to replenish water in the middle reaches of the Huaihe river, mainly for the purpose of ensuring urban and rural water supply, developing Jianghuai shipping and improving river and lake ecology.In this paper, we predict the four aspects water demand including domestic, industrial, agricultural and ecological water demand in the river channel in 2030. According to the rules of lakes and water supply projects, we can obtain the water supply under different typical years. The balance of water supply and demand is analyzed for different circumstances. It is concluded that the average annual total water demand in the planned area in 2030 is 18.335 billion m^3, the total water supply is 15.356 billion m^3, and the water shortage is 2.979 billion m^3.

1. Introduction

The Changjiang river and Huaihe river project connects the Changjiang River and Huaihe River basins and belongs to one of the 172 major water-saving and water supply projects determined by the State Council and started construction in 2016[1].The water supply scope of the project covers 12 cities in anhui province and 2 cities in henan province, covering a total of 55 districts and counties, covering an area of 70,600 km^2, benefiting 41.32 million people.The scale of the Changjiang River and Huaihe River project is 300m^3/s, the scale of the Huaihe River is 280m^3/s, and the total irrigation area of the Huaihe River Basin is 10.85 million mu.In 2030, the average annual water diversion of the Changjiang River Estuary is 3.302 billion m^3; the water input into the Wapu Lake is 2.136 billion m^3.It is an urgent problem to be solved whether water resources can be developed and utilized according to the principle of sustainable development, and provide services for the life, industry, agriculture and ecological development in the planned areas so as to obtain the maximum socio-economic and environmental benefits.The rational allocation of water resources is the main means to solve the shortage of water resources, maintain the balance between supply and demand, and water ecology and environmental protection[2]. To this end, this paper needs to conduct a rational allocation study of water resources to achieve the purpose of alleviating the contradiction between supply and demand of water resources.

IOP Conf. Series: Materials Science and Engineering **794** (2020) 012060 doi:10.1088/1757-899X/794/1/012060

2. Available water supply analysis

Available water supply analysis is to sum up the available water supply of a calculated area in the same planning level year and the same guarantee rate.[3].The analysis of water supply capacity includes surface water availability analysis, groundwater availability analysis, water reuse for water reuse, and external water transfer analysis for the South-North Water Transfer Project[4].

2.1Surface water availability analysis

Ground surface water calculation formula:

$$W_S = \sum \min (Q_i, Y_i) \tag{1}$$

Formula: W_S is the water supply quantity of diversion project; Q_i is the amount of water that can be transferred to meet the water quality requirements in period I; Y_i is the project water pumping capacity[5];In general, the amount of available surface water should be less than or equal to the amount of available surface water[6].

Finally, according to the long series of adjustments, the planned annual average water supply in 2030 is 8.709 billion m^3.

2.2Groundwater availability analysis

Groundwater availability can be divided into shallow and deep groundwater availability[7].The average annual shallow groundwater supply was 2.329 billion m^3in 2010.By 2030, the average annual water supply of shallow groundwater is 2.381 billion m^3.At the same time, the deep groundwater will be used as an emergency water source and will not participate in the water allocation. Therefore, the influence of deep groundwater is not considered in the calculation of available water supply[8].

2.3Analysis of available supply of reclaimed water

According to the proportion of water and drainage in recent years, the urban sewage discharge rate is determined to be 75%, in which all domestic sewage is collected by the pipe network.According to the requirement of reuse rate of industrial water in the planned year, 80% of the total amount of industrial sewage discharged shall be recycled within the plant, and the remaining 20% shall be discharged into the urban sewage pipe network.In the planning scope of 2030, water consumption is 559 million m^3.

2.4Analysis of available water supply of external mixed water

The water supply scope of the Changjiang river and Huaihe river projectand the water supply scope of the East Route of the South-to-North Water Transfer Project are duplicated in Huaibei and Suzhou city. The water volume of the East Route of the South-North Water Transfer Project will be deducted from the scope of this plan. The East Route of the South-to-North Water Transfer Project plans to supply 254 million m^3 of water to Huaibei and Suzhou in 2030.

3. Water demand forecast

The water demand can be divided into two types of water demand in the river channel and outside the river channel. The water demand in the river channel is the average annual water volume of a specific section. The water demand outside the river channel needs to further distinguish the socio-economic water demand and the artificial ecosystem water demand, and the quota method is used for prediction [9].

Water demand of social economy is divided by life, industry and agriculture. Living water needs include urban life and rural life. Industrial water requirements include the power industry (non-hydropower) and the non-power industry[10].Agricultural water needs include farmland irrigation and forestry, animal husbandry and fishing.

3.1Forecast of domestic water demand

Living water demand mainly includes urban and rural residents' water demand, which is predicted by quota method. Estimates of their total domestic water needs need to take into account population and quotas. The domestic water demand of residents is predicted by the per capita water quota method, and the total domestic water demand of the year is predicted by the population in 2010 and the water quota. The domestic water needs of urban residents and rural residents are calculated separately, the calculation formula is shown as follows:

$$W_{ji}^{t} = P_{i}^{t} \times L_{i}^{t} \times 365/1000 \tag{2}$$

$$W_{mi}^{t} = P_{i}^{t} \times L_{i}^{t} \times 365/1000/\eta_{i}^{t} \tag{3}$$

Formula: i is the user classification number, i=1 is the town and i=2 is the countryside; t is the planning year serial number; W_{ji}^{t} is the net water demand of residents in year t; P_{i}^{t} is the number of water users in year t; L_{i}^{t} is the net living quota (liter/person. day) of residents in the t year; W_{mi}^{t} is gross water demand of residents in year t (Ten thousand m³); η_{i}^{t} is the domestic water utilization coefficient of residents in year *t*.

Total population in planning year can be predicted by quota method:

$$p = p_0(1+k)^n \tag{4}$$

Formula: p is the total population in the planning year; p_0 is the total population in 2010; k is the natural growth rate of population during the forecast period; n is the predicted years.

3.2Forecast of industrial water demand

The prediction of industrial water demand can be divided into general industry and thermal power industry, the calculation formula is shown as follows:

$$Q_{i}^{t_2} = Q_{i}^{t_1} \times (1 - r_{i}^{t_2})^{t_2 - t_1} \tag{5}$$

Formula: $r_{i}^{t_2}$ is the annual decline rate of water intake quota of industry i in year r_2 and r_1 (%);i is the general industry or thermal power industry; $Q_{i}^{t_1}$ and $Q_{i}^{t_2}$ is the water intake quota of industry i in year t_1 and t_2,the calculation formula is shown as follows:

$$W_{j}^{t} = \sum_{t=1}^{n} (X_{i}^{t} \times Q_{i}^{t}) \tag{6}$$

$$W_{m}^{t} = \frac{W_{j}^{t}}{\eta_{i}^{t}} \tag{7}$$

Formula: W_{j}^{t} is the industrial net water demand in year t; X_{i}^{t} is the industrial development index of the i industry sector in year t; W_{m}^{t} is the industrial gross water demand in year i; η_{i}^{t} is the industrial water supply utilization factor of industry sector i in year t.

3.3Forecast of agricultural water demand

(1)Forecast of irrigation water demand

According to the surface irrigation area and groundwater irrigation area, the effective irrigation area and water consumption quota are predicted respectively. The comprehensive irrigation quota for farmland is formulated according to the factors such as net irrigation quota for individual crops, current irrigation water level, crop planting structure and utilization coefficient of irrigation water, etc from 1956 to 2010.

Calculate the net quota of comprehensive irrigation:

$$Q_{j} = \sum_{i=1}^{n} Q_{i} \times A_{i} \tag{8}$$

Formula: Q_j is net quota of irrigation (m^3/mu); Q_i is the net quota of irrigation for crops(m^3/mu); A_i is the percentage of crops planted (%)

According to the comprehensive net irrigation quota, calculate the net water requirement for farmland irrigation:

$$W_n^t = \sum_{j=1}^{3}(S_j^t \times Q_j) \tag{9}$$

Formula: W_n^t is the net water demand for farmland irrigation in year t (10,000 m^3); j is the type of irrigation area; S_j^t is the irrigated area of category j in year t (10,000 mu)

Calculation of irrigation gross water requirement:

$$W_g^t = \sum_{j=1}^{3}(S_j^t \times Q_j / (\eta_t^t - \eta_q^t) - W_n^t / (\eta_t^t - \eta_q^t)) \tag{10}$$

Formula: W_g^t is the gross irrigation water demand in year t (10,000 m^3); η_t^t is the utilization coefficient of irrigation water in year *t*.

(2)Water demand forecast for forest, animal husbandry, fish and livestock

The calculation method of water demand of forest, animal husbandry, fish and livestock is the same as above.

3.4Ecological water demand prediction

Ecological water demand refers to the amount of water needed to maintain ecological environment functions.

(1) eco-environmental water demand inside the river channels

The water demand of ecological environment in river is the amount of water needed to maintain river ecosystem. Ecological water demand in river course is not included in supply and demand analysis.

(2) The ecological environment outside the river needs water

The water demand for the ecological environment outside the river course refers to the water demand for the protection, restoration or construction of the ecological environment in a region and the replenishment of water for the maintenance of lakes, marshes and wetlands with a certain surface. The urban ecological environment water demand is predicted separately according to greening, river and lake water supply and environmental sanitation, and the water demand per unit area is multiplied by the corresponding area forecast. Calculation steps are as follows:

$$W = S_G \times q_G + S_C \times q_C \tag{11}$$

Formula: S_G is the green area(hm^2); q_G is green irrigation quota(m³/ hm^2); S_C is urban area(m^2); q_C is the amount of water needed for sanitation per unit area(m^3/m^2).

4. Consider the regulation and storage of three lakes

4.1Basic characteristics of lake

After the implementation of the Changjiangriver and Huaihe river project, the project relies on caizi lake, chaohu lake and wabu lake as storage areas. During the flood season, the lake stores a large amount of water, thus reducing the river water level. During the dry season, the water stored in the lake flows into the river, thus supplementing the river water source. Storage plays an important role in the allocation of water resources. In this regard, it is necessary to consider the regulation and storage of the three lakes, where the simple storage rate can be used for simple optimization. The refill index refers to the ratio between the available water supply and the effective storage capacity of a lake. When the storage capacity of a lake is small, the inflow can make the effective storage capacity be refill for multiple times, and the refill index can be greater than 1.When the reclosing index is 1, the flows of caizi lake, chaohu lake and wabuhu are respectively 31.25m^3/s, 58.64m^3/s and 7.72m^3/s.

4.2Consider the local water supply for three lakes

According to the above lake regulation and storage rules, water resource regulation allocation was carried out and the lake water storage capacity was V_{t+1} through water balance analysis.

$$V_{t+1} = V_t + I_t + W_t - D_t \tag{12}$$

Formula: $V_{t+1}V_t$ is the water storage capacity of the lake at the beginning and end of the t period; I_t is the natural amount of water; W_t is the amount of water introduced at time t; D_t is the water demand of the region in time period t.

Combined with the above equation, the water storage situation at the end of the period was analyzed and decided.

When $V_{t+1} < V_{t,\ min}$,it is necessary to maintain V_{t+1} not lower than $V_{t,\ min}$ on the basis of ensuring water supply in the water demand area, and the discharge amount during this period is:

$$Q_t = 0 \tag{13}$$

Formula: $V_{t,\ min}$ is the lower limit control reservoir capacity of the lake in the t period; Q_t is abandoned water;

When $V_{t,\ min} < V_{t+1} < V_{t,max}$,It is necessary to ensure that V_{t+1} is not higher than $V_{t,\ max}$, and the discharge during this period is:

$$Q_t = 0 \tag{14}$$

Formula: $V_{t,\ max}$ is the upper storage capacity of the lake in the t period.

When $V_{t+1} \geq V_{t,\ max}$, V_{t+1} is equal to $V_{t,\ max}$, and the discharge during this period is:

$$Q_t = V_{t+1} - V_{t,\ max} \tag{15}$$

5. Water supply and demand balance analysis

The Changjiang river and Huaihe river project covers a wide range which spans the Changjiang river and Huaihe river.The balance of water supply and demand is calculated monthly in a long series, with 1970, 1992 and 1978 as the typical years with 50%, 75% and 95% guarantee rates, respectively.The analysis results of supply and demand balance in 2030 within the planning scope of the project are shown in table 1[11].

This study uses a long series of adjustment calculation methods to calculate the balance of water supply and demand within the scope of the Changjiang river and Huaihe river planning in 2030.It is 18.335 billion m^3, the total water supply is 15.356 billion m^3 and the water shortage is 2.981 billion m^3.According to the spatial distribution situation of water shortage, the water shortage situation is relatively serious in the area west of anhui vortex river, east of anhui vortex river, west of henan vortex river and east of henan vortex river[12].To alleviate the increasingly sharp contradiction between supply and demand of water resources .In the long run, cross-basin water transfer projects must be implemented[13].

Table 1Table of analysis results of supply and demand balance within the planning scope of jianghuai project in 2030(Unit: billion m^3)

Calculating area	Year	Water demand					Water supply					lack of water
		Live	industrial	Agricultural	Ecological	Total	Local surface water	Lakes to regulate	Shallow groundwater	Water reuse	External water	
Changjiang	50%	8.76	14.15	20.35	2.49	45.75	43.91	0	1.84	0	45.75	0

River	75%	8.76	14.15	29.75	2.49	55.14	49.74	0	1.84	0	51.58	3.56
	95%	8.76	14.15	41.7	2.49	67.1	51.4	0	1.84	0	53.24	13.86
	multi-year average	8.76	14.15	26.95	2.49	52.35	47.49	0	1.84	0	49.33	3.02
Along Huaihe and Huainan	50%	5.26	16	13.27	0.88	35.41	29.11	1.21	1.31	0	31.63	3.78
	75%	5.26	16	21.42	0.88	43.56	31.74	1.42	1.31	0	34.46	9.1
	95%	5.26	16	32.45	0.88	54.59	30.18	2.04	1.31	0	33.52	21.07
	multi-year average	5.26	16	17.66	0.88	39.81	32.62	1.26	1.31	0	35.18	4.63
West of guohe, Anhui	50%	6.81	10.88	21.64	0.97	40.29	22.51	4.94	1.08	0	28.53	11.76
	75%	6.81	10.88	30.61	0.97	49.26	26.81	6.03	1.08	0	33.92	15.34
	95%	6.81	10.88	38.78	0.97	57.43	19.25	8.43	1.08	0	28.77	28.66
	multi-year average	6.81	10.88	25.16	0.97	43.81	25.73	5.56	1.08	0	32.37	11.44
East of guohe, Anhui	50%	3.02	6.99	11.66	0.49	22.17	7.13	6.36	0.66	2.54	16.69	5.48
	75%	3.02	6.99	10.53	0.49	21.03	9.11	3.9	0.66	2.54	16.22	4.81
	95%	3.02	6.99	14.86	0.49	25.36	5.58	7.61	0.66	2.54	16.4	8.96
	multi-year average	3.02	6.99	10.63	0.49	21.14	8.16	5.01	0.66	2.54	16.37	4.77
West of guohe, Henan	50%	3.04	2.48	10.52	0.39	16.43	3.61	8.97	0.42	0	12.99	3.44
	75%	3.04	2.48	10.24	0.39	16.15	4.33	8.32	0.42	0	13.07	3.08
	95%	3.04	2.48	13.34	0.39	19.25	2.85	9.29	0.42	0	12.56	6.69
	multi-year average	3.04	2.48	10.16	0.39	16.07	4.6	7.73	0.42	0	12.75	3.32
East of guohe, Henan	50%	1.74	2.57	5.88	0.34	10.52	2.63	5.03	0.29	0	7.94	2.58
	75%	1.74	2.57	5.48	0.34	10.13	3.38	4.3	0.29	0	7.97	2.16
	95%	1.74	2.57	7.17	0.34	11.82	1.69	5.16	0.29	0	7.14	4.68
	multi-year average	1.74	2.57	5.54	0.34	10.19	3.02	4.26	0.29	0	7.56	2.63

6. Conclusions and Suggestions

(1) In 2030, 50%, 75%, 95% and perennial average water shortages are 2.702 billion m^3, 3.806 billion m^3, 8.392 billion m^3 and 2.981 billion m^3 respectively. Even after considering the water supply of the eastern route of the south-to-north water transfer project and increasing the utilization of reclaimed water, the increased water supply still cannot meet the demand of the planning scope and the water shortage situation is still severe.[14].

(2) This study is mainly based on the reasonable allocation of water resources from the Changjiang river to the huaihe river. However, the optimal allocation of water resources on how to increase the guarantee rate of water resources by improving the utilization efficiency of local water and water from the Changjiang river remains to be further studied.

Acknowledgments

This work is funded by the National Key R&D Program of China (2016YFC0402203), the National Natural Science Foundation of China (No. 51779013, 51509009), National Public Research Institutes for Basic R&D Operating Expenses Special Project (No.CKSF2019478, CKSF2017061), Water Conservancy Science and Technology Innovation project of Guang Dong Province(2017-03), Natural Science Foundation of Hubei Province(2017CFB606, 2018CFB655).

References

[1] Jun Huang, Guoming Xu. Preliminary discussion on water resources allocation in the north section of river water diversion project[J].Harnessing the huaihe river,2018.
[2] Anhui water resources and hydropower survey and design institute. Report on comprehensive planning of huaihe river basin in anhui province[R].2015
[3] Nanxiang Cheng, Mingming Jia, Water resource simulation allocation model based on rules. Journal of irrigation and drainage,2005.
[4] Qinxiao Zeng, Qiang Wei. Dynamic simulation analysis method of water supply and demand balance. Urban environment and urban ecology,1996.
[5] Xianwei Cheng. The study of the available water resources concept and calculation method both athome and abroad [J]. Express Water Resources & Hydropower Information, 2007,28(2):7-10.
[6] Linlong Bai. Calculation and analysis on available surface water resources in upper Huaihe river [J]. Yangtze River . 2013,44(17):45-48.
[7] Hanxue Du, Guochun Chang, Qiaosheng Zhang .Preliminary understanding of using underground reservoir to store water[J].Progress in water science,2002,13(5)： 618-622.
[8] Fang Hongyuan, Ma Ruichen. Research method on reservoir operation performance in drought[J]. Northwest Water Resources and Water Engineering, 2003, 14(1): 21－23.
[9] Chunquan Li ,Tienan Niu. The application of regression analysis in urban water demand forecasting. Water supply and drainage.1989(4).
[10] Deqian Xu. Medium and long term urban industrial water use forecast.China water supply and drainage.1992,(1): 38-41.
[11] Chuangjiang Wei. Analysis of ecological water consumption system in water resource allocation[J].Journal of Chinese academy of water resources and hydropower.2006,4(4):282-286.
[12] Bin Chen. Discussion on several problems in the stage of integrated planning and allocation of water resources[J]The people of the pearl river.2006,3: 1
[13] Zongzhi Wang,Siyi Wang, Yintang Wang. Initial Water Right Allocation and Joint Regulation of Water Quantity and Quality in a Basin [M]. Beijing: Science Press, 2011.
[14] Zhizhang Wang. Research on water resources optimization allocation model technology and application[D].Beijing: China institute of water resources and hydropower research,2007.

IOP Conf. Series: Materials Science and Engineering **794** (2020) 012061 doi:10.1088/1757-899X/794/1/012061

A numerical investigation of seismic performance for an RC frame with latticed concrete infill walls

Xinyu Cao[1], Xiaojun Li[1,2*]

[1]College of Architecture and Civil Engineering, Beijing University of Technology, Beijing 100124, China

[2]Institute of Geophysics, China Earthquake Administration, Beijing 100081, China

[*]beerli@vip.sina.com

Abstract: Expansive polystyrene granule cement (EPSC) latticed concrete wall can be used as an infill wall. After the expanded polystyrene granule cement form has been demolished, the part that remains is a latticed concrete infill wall comprising lattice beams and columns whose cross section is circular with a diameter of 0.12 m. The horizontal and vertical rebar are arranged within the beams and columns. The dynamic responses of a full-scale single-story reinforced concrete (RC) frame with latticed concrete infill walls subjected to earthquake-induced ground motion are simulated using the nonlinear finite element ABAQUS software. This study aims to investigate the effects of different types of earthquakes on the structure and the acceleration responses of the structure. The results obtained indicate that the structure has excellent seismic behavior, although the structural responses vary between earthquakes. This research may help in promoting the design and application of RC frames with latticed concrete infill walls.

1. Introduction

The expansive polystyrene granule cement (EPSC) latticed concrete wall is a composite wall made of expanded polystyrene granule cement forms and concrete lattice beams and columns[1]. A latticed concrete infill wall serves as the skeleton of the EPSC latticed concrete wall. EPSC latticed concrete walls can be used in either load-bearing or non-load-bearing walls, and they are suitable for both interior and exterior wall applications. Using these walls provides good economic and environmental benefits.

Latticed concrete walls have been the subject of a number of studies. For example, cyclic lateral loading tests to investigate the seismic behavior of the latticed concrete wall have been conducted by various researchers. Most of these researches have focused on the shear span ratio of the wall. Sun et al.[2] conducted a series of cyclic lateral loading tests on 8 latticed concrete walls with a shear span ratio of 0.77, 0.84, and 1.0. Cai et al.[3] also studied the seismic behaviors of 6 latticed concrete walls with different shear span ratios. Both these studies found that the latticed wall has good deformability. Zhou et al.[4, 5] studied the influence of aspect ratio on seismic performance and found that both common concrete and recycled concrete lattice walls have high load-bearing capacity. Dusicka et al.[6] studied the in-plane lateral performance of full-scale grid walls with two different aspect ratios and two different vertical loading conditions. They found that the vertical loading has an effect on the strength and effective stiffness of the wall. Additionally, a few researchers have performed shaking table tests. Tang et al.[7] conducted a series of shaking table tests to investigate the seismic behavior of a full-scale RC frame with an EPSC latticed concrete infill wall. They found that the designed RC frame with the EPSC latticed concrete infill wall has satisfactory seismic performance.

Experimental investigations provide invaluable opportunities for researchers to study the seismic behaviors of the latticed concrete walls. Since the cost of these tests is high, some researchers have conducted their investigations through numerical simulations. Cao et al.[8] studied the seismic behavior of a twelve layer frame-latticed concrete wall in a high-rise residential building using ANSYS. They found that the door and window openings of the wall were the weakest parts of the structure. Asadi et al.[9] conducted a numerical study to determine the ductility and response modification factors for a latticed concrete wall in ABAQUS. They found that the latticed concrete wall has an acceptable ductility for use in high seismic-risk zones.

A new type of latticed concrete infill wall was used in this study. The diameter of the concrete lattice beams and columns is 120 mm. The application of this new type of latticed concrete infill wall faces great challenges due to the lack of historical earthquake performance, experimental data, and numerical studies. In this research, a numerical investigation of the seismic behavior of an RC frame with latticed concrete infill walls was conducted, and the numerical simulation data are analyzed in this paper.

2. Structure and Finite Element Model

This study considers a 3.0 m high, one-story building composed of an RC frame and latticed concrete infill walls. The dimensions are shown in figure 1[7]. The cross section sizes of the rectangular columns and beams are 0.3 × 0.3 m and 0.25 × 0.3 m, respectively[7]. The thickness of the slab is 0.06 m[7]. The sizes of the door and window openings are 0.9 × 2.1 m (width × height) and 1.5 × 1.8 m (width × height), respectively[7]. The strength grade of the concrete used in the RC frame and slab is C30.

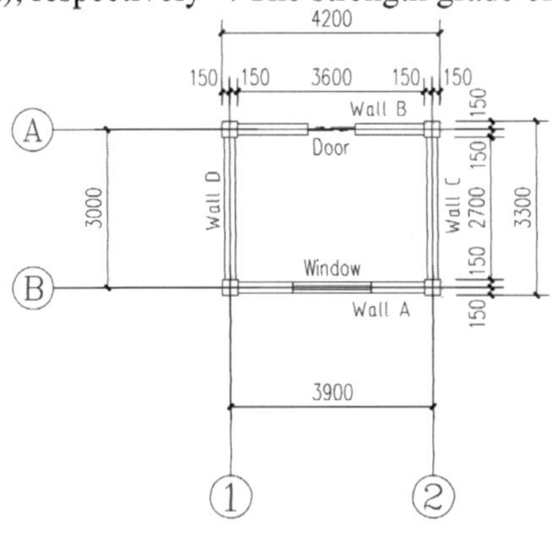

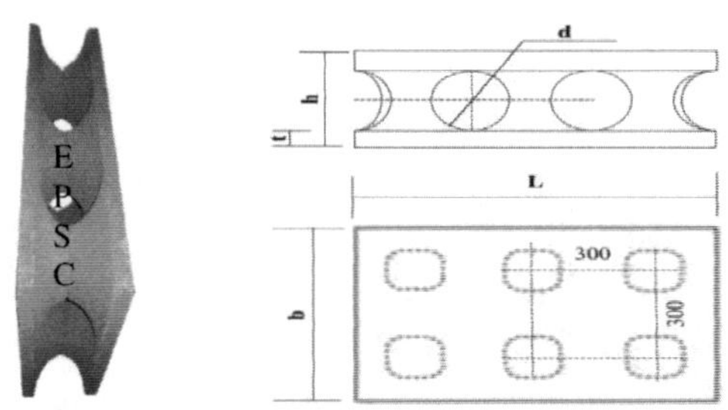

Figure 1.The dimension data of the structure[7] Figure 2.Specification and dimension of EPSC.

The dimensions of the EPSC are 0.9 × 0.6 ×0.21 m(L × b × h). The cross section of the concrete lattice beams and columns is circular with a diameter of 0.12 m (t = 0.045 m), which is smaller than that used by Tang et al. (diameter = 0.16 m)[7]. The spacing of the concrete lattice beams or columns is 0.3 m, as shown in figure 2. The EPSC was removed before conducting the numerical simulations so that only the concrete lattice beams and columns remained. The strength grade of the concrete used in the latticed concrete infill walls is C20.

The plastic damage constitutive model was adopted for concrete in ABAQUS software. The concrete was modeled using solid element C3D8R. The finite element model is shown in figure 3. An additional mass of 1500 kg was applied in the numerical model[7].

Figure 3. Finite element model of RC Frame with Latticed Concrete Infill Wall.

3. Numerical Simulation Resultsand Analysis

The seismic response of a structure will vary greatly with the input seismic waves. To conduct the seismic analysis of a structure, the three elements of the ground motion parameters must be fully considered when selecting typical strong seismic waves from historical records. These elements are amplitude, frequency spectrum, and duration. Two typical strong earthquake records were selected for this study: the El Centro NS record (1940) and the Taft EW record (1952). The acceleration time histories from 0.00-30.00 s of the two input ground motions are shown in figure 4. The peak ground acceleration values of the records were adjusted to 0.3 g. The adjusted seismic waves were then input along the y-direction (shown in figure 5) of the structure. The acceleration measurement location arrangement is shown in figure 5.

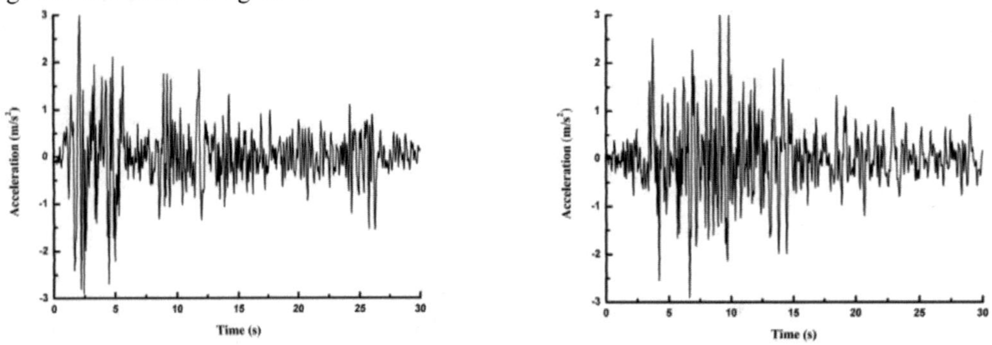

(a) El Centro Record (Adjusted) (b) Taft Record (Adjusted)

Figure 4. Acceleration time histories of the two input ground motions.

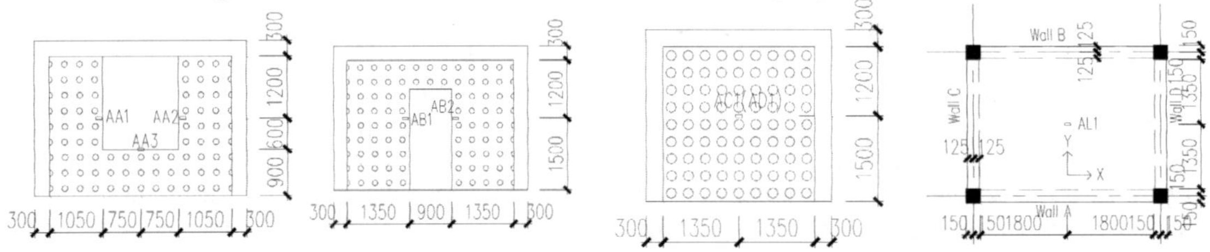

(a) (b) (c) (d)

Figure5.Acceleration measuring point arrangement:(a) Wall A; (b) Wall B; (c) Wall C (D); (d) Slab.

Partof the acceleration response results of the numerical simulations are shown in figure 6 and 7; the figures show the acceleration time histories at the various measurement locations for the El Centro NS record (adjusted) and the Taft EW record (adjusted), respectively. The peak acceleration magnification factors $a_{1\ max}$ and $a_{0\ max}$, where $a_{0\ max}$ is the peak acceleration of the input acceleration record at the bottom of the wall and $a_{1\ max}$ is the peak acceleration at the acceleration measurement location, along the height of the structure are shown in figure 8. In figure 8, the average of the AA1 and AA2 peak acceleration magnification factors is taken as the peak acceleration magnification factor at a height of 1.5 m for wall A; the same is true with AB1 and AB2 for Wall B, while AC1 and AD1 are used for walls C and D, respectively.

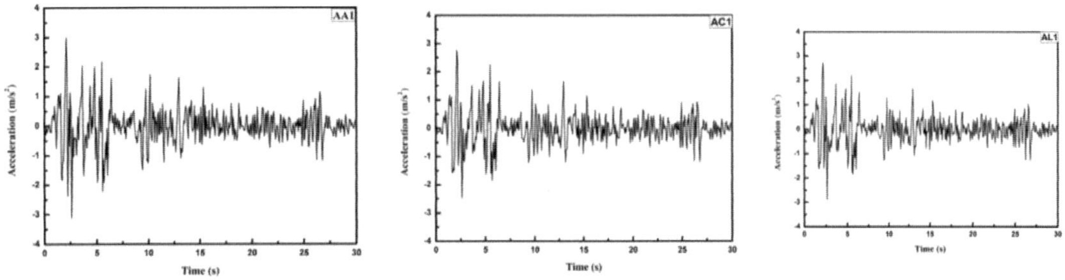

Figure 6.Acceleration time histories for El Centro NS record (Adjusted).

At the bottom of the wall, the peak acceleration for the El Centro NS record (Adjusted) is a little less than 0.3g. The seismic responses are stronger for the walls with openings (walls A and B) than for those without openings (walls C and D).

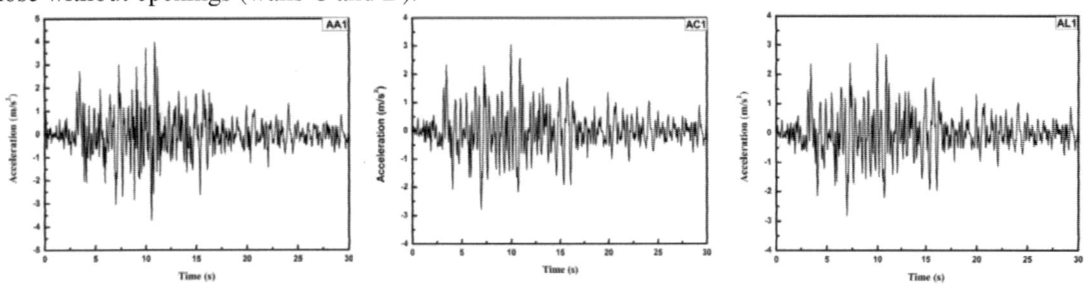

Figure 7.Acceleration time histories for Taft EW record (Adjusted).

Given the Taft EW record (adjusted) as the input motion, the peak acceleration at the bottom of the wall is a little less than 0.3g. The seismic responses are stronger for the walls with openings (A and B) than for those without (C and D), which is similar to the El Centro NS record (Adjusted) results.

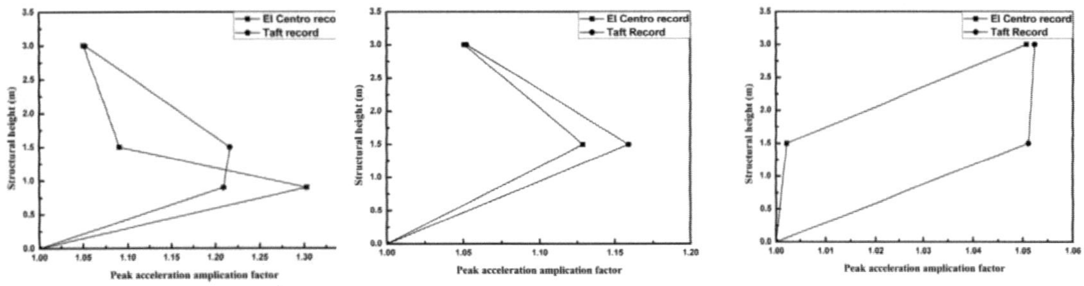

(a) (b) (c)

Figure 8. Peak acceleration magnification factors of the structure: (a) Wall A; (b) Wall B; (c) Wall C (D).

Peak acceleration magnification factors of the structure are shown in figure 8. The peak acceleration magnification factor for walls with openings (walls A and B) is larger at the acceleration measurement locations around the openings (AA1, AA2, AA3, AB1, and AB2), compared with that at the slab (AL1) for both of the adjusted records. The peak acceleration magnification factors are different for different input motions for the same wall. For the walls without openings (C and D), the peak acceleration magnification factors increased along the height of the wall for both the El Centro NS record (adjusted) and the Taft EW record (adjusted). It is worth mentioning that the magnification factor for walls without openings is close to 1.05.

4. Conclusions

A numerical investigation into the seismic performance of an RC frame with latticed concrete infill walls was performed, and the results show that the whole structure has an excellent seismic behavior. Seismic performance can vary among walls. For example, walls without openings have better seismic resistance in comparison with walls with openings. The peak acceleration magnification factors will also vary with the input motion. For walls with openings, strengthening those openings might be considered when it is necessary to improve the seismic resistance of the structure.

Acknowledgments

This work was supported by the National Key Research and Development Program of China (No. 2017YFC1500400) and the National Natural Science Foundation of China (No. 51421005).

References

[1] CECS 173, (2004) Technical specification for dwelling houses with EPSC form latticed concrete wall Tech. Rep., China Planning Press, Beijing, China.

[2] Sun, J., Qian, J., Fang, E., et al. (2002) Experimental study on seismic behavior of concrete grill wall with small shear span ratio Journal of Building Structures, vol. 23, no. 2, pp. 19-26.

[3] Cai, X., Shi, C., Chen, J., et al. (2010) Experimental and analysis of lateral resistant performance of expansive polystyrene granule cement latticed concrete walls Chinese Journal of Solid Mechanics, vol. 32, no. S, pp. 361-364.

[4] Zhou, J., Wang, A., Meng, X., et al. (2012) Influence of aspects ratios on seismic behavior of recycled concrete lattice walls Journal of Shenyang Jianzhu University (Natural Science), vol. 28, no. 2, pp. 241–248.

[5] Zhou, J., Wang, P., Meng, X., et al. (2011) Experimental study on seismic behavior of concrete grillage walls with different shear span ratio Journal of Earthquake Engineering and Engineering Vibration, vol. 31, no. 3, pp. 127–133.

[6] Dusicka, P., Kay, T. (2011) In-Plane Lateral Cyclic Behavior of Insulated Concrete Form Grid Walls Journal of Structural Engineering, vol. 137, no. 10, pp. 1075-1084.

[7] Tang, B., Li, X., Chen, S., et al. (2017) Shaking Table Test of a RC Frame with EPSC Latticed Concrete Infill Wall Shock and Vibration, vol. 2017, Article ID 7163560, 18 pages.

[8] Cao, Q., Xie, R., Shen, Y. (2017) Seismic response analysis of high-rise EPS lattice type concrete Wall Earthquake Resistant Engineering and Retrofitting, vol. 39, no. 5, pp. 18-33.

[9] Asadi, P., Madandoust, R., Zahrai, S. M. (2017) "Response modification factor due to ductility of screen-grid ICF wall system in high seismic risk zones" KSCE Journal of Civil Engineering, vol. 21, no. 1, pp. 258-264.

The statistical distribution of curvatures of natural rivers meanders

Jun Feng[1], Kai Li [2], Liang Ma[3*], Yu Tian[1]

[1]China Insititute of Water Resources and Hydropower Research, Beijing 100044, China

[2]Changjiang River Scientific Research Institute of Changjiang Water Resources Commission, Wuhan 430015, China

[3]China Renewable Energy Engineering Institute, Beijing 100120, China

[*]Corresponding author's e-mail: 252944344@qq.com

Abstract: The rivers meanders are the most common form of natural rivers. In this paper, the shapes of meanders may be characterized in terms of the statistical distribution of curvatures measured along their contours. Deviation degree and span are proposed, in which deviation degree represents the closure degree of the bend and span represents the roundness degree or even degree of curvature distribution of the bend. Using these two indicators, we can effectively distinguish the state of the development of the river bend and thus determine different river regulation schemes. Finally, the method of deducing the reliable curvature of the bend from the satellite pictures of the river is discussed.

1.Introduction

The rivers meanders are the most common form of natural rivers. Its geometric shape of twists and turns and the dynamic characteristics of creeping creep attract the attention of scholars at home and abroad. Many papers have carried out in-depth research on the geometry characteristics of rivers meanders [1-4], self-organization process [5-6] and rivers meanders evolution simulation [7-8], and obtained a series of important research results.

However, in the classification and description of the fractal and developmental state of rivers meanders, they are all qualitative methods, such as the early stage of development and the end stage, which cannot be analyzed more finely. In order to capture the statistical characteristics of the natural development process and evolution of rivers meanders, it is necessary to establish a detailed method for describing the shape of curvatures. This method should be quantified and more closely related to the evolution process. In this paper, we present a key microscopic variable, curvature, using the curvatures distribution of a two-dimensional image of the rivers meanders a tool to describe its shape characteristics.

This paper will detail this method and apply it to 30 natural or artificially rehabilitated river meanders. The first part begins with a quantitative study of two-dimensional shape features, summarizing our curvature-based approach, and details the method of extracting local curvature from river meander satellite photos. The second part is a quantitative description of the application of this method to the development of river meanders, and proposes the possibility of using this method to guide river regulation.

2.Quantification of the two-dimensional shape of river meanders

The shape of the river meanders has long been a concern in the fields of geomorphology, hydraulics and riverbed evolution. There are two basic methods that are often used now. The first method is described using the radius of curvature and the central angle of the river meanders (Fig. 1). In this description, R is the radius of curvatures of the river meander top. There are two problems here: First, the position of the river meanders top is difficult to determine. Second, because the two-dimensional shape function of rivers meanders is unknown, the radius of curvature of the apex is difficult to determine, so this method has certain subjectivity. The second method is to use the degree of curvature to judge [9] (Fig. 2). The degree of curvature C=L/B is more effective in judging the development period of the river meander, but it is described in the same development period of the 2D shape does not apply.

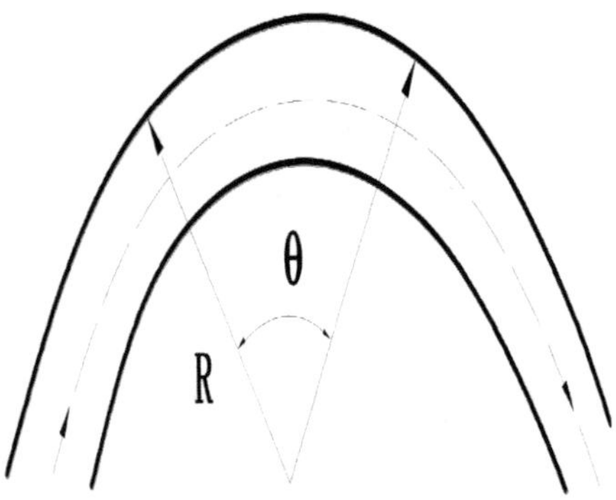

Fig.1 Definition of shape parameters in method one

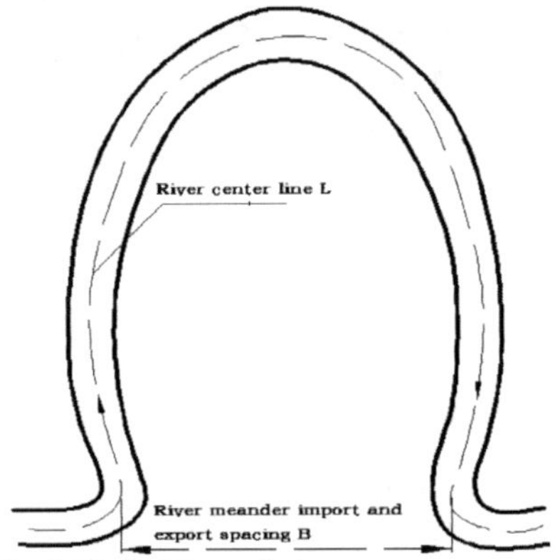

Fig.2 Definition of shape parameters in method two

The plane evolution of a river is essentially the constant change of the curvature of each point on the river. The curvature is the rotation rate of the tangential direction of the point on the curve to the length of the arc, which can be given by $K = d\alpha/ds$. A more vivid understanding is that the curvature is the reciprocal of the radius of the inscribed circle at a point on the curve. Most of the time it is customary to define a contour as convex when K>0 and a concave when K<0.

To describe the shape of the river meander the probability density function of the curvature distribution is $\rho(K)$. $\rho(K)dK$ represents the probability that the curvature is between $K \sim K+dK$. In practice, to distinguish between different distributions, we use the cumulative distribution function of curvature.

$$f(K) = \int_0^K \rho(K')dK'$$

Where $f(K)$ is the ratio of the length of the curve whose curvature is less than K to the total length. When K increases from $0 \sim \infty$, $f(K)$ from 0 to 1. When K is the minimum value, $f(K)$ is 0, when K is the maximum value, $f(K)$ is 1, and when K is the median, $f(K)$ is 1/2. Unlike $\rho(K)$, $f(K)$ does not need to know the function expression of the curve or measure the curvature of all points to derive, because the curve can be discretized and the curvature of the discrete points can be obtained, and they are arranged in order from small to large and record less than this curvature. Arrange them from small to large and record the ratio of the length of the arc segment less than this curvature to the total length. In this paper, we use $f(K)$ and $<K>/K$ as the shape function of the river meanders. The $<K>/K$-average curvature is dimensionless compared to the curvature of a point, so that the shapes of different sizes of river meanders can be compared. The reason for using $<K>/K$ is to make the data point fall in a larger space on the right side of the horizontal coordinate "1", and the image is clearer. To build the formula, $f(K)$ for some simple shape examples will be given below, and lead to two important indicators of the $f(K)$ graph. Prior to this, we will discuss the method of extracting local curvature by relying on the images of the river meander satellite.

There are two common methods for finding the curvature of each point on the curve. One is to use the points on the curve for interpolation, to fit a similar curve, and then calculate the curvature of the point on the interpolation curve instead of the curvature of the original curve point. The advantage of this type of method is that the calculated curvature is relatively accurate and the error is smaller. However, it is difficult to determine the distribution of the interpolation points. The distribution of the points may cause the cumulative distribution function of the curvature to be greatly deformed. Generally, it cannot be used in areas where the curvature changes more severely, especially in areas where the unevenness changes, otherwise there will be a larger Error and large amount of calculation. The second method is to use the known curve to fit the original curve by adjusting the curve parameters to adjust the shape of the curve. The principle is to match as many points as possible on the curve to be sought. This type of method has a small amount of calculation, and the error is visually visible on the graph. It is possible to artificially select a more accurate point, so that the overall error is small. However, the local accuracy is poor, and the adjustment of the fitting curve is subjective. In this paper, two methods are used in combination, and the curve to be obtained is fitted using a known curve. Interpolation fitting is preferred in places where the local error is large.

Systematic errors can be minimized by using cubic spline interpolation to fit the original curve where local errors are large. In order to convert the fitted Cartesian coordinate function to polar coordinates, make sure $y = y(x)$ is the single-valued function. Therefore, should choose the right origin of the Cartesian coordinate system and coincide with the origin of the polar coordinates. After $y = y(x)$ is obtained by cubic spline interpolation, $y = y(x)$ can be converted to $r = r(\theta)$ directly by $x = r\cos\theta$ and $y = r\sin\theta$. Among them $r = \sqrt{x^2 + y^2}$, $\theta = \tan^{-1}(y/x)$. When converted to polar coordinates, you can find the curvature of the point on the curve by

$$K = \frac{r^2 + 2r_\theta^2 - rr_{\theta\theta}}{\left(r^2 + r_\theta^2\right)^{3/2}}$$

Where $r_\theta = dr/d\theta$, $r = d\,r/d\,\theta$ [10].

The method of fitting a shape of a river meander using a known function in a smoother part. In this paper we choose the elliptic function $x^2/a^2 + y^2/b^2 = 1$, where a is the long axis and b is the short axis. By adjusting the long and short axes to change the shape of the ellipse, the known river meander is fitted to match as many points as possible on the contour of the river meander. Then you can find the curvature of any point $P(x, y)$ by

$$K = \frac{ab^4}{\left[\left(a^2 - b^2\right)y^2 + b^4\right]^{3/2}}$$

According to the specific shape of the contour of the river meander, the curvature of each point on the contour curve can be obtained by using these two methods. The following two methods are used to analyze the two practical problems to further explain the curvature distribution method.

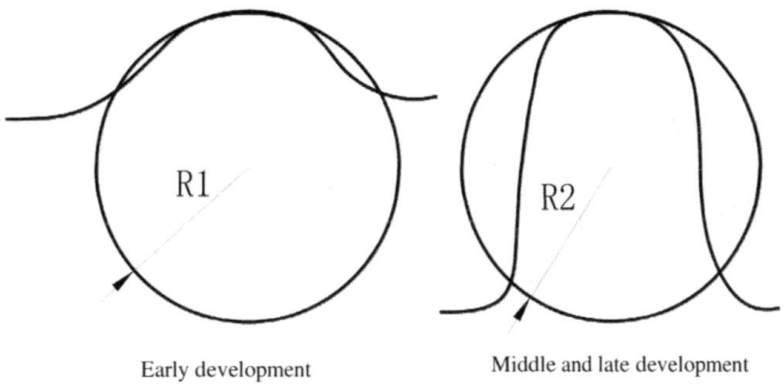

Fig.3 Two meanders in different evolution stages, but they have the same curvature in the apex.

Figure 3 is a two-dimensional plane figure of the two rivers meanders on the map scaled down. Currently, it is obviously not advisable to describe the shape of the river meanders according to the above-mentioned method with the curvature radius of the apex of the river meanders as the characteristic value. Because the two river meanders have the same curvature radius R_1=R_2 of the river meanders, but at different stages of development. The curvature distribution method is used to obtain the curvature distribution of two rivers, as shown in Fig. 4.

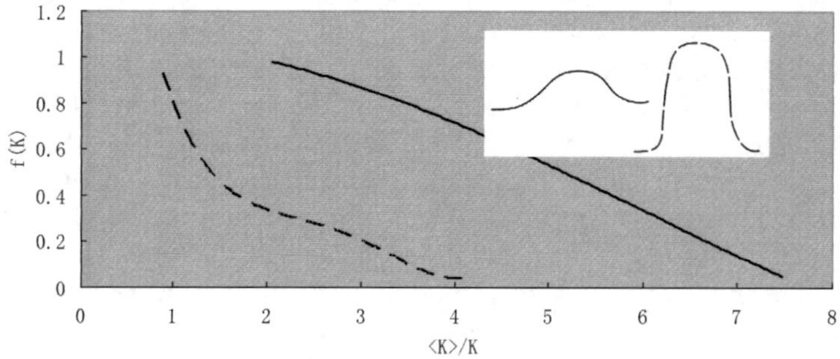

Fig.4 The statistical distribution of curvatures of two meanders in different evolution stages.

Here, an index deviation γ is defined, which is the degree to which the x coordinate value corresponding to the graph deviates from the scale "1" when a = 1/2. This index represents the degree of closure of the river meanders, and the larger the value of the deviation γ is the lower the closure of this river meanders. In this case, it can be seen that the development of the river meanders 1(black solid line) is low and the degree of closure is low, so the deviation degree γ=4.8 is larger. When the river meander is completely closed to form the oxbow lake, its deviation γ reaches a minimum value, which is close to 0. That is to say, as the river meander continues to develop, its deviation γ is continuously reduced. When the river bends straight, the deviation γ suddenly increases from the minimum value, forming a new round of river meander evolution.

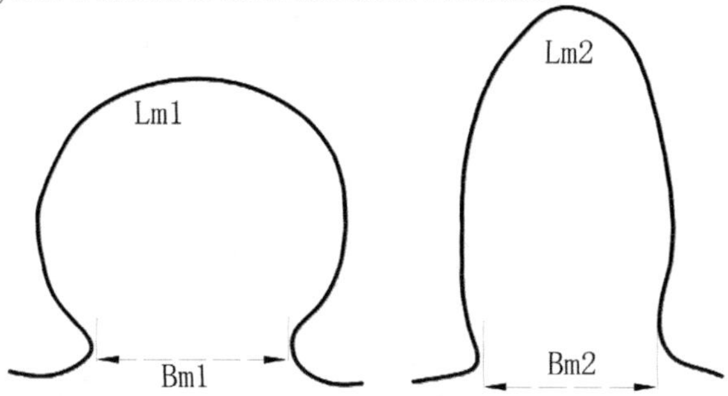

Fig.5 Two meanders which in same evolution stage have the same C=L/B, and their shapes are completely different.

Figure 5 is a two-dimensional plan image of two rivers on the map scaled down according to different scales. It can be clearly seen that the two rivers and meanders are in the middle and late stages of development, with the same curvature, $C_1=L_{m1}/B_{m1}=C_2=L_{m2}/B_{m2}$, but the shape of the river meander is very different, and the shape of the first river meander is closer to a circle. The curvature distribution method is used to obtain the curvature distribution of two river meanders, as shown in Fig. 6.

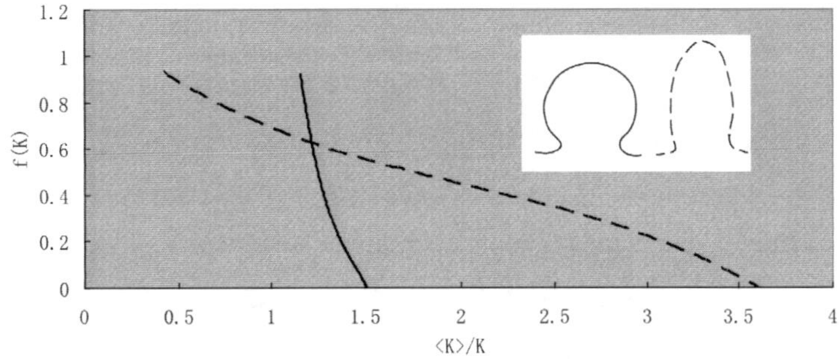

Fig.6 The statistical distribution of curvatures of two meanders which all in the later stage and have completely different shapes.

Here this paper define another indicator span *β*, which is the span of the graph on the abscissa x. This indicator represents the roundness of the river meanders. The smaller the span β is, the more uniform the curvature distribution is, and the shape of the river meander is closer to a part of a circle or a circle. It can be seen from Figure 7 that although the two rivers are in the middle and late stages of development, their spans are significantly different, and the span of the circular river meander (black solid line) is much smaller than that closer to the ellipse river meander (black dotted line).

3.Two-dimensional graphic curvature distribution of natural rivers

In order to test whether the method of curvature distribution can find some rules in natural rivers to distinguish different river meander shapes, this paper selects ten river meanders at different scales in the early, middle and late stages of development as statistical sample space. The Google Earth satellite map is combined with AutoCAD to extract the geometric parameters of the selected river meander. In order to be able to be used as a standard and more clear, we should try to choose a typical river meander, avoiding the controversial transition period, but one thing to be clear is that river meander development is a continuous process, there is no obvious node.

First, select ten river meanders in the early and upper-middle-lower reaches of the Yangtze River. For the sake of clarity, take the first five to draw their curvature distribution (Figure 7), and the other river meander's deviation and span list (table 1).

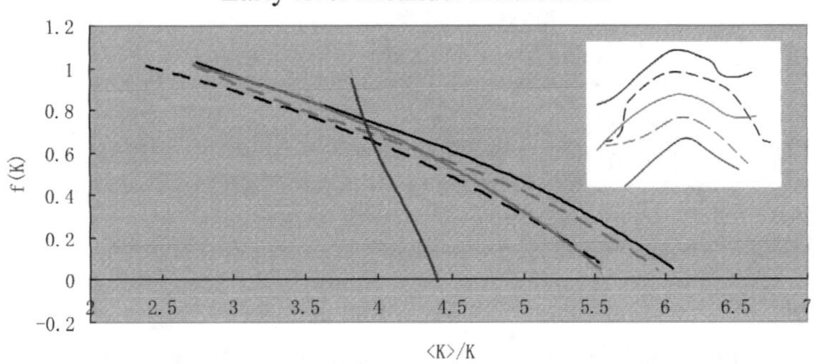

Fig.7 The statistical distribution of evolution of meanders in initial stage

Figure 7 shows the curvature distribution of five rivers at the beginning of development. The deviation γ is in the range of 2.5~3.5, which is far from the scale "1", indicating that the degree of closure is relatively low. The spans of the first four rivers are similar, with a length of 3 units. The span of the blue river meander is smaller, indicating that the curvature of the blue river meander is more evenly distributed. The shape of the river meander is closer to a circle or arc.

In the upper, middle and lower reaches of the Yangtze River, ten river meanders of the Yangtze

River are selected respectively. For the sake of clarity, use the first five to plot its curvature distribution (Figure 8), and the other river meanders are shown in the deviation and span list (Table 1).

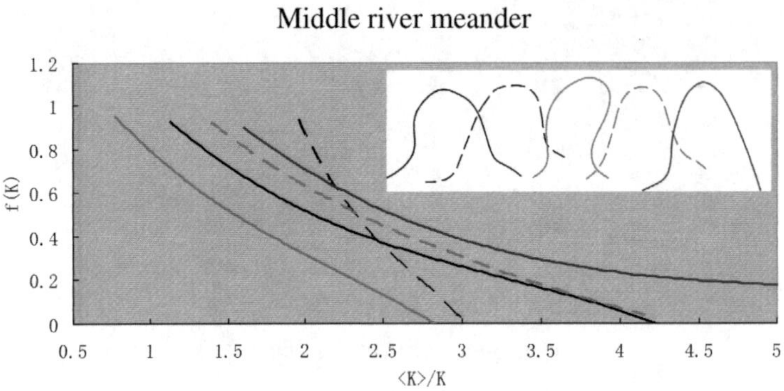

Fig.8 The statistical distribution of evolution of meanders in mid-term stage

Figure 8 shows the curvature distribution of five rivers in the middle of development, which the deviation γ is in the range of 0.5~1.5. The deviation γ of the river meander relative to the early development is closer to the scale "1", indicating that the degree of closure in the mid-development river meander is much higher than that in the early development. However, the span β of the river meander in the mid-development period is irregular and not comparable to the early river meander. It is indicated that the span β is an index for judging roundness, and it can be judged which river meander curvature distribution is more uniform, and it is impossible to judge the development of the river meander.

Finally, ten river meanders in the middle and lower reaches of the Yangtze River were selected. For the clarity of the map, the first five strips were taken to plot their curvature distribution (Fig. 9), and the other river meanders were shown in the deviation and span list (Table1). Figure 9 shows the curvature distribution in the late development of the river meander. The deviation γ is between 0 and 0.5, which indicates that it is closer to the scale "1" than the former two groups of river meanders, and the closure degree is much higher in the late development river meander. It is also easy to see from the two-dimensional shape of the river meander.

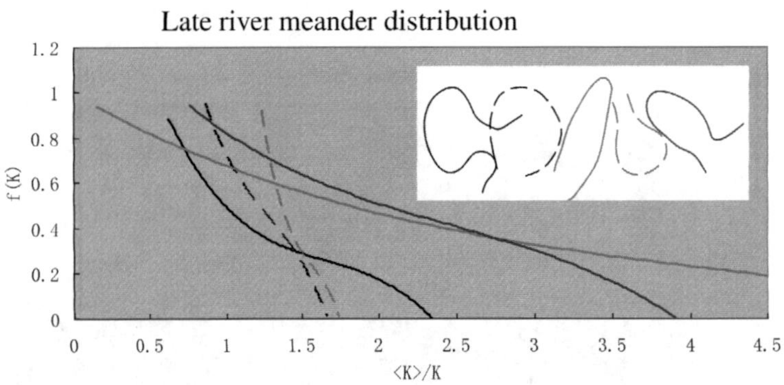

Fig.9 The statistical distribution of evolution of meanders in later stage

Thirty different river meanders are listed in Table 1, all located on the Yangtze River and its tributaries. The statistics of these thirty rivers further indicate that the developmental stage of the river meander can be distinguished from the deviation γ. The early stage of development, the deviation of γ is located between (2.5, 3.5) the deviation of γ in the middle of development is located between (0.5, 1.0), and the deviation of γ in the late development is located between (0, 0.5). However, the data of span β has no obvious law, but it can be seen that the span β of the river meander which is close to a circle or arc is small, and the span β is a sign of whether the curvature distribution is uniform.

Therefore, the span β can judge the shape of the river meander at the same development stage and distinguish it.

Table1 The evolution stage of the meanders which in Yangtze river or its tributary.

River name	River meander location	Developmental stage	Deviate degree	Span	River name	River meander location	Developmental stage	Deviate degree	Span
Qingjiang ririver	Hejiaping	Early stage	3.49	3.25	Yangtze	Xingoudukou	Middle	1.53	3.15
Yangtze ri	jianglingxian	Early stage	3.44	3.10	Xiangjiang	Qixian beach	Middle	1.15	2.51
Yangtze	Luxi town	Early stage	3.57	2.65	Zishui	Baimasi town	Middle	0.67	2.21
Yangtze	Xintan town	Early stage	3.62	3.22	Mangdao	Jiaxi	Middle	1.34	2.12
Jinshui rive	Huanglong tan	Early stage	3.12	0.60	Qinggejiang	Xiaojiawan	Middle	0.85	1.24
Yangtze	Yanji town	Early stage	2.95	2.55	Yangtze	sanjiatan	Later stage	0.02	1.70
Yangtze	Sanhua town	Early stage	3.05	2.75	Yangtze	Tianetu	Oxbow Lake	0.21	0.58
Yangtze	Jingan town	Early stage	3.45	3.07	Yangtze	Jicheng village	Oxbow Lake	0.47	4.51
Yangtze	Zhenjiang city	Early stage	2.50	1.20	Yangtze	Laojianghe	Oxbow Lake	0.38	0.25
Jiajiang	Hongqiaotown	Early stage	3.43	1.05	Yangtze	Baotazhou village	Later	0.25	0.31
Qingjiang	Gaoba city	Middle stage	1.23	3.10	Huangbai	Xujiawo	Later	1.06	2.95
Yangtze	Jingzhou city	Middle	1.33	0.92	Mangdao	Xujiayuan	Later	0.21	0.26
Yangtze	Xiaohekou town	Middle	0.54	2.05	Xiushui	Wujiawang	Later	0.35	3.45
Yangtze	Rongchengtown	Middle	1.35	2.65	Qingshang`	bajiacun	Later	0.41	0.78
Yangtze	Liangjiamen	Middle	1.52	3.25	Yangtze	Wainiutan	Later	0.97	3.05

Many of the rivers have been analyzed using the curvature distribution method above, but why should we choose the curvature distribution for the river meander fractal? How is the curvature distribution of the river meander and the river meander development process linked together? It is only when the curvature distribution can reflect the development mechanism of the river meander that the curvature distribution is appropriate to describe the shape of the river meander. The meaning of curvature in mathematics is the rate of rotation of the tangential direction of a point on the curve to the length of the arc. If it is used in the development of the river meander, assuming that the flow velocity of the river is v, the curvature can be described as the rate at which the angle of the river meander changes. Because if the flow rate is assumed, the time and the arc length can be linked together ($t=L/v$), and the rotation rate of the tangential direction angle to the arc length can be converted into the tangential direction angle versus time, that is, the tangent the rate of change of the direction angle. At the same flow rate and flow rate, the greater the curvature, the greater the rate of change of the tangential direction angle, indicating that the greater the acceleration perpendicular to the flow direction, so the greater the pressure perpendicular to the river bank received by the bank. This explains why the straightening of the curve occurs where the curvature is greatest.

River meander curvature distribution and river management are also closely linked. Above we have linked the curvature to the force per unit volume of water. If we know the flow and the pressure on the river bank, we will naturally think that the actual pressure on the river bank is equal to the pressure on the river bank. At this time, this curvature is the ideal curvature of the river meander. In fact, we have some empirical understandings in this respect. For example, the lower reaches of the Yellow River proposed to rectify the river meander with continuous curves. The purpose is to make every point on the river meander at the best curvature, so that the river meander is at in this relatively balanced state, research in this area will be carried out in the future.

IOP Conf. Series: Materials Science and Engineering **794** (2020) 012062 doi:10.1088/1757-899X/794/1/012062

4.Discussion and conclusions

In this paper, thirty river meanders in different development stages of the Yangtze River and its tributaries were analyzed using a method describing the curvature distribution of the river. It is found that for the complex alluvial river meander, the traditional method cannot distinguish its shape and determine its development state, but the method of curvature distribution can find obvious laws. As the degree of development increases, the degree of deviation γ decreases continuously, approaching zero. The γ change interval of the river meander at the beginning of development is [2.5, 3.5], the deviation range of γ deviation in the mid-development period is [0.5, 1.0], and the deviation range of γ in the late development is [0, 0.5]. In the same developmental period, the contours of the river meander are still diverse, some like ellipse, and some similar to a circle.

Because the time scale of river meander evolution is large, this paper only analyzes the different river meander patterns at the same time, but cannot describe the evolution of a river meander, or explain the change process of curvature distribution. In the subsequent research, many experiments will be carried out, and the longitudinal analysis of multiple river meanders will be carried out at the same time. Further study the complexity and variability of the river meander geometry in the evolution of the river meander.

Acknowledgments

This work is funded by the National Key R&D Program of China (2016YFC0402203), the National Natural Science Foundation of China (No. 51779013, 51509009), National Public Research Institutes for Basic R&D Operating Expenses Special Project (No.CKSF2019478, CKSF2017061), Water Conservancy Science and Technology Innovation project of Guang Dong Province(2017-03), Natural Science Foundation of Hubei Province(2017CFB606, 2018CFB655). Special thanks are given to the anonymous reviewers and editors for their constructive comments.

References

[1] CONSTANTINE J A，DUNE T. Meander cutoff and the controls on the production of oxbow lakes ［J］.Geology，2008，36(1):23-26.
[2] SWAMEE P K, PAEKASH B, THOMAS JV, et al. Changes in channel pattern of river Ganga between Mustafabad and Rajmahal, Gangetic Plains since 18th century [J]. International Journal of Sediment Research, 2003, 18(3):219-231.
[3] WILLI H H. Fargue, founder of experimental river engineering [J]. Journal of Hydraulic Research, 2003, 41(3):227-233.
[4] Nan-Shang, Huang Zheng-wen, et al. Meanders of the Jialing River in China: Morphology and formation [J]. Chinese Science Bulletin, 2007, 52(22):2671-2682.(in Chinese)
[5] STOLUM H H. River meandering as a self-organization process [J]. science, 1996, 271(22): 1701-1712.
[6] Wang Fu-quan, Cao Shu-you, Ding Jing. Self-organization and stability of river bends [J]. Advance in Water Science, 2001, 12(1): 7-16. (in Chinese)
[7] NICOLL T J, HICHIN E J. Platform geometry and channel migration of confined meandering rivers on the Canadian prairies [J]. Geomorphology, 2010, 116: 37-47.
[8] GUNERALP I, Rhoads B L. Empirical analysis of the platform curvature-migration relation of meandering rivers [J]. Water Resources Research, 2009, 45, W09424: 1-15.
[9] Constantine J. A., Dune T. Meander cutoff and the controls on the production of meandering river [J]. water Resources Research, 2009, 45.
[10] E. W. Weisstein, The CRC concise Encyclopedia of Mathematics (CRC Press, New York, 1992).

IOP Conf. Series: Materials Science and Engineering **794** (2020) 012063 doi:10.1088/1757-899X/794/1/012063

Research progress and development direction of drability assessment methods for CFRP reinforced concrete structural members

LiuYan Nian[1], WangLei[2], Zhao XiaoYong[3*]

[1]Sichuan Institute of building Reserch Co., Ltd., Chengdu 610000, China

[2]School of Civil Engineering, Southwest Jiaotong University, Chengdu 610000, China

[3*]Chengdu Municipal Waterworks Co., Ltd., Chengdu Environment Investment Group Co., Ltd., Chengdu 610000, China

[*]Corresponding author's e-mail:Mrzhao@vip.126.com

Abstract. The construction industry is an important part of China's national economic development, in recent years with the economic development is to increase the investment in infrastructure. As time goes on, there be the phenomenon of building use function degradation, in addition, in many reinforced concrete structure, and part of the building is in a bad environment conditions, caused by a lack of concrete durability, reduce safety performance and the structure performance degradation, causing a large number of buildings can not reach the expected service life or the desired function. With the rapid development of reinforcement technology, the theoretical basis and practical engineering experience of concrete reinforcement technology in China have been gradually accumulated and improved, and a lot of achievements have been made. Based on the theoretical research and engineering practice of concrete structure reinforcement, the state and local governments have promulgated a series of codes and standards for the reinforcement and renovation of existing buildings, which provide a good technical guarantee for the reinforcement and renovation of concrete structures. Concrete structural members reinforced by adhesive still need to be affected by external environment in the use process, such as chemical corrosion environment, chlorine corrosion environment, freeze-thaw environment damage and long-term load. Therefore, the test and evaluation of the durability of the members of the load-bearing structure reinforced by adhesive plays an important role in the reinforcement of concrete structure system and the service process after the reinforcement of members. How to evaluate the durability of reinforced concrete structures is of great significance to the durability monitoring of reinforced concrete structures.

1. Introduction

Since the reinforced concrete structure construction methods are widely used in building structure reinforcement and reconstruction project, the reinforcement method has played an important role in the field of structure strengthening, at the same time based on the theory of reinforced concrete structure and related research and engineering practice has gained popularity, paste CFRP reinforced concrete member in reinforcement applications accounted for a larger proportion. Along with the national standard "reinforced concrete structure design code" GB50367-2013 and the construction quality acceptance of building structure reinforcement engineering GB50550-2010, the engineering structure reinforcement material technical specification for safety appraisal GB50728-2011 promulgation and implementation of system specification for concrete structures strengthening engineering design and construction quality inspection and reinforcement material safety appraisal of each link. After the reinforcement of the system and components, the strength, stiffness and integrity of the structural system have been improved, but it is still necessary to consider that the reinforcement of concrete components is still affected by wet and dry conditions, temperature conditions, chemical corrosion environment and long-term load. With the increasing popularity of reinforcement construction, the durability of the structure after reinforcement has become the focus of attention in the process of engineering application. Therefore, it is necessary to consider whether the performance of the structure after reinforcement is degraded and whether the subsequent service life can reach the expected target in the process of putting it into use.

Adhesive reinforcement in concrete structures strengthening methods due to the construction of the paste CFRP concrete member in convenient operation, the craft is mature, CFRP performance is good, is widely used in the reinforcement engineering, paste CFRP concrete component in the structure reinforcement engineering, need to consider include: the factors which influence the durability of the temperature and humidity effect, load impact resistance, resistance to chemical corrosion and ultraviolet radiation, etc. In this paper, the factors influencing the durability of CFRP reinforced concrete members are summarized based on the above environment.

2. influencing factors of durability

2.1 influence of temperature on CFRP materials and reinforced concrete structures

In order to verify whether the performance of CFRP sheet material under temperature cycling is affected, Hutchinson et al. conducted mechanical performance tests on CFRP sheet material and adhesive under temperature cycling at -20~50°C. Every 24h is a cycle, in which temperature is increased for 5h, temperature is decreased for 17h, and temperature and humidity at each end are kept for 1h.[1]The test results show that the elastic modulus, tensile strength and ultimate strain of CFRP sheet under the action of temperature cycling do not decrease, but increase correspondingly.

Houssam, Toutanji, Balaguru et al. conducted representative tests on the mechanical properties of reinforced concrete columns coated with CFRP sheets under freezing-thawing cycles.[2]-[5]The following conclusions are obtained: under freezing-thawing cycle, the strength of concrete columns reinforced with CFRP decreases. The freezing-thawing cycle has no effect on the stiffness of CFRP reinforced columns, and the ductility decreases obviously. Compared with CFRP strengthened columns under room temperature and wet and dry cycling conditions, brittle failure is more likely to occur under freezing-thawing cycling conditions.

Wang fengxia et al. studied the properties of CFRP reinforced concrete under different environmental effects.[6]The results show that the freezing-thawing cycle has less effect on the compressive strength of CFRP reinforced concrete rectangular columns, and more on the tensile bond strength of reinforced concrete cubes.The theoretical analysis of the fiber of CFRP reinforced concrete column at the middle corner of the column height is obtained.

The strength of the carbon fiber sheet (CFS) under different temperature conditions is not affected, aci committee, Wang Jing, xiao-dong zhang for CFRP under the condition of environment related durability test, use black board compared to 63 + 3 °C temperature spray circulation, 18 min / 2 h using daylight natural aging tester for 2000 and 10000 h CFRP and rapid exposure test (equivalent to 10 a and 50 a natural exposure time) exposed to the atmosphere environment for a long time to test the tensile strength of carbon fiber cloth, the experimental results showed that: not found strength reduce phenomenon;[10]The tensile test of CFRP was carried out under the conditions of -54°C, 23°C and 82°C to test the dependence of CFRP on temperature. The test results showed that no strength reduction was found under the set temperature.

In recent years, the research on the freezing-thawing durability of concrete structures strengthened with FRP has accumulated some achievements at home and abroad. The performance characteristics are summarized as follows: 1. Freezing-thawing circulation has a negative effect on the performance of the bond between FRP and concrete, and the degree of influence is related to the strength and frost resistance of concrete; 2. 2. The freezing-thawing cycle has a significant adverse effect on the bonding section performance of ordinary concrete members used for FRP reinforcement; 3. For concrete structures strengthened with FRP materials with high frost resistance, the weakening effect of freezing-thawing on bond section strength is not obvious, but it can be seen from the change of failure mode that freezing-thawing has a certain adverse effect on bond interface. Furthermore, the mechanism of resistance attenuation of the whole structure was further discussed, and relevant theoretical models were established. On the premise of having a large amount of test data, the corresponding reduction coefficient of anti-freeze and thawing durability design is proposed to provide theoretical basis for the establishment of relevant durability specifications.

2.2 influence of humidity on CFRP materials and reinforced concrete structures

CFRP sheet and binder were exposed to a humid environment with a relative temperature of 30°C and a humidity of 100% to verify the effect of humidity conditions on their strength.[1]The test results show that the tensile strength, elastic modulus of tensile strength and ultimate elongation of CFRP sheets do not decrease under the wet environment.

Houssam and Toutanji et al. conducted tests on CFRP sheets under wet and dry cycling conditions. The cycling test conditions (adding 35g salt in 1L of water to simulate seawater) lasted for 4h and the dry conditions (average temperature 35°C and humidity 90%) lasted for 2h, lasting a total of 75d.[9]The experimental results show that the load carrying capacity of FRP strengthened beams under dry/wet cycling conditions is significantly higher than that of beams not strengthened with FRP at room temperature, but the load carrying capacity of FRP strengthened beams at room temperature is decreased.

Toutanji and Houssam et al. conducted experimental studies on concrete columns reinforced with CFRP, and preserved 36 specimens in a humid environment (temperature 25°C, relative humidity 90%) for 58d.[4]After that, the dry and wet cycle was carried out for 18 months, and the cycle times were 300 times. Each cycle under wet conditions (adding 35g salt to 1L of water to simulate seawater) lasted for 4h, and the dry conditions (temperature 35°C, relative humidity 90%) lasted for 2h. The cycle lasted for 75d.The other 18 times were placed at room temperature for 75d as a comparison specimen, and the pulse rate and mass of time were measured once every 30 times through the dry-wet cycle. The results show that the compressive strength of CFRP reinforced columns has no significant effect on the epoxy resin used as the matrix material. The ductility of concrete columns strengthened with CFRP does not decrease under wet and dry cycling conditions.

2.3 influence of seawater on CFRP materials and reinforced concrete structures

The water spray test was carried out at the test temperature of 35±2°C by means of a nozzle brine spray device to test the seawater resistance of the CFRP fabric. The test time was 1a. The test results showed that there was no trend of decrease in tensile strength.[8]

The tensile properties of CFRP were tested by soaking CFRP in sulfate solution, and the bonding properties of CFRP and concrete were also studied. The mechanical properties and failure modes of CFRP reinforced concrete beams were also studied after accelerated deterioration in the severe environment of temperature and humidity alternating cycle and sodium sulfate salt fog.[10]1. Sodium sulfate solution has no effect on the tensile strength of CFRP and the elastic modulus of concrete, but the compressive strength of concrete is related to the soaking time; 2. 2. The bond strength between CFRP and concrete was not affected by soaking in sodium sulfate solution; 3. Through the experimental study on the flexural performance of 9 reinforced concrete beams strengthened by CFRP sheets in the two erosion environments of alternating cycle of temperature and humidity and dry and wet cycle of sodium sulfate spray, the influence of accelerated environmental cycles on the flexural strength, crack development, ductility and failure model of CFRP strengthened beams was analyzed. The results show that the performance of reinforced concrete beams strengthened with CFRP decreases significantly under both accelerated degradation environments.

The strength and ductility of concrete columns strengthened with FRP are decreased after the dry-wet cycling of salt solution, and the dry-wet cycling resistance of CFRP columns is better than that of GFRP columns.While while, jean surplus, li zhao by carbon fiber cloth (CFRP) and glass fiber cloth (GFRP) all gummed reinforced concrete columns wrapped paste reinforcement and stripe intervals in the solution of salt dry-wet circulation under the action of axial compression test, analysis of environmental effect, the types of fiber cloth, fiber cloth paste way affect the performance of concrete columns axial compression test and research.[12]In this paper, it is suggested that in the dry and wet circulation environment of salt solution, the concrete structure can be strengthened by FRP bonding, which can not only improve the mechanical performance of concrete structure, but also delay the erosion of concrete by FRP bonding and prolong the life of concrete structure.

As for the influence of seawater environment conditions on the interfacial properties of cfrp-concrete, wangjizhong used 42 CFRP sheets to conduct the physical and mechanical properties test of CFRP materials, and 72 single shear specimens of concrete were used to test the interfacial bonding properties of cfrp-concrete under seawater environment.[13]The results show that: the durability of CFRP material is good, and the physical properties of CFRP material are not affected by long-term seawater corrosion.The bonding strength of cfrp-concrete decreases with the increase of corrosion time. Long-term seawater corrosion has a negative effect on the interfacial ductility, but has no significant effect on the interfacial strain development law and the bond slip relationship.

2.4 influence of acid-base environment on CFRP materials and reinforced concrete structures

The compressive strength of CFRP reinforced concrete columns under acid environment is lower than that under other environments, and the degree of reduction is related to the strength of concrete and the duration of environmental action. The tensile strength of CFRP reinforced concrete cube under acid environment is lower than that under other environment.[6]Under the action of aqueous solution -- water environment, acid environment and alkali environment, the bonding failure of positive tensile specimen is more likely to occur.

CFRP was impregnated in 50°C, 50% strong acid and NaOH solution, and various concentrated acid and liquid bromo solution at room temperature to test its acid-base resistance. The impregnation time was 250d. The test results showed that the tensile strength and elastic modulus of CFRP were not decreased.[8]

In order to study the durability of frp-concrete interface bond performance under the corrosive environment, wuyutao conducted double-sided shear tests on frp-concrete bond specimens eroded by hydrochloric acid solution, sodium hydroxide solution and sodium sulfate solution. The conclusions are as follows :(1) when the load is small, the slip value increases slowly;[13]When the load reaches about 20% of the ultimate load, cracks begin to appear at the bonding interface. The cracks gradually develop to the loading end, and the slip value increases rapidly. When the load reaches 80% of the ultimate load, the bond slip curve generally appears gentle section. As the load continued to increase, the specimen showed obvious signs of one-time failure. (2) the strain of the FRP sheet is transferred from the loading end to the free end gradually, and the strain distribution curve changes from concave to convex. Using equivalent conversion method, the average shear stress of the specimens in the limit state is generally 2~ 3.5mpa. (3) the bearing capacity of bonded specimens was decreased most seriously after being eroded by hydrochloric acid solution, followed by sodium sulfate solution, and sodium hydroxide solution had almost no effect.

2.5 influence of stress on CFRP reinforced concrete structures

In order to grasp the effect of CFRP on the performance of concrete components under long-term load, Yamagrchi et al. conducted relevant experimental studies, pointing out that under the duration equivalent to 50a, the final strength of CFRP was inferred to be 93% of the initial strength.[14]The test was conducted at room temperature, and the linear relationship was obtained based on the test data of 100h, and the creep fracture stress of CFRP after 50a was inferred to be 79% of the initial strength.

In addition, Tavakkolizadek et al. tested the fatigue resistance of beams repaired by CFRP sheets under repeated loads. According to the test results, the fatigue life of beams strengthened with CFRP sheets externally is 3.5 times longer than that of beams not reinforced with cracked beams.[15]In the logarithm and stress curve (s-n curve) of CFRP load cycles, the average strength decline rate is 5%~8% every 10 years. After one million cycles, the fatigue strength is 50% to 70% of the initial strength and is unaffected by the humidity and temperature of the exposed concrete structure. The study of 10 million load cycles shows that the s-n curve has a downward trend of 5%~8% every 10 years (Critis P.T).

2.6 influence of ultraviolet radiation on CFRP reinforced concrete structures

Yang yongxin et al. conducted experiments on the bond strength of CFRP reinforced concrete beams under ultraviolet conditions. The results showed that the bond strength between CFRP and concrete did not decrease after 2000h of ultraviolet irradiation. In the cut-and-paste test, the failure of the specimens occurred in the concrete and the bond strength did not decrease.[16]

According to above analysis and research results of each part, can be summarized as the following aspects of CFRP paste characteristics on the impact of the durability of concrete structural components: 1, the influence of temperature factor: through the paste CFRP reinforced concrete structures in high and low temperature alternating, high temperature and low temperature environment under the condition of mechanical properties test analysis shows that temperature had no effect on the performance of CFRP materials, the effect on the overall performance of CFRP reinforced concrete structure is not obvious. 2. Influence of humidity factors: the experimental study of cfr-bonded concrete structural members under humidity conditions and dry-wet alternating circulation conditions shows that the effect of humidity environment has no influence on CFRP sheet metal, and the effect of humidity environment has no significant influence on the bearing capacity of CFRP sheet metal reinforced concrete structural members. 3. Influence of medium environment: the experiments of CFRP sheet and CFRP bonded concrete structural members in seawater and acid-base environment show that CFRP sheet is not affected in seawater environment, and the performance of CFRP bonded concrete structural members in seawater environment is affected to some extent; Under the condition of acid and base, CFRP sheet is not affected, and the bearing capacity of CFRP sheet is affected. 4. Under stress and external irradiation, CFRP bonded concrete structural members are affected to a certain extent.

3.existing problems

CFRP reinforced concrete members are widely used in domestic reinforcement and transformation projects, and the durability of CFRP reinforced concrete structural members at home and abroad is becoming more and more mature and perfect. In the engineering structure reinforcement material technical specification for safety appraisal GB50728-2011, for the engineering structure with the durability of reinforcement materials also put forward the corresponding technical indicators, to reinforce the long-term performance of the adhesive for resistance to wet and heat aging, thermal aging, resistance to freezing and thawing ability, long-term stress ability, resistance to fatigue stress ability, resistance to corrosion medium of salt fog, and the roles of seawater immersion, alkaline medium, and the roles of acid medium and other technical indicators put forward clear requirements, but this kind of experimental study on the most complete in the laboratory, and to accelerate the simulation environment test,Due to the lack of durability analysis of reinforced concrete members under real conditions, as well as the lack of experimental studies under a variety of external influence conditions, and the lack of relevant data, it is still unable to provide sufficient theoretical data basis for the durability research of CFRP bonded concrete members.

4. research and development trends

In construction structure reinforcement and reconstruction process, research on the durability of CFRP reinforced concrete member and large amount of research and analysis to improve the verification, evaluation for the durability of building structure after reinforcement is a blank, the current direction of research, set up after the durability of CFRP reinforced concrete structures system evaluation, establish the durability assessment method, calculation theory has important significance. In the process of structural reinforcement design, it is necessary to consider not only the integrity, strength and stiffness of the structural system, but also the durability after reinforcement. In the comprehensive design concept of reinforcement design, the durability factors of materials and the combination of materials and concrete body are taken into account, and the factors affecting the durability of materials such as the normal service life of materials and the performance of the combination of materials and concrete body and the environmental conditions used are also taken into account.

To protect the factors that affect the durability of CFRP reinforced concrete structural members, it is also a focus worthy of further study in the future related inspection and evaluation work. While improving the durability test of reinforced materials, it is necessary to study the durability of reinforced structures, design, evaluate and maintain the durability, and take comprehensive improvement of durability as an important goal of reinforced structures.

References

[1] Hutchinson A R,Hollaway L, c. (1999)emvironmental reinforcement concrete structures with bonding fiber- rc composites[M].Cambridge, UK: Wondhed Publishing Limited.
[2] HOUSSAM A t. (1999)durability characteristics of concrete columns confined with advanced composite materials[J].Composite Structures,44:155-161.
[3] HOUSSAM A T,YONG DENG. (2002) Strength and performance of concrete axially loaded members confined with AFRP composite sheets[J].Composites: Part B,33:255-261.
[4] TOUTANJI H,BALAGURU p. (1998)durability characteristics of concrete columns wrapped with FRP tow sheets[J].Journal of Material in Civil Engineering,10(1):52-57.
[5] QIAO Pi - zhong, XU Ying - wu. (2004)The effects of freeze - thaw and dry - wet conditionings on the mode I fracture of FRP - concrete interface bonds [C]. MALLA R B, MAJI A.E arth&Space 2004: Engineering, Construction, and Operations in Challengine Environments - Ninth Biennial Conference of the Aerospace Division. Houston, USA, ASCE, 601-608.
[6] Wang Feng-xia. (2007) Experimental study on durability of CFRP reinforced concrete [D]. Nanjing: hohei university. (in Chinese)
[7] Zhu hong, Shi Jia-wei, Wu gang, Wu Zhi-shen. (2007)Research review on freeze-thaw cycle durability of FRP reinforced concrete structures [J]. Building structures. (in Chinese)
[8] Yang Xin-yong, Wang jing, Zhang Xiao-dong. (2002) Preliminary study on durability of CFRP reinforced concrete structures.[J]. Port engineering technology. (2): 25 to 27(in Chinese)
[9] HOUSSAM a. T, WILLIAM Gomez, (2007) Durability characteristics of concrete beams externally bonded with FRP composite sheets [J]. J cememt and concrete composites): 351-358.
[10] Yan Zhi-liang. (2007)Experimental study on mechanical and durability of reinforced concrete members strengthened with CFRP [D]. Shenzhen: shenzhen university. (in Chinese)
[11] Li Chen-chen, Gaodan-ying, Zhao jun. (2009.11) experimental study on durability of reinforced concrete columns reinforced with FRP and strip spacing in wet and dry environments [J]. Chinese journal of civil engineering. (in Chinese)
[12] Wang Ji-zhong, Dong Xiao-song, Zhang jian.(2014.10)Durability test research of CFRP reinforced concrete structures [J]. Concrete. (in Chinese)
[13] Wu Yu-tao. (2012) Study on bond and durability of frp-concrete interface under erosion environment [D]. Harbin: Harbin Institute of Technology. (in Chinese)

[14] Yamaguchi T, KatoY, NishimuraT. Uamotot. nakamotofibers of FRP rods made of aramid, carbon and glass fibers[C]. (1997) Proceedings of the Third International Symposium on the Netherlands: Springer Netherlands, (2): 179-186.
[15] TavakkolizadehM, Saadatmanesh h. gealvanic corrosion of carbon and steel in aggressive environments[J]. (2001) Journal of Camposites of Structural Engineering ,5(3):200-210.
[16] Yang Yong-xin, Guo Chun-hong, Caipeng, et al. Effects of ultraviolet radiation on the bonding properties of CFRP and concrete [J]. Industrialarchitecture, 2006,36 (8) 18-20. (in Chinese)

IOP Conf. Series: Materials Science and Engineering **794** (2020) 012064 doi:10.1088/1757-899X/794/1/012064

Application of ABAQUS in Numerical Simulation of Plant Roots Protective Effects on Slopes

Xiao-Lei JI[1]

[1]College of Construction Engineering, Jiangsu Open University, Nanjing, Jiangsu 210019, China.

Abstract: In order to study the slope protection effect of plant roots, the parameters of soil samples collected in the field and oleander roots drawn from laboratory tests were put in the numerical simulation of rooted soil with a 3D finite element software ABAQUS to identify the effects of root morphology and distribution density on the slope soil displacement. The results showed that planting oleander can effectively reduce the displacement of the topsoil and protect the slope. The more the number of fibrous roots and the greater the distribution density, the less the displacement of the slope soil. With the growing plant density, the displacement decreased significantly as well. The results of numerical simulation provided a theoretical basis for the application of plant roots in slope protection engineering.

1.Introduction

During the rapid construction of expressways, a large number of slopes are inevitably generated. These slopes are prone to soil erosion without the protection of natural plants, and geological disasters such as mudslides in severe cases [1-3]. The soil erosion on the slope is mainly due to rainfall. Rainfall causes the water content of slope soil to increase suddenly in a short time, which leads to the increase of soil weight and the decrease of shear capacity of slope soil, thus leading to a wide range of soil displacement and geological disasters easily [4-6]. Plant protective effects on slope are mainly divided into hydrological effect and mechanical effect, and the slope protection by plant roots is related to the mechanical effect [7-8]. The mechanical effect of slope protection is different for different plant roots. The diameter of woody plant roots is large, and the root strength is relatively high, which makes the woody plant roots play a role of anchoring in the slope protection. However, the diameter of the herbaceous plant roots is small, and the roots are buried shallow in the soil. Besides, the tensile strength of the herbaceous plant roots is relatively high, which makes the herbaceous plant roots play a role of reinforcement in the slope protection and resist the external load together with the slope soil in form of root-soil composite [9]. The application of plant roots in slope protection engineering is wide at present. However, the related theory is still lacking. Existing studies on lab shear strength and root strength are mainly conducted considering the roots as reinforced materials to form the root-soil composite, and the influence of root morphology on slope protection effect is not considered in practical engineering [10]. The root morphology is complex, and thus the protective effects on slopes can vary greatly for different kinds of roots. At present, the research on the influence of root morphology on the protective effect on slopes remains lacking [11].

In the present study, a 3D numerical simulation software ABAQUS was used to identify the effects of lateral root number of oleander roots on the slope soil displacement, and quantitatively analyze the plant roots protective effects on slopes, which provided a certain reference value for ecological slope protection engineering [12-14].

2.Finite element model of root-soil composite

2.1 Selection of material parameters

A finite element software ABAQUS was used to conduct relevant numerical simulation. Firstly, the relevant parameters were selected. Research objects soil and oleander roots were taken from the slope of Ningchang Expressway in the present study. The physical parameters of the root model were obtained with the 3-year-old oleander roots through laboratory tests, and the physical parameters of the soil were obtained by GDS triaxial test. Specific parameters are shown in Table 1 and Table 2.

Table 1Material parameters of soil

Material	Modulus of Elasticity(MPa)	Poisson's Ratio	Cohesive Strength C (kpa)	Internal Friction Angle(°)	Unit Weight (KN.m^{-3})	Water Content(%)
Soil	10	0.3	8.3	16	18	26.6

Table 2 Material parameters of oleander root

Material	Modulus of Elasticity(MPa)	Poisson's Ratio	Diameter (mm)	Length (mm)
Root	10	0.3	10	60

2.2 Establishment of finite element model

Through the software ABAQUS, the root-soil composite model was regarded as a three-dimensional separation model, and the soil material was regarded as an ideal elastic-plastic material which adopted the Mohr Coulomb failure criterion. The slope was simulated by C3D8 unit, and the root was simulated by T3D2 unit. Embedded contact was adopted for root and slope contact. It was assumed that the displacement coordination was achieved automatically when the roots were connected with the soil, and the slip and displacement between the contact surfaces between the roots and the soil were not considered in the present study. The original material characteristics of the root and soil were adopted. Slope protection by plant roots is only suitable for slopes with relatively small slope angle. Therefore, the slope model with the slope length of 2 m and the slope angle of 30°was selected for calculation and analysis. The specific simplified model is shown in Fig. 1.

2.3 Calculation model and boundary conditions

Three-dimensional strain model analysis was carried out according to the three-dimensional strain problem. The model boundary adopted the free boundary. The left and right direction of the model adopted the Z direction constraint, while the bottom surface adopted the XYZ three direction constraint. In order to achieve the displacement coordination of the roots and the soil, each root node in the model was coupled to the node on the soil closest to it. The diagram of the calculated model was divided by a grid size of 0.25 m. The finite element grid map of a slope is shown in Fig. 2, and the soil has a total of 27040 grid cells.

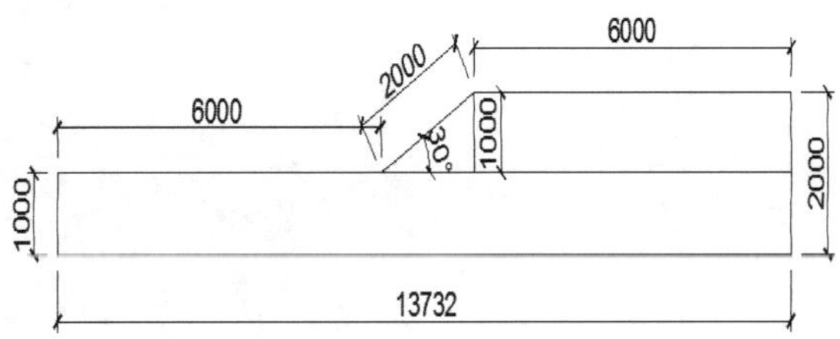

Fig. 1 Schematic diagram of the simplified slope model

IOP Conf. Series: Materials Science and Engineering **794** (2020) 012064 doi:10.1088/1757-899X/794/1/012064

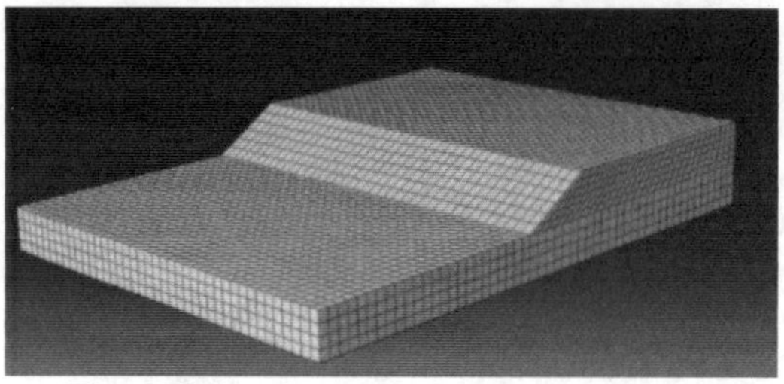

Fig. 2Finite element grid map of a slope

2.4 Numerical simulation of working conditions

The growth morphology of natural oleander roots is very complicated and cannot be strictly simulated according to the actual situation. Therefore, the natural root was simplified according to the morphology of oleander roots. Forty representative oleanders were selected from the ecological slope of Ningchang Expressway, and the parameters on root morphology were analyzed. The oleander roots were simplified as follows. The main root and the fibrous root were of equal section, and the diameter was taken as 10 mm. The length of the main root and the fibrous root was600, and 200 mm, respectively. The angle between the main root and the fibrous root was 30°, and the fibrous roots were divided three types including4, 8, and 12 roots, as shown in Fig. 3. Two rows of plants were laid out on the slope, and the spacing was 0.5, and 1 m for each row of plants respectively, as shown in Fig. 4. Due to the small diameter and light weight of the oleander root, the self-weight of the root was not considered in the load calculation in the present study. Rainfall is the main reason for soil erosion on slopes. The rainfall erosion is related to the size and flow velocity of raindrops and the accumulated water depth on the slope surface. The diameter of raindrops of 6mm, the flow velocity on the slope surface of 10m/s, and the condition of 120mm water accumulation was considered in the numerical simulation of rainfall environment. The self-weight of the slope soil was taken as the imposed load to study the influence of slope protection by plant roots on the displacement field.

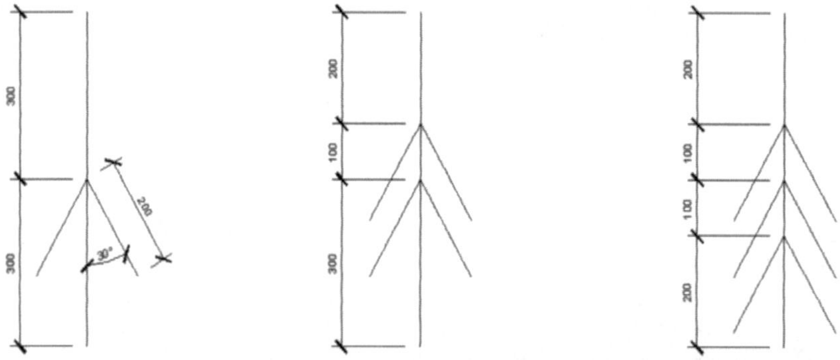

Fig. 3 Schematic diagram of root morphology distribution (mm)

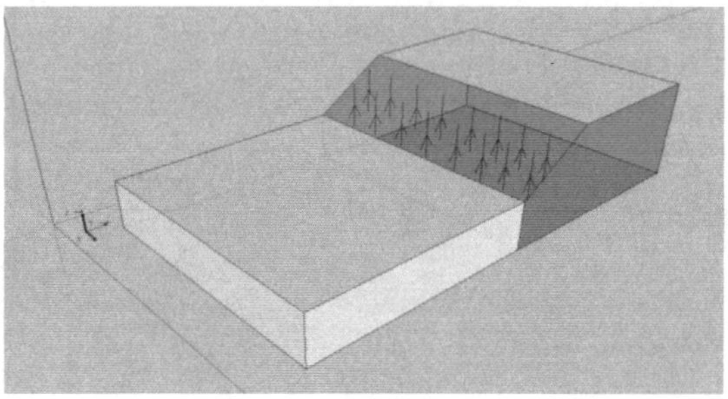

Fig. 4 Schematic diagram of slope root distribution

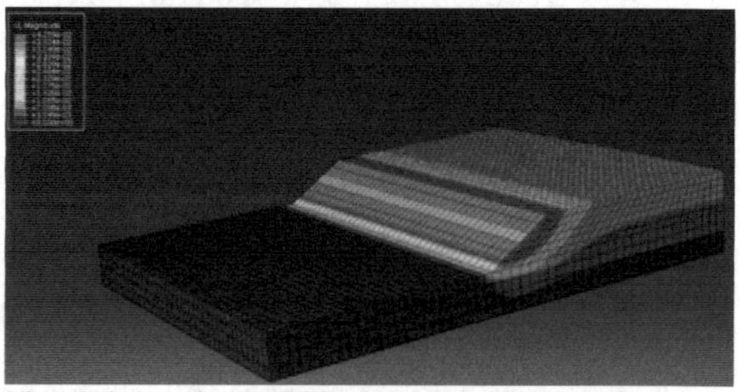

Fig. 5Finite element displacement field nephogram

IOP Conf. Series: Materials Science and Engineering **794** (2020) 012064 doi:10.1088/1757-899X/794/1/012064

3.Numerical simulation results and analysis

According to the finite element numerical simulation, the influence of the number of fibrous roots on the displacement of the slope was studied. The finite element displacement field nephogram of the slope is shown in Fig. 5.

According to the calculation results of finite element software, the maximum horizontal, vertical, and total displacement of the slope with outplant roots was 4.058, 3.452, and 4.259mm, respectively. The displacement field of the slope changed after the layout of the roots with different numbers of fibrous roots, as shown in Table 3.

Table 3Calculation results of the displacement of the slope soil

Number of Fibrous Roots	Distribution Density (root number/m)	Horizontal Displacement(mm)	Vertical Displacement (mm)	Total Displacement(mm)
4	1	3.879	3.330	4.089
	2	3.563	3.113	3.791
8	1	3.749	3.240	3.965
	2	3.311	2.937	3.547
12	1	3.656	3.177	3.879
	2	3.119	2.806	3.369

In order to effectively analyze the influence of the oleander root on the slope displacement, relationship between the displacement of the slope soil and the number of fibrous roots was plotted according to the data in the table, as shown in Fig. 6.

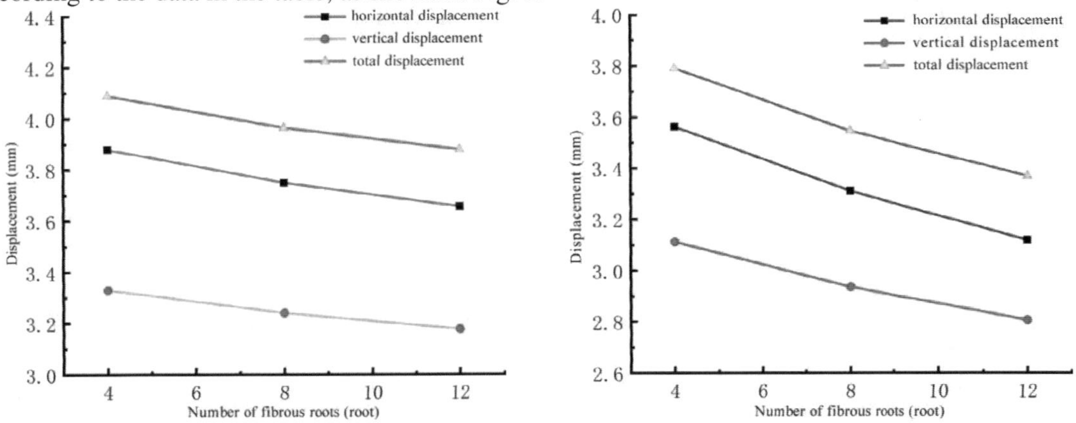

(a)distribution density of 1 root/m (b) distribution density of 2 roots/m

Fig. 6 Relationship between the displacement of the slope soil and the number of fibrous roots

As shown in Fig. 6, the oleander roots on the slope changed the displacement field of the shallow surface soil. The strength of the oleander root is significantly higher than that of the slope soil because the reinforced soil is formed when the roots are planted in the soil. The reinforced soil has stronger bearing capacity compared with that of the slope soil, which reduces the stress of the surface soil, thus reducing the displacement of the surface soil and improving the stability of the surface soil of the slope. According to the curves in Fig. 6, the horizontal, vertical, and total displacement of the shallow surface soil of the slope decrease with the increase of the number and distribution density of the oleander roots. For the slope with large distribution density of roots, the trend of decrease in soil displacement is more obvious and the effect of slope protection by plant roots is more effective with increasing number of fibrous roots.

3.1 Influence of the number of fibrous roots on the horizontal displacement of the slope soil

The effect curve of the number of fibrous roots on the horizontal displacement of surface soil was obtained from the results in Table 3, as shown in Fig. 7.

As shown in Fig. 7, the horizontal displacement of the slope with the distribution density of 2 roots/m containing 4, 8, and 12 fibrous roots decreased by 8.86%, 13.2%, and 17.2% compared with that of the slope with the distribution density of 1 root/m, respectively. The horizontal displacement of the slope with the distribution density of 1 root/m containing 8, and 12 fibrous roots decreased by 3.35%, and 5.74% compared with that of the slope containing 4 fibrous roots, respectively. The horizontal displacement of the slope with the distribution density of 2 roots/m containing 8, and 12 fibrous roots decreased by 7.07%, and 12.5% compared with that of the slope containing 4 fibrous roots, respectively.

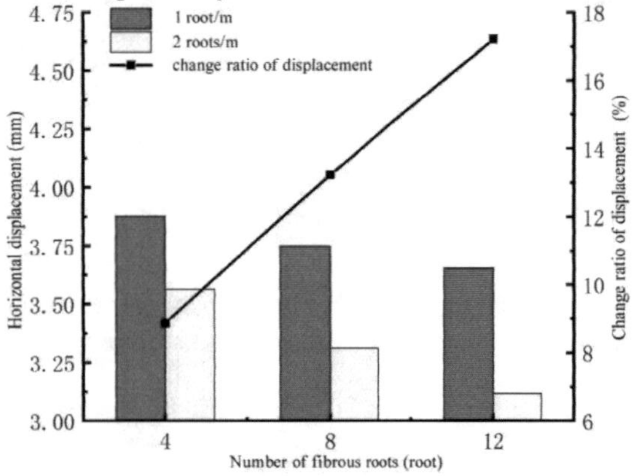

Fig. 7Effect curve of the number of fibrous roots on the horizontal displacement of surface soil

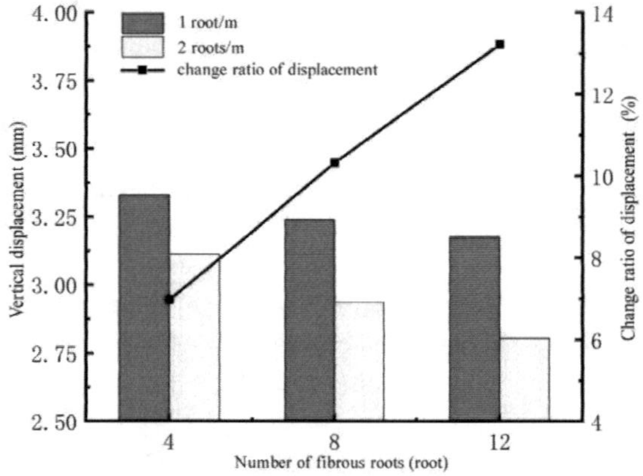

Fig. 8 Effect curve of the number of fibrous roots on the vertical displacement of surface soil

3.2 Influence of the number of fibrous roots on the vertical displacement of the slope soil

The effect curve of the number of fibrous roots on the vertical displacement of surface soil was obtained from the results in Table 3, as shown in Fig. 8.

As shown in Fig. 8, the vertical displacement of the slope with the distribution density of 2 roots/m containing 4, 8, and 12 fibrous roots decreased by 6.97%, 10.32%, and 13.22% compared with that of the slope with the distribution density of 1 root/m, respectively. The vertical displacement of the slope with the distribution density of 1 root/m containing 8, and 12 fibrous roots decreased by 2.7%, and 4.59% compared with that of the slope containing 4 fibrous roots, respectively. The vertical displacement of the slope with the distribution density of 2 roots/m containing 8, and 12 fibrous roots decreased by 5.65%, and 9.86% compared with that of the slope containing 4 fibrous roots, respectively.

3.3 Influence of the number of fibrous roots on the total displacement of the slope soil

The effect curve of the number of fibrous roots on the total displacement of surface soil was obtained from the results in Table 3, as shown in Fig. 9.

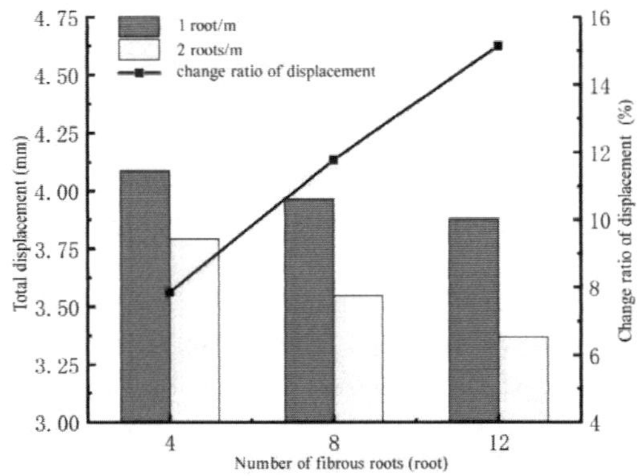

Fig. 9 Effect curve of the number of fibrous roots on the total displacement of surface soil

As shown in Fig. 9, the total displacement of the slope with the distribution density of 2 roots/m containing 4, 8, and 12 fibrous roots decreased by 7.86%, 11.78%, and 15.13% compared with that of the slope with the distribution density of 1 root/m, respectively. The total displacement of the slope with the distribution density of 1 root/m containing 8, and 12 fibrous roots decreased by 3.03%, and 5.14% compared with that of the slope containing 4 fibrous roots, respectively. The total displacement of the slope with the distribution density of 2 roots/m containing 8, and 12 fibrous roots decreased by 6.44%, and 11.13% compared with that of the slope containing 4 fibrous roots, respectively.

4.Conclusions

(1) The finite element numerical calculation results showed that the nonlinear constitutive relation of the slope soil was taken into consideration in the finite element numerical analysis method, which was viable for studying the plant roots protective effects on slopes. The plant roots could effectively change the displacement field of the surface soil of the slope. The soil of the slope interacted with the plant roots, which caused friction between the plant roots and the soil, thus limiting the displacement of the soil and making an effect on slope protection.

(2) With the increase of the number of fibrous roots, the displacement of the surface soil of the slope was smaller, and the plant roots protective effects on slopes were more obvious. The horizontal, and vertical displacement referred to the anterior-posterior, and up-and-down direction of the slope surface, respectively. Test results showed hat the influence of the roots on the horizontal displacement of the soil is greater than that on the vertical displacement. Therefore, the roots can effectively prevent the slope from sliding to the front and avoid soil erosion. In the numerical simulation process, the morphology of oleander root was simplified. However, the morphology of natural oleander root was more complicated in reality, thus the plant roots protective effects on slopes will be more obvious.

(3) Plant roots with different distribution densities had different slope protection effects. There was a certain distance between the slope protection plants roots. During the rainfall process, the soil in the upper part of the slope increased the pressure on roots and the soil around the roots, thus forming the stress arch around the roots. The effect of stress arch on the roots was different for roots with different distribution densities. Test results showed that the trend of decrease in soil displacement is more obvious with increasing distribution density of plant roots. In order to give full play to the soil arching effect between roots, the distribution density of slope protection plants needs to be combined with local climate and soil conditions to protect the slope soil more effectively.

(4) The plant root slope protection is only applicable for the soil erosion in the shallow surface of the slope under rainfall conditions. For the high slope and the slope with deep slide, the slope protection by plant roots needs to be combined with engineering protection measures to ensure the safety and stability of the slope.

Acknowledgments
This research was supported by Science and Technology Project Plan of Ministry of Housing and

Urban-rural Development of the PRC [2016-K4-020]; "Qing Lan Project" of universities in Jiangsu province; Jiangsu province "Six Talents Peak" project[JZ-018];Natural Science Foundation of the Jiangsu Higher Education Institutions of China [18KJB220001].

References

[1] ADEKALU KO, OLORUNFEMI I A, OSUNBITAN J A. Grass Mulching Effecton Infiltration, Surface Run off and Soil Loss of Three Agricultural Soils in Nigeria[J] Bioresource Technology, 2007, 98(4):912-917.

[2] SUN TT, LI B, LIU H B. Study on soil and water conservation effect of three kinds of ecological slope protection [J].Soil and Water Conservation in China, 2017(2):10-12.

[3] RONG H, SHAN D, LIU Y P, et al. Effects of Vegetation Restoration Models on Soil and Water Conservation in Engineering Erosion Area of Grassland [J] Research of Soil and Water Conservation, 2017, 24(3):24 - 28.

[4] AN R, CHAI J R, QIN Y, et al. Analysis on effect of vegetation root-system morphology on slope stability [J]. Water Resources and Hydropower Engineering, 2018, 49(3): 150-156.

[5] YANG Y, YANG J Y, ZHAO T, et al. Ecological restoration of highway slope by covering with straw-mat and seeding with grass-legume mixture[J]. Ecological engineering, 2016, 90(4): 68-76.

[6] ROMANO N, LIGNOLA G P, BRIGANTE M, et al. Residual life and degradation assessment of wood elements used in soil bioengineering structures for slope protection[J]. Ecological engineering, 2016, 90(3): 498-509.

[7] ZHAO B Q, XIA Z Y, XU W N, et al. Review on research of slope eco-restoration technique for engineering disturbed area [J]. Water Resources and Hydropower Engineering, 2017, 48(2): 130-137.

[8] HU P, SONG S G, WU D G. Expermental research on reinforcement mechanism of expressway slope protection with greensward [J]. ROCK AND SOIL MECHANICS, 2008, 29(2): 442 - 444.

[9] HU P, SONG X G, WU D G. Experimental research on reinforcement mechanism of expressway slope protection with greensward[J]. Rock and Soil Mechanics, 2008, 29(2): 442 - 444.

[10] SHEN Y, HE B H, XIANG M H, et al. Effects of Hedgerow on Soil and Water Conservation in Sloping Cropland of the Purple Soil [J]. Journal of Soil and Water Conservation, 2013, 27(2): 47-52.

[11] TARDO G, MICKOVSKI S B. Method for synchronisation of soil and root behaviour for assessment of stability of vegetated slopes[J]. Ecological engineering, 2015, 82: 222-230.

[12] JI J N, ZHANG Z Q, GUO J T, et al. Finite element numerical simulation of Black Locust (Robinia pseudoacacia) and Arborvitae (Platycladus orientalis) roots on slope stability on Loess Plateau of China [J]. Transactions of the Chinese Society of Agricultural Engineering, 2014, 30(19): 146-154.

[13] XIAO B L, LUO S L, CHEN J, et al. Finite element analysis of eco-protection slope through roots [J]. ROCK AND SOIL MECHANICS,2011,32(6):1881-1885.

[14] LIU S L. Numerical Simulation of the Effect of Plant Roots on Slope Reinforcement [J]. Pearl River, 2018, 39(12): 147-150.

Study on Seismic Behaviour of Cold Formed Steel Structure Based on Simplified Mechanical Model

Lei Liu[1*], YongYao[1], YunpengChu[1] and Jiqiang Liu[2]

[1]Civil engineering, Southwest University of Science and Technology, Mianyang, Sichuan Province, 621010, China

[2]Sichuan College of Architecture Technology, Deyang, Sichuan Province, 618000, China

[*]Lei Liu , 568404032@qq.com

Abstract. Cold-formed thin-walled steel buildings adopt the process of layered assembly, which is easy to disassemble and assemble and conforms to the requirements of industrial construction. But there is no clear method of seismic design. Taking a cold-formed thin-walled light steel structure house in Dazhou city as an example, the Pivot connection unit that can characterize the structural damage and the finite element software SAP2000 are used to establish a model, then this model is used to analyse the basic dynamic characteristics of structures and to analyze the time-history response of structures under seismic action. The results show that the simplified modeling analysis of the structure has a strong feasibility, and the lateral stiffness of the structure can resist the action of multiple earthquakes. Under the action of rare earthquake, the displacement between the layers of the structure is lower than the limit of the code, and the hysteretic performance of the structure is good, which can ensure the safety and reliability of the structure.

1. Introduction

Cold-formed thin-walled steel structure has the characteristics of high recovery rate of components and short construction period. It is mainly used for three-story and below civil residence and two-story to three-story villa. The utilization rate of steel is high, and it can meet the requirement of sustainable development. All the structural parts and components are manufactured in the standardized workshop, and assembled on site, so the processing quality is guaranteed. Site construction has little impact on the environment, and it has the advantages of low comprehensive cost and short construction period, therefore it has the value of promotion.Taking a cold-formed thin-wall steel structure building in Dazhou as an example, we study the structure design method of the project. At the same time we use the finite element software SPA2000 to calculation and analysis. Through the established model, we analyse the basic vibration information of the structure. And then we study the linear analysis of structure under the frequent earthquake and the nonlinear analysis of structure under the rare earthquake, through the structural time-history response analysis.Through our research, we hope to provide reference for the engineering application of this kind of structure.

2.Project profile

This project is located in Dazhou, Sichuan province. It adopts cold-formed thin-wall steel structure and adopts new honeycomb wallboard material. The building is a small three-story residential house

IOP Conf. Series: Materials Science and Engineering **794** (2020) 012065 doi:10.1088/1757-899X/794/1/012065

with a building area of 649.24m2 and a total building height of 10.86m. Seismic fortification intensity is 6 degree, and design seismic acceleration is 18cm/s2, and seismic grouping is one group. Site soil category for II class, site characteristic period of 0.35s.

The main frame of the structure is composed of 100×100×2mm, 120×120×2mm, 120×100×3mm square steel tubes. The support adopts C section steel of 120×50×20×1mm, 120×50×20×3mm and square steel tube of 120×60×2mm, 100×50×2mm. The wall material adopts double layer ousong plate sandwich honeycomb com-pound wall paper material, and the floor is composed of pressed steel plate, and the roof is square steel pipe slope roof frame plus C steel purlin ousong plate. The layout of cold-formed thin-walled steel wall is shown in Fig. 1.

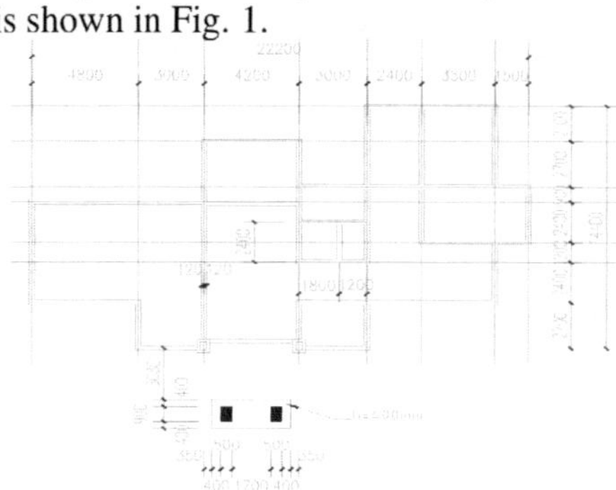

Figure 1.Structural wall layout.

3. Finite element model

The cold-formed thin-walled steel structure residential steel is made of Q235 steel with yield stress of 235MPa, limit stress of 375MPa, elastic modulus of 2.06×105MPa and poisson's ratio of 0.3.

Due to the complexity of structure and connection of cold-formed thin-walled steel walls, there is no theoretical calculation of shear bearing capacity of such walls at home and abroad. The shear performance index is obtained through full scale test [1]. According to the nonlinear simplified analysis method of wall in reference [1], cold-formed thin-walled steel wall can be approximately equivalent to equal-generation pull rod model, in which the pull rod is replaced by Pivot connection unit in SAP2000, and the simplified wall is assembled accordingly according to the shaking table model to obtain a simplified calculation model of the house.

In recent years, it is proposed to simplify the shear bearing capacity of cold-formed thin-walled steel wall by equal-generation pull rod method [2-5]. There are still the following problems to be solved :(1) the nonlinear finite element analysis model of the wall is complex, and the modeling process is tedious and inefficient, and it is difficult to be popularized in practical engineering applications.(2) the calculation model of this method is only applicable to the elastic stage, and nonlinear analysis is no longer applicable.

3.1Wall body modeling

The modeling in this paper [1] refers to the experimental data and modeling method of Huang[1] and the idea of equal-generation pull rod method. Based on the hysteretic rule of line segment plastic Pivot connection ele-ment in SAP2000, this connection element is used to simulate cold-formed thin-walled steel wall according to the experimental data, so as to realize reasonable simplification of the wall, as shown in Fig. 2. In this paper, a degenerate four-line model is adopted to simulate the skeleton curve of the wall, as shown in Fig. 3. The stiff-ness of each section can be calculated according to equation (1).

$$
\begin{aligned}
K_1 &= \frac{|+P_e|+|-P_e|}{|+\Delta_e|+|-\Delta_e|} \\
K_2 &= \frac{|+P_y|+|-P_y|-|+P_e|-|-P_e|}{|+\Delta_y|+|-\Delta_y|-|+\Delta_e|-|-\Delta_e|} \\
K_3 &= \frac{|+P_u|+|-P_u|-|+P_y|-|-P_y|}{|+\Delta_u|+|-\Delta_u|-|+\Delta_y|-|-\Delta_y|} \\
K_4 &= \frac{|+P_f|+|-P_f|-|+P_u|-|-P_u|}{|+\Delta_f|+|-\Delta_f|-|+\Delta_u|-|-\Delta_u|}
\end{aligned}
\tag{1}
$$

As the displacement amplitude increases, the unloading stiffness of the curve decreases gradually, as shown in Fig. 4. The unloading stiffness of specimens at each stage can be calculated according to equation (2). When the characteristic points of specimens are unloaded, the unloading stiffness coefficient can be obtained through displacement difference.

$$
\begin{aligned}
&(1-2): \frac{K}{K_1} = \left(\frac{\Delta}{\Delta_e}\right)^{\lambda_1} \quad \Delta_e < |\Delta| < \Delta_y \\
&(2-3): \frac{K}{K_1} = \left(\frac{\Delta}{\Delta_e}\right)^{\lambda_2} \quad \Delta_y < |\Delta| < \Delta_u \\
&(3-4): \frac{K}{K_1} = \left(\frac{\Delta}{\Delta_e}\right)^{\lambda_3} \quad \Delta_u < |\Delta| < \Delta_f
\end{aligned}
\tag{2}
$$

There, K- unloading stiffness of wall, - displacement of the wall during unloading, - unloading stiff-ness coefficient.

Steel tension strips are set between walls on different floors, and rigid supports and horizontal bracing bars can effectively improve the lateral stiffness and integrity of the combined wall [6]. The tension belt in the wall can only bear tension, while the solid unit in SAP2000 can bear both tension and pressure. In the modeling process, the steel tension belt is simulated by Hook unit. The mechanical behavior of Hook unit is described as equation (3)

$$
f = \begin{cases} k(d-open) & d > open \\ 0 & d \le open \end{cases}
\tag{3}
$$

There, k - spring constant, D - element displacement, Open - initial crack, Open=0 in the formula can simulate the single pull component.

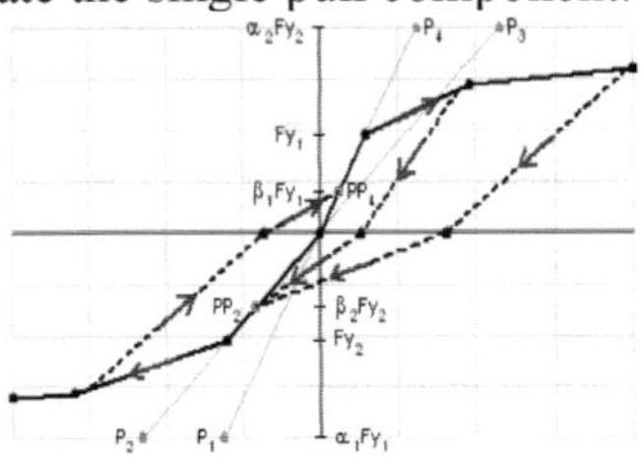

Figure2.Pivot hysteretic rule of wall body.

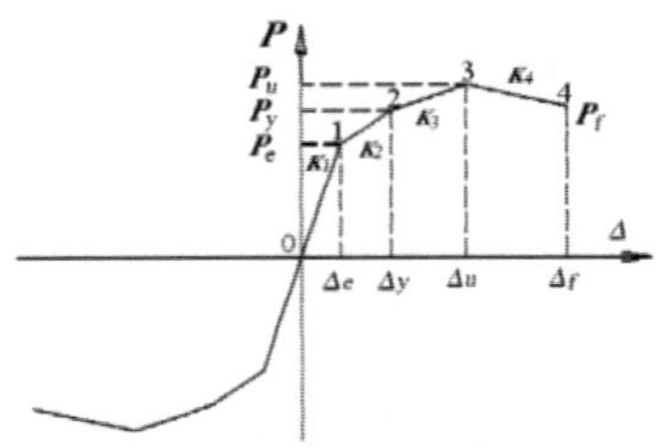

Figure3.Skeleton curve of wall.

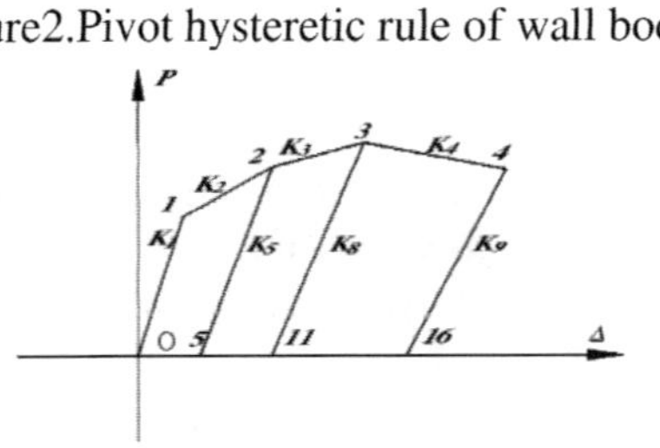

Figure4.Unloading stiffness of wall body.

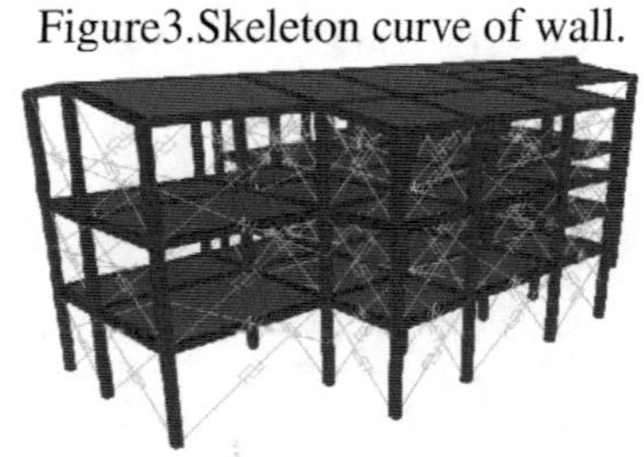

Figure5.Finite element model diagram.

3.2 The load arrangement

The dead load of the first and second floor composite floor is set as 1.42 Kn/m2, and the live load is set as 0.5kn /m2 and 4.0kn /m2 respectively according to the room function and the building load code.The dead weight of the outer wall of cold-formed thin-walled steel wall shall be 1.0kn/m2, and the dead weight of the inner wall shall be 0.5kn /m2.The simplified finite element structure model is shown in Fig.5.

3.3 Basic dynamic characteristics of the structure

Since the y-direction stiffness of the structure is relatively small, the first mode is translational motion in the Y direction, and the second mode is translational motion in the X direction, and the third mode is torsion around the Z-axis. The basic vibration information of the structure is shown in Table 1, and the first three modes are shown in Fig. 6.According to the American approximate formula of the basic natural vibration period (4), the basic vibration period of the structure is calculated as 0.2992s.

$$T = 0.05H^{\frac{3}{4}} \tag{4}$$

The difference between the results of structural modal analysis and this value is 5.24%, which indicates that the results of structural modal analysis are reliable.According to the basic information of structural vibration, the lateral stiffness of the structure in the Y direction is small, so the first mode of vibration of the structure is the translation in the Y direction.Although the plane arrangement of the structure is irregular, the torsion effect can be effectively restrained by the reasonable arrangement of the stiffness.

Table1.Basic vibration information of structure.

Structural vibration mode	T(s)	Mass participation coefficient		
		UX	UY	RZ
The first mode	0.31488	0.01195	0.85015	0.29699
The second order modal	0.27979	0.77242	0.03852	0.14332
The third order vibration mode	0.24734	0.14413	0.03614	0.48712

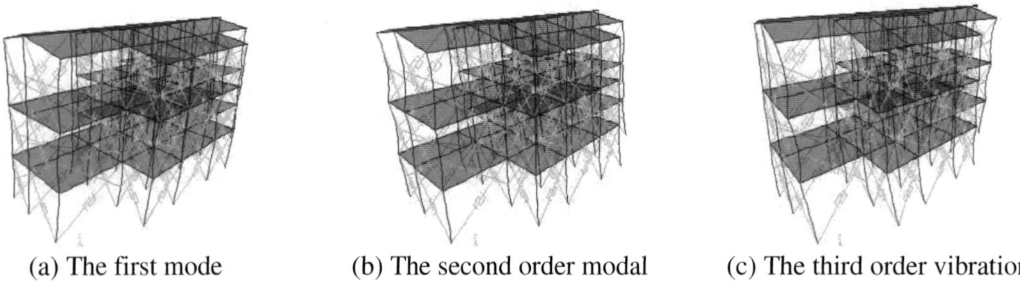

(a) The first mode (b) The second order modal (c) The third order vibration mode

Figure6.The first three vibration modes of the structure.

4. Structural time-history response analysis

4.1 Seismic oscillation

Structure in region belongs to II. The EL-centro wave is selected to simulate the three elements of earthquake. The continuous energy of the seismic wave is uniformly distributed in time-frequency space. Its energy is mainly distributed between [1,5.5]Hz, with a wide distribution range. The energy of [1.5,2.5]Hz frequency band is concentrated between [1,2]s and [4,5]s. The energy of [2.5,5]Hz band is concentrated in the narrower time domain of [2,2.5]s. The peak acceleration of seismic wave is 341.7cm/s^2.

4.2 Structural linear analysis under frequent earthquakes

The building is located in Dazhou city, Sichuan province, and its fortification intensity is 6 degree. For the analysis of the structure under frequent earthquakes, it adopts 6 degree and its peak acceleration is 18gal.

Under the action of frequent earthquakes, the maximum inter-story displacement angles in both directions of the structure appear in the first floor, because the structural deformation is mainly shear deformation, which is inconsistent with other structural systems appearing in the second floor.

As shown in Fig. 7, the envelope curve of inter-layer displacement Angle of the structure can be seen as follows :(1) the maximum inter-layer displacement Angle of the structure in two directions is 2.27×10^{-4} and 2.82×10^{-4}.(2) although the single cold-formed thin-walled steel wall is relatively soft, it is different from the general frame structure. In the stiffness calculation, the contribution of internal partition wall to lateral stiffness should be considered.(3) due to the structure's light weight, the dynamic load generated by acceleration is small.(4) since the connection between components is the response of the structure under the action of earthquakes more than 6 degree, linear analysis is adopted.

The displacement of the top layer of the structure is shown in *Fig. 8*. It can be seen that :(1)the maximum displacement in both directions of the top of the structure is more than 2mm, which indicates that the lateral stiffness of cold-formed thin-walled steel wall structure can meet the application requirements of low-intensity areas.(2) The displacement in Y direction of the top of the structure is greater than that in X direction, the damping ratio in both directions is the same, and the dynamic amplification coefficient of the structure is the same, so the lateral displacement stiffness in X direction of the structure is relatively large, which is consistent with the results of modal analysis.

The time-history curve of acceleration at each elevation of the structure is shown in *Fig. 9*, and the maximum acceleration at each elevation of the structure is shown in *Table 2*. It can be seen that :(1) the maximum acceleration in X direction and Y direction at the elevation of the first layer are 21.572gal and 28.492gal, with a magnification of 1.198 and 1.862 times.(2) the maximum acceleration in X direction and Y direction at the elevation of the second floor are 31.551gal and 40.328gal, and the amplification is 1.753 times and 2.636 times.(3) the maximum acceleration in X and Y directions at the top elevation is 39.266gal and 42.733gal, and the amplification is 2.181 times and 2.793 times, which are similar to the structures of brick and concrete. The magnification factor of acceleration of the structure is related to the soil condition, structural quality, height of the structure floor and seismic intensity, etc. If the self-weight of the structure is reasonably controlled, the magnification factor of acceleration can be effectively controlled.

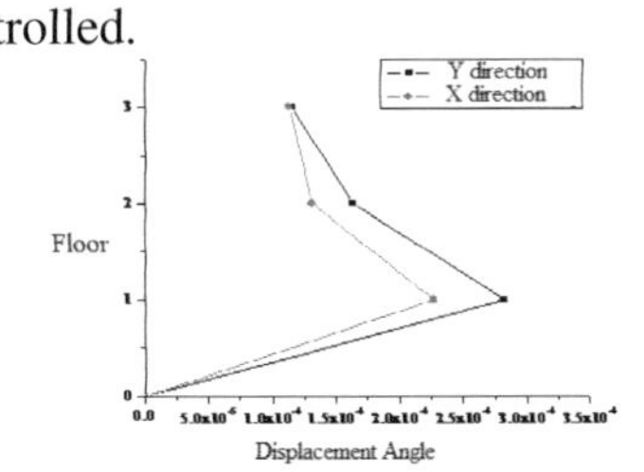

Figure7.Displacement angle envelope of the lower layer under more frequent earthquakes.

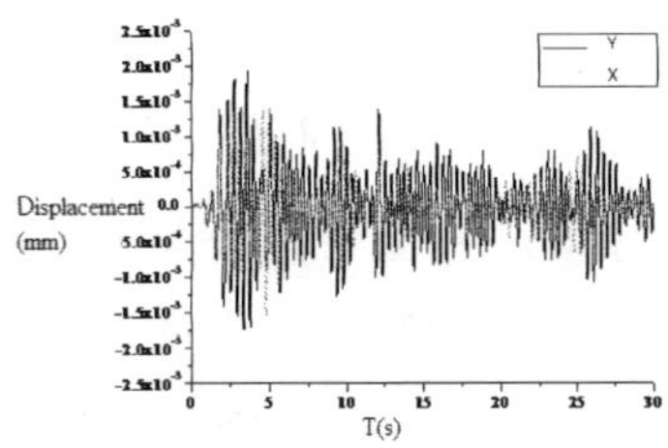

Figure8.Displacement of the top layer under more frequent earthquakes.

IOP Conf. Series: Materials Science and Engineering **794** (2020) 012065 doi:10.1088/1757-899X/794/1/012065

Table2.Maximum acceleration information of each level elevation of the structure.

structure floor	maximum acceleration (gal)		magnification times	
	X	Y	X	Y
One level	21.572	28.492	1.20	1.58
Second level elevation	31.551	40.328	1.75	2.24
Top level	39.266	42.733	2.18	2.37

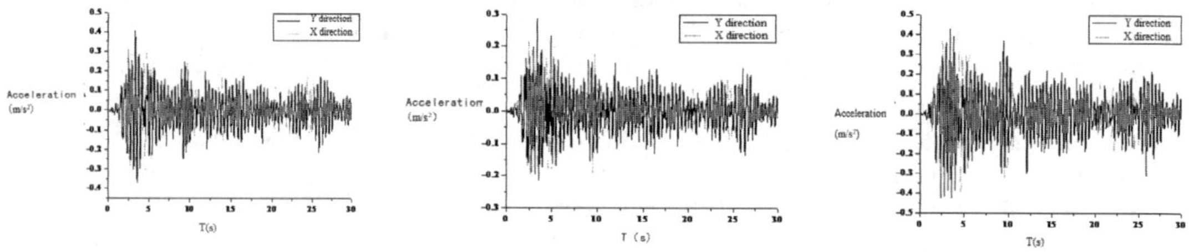

(a) One level (b) Second level elevation (c) Top level

Figure9.Time-history curve of acceleration at floor elevation.

4.3 Structural analysis of rare earthquake

Under the effect of rare earthquakes, the structures are subjected to 6th intensity rare earthquake (peak acceleration 125gal) and 7th intensity rare earthquake (peak acceleration 220gal), respectively, for nonlinear analysis. According to the structural displacement, the acceleration of each layer and the hysteretic curve of the base shear - top layer displacement, the interlayer stiffness and energy dissipation capacity of the structure are analyzed.

As shown in Fig. 10, the envelope curve of inter-story displacement angle of the structure under the action of rarely encountered earthquake is as follows :(1) under the action of 6th intensity rare earthquake, the maximum inter-story displacement Angle in X direction and Y direction appears at the elevation of the first floor of the structure, which are 1/245 and 1/189 respectively.(2) under the action of 7th intensity rare earthquake, the maximum inter-layer displacement angles in X direction and Y direction are 1/119 and 1/118.

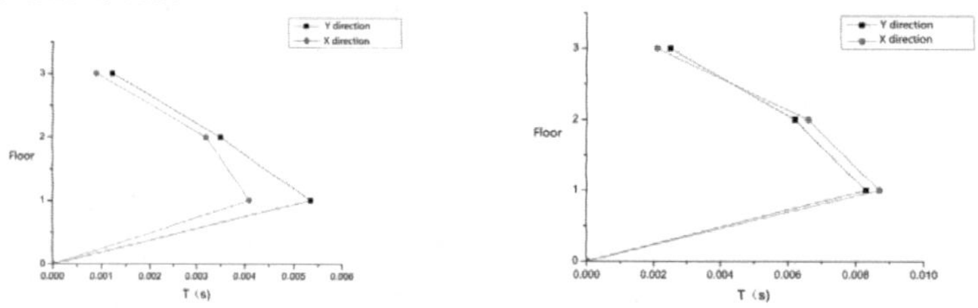

(a)6th intensity rare earthquake (b)7th intensity rare earthquake

Figure10.Structure inter-layer displacement Angle envelope under rare earthquake action.

According to article 5.5.5 of building seismic code (GB50011-2010), the inter-story displacement Angle of the structure under rare earthquake action is no more than 1/50, and the multi-story cold-formed thin-walled steel structure can meet the requirements of the code under the action of 6th intensity rare earthquake and 7th intensity rare earthquake, so it can meet the requirements of stiffness in high-intensity areas. Under the action of a rare earthquake of 7 degree, when time history analysis is carried out, the stiffness in the X direction is greater, the maximum displacement Angle between layers is greater than that in the Y direction, and the X and Y direction acceleration ax: ay=1:0.85. With the increase of seismic intensity, the difference of acceleration amplitudes in the two directions increases, that is, the displacement angle difference between the two directions depends on the difference of seismic amplitudes under the action of 7th intensity rare earthquake.

The time-history curve of displacement of the top layer of the structure is shown in Fig. 11. Under the action of 6th intensity rare earthquake, the maximum displacement of the top layer of the structure in two directions is 0.028m and 0.031m respectively. Under the action of 7th intensity rare earthquake, the maximum displacement of the top of the structure in two directions is 0.057m and 0.060m respectively. In the time history analysis, the displacement in the X direction is greater than that in the Y direction, and the acceleration in the X direction and Y direction ax: ay=1:0.85. With the increase of seismic intensity, the amplitude difference of the acceleration input in the two directions increases. In other words, the displacement difference between the two directions is mainly controlled by the amplitude difference of the earthquake amplitude when the earthquake of 7 degree is rarely encountered.

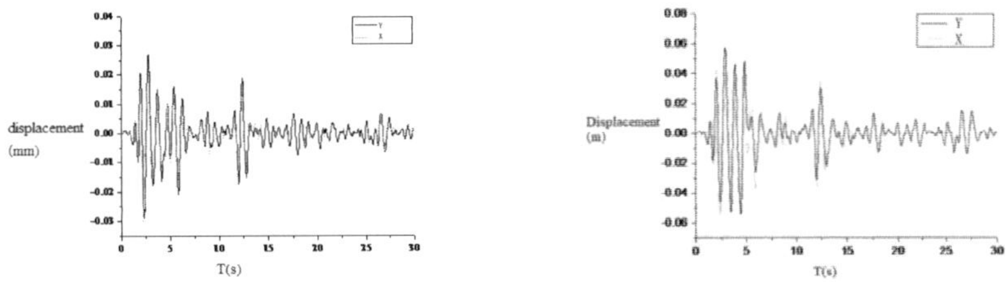

(a) 6th intensity rare earthquake (b)7th intensity rare earthquake
Figure11.Top displacement of the structure under the action of rare earthquake.

Under the action of 7th intensity rare earthquake, the time-history curve of acceleration at the elevation of each layer of the structure is shown in Fig. 12. The maximum acceleration and acceleration magnification in the earthquake process are shown in table 3. According to article 5.1.4 of building seismic code (GB50011-2010), acceleration response spectrum method is adopted, as shown in equation (5).

$$\alpha\,(T)=k\times\beta\,(\mathrm{T}) \tag{5}$$

There, k-the ratio between the peak acceleration of ground motion and acceleration of gravity, β-acceleration magnification.

The energy dissipation capacity of the structure is mainly evaluated by the hysteretic behavior of the shear - top displacement of the base. The hysteretic curve of base shear-top displacement under the action of earthquakes of 6 degree and 7 degree is shown in *Fig. 13*. Before the displacement reaches the elastic limit, the hysteretic curve is basically linear. As the top displacement increases, the hysteresis loop gradually becomes full and dissipates more energy. Maximum acceleration information of each level elevation of the structure is shown in Table 3.

On the whole, although the hysteresis curves are pinched, the hysteresis loops are full and the structure has a strong energy consumption capacity. By comparing the figures in Fig. 13, it is found that the structure pinching phenomenon is more serious under the action of 7th intensity rare earthquake, but the hysteresis loop is more full. Generally speaking, the cold-formed thin-walled steel wall structure system can show good nonlinear performance and has a strong energy dissipation capacity under rare earthquake action.

Table3.Maximum acceleration information of each level elevation of the structure.

		6th intensity rare earthquake			7th intensity rare earthquake		
		One level	Second level elevation	Top level	One level	Second level elevation	Top level
maximum acceleration (gal)	X	217.721	275.216	306.816	303.642	438.496	412.038
	Y	170.708	192.559	273.662	277.540	352.472	407.315
magnification times	X	1.74	2.21	2.45	1.52	2.19	2.06
	Y	1.61	1.81	2.58	1.63	2.07	2.39

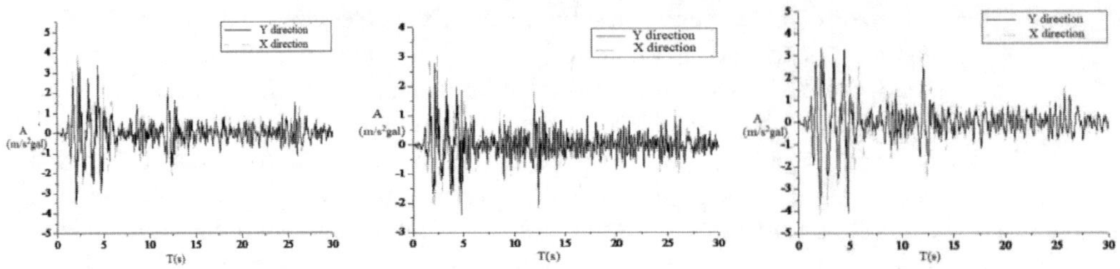

(a) One level (b) Second level elevation (c)Top level

Figure12.Time-history curve of acceleration at floor level (rarely seen at 7 degree).

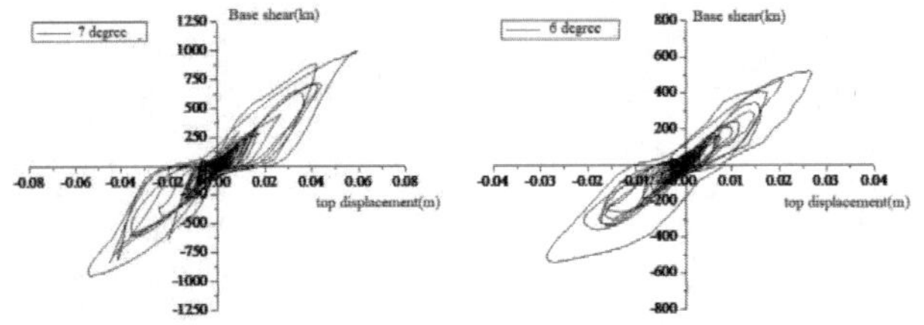

(a)X direction (6 degree) (b) Y direction (6 degree)

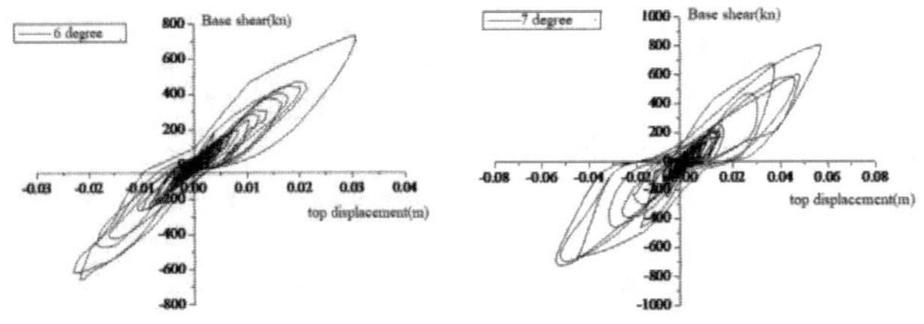

(c)X direction (7 degree) (d) Y direction (7 degree)

Figure13.Hysteresis curve of base shear force-top displacement under the action of 6 degree or 7 degree rare earthquake.

5. Conclusion

According to the structural mode information, the simplified structural modeling has a certain degree of reliability. The stiffness and mass of the structure are evenly distributed.

(1) Under the action of frequent earthquakes, the lateral stiffness of cold-formed thin-walled steel structure is enough to resist the action of frequent earthquakes with low intensity under the action of horizontal loads. The deformation mechanism is different from other structures.

(2) Under the action of a rare earthquake of 6 degree, the maximum inter-layer displacement angles in X direction and Y direction are 1/245 and 1/189.Under the action of a rare earthquake of 7 degree, the maximum inter-story displacement angle is 1/119 and 1/118 respectively, that is, the inter-story displacement of the structure is lower than the limit value, and the inter-story stiffness of the structure meets the requirements. The magnification factor of structural acceleration is equivalent to other structural forms.

Due to the light structure mass, the seismic action of the structure is smaller when the intensity is equal. The hysteretic performance of the structure is good, and the energy dissipation capacity is strong.

Acknowledgement

This research was financially supported by Sichuan Provincial Department of Science and technology (Grant NO. 2019YJ0322) and Deyang "Thirteenth Five-Year Plan" in 2019 (Grant No. DY19C050).

References

[1]Huang, Z.G. (2011) Study on seismic behavior of low-rise cold-formed thin-walled steel buildings. Xi 'an university of architecture and technology, Xi an.

[2]Nie, S.F., Zhou, T.H., Zhou, X.H. et al.(2017) Compression test study and finite element analysis of double-limb cold-formed thin-walled steel box column. Journal of architectural structure, 38(10): 10-20.

[3] Wang, X.L., Chu, Y.P., Yao, Y. et al. (2015) Experimental study on seismic behavior of super thin-walled cold formed steel C frame column-floor beam connection. Journal of civil engineering ,48(7)： 51-59.

[4]Zhou, X.H., Guan, Y., Shi, Y. (2017) Seismic performance analysis of multi-storey cold-formed thin-walled steel housing. Progress of building steel structure,19(6): 11-15.

[5]Chu, Y.P., Yao, Y. (2016) Experimental study on seismic behavior of super thin-walled cold formed steel frame with rectangular cross section between columns and floor beams. Journal of architectural structure,37(7)： 46-53.

[6] Zeynalian, M., Shelley, A.,Ronagh, H.R. (2016)An experimental study into the capacity of cold-formed steel truss connections. Journal of Constructional Steel Research,127:176-186.

IOP Conf. Series: Materials Science and Engineering **794** (2020) 012066 doi:10.1088/1757-899X/794/1/012066

Analysis of slope stability considering the mechanism of the bolt-beam-root pile composite retaining structure

Zhang Yan Hong[1], Sun Bin[2*], Jose Luis Chavez Torres[3]

China Railway (Shanghai) Investment Group Co. LTD, Shanghai, 200135, China

Geotechnical Engineering Institute, Hohai University, Nanjing, Jinagsu, 210024, China

Geotechnical Engineering Institute, Hohai University, Nanjing, Jinagsu, 210024, China

1684985220@qq.com,*191304010069@hhu.edu.cn,jlchavez3@utpl.edu.ec

Abstract. This project focuses on the development of a new type of retaining structure, namely Bolt-Beam-Pile Composite Retaining Structure. The proposed structure is a technical and economically feasible method enhancing the slope stability. A parameter sensitivity analysis carried out to evaluate the behavior of the structure under different scenarios, and the variation of the soil parameters. Finite element method was utilized to stimulate the model and perform the analyses. The results demonstrated that, compared with other three typical retaining structures the new Composite Retaining Structure has larger safety factor and smaller displacement.

1. Introduction

A retaining structure is also known as a support or enclosure structure. The bolt-beam-pile composite retaining structure investigated in this paper is a new type of composite retaining structure formed by the combination of two rows in parallel vertical tree root piles, and anchor with the continuous beam at the top of the pile.

From the structural analysis, the double-row root piles are embedded in soil as portal frames explaining that the structure has greater lateral stiffness than conventional anti-slide piles. The tensile properties of the soil anchor bolts can efficiently limit lateral structural deformation [1].Furthermore the stress condition of the whole structure has a substantial spatial effect, which mobilizes the self-stabilization performance of soil [2].Finite element methods (FEM) is used to study the composite retaining structure by utilizing the software package MIDAS [3].

The methodology used in this paper is PE is based Strength Reduction Method (SRM). It is a finite element method specifically developed to simulate the actual slope failure mechanism and determines both factors of safety and failure behavior. MIDAS is a FE-based software for the analysis of geotechnical projects. Accessing the failure zone of the element, the factor of safety can be consequently calculated. For this purpose, a detailed model can be established through numerical analysis, which can provide sufficiently reliable results [4]. Moreover, elastic-plastic finite elements method for slope stability analysis has the advantages of high accuracy, robustness, and simple calculation process over conventional analysis which are typically used by practicing engineers

[5].The high-pressure grouting technology during the construction of the root piles, and the composite soil around the piles formed by split grouting can make the piles work better. The body is reliably anchored in the soil leading to an increase in the strength of the soil [6].

2. Finite Element Method Analysis of the Bolt-beam-pile structure

In this study, the factors affecting the slope performance such as the shape of sliding surface, the distribution of foundation materials and the rigidity of sliding surface, are taken into account in the stability analysis of slope. Moreover, FEM analysis is carried out by modeling the structure and soil as whole to get better results. The analysis of CRS mechanism was developed in the FEM software.The appropriate model selected to simulate the soil behavior in MIDAS has confirmed the importance significance, and sensitivity of the soil material properties when compared with experimental data. The resulting data from this research was used to validate output mathematical models using a Mohr-Coulomb Plasticity model. The use of software is essential to perform nonlinear analyses by FEM. Furthermore, the software can model more general geometries and soil property variations. As such, such simulations can be applied to complex slope configurations and soil deposits in two or three dimensions to model virtually all types of mechanisms [5].

2.1Definition of ground and structural materials

The constitutive soil model carried out in the present research is Mohr-Coulomb model because this research mainly focuses on the practical application.

Ground materials are defined as below:

Table 1. Ground Materials Properties

Name	Ground material 1	Ground material 2	Ground material 3
Material	Isotropic	Isotropic	Isotropic
Model Type	Mohr-Coulomb	Mohr-Coulomb	Mohr-Coulomb
Elastic modulus (E) (kN/m^2)	10000	20000	30000
Poisson's ratio (Nu)	0.3	0.3	0.3
Unit weight (kN/m^3)	20	22.5	22.5
K0	0.5	0.5	0.5
Cohesion (c) (kN/m^2)	18	22	25
Frictional angle (Phi)	17	21	22

Table 2. Structural features properties

Name	Pile	Beam	Anchor
Model Type	Elastic	Elastic	Elastic
Elastic modulus (E) (kN/m^2)	20500000	20500000	20500000
Poisson's ratio (Nu)	0.167	0.167	0.167

The figure 1 represents an overall view of the two dimensional model of the current research.

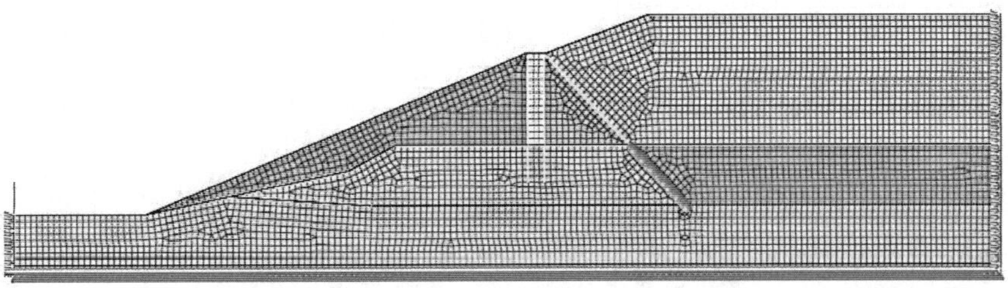

Figure 1. The two dimensional model

2.2Analysis of the bolt-beam-pile CRS with fixed constraints

Table 3. Total displacement CRS/ Fixed Constraints (m)

Node location	Value (m)
11075	0.06
6730	0.05
15815	0.05
13583	0.05
17776	0.05
11207	0.05

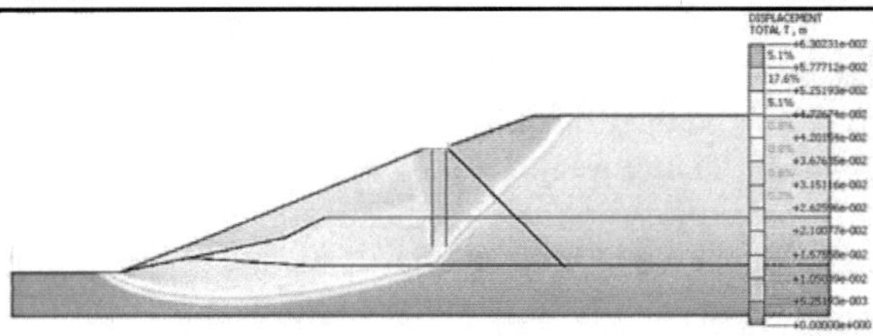

Figure2. Total displacement CRS/ Fixed Constraints

Figure 2 shows that the presence of the structure changes the shape of the sliding arc failure. The increase of the safety factor and the decrease of the displacement demonstrate the technical feasibility of the CRS.

Figure 3. Anchor Axial Forces (kN)Figure 4. The axial force on the beam CRS/Fixedconstraints (kN)

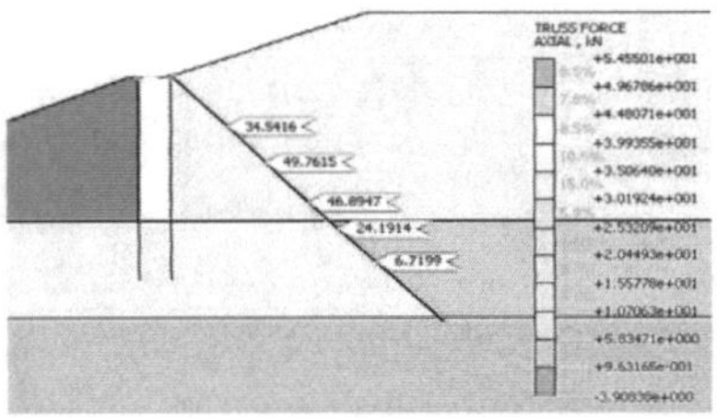

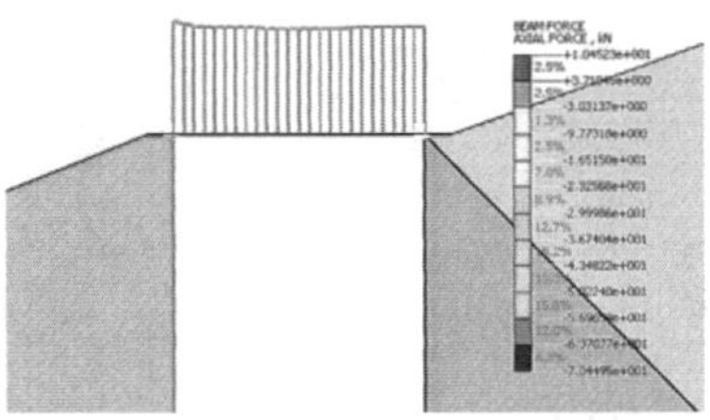

Figure 5. Constraints results

Connections	Scheme	Safety Factor	Decrease of displacement	Piles Axial force (kN)		Piles Shearing forces (kN)		Piles bending moment (kN.m)		Piles deflection (m)		Anchor Truss Axial force (kN)	Beam Axial force (kN)	Beam Bending moments (kN.m)	Beam Deflection (m)
				Left	Right	Left	Right	Left	Right	Left	Right				
Roller		1.91	58.00%	11.07	11.23	11.2	11.12	31.8	60.57	0.0005	0.0017	57.11	11.2	70.1	0.0009
Hinge	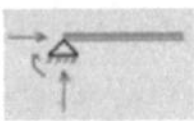	1.78	70.00%	10.36	11.02	2.25	10.99	19.34	60.92	0.0003	0.0018	51.86	9.74	50.28	0.0006
Fixed	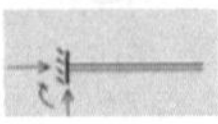	1.86	67.00%	11	11.14	10.4	11.07	19.79	62.94	0.0002	0.0019	49.76	11.2	62.94	0.0006

This comparison showed that the hinge constraint is the optimal choice, because it has smaller inner forces and bending moments, and reasonably reduce the displacement of the initial state of the slope. The appropriate constraint for the bolt-beam-pile CRS, and the pertinence of the implementation of the new CRS structure were in contrast with the typical retaining structure.

3. Conclusions

The research work presented in this paper analyzed the interaction between soil and the new composite retaining structure. after the simulation of the Bolt-Beam-Pile retaining structure by finite elment software, the following conclusions and recommendations are obtained.

- The safety factors and displacement under the three types of constraints are significantly different. The hinge constraint is the optimal displacement constraint.

IOP Conf. Series: Materials Science and Engineering **794** (2020) 012066 doi:10.1088/1757-899X/794/1/012066

- The hinge constraints induce the minimum demand of axial and shearing force, bending moments, and deflections on the structural features. Hence, the hinge is the optimal constraint for the Bolt-Beam-Pile Composite Retaining Structure.
- The practical project is technically feasible because the finite element analysis confirmed the construction pertinence of the Composite Retaining Structure for slope stability.

References

[1] Liao, H.-J., Cheng, S.-H., & Chen, R.-D. (2019). Remedial Measures for Existing Anchored Slopes in Taiwan. ASCE, 1943-2000.

[2] Liu, H., Liant, R., Yang, K., & Nusairat, J. (2011). Slope Stability and Earth Retaining Walls. ASCE, 9-16.

[3] Shi, J., & Yonghui, C. (2015). Three- dimensional numerical parametric study of the influence of basement excavation on te existing tunnel. Journal of Computers and Geotechnics, 146-158.

[4] Cai, F., & Ugai, K. (2004). Numerical Analysis of Rainfall Effects on Slope Stability. ASCE, 68-78.

[5] Midas Elite Center. (2013). Slope Stability Design for Dams and Embankments. Geotechnics - Finite Element Analysis Training. London: Midas UK.

[6] Comodromos, E., Papadopoullou, M., & Laloui, L. (2016). Contribution to the design methodologies of piled raft foundations under combined loadings. Canadian Geotechnical Journal, 559-577.

IOP Conf. Series: Materials Science and Engineering **794** (2020) 012067 doi:10.1088/1757-899X/794/1/012067

Research Status of 1-D Soil Layers Seismic Response

Liu Bingbing[1*], Wang Haiyun[1], Jiang Weiping[1], Zhang Xu[1]

[1]Institute of Engineering Mechanics, China Earthquake Administration, Harbin, 150080, China

[*]Corresponding author's e-mail:liubingbing0925@126.com

Abstract. The local site effect of vibration propagation during earthquakes is usually considered by one-dimensional site response analysis, while the ground motion of the site is the result of the combined effect of the source, the propagation path and the local site conditions. In the study, the influence of ground motion input, soil layer properties and analysis methods need to be considered. Based on a large number of related literatures, this paper briefly summarizes the research status of seismic response of one-dimensional soil layer sites through the determination of bedrock ground motions, soil dynamic constitutive models, and analysis methods. In the future, the development of seismic response of soil sites is prospected and summarized.

1. Introduction

The analysis of soil layer seismic response is one of the important contents in the research field of geotechnical earthquake engineering. Since the 1970s, seismic engineers and geotechnical engineers at home and abroad have begun research in this field and have achieved a series of valuable research results. The earlier research in this area was Wood [1-2]. When collating the seismic damage data of the San Francisco earthquake, he first realized that the site conditions had an impact on the ground motion. Therefore, people began to pay attention to the research on the impact of site conditions on earthquake ground motion. Duck [3] analyzed the collected data and found that there is a certain relationship between the site conditions and the degree of earthquake damage. Wiggins[4] applied this research earlier using numerical methods. In 1964, he began to quantitatively analyze the relationship between ground pieces and ground motion from strong earthquake records.

With the continuous deepening of theoretical research on soil seismic response analysis technology and the continuous development of test technology, the seismic response analysis technology of soil layer has also been continuously developed and perfected, various methods have been developed. For example linear elastic analysis methods and non-linear integration methods, linear total stress method and nonlinear finite element methods. The methods have been developed from simple to complex. This paper discusses and summarizes one-dimensional aspects from the determination of bedrock ground motions, soil dynamic constitutive models, and analytical methods.

2. Determination of bedrock ground motion

The site ground motion is the result of the combined action of the source, propagation path and local site conditions. As input parameters for analysis, the determination of bedrock ground motion is particularly important. Wang Yuqing had summarized three methods to obtain bedrock ground motion: (1)selecting deep well bedrock records;(2)appropriately adjusting bedrock ground motion;(3)use artificial seismic waves that meet the design response spectrum.

At present, the method using bedrock records is more mature than other methods, and the existing strong earthquake records cannot meet the needs of engineering and research. This requires adjustment of the actual bedrock ground motion. Method, by multiplying the recorded ground motion by an appropriate scale to make the amplitude and the period of excellence meet the calculation requirements. Although the scale method is simple and clear in use, it has certain limitations on the scaling ratio. WenqiDu[5] gave a scale limit range of 3-5 times for general use by studying a large amount of ground motion data, but Since the ground motion records used in the analysis are all selected from the western United States, no general research has been performed. Therefore, the use of the proportional method of ground motions at different stations remains to be studied.

At present, the methods for simulation of strong vibrations can be divided into three categories: seismic methods and engineering methods.

2.1 Seismological methods

Based on different source models, propagation functions, and site models, the seismological methods are divided into deterministic, stochastic, and hybrid methods.

The deterministic method is to represent the ground motion as a convolution of the source function and the Green's function, which is based on the representation theorem of Aki et al.[6]Deterministic methods can better simulate low-frequency ground motions, but it is difficult to describe high-frequency ground motions. The stochastic methods are mainly divided into two categories: the empirical Green's function method based on small earthquake records and the random vibration method based on Gaussian band-limited white noise. Both methods can better simulate high-frequency ground motions. The random vibration method can well fit the actual recorded response spectrum characteristics, but there is a certain gap in the fitted waveform.

In order to simulate wide-band ground motions, Kamae Proposed to use Green's function method to simulate low-frequency ground motions and random methods to simulate high-frequency ground motions, and then used the mixed wide-band small earthquake time history as an empirical Green function to synthesize large earthquake time history method. Irikura[7]proposed the idea of predicting future ground motions using the empirical Green function method, summarized the general steps, and simulated a large number of ground motions. This method combines the advantages of the Green's function method and the stochastic method, and is the main method in current seismological methods.

At present, the understanding of the complexity of the source and the propagation medium is not complete, so there is still some uncertainty in the seismic method, and its accuracy needs to be improved. In general, seismic methods have clear physical meaning and are one of the main directions for the development of ground motion simulation.

2.2 Engineering method

The engineering method emphasizes the mathematical description of ground motion parameters, which is the empirical statistical relationship between stochastic process theory and ground motion parameters. As early as 1947, Housner proposed the use of random pulse superposition to simulate ground motions. However, this process is assumed to be unreasonable, So Kanai improved the stationary random process method and proposed a filtered white noise model. In order to reflect the non-stationary nature of the ground motion, Bolottin proposed to multiply the stationary Gaussian process by the envelope function to simulate the ground motion. Based on Bolotin, Scanlan proposed the use of triangular series to simulate strong ground motion, this method can consider the probability of occurrence and the seismic intensity background of bedrock ground motion and can be widely used to this day. In order to solve the problems of fitting accuracy and convergence, many scholars have improved this. Hu Yanxian improved the method of modifying the amplitude spectrum so that the frequency component near the control point of the response spectrum can be distinguished during the fitting, which speeds up the convergence speed and improves the fitting accuracy. Zhai Ximei proposed to use the method of joint modification of power spectrum and amplitude spectrum to improve the fitting accuracy of response spectrum and target spectrum in triangle series method simulation. Cheng Wei proposed a method of directly fitting the response spectrum to avoid the error caused by the target spectrum to power spectrum conversion.

The engineering method is affected by the target response spectrum and the intensity envelope function, so it have certain uncertainty. At the same time, the seismic wave fitted by this method lacks the completely non-stationary characteristics of natural ground motion. The engineering method is simple in theory and easy to apply, at present, the engineering method is still the main method for synthesizing artificial ground motion in engineering construction.

3. Dynamic Constitutive Model

The earliest development of the soil dynamic constitutive model is the linear elastic model. During this period, the soil was considered as a linear elastic material. After the online elastic model, non-linear elastic models appeared, such as Duncan-Chang model. With the understanding of soil nonlinearity, various dynamic constitutive models have emerged, which can be divided into viscoelastic models and elastoplastic models.

3.1 Viscoelastic model

The viscoelastic model introduces a viscous term into the elastic model, which can consider the energy dissipation characteristics of the soil under dynamic loading. The viscoelastic model is divided into an equivalent linear model and the Masing type.

3.1.1 Equivalent Linearization Model

The equivalent linearization model is based on the equal overall dynamics effect, and uses equivalent shear modulus and damping ratio to replace the shear modulus and damping ratio under all strain amplitudes to transform the nonlinear problem into linear The problem is solved in the frequency domain using the principle of superposition[8]. For example, Hardin Drnevich model[9], Ramberg Osgood model, hyperbolic model, etc.

In recent years, many scholars have improved the equivalent linear model. Chen Guoxing introduced the idea of segmentation to reduce the "virtual resonance effect" produced by the equivalent linear model. The combination of equivalent linearization thought and stochastic theory is another direction for the development of equivalent linearization models. In this regard, Wu Zaiguang, Lin Ye, and Koreatown have done a lot of work. Chen Yuan et al. Also established site earthquakes that consider geotechnical nonlinearity Dynamic random numerical simulation method. In addition, the equivalent linearization model has also made breakthroughs in the calculation of soft and frozen soils.

The equivalent linearization model is a breakthrough achievement. It has a simple concept, few calculation parameters, and a simple form. It is widely used in engineering. But after all, it is a simplified method, and it also exposes some shortcomings, such as unreasonable calculation results when strong earthquakes or soft soil layers, and large calculation errors at high frequencies.

3.1.2 The Masing Model

The difference between the Masing model and the equivalent linearization model is that the Masing model directly gives specific expressions of the dynamic stress-strain relationship according to different loading, unloading, and reloading conditions. Is the backbone curve, and then the soil hysteresis curve equation is constructed according to Masing's law. Since the Masing model was proposed in 1926, many scholars have improved and developed Masing-type nonlinear models, such as the Pyke model[10]; Iwan model[11]; Wang Zhiliang, etc., introduced damping ratio degradation coefficients and proposed generalized Masing models. In addition, Li Xiaojun proposed a "dynamic skeleton curve" model based on the Masing model; Qi Wenhao and others constructed a UE model suitable for asymmetric loads according to Masing's law; Zheng Datong developed the Iwan model A new model that can consider work hardening and work softening is proposed.

At present, the viscoelastic model is one of the mainstream dynamic constitutive models. Its simple form and easy-to-measure parameters play an important role in the current seismic response analysis of soil layers. However, it has many deficiencies, such as the failure to consider the strain accumulation of the soil, the influence of the stress path, and the problem of strain softening.

3.2 Elastoplastic model

Many scholars have established many elastoplastic models based on different yield conditions, flow rules, and hardening laws. Li Xiaojun summarized the elastoplastic model, and divided the elastoplastic model into a single yield surface model, a multiple yield surface model, a boundary surface model, an internal time model, a rate model, a nonlinear incremental model, and an empirical relationship model.

The single yield surface model was first developed in the late 1960s. Its stress-strain characteristics are determined by the position of the soil stress state relative to the yield surface. Considering the hysteretic characteristics of the stress-strain relationship of the soil, Concepts such as homosclerosis and kinematic hardening have also been introduced into plastic single yield surface models, such as the hat model proposed by Shen Zhujiang and others.

Because the single-yield-surface model is difficult to reflect the continuous change of soil stress-strain, Iwan proposed a multi-yield-surface model in 1967. The multi-yield surface model reflects the nonlinear hysteresis of the soil through the movement of the yield surface and the isotropic expansion rule. Compared with the single-yield surface model, it can more reasonably simulate the continuous soil stress-strain obtained from the experiment alternative relation. At the same time, plastic multi-yield surface models have emerged. The main differences are the shape of the boundary surface and its moving rules, and the study methods of the hardening modulus field.

After the multi-yield surface model, Bardet et al. Proposed a boundary surface model, which includes a single-sided model and a two-sided model. The two-sided theory was first proposed independently by Dafalias and Krieg. Later, Mroz [77] and others used this theory to describe the constitutive characteristics of soils under round-trip loads and established their own plastic models. The boundary surface model can better simulate the response of soil under dynamic action.

4. Analysis method of siteseismic response

Site seismic response analysis methods are mainly random vibration analysis methods and deterministic dynamic analysis methods. At present, the analysis of random seismic response is immature and there is still a certain distance to apply it to engineering practice. The deterministic method still occupies a dominant position in the application. The earlier deterministic method is the linear elastic wave analysis method, which considers that the soil stress-strain relationship changes linearly. With the accumulation of actual seismic observation data and the in-depth study of soil dynamic characteristics, the equivalent linear method and nonlinear method have been gradually developed.

4.1 Equivalent linearization method

The equivalent linearization method appeared earlier and is widely used at home and abroad. It uses the relationship between dynamic shear modulus, damping ratio and strain to determine the equivalent dynamic shear modulus and damping through multiple linear calculations. The ratio and the strain level corresponding to the equivalent parameter are matched with the effective strain level under the action of the earthquake, thereby transforming the dynamic solution of the complex nonlinear system into a frequency domain iterative solution of the wave equation. Seed[8] first proposed the equivalent linearization method and compiled the corresponding program SHAKE. The proposed equivalent linearization method is a breakthrough in seismic response to soil layers. The method is simple, easy to understand, and easy to operate. Application and promotion. However, the traditional equivalent linearization method also has many problems that are difficult to overcome. The more serious problem is that when the amplitude of the soil strain is large, the calculated high-frequency components of the ground motion are usually lower than the actual strong vibration observations. Based on the above problems, different scholars have proposed various frequency-dependent equivalent linearization methods. Sugito et al. Used frequency-dependent equivalent shear strain to replace the equivalent shear strain originally calculated with a fixed coefficient (such as 0.65), and then equivalent The effective linearization method was improved; Yoshida further analyzed that the high-frequency frequency response magnification calculated by SHAKE was lower than the actual one, and divided the frequency into three sections. Each frequency section used a different formula to calculate the equivalent shear strain; Kausel et al. Equivalent linearization does not consider the influence of high-frequency and small-vibration components, so a relationship between equivalent shear strain and frequency is suggested; Jiang Tong et al. Used a spectral smoothing method to fit the standardized strain spectrum curve to obtain the frequency correlation. New method.

The introduction of the equivalent linearization method is a major breakthrough in the analysis of site seismic response, which has extremely important significance. However, this method converts complex non-linear problems into approximate solutions to linear problems. Problems still occur during the use, such as Problems such as infinite loop phenomena, weak sites, and unreasonable calculations under large earthquakes. In order to improve and improve the existing methods, traditional iterative methods can be replaced by introducing new stable algorithms.

4.2Nonlinear methods

In addition to determining the appropriate constitutive model, the nonlinear method also includes solving dynamic equations and establishing artificial boundary conditions.

4.2.1 Solving the equation of soil motion

The numerical calculation methods of the motion equation mainly include the finite element method and the finite difference method. The finite difference method has strict requirements on the mesh shape of the solution domain, so the finite element method is commonly used.

Kim et al. Proposed a direct finite element method that can consider the anisotropy of the soil and the transmission conditions of any boundary shape, and Song et al. Combined the advantages of the finite element method and the boundary element method and proposed a mapping boundary finite element method Also known as the uniform infinitesimal finite element method. Li Xiaojun combined with an explicit differential numerical integration method to provide a method for analyzing the nonlinear seismic response of soil layers. Methods: Liao Heshan et al. Studied the nonlinear seismic response of the soil layer based on the one-dimensional composite strain wave theory and the characteristic line difference method. Jin Xing et al. Proposed an explicit finite element method for nonlinear analysis of seismic response of horizontally layered sites based on the incremental form of the discrete dynamic balance equation of Li Xiaojun's integral format.

At present, the finite element method is often used, but it also has limitations. The more the finite element mesh is divided, the larger the calculation amount. Therefore, many scholars are still looking for a high-precision, unconditionally stable algorithm.

4.2.2 Establishing boundary conditions

At present, the most commonly used boundary conditions are global artificial boundary, approximate local artificial boundary, and local artificial boundary. Among them, the local artificial boundary conditions are more stable and have a smaller amount of calculation when applied to the nonlinear calculation of complex media, which is convenient to implement.

Local artificial boundary conditions have multiple forms of boundary expression. The viscous boundary was first proposed by Lysmer. It is a boundary condition with low accuracy, which is only the first-order calculation accuracy, but the processing method is simple, the physical concept is clear, and it is widely used in engineering. Smith et al. Proposed and developed superimposed boundaries. It mainly solves the wave problem by superposing the solutions of two problems, fixed boundary and free boundary, which will increase the amount of calculation. Wolf et al. Proposed a double asymptotic Higdon boundary by combining the double asymptotic boundary and the Higdon boundary and developed a viscoelastic boundary. Liao Zhenpeng et al. Proposed and developed a local artificial boundary of the first-order transmission formula, and demonstrated its reliability in the time domain.

These boundaries are easily combined with finite element methods. Among them, the viscoelastic boundary itself is relatively stable and easy to implement, but the accuracy is poor; the transmission boundary is complicated and there are stability problems. Therefore, although many artificial boundary conditions have appeared, there is no recognized optimal method currently affected by stability, accuracy, and ease of implementation.

5. Conclusion

At present, the seismic response analysis methods of soil sites have developed more maturely. However, a lot of work is still needed to truly and reasonably calculate site seismic responses. Starting with the key issues, the research status of the seismic response analysis of the soil layer site is briefly described, and the future research directions are proposed.

1.The best way to determine the bedrock ground motion is to use the measured bedrock ground motion as input, but the number of records currently available is relatively small. Among the methods for considering the simulation of ground motion, the seismological method can consider the physical nature and is worthy of vigorous development.

2.None of the existing soil dynamic constitutive models can completely reflect the nonlinear dynamic characteristics of soils. Therefore, it is still the direction of future efforts to build a soil dynamic constitutive model that is practical, easy to measure and accurate.

3.The equivalent linearization method has some limitations in practical applications. In order to solve this problem, in-depth study of the time-domain nonlinear method is a better development direction. For example, the establishment of high-precision, high-efficiency boundary conditions and the search for a stable solution algorithm.

References

[1] Richter C F. Elementary Seismology[M]. San Francisco, 1950.
[2] Bo JS, Li XJ, Li SY. Some progress of study on the effect of site conditions on ground motion[J],World Earthquake Engineering,2003,19(2):11~15
[3] Li XJ. Analytical method of influence of site soil layer on earthquake ground motion[J]. World Earthquake Engineering,1992,8(2):49~60.
[4] Wiggins, J. H. (1964). Effect of site conditions on earthquake intensity, Proc. ASCE, 90, 279-312
[5] W Du, CL Ning, G Wang. The effect of amplitude scaling limits on conditional spectrum-based ground motion selection[J]
[6] Aki K, Richards P G. Quantitative Seismology: Theory and Methods[M]. NewYork: W H Freeman & Co Ltd, 1980.
[7] Irikura K. Semi-empirical estimation of strong ground motions during large earth quakes[J]. Bulletin, Disaster Prevention Research Institute, Kyoto: Kyoto University, Japan, 1983(33):63-104.
[8] Martin PP, Seed HB. One-dimensional dynamic ground response analysis. Journal of Geotechnical Engineering Division ASCE{1982; 108:935}952
[9] Hardin B O, Drnevich V P. Shear modulus sand damping in soils: design equations sand curves[J]. Journal of Soil Mechanics and Foundations, ASCE, 1972, (7):667–692.
[10] Pyke, R. M. (1979). "Nonlinear soil models for irregular cyclic loading." J. Geotech. Engrg. Div. 105(6), 715-726.
[11] Iwan W D. On a class of models for yielding behavior of continuous and composite systems. ASME Journal of Applied Mechanics, 1967, 34(3): 612617

Study on strength and durability of fair-faced concrete in underground station of urban rail transit

Sheng Ding[1], Weixing Wang[2], Xuanming Yang[1], Shuxian Wang[1], Li Chen[3*]

[1]Shanghai Mechanized Construction Group Co., Ltd, Shanghai 200072;

[2]Jiangsu Yanghu Construction Project Administration Co., Ltd., Changzhou 213159;

[3]Nanjing Institute of Technology, Nanjing 211167

*Corresponding author's e-mail: alexandermacedonia@163.com

Abstract: Concrete is still in the developing stage in China, only some design and construction units can master its construction technology. The side walls and middle columns of underground station of Changzhou Metro Line 2 are made of fair-faced concrete. In this paper, the material properties of the underground station are studied based on the project. In this paper, the working performance, compressive strength, impermeability, sulfate corrosion and carbonization resistance of fair-faced concrete are mainly studied by drawing up the base mix ratio. The research results have reference value for the future development of fair-faced concrete engineering.

1. Introduction

The strength and durability of concrete are important for concrete structures. By carefully controlled construction technology, fair-faced concrete structure takes the original poured surface after concrete pouring as the external surface, and achieves the architectural aesthetic requirements with its own good texture [1, 2]. The smooth surface and clear texture of fair-faced concrete after forming bring a feeling of returning to nature. The overall effect is natural, thick and elegant [3]. At the same time, the steps of plastering and decoration are omitted, the cost is reduced and the time of decoration stage is saved. It is an excellent green building material. At present, the technology of fair-faced concrete is in the development stage in China, which is only a little higher than the ordinary concrete in the performance requirements. The research on fair-faced concrete is not mature, lacking theoretical basis, experimental demonstration and practical engineering verification [4, 5]. There are still many problems in the mix design and construction of fair-faced concrete, especially in the performance of fair-faced concrete. The surface quality of fair-faced concrete is required to be high, so most of the researchers and constructors improved its surface quality from the formwork, but less from the aspect of mix proportion. The formwork is very important to its surface quality, but it is also of great significance to optimize the concrete mix proportion and improve the concrete performance to meet the quality requirements of the concrete surface. In practical engineering, when the surface of fair-faced concrete is directly exposed to the external environment, its durability is particularly important, which not only affects its surface beauty, but also is very important for its corrosion resistance and structural safety. In order to improve its durability, many scholars have conducted a lot of research on its mineral admixtures and additives.

The sidewalls and middle columns of Changzhou Metro Line 2 underground station are made of C35 grade and C45 grade fair-faced concrete respectively. The fair-faced concrete structure of the underground station needs to show its inherent feeling of clear texture, smooth surface and returning to nature, which has spiritual and sensory enjoyment for the waiting passengers, so there are certain requirements for its appearance quality. As an important part of the supporting structure, the sidewalls and middle columns of the underground engineering are generally required that the fair-faced concrete material of which to have preferable bearing capacity and durability, and to meet the requirements of modern buildings. Combing with this project, the working performance, strength and durability of the fair-faced concrete with different mix were mainly studied in this paper, which has reference value for the application of fair-faced concrete.

2. Experiment

2.1 Test materials

The raw materials used for the fair-faced concrete in Changzhou Metro Line 2 project were selected for specimen preparation to ensure the consistency of the concrete material. The selected raw materials were:

a. Cement: Yangzi P·O42.5;

b. Sand: medium sand in zone II;

c. Gravel: 5.0-25.0 two-gradation, the proportion of 5.0-16.0 gravel to 16.0-25.0 gravel is 25%: 75%;

d. High performance water reducing agent: sobote pca-i;

e. Fly ash: Changzhou Guodian Grade I;

f. Mineral powder: Grade S95;

g. Anti crack and waterproof agent: 5% admixture.

2.2Test scheme

The mix proportion of the fair-faced concrete for the sidewalls and the middle columns of the underground station was designed according to the code [6]. By changing the content of the mineral admixture, the mix proportions of the specimens were adjusted. The details of the mix proportions are shown in Table 1 and Table 2. The workability of the fair-faced concretes with different mix proportions were tested according to the slump method in the code [7]. The compressive strengths of the concretes were tested with 150mm×150mm×150mm specimens after standard curing 28 days according to the code [8]. The disabilities were tested according to the requirements about the impermeability, the sulfate resistance and the carbonation resistance in the code [9], in which the method of step-by-step pressure was used for the impermeability test, dry and wet cycle method was used for sulfate resistance test, and fast carbonation method was used for the carbonation resistance test.

Table 1. Mix proportion of C35 fair-faced concrete for the sidewalls

Label	Cement	Fly ash	Mineral powder	Fine aggregate	Crack resistant agent	Coarse aggregate		Water	Additive
	P.O 42.5	Grade I	S95	Medium sand	HME-III	5-16mm	16-25mm	tap water	PCA-I
P1	276	55	37	760	18.4	263	788	160	2.9
P2	239	74	55	760	18.4	263	788	160	2.9
P3	313	37	18	760	18.4	263	788	160	2.9
P4	276	55	37	760	18.4	263	788	160	3.3
P5	276	55	37	760	18.4	263	788	160	2.6

Table 2. Mix proportion of C45 fair-faced concrete for the middle columns

Label	Cement	Fly ash	Mineral powder	Fine aggregate	Coarse aggregate		Water	Additive
	P.O 42.5	Grade I	S95	Medium sand	5-16mm	16-25mm	tap water	PCA-I

P6	307	66	66	705	264	793	155	3.7
P7	263	88	88	705	264	793	155	3.7
P8	351	44	44	705	264	793	155	3.7
P9	307	66	66	705	264	793	155	3.3
P10	307	66	66	705	264	793	155	4.2

3. Results and discussion

3.1Workability

The test results of concrete workability showed that the slump of each group of concrete is 170-180mm and the loss of concrete is 10-15mm after 1 hour. This was obtained by adjusting the dosage of admixture under a constant water binder ratio. It should be noted that the fluidity of P4 and P10 groups of concrete was good, but there was slight bleeding and bottom catching phenomenon. The cohesion, water retention and fluidity of the other groups of concrete were good, which met the requirements of construction. It could be seen that too much water reducing agent will cause bleeding of concrete and affect the working performance of concrete. Therefore, attention should be paid to the dosage of water reducing agent, during the mixing and preparation of fair-faced concrete, in order not to affect the workability of concrete.

3.2Strength

Figure 1 shows the compressive strength of concrete with different mix proportions. The sidewall C35 grade concretes with the mix proportions P1-P5 and the middle column C45 grade concretes with the mix proportions P6-P10 all met the strength requirement for each grade. In the C35 concretes of the sidewall, the compressive strength of P2 group is the lowest, and that of P3 group is the highest, but not significantly higher than those of P1, P4 and P5 groups. In the middle column C45 grade concretes, the compressive strength of P7 group is the lowest, and that of P8 group is the highest, but not significantly higher than those of P6, P9 and P10 groups. The strengths of P3 and P8 group concrete developed well, and separately reached C35 and C45 strength grades at 7 day age.

Comparing P2 group and P7 group with the reference groups, it can be found that though the compressive strength of concrete decreased with the increase of fly ash and mineral powder, the concrete strength decreased to a small extent. This may be due to the fact that the variation of the mineral powder admixture was small, thus has little influence on the concrete strength. Some researches showed that adding a large number of mineral admixture could reduce the strength of concrete, so the amount of mineral admixture should not be too much in actual project, in order to avoid the concrete strength not meeting the requirement and affecting the structural safety.

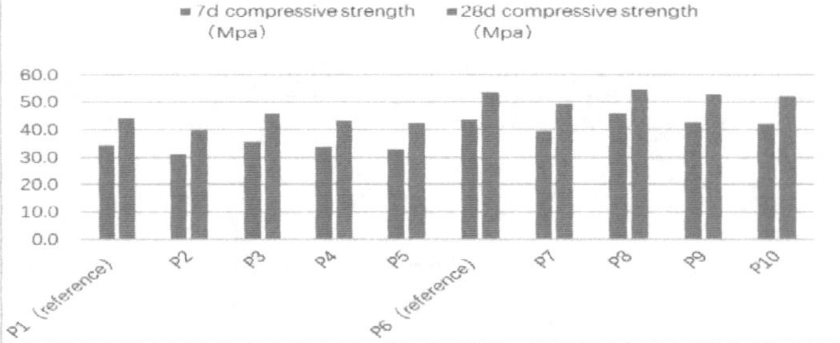

Figure 1. Compressive strengths of the fair-faced concretes with different mix proportions

3.3Impermeability

Table 3 shows the impermeability pressure and grade of the concrete with each mix proportion. The sidewall C35 grade concretes with the mix proportions P1-P5 and the middle column C45 grade concretes with the mix proportions P6-P10 all met the impermeability requirement for each grade. The impermeability grades of concretes are between p15-p17, which met the requirement of the underground station concrete impermeability. The change of mineral admixture content had no significant effect on the impermeability of concrete. In addition, the impermeability of C45 concrete was not higher than that of C35 concrete.

Table 3. Impermeability grade of the fair-faced concretes with different mix proportions

Label	P1	P2	P3	P4	P5	P6	P7	P8	P9	P10
Water pressure (MPa)	1.7	1.6	1.6	1.8	1.7	1.7	1.6	1.6	1.8	1.6
Impermeability grade	P16	P15	P15	P17	P16	P16	P15	P15	P17	P15

3.4 Sulfate resistance

Figure 2 shows the appearance of P8 concrete specimen after 90 cycles of sulfate drying and wetting. The results show that under the action of sulfate drying and wetting cycles, cracks appeared on the concrete surface, even with aggregate peeling off. Table 4 shows the corrosion resistance coefficients of compressive strength of the concretes with different mix proportions after 90 cycles of sulfate drying and wetting. The results show that although the compressive strength of each group of concrete decreased to different degrees, the corrosion resistance coefficients of compressive strength of the other group concretes are higher than 80% except for P3 and P8 concrete specimens. Through the analysis on P1-P3 group concretes and P6-P8 group concretes, it is found that mineral admixtures can improve the sulfate resistance of concrete, which may be due to the combination reaction of mineral admixtures and sulfate ions, so as to improve the corrosion resistance.

Figure 2 Appearance of a specimen of P8 concrete after 90 cycles of sulfate drying and wetting

Table 4. Corrosion resistance coefficient of compressive strength of the concretes after 90 cycles of sulfate drying and wetting

Label	P1	P2	P3	P4	P5
Corrosion resistance coefficient of compressive strength (%)	83.09	87.54	77.24	85.45	91.25
Label	P6	P7	P8	P9	P10
Corrosion resistance coefficient of compressive strength (%)	82.49	97.61	74.90	87.48	98.04

3.5 Carbonation resistance

Table 5 shows the carbonation depth of the concrete specimens with different mix proportions. The results show that the carbonation resistance of C45 concrete is stronger than that of C35 concrete. That is, the higher the strength, the stronger the carbonation resistance. Figure 3-Figure 5 are concrete specimens with different carbonation times. The results show that the carbonation of concrete is deeper with the increase of carbonation time. The carbonation resistance of concrete can be improved by adding some mineral admixtures.

Table 5. Carbonation depth of different mix proportion concrete (mm)

Carbonation time	Label of specimen				
	P1	P2	P3	P4	P5
7d	4.4	5.5	4.3	5.3	4.2
14d	10.4	10.7	13.1	8.3	5.2
28d	12.8	12.2	14.4	9.3	8.1
Carbonation time	Label of specimen				
	P6	P7	P8	P9	P10
7d	0	0	0	0	0
14d	0	0	0	0	0
28d	0	0	0	0	0

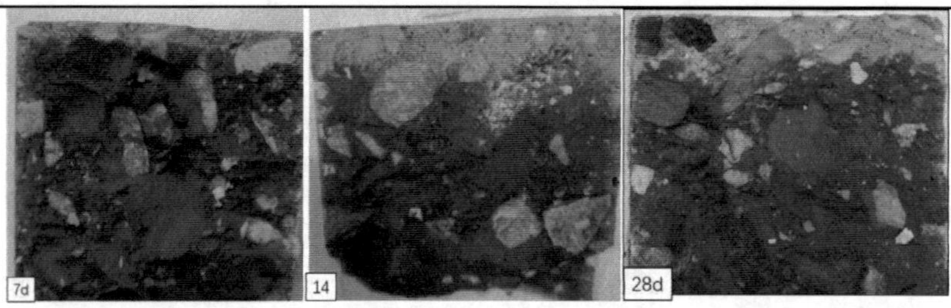

Figure 3. P1 Concrete specimens with different carbonation times

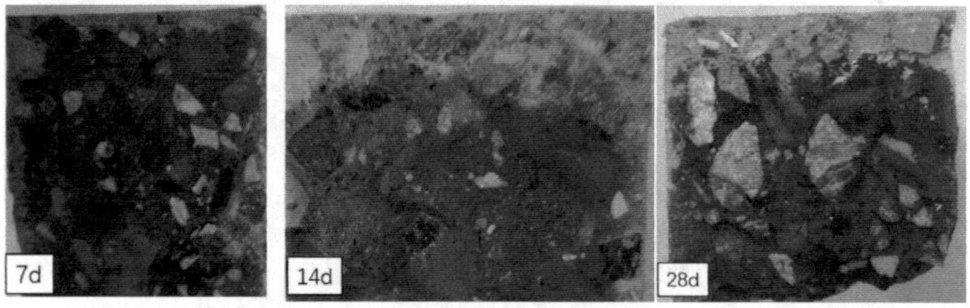

Figure 4. P3 Concrete specimens with different carbonation times

Figure 5. Concrete specimens with different mix proportion with 28d carbonation time

4. Conclusions

Based on the study of the working performance, compressive strength, impermeability, sulfate corrosion resistance and carbonation resistance of the fair-faced concrete materials used in the sidewalls and middle columns of the underground station of Changzhou Metro Line 2, the following conclusions can be drawn from the above results:

a. In the C35 concretes for the sidewalls and C45 concretes for the middle columns with the mix proportions in this paper, except for P4 and P10 groups of concrete, the cohesion, water retention and fluidity of the other groups of concrete are good, which meets the construction performance requirements.

b. The mix proportion of each group of concrete meets the strength requirements, and the amount of mineral admixture should not be too much, otherwise the strength will be reduced.

c. The impermeability grades of C35 concrete for the sidewall and C45 concrete for the middle column are between p15-p17, and are greater than p15, which meets the impermeability requirements. Except for P3 and P8 concrete specimens, the corrosion resistance coefficients of compressive strength of the other groups of concrete specimens after 90 cycles of sulfate drying and wetting are higher than 80%. The higher the strength, the stronger the carbonation resistance. Moreover, with the extension of carbonation time, the deeper the concrete carbonates. Adding a proper amount of mineral admixture can improve the sulfate resistance and carbonation resistance of concrete.

Acknowledgments

Supports from Construction System Science and Technology Guidance Project of Jiangsu (2017ZD131, 2017ZD132, 2018ZD182) are gratefully acknowledged.

Reference

[1] Jun Zhang. Research on the construction technology of as cast finish concrete in housing construction engineering [J]. Construction engineering technology and design, 2018 (8): 704.
[2] Yunhui Pu, Wenyuan Li. Enlightenment of foreign green building promotion on China [J]. Building technology, 2017,48 (7): 678-680.
[3] Guofu Wang, Limin Zhou, Linhai Lu, et al. Research on application technology of as cast finish concrete for rail transit [J]. Concrete, 2017 (03): 146-149.
[4] Zhanye Luo, Min Hou, Tao Li, et al. Study on the construction technology of high performance as cast finish concrete for ASI River Super Large Bridge of Mombasa Nairobi standard gauge railway project in Kenya [J]. Highway transportation technology (application technology version), 2017 (01): 95-98
[5] Wei Guo, Linhai Lu, Longzhi Wang, et al. Overview of concept, research status, existing problems and mix proportion design method of fair faced concrete [J]. Concrete and cement products, 2016 (10): 23-27
[6] China Academy of Building Sciences. JGJ55-2011 Specification of mix promotion design of coordination concrete [S]. Beijing: China Construction Industry Press, 2011.
[7] China Academy of building science. GB/T 50080-2016 Standard for test method of performance on Coordination fresh concrete [S]. China Building Industry Press, 2016.
[8] China Academy of Building Sciences. GB/T 50081-2002 Standard for test methods of mechanical properties of ordinary concrete [S]. Beijing: Building Industry Press, 2003.
[9] China Academy of Building Sciences. GBT 50082-2009 Standard for test methods of long-term performance and durability of ordinary concrete [S]. Beijing: China Planning Press, 2009.

Analysis of Characteristics of Temperature Changes in Heilongjiang River Basin

Bo Yan[1,2], ZheYuan[1,2], JiqiongLi[3], Liqiang Yao[1,2], LiSong[1,2], Yanyi Liu[4], Xiao Zhang[5], Ligang Zhang[6]

[1]Changjiang River Scientific Research Institute, Changjiang Water Resources Commission of the Ministry of Water Resources of China, Wuhan 430010, China

[2]Hubei Key Laboratory of Water Resources & Eco-Environmental Sciences, Wuhan 430010, China

[3]Wuhan Changjiang Kechuang Technology Development Co.Ltd, Wuhan430010, China

[4]Jiangxi Provincial Water Conservancy Planning Design and Research Institute, Nanchang330029, China

[5]Bureau of Hydrology, Changjiang Water Resources Commission of the Ministry of Water Resources of China, Wuhan 430010, China

[6]School of Natural Resources, University of Nebraska-Lincoln, Lincoln, NE 68588, USA

*correspondence: yuanzhe_0116@126.com(Z Yuan)

Abstract. As an international river, the climate change research in Heilongjiang River Basin can provide references for sino-Russia cooperation in tackling climate change. Using the long-term daily measured data from 44 meteorological stations in the Heilongjiang River Basin, the spatial-temporal characteristics of the temperature in the past 59 years were analyzed. The results show that the temperature in the Heilongjiang River Basin has increased significantly and passed the significance test of 0.99. The abrupt change occurred in 1986. The climate change also showed multiple cyclical changes of 16-years, 8-years, and 4-years. The spatial distribution of the annual average temperature of the river basin is generally gradually decreasing from the south to the north as the latitude rises. There is also a spatial difference in the annual temperature increase, but all of the basin have passed the significance test of 99%.

1. Introduction

Heilongjiang is a world-famous international river, and its watershed is located in the mid-high latitudes and cold regions. The climate of the watershed is extremely sensitive to global climate change[1]. In recent years, there have been many studies on the characteristics of temperature changes, but articles on international rivers are Relatively few. Few scholars have studied climate change in Heilongjiang River Basin. A related report from IPCC states that global warming not only affects extreme temperatures and extreme precipitation, but also accelerates the change of the frequency of extreme events such as high temperature, flood, drought, heavy rain, etc[2]. Against the backdrop of global warming, the impact of climate change on water resources in the basin is also particularly significant. Therefore, it is necessary to analyze the characteristics of climate change in Heilongjiang river basin under global warming..

2. Study area, data and methodology

2.1Study area

Heilongjiang River Basin is located in the cold temperate zone and temperate zone, and the monsoon climate is obvious. In winter, the dry and cold air from Siberia brings sunny and dry weather with strong frost. In summer, the warm and humid sea breeze brings heavy rain to improve the water level of the basin and its main tributary. The Heilongjiang River Basin involves China, Russia, Mongolia, and North Korea, and is an important international river

2.2Study data

Daily temperature data series of 44 meteorological stations in the Heilongjiang River Basin from 1954 to 2012 was collected.The complete long-term data is very important for the statistical analysis of the trend, breakout and periodic characteristics of climate change in Heilongjiang River Basin. Data in this paper have all passed the uniformity test of RHtestV.3 [3, 4] software. This software was developed by Environmen Canada's Climate Research Center. It includes the maximum penalty F test (PMFT) and maximum penalty T test (PMT). Considering the lagging first-order autocorrelation of time series, it can be used for the uniformity test of time series on different time scales. Location of 44 stations selected in the temperature analysis Acan be seen in Figure 1。.

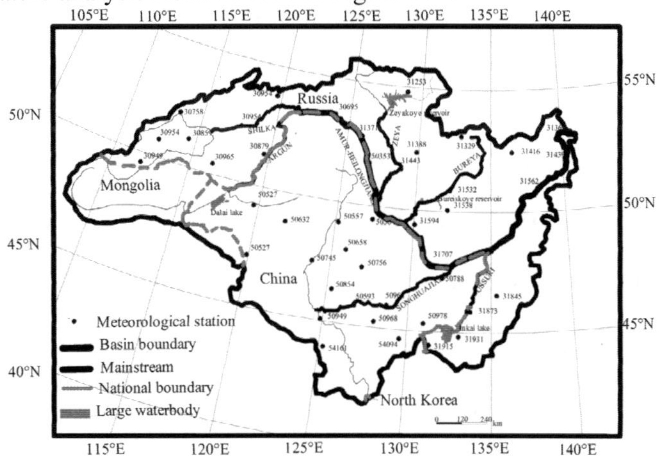

Figure 1.Location of selected meteorological stations for temperature analysis in Heilongjiang River Basin

2.3Methdology

As a common time series trend assessment and breakout diagnosis method, the Mann-Kendall(M-K) test is commonly used in meteorology and hydrology[5-7]. This paper uses the M-K test to quantitatively evaluate the temperature trends and breakout diagnosis in the Heilongjiang River Basin. The Morlet wavelet analysis method [8]is used to analyze the periodicity of temperature changes. the inverse distance weighting method[9]is employed to display the spatial distribution of temperature changes.

3. Results and discussion

3.1Temporal variation characteristics of temperature

3.1.1 Trend analysis of temperature changes

The process of annual mean temperature anomalychanges in Heilongjiang River Basin can be seen in Figure 2. It can be seen from the figure that the annual mean temperature of Heilongjiang River Basin showed an obvious upward trend from 1954 to 2012 and it can be estimated that its upward range is 0.34°C/10a according to the linear trend. Although 0.34 is not a big number, an increase of 0.34°C per decade is also a quantity that cannot be ignored for annual mean temperature and it is higher than the global and northern hemisphere average temperature growth rates. The results of M-K test shows that its upward trend has passed the significance test of 0.99. The M-K statistic of annual mean temperature of Heilongjiang River Basin in China is 5.45 and it has passed the significance test of 99% reliability by M-K test, which has a significant upward trend. This significant upward trend of annual mean temperature will inevitably affect the water cycle process in its basin and it also has a certain impact on the utilization of water resources and the water ecological environment in the basin.

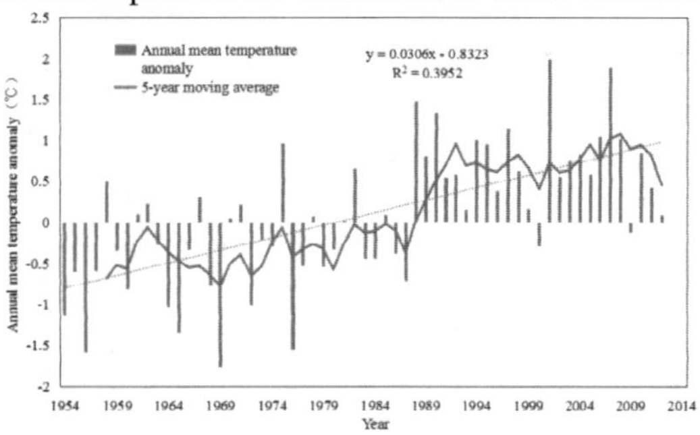

Figure 2. The process of annual mean temperature anomalychanges in Heilongjiang River Basin

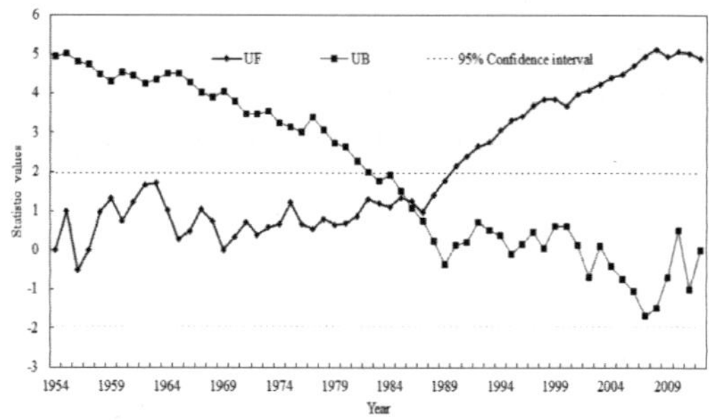

Figure 3. Breakout Diagnosis of annual mean temperature in Heilongjiang River Basin by M-K test

3.1.2 Breakout diagnosis of temperature changes

Figure 3 is a breakout diagnosis chart of annual mean temperature in Heilongjiang River Basin from 1954 to 2012, it is found that 1986 was a change point of the average temperature series by the M-K test, which means that the temperature trend has changed the most in the years around 1986 and the breakout has reached a significance level of 95% (in fact, it has exceeded the significance level of 99%). Therefore, there was a significant temperature change in the late 1980s ,which means that, it has experienced a transition from cold to warm. This sudden change in temperature has brought the annual mean temperature in Heilongjiang River basin to a new level since the late 1980s, which will inevitably affect each link in the hydrological cycle in the basin, thus affecting ecological hydrology situation in the basin.

3.1.3 Periodic analysis of temperature changes

The annual mean temperature series of 59 years from 1954 to 2012 in Heilongjiang River Basin was used as the periodic analysis data of temperature changes. The data was preprocessed by symmetric extension method to eliminate the boundary effect of wavelet transform.

IOP Conf. Series: Materials Science and Engineering **794** (2020) 012069 doi:10.1088/1757-899X/794/1/012069

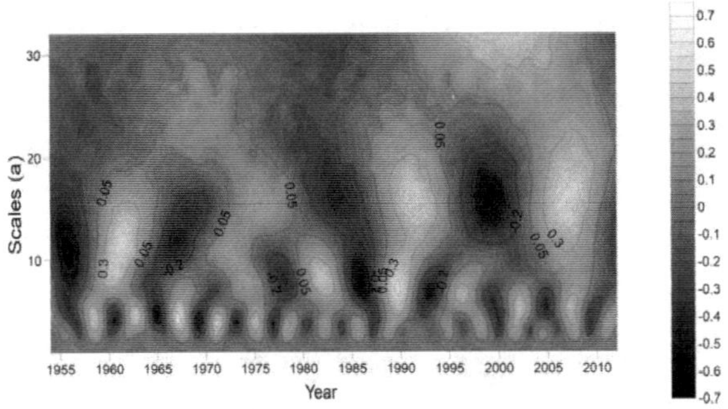

Figure 4. Real part of complex Morlet wavelet coefficients of annual mean temperature in Heilongjiang River Basin

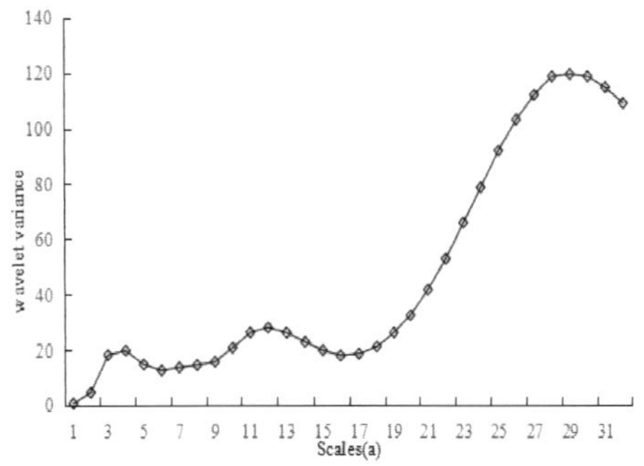

Figure 5. Complex Morlet wavelet variance of annual mean temperature in Heilongjiang River Basin

Figure 4 is a diagram of the real part of wavelet coefficients by wavelet transform using the annual mean temperature series in Heilongjiang River Basin. It can be clearly seen from Figure 4 that the periodic oscillations and mutation characteristics of the temperature in the region over the past 59 years on different time scales. The strength of the signal oscillations in the figure is represented by the size of the gray scale, and the larger the gray scale, the lower the temperature is. The smaller the gray scale, the higher the temperature is. In general, the annual mean temperature in the basin has a small scale of about 4-years, a mesoscale of about 8-years and a large scale of about 16-years. There is a strong signal of the change throughout the period on the scale of about 16-years. It clearly exists that the cycle process of low-high-low-high-low-high-low-high in the temperature change from 1954 to 2012, and it shows a slight increasing trend in large-scale cycles, which is very stable on the whole. The cycle process of high-low-high-low-high-low-high is obvious on the mesoscale of about 8-years from 1971 to around 1995, and it shows an increasing trend in the scale change. Mesoscale cycles of about 8 years disappear in the signal after 1995. For small scales of about 4-years, the changes are more complicated, and they constantly shake around 4-years. In summary, there is a complex nested structure on multiple time period scales in the temperature series in Heilongjiang River Basin.The small scale change is nested in the change on the larger scale.

Figure 5 is a diagram of wavelet variance of the annual mean temperature series in Heilongjiang River Basin.It can be clearly seen from figure 5 that there are three peaks, which are 16, 8 and 4,respectively, indicating that the oscillation periods in the region are 16-years, 8-years and 4-years respectively and the main oscillation period is 4-years. It can be seen from the main oscillation period that the higher temperature isoline is about to close, indicating that it may be in a period of lower temperature in the future.

Based on the analysis of the breakout and periodicity of the temperature time series, despite the periodic fluctuations in the temperature series, it is still difficult to conceal the fact that the overall temperature has risen.

3.2Spatial change characteristics of temperature

Figure 7 is the diagram of the spatial distribution of the annual mean temperature, and Figure 8 is the diagram of the spatial distribution of the annual mean temperature trend change in Heilongjiang River Basin.The annual mean temperature in Heilonjiang River Basin decreases roughly from south to north as the latitude rises in space and the regional differences are obvious. The annual mean temperature in the basin is generally -5.03 °C ~ 5.61 °C. The temperature in the northern and western parts of the basin is lower,which is below 0 °C. The annual mean temperature in the Shilka River, the Argun River, the upstream of the Heilongjiang River, the Zeya River, the Bureya River, the upstream of Nenjiang River and other river basins is lower, which is below 0 °C. Among them, the temperature in the Shilka River, the area where the Argun River flows into the main stream of Heilongjiang River and the upstream areas of the Zeya River and the Bureya River is the lowest, ranging from -5.03 °C to -

3.42 °C. The temperature is higher in the southern part of the basin, ranging from 1.17 °C to 5.61 °C.

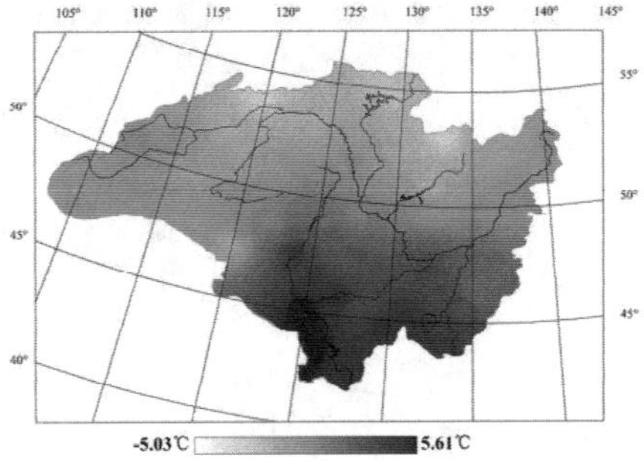

Figure 6. Spatial distribution of annual mean temperature in Heilongjiang River Basin

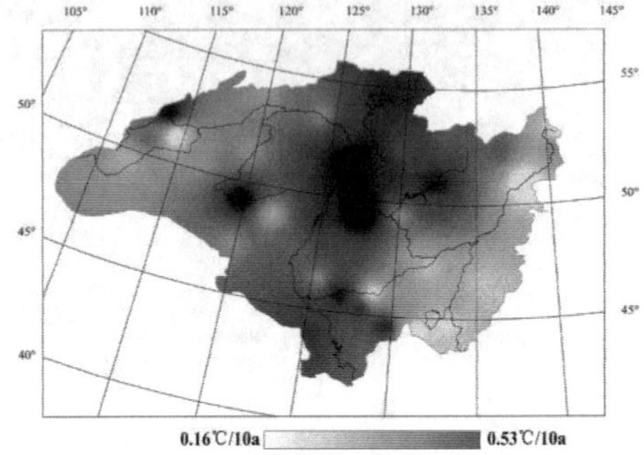

Figure 7. Spatial distribution of annual mean temperature trend change in Heilongjiang River Basin

In order to analyze the change trend of the annual mean temperature in different regions of the basin, a linear trend distribution diagram of the annual mean temperature was drawn (Figure 7). It can be seen that the annual mean temperature of the entire basin has been increasing in the past 59 years, but there is a spatial difference in the annual temperature increase in various regions of the basin.The annual temperature increase in various regions of the basin is 0.16°C~ 0.53°C/10a. In the southwestern regions, central regions and northern regions, the temperature increases greatly, with an increase of 0.47°C to 0.53°C/10a. The M-K method shows that the Z value of the annual temperature in each area of the basin is 3.114 ~ 5.180, which have passed the significance test of 99% , indicating that there is a extremely significant upward trend of the annual temperature all of the basin.

4. Conclusions

By analyzing the trend, breakout and periodicity characteristicsof temperature changes in the Heilongjiang River Basin, and the analysis of the spatial distribution, the following conclusions can be drawn:

a. The temperature increase in the Heilongjiang River Basin was obvious, and a significant abrupt change occurred around 1986.

b. Through wavelet analysis, there are multi-scales periodic changes in the temperature of the Heilongjiang River Basin (16-years, 8-years, and 4-years), and large-scale nesting of small-scales presents more complex periodic changes.

c. The Heilongjiang River Basin has a relatively regular phenomenon with the increase of latitude and the annual average temperature decreases. The whole river basin also shows a significant temperature increase trend, but the temperature increase is irregular in space, showing a phenomenon of high and low staggering.

d. The phenomenon of increasing climate in the Heilongjiang River Basin is in line with the background of global warming. However, as an international river in the high latitudes and cold regions, how to deal with climate change and potential changes in water resources is a problem that countries in the basin, especially China and Russia, must face.

Acknowledgments

This work is funded by the National Key R&D Program of China (2017YFC1502404, 2016YFC0502201), National Public Research Institutes for Basic R&D Operating Expenses Special Project (CKSF2017061, CKSF2017057), Jiangxi Provincial Natural Science Foundation Project (20181BAA208043).Special thanks are given to the anonymous reviewers and editors for their constructive comments.

References

[1] Bo Y , Ziqiang X , Feng H , et al. Climate Change Detection and Annual Extreme Temperature Analysis of the Amur River Basin[J]. Advances in Meteorology, 2016, 2016:1-14.

[2] Henderson-Sellers A , Zhang H , Berz G , et al. Tropical Cyclones and Global Climate Change: A Post-IPCC Assessment[J]. Bulletin of the American Meteorological Society, 1998, 79(1):19-38.

[3]Wang X L . Accounting for autocorrelation in detecting mean shifts in climate data series using the penalized maximal t or F test[J]. Journal of Applied Meteorology & Climatology, 2008, 47(9):2423-2444.

[4] Wang X L . Penalized Maximal F, Test for Detecting Undocumented Mean Shift without Trend Change[J]. Journal of Atmospheric & Oceanic Technology, 2008, 25(3):368-384.

[5] Lins, Harry F, Slack, James R. Streamflow trends in the United States[J]. Geophysical Research Letters, 26(2):227-230.

[6] Ullah S , You Q , Ullah W , et al. Observed changes in precipitation in China-Pakistan economic corridor during 1980–2016[J]. Atmospheric Research, 2018, 210:1-14.

[7] Malakar S, Goswami S, Chakrabarti A. An Online Trend Detection Strategy for Twitter Using Mann–Kendall Non-parametric Test[J]. 2018..

[8] Bolton, Edward W, Maasch, Kirk A, Lilly, Jonathan M. A wavelet analysis of Plio-Pleistocene climate indicators: A new view of periodicity evolution[J]. Geophysical Research Letters, 22(20):2753-2756.

[9] Zimmerman D , Pavlik C , Ruggles A , et al. An Experimental Comparison of Ordinary and Universal Kriging and Inverse Distance Weighting[J]. Mathematical Geology, 1999, 31(4):375 to 390.

IOP Conf. Series: Materials Science and Engineering **794** (2020) 012070 doi:10.1088/1757-899X/794/1/012070

Research on optimal operation of Chaohu Lake Based on the Particle Swarm Optimization algorithm

KaiLi[1,2], JijunXun[1,2*], YongqiangWang[1,2]

[1]Changjiang River Scientific Research Institute, Wuhan, 430010, China.

[2]Hubei Key Laboratory of Water Resources & Eco-Environmental Sciences,Wuhan,430010,China.

[*]Corresponding author's e-mail: 499136963@qq.com.

Abstract. There is a shortage of water resources in the Huaihe River area. With the development of the economy and society, the shortage of water resources is becoming more and more serious. Among them, the Yangtze to Hauihe Water Division is a strategic backbone project to solve the contradiction between water supply and demand in the Huaihe River Basin and the Jianghuai Water Shortage Area, which can improve the water resources allocation capacity in the basin. Chaohu Lake is an important storage place for the Yangtze to Hauihe Water Division. This paper takes the Chaohu optimal operation of the Yangtze to Hauihe Water Division as an example and uses the partical swarm optimization algorithm to solve the problem. When P=95%, through the comparison of conventional operation, after the optimal operation, the water diversion was reduced by 0.071 billion m^3. When P=50%, through the comparison of conventional operation, after the optimal operation, the water diversion was reduced by 0.03 billion m^3. The results show that the optimal operation can reduce the water diversion of the Yangtze to Hauihe Water Division River project more effectively.

1. Introduction

The water transferring line of Yangtze to Hauihe Water Division have some natural lakes such as Caizi Lake, Chaohu Lake, Wabu Lake. Among them, Chaohu Lake is one of the five major freshwater lakes in China[1]. It is located in the lower Yangtze river and in central Anhui Province. The drainage area is 13486 km^2 and the normal water surface is about 780km^2[2]. It is an important of regulation and storage of antze to Hauihe Water Division. Natural runoff is injected into Chaohu lake through the Fengle River, Hangyi River, NanfeiRiver, Baishitian River, which to supply the needs of users in Chaohu lake basin.

Chaohu regulsting storage capacity up to 608 million m^3, which makes Chaohu Lake particularly important in the Yangtze to Hauihe Water Division. On the one hand, it can coordinate the allocation of water resources in the river and realize the mutual adjustment of water resources within the system. On the other hand, it can form a water regulation system for water transmission lines along the line, which has various practical uses such as realizing pre-adjustment of water resources and rational use of flood water resources. However, how to scientifically and rationally use the water storage in Chaohu Lake to greatly improve the water resources allocation capacity is a difficult problem.

About the optimization of the lake was first started in the South-to-North Water Transfer Project. The optimal operation of the lake was first started in the South-to-North Water Diversion Project. The self-optimization simulation model of joint operation was used by five lakes in the first phase of the South-to-North water diversion project in paper of Leishenglong[3]. Lei has presented self-optimization simulation model of joint operation which was used in the first phase of the South-to-North water diversion project. In this paper, taking the minimum amount of pumping as the target, considering the constraints such as supply and demand water and reservoir control water level meter scheduling rules, a low-energy water diversion scheme that can basically meet the water demand of the system is obtained. On that base, YuFengchun[4] combining the characteristics of the South-to-North water diversion project, according to the three lakes in the Eastern Jiangsu Province, a multi-objective optimal operation model with the minimum of two targets is constructed, and the NSGA-II algorithm is used to solve the model. Therefore, the optimal water allocation scheme of the lake group system for the East line of the South-to-North Water Diversion Project is obtained. Fuhansheng[5] based on the analysis and generalization of Jiangsu project in the South-to-North water diversion project, the water resources scheduling scheme in Jiangsu water-receiving area is optimized by combining basic dynamic programming with simulated annealing Algorithm. The model results are good and superior to the conventional scheduling method of the system. In the paper of Shizhengtong[6], in order to solve the problem of joint operation of reservoir groups, this paper puts forward the aim of minimizing the lack of water in the whole system during the calculation period, and the decision variable of the water supply of each reservoir during each period. The optimization and dispatch model of water resources in several lakes with water balance and storage capacity of each reservoir as constraints, and the "Hongze Lake-Luoma Lake-Nansi Lake" reservoir group system in the first phase of the South-North Water Diversion Line as an example, using the large system-dynamic planning polymerization algorithm to solve the model, Results Compared with the conventional scheduling, the water supply of the system has been improved accordingly, and the water shortage and total water disposal have been reduced accordingly.

Summarizing the results of the current research, it is found that most of the optimal operation of lake at this stage is involved in the South-to-North Water Diversion. However, the Yangtze to Hauihe Water Division, as another trans-basin water transfer project carried out in recent years, has little research on it[7]. So, as an important storage place for the Yangtze to Hauihe Water Division, Chaohu Lake is particularly important to study its optimal operation.

This paper takes the Chaohu optimal operation of the Yangtze to Hauihe Water Division as an exampleand uses the partical swarm optimization algorithm to solve the problem. In this way, the plan for the minimum water shortage of water users at various times is obtained and formulate the scheme of how much water to draw from the Yangtze river.

2. Study area

2.1. Case study

In the preliminary design report of the Yangtze to Hauihe Water Division (*initial design*), the scope of the Yangtze River and Huaihe River project will be divided into six regions, the Yangtze River, Huaihe and Huainan, and four areas north of the Huaihe River. In this paper, only takes the Yangtze River area as the research object (figure1), and only considered to the regulation and storage capacity of Chaohu Lake in *initial design*.

Located in Hefei, Anhui Province, Chaohu Lake is one of the five largest freshwater lakes in China[1]. It is an important regulation and capacity sites for the Yangtze to Hauihe Water Division and drainage area is 13,486 square kilometers, which accounting for 93% of the total area of Anhui Province. The highest lake water level was 12.93 m (August 31, 1954, before wuwei major dam collapse) in Chaohu Lake. After the completion of the Chaohu Gate in 1962, the highest water level in Chaohu Lake was 12.71m(July 14, 1991) and the lowest water level was 6.47 m.(November 9, 1978), Chaohu has an annual average water level of 8.33 m. In the Yangtze to Haihe Water Division, the lowest operating water level of Chaohu Lake is 5.8m, the highest operating water level is 6.6m, and the average annual runoff is 3.03 billion m^3. As a multi-year regulating reservoir, its annual regulating water volume is 608 milion m^3.

IOP Conf. Series: Materials Science and Engineering **794** (2020) 012070 doi:10.1088/1757-899X/794/1/012070

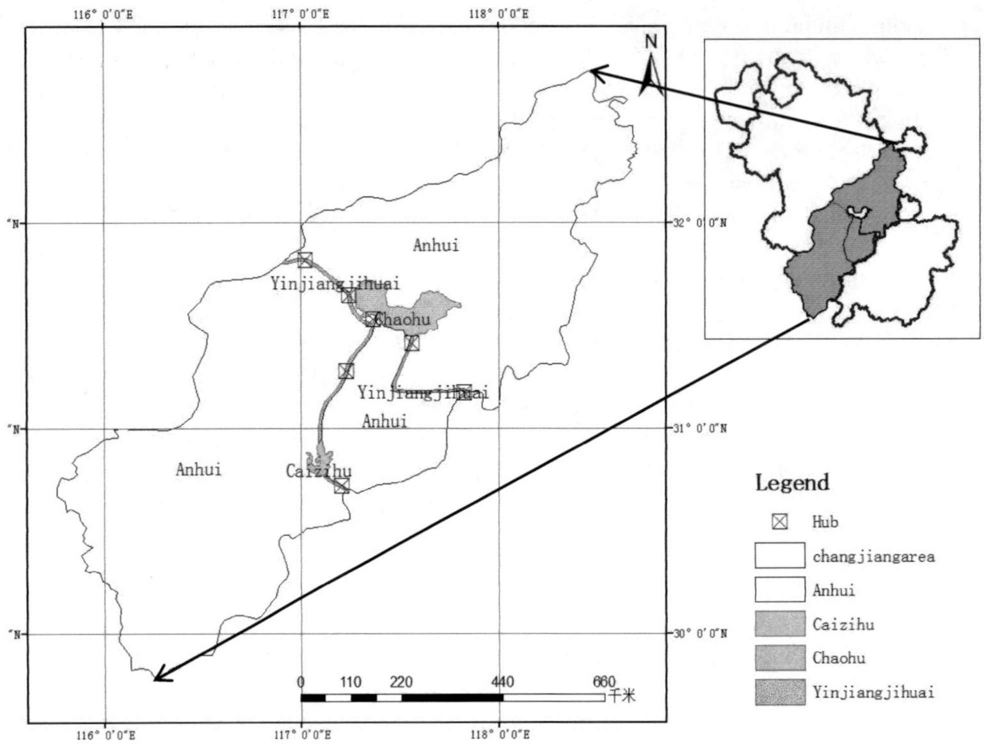

Figure1. Yangtz river area

2.2. Lake Optimization modle

2.2.1. Objective function
This paper uses the minimum water shortage model of water users in the Yangtze River area as the research. The objective function

- $W = \min \sum_{t=1}^{N} W_t^x - X_t$

In the above formula, W is volume of water shortage, and W_t^x is water demand of *t* times, X_t is quantity of water supply of *t* times .

2.2.2. Constraint condition
A Water budget constraint.

- $V_t = V_{t-1} + I_t - X_t$

Where,

V_t is the storage capacity of *t* times.

V_{t-1} is the storage capacity of *t-1* times.

I_t is the discharge into reservoir of *t* times.

B restriction of water level constraint

- $Z_t^{\min} \le Z_t \le Z_t^{\max}$

Where,

$Z_t^{\min}$ maximum water level.

$Z_t^{\min}$ minimum water level.

Z_t dynamic water-level.

C Discharge flow constraint

- $Q_t^{min} \le Q_t \le Q_t^{max}$

Where.

Q_t^{min} maximum discharge flow.

Q_t^{max} minimum discharge flow.

Q_t dynamic discharge flow.

D Water supply and abandoned water constraint

- $$\begin{matrix} V_t > V_t^{max} & C = V_t - V_t^{max} \\ V_t \le V_t^{max} & C = 0 \end{matrix}$$

Where.

V_t is storage capacity of t times.

V_t^{max} maximum storage capacity.

C is abandoning water.

3. Methodology

3.1 Particle Swarm Optimization

When applying PSO to solve problem of optimization, the solution of the problem corresponds to the "particle" in the search space[8-12]. Each particle has its own position and velocity (determining the direction and distance of flight) and a fitness determined by the optimization problem correspondence function. The process of particle optimization is the process of particle memory, following the current optimal particle, and constantly searching in the solution space. Each iteration of the particle is not completely random. If a better solution is found, save it and use it as a basis to find the next solution. When PSO solves the optimization problem, it first initializes to generate a group of random particles (corresponding to a set of random solutions to the problem). In each iteration, the particles update themselves by tracking two *extreme values*. One is the optimal solution found by the particle itself so far, namely the pbest, and the other is the optimal solution found by the entire population so far, namely the gbest. After finding this solution, the particles update their speed and position according to equations (1) and (2). Let the particle velocity information, position information, etc. be represented by an N-dimensional vector. Position vector $X^k = (x_1^k, x_2^k, ..., x_N^k)$ .Velocity vector $V^k = (v_1^k, v_2^k, ..., v_N^k)$ The current optimal position of the particle (local optimal solution to the problem) The $X_p^k = (x_{1.p}^k, x_{2.p}^k, ..., x_{N.p}^k)$ current optimal position of the particle (Global optimal solution corresponding to the problem) $X_g^k = (x_{1.g}^k, x_{2.g}^k, ..., x_{N.g}^k)$ Then the speed and position update formula is as follows.

- $v_n^{k=1} = w \times v_n^k + c_1 + Rnd_1 \times (x_{n.p}^k - x_n^k) + c_2 \times Rnd_1 \times (x_{n.g}^k - x_n^k)$ (1)
- $x_n^{k+1} = x_n^k + v_n^{k+1}$ (2)

Where

w is the inertia weight.

v_n^{k+1} is the speed of the n dimension of the particle in the k+1 iteration.

x_n^{k+1} is the position of the n dimension of the particle in the k+1 iteration.

c_1 and c_2 are acceleration factors.

rand1 and *rand2* are random number between(0,1).

3.2 *Solution procedure*

Steep1: Initialize particle population let k =0, randomly generate L particles according to specific research questions $P^k = (X^k(l), V^k(l))(l = 1, 2, \cdots, L)$ among them, is the position and speed.

- $X^k(l) = (x_1^k(l), x_2^k(l), \cdots, x_N^k(l))$
- $V^k(l) = (v_1^k(l), v_2^k(l), \cdots, v_N^k(l))$

Stept2: Calculate the current fitness of each particle $F^k(l), F^k(l) = f(X^k(l))$

Stept3: Update local best individuals and global best individuals. Find the best fit for each individual so far, recorded $F_p^K(l)$ as and remember that its corresponding position vector is $x_p^k(l)$. Find out the $X_p^k(l) = (x_{1,p}^k(l), x_{2,p}^k(l), \ldots, x_{N,p}^k(l))$ best fitness values for the group so far as F_g^k and remember that its corresponding position vector is X_g^k, $X_g^k = (x_{1,g}^k(l), x_{2,g}^k(l), \ldots, x_{N,g}^k(l))$.

Stept4: The individual's position vector and velocity vector are updated to produce the next generation particle swarm. Update each speed element and position element in all individuals using the following formula:

- $v_n^{k+1}(l) = w \times v_n^k(l) + c_1 \times Rnd_1 \times (x_{n,p}^k(l) - x_n^k(l)) + c_2 \times Rnd_2 \times (x_{n,g}^k(l) - x_n^k(l))$
- $x_n^{k+1}(l) = x_n^k(l) + v_n^{k+1}(l)$

Stept5: Verify that the algorithm iterative termination condition is satisfied, and if it is satisfied, stop the iteration and output the optimal result; no k=k+1, turn to be. The iteration termination condition is *k*=*K* (K is the preset maximum number of iterations) or

- $\frac{\left|F_g^{k+1} - F_g^k\right|}{F_g^k} \le \varepsilon$ （ε *is the accuracy given in advance*）

4. Results and discussion

According to 2030 p=50%, p=95% of incoming water frequency, water demand, local water supply volume to solve the water shortage in the Yangtze River area of the Yangtze to Hauihe Water Division. The PSO algorithm is used to establish the optimal operation model calculation, and the calculation results are compared with the rule scheduling results in this paper.

Table1 .P=95% optimal operation results

Months	Water demand （10^8m^3）	water supply （10^8m^3）	water diversion （10^8m^3） conventional operation	water diversion （10^8m^3） Optimal operation
1	4.74	2.17	2.57	2.57
2	4.74	2.17	2.57	2.57
3	4.74	2.17	2.57	2.57
4	4.74	2.19	2.55	2.55
5	6.28	6.15	0.13	0.13
6	8.33	8.83	0.00	0.00
7	8.33	11.73	0.00	0.00
8	6.28	10.54	0.00	0.00
9	4.74	6.18	0.00	0.00
10	4.74	3.39	1.35	0.64
11	4.74	1.90	2.84	2.84
12	4.74	1.78	2.96	2.96
Amount	67.14	59.2	17.54	16.83

Table2. P=50% optimal operation results

Months	Water demand （10^8m^3）	water supply （10^8m^3）	water diversion （10^8m^3） conventional operation	water diversion （10^8m^3） Optimal operation

1	3.82	2.11	1.71	1.71
2	3.82	2.11	1.71	1.71
3	3.82	2.16	1.66	1.66
4	3.82	2.18	1.64	1.64
5	4.8	9.24	0.00	0.00
6	6.12	6.76	0.00	0.00
7	6.12	8.89	0.00	0.00
8	4.8	8.78	0.00	0.00
9	3.82	6.05	0.00	0.00
10	3.82	4.98	0.00	0.00
11	3.82	3.96	0.00	0.00
12	3.82	2.14	1.68	1.38
Amount	52.4	59.36	8.40	8.10

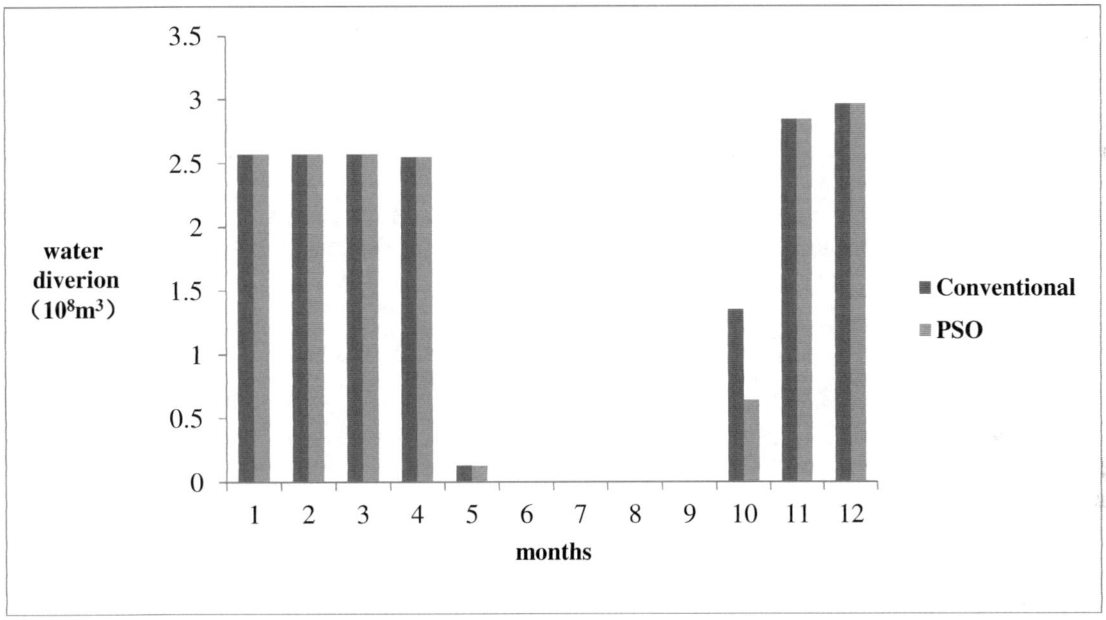

Figure2 .P=95% compare results

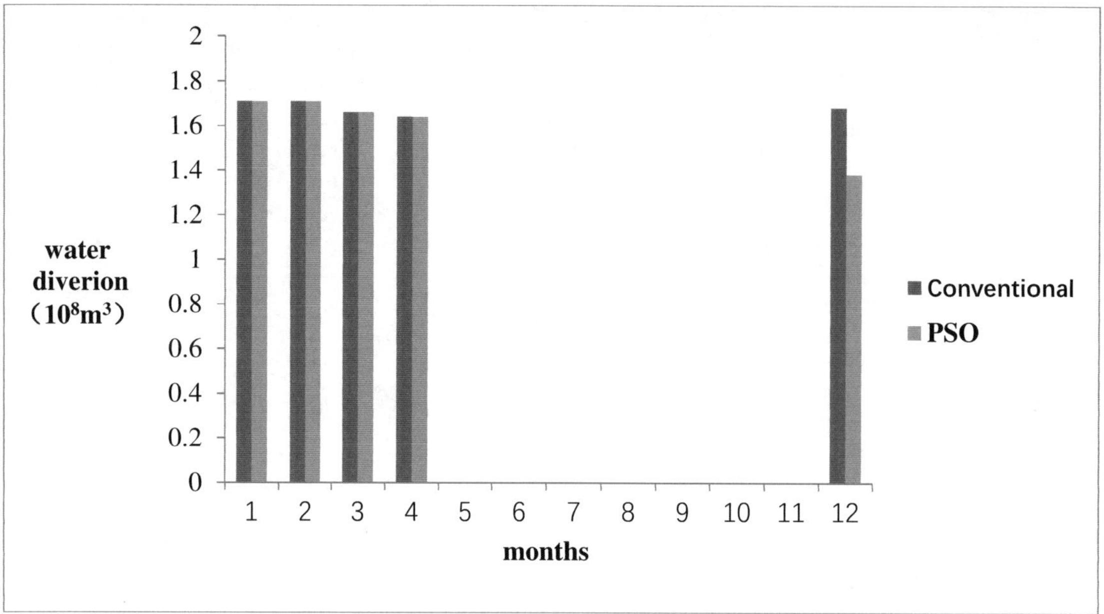

Figure3. P=50% compare results

The PSO-based Chaohu optimal operation model uses 2010 as the current base year to predict the monthly water demand in the near-term planning year of 2030. According to the runoff data of the Yangtze River region from 1976 to 2010, the P-III curve was established, and the P=50% and P=95% incoming water frequency were taken for modeling analysis. It can be seen from the above table that under the conventional operation, at the water frequency of p=95%, the 2030 Yangtze River Basin needs to transfer water by 1.751 billion m^3 through the Yangtze to Hauihe Water Division. After the optimal operation, the Yangtze river region needs to transfer water to 1.683 billion m^3. Through the comparison of conventional operation, after the optimal operation, the water diversion was reduced by 0.071 billion. In 2030, when P = 50% of the incoming water frequency, 0.84 billion m^3 of water needs to be diverted through conventional operation. However, after the optimal operation, the water diversion needs to be 0.81 billion m^3, and the optimal operation can effectively reduce the project water transfer by 0.03 billion m^3.

5. Conclusions

Affected by climate change and human activities, the runoff in the Huaihe River Basin has been declining in the past 10 years. With the rapid development of economic construction, the increasing population and the rapid development of urbanization, the shortage of water resources is becoming more and more obvious. In the contradiction between economic and social development and ecological environmental protection, it is impossible to fundamentally solve the problem of water shortage in the region by relying on water conservation and tapping the potential of local water resources. To alleviate the increasingly sharp contradiction between supply and demand of water resources, in the long run, cross-basin water transfer projects must be implemented[13]. Among them, the Yangtze to Hauihe Water Division is a strategic backbone project to solve the contradiction between water supply and demand in the Huaihe River Basin and the Jianghuai Water Shortage Area, which can improve the water resources allocation capacity in the basin.

As one of the three important storage lakes for the Yangtze to Hauihe Water Division, Chaohu Lake has functions such as flood control, water supply, irrigation and shipping. In this paper, Chaohu Lake is taken as the research object, which is an annual regulation lake. The minimum water demand (water diversion) needs to be the objective function. The local water supply of the reservoir in each period is the decision variable, and the reservoir capacity is the state variable in each period. Based on the reservoir water balance and the reservoir water level, the Chaohu optimal scheduling model was established and solved by particle swarm optimization (PSO). The results were compared with the conventional operation, and the results were compared and analyzed. The results show that the optimal operation can reduce the water diversion of the Yangtze to Hauihe Water Division River project more effectively.

Acknowledgements
This work is funded by the National Key R&D Program of China (2016YFC0402203), the National Natural Science Foundation of China (No. 51779013, 51509009), National Public Research Institutes for Basic R&D Operating Expenses Special Project (No.CKSF2019478,CKSF2017061), Water Conservancy Science and Technology Innovation project of Guang Dong Province(2017-03), Natural Science Foundation of Hubei Province(2017CFB606, 2018CFB655). Special thanks are given to the anonymous reviewers and editors for their constructive comments.

References
[1] DOU,H,S.Jiang,J,H.(2003) China's Five Major Freshwater Lakes . Press of University of Science and Technology of China. Hefei.
[2] Wang,W. Chen,C,P.(2011) Study on the scale of water transfer in the Yangtze to Chaohu Lake Water Diversion .J. China Rural Waterand Hydropower. 2011(08):13-15+19.
[3] Cui,S,L. Lei,S,L.Bai,X,T. Liu,S,C. (1996)Application of Self-optimizing Simulation Technology in Multi-objective Reservoir Optimization Scheduling.Water Resources and Power.J.1996(04):30-36.
[4] Yu,F,C. Fang,G,H. Wang,W,G. Wu,X,W. Wen,X.(2016)Research on Optimal Grouping of Lake Groups in the East Route of South-to-North Water Transfer Project Based on Multi-Objective Genetic Algorithm.J.Journal of Irrigation and Drainage.2016,35(03):78-85.
[5] Fu,H,S. Cheng,J,L. Fang,H,Y.Lu,X,W. (2013)Research on Optimal Allocation of Water Resources for River-Lake-Stall Pumping Station System Based on Dynamic Programming and Simulated Annealing Algorithm.J.Journal of Hydraulic Engineering.2013,44(01):91-96.
[6] Shi,Z,T.(2015)Study on "Single-station-single-station" and "Multi-database" Water Resources Optimization Scheduling Methods in the Jiangsu Section of the South-to-North Water Transfer Project.D.Yangzhou University.2015.
[7] Shen,H,S. (1994)Necessity and favorable conditions for the Yangtze River and Huaihe River project.ZhiHuai.J. 1994
[8] Tim, H. (2008) The Particle Swarm Algorithm. Springer Berlin Heidelberg. Germany
[9] James, K. Russell, E. (1995) Particle swarm optimization. Proceedings of ICNN'95. *In: International Conference on Neural Networks*. IEEE.pp.83-100.
[10] Russel ,E.J, Kennedy. (2002)A new optimizer using particle swarm theory,MHS'95. Proceedings of the Sixth International Symposium on Micro Machine and Human Science. C. IEEE.
[11] Ali,Z. Sayed,F,M. Madjid, E,G. Hojat, K. (2019)Optimal Reservoir Operation Using Bat and Particle Swarm Algorithm and Game Theory Based on Optimal Water Allocation among Consumers.J. Water Resources Management,2019,33(9).
[12] Li ,Y. Huang T,L.Tan,X,L. Zhou,Z,Z.Ma,W,X.(2018) Destratification and Oxygenation Efficiency of a Water-lifting Aerator System in a Deep Reservoir: Implications for Optimal Operation.[J]. Journal of environmental sciences (China),2018,73.
[13] Shen,H,S.(2009) Discussion on some issues in the planning of the Yangtze to Huaihe River Diversion. Jianghuai Water Resources Science and Technology.J. 2009(01):7-9.

IOP Conf. Series: Materials Science and Engineering **794** (2020) 012071 doi:10.1088/1757-899X/794/1/012071

A survey on the satisfaction of living environment in old urban residential areas from the perspective of public infrastructure

YuchenShen[1*], YunWu[1]and WeimiaoFang[2]

[1]Department of Architecture, Zhejiang A&F University, Hangzhou, ZheJiang, 310000, China

[2]The Architectural Design & Research Institute of Zhejiang University, Hangzhou, ZheJiang, 310000, China

[*]Corresponding author's e-mail:mrshenych@foxmail.com

Abstract. Under the circumstance of constantly accelerated urban development process, the old residential areas in urban areas have a lot of problems, such as earlier construction year, longer construction time, aging public infrastructure and lack of public services, etc., which directly affect the satisfaction of residential residents with the living environment. In this paper, a simple data analysis method is used to conduct a questionnaire survey on the residents of Chengnan Garden Residential District, Jiaxing City. The survey results are quantified and analyzed to understand the residents' satisfaction with the public infrastructure of the old residential district. In this paper, the author analyzes the reasons for the satisfaction, and puts forward transformation suggestions for these problems.

1. Introduction

Under the circumstance of constantly accelerated urban development process, the old residential areas in urban areas have a lot of problems, such as earlier construction year, longer construction time, aging public infrastructure and lack of public services, etc., which directly affect the satisfaction of residential residents with the living environment[1]. Aiming at solving the practical problems in urban and rural construction, human settlement environment attempts to establish a new scientific group centering on the coordination between man and nature. It takes the living environment as the research object, and its theoretical basis is derived from the "Ekistics" of C.A.Doxiadis [2]. Public facilities are one of the basic factors for the evaluation of human settlements [3]. This study will investigate the satisfaction of community residents with the public infrastructure of the communityand provide the direction and focus for the reconstruction of the old community. This investigation content is mainly the satisfaction with public infrastructure in residential area, and based on the existing research[3-5] and by combining with the actual survey, the community public infrastructure is divided into seven factors, including "medical and health facilities", "public security and fire safety", "water and electricity supply", "recreational activities", "physical fitness facilities", and "vehicle management". The questionnaire adopts Llikert five-level scale, and the evaluation of each indicator is divided into five grades, including "very satisfied", "satisfied", "average", "dissatisfied", "very dissatisfied". For quantitative analysis, the five factors are assigned with the rating of 5, 4, 3, 2 and 1, respectively.

2. Research methods and data sources

2.1Research object
Jiaxing City is in the northeast of Zhejiang Province. The object of this study is Chengnan Garden Residential Area in Nanhu District, Jiaxing City. Completed in 1999, the residential covers a total floor area of about 80,000 square meters, with a plot ratio of 1.2 and a greening ratio of 36%.

2.2Research method
Among the methods to evaluate human settlements, there are five relatively mature methods, including simple data analysis method, linear weight method, fuzzy comprehensive evaluation method, GIS analysis method and BP neural network model method[6]. The main research content is the satisfaction of residents in Chengnan Garden Community with the public infrastructure. Public infrastructure is a single part of the human settlement environment research, and its study subject is a single residential area, and the investigation scope and sample size are limited, so in this paper, the simple data analysis method is adopted to analyze the data.

2.3Data resources
Table 1. The questionnaire survey.

Total number	Valid number	Effective rate
80	75	93.75%

The survey data of this study was from the residents of this community. During the investigation process, a total of 80 questionnaires were issued, of which 75 were valid and the effective rate was 93.75% (Table 1).

2.4Reliability analysis
Cronbach coefficient α is often used as an index to evaluate internal reliability. If a value is greater than 0.7, it indicates a high reliability, and if it is between 0.35 and 0.7, it indicates a fair reliability, and if it is less than 0.35, it indicates a low reliability [7]. The software SPSS is used to calculate the reliability of 7 factors in the questionnaire, and it is obtained that the Cronbach coefficient α is 0.802, so the data of the questionnaire has a high reliability.

3. Results and analysis

3.1Survey information statistics
Table 2. Gender statistics of interviewed households.

	Man	Woman
People number	40	35
Rate	53.33%	46.67%

In this survey, 40 men and 35 women were interviewed by questionnaires, accounting for 53.33% and 46.67%, respectively (Table 2). The proportion of men and women in this survey is relatively balanced, which can better reflect the comprehensiveness of the survey results.

Table 3. Statistics of the age distribution of the interviewed households

Age	10-19	20-29	30-39	40-49	50-59	Over 60
People number	1	8	13	21	12	20
Rate	1.33%	10.67%	17.33%	28.00%	16.00%	26.67%

In this investigation, the age grades who are investigated are mainly as follows: over 40 years old, accounting for 70.67%, 17.33% for 30-39 years old, and 12% for under 30 years old, 44.00% for 40 to 59 years old, who are not only the main source of income in the family but also undertake the responsibility of taking care of the elderly. The elderly over 60 years old accounts for 26.67%, who often take a leisurely walk in the community. As the main population in the community, they can accurately reflect the public infrastructure of the old community (Table 3).

Table 4. Statistics on the residence years of the interviewed households

Years of residence	<1	1-5	6-10	11-20	>21
People number	2	12	30	21	10
Rate	2.67%	16.00%	40.00%	28.00%	13.33%

In this investigation, the residents surveyed are mainly as follows: those who have lived for more than 10 years account for 41.33%; those who have lived for 6-10 years account for 40.00%, and those who have lived less than 6 years account for 18.67%. The surveyed households generally have lived for a longer period, so they can more accurately reflect the specific conditions of the community's public infrastructure and increase the accuracy of the survey results.

3.2Analysis on the satisfaction of public infrastructure in Chengnan Garden Community

Table 5. Mean value and standard deviation of public infrastructure satisfaction in Chengnan Garden Community

Factor	Mean	Standard deviation
Medical and Health facilities	3.0533	0.8778
Public security and Fire safety	3.6667	0.8055
Water and Electricity supply	3.5467	0.8049
Recreational activities	3.2267	0.8575
Physical fitness facilities	3.4000	0.8641
Vehicle management	2.6800	1.1095
Supporting facilities for education	3.3200	0.8352
Total	3.2705	

In this survey research, the evaluation of 7 factors is divided into 5 grades, and values are assigned to them with "1-5". The higher the satisfaction is, the higher the score will be. According to the questionnaire survey, the data as shown in table 5 is obtained. The overall satisfaction is 3.2705, and residents are generally satisfied with the public infrastructure of Chengnan Garden Community. Among them, the satisfaction degree of residents in the community with medical and health facilities, recreational activities, public construction facilities and educational supporting facilities is between 3 and 3.5, with a moderate satisfaction. "Medical and health facilities" mainly include the medical facilities of the community. There is a community hospital in the community, which is only 300 meters away from Jiaxing People's Hospital. The survey result of this term is 3.0533, which is relatively satisfactory. In terms of "public security and fire safety", the monitoring system covers the community, and there are often fire drills and public security lectures. The score of this survey result is 3.6667, which is relatively satisfactory. In terms of "water and electricity supply", there hasn't been water and electricity supply failure in the community in the past year. There were a few problems in some residents and professional maintenance team is also available in time. The score of this survey result is 3.5467, which is relatively satisfactory. In terms of "recreational activities", the community has an "association for the elderly", which regularly holds recreational activities such as "table tennis competition" and "Mid-Autumn festival evening party" and so on. The score of this survey result is 3.2267, which is relatively satisfactory. In terms of "physical fitness facilities", there are public fitness equipment and a small basketball court in the community. The score of this survey result is 3.4, which is relatively satisfactory. In terms of "supporting facilities for education", there is a kindergarten in the community and Jiaxing Experimental Junior Middle School is less than 2 kilometers away, with perfect educational facilities. The score of this survey result is 3.32, which is relatively satisfactory. However, the satisfaction of "vehicle management" in the community is less than 3, which is not satisfactory. Therefore, a further analysis and research will be conducted to vehicle management.

Table 6. Vehicle management

	Upground Park	Underground Park	Car
Number	About 100	0	About 360

Table 7. Parking Modes and Disadvantages of Vehicles

Mode	Parke on both sides of the road	Parke in public square	Parked in front ofbuilding
Rate	About 72%	About 23%	About 5%
Disadvantages	• Reduce community traffic area • Prone to traffic accidents	• Reduce community public area • Cause congestion and inconvenience for residents	• Congestion at the entrance of residential buildings, causing inconvenience to the entire population • Inconvenience to passing vehicles

Chengnan Garden Community, completed in 1999, is an old community. Therefore, under the circumstance that most families have private cars, the parking spaces in the community are in short supply, and motor vehicle parking spaces are far from enough (Table 6). As the overall lack of vehicle parking spaces, there are a variety of vehicle management problems in the community: parking wars, parking disorder and other phenomena emerge in endlessly. The main parking ways in the community are divided into three situations: parking on both sides of the community road; Parking in a public square; Parking at the gate of the residential building (Table 7).

In the community, many vehicles are parking on both sides of the community road, so the road becomes very narrow, seriously affecting the convenience of residents' life and walking. In the rush hour, there is usually a phenomenon that two cars are opposite, which also leads to the phenomenon of traffic jam in part of the community. The roads of the community originally bear the traffic function. However, due to the parking of vehicles, the traffic capacity is reduced, which is also easy to cause traffic accidents. In addition, it will also lead to the deterioration of the surrounding environment and landscape of the community. Both sides of the road cannot meet the demand for parking, and the entrance of small squares and public fitness equipment are also occupied by vehicles, which also seriously affects residents' daily outdoor fitness activities, greatly reduces the public activity space, and declines their sense of life happiness to a certain extent.

4. Conclusions

In the research on urban living environment of older communities from the perspective of public facilitiessurvey on the satisfaction of residents living in Chengnan Garden Community in Jiaxing, the community residents' satisfaction with medical and health facilities, cultural activities, construction of public facilities, and education facilities is relatively good, but they aren't satisfied with vehicle management, and this is mainly because the old village parking space cannot meet the increasing number of vehicles. Therefore, in order to solve such problem, the following several suggestions are proposed.

4.1Improve road organization in old community

Due to the lack of popularity of cars in the early years in China, "people and cars diverge" wasn't considered in the old residential areas, and in many old residential areas, pedestrians and vehicles share the same road system. With the increase of private cars in the community, this kind of road organization has caused safety risks for residents to travel. Therefore, improving the road organization in the old community has become one of the factors to be considered in the transformation process of the old community. For example, the sidewalks should be extended on the main roads of the main communities. Meanwhile, footpaths can be constructed in the green landscape of the communities to effectively separate pedestrians from vehicles and reduce safety risks.

IOP Conf. Series: Materials Science and Engineering **794** (2020) 012071 doi:10.1088/1757-899X/794/1/012071

4.2Build new parking lots

The government can choose sites around the old residential areas and build new public parking spaces to meet the parking needs of residents in the old residential areas nearby. Meanwhile, it can also build three-dimensional parking spaces, multi-storey parking spaces, or special parking buildings. At the same time, the current science and technology can be used for the unified, scientific, efficient and standardized management of parking lots. Through building new parking lots, the parking difficulty and parking chaos in old residential areas can be improved.

4.3Share parking spaces

There is a large commercial centre near Chengnan Garden Community, and Wal-Mart is 300 meters away, Huafu Square is 500 meters away, and Yaohan is 1.5 kilometers away, so the parking spaces of the commercial centre can be fully used and the parking facilities can be shared, so as to effectively enhance the utilization rate of the parking spaces, improve the urban travel environment to a certain extent, and alleviate the parking problems.

4.4Strengthen parking management

In view of the problem of parking chaos in the old residential areas, the government shall formulate corresponding rules and regulations and improve the level of property management. Meanwhile, it shall manage and timely rectify the phenomenon of disorderly parking in the residential areas and may give corresponding legal penalties to vehicles that don't meet the regulations. In addition, the management of access control at the entrance of the community shall be enhanced, external vehicles shall be registered and charged, etc.

References

[1] Chan, H. (2019) Withurban renewal as the goal, actively promote the comprehensive transformation of old communities. The People's Congress of China,20: 45-46.

[2] Wu, L. (2001) Introduction to Sciences of Human Settlements. China Architecture & Building Press, Beijing.

[3] Liu, L. (2017) Research on the Evaluation Index System of Urban-ruralIntegration in China-Based on the Comprehensive Study of theReliability Analysis and Panel Data Analysis. Hebei Normal University, Shijiazhuang.

[4] Li, X., Yi, B., Yang, J., et al. (2008) The research on the satisfaction to community settlement: taking Dalian City for example. Urban Problems, 1: 56-63.

[5] Li, H., Xu, R., Gao, Z., et al. (2005) Quality Evaluation of Human settlements in a City Scale—A case study on Nanjing City. Human Geography, 1: 1-5.

[6] Zhou, W., Zhang, X., Li, X. (2013) Overview of the Evaluation Methods of Chinese Human Settlements. Safety and Environmental Engineering, 20: 14-18.

[7] Chen, G., Peng, W., Ye, Y., et al. (2011) Communication Research Methods. Fudan University Press, Shanghai.

IOP Conf. Series: Materials Science and Engineering **794** (2020) 012072 doi:10.1088/1757-899X/794/1/012072

Investigation on an application of emergency rescue in the initial natural earthquake by wavelet analysis

MajunPeng, Cheng Wei*

China Nonferrous Metal (Guilin) Geology And Mining Co. Ltd, Guilin, 541004, China

*Corresponding author's e-mail:blackfire513@126.com

Abstract. Earthquake brings great harm to humans. In order to achieve the desired effect and mitigation effectiveness of early warning, it must reduce the time of rapidly assessing earthquake location and processing seismic and other data. Seismic phase identification based on the wavelet analysis is a powerful and widely spread method in identifying the initial earthquakes. We can get the signals' recognition factor of S wave and P wave in different scale. Then according to these recognition factors, we can determine the positioning function of primary earthquake of S wave and P wave. Next through the positioning function, we can easily pick up the first arrival of P wave and S wave. The wavelet packet analysis is a more precise analysis method of signal, which makes the frequency band multi-level division, and further decomposes the high-frequency part. Also, the wavelet analysis does not subdivide, and according to the characteristics of analysis-signal, the adaptive-select the corresponding frequency band is matched by the frequency spectrum of the signal, improving the time-frequency resolution. This paper mainly investigated the applications of initial earthquakes identification on emergency rescue, and it can be a useful theoretical reference for further relevant research.

1. Introduction

The identification of seismic waves in the initial earthquakes is a great significant problem encountered in the research of modern seismology. According to the principle of identification, initial earthquakes identification consists of two parts, manual identification and automatic identification[1]. However, traditional manual identification often depends on the experience of the one who makes the judgment and identification. It is not only time-consuming but also has artificial errors, reading errors, and so on, which will all affect the identification precision. Thus, it is necessary to develop and research for obtaining a stable, practical and efficient automatic identification method and system[2].

Wavelet theory is capable of application on the earthquake emergency and has been developed rapidly during the past several years. In this theory, energy is decomposed based on scales from time series. The multiresolution analysis decomposes the signals into different frequency bands, each of which is still a signal changed with time. Therefore, it is the excellent time-frequency characteristic of the wavelet transform that makes it a powerful tool to analyse the mutational signals and more suitable for detecting the arrival time of abnormal waveform[3]. In this paper, the author comprehensively analysed the natural seismic signals by the application of wavelet packet transform and wavelet transform, and verified the validity and accuracy of this method for the identification of actual natural earthquake, which demonstrates that the method provides a reference for the effective identification of initial natural earthquakes.

2. Theory of initial earthquake identification based on wavelet packet transform signal

The identification of the first arrival of the signals by using the wavelet packet transform is quite useful. Wavelet packet transform $c_{n,j}^k$ refers to a signal as the superposition of the basic unit $2^{k/2}W_n(2^k t - j)$. Just as the Fourier transform regards the signal as the superposition of trigonometric functions, any wavelet packet basis function $2^{k/2}W_n(2^k t - j)$ can be obtained by binary translation, expansion and contraction of $W_n(t)$. The window centre of $W_n(t)$ is (t_n, ω_n). Δt_n and $\Delta\omega_n$ respectively, represent the time window and frequency window width. It can be proved that the window corresponding to the wavelet packet basis function is $\left[2^h(t_n - \Delta t_n + j), 2^h(t_n + \Delta t_n + j)\right] \times \left[2^{-h}(\omega_n - \Delta\omega_n), 2^{-h}(\omega_n + \Delta\omega_n)\right]$. The wavelet packet decomposition transform represents that the size of the window area equals the internal energy of $4\Delta\omega_n \bullet \Delta t_n$, where Wn (t) is the generating wavelet function, k is the scale parameter, and j is the translation factor. With the change of k, the window area does not change, while its time window length and frequency window height are continually adjusting. When k becomes higher, the time window becomes narrower, that is, the time resolution increases [4]. This characteristic contributes to demonstrating the local features of mutational or gradient earthquake signals resulted from the generation of new seismic phase. Therefore, it is feasible to use wavelet packet decomposition transform to identify the initial earthquake of unsteady seismic signals, which is the basis for identifying the initial earthquake of signals [5].

For a N-point discrete signal, the phase plane can be used according to the time-frequency characteristics of the signal since the window in the phase plane corresponds to the coefficient $c_{n,j}^k$ obtained by wavelet packet decomposition. When the energy of signal in the window is relatively strong, it is represented by a high chromaticity in the phase plane; on the contrary, it is represented by a lower chromaticity. The position of the window along the frequency axis is determined by j [6]. If the signal amplitude of each decomposition coefficient filled in the corresponding window with the corresponding chromaticity, a phase plane chromaticity map can be obtained. When the new seismic phase arrives, it will cause a mutational or gradient change of the original signal, and the frequency and energy of the signal will change abnormally in high values. Therefore, the mutational change characteristics of the time-frequency distribution of the signal can be introduced by the distribution of the phase plane chromaticity diagram. Then the time of the seismic phase arrival can be obtained [7, 8].

In conclusion, since different scales own different time and frequency resolutions under wavelet decomposition, wavelet decomposition separates signals contained in different frequency intervals, combining with the time-frequency intuitive diagramof the signal the method used can identify the seismic phase at the corresponding frequency.

3. Verification of analogue signal identification

To verify the signal identification method mentioned before, a more characteristic analogue signal is used for the following research. By processing the analogue signal, the corresponding first-arrival position is identified and compared with the real signal characteristics, and then verifying the validity and accuracy of the identification method.

3.1 Gradient analogue signal and identification results

The gradient signal can be simulated by a combination of functions. In this paper, the simulation of the gradual signal is composed of several sine functions. It is assumed that there are five sine waves with the form of $\sin\frac{\pi}{20}(t-t_{01})\bullet s\bullet H(t-t_{01})$, $\sin\frac{\pi}{10}(t-t_{02})\bullet s\bullet H(t-t_{02})$, $\sin\frac{\pi}{5}(t-t_{03})\bullet s\bullet H(t-t_{03})$, $\sin\frac{2\pi}{5}(t-t_{04})\bullet s\bullet H(t-t_{04})$, and $\sin\frac{\pi}{2}(t-t_{05})\bullet s\bullet H(t-t_{05})$, and all the amplitude is 1, which does not change with time. The length of the discrete data is 512 points, and the sampling rate is s = 40 points/second, where H is step function, t_{01}= 0s, t_{02}= 2.5s, t_{03}= 5s, t_{03}= 7.5s, and t_{04}= 10s .The sine waves as mentioned above are successively superimposed to form the gradient analogue signal used in this paper [2]. If these signals are discretely considered as points, then the first signal always exists from point 0, while the second signal emerges from 100, and the third, the fourth and the fifth signal appears from 200, 300, and 400, respectively.

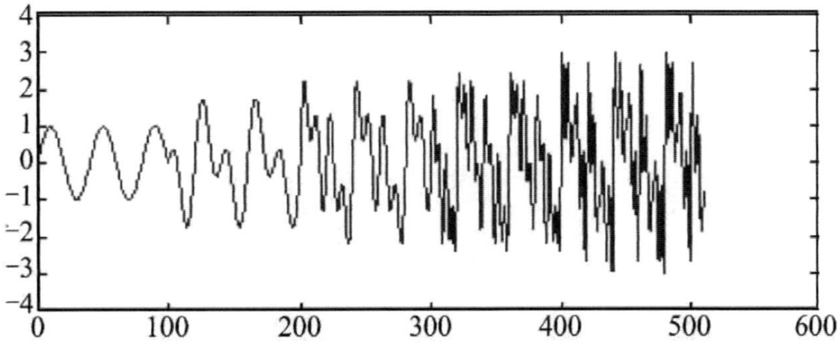

Fig.1 The gradual change

Figure 1 shows an analogue gradient signal formed by the superposition of the above five sine waves. In Figure 1, the x-coordinate represents the number of points, and the y-coordinate represents the amplitude. From the figure, we can see the gradient characteristic of the signal.

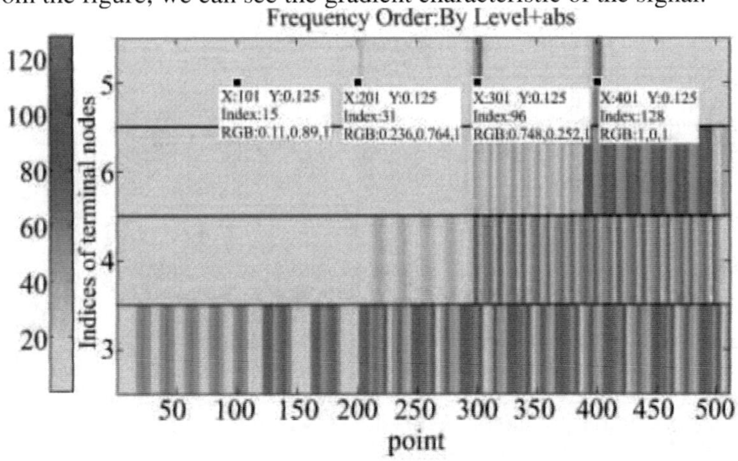

Fig.2 Initial motion recognition of gradual change analog signals in two layers decomposed by db6 wavelet

Figure 2 presents chromaticity diagram of the phase plane generated by conducting two-layer decomposition of the db6 wavelet packet on the signal. The best wavelet base uses logarithmic energy entropy. In this diagram, the x-coordinate is the points of the signal, and the coloured bar represents the frequency of the signal. It can be seen from the distribution of the coloured chromaticity with the points in the figure that when each new signal arrives, there will appear a mutational signal with high frequency. The first arrival position of the signal is the coordinate position of the four black points in the figure, which are 101, 201, 301 and 401 respectively. The position of the first black spot is the initial moving position of the second signal, the position of the second black spot is the initial moving position of the third signal and, so forth. The identification result is basically the same as the actual initial position of the signal (100, 200, 300, 400). The corresponding initial movement times were

IOP Conf. Series: Materials Science and Engineering **794** (2020) 012072 doi:10.1088/1757-899X/794/1/012072

2.525s, 5.025s, 7.525s, and 10.025s, which are essentially in agreement with the actual initial movement times.

3.2 Mutational analogue signal and identification results

The gradient signal described above is the superposition of five sine waves, while the mutational analogue signal is changing the sine wave to the corresponding cosine wave, and then superimpose the five cosine waves one after another, as shown in Figure 3. The following is the method used to identify the gradual signal and here to identify the mutational signal and verify the identification effect of the method on the mutation signal.

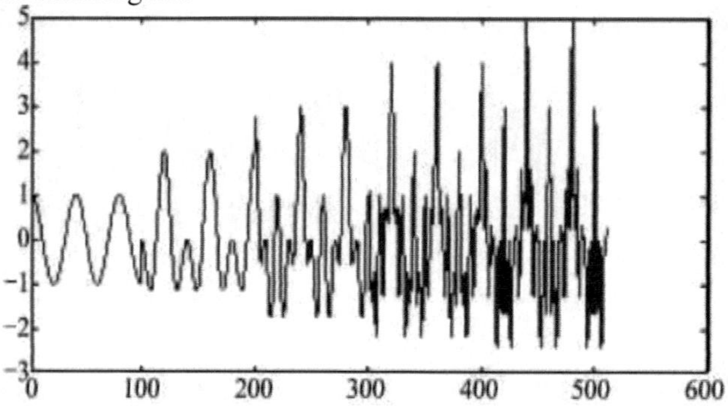

Fig.3 Mutational analog signals

As can be seen from figure 4, the initial moving positions of the mutation signal are 101, 201, 301 and 401, respectively, and the corresponding first arrival time is 2.525s, 5.025s, 7.525s, and 10.025s, respectively, which are basically consistent with the real initial motion time.

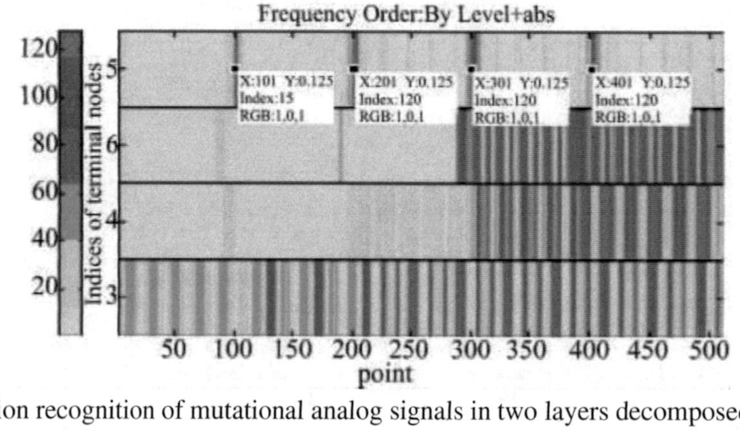

Fig.4 Initial motion recognition of mutational analog signals in two layers decomposed by db6 wavelet

In terms of the identification effect of analogue signals, the combination of wavelet packet transform and discrete wavelet transform can identify the first arrival time of non-stationary signals more accurately. It shows that this method can be applied to the identification of initial earthquake waves.

4. Identification of initial earthquake in actual seismic signals

It is found that wavelet packet transform combined with wavelet transform is an effective method to recognize the initial motion of the signal through previous analysis. Based on the experience and knowledge of analogue signal recognition, we used the method mentioned to identify the actual natural seismic signals of the aftershock of the Wenchuan earthquake. The identification results are as follows:

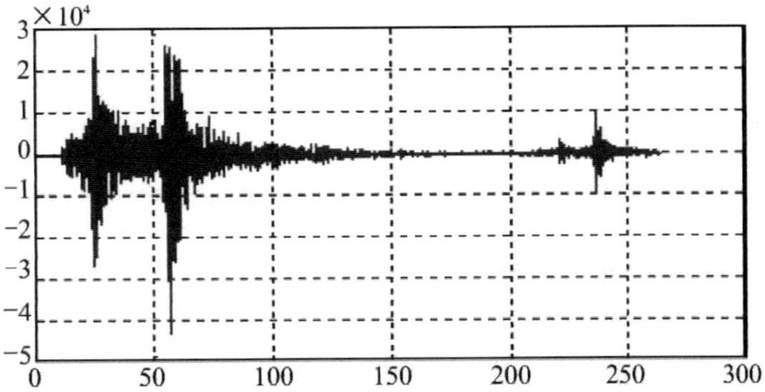

Fig.5The south-north seismic records of Wenchuan earthquake

The south-to-north natural seismic record of the aftershocks during the Wenchuan earthquake is shown in Figure 5, in which the sampling rate of 50 Hz. This paper took the actual natural seismic data as an example to identify the initial earthquake

Due to the complexity of seismic signal components, there exits plenty of noise in the recording process, and thereby, the initial seismic signal identification need to deal with several scales of coefficient energy distribution for the overall analysis.

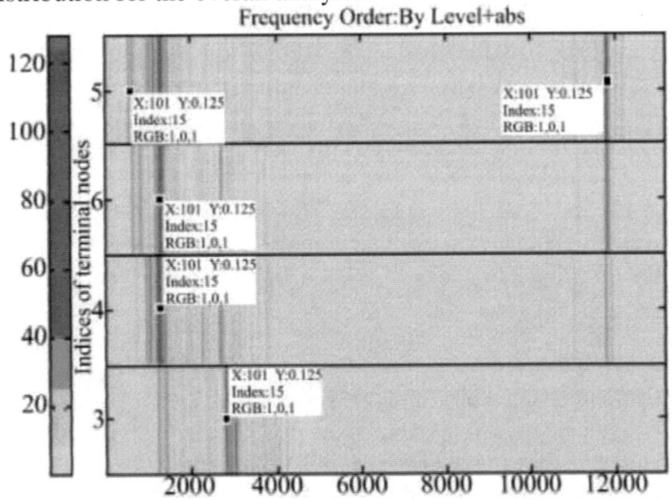

Fig.6 Initial motion recognition of Natural Earthquakes signals in two layers decomposed by db6 wavelet

It can be seen in Figure 6 that the coordinate of the first initial motion position of the seismic signal is at point 591, and the corresponding time is 11.82s. The coordinate of the second initial motion position of the seismic signal is at the point 1271, and the corresponding time is 25.42s. The coordinate of the third initial motion position of the seismic signal is at the point 2811, and the corresponding time is 56.22s. The coordinate of the fourth initial motion position of the seismic signal is at the point 11830, and the corresponding time is 236.6s. Thus, the location and time of several initial seismic points in the seismic signal can be captured.

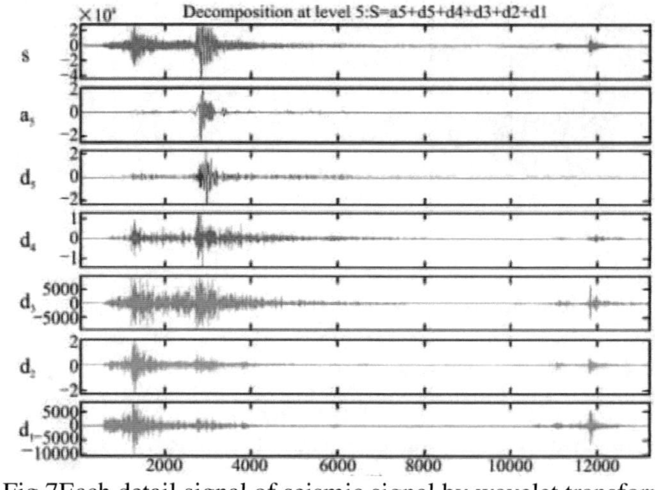

Fig.7Each detail signal of seismic signal by wavelet transform

Figure 7 is the detailed signal distribution diagram of the seismic signal in five layers decomposed by the db6 wavelet transform. The distribution of the highest frequency in the signal is d1, and the higher up the value, the lower the frequency. It can be seen from the frequency distribution in the figure that the highest frequency part of the seismic signal arrives first. The distribution of the lower frequency part of the seismic signal is represented by d5 and a5. According to the acknowledge we known, the first arrived is the P wave of the high-frequency component, followed by the S wave of the intermediate frequency part, then the surface wave (Love Wave) of the low-frequency part, and finally the aftershock.

Combined with the results of the two analyses, the first arrival time of P wave and S wave are 11.82s and 25.42s, respectively. The first arrival time of surface wave (Love Wave) is 56.22s. The initial arrival time of the aftershock is 236.6s.

5. Conclusion and suggestion

This paper discussed applying wavelet analysis technology to the emergency rescue for the initial natural earthquake. The main conclusions obtained are as follow:

1) The wavelet packet decomposition transform possesses a "mathematical microscope" function, which can automatically use different resolution time windows according to different frequency components of the signal. When a specific type of seismic wave arrives, the recorded waveform will make gradient change or mutational change; in other words, the abnormal high-frequency component will emerge. Wavelet packet decomposition will automatically use a short window to improve the temporal resolution, which means the method will focus on the new signal to accurately identify and measure the new seismic waves.

2) At low frequency, the wavelet functions with large-scale value and long period are used while small-scale values and short-period wavelet functions are used at high frequencies. The above transformation characteristics are particularly suitable for the demonstration and processing of seismic signals.

3) There are abundant high-frequency components when the initial of the signal arrives. Thus, it is obviously inappropriate to use low-frequency wavelet to detect the initial arrival of the signal, and in this case, high-frequency wavelet should be employed.

4) The maximum amplitude of waveform recorded is usually not the arrival of the seismic phase. In fact, the corresponding actual arrival of the maximum amplitude should be ahead of the seismic phase. For situations with high signal noise, picking up arrival time according to the maximum amplitude will not cause high errors, but for cases with low signal noise, this treatment will bring large errors. The research in this paper reveals that wavelet packet picking of seismic phases is a quite effective method and can be applied for earthquake emergency rescue and relevant work.

Hence, according to the characteristics of the earthquake-affected area, the following emergency rescue measures are recommended:

(a)Distribute the relevant documents of emergency rescue work to the residents threatened by thenatural earthquake;

(b) Divide the influence area of groundwater activity and set up warning zones and warning signs around the earthquake (and the affected) area;

(c) Monitor the water level changes and earthquake activity characterization phenomena, especially the signs and changes of the existing phenomena and the scope of influence, so as to exercise emergency evacuation and avoidance for the residents within the potential affecting area of the earthquake;

(d) Notify the electric power authority the situation about the earthquake threatening the safety of transmission, and ask the authority to make corresponding protection measures and emergency plans;

References

[1] Y.J. Zhou, Y.S. Pan. (2007) P and S Wave Identification of Rock-bust Wave Based on Wavelet Trans-form. Coal Mining Technology, 12(6): 1–4.

[2] D.P. Wu. (2004) Research on the Real-time Processing Method of Seismic Phase Identification. Master's Degree Dissertation, Institute of Engineering Mechanics China Earthquake Administration.

[3] Z.Z. Zhen, P. Shen. (2001) Wavelet Transform and its application with MATLAB. Seismological Press, Beijing.

[4] X.Q. Liu, H.L. Zhou, Z.Z. Zhen, P. S, X.H. Yang, Y.L. Ma. (1998) Identification Method of Weak Seismic Phase Based on Wavelet Package Transform. ACTA SEISMOLOGICA SINICA, 20(4): 373-380.

[5] X.Q. Liu, H.L. Zhou, W.H. Cao. (2002) Gaussian Chirplet Transform and its Application on Seismic Phase Identification. ACTA SEISMOLOGICA SINICA, 24(6): 607-616.

[6] P.X. Yang, C.H. Deng, X.Q. Liu, Y. Ren, Q.Z. Yam. (2004) Research on the method of automatic identifi-cation of seismic phase in digital seismic record. JOURNAL OF SESMOLOGICAL RESEARCH, 27(4): 308-313.

[7] D.K. Wang, J.Y. Peng. (2006) Wavelet Analysis and its Application in Signal Processing. Publishing House of Electronics Industry, Beijing.

[8] Q.Q. Qin, Z.K. Yang. (1995) Practical Wavelet Analysis. XIDIAN University Press, Xi'an.

IOP Conf. Series: Materials Science and Engineering **794** (2020) 012073 doi:10.1088/1757-899X/794/1/012073

Application of One-Dimensional and Two-Dimensional Coupled Overtopping Embankment Flood Model in Wuyi River

Teng Hui[1,2], Wang Lirong[1,2], LiHanze[1,2], ZhengHongri[1,2]and Li Qian[1,2]

[1]Zhejiang Institute of Hydraulics & Estuary, Hangzhou 310020, Zhejiang Province, China

[2]Zhejiang Provincial Key Laboratory of Hydraulic Disaster Prevention and Mitigation, Hangzhou 310020, Zhejiang Province, China

E-mail: forzath@qq.com

Abstract: Numerical simulation is an effective method to study the law of flood evolution. Adopting a one-dimensional two-dimensional coupling approach, the river flood evolution uses a one-dimensional model and the flood protection areas on both sides of the river adopt a tow-dimensional model. The two-dimensional model is laterally connected to the one-dimensional model, which can simulate the flood evolution process in the two-dimensional flood protection area after the flood of one-dimensional river overtopping the dam. A one-dimensional and two-dimensional coupled overtopping dam flood model in Wuyij River basin is established. The model calculation accuracy is relatively high and the results are overall reasonable.

1.Introduction

In order to resist flood attack and reduce flood damage, it is of great importance to study flood evolution law for the sake of flood control work. Based on the hydrodynamic theory, numerical simulation of flood evolution is an effective method to study the law of flood evolution.

The one-dimensional hydrodynamic model can quickly simulate the evolution process of flood such as in long river sections, complex river networks and has relatively high calculation efficiency. The two-dimensional hydrodynamic model can better simulate the evolution process of flood after flowing out of the river and provide detailed information such as the extent of flood inundation. The flood evolution processes of one-dimensional river channels and two-dimensional flood protection areas affect each other. Coupling the one-dimensional model with the two–dimensional model can simulate the flood evolution process in the two-dimensional flood protection area after flood overtopping the embankment of the one-dimensional river channel. Compared with the separate one-dimensional model and two-dimensional model, it has higher simulation accuracy and faster calculation speed.

This paper adopts a one-dimensional and two-dimensional coupling approach to establish the flood model in Wuyi River basin in Wuyi County, Zhejiang Province and analyses the law of flood evolution.

2. Model theory

2.1 One-dimensional model of the river channel

The equation of the one-dimensional model is the one-dimensional open channel non-constant flow equation. The difference format uses a sin-point central implicit difference format and uses the "catch-up method".

continuity equation:

$$\frac{\partial A}{\partial t}+\frac{\partial Q}{\partial x}=q \tag{1}$$

motion equation:

$$\frac{\partial Q}{\partial t}+\frac{\partial}{\partial x}\left(\frac{Q^2}{A}\right)+gA\frac{\partial z}{\partial x}+gA\frac{Q^2}{K^2}=0 \tag{2}$$

where: x and t are respectively spatial and temporal coordinates; $A$is the area of the river cross section; Q is the discharge of the cross section; q is lateral inflow; z is water level; K is discharge modulus and g is gravitational acceleration.

2.2Two-dimensional model

The two-dimensional model uses a two-dimensional non-constant flow equations describing the motion of the plane water flow, including three equations: the flow continuity equation, the momentum equation in the x direction and the momentum equation in the y direction. The form is as follows:

flow continuity equation:

$$\frac{\partial z}{\partial t}+\frac{\partial p}{\partial x}+\frac{\partial q}{\partial y}=s \tag{3}$$

momentum in the x direction:

$$\frac{\partial p}{\partial t}+\frac{\partial}{\partial x}\left(\frac{p^2}{h}\right)+\frac{\partial}{\partial y}\left(\frac{pq}{h}\right)+gh\frac{\partial z}{\partial x}+\frac{gp\sqrt{p^2+q^2}}{h^2c^2}-\Omega q-E\left(\frac{\partial^2 p}{\partial x^2}+\frac{\partial^2 q}{\partial y^2}\right)=s_{ix} \tag{4}$$

momentum in the y direction:

$$\frac{\partial p}{\partial t}+\frac{\partial}{\partial y}\left(\frac{q^2}{h}\right)+\frac{\partial}{\partial x}\left(\frac{pq}{h}\right)+gh\frac{\partial z}{\partial y}+\frac{gq\sqrt{p^2+q^2}}{h^2c^2}-\Omega p-E\left(\frac{\partial^2 q}{\partial x^2}+\frac{\partial^2 p}{\partial y^2}\right)=s_{iy} \tag{5}$$

where: z is water level; p and q are respectively unit discharges in the x and y direction; h is water depth;s is the source sink term; s_{ix} and s_{iy} are the components of the source sink term in the x direction and y direction; c is Chézy resistance coefficient; Ω is Coriolis force and E is turbulent diffusion coefficient.

According to the above equations, the finite volume method can be used to obtain the water level and water depth at (x, y) and velocity in the x and y directions at each moment.

2.3One-dimensional model coupled with two-dimensional model

The two-dimensional model is laterally connected to the one-dimensional model to simulate the process of mutual exchange and common evolution of water between the one-dimensional river and the two-dimensional protection area. When the water level of the one-dimensional river is higher than the top of the embankment, the flood overflows the river and enters the two-dimensional area. The one-dimensional model provides the overtopped channel side water level and the two-dimensional model provides the water level behind the overtopped embankment. Then weir flow formula is used to calculate the water flow through the lateral connection.

$$Q=\begin{cases} C_1W\left(H_{US}-H_W\right)\sqrt{H_{US}-H_W} & \left(H_{DS}-H_W\right)/H_{US}<2/3 \\ C_2W\left(H_{US}-H_W\right)\sqrt{H_{US}-H_{DS}} & \left(H_{DS}-H_W\right)/H_{US}\geq 2/3 \end{cases} \tag{6}$$

$$C_2=3\sqrt{3}C_1/2 \tag{7}$$

where: $Q$is the discharge overtopping the embankment, m^3/s; C_1 and C_2 are weir flow coefficients; W is weir crest width, m; H_{us} is upstream water level, m; H_{DS} is downstream water level, m; H_W is weir crest height, m.

3. Model application

3.1Model construction

3.1.1One-dimensional model

Taking urban areas and important towns in Wuyi County as protection objects and considering potential flood threats, the main stream of Wuyi river and its tributary Shu river are selected to establish a one-dimensional generalized model. Wuyi river is about 38.8km and 66 sections are arranged. The tributary Shu river is about 10km and 19 sections are arranged. 45 water-related buildings are generalized including 9 weirs, 1 sluice and 35 water-blocking bridges in the model.

The upstream boundary of the model adopts the corresponding flood discharge hydrograph and the downstream one adopts the relationship between water level and discharge. The inflows of other small rivers in the basin are treated as lateral centralized inflows of the main channel.

3.1.2Two-dimensional model

Both sides of the river channel are generalized and simulated using a two-dimensional model. According to the field investigation, the calculation extent should be appropriately expanded to ensure that the main towns and villages on both sides of the Wuyi river basin in Wuyi County are included in the scope of the two-dimensional model. There are 11 generalized protected areas with $128km^2$. The model is generalized with an unstructured triangular grid to accommodate the complex terrain changes in Wuyi river basin. The grid length on the riverside is 20m and the grid length on the non-riverside is 60m. A total of 135273 grids are divided.

3.1.3Model coupling

The lateral connection is used to couple the one-dimensional model and the two-dimensional model. The model is provided with 13 lateral connections and 2066 coupling units.

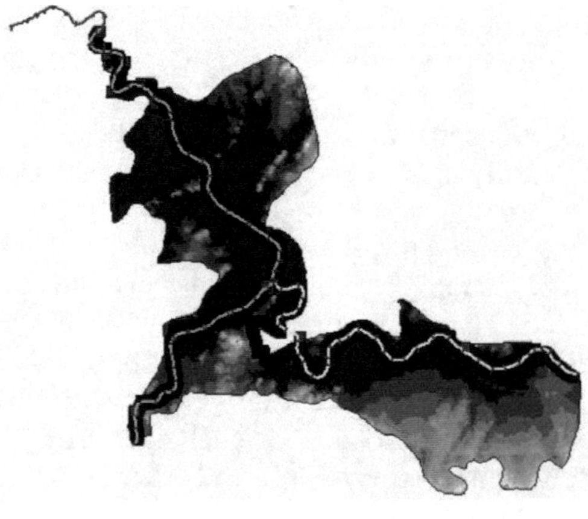

Figure 1.Sketch map of the model

3.2Calibration and verification

In the hydrodynamic model, the parameter to be calibrated is mainly the roughness coefficient, and the roughness is a comprehensive coefficient that reflects the degree of resistance when the water flows. In this paper, the flood on 8.20, 2014 is used for calibration and the flood on 6.25, 2017 is used for verification. The roughness values are shown in table 1.

Table 1.Roughness values

Category	Straight channel	Curved channel	River beach	Settlement place	Forest	Dry land	Paddy field	Clearing land
Roughness (n)	0.025-0.035	0.035-0.04	0.04-0.06	0.07-0.12	0.065	0.06	0.05	0.035

There is Wuyi hydrological station in the basin. The measured water level and discharge hydrograph are collected and compared with the calculated ones and the results are shown in table 2 and figure 2-3. It can be seen that the water level error is less than 0.1m and the discharge error is with 10%. The calculation accuracy is high and the water level and discharge at the flood peak are basically consistent with the measured ones.

The calculated inundation situation is compared with actual survey situation to analyze the rationality of the calculation of the two-dimensional model. According to the information of the local water conservancy bureau, the calculated flood and the actual one are basically consistent in terms of the location of the overtopped embankment and the inundation area. The inundation area is shown in figure 2. Therefore, the flood analysis model and its parameters are reasonable.

Table 2.Results of calibration and verification

Event	Maximum water level at Wuyi hydrological station (m^3/s)			Maximum discharge at Wuyi hydrological station (m)		
	Observation	Calculation	Relative error(%)	Observation	Calculation	Error(m)
Flood on 8.20,2014	2030	2184	7.5	67.4	67.36	-0.04
Flood on 6.25,2017	1580	1608	1.7	66.55	66.49	-0.06

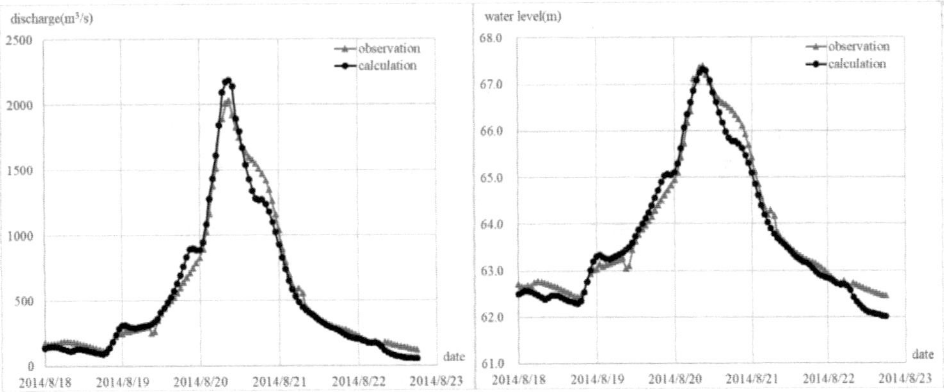

Figure 2.Calibration of the flood on 8.20,2014

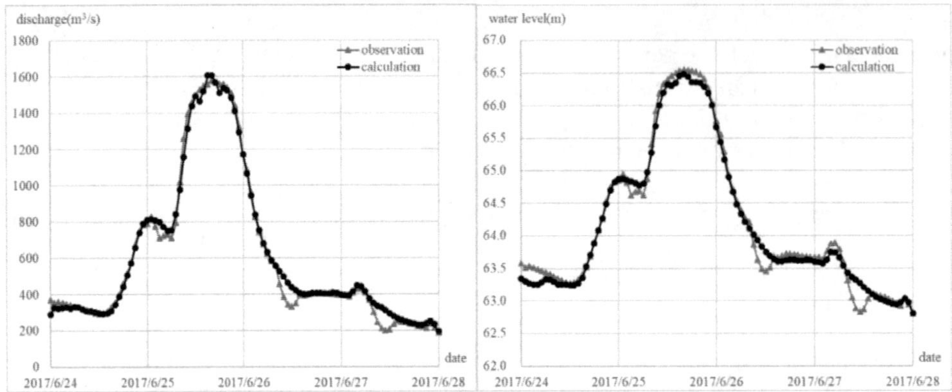

Figure 3.Verification of the flood on 6.25,2017

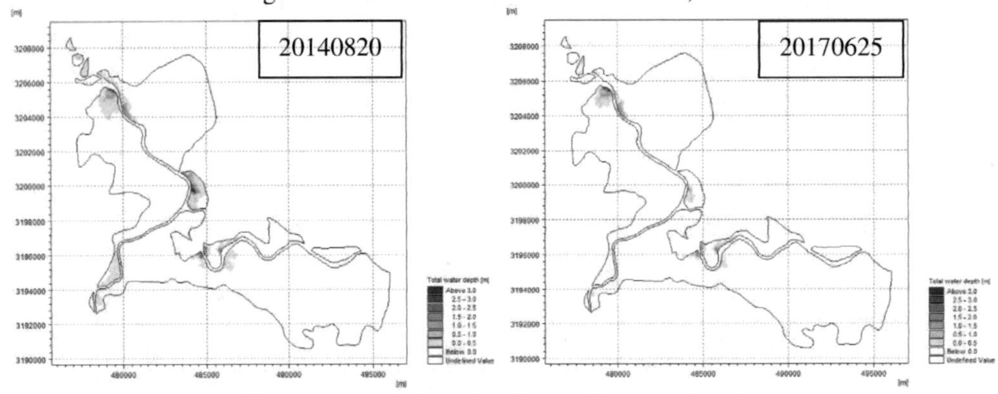

Figure 4.Inundation situations of the two floods

4. Conclusions and suggestions

(1) In this paper, a one-dimensional and two-dimensional coupled overtopping embankment flood model is established in Wuyi river basin. The one-dimensional river channel model uses the one-dimensional open channel non-constant flow equation and the two-dimensional model of the protection area uses the two-dimensional non-constant flow equation. The latter is laterally connected to the former and the former provides water level value and inflow discharge for the latter in real time.

(2) After calibration and verification, the errors between the calculation values and the observation values are small and the calculated water level and discharge hydrograph during the peak are basically consistent with the measured process. The calculated location of the overtopped embankment and inundation area are also basically consistent with the actual flood investigation. Thus, the model calculation accuracy is high, the results are generally reasonable and can provide a scientific basis for flood control in this region.

Acknowledgement

Zhejiang Provincial Research Institute Support Fund(2018F10027)

References

[1] HUAI Wenxin, ZHAO Zhenwu, TONG Hanyi, ZHANG Changjiang. (2003) Numerical simulation on flood in downstream of Weihe Rver andflooded area (I): Mathematic model and its calculation method[J]. Engineering Journal of Wuhan University, Vol.36 No.4:10-14.

[2] ZHANG Huiping, TIAN Fuchang, GENG Qingzhu, WANG Jianjia. (2018) Coupled 1D-2D Hydrodynamic Model for Overflow Flood Considering Complex River Buildings[J]. Water Resources and Power, Vol.36 No.9:81-85.

IOP Conf. Series: Materials Science and Engineering **794** (2020) 012073 doi:10.1088/1757-899X/794/1/012073

[3] FU Chengwei, YUAN Ximin, YANG Min. (2013) A real-time dynamic coupling model forflood routing and its application to flood risk charting[J]. Hydro-Scienceand Engineering,2013 No.5:32-38.
[4] CHEN Wenlong, SONG Lixiang, XING Linghang, ZHANG Wenming, ZHOU Jianzhong. (2014) A 1D-2D coupled mathematical model for numerical simulating of flood routine in flood protected zone[J]. Advances in Water Science, Vol.25 No.6:848-855.
[5] YUAN Ximin, XUE Wenyu, FENG Guona, LI Changyue. (2016) A coupled one and two dimensional hydrodynamic model for analysis of levee-breach flood and its application[J]. Advances in Science and Technology of Water Resources, Vol.36 No.4:53-58.

Research on influencing factors of non-stress gauge measurement on mesoscopic scale

Yongrong Qiu[1], Guoxin Zhang[1*]

[1]State Key Laboratory of Simulation and Regulation of Water Cycle in River Basin, China Institute of Water Resources and Hydropower Research, Beijing 100038, China.

[*]Corresponding author's e-mail:zhanggx@iwhr.com

Abstract: Concrete stress is an important item of dam safety monitoring. The stress of dam body can be mastered and the safety degree of dam can be estimated through systematic monitoring and analysis of the stress and strain of concrete. Non-stress gauge, as well as a strain gauge group, is used to monitor the stress and strain of concrete dams. The stress-free strain of concrete can be obtained by utilizing a non-stress gauge, which is the basis for stress analysis. Results of stress-free monitoring are influenced by surrounding environments, instrument structure, and installation method. Finite element numerical simulation and meso-mechanical methods are used to study factors, such as non-uniform temperature field around the gauge, stiffness of a bucket, and aggregate state of concrete. Results show that the stiffness of bucket and non-uniform temperature field slightly influence the results of non-stress strain measurements. In addition, when the aggregate content is the same, the influence of the aggregate distribution is less. However, extra-large aggregate removal method significantly influences the monitoring results. The influence of the removal method on the stress analysis to monitor the data of the strain gauge group is evaluated using an example. Recommendations to deal with extra-large aggregate in the installation of non-stress gauges are presented.

1. Introduction

Concrete stress is an important item of dam safety monitoring. The observation results of concrete stress state are the main basis for mastering the stress state, estimating the safety degree, checking the design and construction quality of a dam body, and improving dam engineering technology. However, directly measuring the stress of dams is difficult. **Carlson** [1] invented the stress meter to assess stress levels. However, this instrument has limitations. For example, the instrument cannot measure tensile stress. Since there is still no good stress meter, the indirect method is widely used in every concrete dam project in the world, which determines stress by measuring strain and the conversion of strain to stress. Concrete elastic modulus, creep and temperature strain, autogenous volume strain, and other basic data must be obtained to determine stress. Elastic modulus and creep strain can be identified by laboratory testing, whereas temperature strain and autogenous volume strain are measured using a special instrument called non-stress gauge. Non-stress strain observation is the basis of observing stress in dams.

Therefore, the three main factors that affect the stress observation results include the accuracy of strain gauge measurements, the accuracy of non-stress gauge measurements, and the accuracy of elastic modulus and creep parameters of measured concrete. Various factors influence the accuracy of strain gauge measurement, such as temperature compensation [2], length [3], mechanical properties of gauge [4], and method of installation [5]. **Calderón et al**. [4] argued that the two most important conditions must be met to ensure the accuracy of strain gauge measurement: (1) the good interaction between the sensor and the measured structure must be guaranteed; and (2) the presence of the sensor must not perturb the measured structure; the result indicated that the accuracy of measurements is mainly dependent on the goodness of the interaction (strain transfer) between a host material and the anchor pieces of a sensor. Several scholars conducted related research. **Azenha et al**. [6] studied the effect of strain gauge stiffness on measurements and determined that the effect of strain stiffness on measured values is significantly higher at the early stage of concrete than at the later stage. Torres et al. [7] examined the effect of material properties on measured values and determined that small changes in material properties can also affect measurements to a significant degree. **Kesavan et al**. [8] analyzed the effect of wrapper stiffness on measured values and determined that the effect of stainless steel packaging on measured values is more substantial than that of composite packaging.

Most studies focused on strain gauges and have gained a lot of knowledge, but few examined non-stress gauges. Inaccurate non-stress strain measurements cannot indicate the autogenous volume strain of concrete, and obtaining the correct stress results is difficult even if strain gauge observations are accurate [9]. Non-stress gauge is generally composed of a strain gauge and a bucket. Strain gauge is used to measure concrete strain, whereas bucket isolates concrete from external forces. The concrete in the bucket, which is free from external forces, only indicates the strain of other physical and chemical factors of concrete. Therefore, the strain of concrete in the bucket is mainly composed of temperature strain and autogenous volume strain, which includes volumetric strain caused by hydration and the hardening process of cement and possible alkali–aggregate reaction.

To ensure that a stress-free environment is the same as that of a strain gauge group, a non-stress gauge must be embedded near the strain gauge group at approximately 1 m. However, various indeterminate factors may exist in the measurement of non-stress gauge, which leads to uncertainties on whether the non-stress gauge can actually determine stress-free conditions. The main causes of uncertainty that affects the results of stress-free measurement are as follows:

(1) Effect of bucket stiffness. According to **Lv et al**. [10], concrete in stress-free buckets may be subject to the lateral constraints of a bucket wall, thereby resulting in constrain stress. Thus, a stress-free condition is not achieved.

(2) Effect of temperature field. The non-uniform temperature field [11], generally exist near a concrete surface or a water cooling pipe, may cause stress. Thus, stress-free conditions are not met if the non-stress gauge is embedded in a non-uniform temperature field, and the observed values of non-stress strain may contain temperature stress-induced strain.

(3) Effect of aggregate state. When a strain gauge group of a concrete dam is buried, the large aggregate is removed from the backfill and mortar is usually added to facilitate the compaction. The mix ratio of the concrete around the strain gauge may be greatly different with that of actual dam concrete. **Zhao et al**. [12] noted that the effect of the removal of large aggregate on non-stress measurements must be carefully studied.

The three factors mentioned are studied using finite element numerical simulation and meso-mechanics. The effect of the removal of aggregates on non-stress strain gauge measurements was investigated. The influence of measurement error of non-stress strain gauge on the stress analysis of the monitoring data of a strain gauge group was estimated based on the research results.

2. Principle of non-stress gauge

Non-stress gauge measurements are primarily conducted to determine the actual coefficient of thermal expansion and the autogenous volume changes of dam concrete. The coefficient of thermal expansion and the autogenous volume change obtained by a non-stress gauge are used to analyze the strain of the working strain gauge group and calculate the creep stress of concrete. Monitoring the stress of concrete is mainly based on strain gauge group and non-stress gauge. Strain gauge groups measure the strain of concrete, including stress-induced strain and non-stress strain. Based on the assumption that the total strain, stress strain, and non-stress strain in the concrete are ε_m, ε, and ε_0, respectively, the relationship between these types of strains can be expressed as

$$\varepsilon_m = \varepsilon + \varepsilon_0. \tag{1}$$

In the in-situ observation, ε_m and ε_0 can be directly measured, and ε is calculated using Equation (1). Stress-free strain mainly includes thermal expansion strain caused by temperature changes, dry shrinkage strain caused by moisture change, and autogenous volume strain caused by cement hydration. Non-stress strain is expressed as follows:

$$\varepsilon_0 = G_t + \alpha\Delta T + \varepsilon_\omega, \tag{2}$$

where G_t is autogenous volume strain caused by the physical and chemical factors of concrete, $\alpha\Delta T$ is strain attributed to temperature changes, α is thermal expansion coefficient of concrete, ΔT is temperature change, and ε_ω is strain caused by humidity. Mass concrete moisture is almost constant. Thus, strain caused by moisture inside the barrel of concrete can be disregarded.

Various types of non-stress gauge exist in different countries. Two types are generally used in the United States, as shown in Fig. 1. The first type embeds one strain meter vertically and another horizontally near the top of a concrete lift in an isolated cone of concrete, as shown in Fig. 1(a). The second type embeds a concrete pipe with length of 3 or 4 ft (0.9 or 1.2 m) and diameter of 15 in (381 mm), and a strain meter is suspended at the center of a 12 in × 24 in (305 mm × 610 mm) cylinder specimen mold of a type that permits easy stripping, as shown in Fig. 1(b). The non-stress gauge type widely used in China is similar to that in Fig. 1(b) but resembles a cone shape, as shown in Fig. 2. This type was originally obtained from Novosibirsk Hydropower Station and proposed by **Adelman** [13]. Although various types of non-stress gauges have been developed in different countries, these gauges are based on the same design principle.

The non-stress gauge widely used in China is made of a conical bilayer bucket in which concrete is poured in the inner cylinder and a strain gauge is embedded in the central axis. The general structure of non-stress gauge buckets is shown in Fig. 2. Taking a conical bilayer bucket for the strain gauge with length 25cm for example, the outer cylinder, inner cylinder are made of 1.2mm and 0.5mm thick metal sheet respectively, the gap between the inner and outer cylinder is around 10–30 mm and is mostly filled with foam. The inner cylinder is coated with 5 mm-thick asphalt. The concrete in the inner cylinder is not subjected to external force because of the separation of the two layers of the bucket. The concrete is integrally connected to the outer concrete only by the bunghole to maintain the same temperature and humidity as the outer concrete. Thus, the strain of the concrete in the inner cylinder is caused only by temperature, humidity, and autogenous volume changes and not by stress, i.e., the strain measured in the inner cylinder is stress-free strain.

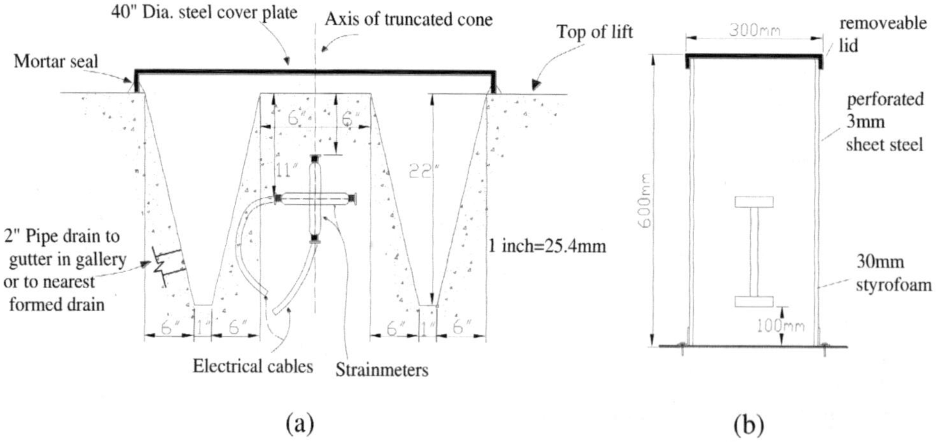

Fig. 1 Diagram of non-stress gauge

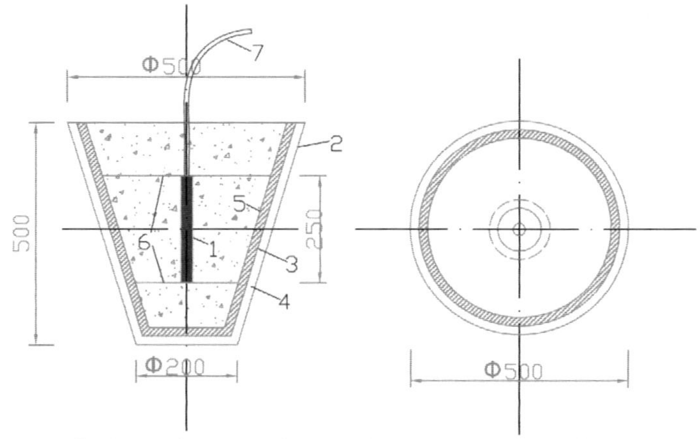

(1 strain gauge, 2 outer cylinder [1.2 mm], 3 inner cylinder (0.5 mm), 4 foam, 5 asphalt layer (5 cm), 6 wire, 7 cable)

Fig. 2 Diagram of non-stress gauge

3. Effect of bucket stiffness

This section aims to examine whether a bucket is laterally bound to the strain of concrete. Thus, the inner concrete is examined to verify whether it is free deformation from the constraint of the bucket. The shape and size of the bucket are assumed to be the same, and only bucket stiffness changes.

The calculation model is shown in Fig. 3(a). The size of the model is set to 1.0 m × 1.0 m. The bucket is composed of double layers of metal with a foamed gap, as presented in Fig. 3(b). The size of the bucket is the same as that in Fig. 2. The calculation model is not subject to external loads but is surrounded by a normal constraint. Concrete is a homogenous material with autogenous volume strain that is assumed to increase linearly with age. The given autogenous strain curve of concrete is shown in Fig. 4. The modulus of elasticity of concrete is set to 35 GPa, and Poisson's ratio is 0.2. The stiffness of the bucket is generally determined by the inner and outer cylinder metal sheet and the middle foam. The elastic modulus of the foam material is less than that of the metal sheet. Therefore, the foam elastic modulus is considered as 50 MPa, and the metal elastic modulus E_S is set as 50, 100, 150, and 200 GPa.

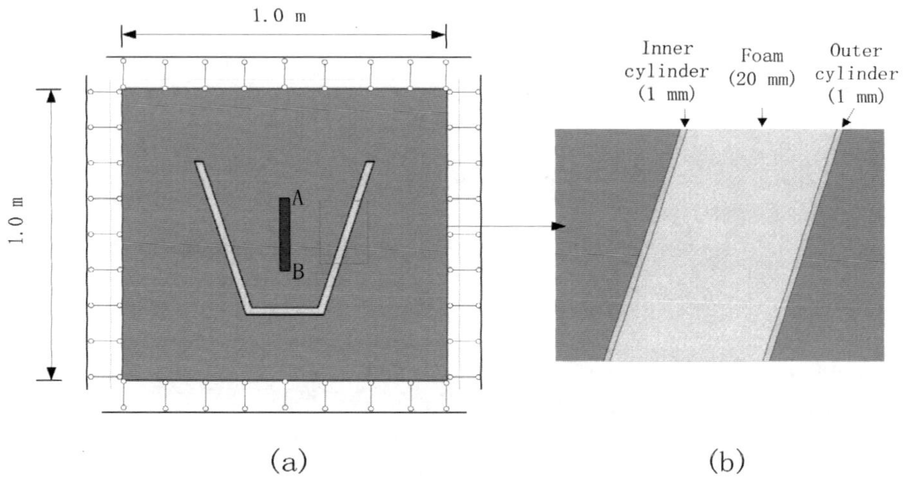

Fig. 3 Computing model

Linear elastic finite element method (FEM) is used for the calculation. The measurement strain of the strain gauge was calculated by dividing the relative displacement of both ends of the strain gauge by the length of the strain gauge, as expressed by

IOP Conf. Series: Materials Science and Engineering **794** (2020) 012074 doi:10.1088/1757-899X/794/1/012074

$$\varepsilon = \frac{L-L_0}{L_0} = \frac{(z_A+d_A)-(z_B+d_B)}{z_A-z_B} = \frac{d_A-d_B}{z_A-z_B}, \tag{3}$$

where ε is the measurement strain of the strain gauge; L_0 is the initial length of the strain gauge; L is the length of the strain gauge after calculation; z_A, z_B, d_A, and d_B are the coordinate height and displacement of the two ends of the gauge, respectively.

The calculation results of the four cases are indicated in Fig. 4. The measurement strain of the gauge of all the cases are basically equal to the given autogenous strain of concrete. The strain difference of the four cases is small. Therefore, the influence of bucket stiffness on the non-stress gauge measurement is minimal.

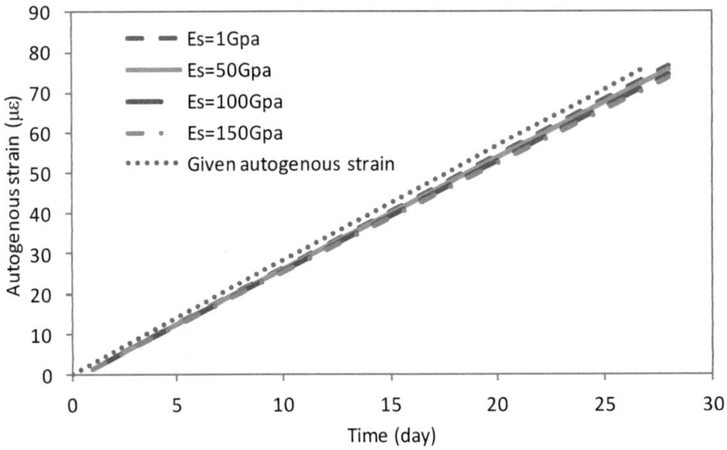

Fig. 4 Results of different bucket stiffness calculations

4. Effect of non-uniform temperature field

Wang et al. [11] noted that the stress-free condition is not satisfied if a non-stress gauge is in a non-uniform temperature field, and the strain measurement may contain a part of the strain caused by the stress induced by the non-uniform temperature. According to thermoelastic theory, for a 3D problem, a temperature field that does not produce a self-constrained thermal stress must satisfy

$$\left.\begin{array}{c}\frac{\partial^2 T}{\partial x^2}+\frac{\partial^2 T}{\partial y^2}=0, \frac{\partial^2 T}{\partial y^2}+\frac{\partial^2 T}{\partial z^2}=0, \frac{\partial^2 T}{\partial z^2}+\frac{\partial^2 T}{\partial x^2}=0 \\ \frac{\partial^2 T}{\partial x\,\partial y}=0, \frac{\partial^2 T}{\partial y\,\partial z}=0, \frac{\partial^2 T}{\partial z\,\partial x}=0\end{array}\right\}. \tag{4}$$

This set of equations is solved as follows:

$$T = a_0 + a_1 x + a_2 y + a_3 z, \tag{5}$$

where a0–a1 are constant coefficients. Equation (5) indicates that the concrete temperature distribution must be linearly distributed to ensure that the concrete in the bucket does not produce thermal stress conditions.

This condition is not always satisfied in a complex and practical engineering structure. Non-uniform temperature fields generally exist near a concrete surface or a water cooling pipe. The temperature field near the water cooling pipe is the most uneven. In this section, the non-uniform temperature field of a cooling pipe is used as an example to study the effect of a non-uniform temperature field on non-stress gauge. The non-uniform temperature field was analyzed to determine if this temperature field disrupts the stress-free condition.

The calculation model is similar to that of the previous section, but the cooling pipe is arranged on four sides. The diameter of the cooling pipe is 20 mm, and the model is free from external constraints. The calculation model is shown in Fig. 5. The initial temperature of the concrete is 30 °C and that of the cooling water is 15 °C. The thermal boundary condition of a pipe wall is set to $T=T_w$. The other boundary is set as an adiabatic boundary, as indicated in Fig. 5. The thermal parameters of all the materials are listed in Table 1. The mechanical parameters of all materials are the same as those in

Section 3. The temperature field of the concrete under the action of water cooling and the thermal stress induced by the temperature field in the non-stress gauge are calculated.

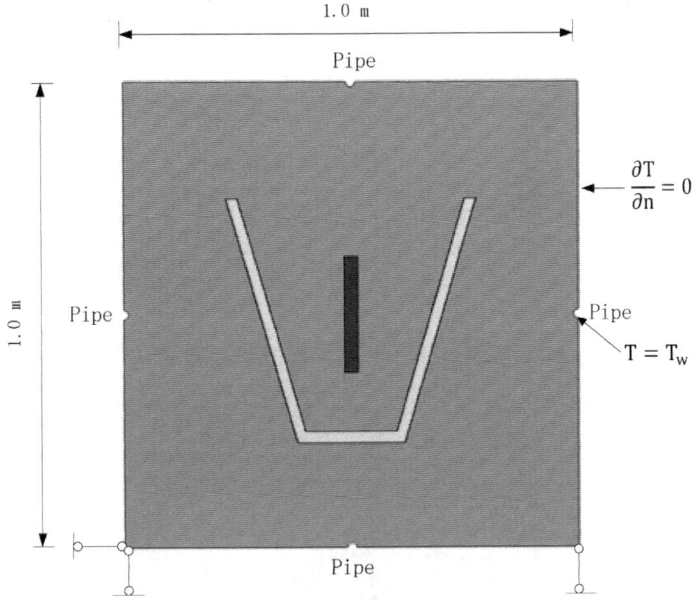

Fig. 5 Computing model

Table 1 Thermal and mechanical parameters of each phase

Phase	Density ($10^3kg/m^3$)	Thermal parameters			Mechanical parameters	
		Coefficient of thermal expansion (/°C)	Coefficient of thermal conductivity (kJ/(m·h·°C))	Specific heat (kJ/(kg·°C))	Elastic modulus (GPa)	Poisson's ratio
Metal sheet	7.5	12.0e-06	100.0	0.46	200	0.3
foam	0.09	50.0e-06	0.15	1.5	0.5	0.2
gauge	2.7	30.0e-06	1.0	0.8	3	0.2
Concrete	1.5	7.5e-06	10.0	1.0	35	0.17

The results are shown in Fig. 6. Fig. 6 (a), (b), and (c) show the temperature distribution after cooling for 3, 7, and 14 days, respectively. Fig. 6 (d), (e) and (f) indicate the stress distribution after cooling for 3, 7, and 14 days, respectively. In the early stages of cooling, the temperature gradient is large and mainly concentrated in the vicinity of the cooling pipes. As the cooling process progresses, the temperature gradient gradually develops inside the model. However, the bucket wall exhibits a strong barrier effect on temperature conduction. Thus, the temperature distribution of concrete in the bucket remains relatively uniform. In the latter stage of the cooling period, the uniformity of the temperature distribution increases. As the temperature gradient of the entire model decreases, the temperature gradient of the concrete in the bucket also reduces.

In the early cooling period, a large temperature gradient near the cooling pipe causes a large stress near the pipe, but the stress in the bucket is still zero. In the mid-cooling period, the stress near the cooling pipe begins to decrease, and the concrete in the bucket produces a certain amount of stress. However, stress remains minimal. In the latter stage of cooling, the stress of the concrete in the bucket decreases further as the overall temperature equalizes.

The results show that the non-uniform temperature field in the non-stress bucket produces a certain amount of stress. However, in actual projects, pipe cooling not only produces an excessively large temperature gradient in the concrete in the bucket because of the heat insulation of the bucket wall and the no-stress gauge is usually installed far from the cooling pipe. The produced thermal stress is not excessively large. Thus, the effect of non-uniform temperature field on no-stress gauge measurements is minimal.

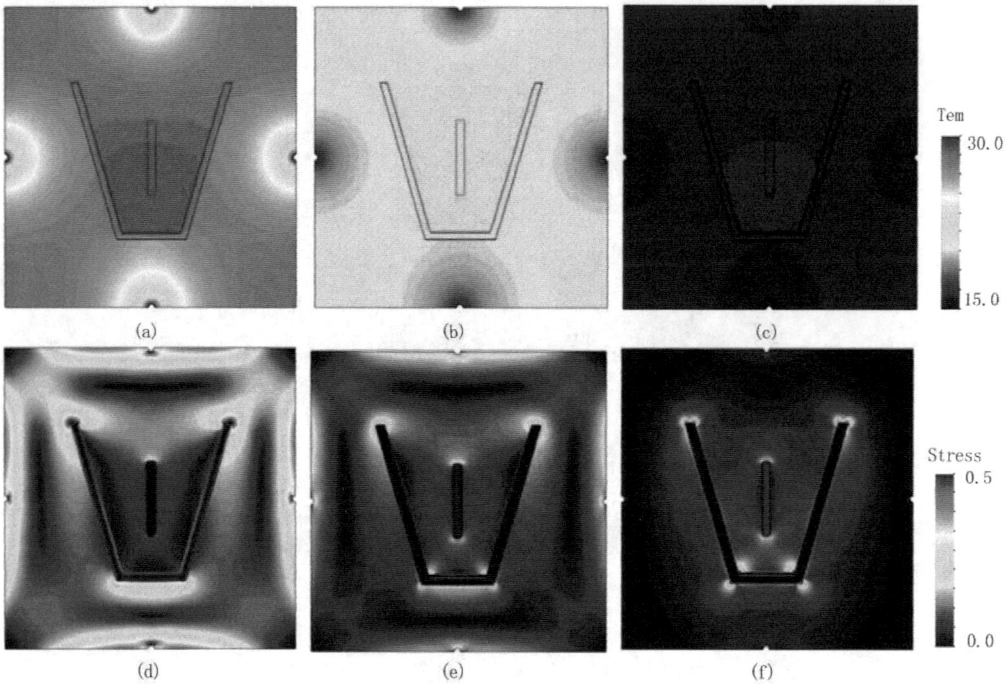

(3 days) (7 days) (14 days)

Fig. 6 Temperature and stress field at specific cooling times

5. Influence of aggregate content and distribution

Concrete is a multiphase composite material that consists of aggregate, mortar, and interface (ITZ) at the meso-scale. The results of studies on concrete meso-mechanics show that aggregate content and distribution significantly affect the properties of concrete. In the installation process of non-stress gauges, the large aggregate in the concrete is removed, which essentially changes the state of the aggregate in the concrete and alters the nature of the measured object. The influence of aggregate state changes on the coefficient of the thermal expansion of concrete and the measurement of autogenous volume strain is crucial for non-stress measurements.

5.1Effect of random distribution of aggregates

This section aims to investigate the effect of aggregate distribution on the stress measurement of non-stress gauges under the same aggregate content and aggregate shape. Four aggregate distribution models are established, as shown in Fig. 7. The aggregate of the four models have the same number and shape, but the distribution differs. The aggregate number of one model is set to 500, and the maximum size of the aggregate is set to 80 mm. The gradation of the aggregate is fit with a full curve.

The calculation parameters of each phase of concrete are shown in Table 2. The calculation parameters of other materials are the same as those in Sections 3 and 4. For computational convenience, mortar autogenous volume strain is assumed to linearly increase with age, as indicated in Fig. 8.

The calculated results of the non-stress strain gauge are shown in Fig. 8. The calculation results show that the measured autogenous volume strains of these four models are basically the same and their differences are minimal. Thus, when the aggregate content is the same, the aggregate spatial distribution on the measurement results of the strain gauges exhibits a minimal effect.

The effect of aggregate distribution on the measurement of the coefficient of thermal expansion has been studied. All of the models are under unit uniform temperature drop, and the measured coefficients of thermal expansion of various models are calculated. The coefficients are 7.39, 7.37, 7.41, and 7.36 με/°C. These results show that the random distribution model slightly influences the coefficient of the monitoring results of thermal expansion.

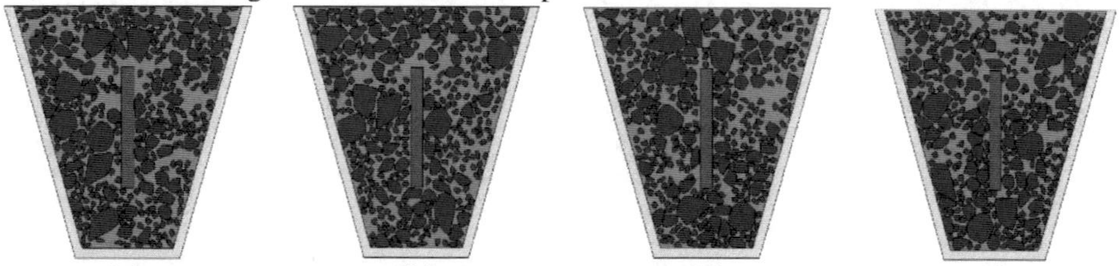

(a) (b) (c) (d)

Fig. 7 Four different aggregate distribution models

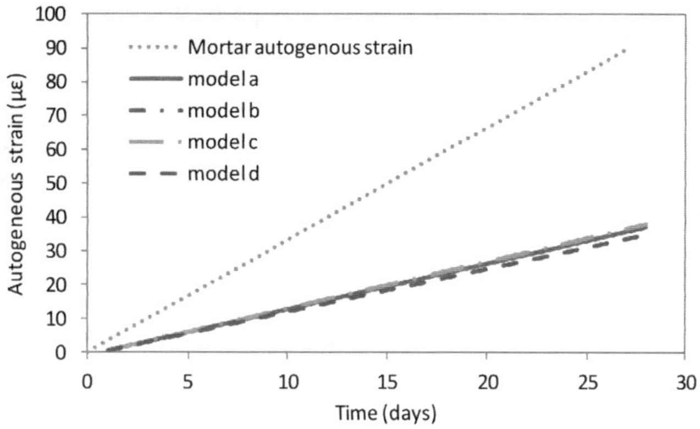

Fig. 8 Calculation results of autogenous volume strain of different aggregate distributions

Table 2 Thermal and mechanical parameters of each phase

Material	Thermal parameters	Mechanical parameters	
	Coefficient of thermal expansion (/°C)	Elastic modulus (GPa)	Poisson's ratio
Aggregate	5.0e-06	50	0.2
Mortar	11.0e-06	20	0.16

5.2Effect of Aggregate Content

The effect of the change in the aggregate content on stress-free measurements is studied in this section, including the effect on the linear expansion coefficient and the autogenous volume deformation. Six meso-scale models with aggregate contents (area fraction) of 0.3, 0.4, 0.5, 0.6, 0.7, and 0.8 are established, as presented in Fig. 9. The aggregate content gradations of the six models are all set to large-sized aggregate (d = 40 to 80 mm): medium-sized aggregate (d = 20 to 40 mm): small-sized aggregate (d = 5 to 20 mm) = 3:2:2.

First, the difference of the coefficient of thermal expansion of various models is studied. In this calculation, the influencing factors of aggregate content and the changes in the elastic modulus of a mortar are considered. Due to the modulus of a hardening material develops with its age, the modulus of the mortar is set in this study as 5, 10, 15, 20, 25, 30, and 35 GPa. The modulus of the aggregate is set as 50 Gpa. The aggregate coefficient of thermal expansion is 5 με/°C. The coefficient of thermal expansion of the mortar is 11 με/°C. The other calculation parameters are the same as those indicated in Section 0.

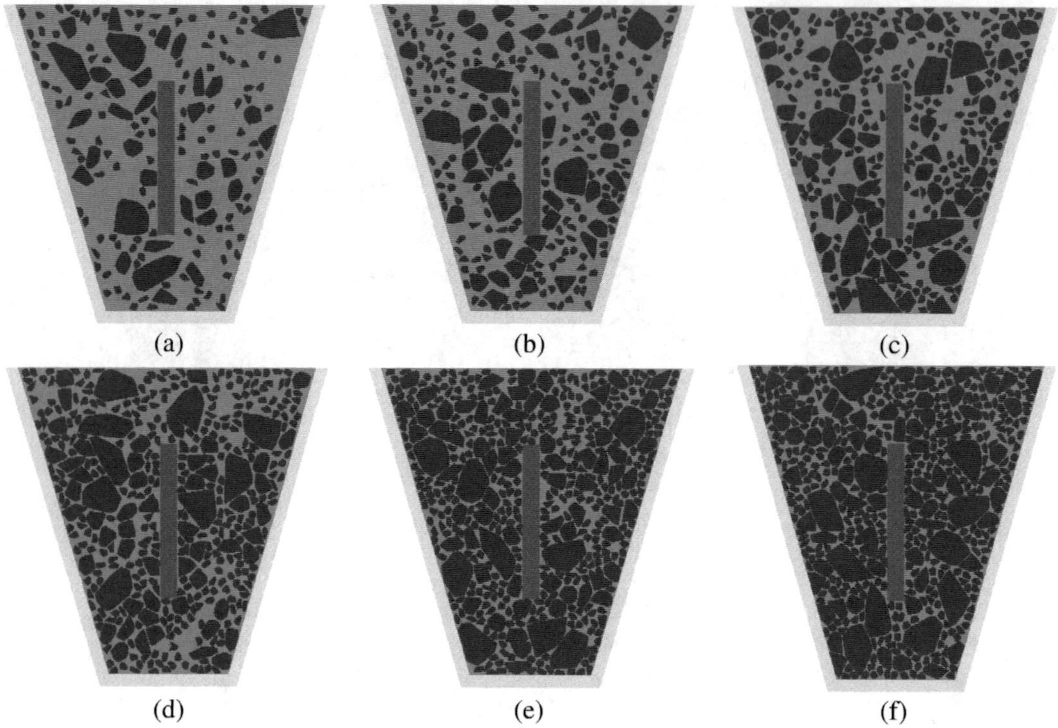

Fig. 9 Six different aggregate content models

Table 3 Effect on coefficient of thermal expansion (με/°C)

Aggregate content	Mortar elastic modulus/aggregate elastic modulus								
	0.1	0.2	0.3	0.4	0.5	0.6	0.7	0.8	0.9
0	11.00	11.00	11.00	11.00	11.00	11.00	11.00	11.00	11.00
0.3	8.37	8.50	8.62	8.72	8.82	8.91	8.99	9.06	9.13
0.4	7.71	7.84	7.96	8.07	8.18	8.27	8.36	8.45	8.53
0.5	7.13	7.25	7.36	7.47	7.57	7.67	7.76	7.84	7.92
0.6	6.61	6.71	6.81	6.91	7.00	7.09	7.17	7.25	7.33
0.7	6.14	6.23	6.31	6.38	6.46	6.53	6.60	6.67	6.74
0.8	5.73	5.78	5.84	5.89	5.95	6.00	6.05	6.10	6.15

The calculation results are listed in Table 3. Taking the 25 Gpa mortar case as example, for these models with the aggregate contents of 0.3, 0.4, 0.5, 0.6, 0.7, and 0.8, the coefficients of thermal expansion are 8.82, 8.18, 7.57, 7.00, 6.46, and 5.95 με/°C, respectively. When the aggregate content increases by 0.1, the coefficient of thermal expansion decreases by roughly 7.0%. The non-stress gauge aggregate content substantially affects the results of the coefficient of thermal expansion measured by the non-stress gauge.

Second, the difference of the autogenous volume strain of various models is studied. The autogenous volume strain of a mortar is set as 10 με. The results are shown in Table 4. Also taking the 25 Gpa mortar case as example, for these models with the aggregate contents of 0.3, 0.4, 0.5, 0.6, 0.7, and 0.8, and the autogenous volume strain is 6.36, 5.29, 4.29, 3.33, 2.43, and 1.58 με, respectively. When the aggregate content increases by 0.1, the autogenous volume strain decreases by around 0.8 με, and the decrease percentage is roughly 16.8% to 34.9%. The aggregate content of concrete in the non-stress bucket significantly influences the measurement of the autogenous volume strain.

Table 4 Effect on autogenous volume strain (με)

Aggregate content	Mortar elastic modulus/aggregate elastic modulus								
	0.1	0.2	0.3	0.4	0.5	0.6	0.7	0.8	0.9
0	10.00	10.00	10.00	10.00	10.00	10.00	10.00	10.00	10.00
0.3	5.62	5.83	6.03	6.20	6.36	6.51	6.65	6.77	6.89
0.4	4.52	4.74	4.94	5.12	5.29	5.45	5.60	5.74	5.88
0.5	3.55	3.75	3.94	4.12	4.29	4.44	4.59	4.74	4.87
0.6	2.68	2.86	3.02	3.18	3.33	3.48	3.62	3.75	3.88
0.7	1.91	2.05	2.18	2.31	2.43	2.55	2.67	2.78	2.89
0.8	1.21	1.30	1.40	1.49	1.58	1.67	1.75	1.84	1.92

6. Effect on stress calculation

6.1Preliminary estimates

The effect of large aggregate removal on the stress analysis can be preliminarily estimated based on the results in Section 5. Table 5 lists the mix proportion of three kinds of concrete in a project. The average aggregate amount is 1831 kg/m^3, and the gradation ratio of aggregates is extra large (d = 80 to 150 mm): large (d = 40 to 80 mm): medium (d = 20 to 40 mm): small (d = 5 to 20 mm) = 3: 3: 2: 2. Based on the assumption that the aggregate density is 2500 kg/m3, the volume ratio of the aggregate is approximately 0.73.

Table 5 Concrete mix proportion

Strength grade	Water-cement ratio	Fly ash content (%)	Sand rate (%)	The amount of material per cubic gauge of concrete (kg/m^3)				
				Water	cement	fly ash	sand	aggregate
$C_{180}40$	0.41	35	23	82	130	70	511	1843
$C_{180}35$	0.45	35	24	82	118	64	538	1833
$C_{180}30$	0.49	35	25	83	110	59	563	1816

Based on the three types of large aggregate treatments, the volume ratio of the aggregate is calculated as follows:

(1) Excluding large aggregate, filled with mortar: the aggregate content becomes 1831 kg/m^3 × (1-0.3) =1281.7 kg/m^3, and the volume ratio is around 1281.7 / 2500= 0.51.

(2) Excluding all concrete, filled with wet-screened concrete: the aggregate content is the same as that of the wet-screened concrete.

The wet-screened concrete has no extra-large aggregate. Based on the assumption that one cube meter concrete excluded extra-large aggregates, the following results are obtained:

Aggregate content is 1831 kg × (1-0.3) = 1281.7 kg.

Concrete volume is 1 m^3 - (1831 kg × 0.3) / 2500 kg/m^3 = 0.65 m^3.

Thus, the aggregate content of one cube meter of wet-screened concrete is determined: 1281.7 kg/0.65 m^3 = 1642.6 kg/m^3.

Thus, the volume ratio of the aggregate becomes roughly 1642.6 / 2500= 0.66.

(3) Excluding large aggregate, filled with medium and small aggregate: the aggregate content remains constant, and the volume ratio is still around 0.73.

Therefore, three treatment methods obtain three volume ratios of 0.51, 0.66, and 0.73 of aggregate. According to the results presented in Table 3 and Table 4, the coefficient of thermal expansion, autogenous volume strain reference value, and relative error can be obtained by interpolation method, as shown in Table 6.

Table 6 Interpolated result of coefficient of thermal expansion and autogenous volume strain of each removal method

Aggregate content	coefficient of thermal expansion		autogenous volume strain	
	error	relative error	error	relative error
0.51	7.51	19.20%	4.19	92.50%
0.66	6.67	5.90%	2.79	28.30%
0.73	6.31	0.00%	2.18	0.00%

The results in Table 6 indicate the effects of stress on the stress analysis, which are preliminary estimated. Based on the assumption that the concrete temperature decrease is 10 °C, the concrete elastic modulus is assumed to be 35 GPa, the coefficient of thermal expansion of the concrete is 10 με/°C, and the relative error of coefficient of thermal expansion is 19.2%. Thus,

$$\Delta\sigma = E\Delta\alpha\Delta T = 35\text{Gpa} \times 10\mu\varepsilon/°\text{C} \times 19.2\% \times 10°\text{C} = 0.67\text{MPa}.$$

Therefore, the coefficient of thermal expansion has a relative error of 19.2%, and thermal stress has an error of approximately 0.67 MPa.

Based on the assumption that the autogenous volume strain is 50 με and its relative error is 92.5%, the concrete elastic modulus is assumed to be 35 Gpa. Thus,

$$\Delta\sigma = E\Delta\varepsilon = 35\text{Gpa} \times 50\mu\varepsilon \times 92.5\% = 1.62\text{MPa}.$$

Therefore, the autogenous volume strain has a relative error of around 92.5%, and the stress has an error of roughly 1.62 MPa. Preliminary assessment indicates that the aggregate content in the non-stress bucket substantially influences the stress analysis.

6.2Project example calculation

The influence on the stress analysis is evaluated based on the calculation results of Section 5, with the actual monitoring data of a certain project as the analysis object. The dam is a concrete double-curvature arch dam with crest elevation of 610 m and maximum dam height of 285.5 m. The measuring point of the 334-m height of dam section No. 16 was taken as the analysis object. The point is embedded with a six-direction strain gauge group and a non-stress gauge. The structure of the six-direction strain gauge is shown in Fig. 10. A six-direction strain gauge consists of six strain gauges, each bound to the bracket, and the bracket frame is a tetrahedron. The location of each strain gauge of the strain gauge group is shown in Fig. 11. Based on the assumption that the strain measurement value of six strain gauges is ε_1, ε_2, ε_3, ε_4, ε_5, and ε_6, the strain of concrete in the Z direction can be calculated by

$$\varepsilon_z = \frac{2\sqrt{3}}{3}(-\varepsilon_1 - \varepsilon_2 - \varepsilon_3 + 3\varepsilon_4 + 3\varepsilon_5 + 3\varepsilon_6). \quad (6)$$

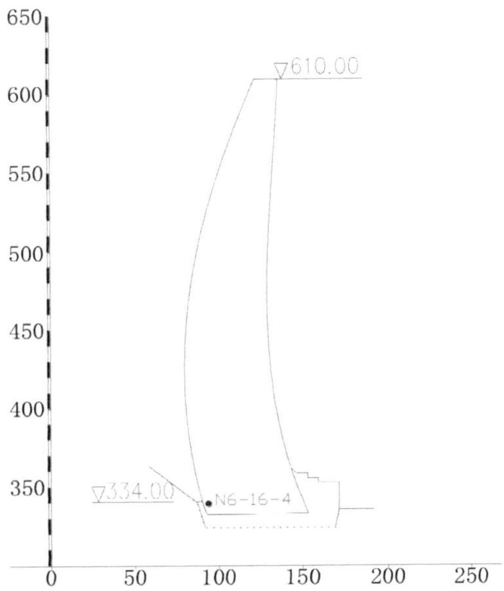

Fig. 10 Sectional drawing of No. 16 dam section and location of measuring points of a certain project

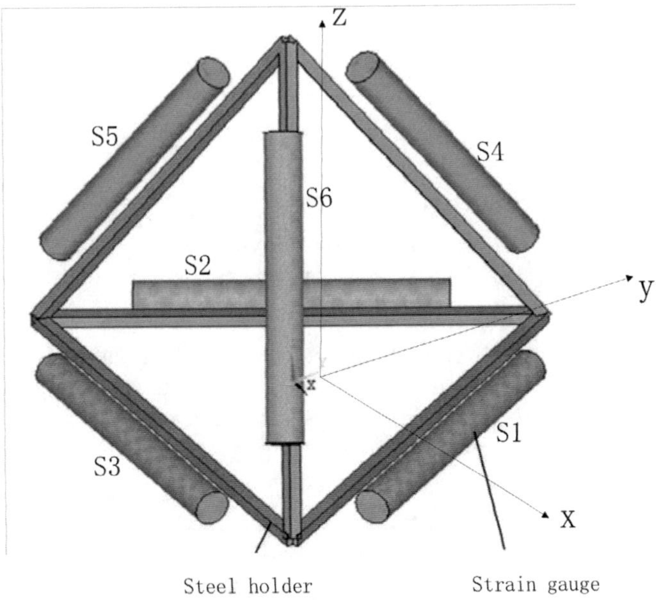

Fig. 11 Structure diagram of a six-direction strain gauge group

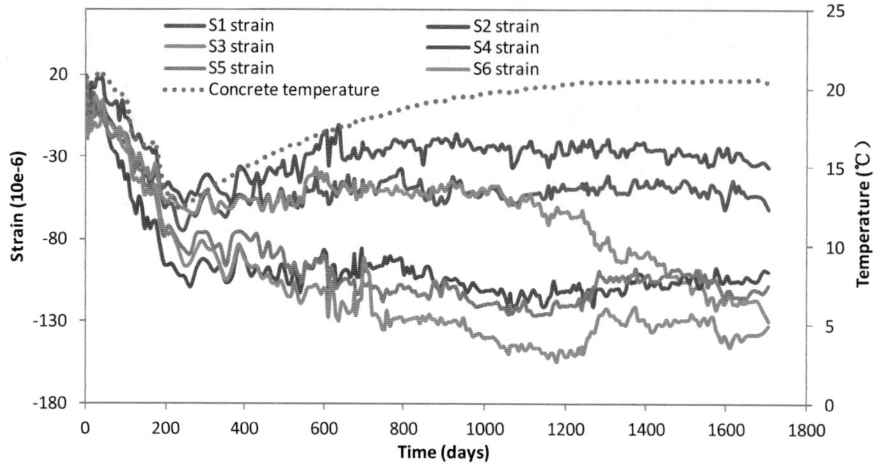

Fig. 12 Measurement data of a strain gauge group

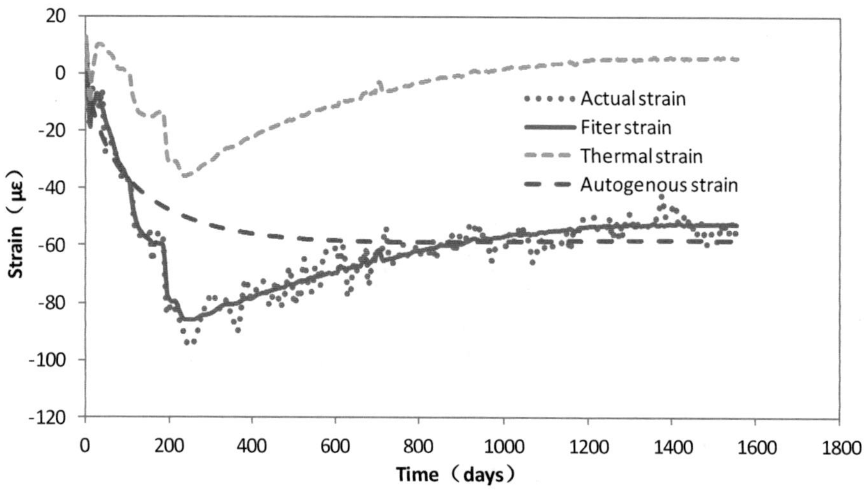

Fig. 13 Measurement data and analysis results of a non-stress strain gauge

Strain measurement data and temperature data of the group of strain gauges are shown in Fig. 12. The non-stress strain gauge measurement data are indicated in Fig. 13. Before the stress analysis, the non-stress strain data must undergo regression analysis, which uses the following formula:

$$\varepsilon_0 = a_1 \Delta T + a_2 \tilde{t} + a_3 \ln(1+\tilde{t}) + a_4\left(1 - e^{k\tilde{t}}\right), \tag{7}$$

where a_1, a_2, a_3, and a_4 are regression coefficients; and ΔT is the temperature difference value that is the temperature minus the initial temperature value $\Delta T = T - T_0$. k is a coefficient, in the current study, k is taken as –0.01. The $\tilde{t}$ *is the* age months *and equals to* age days / 30.

In Formula (7), $a_1 \Delta T$ can be regarded as the thermal strain, and a_1 is the coefficient of thermal expansion obtained by regression. The remaining $a_2\tilde{t} + a_3\ln(1+\tilde{t}) + a_4(1 - e^{k\tilde{t}})$ can be considered as the autogenous volume strain of concrete.

The regression results of the non-stress strain gauge measurement data shown in Fig. 13 are a_1 = 5.154, a_2= –1.370, a_3= –28.983, and a_4 = 315.840. The regression results of thermal strain and the autogenous volume strain of concrete are also shown in Fig. 13.

Stress calculations can be performed based on the results of the regression analysis of the non-stress gauge data. The stress is calculated by employing the incremental method. Based on the assumption that the time is divided into a series of increments $t_1, t_2, \ldots, t_n$, the stress increment can be calculated as follows:

$$\Delta\sigma_i = E_i(\Delta\varepsilon_i - \Delta\varepsilon_i^0 - \Delta\varepsilon_i^c), \quad (8)$$

where $\Delta\varepsilon_i$ is the strain of concrete monitoring, $\Delta\varepsilon_i^T$ is the increment of temperature strain, $\Delta\varepsilon_i^s$ is the increment of autogenous volume strain, $\Delta\varepsilon_i^c$ is the increment of creep strain, and E_i is the concrete elastic modulus at time t_i. The elastic modulus formula of the concrete used in the study is

$$E(t)=E_\infty\left(1-\beta_1 e^{-\gamma_1 t}-\beta_2 e^{-\gamma_2 t}\right), \quad (9)$$

where E_∞ is the final elastic modulus; and β_1, γ_1, β_2, and γ_2 are parameters. In this section, E is considered as $E(t)=44.34\left(1-0.729e^{-0.387t}-0.271e^{-0.017t}\right)$.

$\Delta\varepsilon_i^T$ is calculated by using the following formula:

$$\Delta\varepsilon_i^T = \alpha(T_i - T_{i-1}), \quad (10)$$

where α is the coefficient of thermal expansion using the a_1 result value of regression of Equation (7).

The autogenous volume strain increment is calculated by utilizing the following formula:

$$\Delta\varepsilon_i^s = \varepsilon_i^s - \varepsilon_{i-1}^s, \quad (11)$$

where ε_i^s is the autogenous volume strain of Equation (7), and

$$\varepsilon_i^s = a_2 t + a_3\ln(1+t) + a_4\left(1-e^{kt}\right). \quad (12)$$

The creep strain increment is calculated by the following equation:

$$\Delta\varepsilon_i^c = \varepsilon_i^c - \varepsilon_{i-1}^c, \quad (13)$$

where ε_i^c is the total creep strain for time t_i, which can expressed by the following formula:

$$\varepsilon_i^c = \sum_{j=1}^{i-1}\Delta\sigma_j C(t_i,\tau_j), \quad (14)$$

where $\Delta\sigma_j$ is the stress increment of each time step, and $C(t,\tau)$ is the unit creep of concrete, t is the time (day); and τ is the loading age (day). The creep of the study adopts the following formula:

$$C(t,\tau) = \left(A_1 - B_1\tau^{-C_1}\right)\left(1+e^{-D_1(t-\tau)}\right) + \left(A_2 - B_2\tau^{-C_2}\right)\left(1+e^{-D_2(t-\tau)}\right). \quad (15)$$

The calculation parameter is considered as A_1 = 3.656, B_1 = 5.657, C_1 = 0.219, D_1 = 9.287E-03, A_2 = 1.141E-04, B_2 = 12.676, C_2 = 2.060E-01, and D_2 = 2.269E-01.

Stress increment $\Delta\sigma_i$ for each step is calculated by using Equation (8), and the stress at each step can be calculated cumulatively.

To study the effect of aggregate treatment on stress, the coefficient of thermal expansion and autogenous volume strain of concrete can be deducted. Formulas (10) and (12) are reduced to the following:

$$\Delta\varepsilon_i^T = f_1\alpha(T_i - T_{i-1}), \quad (16)$$

$$\varepsilon_i^s = f_2\left[a_2\tilde{t} + a_3\ln(1+\tilde{t}) + a_4\left(1-e^{k\tilde{t}}\right)\right]. \quad (17)$$

In the aforementioned two formulas, f_1 and f_2 are the reduction coefficients for the coefficients of thermal expansion and autogenous volume strain, respectively.

When the aggregate volume ratio is 0.51 (v = 0.51), f_1 and f_2 are set as f_1 = (1+19.2%) and f_2 = (1+92.5%). When the aggregate volume ratio is 0.66 (v = 0.68), f_1 and f_2 are set as f_1 = (1+5.8%) and f_2 = (1+28.3%). When the aggregate volume ratio is 0.73 (v = 0.73), f_1 and f_2 are set as f_1 = 1.0 and f_2 = 1.0.

The stress calculation results of the three aggregate volume ratio cases are shown in Fig. 14. The figure shows the maximum compressive stresses are -4.34 (v=0.73), -3.71 (v=0.68) and -2.29 (v=0.51) MPa, respectively. Compared with the maximum compressive stresses of v=0.73 case, the relative errors of both v=0.68 and v=0.51 cases are 14.4% and 47.2%. The Fig. 15 shows the stress error of both v = 0.51 and v = 0.66 cases, which were calculated by subtracting the stress of v = 0.73 from the stresses of v = 0.51 and v = 0.66, respectively. The figure shows that for different extra-large aggregate removal method, the difference of the stress calculation can reach roughly 2.2 MPa. The error curves in the figure also show that, at the early stage, the error is small but increases with time. The error gradually stabilizes at the later stage.

Although the above error calculation is related to the specific stress level of measured point, the above results also reflect to a certain extent that the various treatment methods of the large concrete aggregate in the non-stress gauge bucket variably affect the stress analysis of concrete. Therefore, the treatment of large concrete aggregate during the installation of a non-stress gauge must be examined. Third removal method is recommended for large aggregates (excluding large aggregates filled with medium and small aggregates).

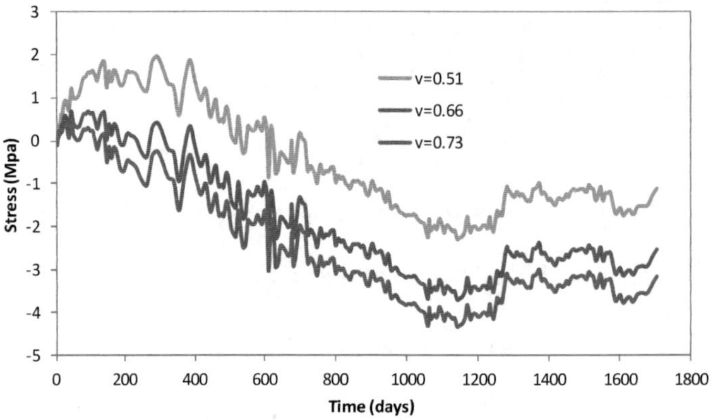

Fig. 14 Stress results of three kinds of removal methods

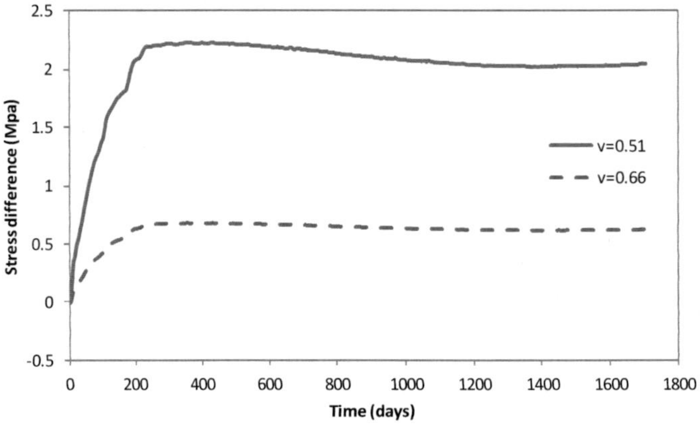

Fig. 15 Stress difference curve of v = 0.59 and v = 0.68 from v = 0.73

7. Conclusions

Non-stress strain gauge measurement is the foundation of investigating dam stress. The measurement plays an important role in strain-stress conversion calculation of concrete. It has various influencing factors, such as bucket effect, temperature field effect, and aggregate content. By using the FEM and meso-mechanical method, this study examines the effects of bucket stiffness, non-uniform temperature field, and aggregate content on measured value of the autogenous volume strain and the coefficient of thermal expansion of concrete. The conclusions are as follows:

(1) The concrete in the bucket can be freely deformed, and the influence of bucket stiffness on the non-stress gauge measurement is small.

(2) The non-uniform temperature field in the non-stress bucket produces a certain amount of stress, but the stress level is low because of the heat insulation in the bucket wall and the distance of the no-stress gauge from the cooling pipe. The effect of non-uniform temperature field on the non-stress gauge measurement results is minimal. In actual projects, the influence of the non-uniform temperature field on the stress-free condition of the non-stress gauge is limited.

(3) When the aggregate content is the same, the influence of the aggregate distribution is less. However, the aggregate content of concrete significantly influences the measured value of the

coefficient of thermal expansion and the autogenous volume strain. When the aggregate content increases by 0.1, for the coefficient of thermal expansion, the decrease percentage can reach 7.0%, and for the autogenous volume strain, the decrease percentage can be up to 16.8% to 34.9%.

(4) Different removal method of extra-large aggregate will cause the different aggregate content of the measured concrete. This factor significantly influences the stress analysis results. Through an analysis example of the monitoring data of a practical engineering, it is revealed that the maximum stress error of the analysis results can reach approximately 2.2 MPa.

In summary, the influence of bucket stiffness and non-uniform temperature field on the measurement result is small, but the aggregate content significantly affects the measurement results. Therefore, in the process of actual instrument installation, the treatment of large aggregates inside the bucket should be controlled strictly.

Acknowledgments
The research was supported by the National Key Research and Development Project of China (Grant NO. 2016YFB0201000), the National Natural Science Foundation of China (Grant NO. 51579252, 51439005).

References

[1] Pirtz D, Carlson R W. (1963) Tests of strain gauges and stress gauges under simulated field conditions. Special Publication, 6: 287-308.
[2] Zeng C. (2014) Research on Influence of Temperature Value on Measured Stress of Concrete Dam. Water Resources & Power.32(2):79-82.
[3] Rodrigues C, Inaudi D. (2010) Laboratory and Field Comparison of Long-Gauge Strain Sensing Technologies.In: European Workshop on Structural Health Monitoring:1289-1294.
[4] Calderón P A, Glisic B. (2012) Influence of mechanical and geometrical properties of embedded long-gauge strain sensors on the accuracy of strain measurement. Measurement Science & Technology, 23(6):65604-65618(15).
[5] Seo Y, Lee J H. (2012) Short- and Long-Term Evaluation of Asphalt Concrete Strain Gauge Installation Methods Applied to the KHCTR. Journal of Transportation Engineering, 138(6):690-699.
[6] Azenha M, Rui F, Ferreira D. (2009) Identification of early-age concrete temperatures and strains: Monitoring and numerical simulation. Cement & Concrete Composites, 31(6):369-378.
[7] Torres B, Payá-Zaforteza I, Calderón P A, et al. (2011) Analysis of the strain transfer in a new FBG sensor for Structural Health Monitoring. Engineering Structures, 33(2):539-548.
[8] Kesavan K, Ravisankar K, Parivallal S, et al. (2010) Experimental studies on fiber optic sensors embedded in concrete. Measurement, 43(2):157-163.
[9] Wang Z, Pan L, Shen H. (2010) Abnormal Phenomenon of Non-stress Strain Gauges in Observation of High Concrete Dams. Hydropower Automation & Dam Monitoring. 34(6), 42-45.
[10] Lv Gaofeng, Wang Yujie, et al. (2015) Analysis of influence of sleeve structure on stress of non-stress meters . dam and safety 2015 (5): 73-78.
[11] Wang Tongsheng. (1978) A basic problem of non-stress gauge in concrete dam. Hydropower Automation & Dam Monitoring, 1978 (1): 66-72.
[12] Zhao Zhiren. (1984) Survey of measured values of autogenous volume change in concrete. Hydropower automation & dam monitoring, 1984 (4): 12-23.
[13] S. Y. Adelman. (1960) Field investigations of concrete hydraulic structures. Moscow, 1960:98-147.

IOP Conf. Series: Materials Science and Engineering **794** (2020) 012075 doi:10.1088/1757-899X/794/1/012075

The Critical Buckling Load of Column under Axial Compressive Force

Kang Xifeng*; GuoXin; Wang Qilei; Liang Qiang;Zhang Fang

Fire Commanding Department, China People's Police University, Langfang, Hebei, 065000, China

*Corresponding author's e-mail: 610590046@qq.com

Abstract. Recent buckling failures of pipelines in permafrost regions reveal that buckling phenomenon is related to the initial displacement of the pipeline due to frost heave. The aim of this study is to investigate whether this relation influent widely the columns applied axial compressive forces. An apparatus was constructed to measure the loads applied on the specimens and the displacements of the specimens. The curve between the two measurement results are shown to lead to an accurate determination of the critical buckling load. Furthermore, a simulation calculation was conducted through finite element method, the numerical results are consistent with the experiment ones. We found that the critical buckling load is inversely correlated to the initial displacement, and further fitting the curves leads to a linear relation.

1. Introduction

Column-liked members which are axially loaded, commonly compressively, are widely used in bridges, aircrafts, vehicles and other industries. Its stability problem has always been an important issue in the research field of machinery and engineering construction. Especially in the field of oil and gas transportation, several accidents of pipeline buckling, such as Canada's Norman-well pipeline buckling accident, Uzbekistan's Tashkent pipeline buckling accident and China's Ge-la pipeline buckling accident, took place.

Figure 1. Buckling accidents in Canada and China

Because the transmission of oil requires a higher temperature, the temperature of the pipe wall during the operation period is higher than that during the construction period, and the temperature difference makes the pipeline expand However, in the axial direction, the continuous expansion of hundreds or even thousands of kilometers of pipeline will cause serious damage: buckling instability. This kind of accidents have made scrapped, causing significant economic losses. According to the investigation of pipeline buckling accident, this kind of accident only occurred in permafrost area, but never took place beyond permafrost area. Further investigations show that every pipeline buckling accident is accompanied by frost heave of soil around the pipeline. Therefore, it is necessary to study the frost heave and buckling of pipes. This kind of problem can be simplified as the buckling problem of compression bar with initial lateral displacement.

The buckling analytical solution of axially compressed structure was given by Euler in 1744, which preliminarily solved the instability problem of fixed length members[1]. Similar research also includes the empirical formula of the relationship between the length and strength of column members proposed by Musschenbroek[2]. With the emergence of the buckling problem of railway and pipeline, Martinet proposed the infinite beam model of winkle foundation, and T. Allan's experiment studied the buckling of thin steel sheet[3-6]. Although these studies promote the understanding of buckling phenomenon to a certain extent, they have great limitations for the buckling of pipes, and fail to solve the nonlinear problems of buckling itself and pipe soil interaction. Musschenbroek ignored the influence of specimen length, and the model of martinet and T. Allan is over simplified, which may lead to a large error with the actual results. With the development of finite element method, the effects of phase transformation characteristics, different constraints, initial geometric defects and load changes on buckling are gradually recognized[7-8]. In the field of petroleum engineering, the buckling studies of sucker rods and pipelines with different materials and working conditions were completed by Yang Meihua, Zhang Xiao, Felipe Franzoni, Falk, etc[9-12]. With an analys is of the public literatures, it is found that there is a lack of test results to verify the simulation study. Given the facts above, this paper investigates the relationship between the initial lateral displacement and the critical buckling load, and carries out small scale test research of buckling.

2.Experiment

Therefore, it is necessary to study the frost heave and buckling of pipes. This kind of problem can be simplified as the buckling problem of compression bar with initial lateral displacement.

2.1apparatus

The experimental apparatus is constructed based on the traditional mechanical test. The initial lateral displacement of the test piece is realized by the middle part of the test piece with the vertical jacking rod with thread. The device is fixed on the stand column of the workbench. The jacking rod and the base of the device are connected by thread. The thread pitch is 1mm. The initial lateral displacement obtained by the test piece is determined by recording the rotation cycles of the jacking rod.

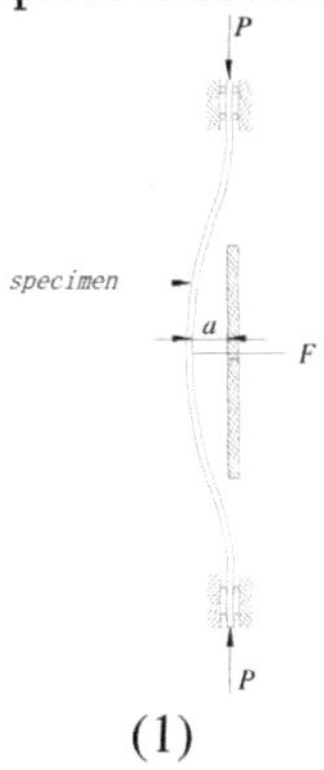

(1)

(2)

Figure 2. Schematic and photo of the experimental system (1) mechanical analysis;(2) photo of the experimental apparatus

In Figure 2. (1), a is the initial displacement, P is the load applied axial on the specimen. After the initial displacement a is generated, the axial load P of the specimen is applied, and the displacement and load changes are measured at the same time. According to the curve between the displacement and the load applied, the critical buckling load is determined, and the relationship between the critical buckling load and the initial lateral displacement is analyzed.

The specimens are straight steel bar with circular cross section, 1 m long and 10 mm in diameter. Defects caused by material are not considered, such as possible micro tissue defects inside.

The specimens are divided into four groups, each group consists of six ones which are applied to the same initial lateral displacement, the values of the initial lateral displacements are 3mm, 5mm, 7mm and 9mm respectively.

2.2experiment results

According to the displacement-load curve based on the test results, the process of loading roughly reveals two stages: the first is the stage of rapid increasing of the load and the second is the stage of buckling. In the former stage, the load increases rapidly, however, the displacement increase slowly; on the contrary, the load in the latter stage decreases gradually after reaching the maximum, but the displacement increases rapidly, indicating that the specimen has buckled. According to the displacement-load curves in Figure 3, all the specimens buckled except for two specimens, and the post buckling load decreased dramatically. The curve of two abnormal test pieces shows that the load starts to decrease immediately after loading, but the displacement increases continuously. This phenomenon is due to the instability of high-pressure oil pump and occasional short-term loss of pressure. Therefore, when the results of buckling critical loads are calculated, the data of two obviously abnormal specimens are ignored.

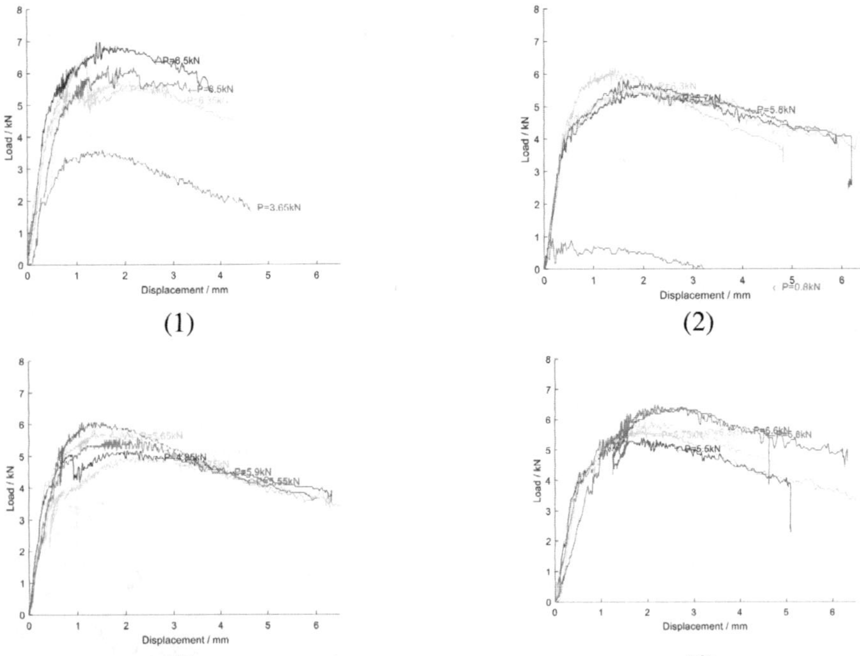

Figure 3. Curves of displacement-load in experiment results
(1) a=3mm; (2)a=5mm; (3) a=7mm; (4) a=9mm

2.3 Result Analysis

The displacement-load curves of each group are compared, and the results are shown in Figure 1. The critical buckling load of each specimen is the highest point of displacement load curve, that is, the point where the load takes the maximum value.

In order to further analyze the relationship between the critical load and the initial lateral displacement and eliminate the accidental error, the average value of critical buckling load of each group of test pieces is calculated, and the results are shown in Table 1.

Table 1. Critical buckling load of specimens

Number of group	Initial Displacement (mm)	Buckling Critical Load(kN)						
		1	2	3	4	5	6	Average
1	3	3.65	6.35	6.7	6.7	6.5	6.15	6.5
2	5	5.5	6.6	5.75	6.8	6.3	6.3	6.31
3	7	6.3	5.7	5.8	6.1	6	0.8	5.73
4	9	5.55	4.75	4.85	5.65	5.9	5.65	5.44

Based on the results of the above table, the fitting curve between the initial lateral displacement (geometric imperfection) and the critical buckling load is determined.

3.Numerical simulation

Through the establishment of the finite element model with the same size as the test specimen, the simulation analysis is carried out,

3.1 The model

The model is established with the beam188 element. The diameter of the simulation specimen is 1cm, and the material is steel. The specific parameters are shown in Table 2.

Table 2. The parameters of the model

Element	Cross section	Prxy(Pa)	ex	R	N	T
Beam188	csolid	2.06e11	0.3	10	16	4

The applied load in the simulation analysis refers to the test load. The maximum value of the buckling critical load in the four groups of specimens is 6.7kN, and the setting of the simulation analysis load is 7kN, slightly higher than the maximum value of the test.

3.2 Numerical results

Figure 4 shows the comparison between the test and simulated buckling critical load when the lateral initial displacement a = 7mm. The blue curve is the experimental result and the red curve is the simulation result.

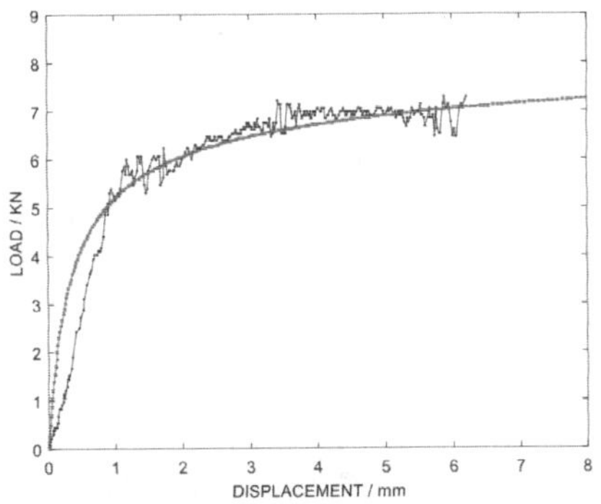

Figure 4. critical buckling load Comparison of experiment and simulation

Comparing with the curves in the above figure, it is found that the curve trend of test results is similar with the simulation results, and the two curves have good consistency. According to the data of each group, as shown in Table 3, the maximum relative error is 8.3%.

Table 3.Relative errors between experiment and simulation results

Initial displacement a(mm)	Experimental load（kN）	Load in simulation （kN）	Error between experiment and

			simulation（%）
3	6.5	6.3	3.07
5	6.31	6.11	3.3
7	6	5.73	8.3
9	5.441	5.446	0.9

4.conclusions

The experimental and simulated displacement responses of specimens with initial lateral displacement show that the smaller the initial lateral displacement is, the larger the curvature of displacement curve changes. There is a significant correlation between the initial lateral displacement and the critical buckling load. The larger the initial lateral displacement is, the smaller the critical buckling load is.The gap between the simulation results and the test results is less than 10%, which shows that the simulation results confirm the test results.

Acknowledgement

The study was supported by the research plan of Hebei province (The critical buckling load of column applied compressive force:172176431)

References

[1] Zhang Xiao, shuaijian. Study on local buckling of X90 pipeline under lateral load [J]. China work safety science and technology, 2018,14 (07): 135-140
[2] Liang Xiaotian. Research on Optimization and control of cable strut tension structure [D]. Zhejiang University, 2017:123-126
[3] Zhang Xinlin. Research on buckling strength of connecting rod of automobile engine [D]. Tsinghua University, 2015:122-123
[4] Felipe Franzoni, Falk Odermann, Dirk Wilckens, Eduards Skuķis, Kaspars Kalniņš, Mariano Andrés Arbelo, Richard Degenhardt. Assessing the axial buckling load of a pressurized orthotropic cylindrical shell through vibration correlation technique[J]. Thin-Walled Structures,2019,137.
[5] Min Jiafeng, He Weiliang, Zhu Haiqing. Buckling analysis method of variable cross-section compression bar based on ANSYS Workbench [J]. Forging equipment and manufacturing technology, 2012,47 (04): 70-72
[6] Zhao Zhenbo, Li Chen, pan Junjie. Study on the influence of local weakening on the stability of compression bar [J]. Mechanical science and technology, 2012,31 (04): 592-596
[7] Zhang Xiaoxia, Zhong Wensheng, Yao yuan. Analysis of the influence of initial deflection and intermediate elastic support on the stability of compression bar [J]. Machine, 2011,38 (06): 1-4
[8] Li Dan. Study on the axial static and dynamic buckling characteristics of TiNi phase change cylindrical shell [D]. China University of science and technology, 2009:96-98
[9] Peng Peng. Numerical simulation of buckling of structures under constraints [D]. Shanghai Jiaotong University, 2009:67-69
[10] Wu Xiangguo. Study on buckling and reliability evaluation method of incomplete structure [D]. Harbin Engineering University, 2006:101-104
[11] Yang Meihua. Study and application of mechanical behavior of sucker rod string in cased production well [D]. Southwest Petroleum Institute, 2005:26-29
[12] Xu Lihui. Calculation of buckling load of steel frame based on ANSYS [a]. Beijing mechanics society. Proceedings of the 24th Annual Meeting of Beijing mechanics society [C]. Beijing mechanics Society: Beijing mechanics society, 2018:3:38-41
[13] Hamaidia Achref, Mohri Foudil, Bouzerira Cherif. Higher buckling and lateral buckling strength of unrestrained and braced thin-walled beams: Analytical, numerical and design approach applications[J]. Journal of Constructional Steel Research,2019,155.

IOP Conf. Series: Materials Science and Engineering **794** (2020) 012076 doi:10.1088/1757-899X/794/1/012076

Study on optimization technology of protective engineering for submerged tube overburden in crossing river tunnel

TU Xiang-yang[1*], TongXiao-lei[1], GAOShi-you[1], LUChen[1]

[1]Pearl River Hydraulic Research Institute, Pearl River Water Resources Commission, Guangzhou 510611, China

*Corresponding author's e-mail: 475958991@qq.com

Abstract: Under the influence of many complex factors, such as flood, tidal power, ship operation, earthquake and construction quality, the stability of the tunnel's underwater overburden layer is crucial to the safe operation of the project. A technical line combined with measured data analysisand physical model testing were utilized to study the features of scouring and sediment for tunnel overburden under several extreme hydrodynamic conditions. In order to ensure the rationtity of anti-scouring stability and economy for tunnel engineering project, optimization design for overburden layer of tunnel was proposed after detailed experiments and analysis. The slurry block stone and stone throwing method can be used to protect the slope and the bottom of the embankment downstream of the tunnel line, which can effectively control the possiblescouring of the concave embankment foot and river bed.The technical line and research methods applied in this paper can be extended to statisfy the engineering design optimization for overburden layer of similar tunnel engineering.

1. Introduction

The anti-scouring stabilityof protection layerfor immersed tunnel engineering is affected by many factors such as water depth, water and sediment dynamics, geological conditions, anti-scouring ability of overburden, construction quality, shipping activity and other human activities[1-2]. Dynamic characteristics of water and sediment, human factors are the main factors which can affect the dangerous scouring of tunnel over burden [3-5]. Usually, two-dimensional flow-sediment dynamic mathematical model and physical model are used to study the maximum scouring depth of River section for tunnel engineering. Aresearch method of erosion protection technology for tunnel overburden can be applied to similar tunnelengineering.

2.Overview of river-crossing tunnel for Guang-Fo line

Crossing Dongping channel 50m upstream of lanshi pier, the river crossing tunnel project is located between Chancheng district and Shunde district in Foshan, China. Map of excavation and backfilling of foundation trench for this river-crossing tunnel is shown in Fig.1.The tunnel overburden is basically backfilled to the original riverbed elevation. Moreover, the tunnel overburden layer of interlocking concrete block for only covers the range of the main channel.

IOP Conf. Series: Materials Science and Engineering **794** (2020) 012076 doi:10.1088/1757-899X/794/1/012076

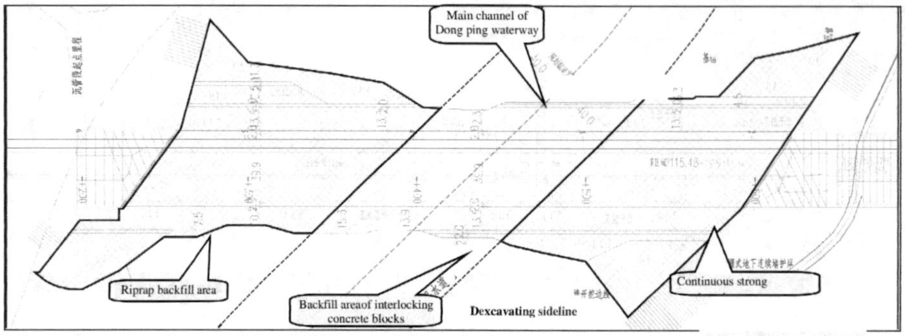

Figure.1Map of excavation and backfilling of foundation trench for river-crossing tunnel of Guangfu Line in Foshan, China.

3.Analysis of River bed Evolution near the Tunnel

Based on the river topographic maps in 1980, 1990, 1998, 2004 and 2009, the variation of riverbed and average scouring and silting intensity at tunnel site were analyzed in details. With a minor amplitude of swing, the dynamic axis of the river is insubstantially stable state. According to the historical evolution and scouring envelope for the cross section of the tunnel's central axis, the elevation of the riverbed on both sides of sinking pipe varies relatively greatly, and the main channel section is relatively stable.

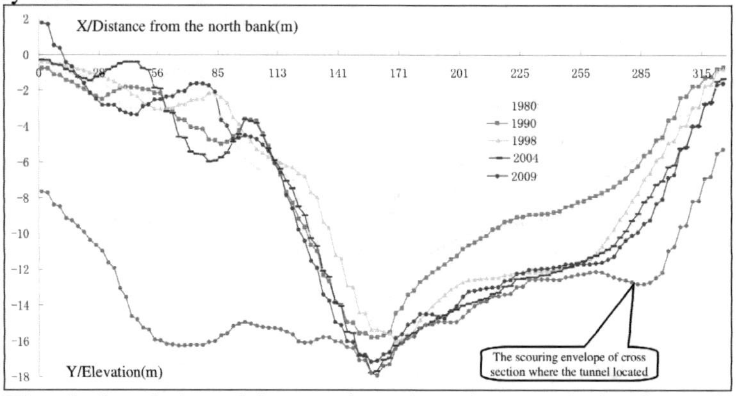

Figure.2 Historical evolution of river section where the centralaxis of tunnel is located

4.Physical model test of eosi on protection for tunnel overburden

4.1 Model design and verification

(1) According to the scope of the river channel and the site conditions, the horizontal and vertical scalesofphysical model are respectively 100 and 50, and the river roughness of the prototype is 0.024-0.026, On the basis of roughness similaritycriterion, the roughness of the model should be 0.018-0.019. The plane layout of physical model for the tunnel of Guang-Fo Line is shown in Fig.3.

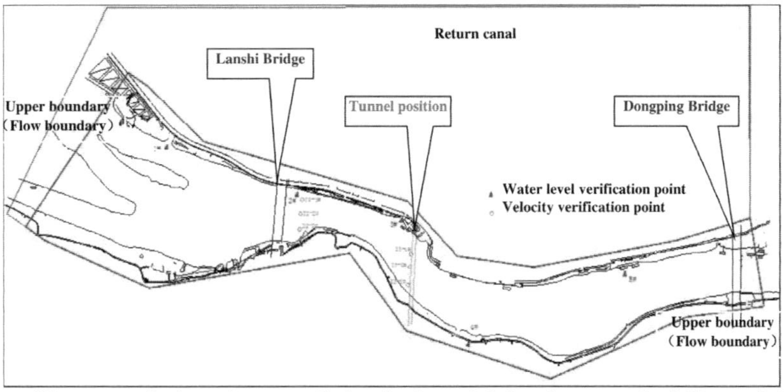

Figure.3Scope and plane layout of physical model for the tunnel of Guangfo Line

(2) The choice of model sand. The critical starting velocity formula proposed by Dou Guoren of the Republic of China is as follows:

$$V_k = m\left(\ln 11\frac{H}{k_s}\right)\sqrt{\frac{\gamma_s-\gamma}{\gamma}gd+0.19\frac{\varepsilon_k+gH\delta}{d}} \quad (1)$$

In the formula(1), m is the coefficient of starting velocity; H is the water depth. According to the results of scouringexperiments in the water tank, the starting velocityof physical sand is between 7.5cm/s and 12.0cm/s under specificwater depths introduced previously, which conforms to acceptable similarity of thestarting velocityfor sediment basically.

Table.1 statistical table for physical model scales

Law of similarity	Scale name	Calculated value
Geometric similarity	Horizontal scale λ_l	100
	Vertical scale λ_h	50
Hydrodynamic Similarity	Velocity scale λ_v	7.07
	Roughness ratio λ_n	1.36
	Scale for flow time λ_t	14.14
	flow ratio λ_Q	35355

(3) Verification for physical model. Under typical hydrological conditions, water level, velocity, scouring and siltingof physical model should be verified in details. The hydrographic datas of water level and velocity measured in the dry season of "2009.2" were used to model calibration.The partialverification results of velocity process of high tide in dry season of "2009.2" were shown in Fig.4.

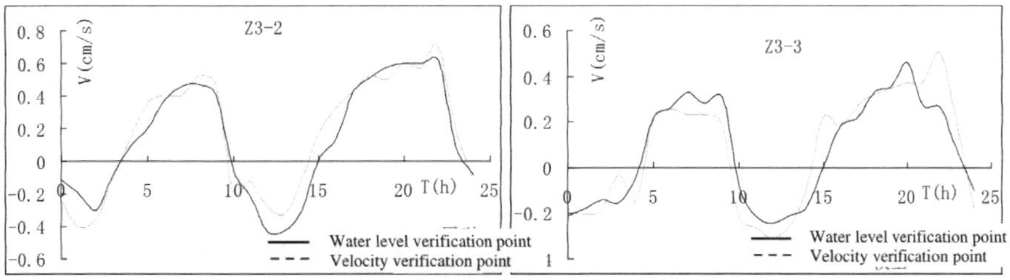

Figure.4 Verification of the "2009.2" spring tide flow rate process

4.2 Model experiments for hydrodynamic characteristic

Under boundary conditions of frequency floodsincluding P=0.33%, 1%, 5% and "99.7" (a middle flood with large tide), the fixed bed current model tests were carried out for hydrodynamic characteristics on the cross-section of tunnel engineering.Before the construction of the project, the effect ofhindrance and transverse-flow the lower reef is evident, and the mainstream leans slightly to the right bank. Considering the main flow will be blocked and deflected by the concave bank on the north side, disadvantage erosion on the riverbed and foot of the embankment by the turbulent flow should be concerned in details.

4.3 Analysis of scouring test results of moving bed

(1) Scouring test of riverbed before theproject construction. After two hours of extreme frequency flood(P=0.33%) scouring in the river channel under topographic conditions of 2009, the result of scouring test is shown in Fig.5. The reef named upper Qu long locates in the mainstream area of the concave bank.

(2) Scouring test of overburden after theproject construction. The riverbed is relatively flat, and the upper and lower reaches of the river are connected smoothlyafter theproject construction. After 2 hours scouringof frequencyflood(P=0.33%), the top and the shoulder of the overburden can remain stability.The area of throwing stones connects smoothly with the upper and lower reaches of the river bed.With few siltationin the north and slight siltationin the south of cross-section of the channel where the tunnel is located, the mixing of bed sand and throwing stones are beneficial to the scouring resistance stability of the tunnel overburden layer.

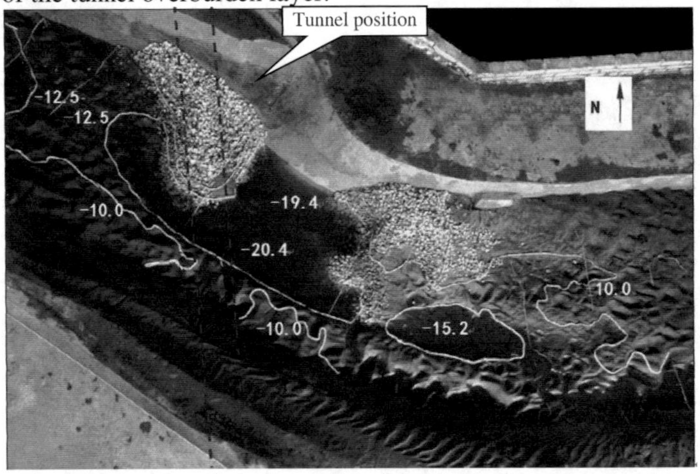

Fig.5The results of souring test before the project construction

(After 2 hours of extremescouring underfrequency flood(P=0.33%) and topographic conditions of 2009)

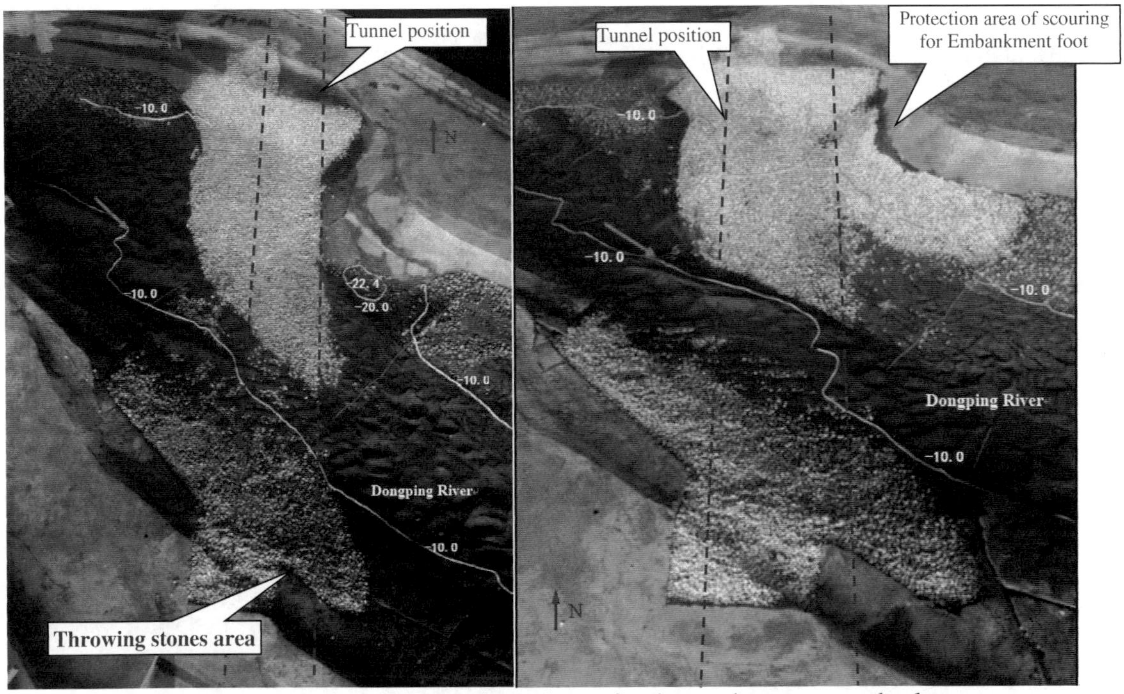

(a)After project construction(b) Protection scheme for the north concave embankment toe.
Figure.6Theresults of extreme scouring test after the tunnel construction
(The frequency flood(P=0.33%) with the low tide level of "05.6 flood")

(3) Scouring test of the protection scheme of the north concave bank embankment. The block stones with a diameter of 30~50cm(50~180kg) were selected, and the throwing thickness is twice the stone particle size (0.6m to 1.0m). The length and width of the filling range are 180m and 25mrespectively. So, the protective area of throwing stone is about $4.6\times10^3 m^2$. The scouring test results of extreme hydrodynamic conditions frequencyflood(P=0.33%) are shown in Fig.6.

5.Conclusion

(1)The rationality and optimization of the scour stability of the overburden of the newly-built river-crossing tunnel project should be carried out by utilizing the advantages of many technical means, such as the analysis of measured hydrological data, the analysis of riverbed evolution, the mathematical model and physical model test. The optimization and demonstration method of the tunnel overburden scour protection project established in this paper can be extended to similar projectsin the future.

(2)For the easily scoured embankment near the tunnel project, the necessary protective engineering should be taken to ensure the stability of the riverbed and the foot of the embankment. The use of masonry slope protection above the low water level and riprap bottom protection at the foot of the embankment can effectively control the adverse scouring of the embankment foot and riverbed on the north concave bank.

(3)The level of construction technology and human activities are main factors which affect the stability of tunnel overburden. Moreover, the destruction of tunnel by human and mechanical forces during operation, the illegal operation of ship anchoringand sand dredger may have adverse effects on the stability of tunnel overburden. Regular monitoring on the safety of underwater terrain, stringent safety monitoring system and emergency plan should be carry out by the tunnel management departments, so as to detect and dispose thehidden dangersof tunnel overburden timely.

IOP Conf. Series: Materials Science and Engineering **794** (2020) 012076 doi:10.1088/1757-899X/794/1/012076

Acknowledgments
The authors would like to thank Professor Xiaoming Wu and Rongli Chen from Pearl River Hydraulic Research Institute, and Pearl River Water Resources Commission for their assistance, and Professor Weiguo He and Yonghui Xing from China Railway Tunnel Survey and Design Institute Co., Ltd., for giving many suggestions and guidance to this paper.

References
[1] Li Zhijun, Wang Qiulin, and Chen Wang, etc. Typical engineering examples of China's sinking-and-tube tunnel and technology innovation and prospect[J], Tunnel Construction (Chinese and English),2018,39 (6):879-894.
[2] Wu Gang, ShenYongfang, Sun Hong. The hydraulic characteristics and stability analysis of the inland floating transport of the sinking tube section[J], Modern tunneling technology,2018,55 (1):170-177.
[3] Chen Shaozhang, Chen Yue, Zhang Mya. Design and construction of immersed tube tunnel[M]. Beijing: Science Press. 2002.05.
[4] HE Wei guo, LYUYang, SONGChaoye. Large Waterway -crossing Immersed Tunnel for Highway and Railway Transportation: Dongping Tunnel in Foshan, China, Tunnel Construction,2018,38(2): 329-336.
[5] The prediction of the scour depth of the extreme flood in the river reach of the river in the estuary of the Qiantang River estuary[J]. Water Scientific Progress,2008,19 (5):685-692.

Study on River Health Assessment Method of the Urbanized Area in Eastern Plain of China

Yifan Ding*, Yongxing Ji, Xiaomei Liu and Jinlong Wu

Shanghai Water Engineering Design & Research Institute, Shanghai, 200061, China

*Corresponding author's e-mail:florawithyou@163.com

Abstract. River health assessment is an important decision-making basis of ecological protection and watershed management. In this study, we established the river health evaluation method of the urbanized area in eastern plain of China. In recent years, Shanghai city has carried out many comprehensive river rehabilitation and ecological restoration projects. Two small rivers with different health condition background were taken as examples to reveal the river health status before and after their rectification. The assessment result shows that the health status of river Arestored from Sick (RHI=0.074) to Sub-healthy (RHI=0.510), and river B from Sub-healthy (RHI=0.585) to Ideal (RHI=0.956). This conclusion is consistent with the actual situation, and the evaluation results are reasonable.

1.Introduction

Ecosystem health of riversis the goal of water environmental management, and river health evaluation is an important bond and basis of water environment and ecosystem management [1].The U.S. Environment Agency (U. S. E. P. A) proposed RBP (Rapid Bioassessment Protocol)incorporated the characteristics of the stream, including width, flow, matrix type and size, into the assessment[2]. ISC(1ndex of Stream Condition) in Australia, RHS(River Habitat Survey) in England, ISG(IndexofStreamGeomorphology) in South Africa, RCE(Riparian, ChannelandenvironmentalInventory) in Sweden, all of them emphasize the significance and importance of river landscapes, river morphology, including river cross-sectional patterns, and the depth ratio of sections, which are important to river ecosystems[3-5].

Since 2010, Chinese government has paid more attention to the ecological protection of rivers and watersheds, and important policies, systems and opinions on the ecological protection and water environment clearly required regular assessment of river health. Somedifferent river health evaluation index systemswereput forward to some important rivers such as The Yellow River, the Yangtze River, and the Pearl River.

However, the domestic and foreign evaluation of river health mainly focuses on large river basins, barely on the urbanized areas or small riversof the eastern plain. The existing index system is not suitable for small urbanized rivers, as the representation and availability of indicators are very different from large watersheds and small rivers. Therefore, it is necessary to carry out research on the rivers in the urbanized areas of the eastern plain in China, which are numerous, insufficient hydrodynamics, dense river networks and high environmental pressure.

In this paper, a river health evaluation system wasputforwardtothisarea, and two smallurbanizedrivers were used as examples to study the riverhealth statuschangesin this area.

IOP Conf. Series: Materials Science and Engineering **794** (2020) 012077 doi:10.1088/1757-899X/794/1/012077

2.Index System and Classification Standards

2.1Index system

Following the principles of scientific, operable, systematic, independent and complete, reference to domestic and foreign literature and the Shanghai River Ecological Governance Design Guide and other guiding documents, an index system of 3 levels,4 dimensions, a total of 9 basic indicators was defined, shown in Table 1.

2.2Weight distribution

Using the expert scoring method, 30 experts answered the questionnaires and thescoreswereanalysed to determine the weight of the indicators, shown in Table 1.

Table 1.Index system and indicator weightsof river health index

Goal layer	Effect layer	Index layer	Meanings of index	Weight	Data source
River Health Index	Water safety	Flood control	Embankment compliance ratio	0.144	Field investigation
	Water environment	Water quality	quality standards for surface water	0.179	Water analysis
		Water Liquidity	Connectivity to near water	0.151	Field investigation
		Transparency	Transparency	0.094	Water analysis
	Water Ecology	Ecological shore protection ratio	Ecological shore protection ratio	0. 106	Field investigation
		Riverside zone vegetation	Vegetation cover in riverbank zone	0.117	Aerial/remote sensing
		Large aquatic plant cover	Total plantscoverage in water	0.129	Aerial/remote sensing
	Human-water relations	Hydrophilic accessibility	Walkwayalongtheriver	0.082	Field investigation
		Resident satisfaction	Resident satisfaction	0.104	Questionnaire

2.3Evaluation criteria

Based on local realities and a large amount of literature, the evaluation criteria for indicators are developed, shown in Table 2.

Table 2. Evaluation criteria of river health index

Index layer	Evaluation value				
	1	0.8	0.6	0.3	0
Flood control /%	[95,100]	[85,95)	[65,85)	[50,65)	[0,50)
Water quality [a]	III and above	IV	V	Inferior V	Black stench
Water Liquidity	connectedon both ends	connected&tube culvert	connected&severed	tube culvert&severed	severedon bothends
Transparency /cm	[80,100]	[50,80)	[30,50)	[10,30)	[0,10)
Ecological shore ratio/%	[90,100]	[70,90)	[50,70)	[30,50)	[0,30)
Vegetation coverage/%	[75,100]	[40,75)	[10,40)	[10,0)	0
Large aquatic plant coverage/%	[75,100]	[40,75)	[10,40)	[10,0)	0
Hydrophilic accessibility	unimpeded both sides	unimpededone side only	partly-throughone side only	only on somespots	completely enclosed

Resident satisfaction/%	[90,100]	[80,90)	[60,80)	[40,60)	[0,40)

[a]Environmental quality standards for surface water, Ministry of Ecological Environment of China

2.4 River Health Grading Standard

River Health index (RHI)is classified as 5 levels: *Ideal*(0.8≤RHI≤1.0), *Healthy* (0.6≤RHI<0.8), *Sub-healthy*(0.4≤RHI<0.6), *Unhealthy*(0.2≤RHI<0.4) and *Sick*(0≤RHI<0.2).

3.Case study

3.1Project Overview

River A (Xihuangtong River) is in Pudong District, amongthe residential areas. River A is about 720m long and 15m wide. The water was turbid and green, the transparency was below 10cm, NH3-N content was as high as 19.1 mg/L, an unpleasant smell was obvious,and river sludge was floating, mixed with oil to oil-stained zones.After the treatment, the NH3-N content is below 2.0 mg/L, the transparency reaches 40cm, and the aquatic plant cover reaches 30%.

River B (Waihuanlindai River) is in the outer ring forest shelterbelt of Baoshan District, about 800m long and 15m wide. The main problem of this river was eutrophication, with a NH3-N content of 2 to 3mg/L. After the measures including shore slope dressing, greening, building culverts to link the near river and walkway transfixion, the flood risk is eliminated, water quality is improved, water connectivity is strengthened, and the river has become a sceniclandscape.

3.2Results and discussions

Basic data was collectedandassessed by the river health index system established above, and the result shows that the health status of river A restored from Sick (RHI=0.074) to Sub-healthy (RHI=0.510), and river B from Sub-healthy (RHI=0.585) to Ideal (RHI=0.956).

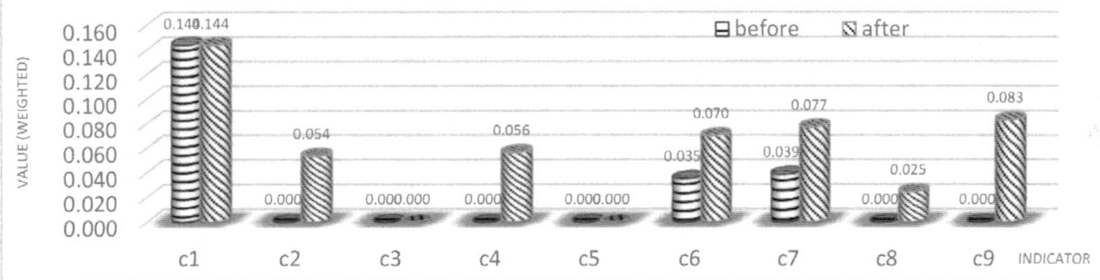

Figure 1. River health evaluation before and after remediation _RiverA

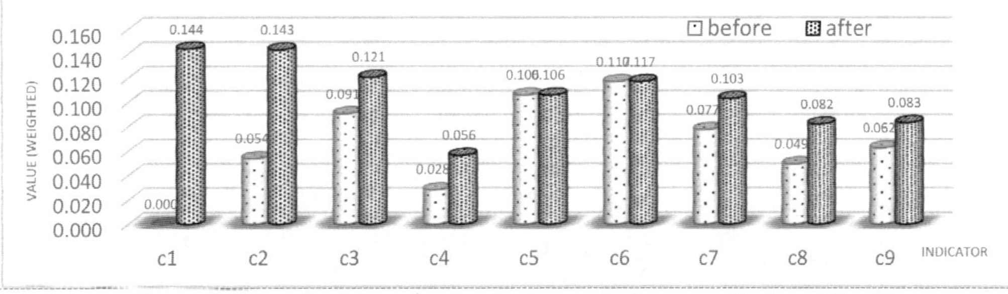

Figure 2. River health evaluation before and after remediation _RiverB

4. Conclusion

In this paper, a complete river health assessment system was established, the method is intuitive and simple. This system was used to assess the health of two small rivers of Shanghai, before and after the

IOP Conf. Series: Materials Science and Engineering **794** (2020) 012077 doi:10.1088/1757-899X/794/1/012077

water environment improvements. And the health of both them was improved significantly. This conclusion is consistent with the actual situation, and the evaluation results were reasonable.

Acknowledgments
Thiswork is sponsored by Shanghai Sailing Program and Self-Leading Talent Program in Huangpu District, Shanghai.

References

[1] Barbour, M. T., Gerritsen, J.,Snyder, B. D. (1999) Rapid Bioassessment protocols for Use in Wadeable Streams and Rivers: Periphyton, Benthic Macroinvertebrates, and Fish. BDS USEPA, California
[2] Cude C G. (2001) Oregon water quality index. Journal of the American Water Resources Association.,37(1):125-137
[3] Y.W. Zhao, & Z.F. Yang. (2005). River health: concept, assessment method and direction. Scientia Geographic Sinica, 25(1), 119-124.
[4] D. PONT, Bernard Hugueny, & C. ROGERS. (2007). Development of a fish-based index for the assessment of river health in Europe: the European fish index. Fisheries Management & Ecology, 14(6), 427-439.
[5] Leah Barclay, Toby Gifford, & Simon Linke. (2017). River listening: acoustic ecology and aquatic bioacoustics in global river systems. Leonardo, 51(3), 1-2.

AUTHOR INDEX